Statistical Physics

This book presents an introduction to the main concepts of statistical physics, followed by applications to specific problems and more advanced concepts, selected for their pedagogical or practical interest. Particular attention has been devoted to the presentation of the fundamental aspects, including the foundations of statistical physics, as well as to the discussion of important physical examples. Comparison of theoretical results with the relevant experimental data (with illustrative curves) is present through the entire textbook. This aspect is facilitated by the broad range of phenomena pertaining to statistical physics, providing example issues from domains as varied as the physics of classical and quantum liquids, condensed matter, liquid crystals, magnetic systems, astrophysics, atomic and molecular physics, superconductivity and many more. This textbook is intended for graduate students (MSc and PhD) and for those teaching introductory or advanced courses on statistical physics.

Key Features:

- A rigorous and educational approach of statistical physics illustrated with concrete examples.
- A clear presentation of fundamental aspects of statistical physics.
- Many exercises with detailed solutions.

Nicolas Sator is associate professor at Sorbonne University, Paris, France. He is a member of the Laboratory of Theoretical Physics of Condensed Matter (LPTMC) and his research focuses on the physics of liquids.

Nicolas Pavloff is professor at Paris-Saclay University, France. He is a member of Laboratoire de Physique Théorique et Modèles Statistiques (LPTMS) and his domain of research is quantum fluid theory.

Lénaïc Couëdel is professor at the University of Sasktchewan, Saskatoon, Canada and researcher at CNRS, France. His research area is plasma physics with a focus on complex plasma crystals.

Statistical Physics

Nicolas Sator, Nicolas Pavloff and
Lénaïc Couëdel

CRC Press
Taylor & Francis Group
Boca Raton London New York

CRC Press is an imprint of the
Taylor & Francis Group, an **informa** business

Designed cover image: Shutterstock_720031951

First edition published 2024
by CRC Press
6000 Broken Sound Parkway NW, Suite 300, Boca Raton, FL 33487-2742

and by CRC Press
4 Park Square, Milton Park, Abingdon, Oxon, OX14 4RN

CRC Press is an imprint of Taylor & Francis Group, LLC

ISBN: 978-1-032-20706-3 (hbk)
ISBN: 978-1-032-22396-4 (pbk)
ISBN: 978-1-003-27242-7 (ebk)

DOI: 10.1201/9781003272427

Typeset in Latin Modern font
by KnowledgeWorks Global Ltd.

Publisher's note: This book has been prepared from camera-ready copy provided by the authors.

Contents

Preface

"According to me, equilibrium statistical mechanics is the most profound and accomplished achievement of science."

D. Ruelle[1], 1991

A fascinating aspect of physics is the new light it sheds on our everyday experience. This is particularly true for statistical physics, which is meant to describe systems with a large number of degrees of freedom, such as those we encounter in our macroscopic world. Another aesthetic and appealing aspect of statistical physics is that it provides an underlying microscopic description of matter: by means of a subtle construction in which renouncement to perfect knowledge is a source of statistical information, it is possible to develop, in an unexpectedly fruitful way, the atomic hypothesis by Democritus. These two aspects guided our presentation of the field: we wanted to highlight the richness of the formal construction of statistical physics – as well as its historical developments – and its vast field of application illustrated by experiments, some of them very recent. The dimensions of this book do not allow to do justice to the extent of the fields in which statistical physics is relevant. The choice of topics was guided by pedagogical considerations, our personal tastes, our research interests and sometimes also by the desire to address aspects not often studied in classical textbooks.

This book is based on the courses (bachelor's and master's degrees in fundamental physics) we taught at Sorbonne Université, at the École Normale Supérieure de Cachan, at the University of Paris-Saclay and at the University of Saskatchewan. The reading of this book requires basic knowledge of thermodynamics as well as of classical and quantum mechanics[2]. The mathematical background required is fairly light and an appendix presents the concepts and techniques that the reader may lack. Each chapter is illustrated by numerous exercises with solutions, inspired by examination and tutorial papers. The first chapters present simple cases (such as the ideal gas) to familiarise students with the new concepts and end with a series of problems presenting more original situations and touching many areas of physics. In the reminder of the book, the exercises are placed at the heart of the chapters to develop technical points or to study particular cases.

Hubert Krivine and Jacques Treiner trained two of us (N. P. and N. S.) and more recently pushed us to start writing this book. We express them our deep and friendly gratitude. The valuable knowledge, constructive criticism and unfailing help of Claude Aslangul and Dominique Mouhanna have been essential. Our warmest thanks to them. Our gratitude also goes to our students, whose stimulating questions have shaped our teaching, as well as to all our colleagues who have helped us to enrich this book and its presentation: Michael Bradley, Xavier Campi, Nicolas Dupuis, Bertrand Guillot, Thierry Hocquet, Glenn Hussey, Thierry Jolicoeur, Giacomo Lamporesi, Pierre-Élie Larré, Mélanie Lebental, Carmelo Mordini, François Ravetta, Guillaume Roux, Gregory Schehr, Marco Tarzia, Christophe Texier, Emmanuel Trizac and Pascal Viot.

---

[1] Quotation from *Hasard et Chaos* (Odile Jacob, 1991). The translation is ours.

[2] The reader will find bibliographical references on these fields of physics thorough the book to complete their knowledge or to review a particular point.

We have consulted many books and articles during the preparation of this book. The following statistical physics textbooks have been particularly useful to us:

- R. Balian, *From Microphysics to Macrophysics: Methods and Applications of Statistical Physics* Volume I (Springer, 2006)

- L. Couture and R. Zitoun, *Statistical Thermodynamics and Properties of Matter* (CRC Press, Taylor & Francis Group, 2000)

- N. Goldenfeld, *Lectures on phase transitions and the renormalization group* (Addison-Wesley, 1992)

- K. Huang, *Statistical mechanics* (John Wiley & Sons, 1987)

- M. Kardar, *Statistical Physics of Particles* (Cambridge University Press, 2007)

- R. Kubo, *Statistical mechanics, an advanced course with problems and solutions* (North-Holland, 1988)

- L. Landau and E. Lifchitz, *Course of Theoretical Physics: Volume 5 Statistical Physics* (Butterworth-Heinemann, 3rd edition, 2013)

- R. K. Pathria and P. D. Beale, *Statistical mechanics* (Academic Press, 2011)

- F. Reif, *Fundamental of statistical and thermal physics* (Mc Graw-Hill, 1985)

- J. P. Sethna, *Statistical mechanics entropy, order parameters, and complexity* (Oxford University Press, 2006)

- R. H. Swendsen, *An introduction to statistical mechanics and thermodynamics* (Oxford University Press, 2012)

Contributors

Nicolas Sator
Sorbonne University
Paris, France

Nicolas Pavloff
Paris-Saclay University
Orsay, France

Lenaic Couedel
University of Saskatchewan
Saskatoon, Saskatchewan, Canada

1 Microscopic Description of a Macroscopic System

"If, in some cataclysm, all of scientific knowledge were to be destroyed, and only one sentence passed on to the next generations of creatures, what statement would contain the most information in the fewest words? I believe it is the atomic hypothesis (or the atomic fact, or whatever you wish to call it) that all things are made of atomslittle particles that move around in perpetual motion, attracting each other when they are a little distance apart, but repelling upon being squeezed into one another. In that one sentence, you will see, there is an enormous amount of information about the world, if just a little imagination and thinking are applied."

R. Feynman[1], 1963

The atomic hypothesis is at the origin of modern physics even before it was validated by the experiments of Jean Perrin[2] in 1908. The objects surrounding us have two aspects: one familiar, observable at our macroscopic scale, the other that escapes our senses, at the microscopic scale, the scale of atoms. The understanding of observable physical phenomena was successfully achieved in the nineteenth century within the framework of theories constructed independently of each other, such as thermodynamics, hydrodynamics or electromagnetism of materials. These theories of continuous media deal with physical quantities defined phenomenologically, based on experimental measurements (pressure, temperature, velocity field of a fluid, magnetic field, etc.) and are based on principles and laws of empirical origin. Besides, at the microscopic scale, the behaviour of atoms is described by the fundamental laws of classical mechanics and, since the beginning of the twentieth century, by the law of quantum mechanics. The atomic hypothesis then offers a fundamental and unifying framework for understanding the many properties of macroscopic matter in terms of its microscopic constituents. However, to meet this challenge, mechanics, whether classical or quantum, is not enough. The change of scale, from the microscopic universe to our macroscopic world, relies on probability theory, a founding element of statistical physics.

This chapter starts by presenting the two descriptions, classical and quantum, of the microscopic state of a system, called a *microstate* (Section 1.1). It is then shown in the framework of classical mechanics how the evolution of microstates (Section 1.2) should, in principle, allow one to determine the system's macroscopic state (Section 1.3). In practice, however, this deterministic approach cannot be successful, and it is essential to resort to probabilities (Section 1.4) and to introduce the concept of *statistical ensemble* (Section 1.5).

1.1 MICROSTATE OF A SYSTEM

In statistical physics, studied systems are composed of many, supposedly elementary, constituents, referred to microscopic scale of the relevant systems. When the number of

---

[1]This quotation from Richard Feynman (1918–1988), American physicist who won the Nobel Prize in 1965 for his work on quantum electrodynamics, is taken from the famous course: *The Feynman lectures on physics, Vol 1*, New Millennium Edition by R. Feynman, R. Leighton and M. Sands (Basic Books, 2011).

[2]Jean Perrin (1870–1942), French physicist, winner of the Nobel Prize in 1926 for proving experimentally the existence of atoms by measuring with different methods the Avogadro number. It was the consistency of these measurements, carried out on phenomena as varied as sedimentation equilibrium or the Brownian motion of small bids suspended in water, that convinced physicists of the reality of atoms. The reader can read with enthusiasm his book, *Atoms* (Ox Bow Press, 1990).

DOI: 10.1201/9781003272427-1

particles N is very large, there is a clear separation between the *microscopic* (particle) scale and the system's *macroscopic* scale (the ensemble of $N \gg 1$ particles), regardless of the size of the particles. This is, of course, the case for atomic systems, whose $N \sim 10^{23}$ particles are molecules, atoms, or so-called *elementary* particles (electrons, photons, etc.). The microscopic scale is then atomic, or even subatomic, that is smaller than the nanometre (10^{-9} m), and the macroscopic scale is larger than the micrometre, typically of the order of a centimetre or a metre. The formalism of statistical physics can be extended more broadly to the study of systems whose particles are themselves observable to the naked eye, without being as numerous as in an atomic system: the grains in a pile of sand ($N \sim 10^9$ to 10^{11} m^{-3}), cars in a traffic jam ($N \sim 10^2$ to 10^3), the stars in a galaxy ($N \sim 10^{11}$), or even the voters in a ballot ($N \sim 10^6$). These very different "particles" have their own characteristics[3], far beyond the scope of the present book limited to atomic systems.

At the microscopic scale, the state of the system, called a *microstate*, is given by the set of physical quantities that completely characterises the state of each of the N particles at a given moment. These microscopic quantities depend on the nature of the particles (atoms, electrons, spins, cars, etc.), the properties under consideration (thermodynamic, magnetic, electronic, etc.) and, of course, the theory used to describe the state of the particles at the microscopic scale (classical or quantum physics). Quantum mechanics is the fundamental theory allowing the understanding of the properties of matter at the atomic scale[4]. The microstate of a system is therefore strictly quantum by nature. However, in many situations it is useful to use classical Newtonian mechanics to describe the microstate of an atomic system. Firstly, because in large domains of thermodynamic conditions, the classical description is a very good approximation, simpler to grasp than the quantum description. This is the case, for example, for most gases and molecular liquids (see Chapter 5). Secondly, because classical mechanics is the fundamental theory on which statistical physics was founded in the nineteenth century, before the discovery of quantum mechanics. Classical mechanics is therefore essential for understanding the work of the founders of statistical physics. Finally, the classical description offers an intuitive approach that allows one to better grasp the probabilistic aspects of statistical physics.

Let us see now how to characterise precisely the microstate of an atomic system, first in the framework of classical physics, then in that of quantum physics.

1.1.1   CLASSICAL FRAMEWORK

In classical Newtonian mechanics, the dynamic state of a particle of mass m without internal structure[5] is completely characterised by its position $\vec{r}$ and by its momentum $\vec{p} = m\vec{v}$, where $\vec{v}$ is the velocity of the particle. Indeed, the knowledge of these two vector quantities specific to the particle at a given time, as well as the force exerted on it by the other particles and possible external fields, allows the exact determination of its trajectory using the fundamental law of dynamics. In the framework of the Hamiltonian formulation[6] of classical mechanics, on which statistical physics is based, the vector quantities $\vec{r}$ and $\vec{p}$ are

---

[3]These characteristics are essentially related to the types of interactions between particles (friction between grains of sand, long-range attraction between stars), to the equilibrium or non-equilibrium state of the system and, in the case of human beings, to the very possibility of modelling their individual behaviour.

[4]Quantum mechanics is presented in the three volumes of *Quantum Mechanics* by Claude Cohen-Tannoudji, Bernard Diu, Franck Laloë (Wiley-VCH, 2019)

[5]The effects due to the internal structure of atoms are neglected when their electronic structure is not taken into account. An asymmetric distribution of electronic charges within an atom, or *a fortiori* a molecule, generates a dipole moment that modifies the interactions with other particles and therefore the dynamic state of each particle.

[6]To familiarise oneself with the basics of analytical mechanics, the reader should consult the book *Analytical mechanics: an introduction* by Antonio Fasano, S Marmi, Beatrice Pelloni (Oxford University Press, USA, 2006).

replaced by scalar quantities, the *generalised coordinates* q_i and their *conjugate momenta* p_i, where $i = 1, 2, \ldots, f$. The number of *degrees of freedom* f is, by definition, the number of generalised coordinates needed to specify the particle's position over time. For example, the position of a particle moving in a three-dimensional space is marked by $f = 3$ coordinates, either Cartesian ($q_1 = x$, $q_2 = y$ and $q_3 = z$) or spherical ($q_1 = r$, $q_2 = \theta$ and $q_3 = \varphi$). In contrast, a particle moving on the surface of a sphere has only $f = 2$ degrees of freedom (the two angles $q_1 = \theta$ and $q_2 = \varphi$). The Hamiltonian approach thus has the advantage of involving only scalar quantities and treating translational and rotational motion in a unified manner, where q_i can denote a length or an angle. In the absence of magnetic field and when the coordinates are Cartesian[7], the conjugate momentum is simply the momentum $p_i = m\dot{q}_i$, where the dot denotes the derivative with respect to time.

The microstate of a system consisting of N particles of mass m (without internal structure) is thus given by the set of N position vectors $\vec{r}_j$ and N momentum vectors $\vec{p}_j$ of each particle $j = 1, 2, \ldots, N$, or equivalently, using the Hamiltonian formulation, by the $f = Nd$ generalised coordinates q_i and the $f = Nd$ generalised momenta p_i, where $i = 1, 2, \ldots, Nd$ and $d = 1, 2$ or 3 is the dimension of the space in which the particles evolve. Note that $2f = 2Nd$ scalar quantities are therefore necessary to determine the system's microstate. In the following, a microstate will be given by:

$$(\mathbf{q}, \mathbf{p}) = \{\vec{r}_j, \vec{p}_j\} = \{q_i, p_i\},$$

where the indices $j = 1, 2, \ldots, N$ and $i = 1, 2, \ldots, Nd$ run over the particles and the set of degrees of freedom, respectively, and where $\mathbf{q}$ and $\mathbf{p}$ are two f-component vectors representing the set of positions and momenta of the N particles, respectively.

In a figurative way, although it cannot be observed in practice, the microstate of an atomic system is a photograph taken at a given moment, sufficiently precise to reveal the positions of all particles and on which the momentum vectors would appear (see Figure 1.1).

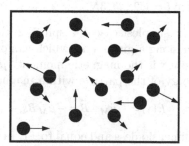

FIGURE 1.1 Schematic representation of a classical microstate in a two-dimensional particle system. The arrows represent the momentum vectors of each particle.

1.1.2 QUANTUM FRAMEWORK

In the framework of quantum mechanics, the microstate of a system is specified at a given time t by the wave function $\Psi(q_1, q_2, \ldots, q_f, t)$. The f degrees of freedom are the generalised coordinates and possibly the spin variables, of purely quantum origin. Spin is an intrinsic quantity of the particles, like the mass and the charge, which plays a fundamental role in statistical physics, as we shall see in Chapter 6.

The quantum nature of a system first manifests by the *quantisation* of certain physical quantities characterising it, in particular the energy, when the particles are confined in a

---

[7]The generalised momentum associated with an angular coordinate is an angular momentum (see Note 24 in Chapter 4).

container or are in bound states[8]. The microstate is thus completely defined by f quantum numbers whose values belong to a *discrete set*. This is an essential difference with the classical description which is based on microscopic variables (positions and momenta) whose real values are *continuous*. Note that in the following examples, the chosen systems have the same number of degrees of freedom in the classical and quantum descriptions, when the particles' spin is not taken into account:

- A spin-less free particle of mass m trapped in a cubic container of side length L. Its quantum state is described by $f = 3$ strictly positive integers, n_1, n_2 and n_3 and its energy is given by:

$$E(n_1, n_2, n_3) = \frac{h^2}{8mL^2} \sum_{i=1}^{3} n_i^2, \qquad (1.1)$$

  where $h = 6.62607015 \times 10^{34}$ J $\cdot$ s is the Planck's constant. The microstate of a perfect quantum gas consisting of N free (non-interacting) particles is therefore characterised by $f = 3N$ integers $n_i > 0$, where $i = 1, 2, \ldots, 3N$.

- A one-dimensional quantum harmonic oscillator. Its quantum state is described by a single positive or zero integer n and its energy is given by:

$$E(n) = \left(n + \frac{1}{2}\right)\hbar\omega, \qquad (1.2)$$

  where $\hbar = h/2\pi$ and ω is the natural angular frequency of the oscillator. It will be seen in Section 4.1.5 that three harmonic oscillators (classical or quantum) describe the oscillation of an atom in a crystalline solid around its equilibrium position (one harmonic oscillator per direction in space). A solid consisting of N atoms can thus be modelled by a set of $3N$ harmonic oscillators. The microstate of such a system is therefore characterised by $f = 3N$ integers, $n_i \geq 0$, where $i = 1, 2, \ldots, 3N$.

- The spin $\vec{s}$ of a particle. In the simplest case of a spin $1/2$, the quantum state is specified by the projection of the spin along a given axis, which can only take two values: $+\hbar/2$ or $-\hbar/2$. Associated with the spin $\vec{s}$ is the magnetic moment[9] $\vec{\mu}_M$, whose projection is $+\mu_M$ or $-\mu_M$. The interaction energy of the particle with a magnetic field $\vec{B}$ is given by:

$$E(s) = -\vec{\mu}_M \cdot \vec{B} = -\mu_M B s, \qquad (1.3)$$

  where the spin variable s is dimensionless and equal to ± 1. In the following, the case where the gyromagnetic ratio $\mu_M > 0$ will be considered. The energy is minimal ($E = -\mu_M B < 0$) when $s = +1$: the spin tends to orient itself in the direction of the magnetic field. To study the magnetic properties of a crystalline solid consisting of N atoms, the degrees of freedom associated with the motions of the atoms are often neglected and only the N spins system is of interest. Its microstate is then characterised by the $f = N$ spin variables $s_i = \pm 1$, where $i = 1, 2, \ldots, N$ (see Figure 1.2). As it will be seen in Section 4.2.1, the spin $1/2$ model has a very general scope that goes beyond the magnetic properties of matter, since it is also used to describe any system in which each particle can only be in *two states*. For example, in a fluid, a position in space is either occupied by a particle or is empty (see the lattice-gas model, page 141). In a binary alloy, a site is occupied by either a type A or a type B atom. In a referendum, a voter votes either "yes" or "no".

---

[8]Particles are in a bound state when energy is required to dissociate them.

[9]The magnetic moment is written $\vec{\mu}_M = g\gamma\vec{s}$, where γ is the gyromagnetic ratio, and g is a dimensionless number, the Landé factor. For a particle of mass m and charge $q \neq 0$, $\gamma = q/(2m)$. For the electron (of mass m_e and charge q_e), $g \simeq 2$ and $\mu_M \simeq \mu_B = q_e\hbar/(2m_e) \simeq 9.27 \times 10{-24}$ J $\cdot$ T^{-1}, where μ_B is the Bohr magneton.

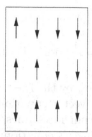

FIGURE 1.2 Schematic representation of a quantum microstate of a system of spins located at the nodes of a square lattice. The "up" and "down" arrows correspond to a spin value of $s = +1$ and $s = -1$, respectively.

When all the degrees of freedom of a system are described by quantum mechanics, one resorts to *statistical quantum physics*, introduced in Chapter 6. Particles' spins play a fundamental role and have major effects on all the physical properties of a system[10]. In addition to the quantisation of physical quantities and the discretisation of microstates, another manifestation of quantum physics is the *indistinguishability* of identical particles (i.e. same mass, charge and spin): one cannot assign an individual state (specified by quantum numbers) to each of them, nor even a label (e.g. a number). Only the collective states of N particles have physical meaning. Thus, the microstate of a quantum gas composed of N identical particles is described by $3N$ quantum numbers (if one ignores the spin), while each particle cannot be assigned an individual set of three quantum numbers.

In contrast, in *classical statistical physics*, each particle has its own position and momentum. In principle one can differentiate the particles and label each of them: the particles are *a priori* distinguishable (see Figure 1.1). This point will be further discussed in Section 2.2. The use of classical statistical physics does not preclude some internal degrees of freedom being described using quantum physics. For example, since the atoms of a crystal are located at the nodes of the lattice, their equilibrium positions can be considered fixed (classical description). Thus, atoms behave as discernible particles with an assigned equilibrium position and therefore are identifiable. Then, the atoms' oscillations around their equilibrium positions can be studied using a classical or quantum description[11]. The spin degrees of freedom are necessarily described by quantum mechanics. However, a system of spins carried by distinguishable particles, as seen in Figure 1.2, is a matter of classical statistical physics, whereas spin is intrinsically quantum. It is therefore necessary to make a difference between the notion of *statistics* (classical or quantum), which is based on the very nature of the particles (distinguishable or indistinguishable), and the notion of *description* (classical or quantum), which indicates the nature of the different degrees of freedom of a system.

1.2 DETERMINISTIC EVOLUTION OF MICROSTATES

Being initially in a specified microstate, a system of N particles will evolve according to the fundamental laws of dynamics. In classical physics, the fundamental principle of dynamics,

---

[10]To summarise rapidly, it is the existence of spin combined with the Pauli exclusion principle that ensures the stability of matter at all scales (see, for example, E.H. Lieb, *The stability of matter*, Rev. Mod. Phys. **48**, 553 (1976)).

[11]Similarly, in Chapter 6, it will be seen that the translational motion of diatomic molecules in an ideal gas can be treated classically, while the degrees of freedom associated with rotations and vibrations, internal to the molecules, are described by quantum mechanics.

applied to each particle j, determines the microstate of the system at any time t:

$$m\frac{\mathrm{d}^2}{\mathrm{d}t^2}\vec{r}_j(t) = \vec{F}_j \quad \text{for} \quad j = 1, 2, \ldots, N, \tag{1.4}$$

where $\vec{F}_j$ is the force exerted on particle j by the $N-1$ other particles (and possible external fields). The solutions of the N coupled vector equations (1.4) give $\vec{r}_j(t)$ (and $\vec{p}_j(t)$) for each particle j. In the framework of quantum mechanics, the evolution of the system's wave function $\Psi(\mathbf{q}, t)$ is governed by Schrödinger's equation:

$$i\hbar\frac{\partial}{\partial t}\Psi(\mathbf{q}, t) = \hat{H}(\mathbf{q}, t)\Psi(\mathbf{q}, t), \tag{1.5}$$

where $\hat{H}$ is the system's Hamiltonian operator. In these two frameworks, classical and quantum, the dynamics of the system is described by a differential equation (1.4) or by a partial differential equation (1.5). The evolution of physical quantities is therefore *deterministic*. In classical mechanics, when knowing the initial microstate, one can, in principle, deduce univocally the future microstates of the system. On the other hand, quantum mechanics is intrinsically probabilistic in nature: while the evolution of the wave function is indeed deterministic, the microstate is only known in terms of probability by reference to a measurement process, the wave packet reduction[12]. It is therefore necessary to distinguish between the probabilities specific to quantum mechanics and the probabilities introduced in the framework of statistical physics

In quantum mechanics, as shown in Equation (1.5), it is the system's Hamiltonian operator $\hat{H}$ that controls the evolution of the wave function. In the framework of classical physics, rather than using the notion of force and the fundamental principle of dynamics (1.4), the unifying formalism of analytical mechanics in its Hamiltonian formulation is chosen. It is based, like quantum mechanics, on the notion of Hamiltonian. In the following, classical analytical mechanics is used to develop the probabilistic approach and to present the basis of statistical physics.

1.2.1 HAMILTONIAN

The Hamiltonian of a classical system is a function of the generalised coordinates and the generalised momenta. It gives the energy of a microstate as the sum of the kinetic energy and the potential energy. For particles with no internal structure and in the absence of magnetic field[13], the Hamiltonian can be written as:

$$H(\{q_i, p_i\}) = \underbrace{\sum_{i=1}^{f}\frac{p_i{}^2}{2m}}_{\text{kinetic energy}} + \underbrace{U_{\text{pot}}(q_1, q_2, \ldots, q_f)}_{\text{potential energy}}, \tag{1.6}$$

where the index i stands for the degrees of freedom of the system. The Hamiltonian takes into account the microscopic characteristics of the system, that is the parameters describing each particle (mass, dipole or magnetic moment, spin for a quantum system...), as well as the interactions between particles and possibly the interactions of the particles with external fields (electromagnetic, gravitational...). The Hamiltonian expression, in particular that of the potential energy, is thus at the heart of the modelling process of a physical system.

---

[12]For example, the probability of observing a particle in the vicinity of point $\vec{r}$ at time t, given by $|\Psi(\vec{r}, t)|^2\mathrm{d}\vec{r}$, evolves deterministically.

[13]In the presence of a magnetic field, the kinetic energy is expressed as a function of the momentum and the vector potential $\vec{A}$ (see L.D. Landau and E.M. Lifshitz, *The classical theory of fields, Course*

In the very common case where the N particles interact via a two-body central potential $u(r)$ (e.g. Lennard–Jones interaction potential discussed in Section 5.1), the Hamiltonian is:

$$H(\{\vec{r}_j, \vec{p}_j\}) = \sum_{j=1}^{N} \frac{\vec{p}_j^2}{2m} + \sum_{i<j} u(r_{ij}), \tag{1.7}$$

where the indices i and j now denote the particles, and r_{ij} is the distance between the particles i and j. The sum over i and j, such that $i < j$, is performed over the $N(N-1)/2$ pairs of interacting particles.

If one can neglect the interactions between particles compared to their kinetic energies or compared to their interaction energies with an external field, the particles can be considered as independent of each other. Under this approximation, the Hamiltonian of the system is:

$$H = \sum_{j=1}^{N} h_j, \tag{1.8}$$

where h_j is the Hamiltonian of a single particle, also called the *one-body Hamiltonian*. This approximation, whose validity depends on the thermodynamic conditions (pressure, temperature...) and the microscopic characteristics of the system, generally allows the analytical calculation of the system's properties and will therefore be used extensively for teaching purposes in this book. The following two cases, the first classical and the second quantum, will be of particular interest:

- The ideal gas is, of course, the most common example: $u(r) = 0$ independently of the distance r between two particles[14]. The Hamiltonian (1.7) is then simply written as the sum of the kinetic energies of N point-like particles:

$$H_{\text{IG}} = \sum_{j=1}^{N} h_j \quad \text{with} \quad h_j = \frac{\vec{p}_j^2}{2m} + u_j, \tag{1.9}$$

where u_j is the internal energy (rotational, vibrational, electronic, dipolar...) of particle j when the particles have an internal structure (molecules, macromolecules...). When the ideal gas is subjected to a gravity field, the potential energy term mgz_j should be added to h_j, where z_j is the altitude of particle j and g is the gravity acceleration (see Section 4.1.5).

- A system of N independent spins–1/2, interacting with an external magnetic field B. Since the interactions between spins are neglected, the Hamiltonian of such a system[15] is, according to Equation (1.3):

$$H_{\text{IS}} = \sum_{j=1}^{N} h_j \quad \text{with} \quad h_j = -\mu_M B s_j, \tag{1.10}$$

where $s_j = \pm 1$.

---

of theoretical physics Volume 2 (Butterworth Heinemann, 1980)). Moreover, if particles have an internal structure, as it is the case for polyatomic molecules, vibrational and rotational energies must be taken into account in the Hamiltonian (1.6) (see Equation (4.41)).

[14]This is, of course, an idealisation. Gas molecules, even a dilute one, interact strongly during collisions, thus changing the microstate of the system. The ideal gas approximation simply assumes that in a given microstate, the interaction energies can be neglected compared to the kinetic energies of the particles (see Section 5.1.1).

[15]Since this system is quantum in nature, its evolution cannot be studied using classical physics.

When interactions between particles cannot be neglected, an extremely important method, called the *mean-field approximation*, allows the study of a system of N interacting particles to be reduced to a model of N independent particles immersed in a mean potential that effectively takes into account the interactions between particles (see Section 5.3 and Chapter 9).

1.2.2 PHASE SPACE DYNAMICS

Once the form of the potential energy U_{pot} is specified, the Hamiltonian (1.6) controls the evolution of the microstates of a classical system through Hamilton's equations[16]:

$$\dot{q}_i = \frac{\partial H}{\partial p_i} = \frac{p_i}{m} \quad \text{and} \quad \dot{p}_i = -\frac{\partial H}{\partial q_i} = -\frac{\partial U_{pot}}{\partial q_i}, \quad \text{for} \quad i = 1, 2, \ldots, f. \tag{1.11}$$

Knowing the initial microstate $(q(0), p(0))$, solving the $2f$ coupled differential Equations (1.11) allows, in principle, to calculate the generalised coordinates and the generalised momenta, $q_i(t)$ and $p_i(t)$, at each time t and for all i. The famous sentence of Pierre–Simon de Laplace (1749–1827), in the introduction to his *Philosophical Essay on Probabilities* (1814), provocatively illustrates this determinism of classical mechanics: "an intelligence which could comprehend all the forces by which nature is animated and the respective situation of the beings who compose it an intelligence sufficiently vast to submit these data to analysis it would embrace in the same formula the movements of the greatest bodies of the Universe and those of the lightest atom; for it, nothing would be uncertain and the future, as the past, would be present to its eyes." He adds, clear-sightedly: "All these efforts in the search for truth tend to lead [the human mind] back continually to the vast intelligence which we have just mentioned, but from which it will always remain infinitely removed."

The time evolution of a system of N particles is naturally represented by the N particle trajectories in real space (a *priori* in three dimensions). However, it is more convenient to represent only one trajectory, that of the system as a whole, at the cost of an effort of imagination: to a given microstate, one can associate a single representative point in an abstract $2f$-dimensional space, called *phase space*[17], each axis of which corresponding to one of the f particles' generalised coordinates or one of the f particles' generalised momenta. The time evolution of the system of N particles is then represented by a single trajectory of the representative point in phase space, called the *phase portrait*.

Consider the example of a single ($N = 1$) classical harmonic oscillator of angular frequency ω in one dimension ($f = Nd = 1$), its Hamiltonian is

$$H(q, p) = \frac{p^2}{2m} + \frac{1}{2} m \omega^2 q^2 = E, \tag{1.12}$$

where E is the energy of the system set by the initial conditions ($H(q(0), p(0)) = E$). The motion of the harmonic oscillator in real space is rectilinear, and its representative point in the $2f$ = two-dimensional phase space is the ellipse shown in Figure 1.3. Note that Hamilton's equations (1.11) do indeed recover the equation of motion: $\dot{p} = m\ddot{q} = -m\omega^2 q$.

---

[16]Since

$$\dot{q}_i = \frac{p_i}{m} \quad \text{and} \quad \dot{p}_i = -\frac{\partial U_{pot}}{\partial q_i} = F_i, \quad \text{for} \quad i = 1, 2, \ldots, f,$$

where F_i is the force associated with the coordinate i deriving from potential U_{pot}, one recovers the more familiar formulation of the fundamental principle of dynamics: $m\ddot{q}_i = F_i$.

[17]In this context, the word "phase" was introduced by Boltzmann in 1872 with reference to the state of a periodic quantity (just like the phase of a sinusoidal function). Maxwell then used it as a synonym for "microstate" and it was finally Gibbs who explicitly associated it with the $2f$ dimensional space (D. D. Nolte, *The tangled tale of phase space*, Physics Today **63**(4) (April 2010) 33-38). "Phase space" should not be confused with "phase diagram", a graphical representation of the physical state (solid, liquid, gas, etc.) of a substance as a function of thermodynamic conditions.

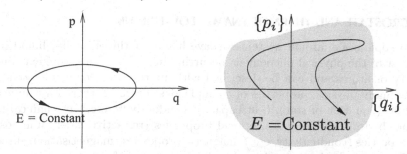

FIGURE 1.3 Phase space trajectory of the representative point of a one-dimensional harmonic oscillator (left) and of any system of N particles in three dimensions (right). The shaded region corresponds to the hypersurface of constant energy E on which the representative point of the system moves. The arrows indicate the direction of the motion.

For N particles moving in a three-dimensional space, it is, of course, impossible to represent graphically the $2f = 6N$-dimensional phase space, but symbolically (and in the two dimensions offered by this page), the set of f spatial coordinates corresponds to the abscissa and the set of f momenta corresponds to the ordinate, as shown in Figure 1.3. The Hamiltonian of an isolated system does not depend explicitly on time and energy is conserved. All the microstates explored by the system in the course of its time evolution therefore have the same energy E and the trajectory of its representative point is located in a $2f - 1$-dimension subspace of the phase space, called an *hypersurface*[18] of energy E. If the N particles are confined in a container, the finite volume reduces the region of phase space accessible to the representative point.

1.3 FROM A MICROSCOPIC DESCRIPTION TO A MACROSCOPIC DESCRIPTION

To understand the physical properties of macroscopic systems from the microscopic behaviour of the particles constituting them, one needs to look at systems made up of a very large number of particles: $N \gg 1$. A macroscopic system is an object that can be directly observed with the human eye, at most with the aid of a magnifying glass or an optical microscope, that is with dimensions greater than one micron (10^{-6} m) and therefore extremely large compared to the atomic scale (10^{-9} m). In macroscopic systems, the number of atoms is therefore typically of the order of Avogadro's number[19]:

$$N \sim N_A = 6.02214076 \times 10^{23} \text{ mol}^{-1}, \tag{1.13}$$

which is by definition the number of particles contained in a mole. For example, under standard conditions of temperature and pressure ($T = 273$ K and $P = 1$ atm), one litre of liquid water contains $N \simeq 3.3 \times 10^{25}$ molecules (the molar mass of water is $M = 18$ g $\cdot$ mol^{-1}) and one litre of ideal gas contains $N \simeq 2.7 \times 10^{22}$ molecules[20].

---

[18]The trajectory of the representative point of a single harmonic oscillator is an ellipse, that is a hypersurface of dimension $2f - 1 = 1$.

[19]Avogadro's number was roughly estimated in the nineteenth century (see Note 38), but it was Jean Perrin who measured it more precisely in 1908, using different experimental approaches. He named the number after the Italian scientist Amedeo Avogadro (1776–1856). According to Avogadro's law (1811), equal volumes of different ideal gases maintained under the same conditions of temperature and pressure contain the same number of molecules.

[20]In a (macroscopic) system consisting of macroscopic particles, $N \ll N_A$. Taking the examples from Section 1.1, the number of (small) grains of sand per m^3 or stars in a galaxy is $N \sim 10^{11}$.

1.3.1   MACROSTATE AND THERMODYNAMIC EQUILIBRIUM

On a macroscopic scale, atomic systems behave like a continuous solid, liquid or gaseous medium. Most of the physical phenomena occurring in such systems were explained before the discovery of atoms and can be described without the need for a microscopic theory. Think of the many physical properties of a simple glass of water: thermodynamic and thermal properties (equation of state, heat capacity, conductivity, etc.), mechanical properties (deformation, hydrodynamics, etc.), optical properties (refractive index, luminosity, etc.), electrical properties (conductivity, etc.), magnetic properties (magnetisation, susceptibility, etc.), etc. The set of physical quantities describing the state of a system on the macroscopic scale is called *macrostate*. Although quite large, macroscopic physical quantities are (infinitely) less numerous than the $6N \gg 1$ microscopic quantities characterising a microstate. Thus, the thermodynamic state of a pure, homogeneous body is completely defined by three thermodynamic quantities – pressure P, volume V and temperature T – linked together by an equation of state, $P(V,T)$, like that of ideal gases. Since the maximum information about a system is given by its microstates, any macroscopic physical quantity can be expressed as a function the particles' generalised coordinates and momenta. This is the case for the total energy E of the system which is given by the Hamiltonian, $E = H(\{q_i, p_i\})$. In other words, the macroscopic state derives from the microscopic state (the reverse being false)[21].

Depending on the conditions in which a system is found during an experiment or an observation, a macroscopic quantity can be of two different natures: either an external parameter or an internal variable. In a statistical physics problem, it is essential to identify and distinguish these two types of physical quantities:

- An *external parameter* is by definition imposed on the system and must therefore be taken into account in the form of a constraint. It remains constant during the evolution of the system and effectively limits the region of phase space accessible to the system to only microstates compatible with the set of external parameters. For example, the volume V of a system confined in a vessel with fixed and rigid walls is an external parameter[22]. Although a function of the microstates (through the Hamiltonian), the energy E of an isolated system is conserved and therefore behaves as an external parameter.

- The value of an *internal variable* depends on the microstate in which the system is found. It is therefore constantly changing on microscopic time scales, typically of the order of $\tau \sim 10^{-12}$ seconds for a liquid or gas, reflecting the underlying molecular agitation. This is, in general, the case for the total kinetic energy and the volume of a gas in a container closed by a mobile piston. In this example, depending on the context, the same quantity (the volume V) can be either an external parameter or an internal variable. Note that an internal variable can be either macroscopic or microscopic, if it characterises a small number of particles, or even a single particle, such as the velocity of a molecule.

During the nineteenth century, the main objective of pioneers of statistical physics was to deduce the thermodynamic properties of a system from the microscopic dynamics of its constituent atoms. Since the objects studied using thermodynamics are macroscopic and *at equilibrium*, their associated thermodynamic quantities are macroscopic and time-independent. By definition, the external parameters are fixed, but a *macroscopic* internal

---

[21]By analogy, in the context of an election, the "microstate" of the electorate is given by the knowledge of the ballot paper cast by each voter (maximum information). A secret ballot reduces (fortunately!) the information on the system. The macroscopic state is then the result of the vote.

[22]The potential energy of the Hamiltonian (1.6) can formally include an interaction term, essentially between the particles and the walls of the container. In general, it is simpler to indicate the region of space (and the corresponding volume V) in which the particles are confined in the form of integration limits.

variable $A(q(t), p(t))$ changes continuously over time, even at equilibrium[23]. In fact, measuring apparatus do not have sufficient temporal (and spatial) resolution to measure the small variations of the internal variables over microscopic timescales, of the order of $\tau \sim 10^{-12}$ seconds. Over the duration of an experiment $t_{\exp} \gg \tau$, the measured value is actually a *time-averaged value*, $\bar{A}$, smoothing out the variations induced by the microscopic evolution of the system:

$$\bar{A} = \frac{1}{t_{\exp}} \int_{t_0}^{t_0 + t_{\exp}} A(\boldsymbol{q}(t), \boldsymbol{p}(t)) \, \mathrm{d}t \quad \rightarrow \quad A_{\mathrm{eq}} \quad \text{at equilibrium.} \tag{1.14}$$

As long as the system has not reached thermodynamic equilibrium, a time-averaged quantity $\bar{A}$ depends on the time t_0 at which the measurement starts. Once the system is in equilibrium, $\bar{A}$ takes on a practically constant value, its equilibrium value A_{eq}, independent of t_0 and $t_{\exp}$, which is by definition a thermodynamic quantity[24]. For example, the temperature of a soft drink taken out of a refrigerator increases during a transient thermalisation period and then reaches a new equilibrium value, which is constant and equal to the outside temperature of the kitchen. In the case of the diffusion of a drop of ink in a glass of water, the density of ink in a 1 cm^3 volume tends towards a constant when the system reaches a new equilibrium. Figure 1.4 shows a succession of non-equilibrium macrostates leading to the equilibrium macrostate.

Once in equilibrium, the system remains in the same equilibrium macrostate, for which $\bar{A} = A_{\mathrm{eq}}$, but continues to explore accessible microstates compatible with the external parameters. The instantaneous value $A(q(t), p(t))$ of the internal variable therefore fluctuates continuously around the equilibrium value A_{eq}. Thus, *the equilibrium macrostate corresponds to a gigantic number of microstates*. Just imagine the number of ways of assigning positions (and momenta) to the air molecules in a room, compatible with the single macrostate corresponding to the room's temperature and pressure.

1.3.2 FAILURE OF THE DETERMINISTIC APPROACH

The deterministic approach to calculate a thermodynamic quantity from the particle dynamics is clear: once the system's Hamiltonian has been written, it is necessary to i) determine the $6N$ initial conditions $(q(0), p(0))$, ii) solve the $6N$ coupled Hamilton differential Equations (1.11), iii) deduce $(q(t), p(t))$ and $A(q(t), p(t))$ at each time t and finally, iv) calculate the time average value $\bar{A}$ of the internal variable (Equation (1.14)) which, after a long time, tends towards the thermodynamic quantity A_{eq}. Since there are $N \sim 10^{23}$ particles!, this approach is impossible to implement for reasons detailed in the following:

---

[23]To simplify the notation, a physical quantity and its value are designated by the same letter A.

[24]The relaxation time, t_{relax}, necessary to reach an equilibrium macrostate, is generally very large compared to the microscopic time ($t_{\mathrm{relax}} \gg \tau$) and depends on the microscopic properties of the considered system (its Hamiltonian), the thermodynamic conditions and its initial macrostate. Three cases can be distinguished according to the duration of the experiment with respect to t_{relax}:

- If $t_{\exp} \gg t_{\mathrm{relax}}$, the system reaches its equilibrium macrostate described by thermodynamics.

- If $t_{\exp} \ll t_{\mathrm{relax}}$, the system does not have time to reach equilibrium and remains trapped in a macrostate, so-called *metastable* state, in which the time-averaged values $\bar{A}$ of the macroscopic internal variables are constant, but different from their equilibrium value A_{eq}. However, it is possible to describe the system using thermodynamics, considering the metastable state as an equilibrium state under effective constraints. For example, hot coffee in a thermally insulated bottle can be considered as an isolated system for a certain period of time, although its temperature ultimately tends towards the outside temperature. This is also the case for a glass, obtained by cooling down a silicate liquid rapidly enough so that its atoms do not have time to form a crystal. The dynamics of this so-called *amorphous* solid are so slow that it can remain in the same macrostate for times that would exceed the age of the universe.

- In contrast, if $t_{\mathrm{relax}} \simeq t_{\exp}$, the system evolves during the duration of the experiment, changing to a non-equilibrium macrostate, which, in principle, cannot be studied with usual thermodynamics.

FIGURE 1.4 Photographs of a drop of ink initially laid in a glass of water (left). Each picture represents a macrostate of the system, first out of equilibrium and then at equilibrium (right). The black square corresponds to a 1 cm^3 volume in which the ink density evolves and tends towards a constant equilibrium value (Source: wikipedia, Blausen, Diffusion `https://commons.wikimedia.org/wiki/File:Blausen_0315_Diffusion.png`).

- Mathematically, the equations of motion cannot be solved analytically for $N > 3$ interacting atoms[25].

- In practice, writing down the $6N \sim 10^{23}$ numerical values associated with a single microstate would require about 10^{20} sheets of paper (or a light year high stack of sheets on the corner of a desk, i.e. 10^{13} km) or a computer memory of the order of 10^{15} gigabytes, which is incommensurately larger than the storage capacity of any computer, regardless of the purchase date.

- Fundamentally, a system of $N \geq 3$ particles can exhibit chaotic behaviour: while the evolution of the system being deterministic, the slightest imprecision on the initial conditions leads to divergence with time and forbid any prediction of future microstates. This is known as "sensitivity to initial condition", as illustrated by the example in Section 1.4.2.

The transition from the microscopic to the macroscopic by a purely deterministic approach must in fact be abandoned: classical (or even quantum) mechanics, which describes the microscopic world, cannot alone explain all the physical phenomena observed at human scale[26]. To achieve this change of scale, it is necessary to introduce a new formalism based on the use of probabilities.

The deterministic approach presented above can nevertheless be applied in the framework of numerical simulations called *molecular dynamics*, at the cost of two approximations due to the use of a computer to process the dynamics of the particles. First, the (differential) equations of motion are solved numerically by successive iterations and time discretisation.

---

[25] More precisely, there are analytical solutions to the two-body problem ($N = 2$) when the interaction potential is of the form $u(r) = kr^n$, where k is a constant, but only for certain values of n (see D. Hestenes, *New Foundations for Classical Mechanics*, (Kluwer Academic Publishers, 1999)), especially for $n = 2$ (harmonic potential) and $n = -1$ (gravitational and Coulomb interactions, the Kepler problem). In the latter case, the analytical solution of the $N \geq 3$-body problem exists (since 1909 for $N = 3$ and since 1991 for $N > 3$), in the form of a series that converges so slowly that it is not used in practice.

[26] This is in contradiction with absolute reductionism, the idea that the properties of any object can be understood only by means of a theory that describes the behaviour of its elementary constituents. The scientific approach, particularly in physics, has developed on the basis of this fruitful idea which has its limits (taken to the extreme, this philosophical position implies that human behaviour can be explained by atomic physics or even by particle physics). Statistical physics is an illustration of this. See P.W. Anderson, *More is different*, Science **177**, 393 (1972).

The calculation is therefore only an approximation. Second, the capacities of current computers limit the number of particles taken into account ($N \sim 10^3$ to 10^6) and the system's evolution duration (typically $\sim 10^{-8}$ s for atomic systems). Despite these restrictions, molecular dynamics is proving to be very effective in determining the properties of physical or biological systems that are too complex to be studied analytically, or subjected to extreme thermodynamic conditions of temperatures and/or pressures, difficult to obtain in laboratory experiments[27].

1.4   THE NEED FOR A PROBABILISTIC APPROACH

The expected failure of the deterministic approach must be put into perspective. The knowledge of microstates would give a plethora of information, impossible to manipulate in practice and ultimately of little interest. Suppose that one knows the magnitude of the velocity of each of the $N \sim 10^{23}$ particles of a gas in equilibrium at a given time t. To make this amount of data intelligible, one would determine the number of particles $N_t(v)$ whose velocity magnitude lies between v and $v + dv$, where dv is the precision on the velocity. By proceeding in this way for a large number of successive time instants, one could, in principle, calculate the time average $\bar{N}(v)$ of the internal variable $N_t(v)$ which, since the system is in equilibrium, would remain constant. Finally, the velocity distribution $P(v)$ could be deduced, giving the probability

$$P(v)dv \simeq \frac{\bar{N}(v)}{N}$$

to find a random particle at a given time with a velocity between v and $v + dv$. The distribution $P(v)$, obtained here from the true velocities of the particles, would provide a sufficiently accurate and usable description of the microscopic quantity v at equilibrium[28]. Thus, only the distributions and time-averaged values of the internal variables (microscopic or macroscopic) are of interest when describing the macrostate of a system in equilibrium, as well as the behaviour of its particles at the microscopic scale.

The central idea of statistical physics is to replace the deterministic approach, based on time average values (1.14) calculated from the laws of mechanics, with a probabilistic approach to microstates, allowing the direct calculation of the probability distributions and the statistical average values of the internal variables, treated as random variables. The use of probabilities in physics was hardly accepted by most physicists of the nineteenth and early twentieth centuries. In principle, the study of the motion of any object, even atoms, was a matter for classical mechanics, a theory that reigned supreme in the natural sciences thanks to its mathematical rigour and predictive power, beautifully illustrated by the understanding of the motion of the planets. Thus, to be fully understood, all physical phenomena – especially thermodynamics – had to be described by deterministic classical mechanics. Faced with the obstacles presented in the previous section, Maxwell[29] in 1860

---

[27]The first two obstacles - mathematical and practical - to the implementation of the deterministic approach are therefore overcome using numerical approximations. On the other hand, as in a real system, the particle dynamics determined using a simulation exhibits chaotic behaviour due to rounding errors. The system's trajectory in phase space obtained by molecular dynamics is therefore very different from the exact trajectory that would be calculated analytically (if this calculation were possible). Thus, the instantaneous value of a variable $A(q(t), p(t))$, can be strongly affected by chaos. It will be seen in Section 2.4 that the latter has no effect on its time average value $\bar{A}$ and thus on the determination of the thermodynamic quantities (see Note 60 in Chapter 2).

[28]By analogy, in the context of an election, the number $N(X)$ and the percentage $P(X)$ of voters who voted for candidate X are of interest.

[29]James Clerk Maxwell (1831–1879) Scottish physicist, first founding father of statistical physics, best known for his famous equations in electromagnetism that unify electricity, magnetism and optics.

and 1867, and Boltzmann[30] in 1868, introduced probabilities into physics, giving rise to statistical physics[31].

Before turning to the formalism of statistical physics itself in Section 1.5, the probabilistic approach will be illustrated for two simple situations that require only a basic knowledge of probability theory.

1.4.1 AN INSTRUCTIVE EXAMPLE: THE JOULE EXPANSION

A closed vessel, thermally insulated from the outside, is divided into two compartments $\mathcal{A}$ and $\mathcal{A}'$ of volumes V and V', respectively, and are separated by a closed valve (see Figure 1.5). The studied system contains one mole of gas, that is $N \simeq 6 \times 10^{23}$ molecules of mass m, initially at equilibrium in compartment $\mathcal{A}'$ while compartment $\mathcal{A}$ is empty. At $t = 0$, the valve is opened (without adding energy into the system) and the gas gradually spreads throughout the vessel, evenly filling both compartments when the final equilibrium macrostate is reached[32].

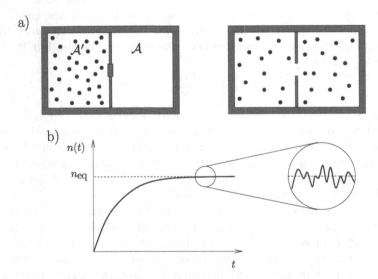

FIGURE 1.5 Joule expansion: a) Compartment $\mathcal{A}'$ of the insulated vessel contains a gas (left). At $t = 0$, the valve separating the two compartments is opened. Once in equilibrium, the gas fills the container uniformly (right). b) The curve $n(t)$, showing the number of molecules contained in compartment $\mathcal{A}$ at time t. Magnification ($\times 10^{12}$) reveals the fluctuations of $n(t)$ around the mean equilibrium value n_{eq}.

In the following chapters of this book, the various thermodynamic aspects of this experiment will be studied using the tools of statistical physics. For the moment, let us focus

---

[30]Ludwig Boltzmann (1844–1906), Austrian physicist, second founding father of statistical physics, who contributed to its development by generalising and proving more formally the results found by Maxwell. His work on the evolution of a system towards equilibrium founded non-equilibrium statistical physics.

[31]For the history of statistical physics with technical aspects, the reader may wish to consult the works of J. Uffink, *Compendium of the foundations of classical statistical physics*, in J. Butterfield and J. Earman (Eds.), *Philosophy of Physics* (Amsterdam: North Holland), 923 (2006), which is easily found on the internet.

[32]This experiment, sometimes called the *Joule-Gay-Lussac relaxation experiment*, was performed by Joseph-Louis Gay-Lussac in 1807, then by James Prescott Joule in 1845. It shows that the internal energy of a (ideal) gas depends only on its temperature.

on only one internal variable, the number of molecules $n(t)$ contained in compartment $\mathcal{A}$ at a given time t, whose evolution as a function of time is shown in Figure 1.5: initially at zero, $n(t)$ increases during a transient regime. It then reaches a constant value, n_{eq}, its equilibrium value. The density of the gas in compartment $\mathcal{A}$, equal to $\rho_m = mn_{eq}/V$, is a measurable macroscopic quantity. If one were able to measure $n(t)$ with a very high accuracy, fluctuations would be observed around the time average value n_{eq}, reflecting the constant change of the microstate of the system in equilibrium, that is molecular agitation of the gas molecules: at all times, molecules enter or leave compartment $\mathcal{A}$ without exact number compensation.

The thermodynamic state, that is the equilibrium macrostate of the system, is characterised by the mean number n_{eq}, which can be calculated by expressing the internal variable $n(t)$ as a function of microscopic quantities:

$$n(t) = \sum_{j=1}^{N} c_j(t), \qquad (1.15)$$

where $c_j(t)$ depends only on the position $\vec{r}_j$ of molecule j. It is equal to 1 if molecule j is in compartment $\mathcal{A}$ at time t, and 0 otherwise. With a deterministic approach, which is here impossible to apply, one would have to solve the equations of motion and then determine $\vec{r}_j(t)$, $c_j(t)$ and $n(t)$ at each time instant, even during the transient regime, and finally calculate the time average value $\bar{n}$ at long times to obtain the value n_{eq}.

The probabilistic approach is here very intuitive and is based on two simple assumptions. First, it can be assumed that the molecules' positions are *independent* of each other. This assumption is false for molecules that are close enough to interact, but in a macroscopic system, each particle is far away from the overwhelming majority of other particles. Then, in equilibrium, molecules occupy both compartments uniformly. Thus, in a given microstate, the probability p that molecule j is in compartment $\mathcal{A}$ (i.e. $c_j = 1$) is equal to the fraction of the volume of the vessel occupied by compartment[33] $\mathcal{A}$:

$$p = \frac{V}{V + V'},$$

and this independently of the other $N - 1$ molecules. The probability that molecule j is in compartment $\mathcal{A}'$, (so that $c_j = 0$), is of course, $1 - p$. The microscopic quantity c_j is then treated as a random variable taking the value 1 with probability p and the value 0 with probability $1 - p$. Its *statistical mean value* is:

$$\langle c_j \rangle = 1 \times p + 0 \times (1 - p) = p.$$

In the following, $\langle A \rangle$ is the statistical mean value of the internal variable A, to distinguish it from its non-probabilistic time-averaged value $\bar{A}$. According to (1.15), the number of molecules n contained in compartment $\mathcal{A}$ can also be seen as a random variable whose statistical mean value intuitively corresponds the equilibrium value n_{eq}:

$$n_{eq} \equiv \langle n \rangle = \sum_{j=1}^{N} \langle c_j \rangle = Np = \frac{NV}{V + V'}.$$

As will be seen later, the substitution of a statistical average for a time-average to estimate the value of physical quantity in equilibrium is the cornerstone of statistical physics. Of

[33] If the trajectory of molecule j were followed for a sufficiently long time, it would be seen to explore the entire volume of the vessel and move many times from one compartment to the other, spending in each compartment a time proportional to its volume. Thus, the molecule would be in compartment $\mathcal{A}$ for a fraction p of the duration of the experiment (see Section 2.4).

course, this approach tells nothing about the system's dynamics and the transient regime that precedes the equilibrium macrostate. However, one can go beyond the calculation of the equilibrium value predicted by thermodynamics and estimate the amplitude of fluctuations around the mean value n_{eq} by calculating the variance of the random variable n. By definition, the variance $\Delta^2_{c_j}$ of the variable c_j is given by[34]

$$
\begin{aligned}
\Delta^2_{c_j} &= \langle (c_j - \langle c_j \rangle)^2 \rangle = \langle (c_j^2 + \langle c_j \rangle^2 - 2c_j\langle c_j \rangle) \rangle \\
&= \langle c_j^2 \rangle - \langle c_j \rangle^2 \\
&= \left(1^2 \times p + 0^2 \times (1-p)\right) - p^2 = p(1-p).
\end{aligned}
$$

Since n is written as the sum of independent random variables c_j, its variance Δ^2_n is the sum of the variances $\Delta^2_{c_j}$, so

$$
\Delta^2_n = \sum_{j=1}^{N} \Delta^2_{c_j} = Np(1-p).
$$

Note that in order to fully characterise the random variable n, one would have to calculate its probability distribution $P_N(n)$, that is the probability that there are exactly n molecules in compartment $\mathcal{A}$ (see exercise 1.2). The standard deviation Δ_n is therefore of the order of $\sqrt{N} \sim 10^{11}$. Thus, there are about $n_{eq} \pm \Delta_n \simeq 3 \times 10^{23} \pm 4 \times 10^{11}$ molecules in compartment $\mathcal{A}$ for $V = V'$ (i.e. $p = 1/2$). The deviations from the mean given by Δ_n seem very large, but with respect to the mean value they are insignificant: the relative fluctuations are of the order of[35]

$$
\frac{\Delta_n}{n_{eq}} \simeq \frac{1}{\sqrt{N}} \sim 10^{-12},
$$

for one mole of gas. As will be seen later, this behaviour of the fluctuations is typical in statistical physics. Thus, the thermodynamic quantities (here n_{eq}) identified with statistical averages, have such a small relative uncertainty (of the order of 10^{-12} for a macroscopic system) that the predictions of the probabilistic approach are certain in practice. The calculation of n_{eq} and Δ_n illustrates well the spirit of this approach, but the formalism of statistical physics is necessary for its generalisation and the determination of all thermodynamic quantities (temperature, pressure, entropy...).

1.4.2 RUDIMENTS OF GAS KINETIC THEORY

At the origins of statistical physics, the kinetic theory of gases[36] was the first atomic theory to try to show that the macroscopic properties of gases are due to the motions of their molecules. It was therefore based on the atomic hypothesis, which was widely contested throughout the nineteenth century. Faced with the complexity of the problem to be solved, Clausius[37] in 1857, Maxwell in 1860 and Boltzmann in 1868, first assumed that the molecules of a dilute gas behave like rigid spheres that only interact at the moment

---

[34]Fluctuations are evaluated by taking the average of the *square* of the deviation from the mean, as the average of the deviation from the mean, $\langle (c_j - \langle c_j \rangle) \rangle = \langle c_j \rangle - \langle c_j \rangle$, is always zero.

[35]In Figure 1.5, the value $n_{eq} \simeq 10^{23}$ corresponds to a length of ~2 cm on the ordinate axis. At this scale, the fluctuations around the mean ($\Delta_n \sim 10^{11}$) are therefore of the order of 10^{-12} cm, much smaller than the line thickness...

[36]Although informative, this section can be skipped at first reading.

[37] As early as 1738, in a poorly developed theory, Daniel Bernoulli (1700–1782) correctly interpreted the pressure exerted by a gas as the resultant of the collisions of the molecules against the walls of a container and showed that $PV \propto \bar{v}^2$, where $\bar{v}$ is the characteristic velocity of the molecules. It was not until 1857, under the scientific authority of the German physicist Rudolf Clausius (1822–1888) whose contribution to the foundation of thermodynamics was well known, that the kinetic theory of gases started to be really developed. It achieved its first success with Maxwell in 1860.

of a collision (hard sphere gas). The difficulty is then to correctly treat the collisions they undergo in order to express the gas thermodynamic quantities (pressure and temperature) as a function of the microscopic quantities characterising the molecules (mass, diameter and velocity). To this end, gradually abandoning a purely deterministic approach, the pioneers of the kinetic theory of gases consider the position and velocity of the particles as random variables and introduce hypotheses on their probability distribution function (as it was done in Section 1.4.1).

In the spirit of the Clausius' work (1857), one mole of gas, that is $N = N_A$ molecules, contained in a cubic vessel of side L is considered. Pressure, volume and temperature are P, $V = L^3$ and T, respectively. According to thermodynamics, for a diluted gas, its equation of state is approximated by the ideal gas equation of state. For one mole of gas:

$$PV = RT, \qquad (1.16)$$

where $R \simeq 8.314 \ \text{J} \cdot \text{mol}^{-1} \cdot \text{K}^{-1}$ is the ideal gas constant. With the gas maintained at standard conditions of temperature and pressure ($T = 273$ K and $P = 1$ atm), the volume occupied by one mole of gas is, according to Equation (1.16), $V = 22.4$ litres (or $L = 28$ cm). The particle number density, that is the number of particles per unit volume, is therefore[38]:

$$\rho = \frac{N}{V} \simeq 2.7 \times 10^{25} \ \text{m}^{-3}. \qquad (1.17)$$

The microscopic description of the gas is based on the following assumptions:

- Molecules behave as hard spheres of diameter σ and mass m.

- The gas is sufficiently diluted to neglect the interactions between molecules, except during collisions, which are assumed elastic (kinetic energy is conserved).

- All molecules have the same constant velocity in magnitude, written as $\tilde{v}$.

These simplifying assumptions are not necessarily realistic (especially the third one), but they allow one to easily determine approximate expressions for pressure and temperature as a function of σ, m and $\tilde{v}$.

Pressure

The pressure exerted by a gas on its container walls is the force per unit area resulting from the numerous collisions between the molecules and the walls. Let us start by evaluating the number of collisions experienced by a wall of area $\mathcal{A} = L^2$ during a very short time Δt. The gas is considered sufficiently diluted so that all molecules have uniform rectilinear motion and do not experience any collision with each other. During Δt, a molecule hits the wall if it meets the following two conditions:

- Its velocity vector is oriented towards the wall, which is the case with a probability $p_{r_1} \simeq 1/6$, since space is isotropic (for simplicity, deviations from the 6 equiprobable directions of space are neglected).

- The molecule must be close enough to the wall to reach it during the time Δt. The distance from the wall must therefore be less than $\tilde{v}\Delta t$, inside the volume $\mathcal{A}\tilde{v}\Delta t$ shown in Figure 1.6. At equilibrium, with molecules uniformly distributed in the vessel, a molecule can be found in this region with the probability $p_{r_2} = \mathcal{A}\tilde{v}\Delta t/V$.

---

[38]This number, called the *Loschmidt constant*, was first estimated in 1865 by the physicist Johann Josef Loschmidt (1821–1895) using Maxwell's work on the kinetic theory of gases. He found 1.8×10^{24} molecules per m^{-3} and deduced the diameter of an air molecule: about 1 nm (*On the size of the air molecule*, Proceedings of the Academy of Science of Vienna **52**, 395 (1865)).

Since the orientation of the velocity vector is *independent* of the position of the molecule in the container, a molecule fulfils both these conditions with the probability $p_{r_1} p_{r_2}$. Out of the N gas molecules, there are therefore on average $\tilde{N}_{\text{coll}} = N p_{r_1} p_{r_2}$ molecules that will hit surface $\mathcal{A}$ during the time Δt, so

$$\tilde{N}_{\text{coll}} = \frac{1}{6} \rho \mathcal{A} \tilde{v} \Delta t, \tag{1.18}$$

where $\rho = N/V$ is the particle number density.

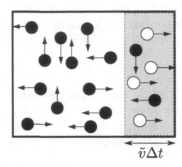

FIGURE 1.6   Only the white molecules, which are in the greyed area of width $\tilde{v} \Delta t$ and whose velocity vector is oriented to the right, will collide with the right wall during the time Δt.

The total force F exerted by these $\tilde{N}_{\text{coll}}$ molecules on the wall during Δt is given by the fundamental principle of dynamics projected on the axis perpendicular to the wall:

$$F = \frac{\mathrm{d}p}{\mathrm{d}t} \simeq \tilde{N}_{\text{coll}} \frac{\Delta p}{\Delta t},$$

where Δp is the change in wall momentum after a single collision. Consider an incident molecule, whose momentum before the collision is $p = m\tilde{v} > 0$. Since collisions are assumed elastic, it bounces off the wall with a momentum $p = -m\tilde{v}$. The change in momentum of the molecule is therefore $-2m\tilde{v}$. The momentum of the system "molecule + wall" being conserved during the shock, the variation of momentum of the wall is: $\Delta p = 2m\tilde{v}$. The pressure exerted by the gas on the wall of area $\mathcal{A}$ is therefore:

$$P = \frac{F}{\mathcal{A}} = \tilde{N}_{\text{coll}} \frac{\Delta p}{\mathcal{A} \Delta t} = \tilde{N}_{\text{coll}} \frac{2m\tilde{v}}{\mathcal{A} \Delta t},$$

or according to Equation (1.18):

$$P = \frac{1}{3} \rho m \tilde{v}^2. \tag{1.19}$$

This equation, obtained by Clausius in 1857, is historically the first relationship between thermodynamic quantities (pressure and mass density $\rho_{\text{m}} = \rho m$) and a microscopic quantity (the velocity $\tilde{v}$ of the gas molecules).

Temperature

Using the ideal gas equation of state (1.16) and Equation (1.19) for pressure, temperature can be written as a function of velocity $\tilde{v}$:

$$T = \frac{N_{\text{A}}}{3R} m \tilde{v}^2 = \frac{M}{3R} \tilde{v}^2, \tag{1.20}$$

where $M = N_{\text{A}} m$ is the gas molar mass. As Clausius showed in 1857, temperature is thus a macroscopic measurement of the thermal agitation of the gas molecules expressed through their velocity $\tilde{v}$. In Chapter 4, it will be shown rigorously using the formalism of statistical physics that relation (1.19) is true in the ideal gas case, provided that the characteristic velocity $\tilde{v}$ is properly defined (see Note 17 in Chapter 4). It will be seen in Section 4.1.5 that Equation (1.20) is even correct in general, when interactions between molecules are taken into account (see Note 22 in Chapter 4).

As the gas is diluted, its potential interaction energy can be neglected in front of its kinetic energy. The total energy E of the gas is therefore equal to the sum of the kinetic energies of each molecule, that is according to Equation (1.20)

$$E = \sum_{j=1}^{N_A} \frac{1}{2} m \tilde{v}^2 = \frac{1}{2} M \tilde{v}^2 = \frac{3}{2} RT, \qquad (1.21)$$

for one mole of gas. Thus, the formula giving the internal energy of an ideal gas as known from thermodynamics is recovered.

These early results were rediscovered by Maxwell in 1860 in a richer and more promising theoretical framework, based on probability theory, which was not very well developed at that time. Particles are no longer characterised by a single characteristic velocity, but by a velocity distribution function $P(v)$, the expression of which was determined by Maxwell using simple symmetry arguments (see exercise 1.3). This demonstration was done by Maxwell in 1867 when he treated collisions using a hypothesis known as "molecular chaos": the velocities of two colliding molecules are independent of each other. In 1868, Boltzmann extended this approach in a more rigorous way by generalising it, laying the formal foundation of statistical physics[39].

Estimation of microscopic quantities

The evaluation of microscopic quantities characterising the gas molecules is now possible. Knowing the value of Avogadro's number (see Equation (1.13)), the mass of a molecule is derived from the molar mass. Foe example:

$$m = \frac{M}{N_A} \simeq 3 \times 10^{-26} \text{ kg},$$

for a water molecule ($M = 18$ g · mol^{-1}). The diameter σ of the molecules can be estimated using the density of a liquid in which all molecules are assumed to be in contact, each occupy a volume of the order of σ^3. For one mole, one gets:

$$\rho_m = \frac{M}{N_A \sigma^3} \quad \text{i.e.} \quad \sigma = \left(\frac{M}{\rho_m N_A} \right)^{\frac{1}{3}} \simeq 3 \text{ Å},$$

where $\rho_m = \rho m$ is the mass density ($\rho_m \simeq 1$ g.cm^{-3} for liquid water at room temperature). The typical velocity of the molecules is according to Equation (1.20),

$$\tilde{v} = \sqrt{\frac{3RT}{M}} \simeq 600 \text{ m} \cdot \text{s}^{-1},$$

or ~2100 km.h^{-1}, for a water molecule under standard conditions of temperature and pressure. This gives the characteristic time for atom motion between two collisions in a fluid $\tau \sim \sigma / \tilde{v} \simeq 10^{-12}$ s.

---

[39]Boltzmann extended his theory to study the approach to equilibrium and in 1872 established an integral-differential equation, the so-called *Boltzmann equation*, which, in principle, allows the calculation of the evolution of $P(v)$ as a function of time in a dilute non-equilibrium gas. The description of this work, which is the basis of non-equilibrium statistical physics, is largely beyond the scope of this book. Please refer to the book by M. Le Bellac, F. Mortessagne, G. Batrouni, *Equilibrium and Non-Equilibrium Statistical Thermodynamics* (Cambridge University Press, 2010).

Throughout its trajectory, a molecule undergoes numerous collisions with other molecules. To calculate approximately the number of collisions per unit of time, one can assume that despite collisions, it remains stationary for a very short time Δt. First consider the incident molecules moving towards the target molecule along a given direction of space, on the axis passing through its centre (see Figure 1.7). It is further assumed that the incident molecules do not collide with each other. During the time Δt, the stationary molecule is hit by molecules whose centre is located in the cylinder of length $\tilde{v}\Delta t$ and radius σ (σ is also the diameter of the molecules) as shown in Figure 1.7. One therefore just needs to count the number of molecules that during Δt hit the area $\mathcal{A} = \pi\sigma^2$, called the *cross-section*. According to Equation (1.18), and taking into account molecules coming from the 6 directions of space, the number of collisions undergone by a molecule per unit of time is thus of the order of:

$$6\frac{\tilde{N}_{\text{coll}}}{\Delta t} = \rho\pi\sigma^2\tilde{v} \simeq 10^9 \text{ s}^{-1}. \tag{1.22}$$

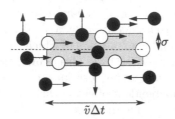

$$\tilde{v}\Delta t$$

FIGURE 1.7 Only white molecules, whose velocity vector is oriented to the right and whose centres are contained in the grey cylinder of length $\tilde{v}\Delta t$ and radius σ, will collide with the target molecule (hatched) from the left during the time Δt.

A molecule therefore experiences a collision every $\delta t = 1/(\rho\pi\sigma^2\tilde{v}) \simeq 10^{-9}$ seconds or so. The *mean free path*, that is the average distance l travelled by a molecule between two successive collisions, is of the order of the distance travelled during the time δt at a velocity $\tilde{v}$:

$$l \simeq \tilde{v}\delta t = \frac{1}{\rho\pi\sigma^2} \simeq 1000 \text{ Å}, \tag{1.23}$$

which is large compared to the average distance between two neighbouring molecules, estimated by $d = (V/N)^{1/3} = (1/\rho)^{1/3} \simeq 30$ Å (i.e. about 10σ), where V/N is the average volume per molecule. Indeed, at low density, the relative motion of two neighbour molecules tends to move them apart from each other, thus avoiding a collision (on the other hand at high density $l \sim d \sim \sigma$). The mean free path was introduced by Clausius (1858) to explain the low diffusion speed of a perfume in the air, observed at human scale, while that the speed of molecules between two collisions is of the order of 1000 km·h^{-1}: the trajectory of a molecule (of perfume) is not rectilinear, but is permanently deflected by collisions with other molecules, thus slowing down its progression in the medium (air).

Validity of the classical description

Since classical mechanics is an approximation of quantum mechanics, the use of classical physics is justified when the quantities of a system having the dimension of an action (i.e. ML^2T^{-1}, e.g. a length multiplied by a momentum) are large compared to Planck's constant, h. Using the average distance d between two molecules and Equation (1.20) for the characteristic velocity $\tilde{v}$, this condition is written as:

$$d\, m\tilde{v} \gg h, \quad \text{thus} \quad d \gg \frac{h}{m\tilde{v}} \simeq \frac{h}{\sqrt{m\frac{R}{N_A}T}}.$$

The right-hand side of the above inequality is the typical quantum wavelength associated with a molecule of mass m in a system at temperature T. Classical mechanics is therefore justified if

$$d \gg \lambda_T, \quad \text{where} \quad \lambda_T \equiv \frac{h}{\sqrt{2\pi m\frac{R}{N_A}T}} \tag{1.24}$$

is, by definition, the *de Broglie thermal wavelength*. The distance d is estimated by taking the cubic root of the volume per molecule $V/N = 1/\rho$, that is $d = (1/\rho)^{1/3}$. The above inequality becomes:

$$\rho\, \lambda_T^3 \ll 1. \tag{1.25}$$

The classical approximation is better the lower the density ρ (or pressure P), the higher the temperature T and the larger the molecules mass m. Under standard conditions of temperature and pressure, according to (1.17), $d \simeq 30$ Å, for all gases assumed ideal. This distance is much larger than $\lambda_T \simeq 0.7$ Å, calculated for the lightest of them, di-hydrogen ($M_{H_2} = 2$ g·mol^{-1}). Therefore, gases at room temperature and atmospheric pressure can be studied using classical physics[40].

Sensitivity to initial conditions

Finally, the chaotic behaviour of an atomic system is illustrated by considering successive collisions of a molecule using a very simple model[41]. Figure 1.8 represents the collision of a spherical molecule moving with a velocity $\tilde{v}$ towards another stationary molecule (for simplicity). Initially, the first molecule has just collided and is at a distance from the second molecule typically equal to the mean free path l. At the instant of the collision, the (contact) distance between the two molecules is equal to their diameter σ. After the (assumed) elastic collisions, the trajectory of the incident molecule is deviated by an angle $\pi - 2\alpha$ from its initial direction (see Figure 1.8).

In the right-angle triangles ABN and $A'BN$, $BN = l\sin\theta = \sigma\sin\alpha$. From Equation (1.23), $l \gg \sigma$, so the angle θ is small. Then,

$$\sin\alpha = \frac{l}{\sigma}\sin\theta \simeq \frac{l}{\sigma}\theta, \quad \text{or by differentiating} \quad \delta\alpha = \frac{l}{\sigma}\frac{1}{|\cos\alpha|}\delta\theta.$$

If the incident angle θ is known with an uncertainty $\delta\theta$, the uncertainty on the direction $\theta' = \theta + \pi - 2\alpha$ taken by the incident molecule after the collision is

$$\delta\theta' = \delta\theta + 2\delta\alpha = \left(1 + \frac{l}{\sigma}\frac{2}{|\cos\alpha|}\right)\delta\theta \geq \left(1 + 2\frac{l}{\sigma}\right)\delta\theta \simeq \frac{2l}{\sigma}\,\delta\theta,$$

since $|\cos\alpha| \leq 1$ and $l \gg \sigma$. Thus, the uncertainty on the direction is multiplied by $2l/\sigma$ after each collision. After n collisions, the uncertainty $\delta\theta^{(n)}$ on the direction of the molecule is:

$$\delta\theta^{(n)} \simeq \left(\frac{2l}{\sigma}\right)^n \delta\theta, \quad \text{with} \quad \frac{l}{\sigma} \simeq 300. \tag{1.26}$$

---

[40] As temperature decreases, λ_T increases. Nevertheless, the classical approximation remains in general quite accurate for ordinary gases. A notable exception is helium, whose liquefaction temperature is extremely low: $T = 4.2$ K (for He^4 at $P = 1$ atm). At this temperature, $d \sim \lambda_T \simeq 4$ Å and quantum effects are important (see Chapters 7 and 8).

[41] This section repeats a calculation presented in Supplement IV.B of B. Diu, C. Guthmann, D. Lederer and B. Roulet *Physique Statistique* (Hermann, 1989, in French).

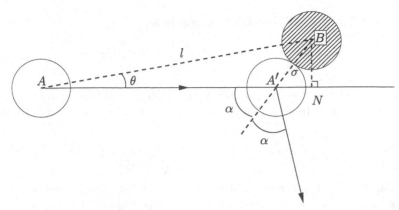

FIGURE 1.8 Collision between two molecules: the first one (white) is in A at the initial time and in A' at the time of collision with the second molecule (hatched) stationary at point B (N is the orthogonal projection of B on the AA' axis).

The smallest inaccuracy $\delta\theta$ is therefore amplified and grows exponentially with n. This inaccuracy may be due to a measurement made during the experiment (or due to numerical rounding in a molecular dynamics simulation) or to an external perturbation. For example, even for a system assumed isolated, the effect of external gravitational fields on the molecules cannot be removed. The inaccuracy due to the influence of the mass of an electron at the other end of the Universe is extremely small[42], $\delta\theta \simeq 10^{-138}$ radians, but completely modifies the trajectory of a molecule in the gas when $\delta\theta^{(n)} = \pi$. According to Equations (1.26) and (1.22), it occurs after only:

$$ n \simeq \frac{\ln\left(\delta\theta^{(n)}/\delta\theta\right)}{\ln(2l/\sigma)} \simeq 50 \quad \text{collisions, or a few tens of nanoseconds!} $$

Although the motion of a molecule is deterministic, any prediction of its trajectory is impossible after only a few nanoseconds when taking into account the numerous perturbations[43].

---

[42] An electron of mass $m' \simeq 10^{-30}$ is at distance $r \simeq 10^{26}$ m from the system. The gravitational interaction energy ($|e_{\text{pot}}| = Gmm'/r \simeq 10^{-92}$ J, where $G \simeq 6.7 \times 10^{-11}$ m^3.kg^{-1}.s^{-2} is the gravitational constant) between this electron and a molecule of mass m in the system is extremely small compared to the kinetic energy of the molecule ($e_{\text{kin}} = m\tilde{v}^2/2 \simeq 10^{-21}$ J). The perturbing electron creates in first approximation a uniform gravity field g since r is much larger than atomic lengths. The parabolic deviation followed by the molecule with velocity $\tilde{v}$ is given by

$$ \delta y = \frac{g}{2}\left(\frac{x}{\tilde{v}}\right)^2, \quad \text{where} \quad g = \frac{Gm'}{r^2}. $$

Between two collisions, after a distance $x \simeq l$, the angle of deflection due to the gravity field g is

$$ \theta(r) \simeq \frac{\delta y}{x} = \frac{Gm'l}{2r^2\tilde{v}^2} \simeq 10^{-105} \text{ rad} \ldots $$

A very small uncertainty, say $\delta r \simeq l$, in the distance r implies an error in the angle θ:

$$ \delta\theta = \left|\frac{d\theta}{dr}\right| \delta r \simeq \frac{Gm'l^2}{r^3\tilde{v}^2} \simeq 10^{-138} \text{ rad}. $$

[43] The estimate of such a perturbation is inspired by an example given by the French mathematician Emile Borel (1871–1956) in 1914 to justify the introduction of statistical mechanics (*Introduction géométrique à quelques théories physiques*, page 98 (Gauthiers-Villars, Paris, in French) (1914)). It is instructive to observe the chaotic behaviour of interacting particles using molecular dynamics software: enter "applet molecular dynamics" in an internet search engine.

1.5 STATISTICAL DESCRIPTION OF A SYSTEM: STATISTICAL ENSEMBLES

The previous section provides a succinct presentation of the spirit of statistical physics. It is now a question of generalising the probabilistic approach in a framework formalised by Gibbs[44] in 1902, in which the time averages of internal variables are replaced by statistical averages, called *ensemble averages*. The kinetic theory of gases and the first approaches of Maxwell and Boltzmann were rooted in the mechanics and focused on treating the individual trajectories of particles, in particular collisions, at the cost of simplifying hypotheses (hard spheres, diluted gas). As early as 1868, Boltzmann applied probability theory to every particle of a system as a whole, described in a very general way using Hamiltonian formalism. He defined the probability of a given microstate as the fraction of time that the system spends in this microstate during its evolution. Later, in 1871, he introduced the concept of statistical ensemble, further developed by Gibbs in 1902, which offers a more abstract conceptual framework, but also drops the use of mechanics for a reasoning over the entire system and no longer on individual particle behaviour. In the 1920s, this generalisation enabled the extension of the scope of statistical physics to systems described by quantum physics.

FIGURE 1.9 The three founding fathers of statistical physics photographed at the age of nearly 24: James Clerk Maxwell in 1855 (left, `https://en.wikipedia.org/wiki/James_Clerk_Maxwell`), Ludwig Boltzmann in 1868 (centre, `https://fr.wikipedia.org/wiki/Ludwig_Boltzmann`) and Josiah Willard Gibbs between 1863 and 1866 (right, `https://en.wikipedia.org/wiki/Josiah_Willard_Gibbs`).

1.5.1 STATISTICAL ENSEMBLES

Consider a system prepared under experimental conditions characterised by given external parameters. Once the system has reached equilibrium, an internal variable A fluctuates around its time-averaged value $\bar{A}$, equal to the equilibrium value A_{eq}. Let A_m be its value at time $t = m\delta t$, where m is an integer and δt a fixed time interval. The time average (Equation 1.14) can be evaluated with the discrete sum[45]:

$$A_{eq} = \frac{1}{M} \sum_{m=1}^{M} A_m \quad \text{with} \quad M \gg 1. \tag{1.27}$$

---

[44] Josiah Willard Gibbs (1839–1903), American physicist, father of statistical physics alongside Maxwell and Boltzmann, whose results he synthesised and systematised. In 1902, he published the first treatise on statistical physics, *Elementary principles of statistical mechanics* (Dover, 2015). He invented the terminology and the name of this new discipline, statistical mechanics.

[45] This is how thermodynamic quantities are calculated in molecular dynamics simulations.

Note that the value of this sum does not depend on the ordering of the microstates resulting from the time evolution of the system. To estimate A_{eq}, the number of microstates corresponding to the equilibrium macrostate just needs to be large enough (i.e. $M \gg 1$).

From a different point of view, it is possible to interpret Equation (1.27), not as a *time average* carried out on a *unique* evolving system (m indicating a given instant), but as an *ensemble average*, carried out on $M \gg 1$ identical systems at a given time instant, in *a priori* different microstates independent of each other (m refers then to a given microstate)[46]. This imaginary ensemble of identical systems, that is characterised by the same Hamiltonian and the same external parameters, is called *statistical ensemble* or *Gibbs ensemble*[47].

Under this formalism, particle dynamics and time are no longer involved. An internal variable A, whose value is, in principle, *determined* by the equations of motion, is treated as a *random variable* characterised by a *probability distribution* $P(A)$. Its time-averaged value $\bar{A}$, defined by Equation (1.14), is then replaced by a *statistical mean value*, called the *ensemble average*, noted $\langle A \rangle$ to distinguish these two averages of very different natures. The probabilistic approach is therefore based on the following hypothesis, which can be considered as the first postulate of equilibrium statistical physics:

$$\boxed{\bar{A} = \langle A \rangle \quad \text{at equilibrium.}} \tag{1.28}$$

There is no formal proof of this equality, which is the cornerstone of statistical physics. Nevertheless, it will be discussed with theoretical arguments in Section 2.4. To conclude, it is the excellent predictions of statistical physics, compared to experimental data, that validate this hypothesis, as it will be seen in the following.

1.5.2 PROBABILITY OF A MICROSTATE AND ENSEMBLE AVERAGES

In order to implement the probabilistic approach, one needs to express the ensemble averages of seemingly accessible microstates, that is compatible with the external parameters. Consider a statistical ensemble defined by the external parameters of the studied system. Depending on whether the description of the microstates is quantum or classical, the used probability theory is discrete or continuous. Since the probabilities associated with discrete events are simpler to conceive – thanks to our long experience with gambling – the discrete description will be described first.

In the framework of quantum physics, among the M replicas of the system, M_m are in a given microstate m. The number

$$p_m = \frac{M_m}{M} \quad \text{for} \quad M \gg 1,$$

is interpreted as the probability of the system to be found in microstate m. The considered statistical ensemble is therefore characterised by a *probability distribution* of the microstates p_m which depends on the external parameters and satisfies the normalisation condition:

$$\sum_m p_m = 1, \tag{1.29}$$

where the sum is over all accessible microstates. The ensemble average of the internal variable A is given by:

$$\langle A \rangle = \sum_m A_m \, p_m, \tag{1.30}$$

---

[46] One can think of the following two experiments: on one hand, a single die is thrown $M \gg 1$ time in a row while recording the obtained number for each throw. On the other hand, one throws (only once) $M \gg 1$ identical dice. In principle, the probability of obtaining a given number are the same in both experiments.

[47] In 1884, Boltzmann introduced the notion of statistical ensemble, called *ergode* and *holode*. In 1902, Gibbs named these ensembles *microcanonical* and *canonical* respectively.

where the sum is taken over all accessible microstates and A_m is the value of the internal variable in the microstate m. Like p_m, the average value of an internal variable therefore depends on the external parameters. In general, the amplitude of the fluctuations of A around its mean value is given by its variance Δ_A^2 defined as:

$$\Delta_A^2 = \langle (A - \langle A \rangle)^2 \rangle = \langle A^2 \rangle - \langle A \rangle^2 \geq 0, \tag{1.31}$$

where Δ_A, the standard deviation of the distribution, has the dimension of the internal variable A. Finally, the internal variable is completely characterised by its probability distribution $P(A)$, the probability that its value is equal to A:

$$P(A) = \sum_{\{m|A_m=A\}} p_m, \tag{1.32}$$

where the sum is performed on all microstates for which the internal variable is equal to A. The average value (and variance) of A can also be calculated using $P(A)$:

$$\langle A \rangle = \sum_A A \, P(A),$$

where the sum is over all possible values of the internal variable.

When a system is studied within the framework of classical physics, its microscopic variables vary continuously and the statistical ensemble is characterised by a probability density function $\rho(\mathbf{q}, \mathbf{p})$ such that the probability dP to find the system in the microstate $(\mathbf{q}, \mathbf{p})$, to within $d^f q$ and $d^f p$, is given by:

$$dP = \rho(\boldsymbol{q}, \boldsymbol{p}) \, d^f q \, d^f p = \rho(q_1, q_2, \ldots, q_f, p_1, p_2, \ldots, p_f) \prod_{i=1}^{f} dq_i dp_i,$$

where the ith coordinate and momentum are between q_i and $q_i + dq_i$ and between p_i and $p_i + dp_i$ respectively. The normalisation of the probability density function implies:

$$\int \rho(\boldsymbol{q}, \boldsymbol{p}) \, d^f q \, d^f p = 1 \tag{1.33}$$

and the ensemble average of the internal variable A is given by:

$$\langle A \rangle = \int A(\boldsymbol{q}, \boldsymbol{p}) \rho(\boldsymbol{q}, \boldsymbol{p}) \, d^f q \, d^f p, \tag{1.34}$$

where in both equations the bounds of the integral correspond to the accessible domain of phase space. The variance and standard deviation are defined by Equation (1.31) and the probability density function $f(A)$, which gives the probability $P(A)$ that the internal variable lies between A and $A + dA$, is by definition:

$$P(A) = f(A)dA = \int_{[A, A+dA]} \rho(\boldsymbol{q}, \boldsymbol{p}) \, d^f q \, d^f p, \tag{1.35}$$

where the bounds of the integral correspond to the domain of the phase space such that the value of the internal variable lies between A and $A + dA$. Note that since a probability is dimensionless, the probability density f has the inverse dimension of the quantity A.

To establish the transition from microscopic to macroscopic, a fundamental task remains: determine, for a given statistical ensemble, the probability distribution (or density function) of the microstates p_m (or $\rho(\mathbf{q}, \mathbf{p})$) that is involved in the definition of the ensemble average of an internal variable, Equation (1.30) (or Equation (1.34)).

1.6 EXERCISES

Exercise 1.1. Kelvin's glass of water

Lord Kelvin (1824–1907) illustrated the huge number of atoms making up matter around us with the following example[48]. Suppose you could mark water molecules in a glass, pour them into the sea and wait long enough for the marked molecules to spread evenly in the world's oceans. Dip the glass into the sea, how many labelled molecules will it contain?

1. First, what is the number N_v of labelled molecules contained in a glass of volume $v = 20$ cl? The mass density density of water is $\rho_m = 1$ g $\cdot$ cm^{-3} and the molar mass of water is $M = 18$ g $\cdot$ mol^{-1}.

2. The oceans cover 71% of the Earth's surface (radius $R = 6400$ km) and their average depth is $h = 3.8$ km. What is the total volume V of the oceans?

3. Finally, deduce the number n of labelled molecules found in the glass after it has been dipped into the sea.

Solution:

1. $N_v = \dfrac{m}{M} N_A = \dfrac{\rho_m v}{M} N_A \simeq 6.7 \times 10^{24}$.

2. $V = 0.71(4\pi R^2)h \simeq 1.4 \times 10^{21}$ *litres.*

3. $n = \dfrac{v}{V} N_v \simeq 1000$ *labelled molecules! Although the volume of the oceans is enormous (compared to glass), it is the huge value of Avogadro's number that gives this surprising result.*

Exercise 1.2. Density fluctuations

A container is divided into two compartments $\mathcal{A}$ and $\mathcal{A}'$ of volumes V and V', which exchange particles (see Figure 1.5). It contains $N = N_A$ molecules uniformly distributed on average. The particles are assumed statistically independent.

1. What is the probability $P_N(n)$ of finding exactly n molecules in compartment $\mathcal{A}$? Check the normalisation of $P_N(n)$.

2. Calculate the average number $\langle n \rangle$ and the variance Δ_n^2 of the distribution $P_N(n)$.

3. To describe small fluctuations around the average value, the variable $s = n - \langle n \rangle$ is introduced. Write the distribution $Q_N(s)$ of variable s as a function of s/N using Stirling's formula given by Equation (A.20) in Appendix A.3. Since $s \ll \langle n \rangle$, treat s as a continuous variable and show that $Q_N(s)$ is a Gaussian distribution. Is this result surprising?

Solution:

1. *Since molecules are uniformly distributed, a given molecule has a probability $p = V/(V + V')$ of being in compartment $\mathcal{A}$ and $q = 1 - p$ of not being there. Molecules are assumed to be statistically independent:*

$$P_N(n) = \binom{N}{n} \underbrace{ppp \dots p}_{n \ times} \underbrace{qqq \dots q}_{N-n \ times} = \binom{N}{n} p^n q^{N-n},$$

---

[48]Based on Erwin Schrödinger's book, *What is life?: The physical aspect of the living cell* (Cambridge University Press, reprint edition 2012).

where $\binom{N}{n}$ is the binomial coefficient, that is the number of ways to choose n molecules among N:

$$\binom{N}{n} = N(N-1)\ldots(N-n+1).\frac{1}{n!} = \frac{N!}{n!\,(N-n)!}.$$

In fact, there are N possibilities for choosing the first molecule, $N-1$ for the second, and $N-(n-1)$ for the last. To disregard the order in which the selected molecules are chosen, one must finally divide by $n!$, the number of permutations between n molecules. This results is the binomial distribution. According to Newton's binomial formula:

$$\sum_{n=0}^{N} P_N(n) = \sum_{n=0}^{N} \binom{N}{n} p^n (1-p)^{N-n} = (p+1-p)^N = 1.$$

In general, it is strongly recommended to start a probability exercise with simple special cases. For example, consider here the case with $N = 4$ and list the 6 configurations for which $n = 2$ molecules are in compartment $\mathcal{A}$.

2. The mean value and variance can be calculated by treating n as a sum of independent random variables c_j (see Section 1.4.1) or use the following trick: treat p and q as two independent variables and at the end of the calculation use $p + q = 1$. Thus:

$$\langle n \rangle = \sum_{n=0}^{N} n\,P_N(n) = \sum_{n=0}^{N} \binom{N}{n} n\, p^n q^{N-n} = p\frac{\partial}{\partial p}\Big(\sum_{n=0}^{N}\binom{N}{n}p^n q^{N-n}\Big)$$

$$= p\frac{\partial}{\partial p}(p+q)^N = Np\underbrace{(p+q)^{N-1}}_{1} = Np.$$

Similarly,

$$\langle n^2 \rangle = \sum_{n=0}^{N}\binom{N}{n}n^2\,p^n q^{N-n} = p\frac{\partial}{\partial p}\Big(p\frac{\partial}{\partial p}\sum_{n=0}^{N}\binom{N}{n}p^n q^{N-n}\Big)$$

$$= p\frac{\partial}{\partial p}\Big(pN(p+q)^{N-1}\Big) = p\Big(N(p+q)^{N-1} + pN(N-1)(p+q)^{N-2}\Big)$$

$$= pN(1+p(N-1)) = pN(pN+q).$$

The variance is given by $\Delta_n^2 = \langle n^2 \rangle - \langle n \rangle^2 = pN(pN+q) - (pN)^2 = Npq.$

3. Trivially[49] $\langle s \rangle = 0$ and $\Delta_s^2 = \Delta_n^2 = Npq$. With the change of variable $n = Np + s$, one gets:

$$Q_N(s) = P_N(Np + s) = \frac{N!}{(Np+s)!(Nq-s)!}p^{Np+s}q^{Nq-s}.$$

Using Stirling's formula (A.20):

$$Q_N(s) \simeq \frac{N^N\,p^{Np+s}q^{Nq-s}}{(Np+s)^{Np+s}(Nq-s)^{Nq-s}}\sqrt{\frac{N}{2\pi(Np+s)(Nq-s)}},$$

so

$$Q_N(s) \simeq \frac{1}{\left(1+\frac{s}{Np}\right)^{Np+s}\left(1-\frac{s}{Nq}\right)^{Nq-s}}\frac{1}{\sqrt{2\pi Npq\left(1+\frac{s}{Np}\right)\left(1-\frac{s}{Nq}\right)}}.$$

---

[49] Given a random variable X, it is easy to show that for two real numbers a and b

$$\langle aX + b \rangle = a\langle X \rangle + b \quad \text{et} \quad \Delta_{aX+b}^2 = a^2\Delta_X^2.$$

Under the square root, one can ignore s/Np and s/Nq since these terms are much smaller than 1. The factor in front of the square root contains terms of the form $(1 + \epsilon)^M$. However, one cannot use a Taylor expansion, $(1 + \epsilon)^M \simeq 1 + M\epsilon$ for $\epsilon \to 0$, since $M \to \infty$. Therefore, $\epsilon M \sim s$ can not be neglected in front of 1. Taking the logarithm, one gets:

$$\ln Q_N(s) \simeq -(Np + s)\ln\left(1 + \frac{s}{Np}\right) - (Nq - s)\ln\left(1 - \frac{s}{Nq}\right) + \ln\frac{1}{\sqrt{2\pi Npq}}.$$

Using the second-order Taylor expansion of $\ln(1 + \epsilon) \simeq \epsilon - \frac{\epsilon^2}{2}$, one obtains:

$$\ln Q_N(s) \simeq -(Np + s)\left(\frac{s}{Np} - \frac{1}{2}\left(\frac{s}{Np}\right)^2\right) - (Nq - s)\left(-\frac{s}{Nq} - \frac{1}{2}\left(\frac{s}{Nq}\right)^2\right) + \ln\frac{1}{\sqrt{2\pi Npq}},$$

so

$$\ln Q_N(s) \simeq -s - \frac{s^2}{2Np} + s - \frac{s^2}{2Nq} + \ln\frac{1}{\sqrt{2\pi Npq}} = -\frac{s^2}{2Npq} + \ln\frac{1}{\sqrt{2\pi Npq}}.$$

Thus,

$$Q_N(s) \simeq \frac{1}{\sqrt{2\pi Npq}}\, e^{-\frac{s^2}{2Npq}}.$$

One finds a (normalised) Gaussian distribution, centred (i.e. with mean $\langle s \rangle = 0$) and variance $\Delta_s^2 = Npq$. This result is a direct application of the central limit theorem. Indeed, according to (1.15) the random variable n (and thus s) being written as the sum of a large number ($N \gg 1$) of independent random variables with finite variance, its probability distribution function tends toward a Gaussian distribution function. The obtained function is extremely peaked, since the fluctuations ($\Delta_s = \sqrt{Npq}$) with respect to the average number of molecules ($\langle n \rangle = Np$) in compartment $\mathcal{A}$ are very small, $\sim 1/\sqrt{N} \sim 10^{-12}$ in relative value.

Exercise 1.3. The Maxwell velocity distribution

To go beyond the first results found by Clausius (see Section 1.4.2), Maxwell determined in 1860 the velocity distribution of gas molecules using very simple assumptions, introducing for the first time in a systematic way the use of probabilities in physics. Since space is isotropic, the velocity $\vec{v}$ components of a molecule have the same probability density function $F(v_\alpha)$, where $\alpha = x, y, z$ (assumption 1).

1. Assuming that the three components of velocity are independent random variables (assumption 2), express as a function of F the probability $P(v_x, v_y, v_z)\, dv_x\, dv_y\, dv_z$ that the three components of velocity lie between v_x and $v_x + dv_x$, v_y and $v_y + dv_y$ and v_z and $v_z + dv_z$. Why is this assumption objectionable?

2. According to Maxwell, after a collision "all scattering directions are equally likely", such that $P(v_x, v_y, v_z)$ depends only on the magnitude of the velocity, or its square $v^2 = v_x^2 + v_y^2 + v_z^2$, and not on the direction of $\vec{v}$ (assumption 3). Deduce that

$$P(v_x, v_y, v_z) = A\, e^{-b(v_x^2 + v_y^2 + v_z^2)} \quad \text{where} \quad A, b > 0. \tag{1.36}$$

3. Express A as a function of b using Gaussian integrals (see Appendix A.2).

4. What is $\langle v_x \rangle$? Calculate $\langle v^2 \rangle$. Derive b as a function of the gas temperature T, assuming that the typical velocity in Equation (1.20) is such that $\tilde{v}^2 = \langle v^2 \rangle$.

Solution:

1. *The three random variables v_x, v_y and v_z being independent:*

$$P(v_x, v_y, v_z)\mathrm{d}v_x\mathrm{d}v_y\mathrm{d}v_z = F(v_x)\mathrm{d}v_x\, F(v_y)\mathrm{d}v_y\, F(v_z)\mathrm{d}v_z.$$

However, after a collision, the three components of the velocity of a particle are clearly not independent. In a Newtonian logic, Maxwell considered hypothesis 2 as "precarious" in his 1867 article. It makes sense in probabilistic terms: the knowledge of one component of the velocity vector of a randomly chosen particle tells nothing a priori *about its other two components.*

2. *Therefore:*

$$P(v_x, v_y, v_z) = P(v_x^2 + v_y^2 + v_z^2) = F(v_x)F(v_y)F(v_z) \quad \text{for all } v_x,\ v_y \text{ and } v_z.$$

In the case of $v_y = v_z = 0$, $P(v_x^2) = \big(F(0)\big)^2 F(v_x)$ and similarly for $P(v_y^2)$ and $P(v_z^2)$ so:

$$P(v_x^2 + v_y^2 + v_z^2) = \frac{1}{\big(F(0)\big)^6} P(v_x^2)P(v_y^2)P(v_z^2).$$

Only the exponential function satisfies this functional equation. This gives Equation (1.36). A normalised probability density P which does not diverge when $v_\alpha \to \pm\infty$ implies $b > 0$. Similarly, the normalisation constant A is strictly positive.

3. *The normalisation of P implies $1 = \displaystyle\int_{-\infty}^{+\infty} \mathrm{d}v_x \int_{-\infty}^{+\infty} \mathrm{d}v_y \int_{-\infty}^{+\infty} \mathrm{d}v_z\, P(v_x, v_y, v_z)$, thus*

$$1 = A \int_{-\infty}^{+\infty} \mathrm{d}v_x\, \mathrm{e}^{-bv_x^2} \int_{-\infty}^{+\infty} \mathrm{d}v_y\, \mathrm{e}^{-bv_y^2} \int_{-\infty}^{+\infty} \mathrm{d}v_z\, \mathrm{e}^{-bv_z^2} = A\bigg(\underbrace{\int_{-\infty}^{+\infty} \mathrm{d}v_x\, \mathrm{e}^{-bv_x^2}}_{\sqrt{\frac{\pi}{b}}}\bigg)^3,$$

from Equation (A.15) giving the integral $I_0(\alpha)$ in Appendix A.2. Thus $A = (b/\pi)^{3/2}$. The probability density P therefore depends only on the parameter b:

$$P(v_x, v_y, v_z) = \left(\frac{b}{\pi}\right)^{\frac{3}{2}} \mathrm{e}^{-b(v_x^2 + v_y^2 + v_z^2)} \quad \text{and} \quad F(v_x) = \sqrt{\frac{b}{\pi}}\, \mathrm{e}^{-bv_x^2}.$$

4. *Since space is isotropic:*

$$\langle v_x\rangle = \langle v_y\rangle = \langle v_z\rangle = \int_{-\infty}^{+\infty} \mathrm{d}v_x\, v_x F(v_x) = \sqrt{\frac{b}{\pi}} \int_{-\infty}^{+\infty} \mathrm{d}v_x\, v_x \mathrm{e}^{-bv_x^2} = 0.$$

In addition $\langle v^2\rangle = \langle v_x^2\rangle + \langle v_y^2\rangle + \langle v_z^2\rangle = 3\langle v_x^2\rangle$ (isotropic space), therefore:

$$\langle v^2\rangle = 3 \int_{-\infty}^{+\infty} \mathrm{d}v_x\, v_x^2 F(v_x) = 3\sqrt{\frac{b}{\pi}} \underbrace{\int_{-\infty}^{+\infty} \mathrm{d}v_x\, v_x^2 \mathrm{e}^{-bv_x^2}}_{\frac{\sqrt{\pi}}{2b^{\frac{3}{2}}}} = \frac{3}{2b},$$

from Equation (A.15) for integral $I_2(\alpha)$ in Appendix A.2. By identifying in Equation (1.20), $\bar{v}^2$ and $\langle v^2\rangle$ (which will be demonstrated in Section 4.1.5) one therefore gets:

$$b = \frac{M}{2RT}, \quad thus \quad P(v_x, v_y, v_z) = \left(\frac{M}{2\pi RT}\right)^{\frac{3}{2}} \mathrm{e}^{-\frac{M}{2RT}v^2}. \tag{1.37}$$

Exercise 1.4. Distribution of the separation length between two neighbour particles

An ideal gas composed of $N \gg 1$ particles is confined in a container of volume V. Particles' positions are statistically independent from each other. In a given microstate, a particle, named j_0, close to the centre of the container is chosen.

1. Give the order of magnitude of the distance d between a particle and its closest neighbour as a function of the particle number density $\rho = N/V$.

2. What is the probability $P_1(r)\mathrm{d}r$ that a given particle is at a distance between r and $r + \mathrm{d}r$ from particle j_0?

3. What is the probability $Pr_{N-2}(r)$ that the other $N - 2$ particles are at a distance larger than r of particle j_0? One writes v for the volume of the sphere of radius r centred on particle j_0.

4. Deduce the probability $P(r)\mathrm{d}r$ that j_0's closest neighbour is at a distance between r and $r + \mathrm{d}r$. Express the probability density function $P(r)$ as a function of ρ for $N \gg 1$ and $V \gg v$. Check that $P(r)$ is indeed normalised.

5. Express the mean distance $\langle r \rangle$ between particle j_0 and its closest neighbour with the help of the Γ-function, $\Gamma\left(\dfrac{4}{3}\right) = \displaystyle\int_0^\infty \mathrm{d}t\, t^{\frac{1}{3}}\,\mathrm{e}^{-t} \simeq 0.89$ (see Appendix A.1). Compare the distance $\langle r \rangle$ to the order of magnitude d estimated in question 1.

Solution:

1. _The mean volume per particle is $V/N = 1/\rho$. If it is assumed cubic for simplification, the cube's side, $d = (1/\rho)^{\frac{1}{3}}$, is an estimation of the mean distance between two neighbour particles._

2. _The position of a given particle ($j = 27$ for example) is between the two spheres of radii r and $r + \mathrm{d}r$ (region with a volume $4\pi r^2 \mathrm{d}r$) with the probability $P_1(r)\mathrm{d}r = 4\pi r^2 \mathrm{d}r/V$, because the particles are spatially uniformly distributed._

3. _The probability that a given particle ($j = 56$ for example) is outside of volume $v = \frac{4}{3}\pi r^3$ is $\frac{V-v}{V}$. For $N - 2$ independent particles: $Pr_{N-2}(r) = \left(1 - \dfrac{4\pi r^3}{3V}\right)^{N-2}$._

4. _One gets $P(r)\mathrm{d}r = (N-1)Pr_{N-2}(r + \mathrm{d}r)P_1(r)\mathrm{d}r = (N-1)\left(1 - \dfrac{4\pi(r + \mathrm{d}r)^3}{3V}\right)^{N-2}\dfrac{4\pi r^2 \mathrm{d}r}{V}$, because there are $N-1$ possibilities to choose the particle that will be the closest particle at a distance r within $\mathrm{d}r$, whereas the other $N-2$ particles are at a distance larger than $r + \mathrm{d}r$. At the first order, for $N \gg 1$ and $V \gg r^3$: $N - 1 \simeq N - 2 \simeq N$ and_

$$P(r)\mathrm{d}r \simeq 4\frac{N}{V}\pi r^2\left(1 - \frac{4\pi r^3}{3V}\right)^N \mathrm{d}r = 4\rho\pi r^2\, \mathrm{e}^{N\ln\left(1 - \frac{4\pi r^3}{3V}\right)}\mathrm{d}r \simeq 4\rho\pi r^2\, \mathrm{e}^{-\frac{4}{3}\rho\pi r^3}\,\mathrm{d}r,$$

where the series expansion $\ln(1 - x) \simeq -x$ for $x \to 0$ was used. Taking $t = 4\rho\pi r^3/3$, one gets indeed $\displaystyle\int_0^\infty \mathrm{d}r\, P(r) = \int_0^\infty \mathrm{d}t\, \mathrm{e}^{-t} = 1$, where the integral upper bound was extended to infinity since the exponential integrand decrease very rapidly with respect to the macroscopic size of the container.

5. _One gets $\langle r \rangle = \displaystyle\int_0^\infty \mathrm{d}r\, r P(r) = 4\rho\pi\int_0^\infty \mathrm{d}r\, r^3\, \mathrm{e}^{-\frac{4}{3}\rho\pi r^3} = \int_0^\infty \mathrm{d}t\left(\frac{3t}{4\rho\pi}\right)^{\frac{1}{3}}\mathrm{e}^{-t} = \left(\frac{3}{4\rho\pi}\right)^{\frac{1}{3}}\Gamma(\frac{4}{3})$, taking $t = 4\rho\pi r^3/3$, that is $\langle r \rangle \simeq 0.55\, d$._

2 Microcanonical Statistical Ensemble

"It can scarcely be denied that the supreme goal of all theory is to make the irreducible basic elements as simple and as few as possible without having to surrender the adequate representation of a single datum of experience."

A. Einstein[1], 1933

From a theoretical point of view, the simplest system is an isolated system since all interactions with the external environment are neglected. The energy E, the number of particles N and the volume V of the system are then fixed. With these chosen external parameters (N, V, E), the statistical ensemble, called *microcanonical ensemble*, is completely defined. The probability distribution function of the associated microstates is then the result of a postulate, known as the "*fundamental*" postulate, which completes the first postulate of equality of the temporal and ensemble averages (Equation (1.28)). The role of the microcanonical ensemble is central in statistical physics. Firstly, because the energy and the number of particles are quantities defined at all scales, which therefore allow the control of the transition from microscopic to macroscopic in the limit $N \gg 1$. Secondly, because the other statistical ensembles, characterised by different external parameters, derive from the microcanonical ensemble.

After stating the fundamental postulate (Section 2.1), the enumeration process of the accessible microstates, a central calculation in the microcanonical ensemble allowing the deduction of the system's thermodynamic properties, will be explained (Section 2.1). Next, it will be shown that a macroscopic internal variable is in general distributed according to an extremely sharp Gaussian distribution. In other words, fluctuations around the average value are negligible (Section 2.3). Finally, the foundations of statistical physics and the justification of the two postulates will be discussed (Section 2.4).

2.1 POSTULATES OF EQUILIBRIUM STATISTICAL PHYSICS

Equilibrium statistical physics is based on two postulates. As discussed in Section 1.5, the first postulate states that in a system in equilibrium, the time averages $\bar{A}$ are equal to the ensemble averages $\langle A \rangle$, characterised by system's external parameters:

$$\boxed{\text{First postulate:} \quad \bar{A} = \langle A \rangle \quad \text{in equilibrium.}} \tag{2.1}$$

Consider now the case of a system contained in a closed vessel with rigid and fixed walls. The system does not exchange work with the outside environment. Moreover, the number of particles and the volume of the system are constant. If, additionally, the walls are adiabatically insulated, no heat is exchanged with the outside environment either: the system is isolated and its energy is conserved.

The microcanonical ensemble is a statistical ensemble of isolated systems that all have the same external parameters: number of particles N, volume V and energy E. The probability that a system is in a given accessible microstate, that is compatible with external parameters, is given by the *fundamental* (or second) postulate of statistical physics:

---

[1] Quoted from *On the Method of Theoretical Physics*, the Herbert Spencer Lecture, Oxford, June 10, 1933.

DOI: 10.1201/9781003272427-2

> For an *isolated* system at thermodynamic equilibrium, the system can be found with *equal probability* in any microstate consistent with the external parameters.

Although there is no proof of these two postulates, theoretical arguments, presented in Section 2.4, show that they are compatible with the fundamental laws of classical mechanics. For the moment, the fundamental postulate will just be qualitatively justified. First (not the strongest argument), the equiprobability of microstates is simple and elegant, as prescribed by Einstein in his scientific formulation of the "Ockham's razor" (see the epigraph to this chapter). Secondly, there is no physical reason to favour one microstate over another when the system is in equilibrium. Finally, it is a pragmatic choice that is remarkably validated by *a posteriori* experiments, as will later be seen in this book[2]. In a more graphical way, the situation is a analogous to that of a player who postulates the equiprobability of the numbers drawn with a six-sided die. If during the game one number comes up more often than the others, the player will conclude that the die is loaded and will change his initial hypothesis. The fundamental postulate has never been questioned.

Let us insist: the fundamental postulate is only valid for a system that is i) isolated and, ii) in equilibrium. This first condition seems restrictive, but it will be seen in Chapter 4 that common situations, such as an object in contact with a thermostat fixing the system's temperature, can be studied within the framework of the microcanonical ensemble. Thus, all results obtained in this chapter follow from the postulate of equality of time and ensemble averages (Equation (2.1)) and the fundamental postulate. The second condition – the system must be in equilibrium – is not binding as long as one is interested in thermodynamic quantities[3].

2.2 COUNTING ACCESSIBLE MICROSTATES

According to the fundamental postulate, in the microcanonical ensemble, all accessible microstates have an equal probability. To express the microstates' *uniform* probability distribution, it is necessary to count them. Let $\Omega(N, V, E)$ be the total number of microstates accessible to an isolated system of energy equal to E and consisting of N particles confined in a volume V. All statistical information[4] about the system and the determination of physical quantities (ensemble average values) characterising it are therefore based on the calculation of $\Omega(N, V, E)$.

2.2.1 UNCERTAINTY ON THE ENERGY OF AN ISOLATED SYSTEM

To count the accessible microstates of an isolated macroscopic system, one must postulate that its energy is not exactly fixed, but lies between E and $E + \delta E$, where the uncertainty $\delta E \ll E$. This assumption is experimentally valid, since a system cannot be perfectly isolated, if only because of long-range interactions with the rest of the Universe (see the discussion on the sensitivity to initial conditions, Section 1.4.2). However, the reason for introducing the uncertainty on the value of energy – and only on the energy – is not from experimental origin. It is only due to the counting of microstates[5]. Indeed, the energy of

---

[2]It was the American physicist Richard Tolman who established in 1938 the equiprobability of microstates as an inherent postulate of the statistical approach. Equiprobability is the only non-arbitrary choice that can be made *a priori* about the shape of the probability density function.

[3]To study non-equilibrium properties, it is necessary to introduce assumptions about the microscopic dynamics of the system (see Note 39 in Chapter 1).

[4]By analogy, the probability of getting a result with an (unloaded) die with Ω faces is $1/\Omega$, so that the probability distribution is normalised.

[5]In practice, macroscopic quantities cannot be known exactly. For example, the volume V of a container is poorly defined at the atomic scale (asperities in the walls, adsorption of molecules on the walls...) and

a macroscopic system can be seen as a continuous quantity in the framework of thermo-dynamics, whereas in quantum mechanics the energy of a confined and isolated system is quantified. Thus, in a strict sense, the number of accessible microstates is a discontinuous function of the energy that goes from zero for energy values lying between energy levels to a value equal to the degeneracies of the accessible levels (see Figure 2.1).

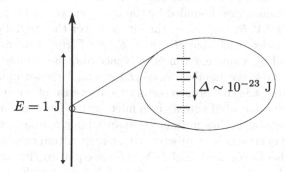

FIGURE 2.1 Energy scale of a macroscopic system. The inset is a zoom on the scale illustrating the discretised energy levels separated by intervals of the order of Δ.

To implement the transition from a discretised system to a continuous system that supports the transition from microscopic to macroscopic, one must count the number of accessible microstates over an energy interval $\delta E \ll E$, but which still contains a large number of energy levels. In other words, $\Delta \ll \delta E \ll E$, where Δ is the typical gap between two energy levels. For a macroscopic system[6], $E \simeq 1$ to 10^3 J and $\Delta \simeq 10^{-30}$ à 10^{-20} J. The number of accessible microstates $\Omega(N,V,E)$ is thus defined as the number of microstates whose energy lies between E and $E + \delta E$, with $\delta E \ll E$. Thus, $\Omega(N,V,E)$ varies almost continuously with E. Note that it is often simpler to calculate the number of accessible microstates with energies less than E, denoted $\Phi(N,V,E)$, and deduce:

$$\Omega(N,V,E) = \Phi(N,V,E + \delta E) - \Phi(N,V,E) \simeq \frac{\partial \Phi}{\partial E} \, \delta E, \qquad (2.2)$$

since[7] $\delta E \ll E$. Although $\Omega(N,V,E)$ depends explicitly on δE, it will be shown that in practice *the physical quantities characterising a macroscopic system are all independent of the arbitrary (small) value of δE*, as it must be in thermodynamics.

---

the number of particles N in a macroscopic system cannot be known to the nearest unit (a very small uncertainty, say for example 0.1 μg of water, corresponds to 10^{15} water molecules!). However, it is not necessary to introduce uncertainties δV and δN on these quantities to count the number of accessible microstates of an isolated system.

[6]The value of Δ is, of course, specific to the studied system. Le us consider three examples:

- A mole ideal quantum gas in three dimensions at the (classical) limit of large quantum number. From Equation (1.1), its energy is given by $E = \epsilon(n_1^2 + n_2^2 + \cdots + n_{3N_A}^2) \simeq 3\epsilon N_A n^2$, where $\epsilon = h^2/8mL^2$, and taking $n_i = n \gg 1$ for all i. The gap between two energy levels is obtained when a single quantum number n_i increases by 1 (n to $n + 1$), thus $\Delta \simeq \epsilon((n + 1)^2 - n^2) \simeq 2\epsilon n$. One deduces $\Delta \simeq \sqrt{4E\epsilon/3N_A} \simeq h\sqrt{RT/4ML^2} \simeq 10^{-31}$ J, since at the classical limit, $E \simeq 3RT/2 \simeq 3$ kJ at room temperature (and with $L \simeq 10$ cm).

- A system of harmonic oscillators describing the oscillations of the atoms of a crystal. From Equation (1.2): $\Delta \simeq \hbar\omega \sim 10^{-20}$ J (see exercise 4.3).

- An electron spin system. From Equation (1.3): $\Delta \simeq \mu_B B \simeq 10^{-23}$ J from a magnetic field $B = 1$ T. (see Note 9 of Chapter 1).

[7]It will be seen in Note 27 that this first order approximation is valid only when trying to estimate $\ln \Omega(N,V,E)$, as it will be the case in the remaining of this book.

2.2.2 DISCRETISATION OF PHASE SPACE IN THE FRAMEWORK OF CLASSICAL MECHANICS

While it is conceivable to count discretised quantum microstates, classical microstates are, in principle, uncountable, since in classical mechanics microscopic quantities change continuously. In phase space, introduced in Section 1.2.2, the set of microstates of energy lower than E corresponds to the region bounded by the hypersurface of energy E and whose $2f$-dimensional volume $\mathcal{V}(N,V,E)$ is given by the integral over the spatial coordinates and the momenta of the N particles, such that $H(\boldsymbol{q},\boldsymbol{p}) \leq E$ (see Figure 2.2). Since a microstate is, in principle, represented by a single *dot* in phase space and the volume $\mathcal{V}(N,V,E)$ contains an infinite number of dots, the number $\Phi(N,V,E)$ should have an infinite value. In order to count the microstates, it is therefore necessary to discretise phase space into *elementary cells* of volume $\mathcal{V}_0$, each associated with a given microstate. The fundamentally discrete nature of quantum microstates is thus introduced in an *ad hoc* manner into the framework of classical physics. Since microstates are discretised, as in quantum mechanics, an uncertainty δE on the energy is also introduced, and $\Omega(N,V,E)$ is equal to the number of elementary cells between the two hypersurfaces of energy E and $E + \delta E$ (see Figure 2.2). To calculate $\Omega(N,V,E)$, Equation (2.2) is used with

$$\Phi(N,V,E) = \frac{\mathcal{V}(N,V,E)}{\mathcal{V}_0} = \frac{1}{\mathcal{V}_0} \int_V \int_{H(\boldsymbol{q},\boldsymbol{p}) \leq E} \mathrm{d}^f q \mathrm{d}^f p, \qquad (2.3)$$

where the integration bounds indicate that the energy $H(\boldsymbol{q},\boldsymbol{p})$ is less than E and that the spatial coordinates of the particles are bounded by the walls of the container of volume V.

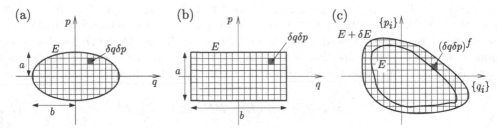

FIGURE 2.2 Volume of phase space corresponding to microstates of energy less than E: (a) for a one-dimensional harmonic oscillator (i.e. $f = 1$ degree of freedom); (b) for a one-dimensional free particle ($f = 1$); (c) for any system of N particles in three dimensions ($f = 3N$). In (c), the two hypersurfaces of energy E and $E + \delta E$ are shown with solid lines.

The presence of $\mathcal{V}_0$ at the denominator of Equation (2.3) is essential for dimensional reasons (Φ is dimensionless), but the choice of its value, on which $\Phi(N,V,E)$ explicitly depends, can be questionable. Writing the elementary volume in the form $\mathcal{V}_0 = (\delta q \delta p)^f$, where δq and δp are the width of the elementary cell associated with each of the f spatial coordinates and the f momenta, respectively. δq and δp can be interpreted as uncertainties on the particles' spatial coordinates and momenta. The number of accessible microstates and the associated derived physical quantities should then depend on the degree of precision on the microscopic quantities. However, in thermodynamics, macroscopic quantities (pressure, temperature, etc.) are measured independently of the knowledge of microscopic quantities one can gather. In fact, it will be seen in Chapter 3 that among all thermodynamic quantities calculated from $\Omega(N,V,E)$, only the entropy and the chemical potential depend on $\mathcal{V}_0$ by means of an additive constant[8]. Here, a loophole in classical physics at the atomic

---

[8]In practice, changes of entropy (and energy) are measured. Thus, the value of the additive constant, which depends on $\mathcal{V}_0$, is generally unimportant.

scale is introduced. It can only be closed thanks to quantum mechanics. According to the correspondence principle, in the limit of large quantum numbers (and at high energies), classical physics is a good approximation of quantum physics[9]. A simple dimensional analysis shows that the product $\delta q \delta p$ is homogeneous to an action and can therefore be written $\delta q \delta p = xh$, where h is Planck's constant. In order to determine the value of the multiplicative factor x, $\Phi(N, V, E)$ can be calculated for a given system, first using quantum formalism and then using classical formalism. Finally, the results obtained in the high-energy limit are compared. To simplify the calculations, two systems with only one degree of freedom ($f = 1$) are successively considered:

- One-dimensional harmonic oscillator. In quantum mechanics, the energy is given by Equation (1.2): $E(n) = (n+1/2)\hbar\omega$, where $n = 0, 1, 2, \ldots$. Taking into account the ground state ($n = 0$), the number of microstates with energies less than E (corresponding to the nth energy level) is:

$$\Phi_{\text{quantum}}(E) = n + 1 = \frac{E}{\hbar\omega} + \frac{1}{2} \simeq \frac{2\pi E}{h\omega}, \qquad (2.4)$$

  in the limit of large quantum numbers ($n \gg 1$ so $E \gg \hbar\omega$). In the framework of the classical physics, the phase space trajectory of a harmonic oscillator at a fixed energy E is an ellipse given by Equation (1.12), whose half-axes are $a = \sqrt{2Em}$ and $b = \sqrt{2E/m\omega^2}$. As illustrated in Figure 2.2, the number of microstates with an energy less than E is thus obtained by dividing the volume of the corresponding phase space (here two-dimensional), that is the ellipse area πab, by the volume of the elementary cell $\mathcal{V}_0 = \delta q \delta p$ (here $f = 1$):

$$\Phi_{\text{classical}}(E) = \frac{\pi ab}{\delta q \delta p} = \frac{2\pi E}{\delta q \delta p \,\omega}. \qquad (2.5)$$

  Comparing Equations (2.4) and (2.5), one immediately sees $\delta q \delta p = h$, $x = 1$.

- Free one-dimensional particle in a box of length L. In quantum mechanics, the particle's energy is given by Equation (1.1): $E(n) = n^2 h^2 / 8mL^2$, where $n = 0, 1, 2, \ldots$ from which one immediately gets the number of microstates n with an energy less than E:

$$\Phi_{\text{quantum}}(E) = n = \frac{2L\sqrt{2mE}}{h}. \qquad (2.6)$$

  In the framework of classical physics, the one-dimensional free particle has an energy given by $E = p^2/2m$, so $p = \pm\sqrt{2mE}$, depending on whether the particle is moving to the left or to the right. Moreover, $-L/2 \le q \le L/2$. The back and forth motion of the particle in the box is described in phase space by a rectangle of sides $a = 2\sqrt{2mE}$ and $b = L$ (see Figure 2.2). The number of microstates with an energy less than E is therefore equal to the area of this rectangle divided by $\delta q \delta p$, that is

$$\Phi_{\text{classical}}(E) = \frac{ab}{\delta q \delta p} = \frac{2L\sqrt{2mE}}{\delta q \delta p}. \qquad (2.7)$$

  Comparing Equations (2.6) and (2.7), one finds again $\delta q \delta p = h$.

The two examples above show that the volume of the elementary cell associated with a microstate must be equal to $\mathcal{V}_0 = h^f$ in order to have a classical count corresponding to the quantum count in the limit of large quantum numbers[10]. Of course, only two particular

---

[9] Provided that quantum irregularities are smoothed out.

[10] This result is in agreement with Heisenberg's inequality: $\Delta q \Delta p \ge \hbar/2$.

cases were discussed, but the very different nature of these two systems suggests the generality of the result, which is indeed confirmed by experimental data such as Sackur and Tetrode results as early as 1912 (see Section 3.2.1). Note for the moment the importance of this discovery in the development of quantum mechanics. In 1900, Planck introduced the constant h, associated with the quanta notion, in order to be able to count the microstates of a photon gas. By measuring the value of h in a mercury gas, Sackur and Tetrode provided one of the first confirmations of the quantum hypothesis and its generalisation to atomic systems. The derivation of the volume $\mathcal{V}_0$ of the elementary cell of phase space from quantum mechanics is explained in Section 6.1.1.

2.2.3 DISTINGUISHABLE OR INDISTINGUISHABLE PARTICLES IN CLASSICAL PHYSICS

When particles' *positions* are treated in the framework of classical mechanics, each particle is assigned its own coordinates. Even identical particles[11] are therefore identifiable (e.g. they can be labelled) and are, in principle, *distinguishable* (see Section 1.1.2). Thus, each phase space axis is linked to the spatial coordinates or momenta of a given particle. As seen in Section 1.1.2, distinguishing atoms located at the nodes of a crystal lattice is legitimate, since each atom can be assigned a permanent and labelled equilibrium position. A microstate is then defined by an ordered list of individual states specifying, for example, the spin or the oscillation degree of freedom around the equilibrium position of each particle. Starting from a specific microstate, swapping particles individual states results in a microstate quite different from the original one[12], as shown in the example in Figure 2.3.

FIGURE 2.3 Exchange of the state of two discernible particles, labelled 1 and 2, located at the nodes of a lattice: the two microstates are different.

However, when identical particles are not localised, such as gas or liquid molecules, they can no longer be assigned a permanent characteristic (such as an equilibrium position) allowing them to be distinguished from each other. The individual state of each particle is then specified, among other things, by its position in the container. Starting from a specific microstate, if we exchange the individual states of the particles (positions, momenta and possibly internal degrees of freedom), the obtained microstate is *similar* to the original one, only the particles' labels have changed, as illustrated in Figure 2.4.

However, these labels play no role in the physical description of a system of non-localised and (supposedly) identical particles. Thus, the macroscopic internal variables, defined over the entire particle system (kinetic and potential energies, velocity distribution...) have the same values in the $N!$ similar microstates, obtained by permutation of the individual

---

[11] At the atomic scale, identical particles have exactly the same mass, charge and spin, such as electrons or given isotopes of a chemical element. On a larger scale, particles cannot, in principle, be strictly identical. Macromolecules (DNA, polymers, etc.), colloidal particles (with dimensions $\leq \mu$m) or sand grains do not have rigorously identical mass, number of atoms, charge or shape. However, for a given level of description, possible or chosen, these particles can be considered identical, if they present the same aspect, and be treated as such in a model (in particular through the Hamiltonian).

[12] More precisely, a Hamiltonian of interacting neighbouring spins is not invariant by permutation of particles spin when particles' positions are fixed. For example, the energy of the system is different in the two microstates shown in Figure 2.3.

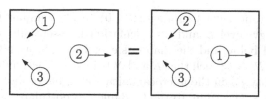

FIGURE 2.4 Permutation of the individual states (positions and momenta) of two particles, labelled 1 and 2, in a fluid (the arrows represent the particles' momenta). If the particles are indistinguishable (i.e. without labels), the two microstates are identical. By analogy, consider lottery balls: seen at close range, their numbers can be seen (and the microstates above are different), seen at a distance they seem identical (microstates are also identical).

states (or labels) of the N particles[13]. The representative points of these similar microstates are in $N!$ different cells of phase space, whereas they should count as only one microstate in the calculation of Ω. In order to count the accessible microstates correctly, one must therefore abandon the labelling of each individual particles, as it is, in principle, allowed by classical physics: one must consider them as "indistinguishable"[14] (unlabelled) particles, such that the system microstate is invariant by permutations of their individual states. The number of accessible microstates calculated for N particles assumed distinguishable using Equation (2.3) must therefore be divided by $N!$ when the particles are considered indistinguishable[15]:

$$\Omega_{\text{Indiscernables}} = \frac{1}{N!}\, \Omega_{\text{Discernables}}. \tag{2.8}$$

To summarise, in classical statistical physics, identical particles are, in principle, distinguishable, but can be treated as indistinguishable particles when the aim is to count the system's number of accessible microstates[16]. It should be remembered that particles of a crystal must be *a priori* considered distinguishable, and fluid particles as indistinguishable. However, since this choice affects the expression of Ω, and therefore the resulting thermodynamic quantities, it must be validated *a posteriori* by ensuring that the calculated physical quantities have the expected thermodynamics properties. It will be seen in Chapter 3 that only the entropy and the chemical potential depend on the supposedly distinguishable or indistinguishable nature of the particles.

2.2.4 $\Omega(N, V, E)$ AND MICROCANONICAL AVERAGES

In the framework of quantum physics, in order to count the number of accessible microstates of a system, one considers in practice a *finite* number of particles at energy E corresponding

[13]Indeed, once the N individual states (positions and momenta) are given, there are N possible states for particle number 1, there are $N - 1$ for particle number 2, $N - 2$ for particle number 3 and so on, giving $N(N-1)(N-2)\ldots 1 = N!$ possible permutations.

[14]Strictly speaking, this is a misuse of language, since indistinguishability is a quantum property (see Chapter 6): the square of the modulus of the wave function associated with a system of identical particles must be invariant by permutation of the particles. One should therefore use another term in classical physics, for example "*alike particles*", to characterise distinguishable particles one does not seek to *differentiate*. In accordance with usage, the term "*indistinguishable*" is nevertheless used in the following.

[15]Equation (2.8) is correct only in the Maxwell-Boltzmann approximation (see Section 6.4.3), when several particles cannot share the same individual state (see Exercises 2.2 and 2.5). This is the case of a dilute gas, or more generally when repulsive forces forbid particles to be located at the same position.

[16]Note that this does not apply to *quantum statistical physics*, since in this case, particles are intrinsically indistinguishable (see Chapter 6). For some authors, the (possible) indistinguishability of particles in classical statistical physics has a quantum origin. For others, on the contrary, it depends only on our ability, or even our will, to distinguish or not the particles of a classical system (see the discussion of Gibbs' paradox, Section 3.5.3). This point, which is still being debated today, is discussed in J. Uffink's article (see Note 31) and in D. Frenkel's article, "*Why colloidal systems can be described by statistical mechanics: some not very original comments on the Gibbs paradox*", Mol. Phys. **112**, 2325 (2014).

to a given energy level (and thus assuming $\delta E = 0$)[17]. The number $\Omega(N,V,E)$ is then the degeneracy of the energy level E and its calculation is essentially a combinational logic problem (see Example 1 below and the chapter's exercises in Section 2.5). The "continuous limit" is obtained when $N \to \infty$ such that E and N behave as continuous variables. Knowing $\Omega(N,V,E)$, the probability p_m in the microcanonical ensemble to find the system in a given accessible microstate m is, according to the fundamental postulate:

$$p_m = \frac{1}{\Omega(N,V,E)},$$ (2.9)

which is indeed independent of the microstate m and respects the normalisation condition (Equation (1.29)). According to Equation (1.30), the average value of an internal variable A is then given in the microcanonical ensemble by[18]:

$$\langle A \rangle = \sum_m A_m \, p_m = \frac{1}{\Omega(N,V,E)} \sum_{m=1}^{\Omega} A_m,$$

where A_m is the value of the internal variable in microstate m and the sum is performed over all accessible microstates.

Using classical physics (Section 2.2.2), the number of microstates with energy less than E is given by

$$\Phi(N,V,E) = \underbrace{\left(\frac{1}{N!}\right)}_{\text{if indistinguishable}} \int_V \int_{H(q,p) \leq E} \frac{d^f q \, d^f p}{h^f},$$

for N (assumed) indistinguishable particles. The factor $1/N!$ is omitted if the particles are (assumed) distinguishable. If the particles are atoms without internal structure that evolve in a $d = 1, 2$ or 3 dimensional space, then $f = Nd$. The number of accessible microstates, $\Omega(N,V,E)$, is derived such that the q-coordinates of the N particles are bounded by the walls of the container of volume V and such that $E \leq H(q,p) \leq E + \delta E$, where $\delta E \ll E$. From Equation (2.2), one obtains[19]:

$$\Omega(N,V,E) = \underbrace{\left(\frac{1}{N!}\right)}_{\text{if indistinguishable}} \int_V \int_{E \leq H(q,p) \leq E+\delta E} \frac{d^f q \, d^f p}{h^f}.$$ (2.10)

---

[17]To take into account the uncertainty δE, one can formally write

$$\Omega(N,V,E) = \int_E^{E+\delta E} dE' \sum_m \delta(E' - E_m(N,V)),$$ (2.11)

where the discrete sum is performed on *all* microstates m. The accessible microstates are selected using the Dirac function.

[18]As in Note 17, average values can be written as an integral taking into account the uncertainty δE:

$$\langle A \rangle = \frac{1}{\Omega} \int_E^{E+\delta E} dE' \sum_m A_m \, \delta(E' - E_m(N,V)).$$ (2.12)

[19]It is sometimes useful to write $\Omega(N,V,E)$ as:

$$\Omega(N,V,E) \quad = \quad \underbrace{\left(\frac{1}{N!}\right)}_{\text{if indistinguishable}} \int_E^{E+\delta E} dE' \int \frac{d^f q \, d^f p}{h^f} \, \delta(E' - H(q,p))$$

$$\simeq \quad \underbrace{\left(\frac{1}{N!}\right)}_{\text{if indistinguishable}} \delta E \int \frac{d^f q \, d^f p}{h^f} \, \delta(E - H(q,p)),$$ (2.13)

since $\delta E \ll E$.

In the microcanonical ensemble, the probability $\mathrm{d}P$ to find the system in the accessible microstate $(\boldsymbol{q}, \boldsymbol{p})$ to within $\mathrm{d}^f q$ and $\mathrm{d}^f p$ is therefore:

$$\mathrm{d}P = \rho(\boldsymbol{q}, \boldsymbol{p}) \mathrm{d}^f q \mathrm{d}^f p = \frac{1}{\Omega(N,V,E)} \frac{\mathrm{d}^f q \mathrm{d}^f p}{h^f}. \qquad (2.14)$$

The probability $\mathrm{d}P$ is completely independent of the microstate $(\boldsymbol{q}, \boldsymbol{p})$ and the probability density $\rho(\boldsymbol{q}, \boldsymbol{p})$ satisfies the normalisation condition[20] (Equation (1.33)). According to Equation (1.34), the average value of an internal variable A in the microcanonical ensemble is given by:

$$
\begin{aligned}
\langle A \rangle &= \int A(\boldsymbol{q}, \boldsymbol{p}) \rho(\boldsymbol{q}, \boldsymbol{p}) \, \mathrm{d}^f q \mathrm{d}^f p \\
&= \frac{\int_V \int_{E < H(\boldsymbol{q}, \boldsymbol{p}) < E + \delta E} A(\boldsymbol{q}, \boldsymbol{p}) \, \mathrm{d}^f q \mathrm{d}^f p}{\int_V \int_{E < H(\boldsymbol{q}, \boldsymbol{p}) < E + \delta E} \mathrm{d}^f q \mathrm{d}^f p},
\end{aligned}
$$

whether the particles are distinguishable or indistinguishable. As mentioned above, apart from entropy and chemical potential, thermodynamic quantities do not generally depend on the distinguishable or indistinguishable nature of the particles, nor on the volume of the elementary cell of phase space $\mathcal{V}_0 = h^f$.

Example 1: Ideal paramagnetic crystal

Consider N spins-1/2 carried by atoms located at the nodes of a crystal lattice and immersed in a magnetic field B. The interactions between spins are neglected. The system is supposed isolated and has a fixed energy E, given by the Hamiltonian (1.10):

$$E = - \sum_{j=1}^{N} \mu_M B s_j \quad \text{with} \quad s_j = \pm 1.$$

The N spins are treated as *independent* and *distinguishable* particles. The external parameters are N, B and E. Since the spin degrees of freedom is of interest, the crystal volume does not intervene in the following. To begin with, consider for example[21] $N = 5$ isolated spins at energy $E = -\mu_M B$.

At this energy, all accessible microstates have three +1 spins and two -1 spins (for example ↑↑↓↓↑). The number of accessible microstates is therefore the number of ways to choose three spins out of five (once chosen, they are assigned the value +1, the remaining two spins having necessarily the value -1), or:

$$\Omega = \binom{5}{3} = \frac{5!}{3! \, 2!} = 10.$$

To be convinced of this, one can easily list the ten accessible microstates:

$$\uparrow\uparrow\uparrow\downarrow\downarrow \quad \uparrow\uparrow\downarrow\uparrow\downarrow \quad \uparrow\uparrow\downarrow\downarrow\uparrow \quad \uparrow\downarrow\uparrow\uparrow\downarrow \quad \uparrow\downarrow\uparrow\downarrow\uparrow \quad \uparrow\downarrow\downarrow\uparrow\uparrow$$
$$\downarrow\uparrow\uparrow\uparrow\downarrow \quad \downarrow\uparrow\uparrow\downarrow\uparrow \quad \downarrow\uparrow\downarrow\uparrow\uparrow \quad \downarrow\downarrow\uparrow\uparrow\uparrow$$

---

[20] The probability $\mathrm{d}P$ differs depending on whether the particles are distinguishable or indistinguishable: $\mathrm{d}P_{\text{indistinguishable}} = N! \, \mathrm{d}P_{\text{distinguishable}}$. When the particles are indistinguishable, the integral over phase space that provides the normalisation of $\mathrm{d}P$ must be divided by $N!$ as in Equation (2.10) giving Ω.

[21] The energy of this 5-spins system can only take the following values: $\pm 5\mu_M B$, $\pm 3\mu_M B$ and $\pm \mu_M B$.

Since the particles are distinguishable, it can be seen that, in general, a permutation of spins changes the microstate[22]. The average value of any spin is given by:

$$\langle s_j \rangle = (+1)\frac{6}{10} + (-1)\frac{4}{10} = \frac{1}{5},$$

since the 10 microstates have the same probability $p_m = 1/10$ and a given spin j (for example the first in the chain) takes the value $s_j = +1$ in six microstates and the value $s_j = -1$ in four.

The generalisation to the case of a system of N independent spins at energy E is straightforward: energy being constant, it depends only on the number N_+ of spins parallel to the magnetic field ($s_j = +1$). The number of spins antiparallel to the magnetic field ($s_j = -1$) is necessarily $N_- = N - N_+$. Thus, one obtains:

$$E = -\sum_{j=1}^{N} \mu_M B s_j = -\Big(+ \mu_M B N_+ - \mu_M B(N - N_+) \Big) = \mu_M B(N - 2N_+).$$

The number of accessible microstates at energy E, comprised between $-N\mu_M B$ and $+N\mu_M B$, is therefore equal to the number of ways to choose N_+ spins $+1$ among N, that is

$$\Omega(N,E) = \binom{N}{N_+} = \frac{N!}{N_+!\ \underbrace{(N - N_+)!}_{=N_-}} \quad \text{where} \quad N_+ = \frac{N}{2}\Big(1 - \frac{E}{N\mu_M B}\Big). \tag{2.15}$$

Example 2: Ideal classical gas

Consider an ideal classical gas consisting of N *indistinguishable* identical particles of mass m with no internal structure, confined in an isolated cubic container of volume $V = L^3$. Its Hamiltonian is given by Equation (1.9):

$$H_{\mathrm{IG}} = \sum_{j=1}^{N} \frac{\vec{p}_j^2}{2m} = \sum_{i=1}^{f} \frac{p_i^2}{2m}, \tag{2.16}$$

where the last sum is over the $f = 3N$ degrees of freedom of the system. Since the particles are *indistinguishable*, the number of microstates of energy less than E is:

$$\Phi(N,V,E) = \underbrace{\Big(\frac{1}{N!}\Big)}_{\text{if indistinguishable}} \int_V \int_{\sum_{i=1}^{f} \frac{p_i^2}{2m} \leq E} \frac{d^f q\, d^f p}{h^f}.$$

Let us start with the case of a single particle ($f = 3$):

$$\Phi(1,V,E) = \frac{1}{h^3} \underbrace{\int_L dq_1 \int_L dq_2 \int_L dq_3}_{=L^3=V} \underbrace{\int\int\int_{p_1^2+p_2^2+p_3^2 \leq 2mE} dp_1\,dp_2\,dp_3}_{=\frac{4}{3}\pi(\sqrt{2mE})^3}$$

$$= \frac{4}{3}\pi \frac{V(\sqrt{2mE})^3}{h^3}.$$

---

[22]If these spins were treated as indistinguishable particles, there would be only one accessible microstate at a given energy. Since different particles can be in the same individual spin state, Equation (2.8) is clearly wrong here.

Indeed, since the particle moves freely in the container, the integral over each of the three generalised spatial coordinates is L. The triple integral over the momenta is the volume of the sphere of radius $\sqrt{2mE}$. This calculation can be extended to the case of N independent and indistinguishable particles

$$\Phi(N,V,E) = \frac{1}{N!}\frac{1}{h^f} \underbrace{\int_L \mathrm{d}q_1 \int_L \mathrm{d}q_2 \cdots \int_L \mathrm{d}q_f}_{=L^f} \underbrace{\int \int \cdots \int_{p_1^2+p_2^2+\cdots+p_f^2 \leq 2mE} \mathrm{d}p_1\mathrm{d}p_2 \ldots \mathrm{d}p_f}_{=\mathcal{V}_f(E)},$$

where the integral $\mathcal{V}_f(E)$ is the volume of the hypersphere of radius $\mathcal{R} = \sqrt{2mE}$ in a f-dimensional space. It is written for reasons of homogeneity as:

$$\mathcal{V}_f(E) = c_f\mathcal{R}^f, \quad \text{where} \quad c_f = \frac{\pi^{\frac{f}{2}}}{\left(\frac{f}{2}\right)!} \tag{2.17}$$

is a dimensionless constant, equal to the volume of the hypersphere[23] of radius $\mathcal{R} = 1$ (see Equation (A.25) in Appendix A.4). Thus:

$$\Phi(N,V,E) = \frac{1}{N!}\, c_f\left(\frac{L\sqrt{2mE}}{h}\right)^f \quad \text{where} \quad f = 3N. \tag{2.18}$$

This gives the number of microstates with an energy lying between E and $E + \delta E$:

$$\begin{aligned} \Omega(N,V,E) &= \frac{\partial\Phi}{\partial E}\,\delta E \\ &= \frac{1}{N!}\, c_f\left(\frac{L\sqrt{2m}}{h}\right)^f \frac{f}{2}E^{\frac{f}{2}-1}\delta E. \end{aligned} \tag{2.19}$$

For an ideal gas in three dimensions $f = 3N \gg 1$), the dependence of $\Omega(N,V,E)$ on V and E is:

$$\Omega(N,V,E) \propto L^{3N}E^{\frac{3N}{2}-1} \simeq V^N E^{\frac{3N}{2}}, \tag{2.20}$$

where in the exponent of E, "1" has been neglected in front of $3N/2$.

2.2.5 DEPENDENCE OF $\Omega(N,V,E)$ ON E

The two calculations of the number of accessible microstates discussed in Section 2.2.4 are for *independent* particle systems. In general, when interactions between particles are not neglected, Ω cannot be calculated exactly. To determine the approximate dependence of Ω on the energy E, a reasoning over a single degree of freedom of energy E_1 can be done. The number of microstates of energy less than E_1, $\Phi_1(E_1)$, is by definition an increasing function of energy, of the order of E_1/Δ at high energy ($E_1 \gg \Delta$), where Δ is the typical gap between two energy levels. Since in general Δ is not fixed, but is a function of the energy, and the energy levels are degenerate, Φ_1 is not exactly proportional to E_1. As a first approximation, one can write:

$$\Phi_1(E_1) \simeq \left(\frac{E_1}{\Delta}\right)^{\alpha_1} \quad \text{where} \quad \alpha_1 \sim 1. \tag{2.21}$$

Thus, for the harmonic oscillator and the one-dimensional free particle, one gets $\alpha_1 = 1$ and $\alpha_1 = 1/2$ from Equations (2.5) and (2.7), respectively. For a macroscopic system of energy E, each of the f degrees of freedom has an energy typically equal to $E_1 = E/f$ and $\Phi_1(E_1)$

---

[23]In the case $f = 3$, the volume of a sphere is recovered with $c_3 = 4\pi/3$, since $(3/2)! = (3/2)(1/2)!$, where $(1/2)! = \sqrt{\pi}/2$.

microstates of energy less than E_1. If the f degrees of freedom were independent, the number of microstates of energy less than E would be proportional to $[\Phi_1(E_1)]^f = (E/(f\Delta))^{\alpha_1 f}$, according to Equation (2.21). Although in general the interactions between particles reduce the number of accessible microstates[24], the dependence of $\Phi(N,V,E)$ on E is given as a first approximation by[25]:

$$\Phi(N,V,E) \simeq \left(\frac{E}{E_0}\right)^\alpha \quad \text{with} \quad \alpha \simeq \alpha_1 f \simeq f, \tag{2.22}$$

where E_0 is the energy of the fundamental level (of the order of $f\Delta$). Thus, the number of accessible microstates increases extraordinarily rapidly[26] with the energy of a macroscopic system (for which $\alpha \simeq f \simeq 10^{23}$). From Equation (2.2):

$$\frac{\Omega(N,V,E)}{\Phi(N,V,E)} = \frac{\Phi(N,V,E+\delta E)}{\Phi(N,V,E)} - 1 \simeq \left(\frac{E+\delta E}{E}\right)^f - 1 = e^{f \ln\left(1+\frac{\delta E}{E}\right)} - 1 \simeq e^{f\frac{\delta E}{E}} - 1, \tag{2.23}$$

since $\delta E \ll E$. And with $f \simeq 10^{23}$, $f\delta E/E \ll f$. To compare the gigantic values of Φ and Ω, it is relevant to calculate the logarithm of Equation (2.23), that is

$$\ln \Omega(N,V,E) \simeq \underbrace{\ln \Phi(N,V,E)}_{\sim f} + \underbrace{\ln\left(e^{f\frac{\delta E}{E}} - 1\right)}_{\ll f},$$

since according to Equation (2.22), $\ln \Phi \simeq f \ln\left(\frac{E}{E_0}\right) \simeq f$ and $\ln\left(e^{f\frac{\delta E}{E}} - 1\right) \ll \ln\left(e^f - 1\right) \simeq f$. Since $N \simeq f$, one shall remember that[27]:

$$\ln \Omega(N,V,E) \simeq \ln \Phi(N,V,E) \simeq f \ln \frac{E}{E_0} \simeq f \simeq N. \tag{2.24}$$

Thus, and this is important in the following, $\ln \Omega$ does not in practice depend on δE (whereas Ω does explicitly). At high energy E, there are as many (or even more) accessible microstates in the small interval $[E, E + \delta E]$, where $\delta E \ll E$, as in the large interval $[E_0, E]$, where $E \gg E_0$. This surprising result is due to the extremely fast growth of $\Omega(E)$ with E and thus to the gigantic value of f. By introducing the density of state $d(E)$ defined by:

$$d(E) \equiv \frac{\partial \Phi}{\partial E}, \tag{2.25}$$

one gets:

$$\Phi(E) = \int_{E_0}^E d(E')\, \mathrm{d}E' \quad \text{and} \quad \Omega(E) = \int_E^{E+\delta E} d(E')\, \mathrm{d}E' \simeq d(E)\delta E.$$

Figure 2.5 shows a graphical representation of the equality given in Equation (2.24).

---

[24] For example, the repulsion between hard spheres limits the number of positions these spheres can take in a given volume.

[25] It will be seen in Section 3.2.1 that this equation is not valid when the energy E has an upper limit, as it is the case for a system of N independent spins interacting with a magnetic field B, whose energy is always less than $N\mu B$ (see Section 2.2.4).

[26] To convince oneself of this, one can plot on the same graph and in logarithmic scale the curves $\Phi(x) = x^f$ with $f = 1, 2, 10, 100$ and 1000. Then the shape of the curve for $f = 10^{23}$ can be guessed.

[27] The approximation (2.2), $\Omega_{\mathrm{ap}} \simeq \frac{\partial \Phi}{\partial E} \delta E$, is valid only when trying to estimate $\ln \Omega$. Indeed, according to Equations (2.22) and (2.23), and with $x = f\frac{\delta E}{E} \ll f$:

$$\frac{\Omega_{\mathrm{ap}}}{\Omega} \simeq \frac{\partial \Phi}{\partial E} \delta E \times \frac{1}{\Omega} \simeq f\Phi \frac{\delta E}{E} \times \frac{1}{\Phi(e^{f\frac{\delta E}{E}} - 1)} = \frac{x}{e^x - 1} \leq 1 \quad \text{with } 0 < x \ll f,$$

but $\ln \Omega_{\mathrm{ap}} \simeq \ln \Omega + \ln(x/(e^x - 1)) \simeq \ln \Omega$, because $\ln \Omega \simeq f$ and $|\ln(x/(e^x - 1))| \leq x \ll f$.

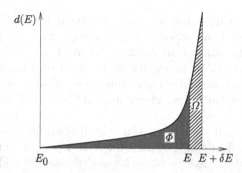

FIGURE 2.5 Schematic density of states $d(E)$ as a function of energy E. The number of microstates, $\Phi(E)$, between E_0 and E (shaded area) is comparable to the number of microstates, $\Omega(E)$, between E and $E + \delta E$ (hatched area) while $\delta E \ll E$ (the figure is not to scale).

To illustrate this geometrically, consider an apple whose skin has a thickness $d\mathcal{R}$, negligible compared to its inner radius. In an f-dimensional space, the inner volume of the apple, assumed to be hyperspherical, is according to Equation (2.17), $\mathcal{V}_{\text{inner}} = c_f \mathcal{R}^f$. The area of a hypersphere is equal to $d\mathcal{V}_{\text{inner}} = f c_f \mathcal{R}^{f-1}$ (as an exercise, check this for $f = 2$ and $f = 3$), the volume occupied by the skin is therefore given by

$$\mathcal{V}_{\text{skin}} = \underbrace{f c_f \mathcal{R}^{f-1}}_{\text{surface}} d\mathcal{R}, \quad \text{thus} \quad \frac{\mathcal{V}_{\text{skin}}}{\mathcal{V}_{\text{inner}}} = f \frac{d\mathcal{R}}{\mathcal{R}}.$$

Although the skin is very thin, say $d\mathcal{R}/\mathcal{R} = 10^{-3}$, its volume $\mathcal{V}_{\text{skin}}$ can be very large compared to the inner volume of the apple if $f \gg \mathcal{R}/d\mathcal{R} \gg 1$, contrary to what is observed in a $f = 3$ dimensional space.

2.3 PROBABILITY DISTRIBUTION FUNCTION OF A MACROSCOPIC INTERNAL VARIABLE

In statistical physics, since internal variables are treated as random variables, they are completely characterised by their probability distribution function. After having seen in Section 2.2.4 how to express the mean value $\langle A \rangle$ of an internal variable A in the microcanonical ensemble, the shape of its probability distribution $P(A)$ as well as the amplitude of the fluctuations around its mean value (estimated by the standard deviation Δ_A) are now of interest.

In the following, notations of the quantum framework are used in all generality. Let A_m be the value of the internal variable in a given microstate m. As with energy, an uncertainty $\delta A \ll A$ is introduced on the value of the internal variable. The probability $P(A)$ to find the internal variable between A and $A + \delta A$ is equal to the probability of the system to be in one of the microstates $\Omega_p(E \mid A)$ at energy E such that $A \leq A_m \leq A + \delta A$, among the $\Omega(E)$ accessible microstates (the dependence in N and V are omitted to simplify the notation). According to the fundamental postulate, all accessible microstates of an isolated system at equilibrium have the same probability $p_m = 1/\Omega(E)$. Thus (see Equation (1.32)):

$$P(A) = \sum_{\{m \mid A \leq A_m \leq A + \delta A\}} p_m = \frac{\Omega_p(E \mid A)}{\Omega(E)}. \tag{2.26}$$

The distribution $P(A)$ is normalised, since by summing $\Omega_p(E \mid A)$ over all possible values of A one recovers the total number of accessible microstates $\Omega(E)$. Equation (2.26) is general and concerns any internal variable of an isolated system at equilibrium. In the case of a *microscopic* variable – that is attached to a particle (or to a small number of particles), such as the velocity or the spin of an atom – the mathematical expression depends on the considered internal variale, in particular its classical or quantum nature. In practice, when a microscopic internal variable is quantum, the calculation of $P(A)$ is essentially a combinatorial analysis problem for which we assume[28] $\delta A = 0$, since the values of A are then discrete.

On the other hand, it will be shown that the probability distribution function of a *macroscopic* internal variable, defined over all the $N \gg 1$ particles of a system as is the case for thermodynamic quantities, nears a *Gaussian distribution law*. The macroscopic internal variables of a quantum (and *a fortiori* classical) system, as the energy, behave as continuous quantities (see Section 2.2.1). In the continuous limit, the probability $P(A)$ that the internal variable lies between A and $A + dA$ is expressed by the probability density function $f(A)$:

$$P(A) = f(A)dA. \tag{2.27}$$

In equilibrium, the thermodynamic quantity associated with the internal variable A is a constant equal to A_{eq}. In statistical physics, this equilibrium value is linked to the ensemble average value: $\langle A \rangle = A_{eq}$. Although the value of the internal variable depends on the system microstate (which is constantly changing), it is observed that, at equilibrium, it remains extremely close to A_{eq}. In order to reproduce this behaviour, the probability density function $f(A)$ must have the following properties: A_{eq} must be the most probable value of A (i.e. the maximum of $f(A)$) and the relative fluctuations of the internal variable around A_{eq} must be very small. In other words, $f(A)$ must show a very sharp peak at $A = \langle A \rangle = A_{eq}$. According to Equation (2.24), $\Omega_p(E \mid A_{eq})$ must therefore be of the order of $\Omega(E)$. And from Equation (2.24), for a macroscopic system[29]:

$$\ln \Omega_p(E \mid A_{eq}) \simeq \ln \Omega(E) \simeq f \gg 1. \tag{2.28}$$

To study the behaviour of $f(A)$ in the vicinity of its maximum, let us write a Taylor expansion of $\ln \Omega_p(E \mid A)$ around $A = A_{eq}$:

$$\ln \Omega_p(E \mid A) = \ln \Omega_p(E \mid A_{eq}) + (A - A_{eq}) \frac{\partial \ln \Omega_p(E \mid A)}{\partial A}\bigg|_{A=A_{eq}}$$
$$+ \frac{1}{2!}(A - A_{eq})^2 \frac{\partial^2 \ln \Omega_p(E \mid A)}{\partial A^2}\bigg|_{A=A_{eq}} + \ldots. \tag{2.29}$$

It will be seen later that terms of order greater than two are negligible. Since $\ln \Omega_p(E \mid A)$ is maximal in A_{eq}:

$$\frac{\partial \ln \Omega_p(E \mid A)}{\partial A}\bigg|_{A=A_{eq}} = 0 \quad \text{and} \quad \frac{\partial^2 \ln \Omega_p(E \mid A)}{\partial A^2}\bigg|_{A=A_{eq}} < 0.$$

---

[28] Thus, in the example of independent spins discussed in Section 2.2.4, the spin is a microscopic internal variable that can only take two values $A = \pm 1$. Therefore, on can assume $\delta A = 0$. The probability $P(s_1 = 1)$ that a given spin, say number "1", takes the value $+1$ is equal to the number of ways to choose $(N_+ - 1)$ spins $+1$ among $(N - 1)$ (since the value of spin "1" is fixed), i.e. $\Omega_p(N, E \mid s_1 = 1) = \binom{N-1}{N_+ - 1}$, divided by $\Omega(N, E)$ given by Equation (2.15). Naturally, $P(s_1 = 1) = N_+/N$ and $P(s_1 = -1) = (N - N_+)/N$.

[29] Even if $\Omega_p(E \mid A_{eq})$ represent only a small fraction of all the $\Omega(E)$ accessible microstates, let us say 10^{-n} with $1 \ll n \ll f$, one gets:

$$\ln \Omega_p(E \mid A_{eq}) = \ln \Omega(E) - n \ln 10 \simeq \ln \Omega(E),$$

since $n \ln 10$ is negligible in front of $\ln \Omega(E) \simeq f \simeq 10^{24}$.

From Equations (2.26), (2.27) and (2.29):

$$P(A) = f(A)\mathrm{d}A = \frac{e^{\ln \Omega_\mathrm{p}(E\,|\,A)}}{\Omega(E)} \simeq \frac{\Omega_\mathrm{p}(E\,|\,A_\mathrm{eq})}{\Omega(E)}\, e^{-\frac{(A-A_\mathrm{eq})^2}{2\Delta_A^2}}, \tag{2.30}$$

where

$$\frac{1}{\Delta_A^2} \equiv -\left.\frac{\partial^2 \ln \Omega_\mathrm{p}(E\,|\,A)}{\partial A^2}\right|_{A=A_\mathrm{eq}} > 0. \tag{2.31}$$

In Equation (2.30), the prefactor $\Omega_\mathrm{p}(E\,|\,A_\mathrm{eq})/\Omega(E)$ is independent of A. Around its maximum, $f(A)$ thus behaves as a Gaussian (normal) distribution of average value $\langle A \rangle = A_\mathrm{eq}$ and variance Δ_A^2. The normalisation of $f(A)$ implies (see the expression (A.15) of $I_0(\alpha)$ in appendix A.2):

$$P(A) = f(A)\mathrm{d}A = \frac{1}{\sqrt{2\pi\Delta_A^2}}\, e^{-\frac{(A-A_\mathrm{eq})^2}{2\Delta_A^2}}\, \mathrm{d}A. \tag{2.32}$$

This almost universal behaviour of $f(A)$ can be interpreted as a consequence of the *central limit theorem*. Indeed, in the very general case where interactions between particles are short-range, a macroscopic system can be considered, to a first approximation, as a set of $M \gg 1$ weakly interacting subsystems. A macroscopic internal variable can then be written as a sum (divided by M), each term being the contribution of one of these many *quasi-independent* subsystems. By describing each of these terms as a random variable (of any but *finite variance*), the central limit theorem implies that the probability density function of the macroscopic internal variable tends toward a Gaussian distribution function[30].

Let us now estimate the order of magnitude of the relative fluctuations, Δ_A/A_eq. To do so, let us distinguish the nature, extensive or intensive, of the considered internal variable. By definition, a *macroscopic* quantity is *extensive* if it is proportional to the quantity of matter contained in the system and it is *intensive* if it does not depend on it. Thus, by bringing together α identical macroscopic systems, that is maintained under the same thermodynamic conditions $(N, V, P, T, \rho = N/V\ldots)$, a new system having a number of particles and a volume multiplied by α is obtained (N and V are therefore extensive). However, the pressure, temperature and density of the new system remain unchanged (P, T and ρ are therefore intensive)[31]. When expressing a macroscopic internal variable A as a function of the number of particles or the number of degrees of freedom $f \gg 1$, it is extensive if $A_\mathrm{eq} \sim f$ or intensive if $A_\mathrm{eq} \sim f^0 = 1$. According to Equation (2.31), the standard deviation of an internal variable,

$$\Delta_A \sim \sqrt{\frac{A_\mathrm{eq}^2}{\ln \Omega_\mathrm{p}(E\,|\,A_\mathrm{eq})}}, \quad \text{where} \quad \ln \Omega_\mathrm{p}(E\,|\,A_\mathrm{eq}) \simeq f \quad \text{from Equation (2.28)},$$

behaves as $\sqrt{f}$, if A is an extensive quantity, or as $1/\sqrt{f}$, if A is an intensive quantity. Finally, the relative fluctuations of a macroscopic internal variable, whether intensive or extensive,

---

[30]This use of the central limit theorem is the core of the statistical physics approach of the Russian mathematician Alexander Khintchin (1878–1959) (*Mathematical Foundations of Statistical Mechanics* (Dover, 1999)). Note that the considered system must be outside the vicinity of a critical point, so that the correlations between particles of different subsystems are weak (see Chapter 11).

[31]Practically, consider a physical quantity $A(x_1, x_2, \ldots, y_1, y_2, \ldots)$ that can be expressed as a function of intensive variables $x_1, x_2\ldots$ and extensive variables $y_1, y_2\ldots$ For any real number α:

- if $A(x_1, x_2, \ldots, \alpha y_1, \alpha y_2, \ldots) = \alpha A(x_1, x_2, \ldots, y_1, y_2 \ldots)$, then A is extensive,
- if $A(x_1, x_2, \ldots, \alpha y_1, \alpha y_2, \ldots) = A(x_1, x_2, \ldots, y_1, y_2 \ldots)$, then A is intensive.

Therefore, a quantity that depends only on intensive quantities or ratio of extensive quantities is necessarily intensive.

are extremely small:

$$\frac{\Delta_A}{A_{eq}} \sim \frac{1}{\sqrt{f}} \simeq 10^{-12} \quad \text{for} \quad f \simeq 10^{23},$$

as illustrated in Figure 2.6. Since the deviations from the mean value are very unlikely, the Taylor expansion (2.29), truncated to the second order, is an excellent approximation[32]. For a Gaussian distribution function, the probability of a deviation (in absolute value) from the mean value greater than $5\Delta_A$ is of the order of 0.00005%.

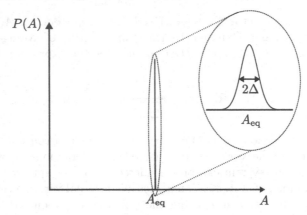

FIGURE 2.6 The probability distribution of a macroscopic internal variable is extremely peaked in $A = A_{eq}$. The inset magnifies the Gaussian behaviour of $P(A)$ and the relative fluctuations of the order of $\sim 10^{-12}$.

It should be remembered that in an isolated macroscopic system in equilibrium, the probability density function of any macroscopic internal variable A (extensive or intensive) behaves like an extremely sharp Gaussian distribution, whose ensemble average $\langle A \rangle$ corresponds to its equilibrium value A_{eq}. In a macroscopic system, the number of particles ($N \simeq N_A$) reduces the fluctuations around the equilibrium value to the order of 10^{-12}, generally beyond the reach of observations or experimental measurements[33]. Thus, *the gigantic value of Avogadro's number fully justifies a statistical approach, its predicted values being almost certain.*

2.4 DISCUSSION ON THE FOUNDATIONS OF STATISTICAL PHYSICS

Since classical (quantum) statistical physics is based on classical (quantum) mechanics, many efforts have been made since the work of the founding fathers to rigorously justify the statistical approach and make the discipline a true branch of mechanics[34]. However, the foundations of statistical physics are still the subject of theoretical studies and discussions, the mathematical and conceptual aspects of which are beyond the scope of this

---

[32]More precisely, in the Taylor expansion given by Equation (2.29), the order $n > 2$ can be neglected since:

$$(A - A_{eq})^n \left. \frac{\partial^n \ln \Omega_p(E \mid A)}{\partial A^n} \right|_{A=A_{eq}} \sim \left(\frac{A - A_{eq}}{A_{eq}} \right)^n \ln \Omega_p(E \mid A_{eq}) \sim \left(\frac{1}{\sqrt{f}} \right)^n f \sim f^{1-\frac{n}{2}},$$

where $(A - A_{eq})/A_{eq}$ is the order of relative fluctuations. For $f \gg 1$, it is perfectly justified to neglect terms of order $n > 2$ in front of the terms of order $n = 2$.

[33]At the formal "thermodynamic" limit, defined by $N \to \infty$ and $V \to \infty$ such that the density $\rho = N/V$ remains constant, relative fluctuations are equal to zero.

[34]The reader only interested in the operative aspects of statistical physics can skip this section at first read. However, it is strongly recommended to come back to it afterwards...

book[35]. An overview is given here, both to clarify the fundamental ideas and to familiarise the reader with the concepts of statistical physics.

Statistical physics of systems in equilibrium is based on two basic assumptions (see Section 2.1): (i) the equality of time and ensemble averages and (ii) the equiprobability of the accessible microstates of an isolated system in equilibrium. In the absence of rigorous and general demonstrations, a first approach is to admit these two hypotheses as postulates, then to calculate the properties of a system and finally to compare them with experimental data. This pragmatic approach, shared by the vast majority of physicists, is justified by the operational power of statistical physics, as it will be seen in the rest of this book. It is nevertheless important to try to justify as rigorously as possible (or even to prove) the two postulates from the fundamental laws of mechanics. Thus, it will be shown that *Liouville's theorem* establishes the compatibility of the fundamental postulate with the deterministic dynamics of the particles in an isolated system. The controversial role in statistical physics of the *ergodic problem*, which aims at proving mathematically the equality of time and ensemble averages will then be discussed. Finally, an approach proposed by Boltzmann, which allows the qualitative understanding of why statistical physics successfully explains the transition from the microscopic world to the macroscopic world, will be presented.

2.4.1 LIOUVILLE'S THEOREM

Without rigorously proving the fundamental postulate, it will here be shown that it is compatible with the laws of classical mechanics. Let us consider an ensemble of $M \gg 1$ isolated systems with the same fixed external parameters N, V and E (for simplicity, δE is neglected in front of E in the remaining of this section). In the $2f$-dimensional phase space, this ensemble of systems is represented at a given time by a swarm of M representative points that evolve according to Hamilton's equations (1.11) and *independently* of each other (see Figure 2.7).

To describe the dynamics of this ensemble, the phase-space density, $\rho(\boldsymbol{q}, \boldsymbol{p}, t)$, is introduced. $M\rho(\boldsymbol{q}, \boldsymbol{p}, t)\delta^f q \delta^f p$ is thus the number of representative points located between $\boldsymbol{q}$ and $\boldsymbol{q} + \delta^f q$ and between $\boldsymbol{p}$ and $\boldsymbol{p} + \delta^f p$ at time t. For the moment, the M systems are not treated as a *statistical* ensemble, so ρ is not a probability density function, but a particle density in a $2f$-dimensional space. Since the representative points move continuously according to the laws of classical mechanics, the density $\rho(\boldsymbol{q}, \boldsymbol{p}, t)$ also evolves deterministically. By analogy with fluid dynamics, the swarm can be seen as a drop of liquid – consisting of M independent "particles" (the M systems)[36] – flowing in a $2f$-dimensional phase space, at velocity $(\dot{\boldsymbol{q}}, \dot{\boldsymbol{p}})$ at point $(\boldsymbol{q}, \boldsymbol{p})$. In the framework of classical mechanics, Liouville's equation is the evolution equation for density $\rho(\boldsymbol{q}, \boldsymbol{p}, t)$.

---

[35]For more information on these issues, see the following articles:

- J. Bricmont, *Science of chaos or chaos in science?*, Annals of the New York Academy of Sciences **775**, 131 (1995).

- S. Goldstein, *Boltzmann's approach to statistical mechanics*, in *Chance in physics*, Lecture notes in physics **574**, 39 (Springer-Verlag, 2001).

- J.L. Lebowitz, *Boltzmann's entropy and time's arrow*, Physics Today **46**, 32 (1993).

- O. Penrose, *Foundations of statistical mechanics*, Rep. Prog. Phys. **42**, 1937 (1979).

- J. Uffink, *Compendium of the foundations of classical statistical physics*, in J. Butterfield and J. Earman (Eds.), *Philosophy of Physics* (Amsterdam: North Holland), 923 (2006).

- L. Sklar, *Physics and Chance, Philosophical issues in the foundations of statistical mechanics* (Cambridge University Press, 1995).

[36]Hamiltonian dynamics dictates that the trajectories of two separated systems in phase space do not intersect: if there were a point of intersection, these systems would share the same (initial) conditions and their future and past trajectories would be identical. The swarm representative points therefore behave like the centres of "hard spheres" whose trajectories never cross (see Figure 2.7).

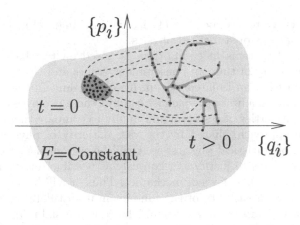

FIGURE 2.7 Evolution in the accessible phase space (E = Constant) of a swarm of representative points: compact at $t = 0$, it evolves by stretching and branching. The trajectories of some of these points are represented by dashed curves.

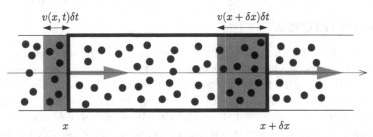

FIGURE 2.8 One-dimensional fluid flow in a tube of cross-section $\mathcal{A}$ moving in the direction of increasing x. The shaded regions on the left and right correspond to particles entering and leaving the volume element of length δx during time δt, respectively.

Demonstration: Firstly, the conservation of the number of representative points during the flow in phase space translates into the continuity equation[37]:

$$\frac{\partial \rho}{\partial t} + \sum_{i=1}^{f} \left(\frac{\partial(\rho \dot{q}_i)}{\partial q_i} + \frac{\partial(\rho \dot{p}_i)}{\partial p_i} \right) = 0. \tag{2.33}$$

---

[37]For simplicity, consider the hydrodynamic flow of a real fluid in a tube of cross-section $\mathcal{A}$ oriented along the x-axis (see Figure 2.8). The particle density and the flow velocity are written $\rho(x, t)$ and $v(x, t) = \dot{x}$ respectively. Between the time instants t and $t + \delta t$, the change δM in the number of particles in the volume element $\mathcal{A} \delta x$ is

$$\delta M = (\rho(x, t + \delta t) - \rho(x, t)) \mathcal{A} \delta x \simeq \frac{\partial \rho(x, t)}{\partial t} \delta t \, \mathcal{A} \delta x. \tag{2.34}$$

This variation is also expressed as the difference between the number of particles entering into the volume element at x with a velocity $v(x, t)$ and the number of particles leaving at $x + \delta x$ at velocity $v(x + \delta x, t)$, during the time δt. The incoming particles were contained in the tube section of length $v(x, t) \delta t$ upstream of the volume element, i.e. $\rho(x, t) v(x, t) \delta t \mathcal{A}$ particles. Reasoning similarly for the outgoing particles, one finds:

$$\delta M = \underbrace{\rho(x, t) v(x, t) \delta t \mathcal{A}}_{\text{entering in } x} - \underbrace{\rho(x + \delta x, t) v(x + \delta x, t) \delta t \mathcal{A}}_{\text{leaving in } x + \delta x} \simeq -\frac{\partial(\rho v)}{\partial x} \delta x \, \delta t \, \mathcal{A}. \tag{2.35}$$

Equating Equations (2.34) and (2.35), one obtains the one-dimensional continuity equation:

$$\frac{\partial \rho}{\partial t} = -\frac{\partial(\rho v)}{\partial x} = -\frac{\partial(\rho \dot{x})}{\partial x},$$

which is easily generalised to the more abstract case of a flow in $2f$-dimensional phase space. In that case, the velocity $\boldsymbol{v} = (\dot{\boldsymbol{q}}, \dot{\boldsymbol{p}})$, distinguishes the axes associated with the coordinates $(q_1, q_2, \ldots, q_f)$ from the axes associated with the momenta $(p_1, p_2, \ldots, p_f)$: $\frac{\partial \rho}{\partial t} + \operatorname{div} \rho \boldsymbol{v} = 0$, i.e. Equation (2.33).

Furthermore, the particles' motion in real space, and thus the evolution of the representative points in phase space, obey the laws of classical mechanics expressed by Hamilton's Equations (1.11). By replacing $\dot{q}_i$ and $\dot{p}_i$ by the partial derivatives of the Hamiltonian H, the continuity equation (2.33) becomes:

$$\frac{\partial \rho}{\partial t} + \sum_{i=1}^{f} \left(\frac{\partial \rho}{\partial q_i} \dot{q}_i + \frac{\partial \rho}{\partial p_i} \dot{p}_i + \rho \left(\frac{\partial \dot{q}_i}{\partial q_i} + \frac{\partial \dot{p}_i}{\partial p_i} \right) \right) = 0$$

$$\frac{\partial \rho}{\partial t} + \sum_{i=1}^{f} \left(\frac{\partial \rho}{\partial q_i} \dot{q}_i + \frac{\partial \rho}{\partial p_i} \dot{p}_i + \rho \underbrace{\left(\frac{\partial^2 H}{\partial q_i \partial p_i} - \frac{\partial^2 H}{\partial p_i \partial q_i} \right)}_{0} \right) = 0$$

$$\frac{\partial \rho}{\partial t} + \sum_{i=1}^{f} \left(\frac{\partial \rho}{\partial q_i} \dot{q}_i + \frac{\partial \rho}{\partial p_i} \dot{p}_i \right) = 0.$$

The left-hand side of this equation is the *total derivative* of the density with respect to time, denoted $\frac{d\rho}{dt}$, that is the time variation of $\rho(q,p,t)$ evaluated by an observer moving with the swarm[38]. The first term, the partial derivative of ρ with respect to t, corresponding to an explicit time dependence of ρ, is the variation evaluated at a fixed point (q,p) of the phase space. The second term describes the implicit variation of the density as a function of time, due to the evolution of the coordinates and momenta. Liouville's equation[39] can be deduced from this:

$$\frac{d}{dt} \rho(q,p,t) \equiv \frac{\partial \rho}{\partial t} + \sum_{i=1}^{f} \left(\frac{\partial \rho}{\partial q_i} \frac{\partial H}{\partial p_i} - \frac{\partial \rho}{\partial p_i} \frac{\partial H}{\partial q_i} \right) = \frac{\partial \rho}{\partial t} + \{\rho, H\} = 0, \qquad (2.36)$$

where the Poisson brackets of ρ with H, written $\{\rho, H\}$, were introduced. In other words, the density is conserved over time, as is the density of an incompressible fluid. Note that this result, which is based only on the continuity equation and on Hamilton's equations, is exclusively a matter of (classical) mechanics. Since the number of representative points is constant, *Liouville's theorem* can be stated as follows: the volume of phase space occupied by the swarm is conserved during its evolution. On the other hand, its shape can change over time: from being compact (if the M systems of the swarm are initially in neighbouring microstates), the swarm can stretch into filaments extending into many regions of phase space while retaining its initial volume (see Figure 2.7).

Consequence for a system in thermodynamic equilibrium

In statistical physics, the phase-space density of representative, $\rho(q,p,t)$, is interpreted as the *probability density function* associated with the microcanonical statistical ensemble of

---

[38] Indeed, in one dimension of space, the total derivative of $\rho(x,t)$ with respect to t is written as:

$$\frac{d\rho}{dt} \equiv \lim_{\delta t \to 0} \frac{\rho(x(t+\delta t), t+\delta t) - \rho(x(t), t)}{\delta t} = \lim_{\delta t \to 0} \frac{\rho(x(t) + \dot{x}\delta t, t+\delta t) - \rho(x(t), t)}{\delta t}$$

$$\equiv \lim_{\delta t \to 0} \frac{1}{\delta t} \left(\rho(x(t), t) + \frac{\partial \rho}{\partial x} \dot{x}\delta t + \frac{\partial \rho}{\partial t} \delta t - \rho(x(t), t) \right) = \frac{\partial \rho}{\partial t} + \frac{\partial \rho}{\partial x} \dot{x}.$$

[39] Joseph Liouville (1809–1882) was a French mathematician and member of the Constituent Assembly in 1848. It was in fact the German mathematician Carl Gustav Jakob Jacobi (1804–1851) who applied to analytical mechanics a generalised Liouville's theorem (established in 1838) on differential equations and proved Equation (2.36) in 1842. Boltzmann then used it in 1871, in the context of the emerging field of statistical physics, and named the resulting theorem after Liouville (see D. D. Nolte, *The tangled tale of phase space*, Physics Today **63**, 33 (April 2010)).

M isolated systems. Intuitively, there is no reason based on the laws of mechanics for the representative points to favour one region of phase space over another, thus supporting the fundamental postulate and the uniform probability density that follows from it. More rigorously, in thermodynamic equilibrium, the ensemble averages, defined by Equation (1.34) are time-independent. The density must therefore be stationary. Let, for any microstate $(\boldsymbol{q}, \boldsymbol{p})$:

$$\frac{\partial \rho(\boldsymbol{q}, \boldsymbol{p}, t)}{\partial t} = 0. \tag{2.37}$$

According to Liouville's Equation (2.36), this is the case if $\{\rho, H\} = 0$. In other words, $\rho(\boldsymbol{q}, \boldsymbol{p})$ must be expressed only in terms of constants of the motion, that is quantities $\alpha_i(\boldsymbol{q}, \boldsymbol{p})$ which depend on time via microscopic variables, but whose total derivative, $\frac{\mathrm{d}\alpha_i(\boldsymbol{q}, \boldsymbol{p})}{\mathrm{d}t}$, is equal to zero[40]:

$$\rho(\boldsymbol{q}, \boldsymbol{p}) = \rho(\alpha_1(\boldsymbol{q}, \boldsymbol{p}), \alpha_2(\boldsymbol{q}, \boldsymbol{p}), \dots) = \rho(\alpha_1, \alpha_2, \dots) = C, \tag{2.38}$$

where C is a constant. Energy is, of course, a constant of motion in an isolated and confined system. If there are no others constants of motion[41], according to Equation (2.38), $\rho(\boldsymbol{q}, \boldsymbol{p})$ depends only on E and thus has the same value in every accessible microstate $(\boldsymbol{q}, \boldsymbol{p})$. Finally, Liouville's theorem shows that once in equilibrium, a *statistical ensemble* of $M \gg 1$ isolated systems is described by a uniform probability density function. However, it was not proved from the laws of mechanics that the microcanonical statistical ensemble describes equilibrium properties of a single isolated macroscopic system[42]. To do so, one would have to prove the equality of time averages and ensemble averages.

2.4.2 ERGODIC PROBLEM, HYPOTHESIS AND THEORY

The *ergodic problem* is a part of the mathematical theory of dynamical systems, which was originally developed in the framework of classical mechanics. Its purpose is to demonstrate,

---

[40] Assume for simplicity that $\rho(\alpha)$ depends on only one constant of motion, α. From Equation (2.36):

$$\frac{\partial \rho}{\partial t} = -\underbrace{\sum_{i=1}^{f} \left(\frac{\partial \rho(\alpha)}{\partial q_i} \frac{\partial H}{\partial p_i} - \frac{\partial \rho(\alpha)}{\partial p_i} \frac{\partial H}{\partial q_i} \right)}_{= \{\rho, H\}} = -\frac{\mathrm{d}\rho}{\mathrm{d}\alpha} \underbrace{\sum_{i=1}^{f} \left(\frac{\partial \alpha}{\partial q_i} \frac{\partial H}{\partial p_i} - \frac{\partial \alpha}{\partial p_i} \frac{\partial H}{\partial q_i} \right)}_{= \{\alpha, H\}} = 0,$$

since $\{\alpha, H\} = \frac{\mathrm{d}\alpha}{\mathrm{d}t} = 0$, α being a constant of motion.

[41] The systems studied in statistical physics are usually confined in an immovable container, with which they exchange momentum and angular momentum when the particles hit the walls. As elastic collisions are assumed, energy is, in principle, the only constant of motion. In the exceptional case where momentum and angular momentum are also conserved, and thus become external parameters of the system, a special treatment is required (see *Course of theoretical physics: Statistical physics*, Volume 5, by L.D. Landau and E.M. Lifshitz (Butterworth-Heinemann, 3rd edition, 2013)).

[42] Liouville's theorem is central to the approach to statistical physics proposed by Lev Landau (1908–1968), winner of the Nobel Prize in 1962 for his theoretical work in condensed matter physics (see *Course of theoretical physics: Statistical physics*, Volume 5, by L.D. Landau and E.M. Lifshitz (Butterworth-Heinemann, 3rd edition, 2013)). Without resorting to the abstract concept of statistical ensemble, the Russian physicist reasoned on a *single* macroscopic system supposed to be made up of a very large number of open subsystems, themselves macroscopic, and considered statistically *independent* and quasi-isolated for a *sufficiently short time* (statistical independence hypothesis). By defining the probability $\rho(\boldsymbol{q}, \boldsymbol{p}) \mathrm{d}^f q \mathrm{d}^f p$ to find a subsystem in a given microstate as the fraction of time – in the limit of *infinite times* – that its representative point spends in the volume element $\mathrm{d}^f q \mathrm{d}^f p$ centred in $(\boldsymbol{q}, \boldsymbol{p})$, the statistical average calculated with probability density function ρ is equivalent to the time average without having to use the concepts of statistical ensemble and ensemble average. Landau then showed that statistical independence and Liouville's theorem imply uniformity of the probability density function (as claimed by the fundamental postulate). Although elegant, the assumption of statistical independence is questionable, and there is little discussion of this approach in the literature (see Penrose's article (1979) (cited in footnote 35) and N. Singh, *On the foundations of statistical mechanics: A brief review*, Mod. Phys. Lett. B **27**, 1330003 (2013)).

for isolated systems, the equality of time averages in the *infinite time limit* and equilibrium microcanonical ensemble averages:

$$\lim_{t_{\exp} \to +\infty} \frac{1}{t_{\exp}} \int_{t_0}^{t_0+t_{\exp}} A(\boldsymbol{q}(t), \boldsymbol{p}(t)) \, \mathrm{d}t \overset{?}{=} \int A(\boldsymbol{q}, \boldsymbol{p}) \, \rho(\boldsymbol{q}, \boldsymbol{p}) \, \mathrm{d}^f q \mathrm{d}^f p, \qquad (2.39)$$

where t_0 is the initial time and $\rho(\boldsymbol{q}, \boldsymbol{p})$ the probability density function given by Equation (2.14). A dynamical system is said to be "ergodic" if it satisfies Equation (2.39). Most of the statistical physics literature gives a fundamental role to the ergodic problem, as its solution seems to reduce statistical physics to a branch of classical mechanics and justify the use of probabilities and statistical ensembles. It is nevertheless important to note that statistical physics is based on the equality $\bar{A} = \langle A \rangle$, where the time average $\bar{A}$ is measured over the *finite* duration $t_{\exp}$ of an experiment (see Equation (1.14)), whereas the ergodic problem is in the formal limit $t_{\exp} \to +\infty$. Two questions then arise: does Equation (2.39) hold for systems studied in statistical physics? And is the infinite time limit relevant given the time scales involved in these systems?

To justify Equation (2.39), physicists often invoke a hypothesis put forward by Boltzmann in 1868 and subsequently called the *ergodic hypothesis*[43]: during its time evolution, an isolated system successively explores *all* accessible microstates. In other words, the phase space trajectory of the system's representative point goes through *all* points of the hypersurface of energy E. If the ergodic hypothesis were correct, not only would the ergodic problem be solved[44], but one could also deduce that the microcanonical probability density function – uniform over the accessible phase space – is the only one that is stationary and therefore able to describe the equilibrium of an isolated system (see Uffink (2006), Note 35). The ergodic hypothesis thus seems to solve the problems posed by the foundations of statistical physics. However, the role of ergodicity has been controversial since its inception and is still debated today (see Bricmont (1995), Goldstein (2001) and Sklar (1995), Note 35).

First, Rosenthal and Plancherel demonstrated in 1913 that the ergodic hypothesis in the strict sense is false[45], as Boltzmann himself thought. Indeed, a system can explore only a tiny part of the *a priori* accessible microstates, if it is initially in very particular given microstates. This is the case of particles perfectly aligned on an axis perpendicular to a wall of a cubic enclosure and with velocity vectors parallel to the same axis. These particles have a permanent back and forth motion between two opposite walls and the system's representative point cannot explore the entire accessible phase space. A weaker hypothesis is then introduced, known as "quasi-ergodic": in the course of its time evolution, an isolated

---

[43]Boltzmann introduced this hypothesis, without making it an essential element of his approach (see Uffink (2006), Note 35). It was Maxwell who gave it its current formulation in 1879 and a central role in the foundations of statistical physics. The term "ergodic" was coined by Boltzmann in 1884 to designate the statistical ensemble, later called *microcanonical* by Gibbs, all of whose systems verify the so-called *ergodic hypothesis* (see Uffink (2006), Note 35).

[44]Consider a volume element of the phase space $\mathrm{d}^f q \mathrm{d}^f p$. On the one hand, according to the ergodic hypothesis, *every* system of the statistical ensemble spends an equal time $\mathrm{d}t$ in this volume element over phase space trajectories followed for a time $t_{\exp}$ assumed long enough so that each system has the time to explore the whole accessible phase space. The cumulative time Δt spent by the $M \gg 1$ systems in this volume element is therefore $\Delta t = M \mathrm{d}t$. Moreover, in equilibrium, there are at each time instant in the microcanonical statistical ensemble $\mathrm{d}M = M\rho(\boldsymbol{q}, \boldsymbol{p}) \, \mathrm{d}^f q \mathrm{d}^f p$ systems whose representative points are located in the volume element $\mathrm{d}^f q \mathrm{d}^f p$. Therefore, $\Delta t = \int_0^{t_{\exp}} \mathrm{d}M \mathrm{d}t = t_{\exp} \mathrm{d}M$. By equalising the two estimations of Δt, one deduces:

$$\frac{\mathrm{d}t}{t_{\exp}} = \frac{\mathrm{d}M}{M} = \rho(\boldsymbol{q}, \boldsymbol{p}) \, \mathrm{d}^f q \mathrm{d}^f p.$$

The probability of finding the studied system at a given time in the considered volume element of phase space is therefore equal to the probability that one of the systems of the microcanonical statistical ensemble is in that same volume. Equation (2.39) can be deduced from this result.

[45]Except in the very special case of systems with a one-dimensional phase space hypersurface of constant energy E (i.e. $f = 1$), such as the one-dimensional harmonic oscillator.

system successively explores *almost all* accessible microstates. Physically, this nuance in the formulation of the ergodic hypothesis does not seem important since microstates leading to this type of non-ergodic behaviour are extremely rare. In addition, weak interactions with the outside world, necessarily present even though the system is assumed isolated, disturb the trajectories of the particles, thus allowing the representative point to explore the accessible phase space. However, these two assumptions are mathematically very different. On the one hand, the ergodic hypothesis, although false, implies Equation (2.39). On the other hand, the quasi-ergodic hypothesis has not been proved for the systems of interest in statistical physics. Moreover, it does not necessarily imply in any case Equation (2.39)[46]. Faced with this impasse, a new hypothesis, called *metric transitivity*, was introduced in the 1930s, giving rise to a branch of mathematics, paradoxically called *ergodic theory*, which is well beyond the scope of this book[47]. According to the von Neumann and Birkhoff ergodic theorem (1931), a system is ergodic (i.e. it verifies Equation (2.39)) if and only if it is metrically transitive. However, until now, there is no proof that realistic systems studied in statistical physics verify this dynamical systems property[48].

Secondly, a system does not have to be ergodic to be very well described by statistical physics. Thus, *independent particles* model systems, such as ideal gases, are clearly not ergodic. In the absence of interactions between particles, they retains their energies over time (as well as the magnitude of their momenta in the case of ideal gases). In other words, the total energy distribution among particles is fixed and the system's representative point remains confined to the small region of the accessible phase space compatible with this unique distribution[49]. Similarly, solids are not ergodic systems. As their atoms are located around the nodes of a crystal lattice, the system's representative point explores only a small region of the accessible phase space (at a given energy E), compatible with the position and the orientation of the lattice[50].

Finally, it is not clear if ergodicity is a relevant property to justify the replacement of time averages by ensemble averages in statistical physics, as an estimation of time scales

---

[46]There are systems that verify the quasi-ergodic assumption, but not Equation (2.39) (see Uffink (2006), Note 35).

[47]A dynamical system has the property of metric transitivity if the accessible phase space cannot be decomposed into regions of *finite* (non zero) volume such that every trajectories initiated within these regions stay confined in them (see Uffink (2006), Note 35). For such a system, the set of phase space points that cannot be explored by a trajectory is effectively negligible.

[48]Systems of hard spheres have been the subject of many mathematical studies since the 1960s. For example, Russian mathematician Yakov Sinai demonstrated in 1963 that under certain conditions a system of (only) two (two-dimensional) hard disks is metrically transitive (see Uffink (2006), Note 35). His complex work has been continued to the present day, in particular by mathematician Nàndor Simányi, but the question of the ergodicity of a system of N hard spheres studied in physics remains open (see *The Boltzmann-Sinai ergodic hypothesis in full generality*, arXiv:1007.1206v2 (2010)). Not to mention more realistic fluid models that take into account attraction between particles.

[49]In ergodic theory itself, the number of particles is arbitrary. In 1949, Russian mathematician Khintchine explicitly took into account the $N \gg 1$ limit for macroscopic systems and demonstrates the following result: in the particular case of *independent particle* systems and internal variables that are written as a sum over all particles ($A = \sum_{j=1}^{N} a_j$, where a_j characterises the particle j), $\bar{A}$ tends *in probability* towards $\langle A \rangle$ in the infinite time limit for $N \to \infty$ (see Uffink (2006), Note 35). It is therefore a weak version of ergodicity.

[50]By abuse of language a liquid is said to be ergodic, in the sense that the system can, *in principle*, explore (almost) any region of the accessible phase space. During a thermodynamic phase transition of solidification, the system moves from a liquid (so-called *ergodic*) to a solid (clearly non-ergodic) equilibrium state. The *ergodicity break* is then due to a symmetry breaking in the system: unlike liquid, crystals are not invariant by translation and rotation symmetry in space. The case of the *glass transition* is different. When a liquid is cooled sufficiently rapidly below its melting temperature, its atoms do not have time to position themselves at the nodes of a lattice and to form a crystal in equilibrium. The resulting system, called *glass*, has the mechanical properties of a solid, but its disordered microscopic structure is typical of a liquid. As the system remains trapped in non-equilibrium states, the ergodicity breakdown here is dynamic in nature: the system does not have time to relax to the equilibrium state (the crystal) for the duration of the experiment (see Note 24 in Chapter 1).

shows. Indeed, in dynamical systems theory, Equation (2.39) is considered in the infinite time limit. For a *macroscopic* system, the characteristic time of ergodicity is typically the time needed for its representative point to explore (almost) the entire accessible phase space. This time, which can only be estimated approximately, as Boltzmann did in 1896, is of the order of $t_{ergo} \simeq 10^{10^{23}}$ years[51]. This gigantic time is far greater than the age of the universe since the Big Bang ($\sim 14 \times 10^9$ years). For a physicist, the infinite limit in Equation (2.39) should be taken literally! The duration of an experiment is generally much larger than microscopic times ($t_{micro} \simeq 10^{-12}$ s), but it is completely negligible compared to the ergodicity characteristic time: $t_{micro} \ll t_{exp} \ll t_{ergo}$. Thus, during an experiment, a system explores only a tiny fraction of the accessible microstates (of the order of $1/10^{10^{23}}$). It is fortunately not necessary to wait until the dawn of time to measure macroscopic quantities (time averages) in equilibrium and compare them (successfully!) with the statistical physics ensemble averages[52].

Although ergodicity is often invoked, its role, and the role of the dynamical systems properties in general, seems to be of little relevance in justifying the foundations of equilibrium statistical physics.

2.4.3 STATISTICAL PHYSICS ACCORDING TO BOLTZMANN

Why can we then replace time averages measured *in equilibrium* during an experiment, in which the system explores only a *minuscule part* of accessible phase space, by an ensemble average calculated over *all* accessible microstates? Boltzmann provided a convincing answer to this question, which has recently been taken up in illuminating articles (see Bricmont (1995), Goldstein (2001) and Lebowitz (1993), Note 35).

An isolated macroscopic system can be in different macrostates at the same energy. For example, consider a drop of ink initially deposited in a glass of water. The ink is assumed to be made up of a single type of molecule and the (water + ink) system is assumed isolated. Photographs taken at different times show a succession of macrostates (see Figure 1.4). For each of these macrostates, there is a gigantic number of microstates, equal to the number of ways to assign positions and momenta to water and ink molecules that are compatible with the shape of the drop in that macrostate. Let Γ_M be the ensemble of Ω_M similar microstates that, on a macroscopic scale, have the appearance and the physical properties of macrostate[53] M: in each microstate $(\boldsymbol{q}, \boldsymbol{p})$ of Γ_M, a macroscopic internal variable $A(\boldsymbol{q}, \boldsymbol{p})$

---

[51] Indeed, according to Equation (2.24), the number of accessible microstates for a N particles system is $\Omega \sim 10^N$. Since the atomic dynamics characteristic time is typically $t_{micro} \simeq 10^{-12}$ s (see Section 1.4.2), a system can be considered to change microstate on average every t_{micro} seconds. For $N \gg 1$, the time required to explore the accessible phase space is therefore of the order of $t_{ergo} \simeq \Omega \, t_{micro} \simeq 10^{(N-12)} \simeq 10^N$ seconds (or in any unit of time and regardless of the estimate of t_{micro}).

[52] Some authors have used ergodicity to prove the very existence of an equilibrium state. However, ergodicity does not imply the existence of a stationary state, as the harmonic oscillator example shows. On the contrary, non-ergodic systems can reach steady states (see Bricmont (1995), Note 35). Other properties of dynamical systems are then invoked, in particular the mixing property. Briefly and without formalism, a system is mixing if an initially compact swarm of representative points spreads over time to cover almost all of the accessible phase space, similar to the diffusion of a drop of ink in water observed in the real world. Since the volume of the swarm is conserved (according to Liouville's theorem), its shape must become extremely branched so that the representative points are more or less uniformly distributed over the accessible phase space (see Figure 2.7). This property is stronger than metric transitivity (see Note 47). It can be intuitively seen that for such a system, an average calculated on this ensemble of representative points is very close to the ensemble average obtained by averaging over *all* accessible microstates. However, as for ergodicity, there is no proof that a realistic system is mixing and this property is also defined in the infinite time limit. The mixing property is therefore not satisfactory to ensure that a system reaches an equilibrium state (see Bricmont (1995), Goldstein (2001) and Uffink (2006), Note 35). Note that Landau's approach, presented in footnote 42, does not invoke the ergodic assumption, but also relies on an infinite time limit.

[53] During the exposure time of a photo representing macrostate M, typically 10^{-2} seconds, the system explores a very large number of microstates (of the order of $10^{-2}/t_{micro} \simeq 10^{10}$, according to the estimate of

has (almost) the same value[54] A_M. Phase space can then be decomposed into domains of macrostates according to the value of A_M, in a relatively arbitrary way, since the boundaries depend on the precision one has on the macroscopic internal variables. This operation, known as "coarse-graining", makes sense in the $N \gg 1$ limit, when relative fluctuations of the macroscopic internal variables are very small and the distinction between micro- and macro-states becomes obvious.

According to Boltzmann, an equilibrium state is by definition a macrostate whose number of associated microstates Ω_{eq} is very large in front of the number of microstates associated with other so-called *non-equilibrium macrostates*. The $\Omega(N,V,E)$ accessible microstates thus decompose into Ω_{eq} *typical* equilibrium microstates belonging to the ensemble Γ_{eq} and $\Omega(N,V,E) - \Omega_{eq}$ atypical (non-equilibrium) microstates, such that[55]:

$$\Omega(N,V,E) = \Omega_{eq} + \underbrace{\sum_{M \text{ non-equilibrium.}} \Omega_M}_{\ll \Omega_{eq}} \simeq \Omega_{eq}. \qquad (2.40)$$

Phase space decomposition is illustrated in Figure 2.10.

To give the order of magnitude of the ratio $\Omega_{eq}/\Omega(N,V,E)$, consider an ideal gas that can be in different macrostates in which the N molecules occupy a region of volume $\alpha V < V$, where V is the volume of the container (see Figure 2.9).

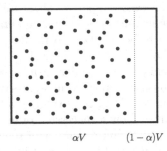

$\alpha V \qquad\qquad (1-\alpha)V$

FIGURE 2.9 Microstate corresponding to a non-equilibrium macrostate: particles are distributed in a volume $\alpha V < V$, where V is the total volume of the container.

In equilibrium, the molecules occupy the whole container ($\alpha = 1$). In contrast, macrostates in which $\alpha < 1$ correspond to non-equilibrium macrostates, if $\alpha \neq 1$. It was discussed in Section 2.2.4 that $\Omega(N,V,E) \sim V^N$. The number of microstates associated with non-equilibrium macrostates therefore behaves as $(\alpha V)^N$. From Equation (2.40), the fraction of the accessible phase space occupied by Γ_{eq} is given by:

$$\frac{\Omega_{eq}}{\Omega(N,V,E)} \simeq 1 - \left(\frac{\alpha V}{V}\right)^N = 1 - e^{10^{23} \ln \alpha} \simeq 1, \quad \text{with} \quad \alpha < 1, \qquad (2.41)$$

for $N = 10^{23}$, whatever the value[56] of $\alpha \neq 1$. This estimate can be obtained by using the Gaussian approximation of the probability density function of a macroscopic internal

t_{micro} presented in Note 51). This number is itself negligible compared to the total number of microstates contained in ensemble Γ_M.

[54]However, a *microscopic* internal variable, such as the velocity of a particle, can considerably fluctuate in ensemble Γ_M.

[55]Equation (2.28) is recovered, with by definition $\Omega_{eq} \equiv \Omega_p(E \,|\, A_{eq})$, which follows from the extremely peaked form of the probability density function of an internal variable around its equilibrium value (see Section 2.3).

[56]An empty region of particles observable on human scale must have a macroscopic volume, say greater than 1 mm³, so $\alpha < (1 - 10^{-9})$ for a container of volume $V = 1$ m³. Therefore, $\Omega_{eq}/\Omega(N,V,E) \simeq (1 - 10^{-10^{14}}) \simeq 1$.

variable[57] (see Section 2.3): the probability of observing a relative deviation to the mean value greater than, for example 10^{-6}, is of the order of $10^{-10^{12}}$. In other words, the ensemble of microstates Γ_{eq} typical of the equilibrium macrostate – sometimes called the *Boltzmann sea* – occupies almost the entire accessible phase space. In addition, in all these microstates, a macroscopic internal variable takes on values (almost) equal (say to within 10^{-12}) to the equilibrium value A_{eq}.

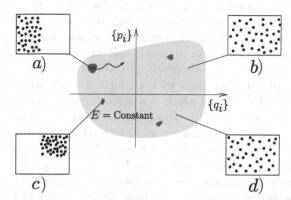

FIGURE 2.10 The accessible phase space (clear grey) is almost entirely occupied by the Boltzmann sea, consisting of typical equilibrium microstates (e.g. b) and d)). Each island (dark grey) groups together an ensemble of atypical microstates (such as a) and c)) corresponding to a non-equilibrium macrostate. Their hypersurfaces are *considerably enlarged to be visible on the figure*. The arrowed curve illustrates the phase space trajectory of the representative point of a gas prepared in the initial non-equilibrium microstate a) (to avoid overloading the figure, the momenta vectors are not drawn).

The vastness of the Boltzmann sea allows to answer – at least qualitatively – the questions raised about the foundations of statistical physics. Firstly, the reach of equilibrium can be understood as follows: starting from an atypical microstate, associated with a non-equilibrium macrostate (e.g. a drop of ink in a glass of water), the system's representative point first evolves in a small region of phase space corresponding to a succession of non-equilibrium macrostates. Sooner or later, it leaves this small region and reaches the gigantic Boltzmann sea Γ_{eq}, in which it will continue its journey (see Figure 2.10)[58]. Note again that it is not necessary for the system to visit all accessible microstates to reach equilibrium as suggested by the ergodic problem[59].

Then, the postulate of equality of time and ensemble (here microcanonical) averages is justified on two grounds. Firstly, in each typical (equilibrium) microstate a macroscopic internal variable has by definition a value $A(\boldsymbol{q}, \boldsymbol{p}) \simeq A_{eq}$. The time average $\bar{A}$ is therefore

---

[57]Indeed, the probability that a Gaussian random variable deviates from its mean value by more than $n = 10^6$ times its standard deviation ($\sim 10^{-12}$) is equal to $\mathrm{erfc}(n/\sqrt{2})$, where the complementary error function is defined as:

$$\mathrm{erfc}(x) \equiv \frac{2}{\sqrt{\pi}} \int_x^{\infty} e^{-u^2} \mathrm{d}u \simeq \frac{e^{-x^2}}{\sqrt{\pi}x} \quad \text{for } x \to \infty,$$

that is a probability $\simeq e^{-(10^6)^2}/10^6 \sim 10^{-10^{12}}$, for $x = 10^6$.

[58]The approach to equilibrium must be studied within the theoretical framework of non-equilibrium statistical physics which takes into account the system's dynamics. In some situations, a system can remain trapped for a long time in atypical microstates: a metastable macrostate. This is the case for glasses (see Note 24 in Chapter 1).

[59]Thus, in the drop of ink in water example, as soon as the colour of the mixture is uniform, equilibrium can be considered as reached, without waiting the "infinite" amount of time necessary for every ink molecule to collide with every water molecule in every point in the container.

measured from instantaneous values that are practically constant during the experiment (see Figure 1.5). Thus, the tiny fraction of typical microstates visited by the system during the experiment is sufficient[60] for $\bar{A}$ to converge rapidly towards the equilibrium value A_{eq}. Secondly, the Ω_{eq} typical (equilibrium) microstates represent almost all the $\Omega(N, V, E)$ accessible microstates. The atypical microstates, in contrast, are so few in number that, although they correspond to internal variables values far removed from equilibrium values, their contribution to the ensemble averages is negligible due to the overwhelming number of $\Omega_{eq} \simeq \Omega(N, V, E)$ typical microstates: equilibrium time averages, estimated on a *small fraction of typical microstates*, can thus be replaced by ensemble averages defined on *all* accessible microstates, whether typical or atypical. And since in the microcanonical ensemble all accessible microstates have the same probability (fundamental postulate), the situation is similar to that of a player who plays with an unloaded die with $\Omega(N, V, E) \gg 1$ faces (microstates) on the vast majority of which is written (almost) the same value[61], A_{eq}.

To conclude, there is no consensus among physicists on the foundations of statistical physics. Despite many attempts to reduce statistical physics to a branch of mechanics and demonstrate rigorously the two postulates with the help of dynamical systems theory (ergodicity, mixing...), it seems that the transition from microscopic systems to macroscopic systems cannot rely only on microscopic, classical or quantum, theories. As Boltzmann's ideas presented in this section and the more formal developments of his successors qualitatively show, macroscopic properties of a system can only be deduced from its microscopic nature if one takes into account the large number $N \gg 1$ of system's particles and uses probability theory (see Penrose (1979), note 35). The success of statistical physics then relies essentially on the extraordinarily large value of N and the resulting vastness of the Boltzmann sea, making probabilistic predictions almost certain.

---

[60]The success of the molecular dynamics simulations, presented in Section 1.3.2, is based on this observation. The computer-simulated trajectories are different from the exact trajectories that would be calculated analytically, but if equilibrium is reached, the small number ($\sim 10^7$ iterations) of typical microstates explored is sufficient to correctly calculate time-averaged values (see D. Frenkel, *Simulations: The dark side*, Eur. Phys. J. Plus **128**, 10 (2013)). Even if the small number of particles ($N \simeq 10^3$ to 10^6) taken into account in a simulation implies large fluctuations around these average values. Note that chaos disturbs the system's trajectory, real or simulated, but has no effect on the determination of thermodynamic quantities, since its representative point evolves anyway in the Boltzmann sea.

[61]According to the fundamental postulate, the four microstates *a*), *b*), *c*) and *d*) shown in Figure 2.10 have the same probability. This may seem surprising, as no gas *in equilibrium* has ever been observed to condense even briefly in one part of a container: imagine the air molecules around you concentrating in one corner of the room where you are standing, leaving you locally in a vacuum..., but typical microstates (such as *b*) and *d*)) are immensely more numerous than atypical microstates (such as *a*) and *c*)).

2.5   EXERCISES

Exercise 2.1.   (Spinless) Ideal quantum gas

Consider firstly a particle of mass m, with no internal structure and no spin, confined in a cubic container of volume $V = L^3$.

1. How can the microstate of the system be characterised in the framework of quantum mechanics? Since the energy of the particle is given by Equation (1.1), what are the energies E_i and the degeneracies d_i of the energy levels $i = 1, 2, \ldots, 15$?

2. In a semi-classical approximation, it can be assumed that the three quantum numbers associated with the particle are distinguishable. Introducing a phase space whose three axes carry the integer values of n_1, n_2 and n_3, give the expression of $\Phi(1, V, E)$, the number of microstates with an energy less than E in the high energy limit.

3. Deduce the number of accessible microstates $\Omega(1, V, E)$ for an argon atom (with $M = 40$ g $\cdot$ mol^{-1}) contained in a can of volume $V = 1$ litre, whose energy is between E and $E + \delta E$, where $E = 6 \times 10^{-21}$ J and $\delta E = 10^{-30}$ J.

4. Consider now N identical non-interacting particles. The energy of this ideal gas is given by

$$E(\{n_i\}) = \epsilon \sum_{i=1}^{3N} n_i^2 \quad \text{where} \quad \epsilon = \frac{h^2}{8mL^2}.$$

   By generalising the result of question 2, calculate the number of microstates $\Phi(N, V, E)$ having an energy less than E (particles are assumed indistinguishable). Compare the obtained expression with Equation (2.18) determined in the framework of classical physics.

Solution:

1. _A microstate is characterised by three strictly positive integer numbers n_1, n_2 and n_3. With $\epsilon = h^2/(8mL^2)$, the energy levels listed in Table 2.1 are obtained. This exercises gets rapidly fastidious... However it is interesting to write a small computer code to numerically calculate d_i until $i = 10000$ and deduce $\Phi(1, V, E_i) = \sum_{j=1}^{i} d_j$. The obtained curve can then be compared to the one given by Equation (2.42) derived in the next question._

2. _$\Phi(1, V, E)$ is equal to the number of triplets (n_1, n_2, n_3) such as $n_1^2 + n_2^2 + n_3^2 \leq R^2 = E/\epsilon$. Since the three quantum numbers are strictly positive integers, the representative points of the accessible microstates are localised in a part of phase space such as $n_{1,2,3} > 0$. The number $\Phi(1, V, E)$ is therefore approximately the number of cubes of side 1 contained in $1/8^{\text{th}}$ of the sphere of radius $R = \sqrt{E/\epsilon} \gg 1$:_

$$\Phi(1, V, E) \simeq \frac{1}{8} \frac{4}{3} \pi R^3 = \frac{4\pi L^3}{3h^3} (2mE)^{3/2}. \tag{2.42}$$

3. _For one argon atom of mass $m = M/N_A$, one obtains with Equation (2.2):_

$$\Omega(1, V, E) \simeq \frac{\partial \Phi}{\partial E} \delta E = \frac{2\pi V}{h^3} (2m)^{\frac{3}{2}} \sqrt{E} \, \delta E \simeq 10^{20}.$$

4. _By definition, $\Phi(N, V, E)$ is equal to the number of combinations $(n_1, n_2, \ldots, n_{3N})$ such as $\sum_{i=1}^{3N} n_i^2 \leq R^2 = E/\epsilon$. Since the $3N$ quantum numbers are strictly positive integers, the representative points of the accessible microstates are localised in the part of phase space such as $n_{1,2,\ldots,3N} > 0$. The number $\Phi(N, V, E)$ is thus approximately the number of hypercubes_

TABLE 2.1 The 15 first energy levels of a particle confined in a cubic box.

niveau i	$n_1\ n_2\ n_3$	d_i	E_i/ϵ
15	4 3 2	6	29
14	3 3 3 and 5 1 1	4	27
13	4 3 1	6	26
12	4 2 2	3	24
11	3 3 2	3	22
10	4 2 1	6	21
9	3 3 1	3	19
8	4 1 1	3	18
7	3 2 2	3	17
6	3 2 1	6	14
5	2 2 2	1	12
4	1 1 3	3	11
3	1 2 2	3	9
2	1 1 2	3	6
1	1 1 1	1	3

of side 1 contained in the fraction $1/2^{3N}$ of the hypersphere of radius $R = \sqrt{E/\epsilon} \gg 1$, in a 3N-dimensions space:

$$\Phi(N,V,E) \simeq \frac{1}{N!}\frac{1}{2^{3N}}c_{3N}R^{3N} = \frac{1}{N!}c_{3N}\left(\frac{L\sqrt{2mE}}{h}\right)^{3N},$$

where c_{3N} is the volume of the hypersphere of radius 1 given by Equation (2.17). The factor $1/N!$ takes into account the indistinguishability of the particles in this semi-classical approximation. Expression (2.18) obtained in the classical physics framework is recovered.

Exercise 2.2. Sorting

Two boxes can contain two objects each. What is the number of ways to arrange two objects in these two boxes, depending on whether the objects are distinguishable or indistinguishable? Does Relation (2.8) hold and why?

Solution:

In terms of particles and individual states, there are two particles (objects) and two different individual states (boxes). Two particles can occupy the same individual state (the same box). Since the two particles are discernible, they are represented as • and ∘ respectively. There are then $\Omega_{Discernables} = 4$ accessible microstates:

| • | ∘ | | ∘ | • | | •∘ | | | | •∘ |.

Since the two particles are indistinguishable, they are both represented by the same symbol •. There are then $\Omega_{Indiscernables} = 3$ accessible microstates:

| • | • | | •• | | | | •• |.

In this example, Relation (2.8) is clearly false, because two particles (objects) can share the same individual state (the same box).

Exercise 2.3. System of independent classical harmonic oscillators

Consider an isolated system consisting of N distinguishable and independent one-dimensional *classical* harmonic oscillators of mass m and angular frequency ω. In this problem, the volume of the system is not involved.

1. Give the expression of the system's Hamiltonian.

2. Express $\Phi(N, E)$, the number of microstates with an energy less than E, as an integral over phase space. To calculate this integral the following change of variables $q_i' = m\omega q_i$ will be performed. Deduce the expression of $\Omega(N, E)$.

Solution:

1. The Hamiltonian is $H(\boldsymbol{q}, \boldsymbol{p}) = \sum\limits_{i=1}^{N} \left(\dfrac{p_i^2}{2m} + \dfrac{1}{2}m\omega^2 q_i^2 \right)$.

2. *Since the harmonic oscillators are distinguishable and since there are $f = N$ degrees of freedom:*

$$\Phi(N, E) = \frac{1}{h^N} \int_{\sum\limits_{i=1}^{N} \left(\frac{p_i^2}{2m} + \frac{1}{2}m\omega^2 q_i^2 \right) \leq E} \mathrm{d}^N q \, \mathrm{d}^N p.$$

With the proposed change of variables, one gets

$$\Phi(N, E) = \frac{1}{(hm\omega)^N} \int_{\sum\limits_{i=1}^{N} \left(p_i^2 + q_i'^2 \right) \leq 2mE} \mathrm{d}^N q' \, \mathrm{d}^N p.$$

The integral is the volume of a hypersphere of radius $\sqrt{2mE}$ in a $2N$ dimensional space. Knowing the volume of the $2N$-dimensional hypersphere of radius 1 (i.e. $c_{2N} = \pi^N / N!$), one gets:

$$\Phi(N, E) = \frac{1}{(hm\omega)^N} c_{2N} (\sqrt{2mE})^{2N} = \frac{1}{N!} \left(\frac{E}{\hbar\omega} \right)^N.$$

Note that the factor $1/N!$ comes from the expression for c_{2N} (the particles are assumed distinguishable). Note that $\Phi \sim E^N$, the typical behaviour given by Equation (2.22) for independent particle systems. Finally:

$$\Omega(N, E) \simeq \frac{\partial \Phi}{\partial E} \, \delta E = \frac{1}{(N-1)!} \left(\frac{E}{\hbar\omega} \right)^{N-1} \frac{\delta E}{\hbar\omega}. \tag{2.43}$$

Exercise 2.4. System of independent quantum harmonic oscillators

An isolated system consists of N *distinguishable* and *independent* one-dimensional quantum harmonic oscillators of angular frequency ω.

1. Give the expression of the energy of the system as a function of the quantum numbers n_i, where $i = 1, 2, \ldots, N$. Deduce that fixing the energy E of the system is equivalent to fixing the total number of vibrational quanta:

$$M = \sum_{i=1}^{N} n_i. \tag{2.44}$$

2. Calculate $\Omega_0(N, M)$, the number of microstates with an energy exactly equal to E ($\delta E = 0$), corresponding to M quanta, first in the special case where $N = 2$ and $M = 3$, then in the general case. The M quanta are considered indistinguishable.

3. Check the obtained expression by directly calculating the degeneracy of the energy levels of a three-dimensional harmonic oscillator ($N = 3$ and any M).

4. Calculate $\Omega(N, E)$, the number of accessible microstates with energies between E and $E + \delta E$, in the high-energy limit (N fixed with $M \gg N$). Compare with the result obtained in the framework of classical physics (Exercise 2.3).

Solution:

1. The energy of N independent oscillators is given by:

$$E = \sum_{i=1}^{N} \hbar\omega(n_i + \frac{1}{2}) = \hbar\omega(M + \frac{N}{2}). \qquad (2.45)$$

Fixing the energy E is therefore equivalent to fixing the number of vibrational quanta, M.

2. $\Omega_0(N, M)$ is the number of ways to distribute M (indistinguishable) quanta among N (discernible) oscillators. For each oscillator, $n_i = 0, 1, 2, \ldots, M$, with Constraint (2.44). For example, there are $\Omega_0(2, 3) = 4$ distributions (n_1, n_2) of $M = 3$ quanta among $N = 2$ oscillators: $(0, 3)$, $(1, 2)$, $(2, 1)$ and $(3, 0)$. In general, the problem is equivalent to distributing M indistinguishable balls into N discernible boxes, separated by $N - 1$ walls, as shown in Figure 2.11.

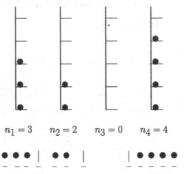

$$n_1 = 3 \qquad n_2 = 2 \qquad n_3 = 0 \qquad n_4 = 4$$

FIGURE 2.11 A distribution of $M = 9$ quanta among $N = 4$ discernible harmonic oscillators. In the lower figure, the | symbol represents a separation between two harmonic oscillators. Counting the accessible microstates is then equivalent to counting the number of ways in which M (•) and $N - 1$ (|) can be distributed over $M + N - 1$ positions (−).

There are therefore M balls to be distributed among $M + N - 1$ positions, so

$$\Omega_0(N, M) = \binom{M + N - 1}{M} = \frac{(M + N - 1)!}{M!(N - 1)!}. \qquad (2.46)$$

One finds again $\Omega_0(2, 3) = 4$.

3. $\Omega_0(3, M)$ can be calculated directly: first, assign n_1 quanta to the first oscillator, that is $0 < n_1 < M$. Then, $M - n_1$ quanta are still available fro the second oscillator, so $0 \le n_2 \le M - n_1$. The number of quanta allocated at the third oscillator is then fixed: $n_3 = M - n_1 - n_2$. Thus:

$$\Omega_0(3, M) = \sum_{n_1=0}^{M} \sum_{n_2=0}^{M-n_1} 1 = \sum_{n_1=0}^{M} (M - n_1 + 1) = (M + 1)^2 - \sum_{n_1=0}^{M} n_1$$

$$= (M + 1)^2 - \frac{M(M + 1)}{2} = \frac{(M + 1)(M + 2)}{2}.$$

This result is also found using the general expression (2.46).

4. In the high energy limit $M \gg N$, Equation (2.46) becomes

$$\Omega_0(N, M) = \frac{1}{(N - 1)!} \underbrace{(M + N - 1)}_{\simeq M} \underbrace{(M + N - 2)}_{\simeq M} \ldots \underbrace{(M + 1)}_{\simeq M}$$

$$\simeq \frac{1}{(N - 1)!} M^{N-1} \simeq \frac{1}{(N - 1)!} \left(\frac{E}{\hbar\omega}\right)^{N-1},$$

since, from Equation (2.45), $E \simeq \hbar\omega M$ at the limit $M \gg N$. Finally, since there are $\delta E / \hbar\omega$ energy levels separated by $\hbar\omega$, between E and $E + \delta E$:

$$\Omega(N,E) \simeq \Omega_0(N,M) \frac{\delta E}{\hbar\omega} \simeq \frac{1}{(N-1)!} \left(\frac{E}{\hbar\omega} \right)^{N-1} \frac{\delta E}{\hbar\omega}.$$

Equation (2.43), obtained using the framework of classical physics, is indeed the high energy limit of the quantum expression of $\Omega(N,E)$. It is, of course, the volume of the elementary cell of the phase space, chosen equal to h^f, that allows this correspondence (see Section 2.2.2).

Exercise 2.5. Distinguishable/indistinguishable particles

Consider a N particles quantum system. To each energy level E_i is associated a number of quantum states g_i (degeneracy). A macrostate of energy E is defined by the ensemble of numbers of particles $\{N_i\}$ assigned to each energy level E_i for $i = 1, 2, \ldots, r$ (see Figure 2.12).

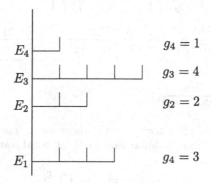

FIGURE 2.12 Energy levels E_i and degeneracies g_i of a quantum system.

1. Express E and N as a sum over the energy levels.

2. Show that the number of microstates $\Omega(N,E)$ associated with a given distribution $\{N_i\} = (N_1, N_2, \ldots, N_r)$ can be decomposed as follows:

$$\Omega(N,E) - \Omega_0 \prod_{i=1}^{r} \Omega_i, \qquad (2.47)$$

where Ω_0 is the number of ways to distribute the N particles over the r energy levels according to a given $\{N_i\}$ distribution and Ω_i is the number of ways to distribute the N_i particles at the energy level E_i among its g_i quantum states.

3. Assume N *distinguishable* particles (numbered from 1 to N). Calculate Ω_0 and Ω_i in the case where any number of particles can be in the same quantum state (*Maxwell-Boltzmann statistics*). Deduce the expression of $\Omega_{\mathrm{MB}}(N,E)$.

4. In the rest of the exercise, the N particles are considered *indistinguishable*. What is now the value of Ω_0?

5. Calculate Ω_i in the case where any number of particles can occupy the same quantum state (*Bose–Einstein statistics*). Deduce $\Omega_{\mathrm{BE}}(N,E)$ and its expression in the limit of low particle densities per energy level ($N_i \ll g_i$).

6. Calculate Ω_i in the case where *at most* one particle can be in a given quantum state (*Fermi-Dirac statistics*). It imposes the condition $0 \leq N_i \leq g_i$. Deduce $\Omega_{\mathrm{FD}}(N,E)$ and its expression in the limit of low particle densities per energy level.

Bose–Einstein and Fermi-Dirac quantum statistics will be introduced in Chapter 6.

Solution:

1. Trivially $E = \sum_{i=1}^{r} N_i E_i$ and $N = \sum_{i=1}^{r} N_i$.

2. *The particles assigned to each energy level are chosen according to the $\{N_i\}$ distribution (Ω_0 choices). Then, for each energy level i, the number of ways to distribute the N_i particles among its g_i quantum states is counted (Ω_i distributions). Since the energy levels are treated independently of each other, Equation (2.47) is obtained.*

3. *Since the particles are distinguishable, Ω_0 is the number of ways to choose N_1 particles from N particles for level 1, and then N_2 particles from the remaining $N - N_1$ for level 2, and so on:*

$$\Omega_0 = \binom{N}{N_1} \binom{N - N_1}{N_2} \binom{N - N_1 - N_2}{N_3} \cdots \binom{N - N_1 - N_2 \cdots - N_{r-1}}{N_r}$$

$$= \frac{N!}{N_1!(N - N_1)!} \frac{(N - N_1)!}{N_2!(N - N_1 - N_2)!} \cdots \frac{(N - N_1 - N_2 \cdots - N_{r-1})!}{N_r! \underbrace{(N - N_1 - N_2 \cdots - N_r)!}_{=0!=1}}$$

$$= \frac{N!}{N_1! N_2! \ldots N_r!}.$$

Then, for each particle of the energy level i, there are g_i possible states, thus $\Omega_i = g_i^{N_i}$. $\Omega(N, E)$ in the so called Maxwell-Boltzmann (MB) classical statistic can be deduced:

$$\Omega_{\mathrm{MB}}(N, E) = N! \prod_{i=1}^{r} \frac{g_i^{N_i}}{N_i!}.$$

4. *Since the particles are indistinguishable, $\Omega_0 = 1$.*

5. *Thus, Ω_i is the number of ways in which N_i (indistinguishable) particles can be distributed among g_i (discernible) quantum states, with no limitation on the number of particles per state. This is the situation encountered in Exercise 2.4, dealing with quantum harmonic oscillators, so from Equation (2.46):*

$$\Omega_i = \binom{N_i + g_i - 1}{N_i}.$$

$\Omega(N, E)$ *in the so-called Bose–Einstein (BE) quantum statistics can be deduced:*

$$\Omega_{\mathrm{BE}}(N, E) = \Omega_0 \prod_{i=1}^{r} \Omega_i = \prod_{i=1}^{r} \frac{(g_i + N_i - 1)!}{N_i!(g_i - 1)!}.$$

At the limit $g_i \gg N_i$:

$$\Omega_{\mathrm{BE}}(N, E) = \prod_{i=1}^{r} \frac{1}{N_i!} \underbrace{(g_i + N_i - 1)}_{\approx g_i} \underbrace{(g_i + N_i - 2)}_{\approx g_i} \ldots \underbrace{(g_i + 0)}_{= g_i}$$

$$\simeq \prod_{i=1}^{r} \frac{g_i^{N_i}}{N_i!} = \frac{1}{N!} \, \Omega_{\mathrm{MB}}.$$

At this limit, the number of individual states (g_i) is very large compared to the number of particles (N_i), and Equation (2.8) is recovers, linking the expressions of Ω for discernible particles (MB) and for indistinguishable particles (BE).

6. *Since the number of particles per quantum state is 0 or 1, it imposes* $0 \leq N_i \leq g_i$. *The number of ways to distribute* N_i *particles among* g_i *states that contain at most one particle is by definition:*

$$\Omega_i = \binom{g_i}{N_i}.$$

$\Omega(N,E)$ *in the so-called Fermi-Dirac (FD) quantum statistics can be deduced:*

$$\Omega_{\mathrm{FD}}(N,E) = \Omega_0 \prod_{i=1}^{r} \Omega_i = \prod_{i=1}^{r} \frac{g_i!}{N_i!(g_i - N_i)!}.$$

At the limit $g_i \gg N_i$:

$$\Omega_{\mathrm{FD}}(N,E) \quad = \quad \prod_{i=1}^{r} \frac{1}{N_i!} \underbrace{(g_i + 0)}_{=g_i} \underbrace{(g_i - 1)}_{\simeq g_i} \dots \underbrace{(g_i - (N_i - 1))}_{\simeq g_i}$$

$$\simeq \quad \prod_{i=1}^{r} \frac{g_i^{N_i}}{N_i!} = \frac{1}{N!} \, \Omega_{\mathrm{MB}}.$$

At this limit, the result obtained in the framework of Bose–Einstein statistics (previous question) is recovered.

3 Statistical Thermodynamics

> "Thermodynamics is a funny subject. The first time you go through it, you don't understand it at all. The second time you go through it, you think you understand it, except for one or two small points. The third time you go through it, you know you don't understand it, but by that time you are so used to it, it doesn't bother you anymore."
> A. Sommerfeld[1], around 1950

The postulates of statistical physics allow the determination of probability distributions and mean values of internal variables of an isolated macroscopic system in equilibrium (see Chapter 2). However, to recover thermodynamics from the microscopic description of a system, usual thermodynamic quantities must be expressed as ensemble average values. To this end, the fundamental postulate will be applied to an isolated system consisting of two interacting macroscopic subsystems. It will thus be seen that thermal equilibrium implies the definitions of statistical temperature (Section 3.1) and statistical entropy, expressed in terms of the number of accessible microstates, Ω, by the famous Boltzmann formula (Section 3.2). Every thermodynamic quantities (temperature, pressure, chemical potential...) then follow from this definition (Section 3.2). Other definitions of statistical entropy, that fall within a more general framework than statistical physics called *information theory*, will then be discussed (Section 3.4). With a microscopic definition of entropy, the second law of thermodynamics will be revisited, and a probabilistic interpretation of irreversibility will be given and we will "interpret in molecular language the most abstract concepts of thermodynamics"[2] (Section 3.5).

Consider an isolated macroscopic system $\mathcal{A}_{\text{tot}}$, of volume V_{tot} and total energy E_{tot}. $\mathcal{A}_{\text{tot}}$ contains $N_{\text{tot}} \gg 1$ particles. The system is divided in two (also) macroscopic subsystems, $\mathcal{A}$ and $\mathcal{A}'$, of volumes V and V', and energies E and E', and contains $N \gg 1$ and $N' \gg 1$ particles, respectively (see Figure 3.1), such that:

$$V_{\text{tot}} = V + V'$$
$$N_{\text{tot}} = N + N'.$$

Depending on the characteristics of the wall separating them, the two subsystems can be isolated (if the wall is adiabatic, fixed and impermeable), can exchange heat (if the wall is diathermic), work (if the wall is mobile) and particles (if the wall is permeable).

First, assume that the wall is adiabatic (no heat exchange between $\mathcal{A}$ and $\mathcal{A}'$), fixed (V and V' are constant) and impermeable (N and N' are constant). In such a case, shown in Figure 3.1(a), the two subsystems do not exchange energy and are both isolated (E and E' are thus constant). The total energy is

$$E_{\text{tot}} = E + E'.$$

The energy E of $\mathcal{A}$ being imposed, it is also an external parameter of the system $\mathcal{A}_{\text{tot}}$. Therefore, the partial number of microstates accessible to $\mathcal{A}_{\text{tot}}$ for this distribution of

---

[1] Cited in S.W. Angrist and L.G. Helper *Order and Chaos – Laws of energy and entropy*, p. 215 (New York: Basic Books, 1967).

[2] As Louis de Broglie spoke about Boltzmann and Gibbs in his presentation, *Reality of molecules and the work of Jean Perrin*, given at the annual awards session on December 17, 1945.

DOI: 10.1201/9781003272427-3

(a)   (b) d

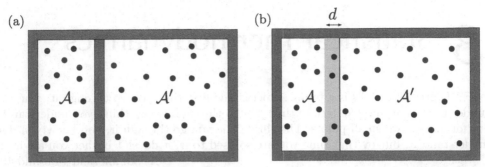

FIGURE 3.1 An isolated system $\mathcal{A}_{\text{tot}}$ consists of two subsystems $\mathcal{A}$ and $\mathcal{A}'$ separated by a fixed and impermeable wall. (a) The wall is adiabatic: $\mathcal{A}$ and $\mathcal{A}'$ are both isolated. (b) The wall is diathermic: $\mathcal{A}$ and $\mathcal{A}'$ exchange heat by conduction through the wall. Only particles of $\mathcal{A}$ that are close to the wall in the region of width d, contribute to the interaction energy between $\mathcal{A}$ and $\mathcal{A}'$ (d is the range of atomic interactions).

energy, $\Omega_{\text{P}_{\text{tot}}}(E_{\text{tot}} \,|\, E)$, depends on E_{tot} and E (in order to ease the notation, the other external parameters, N, N', V, V' and $E' = E_{\text{tot}} - E$ are not mentioned in the parameters of $\Omega_{\text{P}_{\text{tot}}}$). Furthermore, the numbers of accessible microstates of $\mathcal{A}$ and $\mathcal{A}'$ are $\Omega(E)$ and $\Omega'(E' = E_{\text{tot}} - E)$, respectively. For this energy distribution, since the two subsystems are (statistically) independent, they can each be in any accessible microstate. Consequently:

$$\Omega_{\text{P}_{\text{tot}}}(E_{\text{tot}} \,|\, E) = \Omega(E) \cdot \Omega'(E_{\text{tot}} - E). \tag{3.1}$$

A microstate m_{tot} of $\mathcal{A}_{\text{tot}}$ is completely specified by the microstate m of $\mathcal{A}$ and the microstate m' of $\mathcal{A}'$. According to the fundamental postulate, the probability to find the isolated system $\mathcal{A}_{\text{tot}}$ in equilibrium in a given microstate m_{tot} is[3]:

$$p_{m_{\text{tot}}} = \frac{1}{\Omega_{\text{P}_{\text{tot}}}(E_{\text{tot}} \,|\, E)} = \frac{1}{\Omega(E)} \frac{1}{\Omega'(E')},$$

It is the product of the probabilities to find each isolated subsystem, $\mathcal{A}$ and $\mathcal{A}'$, in equilibrium in a given microstate m of energy E and a given microstate m' of energy E', respectively.

Example: Two independent spin subsystems in a magnetic field

Consider, for example, an isolated system $\mathcal{A}_{\text{tot}}$ consisting of $N_{\text{tot}} = 5$ spins-1/2 immersed a magnetic field B and with a fixed total energy, $E_{\text{tot}} = \mu_M B$. This system is separated into two isolated subsystems $\mathcal{A}$ and $\mathcal{A}'$, consisting of $N = 2$ and $N' = 3$ spins, with fixed energies $E = 0$ and $E' = E_{\text{tot}} - E = \mu_M B$, respectively. The interactions between the spins are neglected (see Hamiltonian (1.10)). Here, the accessible microstates can be easily listed in both subsystems: at an energy $E = 0$, $\mathcal{A}$ can be in $\Omega = 2$ microstates: $\uparrow\downarrow$ or $\downarrow\uparrow$. At energy $E' = \mu_M B$, $\mathcal{A}'$ can be in $\Omega' = 3$ microstates: $\uparrow\downarrow\downarrow$, $\downarrow\uparrow\downarrow$ or $\downarrow\downarrow\uparrow$. For this distribution of energy between the two subsystems, system $\mathcal{A}_{\text{tot}}$ can therefore be in $\Omega_{\text{P}_{\text{tot}}}(E_{\text{tot}} = \mu_M B \,|\, E = 0) = \Omega \cdot \Omega' = 2 \times 3 = 6$ microstates.

3.1 TEMPERATURE

In thermodynamics, when two macroscopic systems exchange heat, they are at the same temperature once thermodynamic equilibrium is reached. This equilibrium condition will

---

[3]For simplicity, the framework of quantum mechanics is used.

be demonstrated and a definition of temperature in the framework of statistical physics will be given.

To this end, consider the case, represented in Figure 3.1(b), where the two subsystems $\mathcal{A}$ and $\mathcal{A}'$, are now separated by a *diathermic*, fixed and impermeable wall (N, V, N' and V' are external parameters): $\mathcal{A}$ and $\mathcal{A}'$ thus exchange energy in the form of heat by thermal conduction through the wall. The total energy is then decomposed into[4] $E_{\text{tot}} = E + E' + E_{\text{int}}$, where E_{int} is the interaction energy between $\mathcal{A}$ and $\mathcal{A}'$. For *macroscopic* subsystems and *short-range* interatomic interactions, energy exchanges allow both subsystems to reach thermodynamic equilibrium. However, the interaction term E_{int} can be neglected with respect to E and E'. This is because only particles at a distance from the wall less than the range of interatomic interactions ($d \simeq 1$ nm) contribute to E_{int}. Assume now that the container of $\mathcal{A}$ is a cube of side $L \simeq 1$ cm. The interacting particles are located in a volume $L^2 d$ (see Figure 3.1(b)), and $E_{\text{int}} \sim \epsilon \rho L^2 d$, where ϵ is the energy per particle and ρ the particle number density. In contrast, $E \sim \epsilon \rho L^3$. The ratio of E_{int}/E (and E_{int}/E') is therefore of the order of $d/L \simeq 10^{-7}$. The two macroscopic subsystems are so-called *weakly coupled* and the energy is practically *additive*:

$$E_{\text{tot}} = E + E' + E_{\text{int}} \simeq E + E'. \tag{3.2}$$

As it will be seen, this approximation is essential in statistical physics[5].

3.1.1 THERMAL EQUILIBRIUM BETWEEN MACROSCOPIC SYSTEMS

To determine the condition of thermal equilibrium between the two subsystems, let us focus on subsystem $\mathcal{A}$. As a result of energy exchanges, its energy E fluctuates and behaves as an internal variable described by the probability distribution $P(E)$. Note that E is also an internal variable of the isolated system $\mathcal{A}_{\text{tot}}$.

Let us begin by expressing the number of accessible microstates $\Omega_{\text{tot}}(E_{\text{tot}})$ in the isolated system $\mathcal{A}_{\text{tot}}$. The weak coupling approximation amounts to considering the two subsystems as statistically independent. Thus, for a given partition of the total energy between $\mathcal{A}$ and $\mathcal{A}'$ (E and $E' = E_{\text{tot}} - E$, respectively), the partial number of accessible microstates $\Omega_{\text{ptot}}(E_{\text{tot}} \mid E)$ is given by Equation (3.1). The total number of accessible microstates $\Omega_{\text{tot}}(E_{\text{tot}})$ is thus obtained by summing $\Omega_{\text{ptot}}(E_{\text{tot}} \mid E)$ over all possible values of E compatible with energy conservation (Equation (3.2)):

$$\Omega_{\text{tot}}(E_{\text{tot}}) = \sum_E \Omega_{\text{ptot}}(E_{\text{tot}} \mid E) = \sum_E \Omega(E) \cdot \Omega'(E_{\text{tot}} - E). \tag{3.3}$$

For macroscopic systems, this sum contains a very large number of terms, all positive. Thus, when a constraint is relaxed (replacing the adiabatic wall with a diathermal wall), $\Omega_{\text{tot}}(E_{\text{tot}})$

---

[4]In principle, one should study the interactions of $\mathcal{A}$ and $\mathcal{A}'$ with their common wall. For simplicity, one can consider that the wall has no width (see Figure 3.1(b)) or include it in subsystem $\mathcal{A}'$ and reason about $\mathcal{A}$.

[5]This approximation is no longer valid when at least one spacial dimension of the system is of the order of the interactions range between particles ($L \simeq d$). This is the case for (i) systems of atomic dimensions (atomic nuclei, atomic aggregates, nanoparticles, a thin layer on the surface of a substrate...), and (ii) systems with long-range interactions potential $u(r)$, i.e. such that $u(r) \propto 1/r^\alpha$, where α is less than or equal to the dimension of space (see A. Campa et al., *Physics of Long-Range Interacting Systems* (Oxford University Press, 2014)), such as galaxies (gravitational interactions, $\alpha = 1$) or ferrofluids (magnetic dipole interactions, $\alpha = 3$). The detailed study of systems with long-range interactions is beyond the scope of this book. For an extended discussion, see the book edited by T. Dauxois, S. Ruffo, E. Arimondo, and M. Wilkens, *Dynamics and Thermodynamics of Systems with Long Range Interactions* in *Series: Lecture Notes in Physics* (Springer-Verlag Berlin Heidelberg, 2002). Note that in an electrically neutral medium such as plasmas or electrolytes (electromagnetic or only electrostatic interactions, $\alpha = 1$), the electric field screening due to mobile charges of opposite sign reduces the interactions effective range (see Exercise 5.8).

increases prodigiously going from a single term, $\Omega_{p_{tot}}(E_{tot} \mid E)$, to the sum (3.3). This point will be discussed again in Section 3.5.

The probability $P(E)$ that subsystem $\mathcal{A}$ has an energy E is by definition and according to Equation (3.1):

$$P(E) = \frac{\Omega_{p_{tot}}(E_{tot} \mid E)}{\Omega_{tot}(E_{tot})} = \frac{\Omega(E) \cdot \Omega'(E_{tot} - E)}{\Omega_{tot}(E_{tot})}. \tag{3.4}$$

This result is found by applying the fundamental postulate to the isolated system $\mathcal{A}_{tot}$: in equilibrium, all accessible microstates m_{tot} of $\mathcal{A}_{tot}$, and thus the pairs of microstates m and m' of $\mathcal{A}$ and $\mathcal{A}'$, respectively, have the same probability $p_{m_{tot}} = 1/\Omega_{tot}(E_{tot})$. The probability to find the subsystem in a given microstate m of energy E is thus:

$$p_m = \frac{1 \times \Omega'(E_{tot} - E)}{\Omega_{tot}(E_{tot})}. \tag{3.5}$$

Indeed, the microstate m of $\mathcal{A}$ being imposed so is its energy E. Therefore, subsystem $\mathcal{A}'$ has an energy $E' = E_{tot} - E$ and can then be found in one of its $\Omega'(E_{tot} - E)$ accessible microstates, each having the same probability $p_{m_{tot}} = 1/\Omega_{tot}$. By definition, there are $\Omega(E)$ equiprobable microstates (with the probability p_m) having an energy equal to E (like microstate m). The probability to find the subsystem in a microstate of energy E is therefore $P(E) = \Omega(E)\, p_m$, or Equation (3.4).

Example: Two independent spin subsystems in a magnetic field

Consider again the example of an isolated 5 spins-1/2 system immersed in a magnetic field B and having a total energy $E_{tot} = E + E' = \mu_M B$. The two subsystems $\mathcal{A}$ ($N = 2$) and $\mathcal{A}'$ ($N' = 3$) are in thermal contact. The possible partitions of the total energy E_{tot} between the two subsystems are given in Table 3.1:

TABLE 3.1 Possible distributions of the total energy E_{tot} of an isolated 5 spins-1/2 system immersed in a magnetic field B.

E	Ω	Microstates of $\mathcal{A}$	E'	Ω'	Microstates of $\mathcal{A}'$
$2\mu_M B$	1	$\downarrow\downarrow$	$-\mu_M B$	3	$\downarrow\uparrow\uparrow,\ \uparrow\downarrow\uparrow$ or $\uparrow\uparrow\downarrow$
0	2	$\downarrow\uparrow$ or $\uparrow\downarrow$	$\mu_M B$	3	$\downarrow\downarrow\uparrow,\ \downarrow\uparrow\downarrow$ or $\uparrow\downarrow\downarrow$
$-2\mu_M B$	1	$\uparrow\uparrow$	$3\mu_M B$	1	$\downarrow\downarrow\downarrow$

Therefore, $\Omega_{tot}(E_{tot} = \mu_M B) = 1 \times 3 + 2 \times 3 + 1 \times 1 = 10$ and $P(E)$ can now be derived:

$$P(2\mu_M B) = \frac{\Omega(2\mu_M B)\Omega'(-\mu_M B)}{\Omega_{tot}(\mu_M B)} = \frac{3}{10}$$

$$P(0) = \frac{\Omega(0)\Omega'(\mu_M B)}{\Omega_{tot}(\mu_M B)} = \frac{3}{5}$$

$$P(-2\mu_M B) = \frac{\Omega(-2\mu_M B)\Omega'(3\mu_M B)}{\Omega_{tot}(\mu_M B)} = \frac{1}{10}.$$

For macroscopic systems, the shape of $P(E)$ is easily deduced from the behaviour of Ω as a function of E, given by Equation (2.24). Indeed, according to Equation (3.4), $P(E)$ is the product (to within a constant $1/\Omega_{tot}(E_{tot})$ independent of E) of two positive functions: $\Omega(E) \sim E^f$ which grows very rapidly with E (because $f \gg 1$) and $\Omega'(E_{tot} - E) \sim (E_{tot} - E)^{f'}$, which decreases very rapidly with E (because $f' \gg 1$). The probability distribution $P(E)$ shows thus a very sharp peak centred on a specific value $E = E_{eq}$. More precisely, the energy E of the macroscopic subsystem being a macroscopic internal variable of the isolated system $\mathcal{A}_{tot}$, the analysis presented in Section 2.3 can be used. E can be treated as a continuous variable and its probability density function $f(E)$ is a Gaussian distribution (Equation (2.32)):

$$P(E) = f(E)dE \simeq \frac{1}{\sqrt{2\pi\Delta_E^2}} e^{-\frac{(E-E_{eq})^2}{2\Delta_E^2}} \, dE. \tag{3.6}$$

The mean value of the energy is $\langle E \rangle = E_{eq}$. Its variance Δ_E^2 is defined by Equation (2.31):

$$\frac{1}{\Delta_E^2} = -\left.\frac{\partial^2 \ln \Omega_{Ptot}(E_{tot} \mid E)}{\partial E^2}\right|_{E=E_{eq}}. \tag{3.7}$$

Remember that for $N \simeq 10^{23}$, the relative fluctuations Δ_E/E_{eq} are extremely small (of the order of 10^{-12}). In other words, the energy hardly fluctuates around its most probable value $E = E_{eq}$, interpreted as thermodynamic equilibrium value, that is the system's *internal energy* (often noted U in thermodynamics). Let us now discuss what the existence of the maximum of $P(E)$ at $E = E_{eq}$ implies on the equilibrium condition between the two subsystems. According to Equation (3.4), the value of E_{eq} is such that:

$$\left.\frac{\partial P(E)}{\partial E}\right|_{E=E_{eq}} = \frac{1}{\Omega_{tot}(E_{tot})}\left(\frac{\partial \Omega(E)}{\partial E}\Omega'(E_{tot}-E) + \Omega(E)\frac{\partial \Omega'(E_{tot}-E)}{\partial E}\right)\Bigg|_{E=E_{eq}}$$

$$= \frac{1}{\Omega_{tot}(E_{tot})}\left(\frac{\partial \Omega(E)}{\partial E}\Omega'(E') - \Omega(E)\frac{\partial \Omega'(E')}{\partial E'}\right)\Bigg|_{E=E_{eq}} = 0,$$

since $E_{tot} - E = E'$ and $dE = -dE'$. Therefore,

$$\left.\frac{1}{\Omega}\frac{\partial \Omega}{\partial E}\right|_{E=E_{eq}} = \left.\frac{1}{\Omega'}\frac{\partial \Omega'}{\partial E'}\right|_{E'=E'_{eq}}.$$

The two terms of this equality represent the same quantity, one expressed for system $\mathcal{A}$ (left), the other for system $\mathcal{A}'$ (right). It can be written as

$$\beta(E) \equiv \left.\frac{\partial \ln \Omega(E)}{\partial E}\right|_{N,V}, \tag{3.8}$$

where the partial derivative is calculated at constant N and V, since the volume and the number of particles are in this case external parameters. The quantity β is associated with a given system and depends on its characteristics through the number of accessible microstates. At equilibrium, it leads to

$$\beta(E_{eq}) = \beta'(E'_{eq}). \tag{3.9}$$

This equality obtained in thermal equilibrium suggests a direct link between the quantity β and the thermodynamic temperature of a macroscopic system.

3.1.2 DEFINITION AND PROPERTIES OF STATISTICAL TEMPERATURE

According to Equation (3.8), β has the dimension of the inverse of an energy. To define the statistical temperature T of a macroscopic system in equilibrium from β, the Boltzmann constant, k_B, must be introduced such that

$$\frac{1}{k_B T} \equiv \beta(E_{eq}) = \left.\frac{\partial \ln \Omega(E)}{\partial E}\right|_{N,V}, \tag{3.10}$$

where the partial derivative is calculated in $E = E_{eq} = \langle E \rangle$ and with the system in thermal contact with another one so its energy is an internal variable. In the case of an isolated system, the equilibrium energy is equal to the *fixed and constant* system's energy[6]. Like any thermodynamic quantity, temperature is defined, in principle, for a *macroscopic* system in equilibrium. The Boltzmann constant, whose value is exactly[7]

$$k_{B} = 1.380\ 649 \times 10^{-23}\ \text{J.K}^{-1}, \tag{3.11}$$

is the proportionality factor that links the statistical temperature T to the thermodynamic temperature measured in Kelvin. It is also related to the Avogadro number N_A and to the ideal gas constant $R \simeq 8.314\ \text{J} \cdot \text{mol}^{-1} \cdot \text{K}^{-1}$, by the relation $R = k_B N_A$. By definition, the Boltzmann constant allows the conversion of an energy into a temperature[8]. It is interesting to remember the value of thermal energy $k_B T$ at room temperature:

$$k_{B}T \simeq 1/40\ \text{eV} \quad \text{at } T = 300\ \text{K},$$

where an electron volt is $\sim 1.602 \times 10^{-19}$ J. Thus, at room temperature, chemical, metallic, ionic or covalent bonds (typically from 1 to 10 eV) are not very sensitive to thermal agitation, whereas van der Waals type bonds (~ 0.01 eV), which ensure the attraction between atoms in simple fluids (see Chapter 5), are easily broken.

It remains to verify that the statistical temperature T has the expected properties of thermodynamic temperature. By definition, Equation (3.9) ensures that two macroscopic systems, $\mathcal{A}$ and $\mathcal{A}'$, in thermal equilibrium have the same (statistical) temperature: $T = T'$. Moreover, since for a macroscopic system, $\Omega(E) \sim E^f$ with $f \gg 1$, Definition (3.10) implies:

- T is positive, as expected for an absolute temperature[9].

- T is naturally an *intensive* variable (see Section 2.3), because

$$k_{B}T = \left(\frac{\partial \ln \Omega(E)}{\partial E} \bigg|_{E=E_{eq}} \right)^{-1} \simeq \left(\frac{\partial \ln E^f}{\partial E} \bigg|_{E=E_{eq}} \right)^{-1} \sim \frac{E_{eq}}{f}, \tag{3.12}$$

where E and f are extensive quantities. It can be seen that T is approximately the internal energy E_{eq} per degree of freedom (divided by k_B).

---

[6]To measure the temperature of a system initially isolated at energy E, it is put in thermal contact with a "thermometer". For a reliable measurement, the energy exchanges with the thermometer must be negligible compared to E, so that once equilibrium is reached, $E_{eq} = \langle E \rangle$ remains close to the initial energy E.

[7]As part of the 2019 redefinition of SI base units.

[8]Formally, a temperature could be expressed in joules, but it is an inconvenient unit: imagine, for example, a weather forecast giving a temperature of 4.14×10^{-21} J... Thus, Boltzmann's constant is essentially a conversion factor that is not as fundamental in nature as the speed of light c, the Planck's constant $\hbar = h/2\pi$ and the gravitational constant G. Within the Planck's system of units, these three constants are sufficient to express the three dimensions, length, time and mass, associated with space, time and matter, fundamental concepts of theoretical physics in the process of unification: $l_P = \sqrt{G\hbar/c^3} \simeq 1.6 \times 10^{-35}$ m, $t_P = \sqrt{G\hbar/c^5} \simeq 5.4 \times 10^{-44}$ s and $m_P = \sqrt{\hbar c/G} \simeq 2.2 \times 10^{-8}$ kg. Since three dimensions (for example L, T and M) are sufficient to express the dimensions of all physical quantities, it is not essential to introduce another constant (like k_B). Moreover, c, h and G represent limit values that mark the appearance of new physical phenomena: relativistic, quantum and gravitational respectively (although the status of G is not clear in the absence of a quantum theory of gravitation). This is not the case for Boltzmann's constant: its value does not define the domain of application of statistical physics.

[9]This property is no longer valid when the system's energy has an upper bound. In Section 3.2.1, it is shown that it is possible to decouple the behaviour of the nuclear spins of atoms in a crystal from the kinetic and potential energies. In this case, the number of microstates accessible to the "isolated" spin system can decrease with energy E, leading to negative absolute temperatures, which can be experimentally evidenced. This surprising result, specific to this particular type of system, does not violate the laws of thermodynamics, since it concerns only a portion of the system's degrees of freedom (the spins). Indeed, the entire physical system (the crystal) is not in equilibrium during the measurements.

- According to Equation (3.12), T is, of course, an increasing function[10] of the internal energy E_{eq}.

Heat capacity at constant volume (also called *thermal capacity* or *specific heat*) is the energy that must be supplied to a fixed volume system, to increase its temperature by one degree kelvin[11]. By definition:

$$C_V \equiv \left. \frac{\partial E_{eq}}{\partial T} \right|_{N,V}, \tag{3.13}$$

where E_{eq} is the system's internal energy: $E_{eq} = \langle E \rangle$ if the system is in thermal contact or $E_{eq} = E$ if the system is isolated (fixed energy E). In this last case, by deriving Equation (3.10) with respect to E, one gets:

$$\left. \frac{\partial (1/k_B T)}{\partial E} \right|_{N,V} = -\frac{1}{k_B T^2} \left. \frac{\partial T}{\partial E} \right|_{N,V} = \left. \frac{\partial^2 \ln \Omega}{\partial E^2} \right|_{N,V},$$

leading to

$$C_V = -\left(k_B T^2 \left. \frac{\partial^2 \ln \Omega}{\partial E^2} \right|_{N,V} \right)^{-1}.$$

Since $\Omega(E) \sim E^f$ and $k_B T \sim E/f$, it can be seen that $C_V \sim k_B f$. Heat capacity is an extensive quantity that has the dimension of the Boltzmann constant. It depends on the thermodynamic conditions and on the considered system. Moreover, since T is an increasing function of the internal energy (see Note 10), C_V is positive[12]. Finally, in the case of two macroscopic subsystems $\mathcal{A}$ and $\mathcal{A}'$ in thermal equilibrium ($T = T'$), it can be deduced from Equations (3.1), (3.6) and (3.7) that:

$$\frac{\partial^2 \ln P(E)}{\partial E^2} = -\frac{1}{\Delta_E^2} = \left(\frac{\partial^2 \ln \Omega(E)}{\partial E^2} + \frac{\partial^2 \ln \Omega'(E')}{\partial E'^2} \right) = -\left(\frac{1}{k_B T^2 C_V} + \frac{1}{k_B T^2 C_V'} \right),$$

where C_V and C_V' are the heat capacities of $\mathcal{A}$ and $\mathcal{A}'$, respectively. Therefore:

$$\Delta_E^2 = \langle E^2 \rangle - \langle E \rangle^2 = k_B T^2 \frac{C_V C_V'}{C_V + C_V'} > 0, \tag{3.14}$$

since the variance of a random variable X is necessarily positive (indeed, we have by definition $\Delta_X^2 = \langle X^2 \rangle - \langle X \rangle^2 = \langle (X - \langle X \rangle)^2 \rangle \geq 0$). For two similar macroscopic subsystems, $C_V' \simeq C_V$, and Equation (3.14) imposes $C_V > 0$. Thus, the positivity of heat capacity is associated with the negative sign of the second derivative of $\ln P(E)$ in $E = E_{eq}$, that is with the maximum of $P(E)$ and thus with the stability of thermal equilibrium[13]. It will be seen in the following sections that the calculation of the heat capacity of gases, solids and spin systems has played a great role in the development of statistical physics.

---

[10]In general, when energy is added to a system, its temperature increases, except during a first order phase transition, such as fusion or vaporisation. During a first order phase transition, the energy exchanged by the system with the external medium is used to convert the system from one phase to another (and T remains constant).

[11]According to the first law of thermodynamics (Equation (3.53)), when volume and other external parameters are fixed, energy variations are due to heat exchange only.

[12]Thermodynamics can be extended to the case of systems consisting of a small number of particles (atomic nuclei or atomic aggregates) or to the case of long-range interactions systems (galaxies). These systems can exhibit unexpected macroscopic behaviours, such as negative heat capacity: by adding energy to the system, the temperature decreases (see D. Lynden-Bell, *Negative specific heat in astronomy, physics and chemistry*, Physica A **263**, 293 (1999)). These systems, already mentioned in Note 5, are very particular and do not fall within the scope of this book.

[13]If two macroscopic systems, initially at different temperatures, are brought into thermal contact, the colder system absorbs the energy supplied by the hotter system. Since heat capacities are positive, the temperature of the colder system increases and the temperature of the warmer system decreases until both temperatures equalise in thermal equilibrium.

3.2 STATISTICAL ENTROPY

In thermodynamics, temperature T is defined from the entropy S by the following equation[14]:

$$\frac{1}{T} \equiv \frac{\partial S}{\partial E}\bigg|_{N,V}, \tag{3.15}$$

where E is the internal energy. Equation (3.15) is obviously similar to the statistical temperature definition (Equation (3.10)). This similarity leads to the definition of entropy in statistical physics.

3.2.1 DEFINITION AND PROPERTIES OF STATISTICAL ENTROPY

The statistical entropy S (also called *Boltzmann entropy*) of an *isolated macroscopic system at equilibrium* is defined by the Boltzmann formula:

$$S(N,V,E) \equiv k_B \ln \Omega(N,V,E). \tag{3.16}$$

To establish this formula[15], Boltzmann did not follow the reasoning presented in this chapter. Using an ideal gas model and an argument from combinatorial analysis (see Note 55), he showed in 1877 that $S \propto \ln \Omega$. It was in fact Max Planck who wrote in 1900 entropy as Equation (3.16) in the context of his study of blackbody radiation, intimately mixing electromagnetism and the then controversial statistical physics with the early stages of quantum theory. In one stroke, Planck introduced and evaluated not only the constant k_B, named in homage to the Austrian physicist, but also the fundamental constant h, which bears his name thus giving birth to quantum theory[16].

If the temperature formula is used to deduce statistical entropy, it is indeed Boltzmann's formula (Equation (3.16)) that constitutes the keystone of statistical physics of equilibrium systems, by defining a macroscopic quantity, S, from a quantity characterising the microscopic scale, Ω. Recall that entropy and thermodynamic quantities are defined, in principle, only for macroscopic systems in equilibrium. The entropy $S(N,V,E)$, a function of the external parameters of the microcanonical ensemble (like Ω), is the thermodynamic potential used to describe an isolated system, from which every other thermodynamic properties is deduced, as it will be seen in the following[17].

Properties of statistical entropy

As with statistical temperature, it remains to verify that the statistical entropy S, defined by Equation (3.16), has the expected properties of thermodynamic entropy. For an isolated macroscopic system in equilibrium, $\Omega(E) \sim (E/E_0)^f$ with $f \gg 1$, where E_0 is the energy of the fundamental level (Equation (2.24)). Therfore:

- $S = k_B \ln \Omega \sim k_B f \ln(E/E_0)$ is naturally an extensive quantity. Boltzmann's constant ensures the homogeneity of S and imposes a value of S of the order of $R \ln(E/E_0) \sim 1$ to 100 J $\cdot$ mol^{-1} $\cdot$ K^{-1}, where $R = k_B \mathcal{N}_A$ is the ideal gas constant.

---

[14]Clausius (see Note 37 in Chapter 1) introduced the term "entropy" in 1865, by analogy with the word "energy" from the Greek word $\tau\rho o\pi\eta$, which means "transformation" (from heat to work). He would have noted it as S in honour of the French physicist Sadi Carnot (see Note 62).

[15]Which is engraved on Boltzmann's tombstone.

[16]In the work of Boltzmann and Planck, S is in fact proportional to $\ln W$, to within one additive constant, where W is the probability (*Wahrscheinlichkeit* in German) to find the system in a given macrostate. This definition of S, which is not widely used today, leads to the same entropy expression for a classical ideal gas whether atoms are distinguishable or indistinguishable (R. H. Swendsen, *An Introduction to Statistical Mechanics and Thermodynamics* (Oxford University Press, 2012)).

[17]It was shown in Section 2.2.5 that for a macroscoscopic system $\ln \Omega$ is independent of the uncertainty δE. This is also the case for entropy and the other thermodynamic quantities deriving from it.

- S is also a positive quantity (since $\Omega \geq 1$), which increases with energy, since $T > 0$ (except for the case mentioned in Note 9). In the framework of quantum physics, when temperature decreases, the energy tends towards the fundamental level E_0. When the latter is non-degenerate (ideal crystal, i.e. $\Omega_0 = 1$), $\lim_{T \to 0} S = S_0 = k_B \ln \Omega_0 = 0$. This is the third principle of thermodynamics[18].

- Finally, S is an additive quantity. To establish this, consider the example of a system $\mathcal{A}_{tot}$ divided into two macroscopic subsystems $\mathcal{A}$ and $\mathcal{A}'$. If they are isolated from each other, that is statistically independent, then according to Equation (3.1) and the property of the logarithm function:

$$S_{tot}(E_{tot}) = k_B \ln \Omega_{p_{tot}}(E_{tot} \mid E) = k_B \ln (\Omega(E)\Omega'(E')) = S(E) + S'(E'),$$

where E and E' are the fixed energies of $\mathcal{A}$ and $\mathcal{A}'$, respectively (to ease notations, the other external parameters, N, N', V and V' are not explicitly written). If the two macroscopic subsystems are in thermal contact, in the weak coupling approximation, one obtains according to Equation (3.3):

$$S_{tot}(E_{tot}) = k_B \ln \Omega_{tot}(E_{tot}) = k_B \ln \sum_E \Omega(E)\Omega'(E'). \qquad (3.17)$$

Now according to Equation (3.4), for any value of E,

$$\Omega(E)\Omega'(E') = \Omega_{tot}(E_{tot})P(E).$$

In particular, for $E = E_{eq} = \langle E \rangle$: $\Omega(\langle E \rangle)\Omega'(\langle E' \rangle) = \Omega_{tot}(E_{tot})P(\langle E \rangle)$. Eliminating $\Omega_{tot}(E_{tot})$:

$$\Omega(E)\Omega'(E') = \frac{\Omega(\langle E \rangle)\Omega'(\langle E' \rangle)}{P(\langle E \rangle)}P(E).$$

Equation (3.17) then becomes

$$
\begin{aligned}
S_{tot}(E_{tot}) &= k_B \ln \left(\frac{\Omega(\langle E \rangle)\Omega'(\langle E' \rangle)}{P(\langle E \rangle)} \underbrace{\sum_E P(E)}_{=1} \right) \\
&= S(\langle E \rangle) + S'(\langle E' \rangle) - k_B \ln P(\langle E \rangle),
\end{aligned}
$$

since the probability distribution function $P(E)$ is normalised. Using the Gaussian form, Equation (3.6), of $P(E)$, one has $P(\langle E \rangle) = f(\langle E \rangle)dE \simeq \delta E / \sqrt{2\pi\Delta_E^2}$, so

$$S_{tot}(E_{tot}) \simeq \underbrace{S(\langle E \rangle)}_{\sim k_B f} + \underbrace{S'(\langle E' \rangle)}_{\sim k_B f'} + \underbrace{k_B \ln \frac{\sqrt{2\pi\Delta_E^2}}{\delta E}}_{\ll k_B f, k_B f'} \simeq S(\langle E \rangle) + S'(\langle E' \rangle), \qquad (3.18)$$

---

[18]Similarly, macroscopic systems with a weakly degenerate ground level, such that $\ln \Omega_0 \ll f$, have a negligible entropy at $T = 0$. This is for example the case in the Ising model without external field where $\Omega_0 = 2$ (see Section 9.1.1). In practice, systems can remain trapped at low temperature in non-equilibrium macrostates associated with a very large number of microstates ($\ln \Omega_0 \sim f$), leading to a *residual entropy* at $T = 0$. This is, of course, the case for glasses (see Note 24 in Chapter 1), and also some crystals, such as water ice (see Exercise 3.1) or carbon monoxide, whose residual entropy measured at $T \simeq 0$ is $S_0 = 4.2$ J $\cdot$ mol$^{-1} \cdot$ K^{-1}: each molecule positioned at a node of the crystal lattice has two possible individual states of similar energy depending on its orientation (CO or OC). Assuming these two states are equiprobable, then $\Omega_0 = 2^{N_A}$ for one mole of molecules and $S_0 = k_B N_A \ln 2 = 5.8$ J $\cdot$ mol$^{-1} \cdot$ K^{-1}. This value, greater than the measured one, suggests that the molecules' orientations are not in fact equiprobable.

where to estimate the orders of magnitude of the different terms, $f, f' \gg 1$ and from Equation (3.14), $\Delta_E^2 \sim (k_B T)^2 f f' / (f + f')$. Following the reasoning of Section 2.2.5 for the derivation of Equation (2.24), it can be seen that for any value of δE, the third term in $\ln(\Delta_E / \delta E)$ is negligible in front of $k_B f$ and $k_B f'$. The additivity of entropy (and energy) is thus a consequence of the weak coupling approximation and the tiny fluctuations of the energy around of its average value. As shown by Equation (3.18), once in equilibrium, two systems $\mathcal{A}$ and $\mathcal{A}'$ behave as if they were isolated, at energy $\langle E \rangle$ and $\langle E' \rangle$, respectively. The thermodynamic entropy is then given by the Boltzmann formula for these average energy values, that is $S(N, V, \langle E \rangle)$ and $S'(N', V', \langle E' \rangle)$. This important point will be discussed again in Chapter 4.

If statistical entropy and temperature have the expected thermodynamics properties, it was not shown yet that these two statistical quantities behave quantitatively like thermodynamic entropy and temperature for a given system. Once again, the statistical approach must be validated by comparing theoretical predictions with experimental data.

Is entropy a measure of disorder?

Entropy is often interpreted as a measure of the system's "disorder". This word certainly speaks for itself – it is easy to see what a child's bedroom looks like – but it characterises the appearance of a system too loosely to be identified with entropy. Indeed, entropy has a precise meaning in statistical physics: it is the measure of the number of accessible microstates Ω of an isolated system. Consider a monoatomic system. At $T = 0$ K, the crystal lattice, associated with a single microstate ($\Omega = 1$), does give the image of a perfectly ordered system with zero entropy. On the contrary, at high temperature, the system is in a high entropy gaseous macrostate associated with a very large number of accessible microstates. The distribution of atoms' positions effectively illustrates the disordered aspect of the system.

Regardless of the observation scale, the intuitive notion of disorder can nevertheless be a source of error, as in the following two examples[19]:

- At the microscopic scale, a glass of water containing floating ice cubes is an inhomogeneous system that appears more disordered than liquid water alone (homogeneous). The entropy of the first system (liquid+solid) is, however, lower than the entropy of the second system (liquid only).

- Liquid crystals, which are parts of LCD (Liquid Crystal Display) screens, illustrate well the ambiguity of the relationship between entropy and disorder at the microscopic scale. In a liquid crystal, molecules have an elongated shape. This strong anisotropy at the microscopic scale implies the existence of very particular phases directly observable at human scale. At high temperature (or low density), the system is in an isotropic phase, characterised by a uniform distribution of the positions of the molecules' centres of mass and a uniform distribution of molecules' angular orientations (see Figure 3.2). This is called *translational* and *orientational* disorder.

  By lowering the temperature (or by increasing the molecules number density), an ordered phase appears, called *nematic phase*[20]: molecules tend to align with each other (orientational order), decreasing orientational entropy, S_{or}, associated with the angular degrees of freedom of the molecules. However, as the molecules are almost aligned, their lateral motion with respect to each other is facilitate. This results in

---

[19]Other examples are discussed in the article by F.L. Lambert, *Disorder – A Cracked Crutch for Supporting Entropy Discussions*, J. Chem. Educ. **79**, 187 (2002).

[20]Nematic means "thread" in Greek.

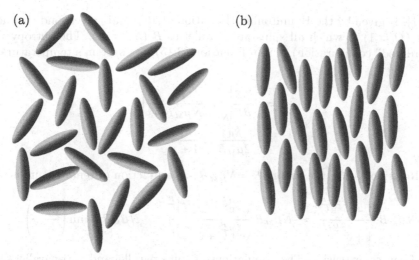

FIGURE 3.2 Liquid crystal in (a) the isotropic phase at high temperature or low density and (b) in the nematic phase at low temperature or high density. The rod-shape organic molecules constituting the crystal are a few nanometers long.

an increase of the translational entropy, S_{trans}, associated to the translational degrees of freedom of their centres of mass (translational disorder). It can be shown that the decrease of S_{or} is more than balanced by the increase of S_{trans}: the nematic phase, which seems the most ordered, has thus a higher entropy ($S_{\text{or}} + S_{\text{trans}}$) than the isotropic phase[21]. The nematic-isotropic phase transition will be discussed in Section 9.3.

Example 1: Ideal paramagnetic crystal and negative temperatures

According to Equation (2.15) giving Ω, the entropy of a system of N independent spins immersed in a magnetic field B is:

$$S(N,E) = k_{\text{B}} \ln \Omega(N,E) = k_{\text{B}} \ln \frac{N!}{N_+! \, N_-!}, \tag{3.19}$$

where

$$N_+ = \frac{N}{2}\left(1 - \frac{E}{N\mu_M B}\right) \quad \text{and} \quad N_- = \frac{N}{2}\left(1 + \frac{E}{N\mu_M B}\right), \tag{3.20}$$

are the number of spins parallel and antiparallel to the magnetic field, respectively. In the case of a macroscopic system, for N, N_+, $N_- \gg 1$, the entropy can be deduced with the use of the Stirling's formula (Equation (A.21)):

$$S(N,E) \simeq k_{\text{B}}\left(N \ln N - N_+ \ln N_+ - N_- \ln N_-\right). \tag{3.21}$$

Taking the dimensionless quantity $x = E/(N\mu_M B)$ and with Equation (3.20), one gets:

$$S(N,x) \simeq -k_{\text{B}}N\left(\frac{1-x}{2}\ln\frac{1-x}{2} + \frac{1+x}{2}\ln\frac{1+x}{2}\right), \tag{3.22}$$

where $-1 \le x \le 1$. The entropy (plotted as a function of x in Figure 3.3) is naturally positive and extensive. Consider now fixed N and B. The ground state energy, $E_0 = -N\mu_M B$

---

[21] This behaviour is typical of entropy-controlled phase transitions described by D. Frenkel in his article *Entropy-driven phase transition*, Physica A **263**, 26 (1999).

(i.e. $x = -1$), is given by the Hamiltonian (Equation (1.10)) and corresponds to the unique microstate ($\Omega = 1$) in which all spins are parallel to B ($N_+ = N$). The entropy[22] is then $S_0 = 0$ ("ordered" configuration). From Equation (3.15), the system's temperature is given by:

$$
\frac{1}{T} = \left.\frac{\partial S}{\partial E}\right|_N = \frac{1}{N\mu_M B}\left.\frac{\partial S}{\partial x}\right|_N
$$

$$
= \frac{k_B}{2\mu_M B}\ln\frac{1-x}{1+x}. \tag{3.23}
$$

Inverting Relation (3.23), the energy $E = N\mu_M Bx$ as a function of temperature is obtained:

$$
E = N\mu_M B\,\frac{1 - e^{\frac{2\mu_M B}{k_B T}}}{1 + e^{\frac{2\mu_M B}{k_B T}}} = -N\mu_M B\,\frac{e^{\frac{\mu_M B}{k_B T}} - e^{-\frac{\mu_M B}{k_B T}}}{e^{\frac{\mu_M B}{k_B T}} + e^{-\frac{\mu_M B}{k_B T}}} = -N\mu_M B\,\tanh\left(\frac{\mu_M B}{k_B T}\right), \tag{3.24}
$$

which is, of course, extensive. The populations of spins parallel and anti-parallel to B given by Equations (3.20) can be deduced:

$$
\frac{N_+}{N} = \frac{e^{\frac{\mu_M B}{k_B T}}}{e^{\frac{\mu_M B}{k_B T}} + e^{-\frac{\mu_M B}{k_B T}}} \quad \text{and} \quad \frac{N_-}{N} = \frac{e^{-\frac{\mu_M B}{k_B T}}}{e^{\frac{\mu_M B}{k_B T}} + e^{-\frac{\mu_M B}{k_B T}}}. \tag{3.25}
$$

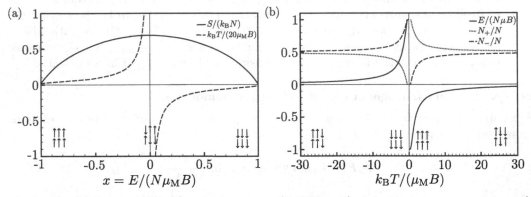

FIGURE 3.3   (a) Entropy ($S/k_B N$) and temperature ($k_B T/20\mu_M B$) of the ideal paramagnetic crystal as a function of energy ($x = E/N\mu_M B$). (b) Energy ($E/N\mu_M B$) and fractions of spins parallel (N_+/N) and anti-parallel (N_-/N) to the magnetic field as a function of temperature ($k_B T/\mu_M B$).

As can be seen in Figure 3.3, as long as $E < 0$, S increases with energy (i.e. $T > 0$) and presents a maximum[23] $S(N,0) \simeq k_B N \ln 2$ for $E = 0$. According to Equation (3.22), the

---

[22]The approximate entropy (Equation (3.22)), obtained for $N_+ \gg 1$, also leads to $S_0 = 0$ in the limit $x \to -1$, since $\lim_{\epsilon \to 0} \epsilon \ln \epsilon = 0$.

[23]Curiously, this maximum value of the entropy (for $E = 0$, i.e. $N_+ = N/2$) corresponds to the total number of microstates (i.e. 2^N) obtained for all possible values of the energy, or, equivalently, of $N_+ = 0, 1, 2, \ldots, N$, since each of the N spins has two possible states. In other words:

$$
2^N = \sum_{N_+=0}^{N} \binom{N}{N_+} > \binom{N}{N/2}.
$$

One term of the sum, $\binom{N}{N/2}$, cannot be equal to its total. In fact, the exact formula for the entropy (Equation (3.19)), given for $N_+ = N/2$ and taking into account the term in $\sqrt{N}$ in Stirling's formula (A.20)

temperature is then infinite: the interaction energy between the spins and the magnetic field is negligible in front of the thermal energy ($\mu_M B \ll k_B T$). Each spin can be equiprobably parallel or anti-parallel to the magnetic field (with the condition $N_+ = N_- = N/2$), giving the system a disordered aspect.

On the other hand, for $E > 0$, entropy decreases with energy: the *absolute temperature (in degree kelvin) of the system is thus negative!* This result, which seems to be in contradiction with the laws of thermodynamics, is due to the *finite number of individual states* (here equal to two) accessible to the particles. This is specific to quantum systems (see exercise 3.4). In such a case, the energy of the system has an upper bound ($E < E_{max}$) given by a weakly degenerate energy level. In the above example, $E_{max} = +N\mu_M B$ (i.e., $x = 1$) when all spins are antiparallel to the magnetic field. So $\Omega = 1$ and $S = 0$. Therefore, the entropy has a maximum for an intermediate energy (here $E = 0$), above which S decreases with E, implying negative temperatures[24]. In general, a system's energy has no upper bound (the number of possible individual states is not limited) and the absolute temperature is necessarily positive. For example, this is the case for quantum harmonic oscillators or classical particles, whose kinetic energy does not have an upper bond.

Negative temperatures were first observed experimentally[25] in 1950 in ionic crystals, such as lithium fluoride. In these materials, the nuclear spins of atoms interact weakly with the other degrees of freedom associated with the crystal lattice (which describe for example the vibrations of the atoms). After a sudden reversal of the applied magnetic field, the spin system, whose relaxation time is very small ($\sim 10^{-5}$ s), can thus be practically isolated and equilibrated for a few minutes at a given, possibly negative, temperature, called *spin temperature*. The latter is deduced from the population ratio, $N_+/N_- = e^{2\mu_M B/k_B T}$, measured by nuclear magnetic resonance (NMR). A negative temperature ($T < 0$) is then associated with a *population inversion* ($N_+/N_- < 1$), that is when there are more particles (spins) in the high-energy state (anti-parallel) than in the low-energy state (parallel), a phenomenon found in lasers. The crystalline lattice is at a necessarily positive temperature, since its energy, in particular its kinetic energy, does not have an upper bound. Thereafter, the whole system (spins and crystal lattice) reaches thermodynamic equilibrium at the same positive temperature. In 2013, stable negative temperatures, associated with atomic (not spin) degrees of freedom, were measured in a quantum gas of ultra-cold potassium atoms trapped in a magneto-optical lattice[26]. In such a system, the kinetic and potential energies can indeed be controlled and an upper bound be imposed[27].

---

is:

$$S(N, 0) = k_B \ln \frac{N!}{\frac{N}{2}! \frac{N}{2}!} \simeq k_B N \ln 2 - \frac{k_B}{2} \ln \frac{\pi N}{2},$$

which is indeed smaller than $k_B N \ln 2$. As it should be for a macroscopic system, the term in $\ln N$ is negligible compared to the term in N.

[24]Note that a system at $T < 0$ is hotter than a system at $T > 0$, since the (positive) energy of the first is greater than the (negative) energy of the second. When put in thermal contact, the first system gives up heat to the second. This is not a paradox, it is simply an inadequate definition of thermodynamic temperature. It would have been defined by the quantity $\theta = -1/T$ (which can be plotted with the help of Equation (3.23)), the "temperature" θ would have been higher as the system gets hotter system, independently of the sign of θ.

[25]E.M. Purcell and R.V. Pound, *A Nuclear Spin System at Negative Temperature*, Phys. Rev. **81**, 279 (1951).

[26]L.D. Carr, *Negative temperatures?*, Science **339**, 42 (2013) and S. Braun et al., *Negative absolute temperature for motional degrees of freedom*, Science **339**, 52 (2013).

[27]The study has led to a controversy over the very definition of entropy in statistical physics. For some authors, entropy should be defined as $S = k_B \ln \Phi$, where Φ is the number of microstates of energy less than E and is a necessarily increasing function of E: temperature would thus always be positive, even when the energy has an upper bound (J. Dunkel and S. Hilbert, *Consistent thermostatistics forbids negative absolute temperature*, Nat. Phys. **10**, 67 (2014) and I. Sokolov, *Not hotter than hot*, Nat. Phys. **10**, 7 (2014)). For usual macroscopic systems, according to Equation (2.24), this definition of entropy is equivalent to

Example 2: Ideal classical gas

According to Equation (2.24) and Expression (2.18) for $\Phi(N,V,E)$, the entropy of an ideal gas made of $N \gg 1$ *indistinguishable* monoatomic particles of mass m contained in a volume $V = L^3$ is written as (with $f = 3N$):

$$S(N,V,E) = k_{\mathrm{B}} \ln \Omega(N,V,E) \simeq k_{\mathrm{B}} \ln \Phi(N,V,E) = k_{\mathrm{B}} \ln \left(\frac{1}{N!} \frac{1}{\left(\frac{3N}{2}\right)!} \left(\frac{L\sqrt{2\pi m E}}{h} \right)^{3N} \right).$$

Using Stirling's formula (A.21), one gets:

$$S(N,V,E) \simeq k_{\mathrm{B}} \left(-N \ln N + N - \frac{3N}{2} \ln \frac{3N}{2} + \frac{3N}{2} + N \ln \left(\frac{L\sqrt{2\pi m E}}{h} \right)^3 \right).$$

By grouping terms together, the Sackur-Tetrode formula can be derived[28]:

$$S(N,V,E) = k_{\mathrm{B}} N \left(\ln \left(\frac{V}{N} \right) + \frac{3}{2} \ln \left(\frac{E}{N} \right) + \frac{3}{2} \ln \left(\frac{4\pi m}{3h^2} \right) + \frac{5}{2} \right), \qquad (3.26)$$

which is, of course, extensive[29], as it should be for the thermodynamic entropy of a homogeneous atomic system. It is the factor $1/N!$ in Expression (2.9) of Ω that allows the extensivity of the entropy[30] by taking into account the indistinguishability of the particles. The gas temperature is then given by:

$$\frac{1}{T} = \left. \frac{\partial S}{\partial E} \right|_{N,V} = \frac{3}{2} k_{\mathrm{B}} N \frac{1}{E} \quad \text{or} \quad E = \frac{3}{2} N k_{\mathrm{B}} T. \qquad (3.27)$$

Thus the energy of the monoatomic ideal gas is recovered, as already seen in thermodynamics and in the kinetic theory of gases (Equation (1.21))[31].

Expression (3.26) of a monoatomic ideal gas entropy is named after two physicists, Otto Sackur and Hugo Tetrode, who determined it independently of each other in 1912 (the latter

---

Boltzmann's definition, and thus has no advantage. Rather, it would not verify some fundamental properties of thermodynamic entropy (D. Frenkel and P.B. Warren, *Gibbs, Boltzmann, and negative temperatures*, Am. J. Phys. **83**, 163 (2015) and U. Schneider *et al.*, *Comment on "Consistent thermostatistics forbids negative absolute temperatures"*, arXiv:1407.4127 (2014)).

[28] When using the exact expression (2.19) of Ω as a function of Φ, a corrective term in $k_{\mathrm{B}} \ln \frac{3N}{2} \frac{\delta E}{E}$ must be added to the entropy (3.26). The dependence of this extra term is in $\ln N$ and is balanced by the terms of higher order coming from Stirling's formula ($-k_{\mathrm{B}} \ln \sqrt{2\pi N} \sqrt{3\pi N}$) and which were neglected. Indeed: $\ln N! \simeq N \ln N - N + \ln \sqrt{2\pi N}$. Sackur-Tetrode formula is therefore exact to within a constant in $k_{\mathrm{B}} \ln \frac{\delta E}{E} \sim 10^{-23}$ – perfectly negligible with respect to $k_{\mathrm{B}} N \sim 1$.

[29] By multiplying the extensive quantities N, V and E by a factor α, entropy is also multiplied by α: $S(\alpha N, \alpha V, \alpha E) = \alpha S(N,V,E)$.

[31] The (incorrect) entropy expression in the absence of the $1/N!$ factor, for (assumed) distinguishable particles of an ideal gas is:

$$S_{\mathrm{Distinguishable}}(N,V,E) = S(N,V,E) \underbrace{-k_{\mathrm{B}} \ln(1/N!)}_{\simeq k_{\mathrm{B}} N \ln N - k_{\mathrm{B}} N} \simeq k_{\mathrm{B}} N \left(\ln(V) + \frac{3}{2} \ln \left(\frac{E}{N} \right) + \frac{3}{2} \ln \left(\frac{4\pi m}{3h^2} \right) + \frac{3}{2} \right), \qquad (3.28)$$

which is not extensive (since V is not divided by N).

[31] Using Equation (3.27), entropy is written as:

$$S(N,V,E) = k_{\mathrm{B}} N \left(\ln \left[\frac{V}{N} \left(\frac{2\pi m k_{\mathrm{B}} T}{h^2} \right)^{\frac{3}{2}} \right] + \frac{5}{2} \right) = k_{\mathrm{B}} N \left(-\ln \rho \lambda_T^3 + \frac{5}{2} \right), \qquad (3.29)$$

where λ_T is the de Broglie thermal wavelength defined by Equation (1.24). From Equation (1.25), when the classical framework is valid ($\rho \lambda_T^3 \ll 1$), the ideal classical gas entropy is, of course, positive.

was then 17 years old!), by discretising phase space of an atomic system using the constant h (see Section 2.2.2). Knowing the latent heat of vaporisation of mercury as well as the heat capacity of its liquid phase and dealing with the vapour as an ideal gas in order to use Equation (3.26), they found[32] the value of h estimated by Planck for a photon gas in 1900. From photon gas to atomic gases to solids (see the heat capacity models of solids of Einstein (1907) and Debye (1912), presented in Exercise 4.3 and Section 7.1, respectively), the quantum hypothesis was widespread in statistical physics more than a decade before the advent of quantum mechanics.

3.2.2 PRINCIPLE OF MAXIMUM ENTROPY AND SECOND LAW OF THERMODYNAMICS

In Section 3.1, the thermal equilibrium condition between two subsystems was established and statistical temperature was introduced using the probability distribution $P(E)$ of the energy E of one of the subsystems. This approach can be generalised to any macroscopic internal variable A of an isolated system of energy E_{tot} (to ease notations, the other external parameters, N_{tot} and V_{tot}, are not mentioned in the following). The probability distribution of an internal variable A is given by the general expression (2.26):

$$P(A) = \frac{\Omega_{\text{p}}(E_{\text{tot}} \,|\, A)}{\Omega(E_{\text{tot}})} = \frac{1}{\Omega(E_{\text{tot}})} \, e^{\frac{S_{\text{p}}(E_{\text{tot}} \,|\, A)}{k_{\text{B}}}}, \tag{3.30}$$

where $S_{\text{p}}(E_{\text{tot}} \,|\, A) \equiv k_{\text{B}} \ln \Omega_{\text{p}}(E_{\text{tot}} \,|\, A)$ is by definition the partial statistical entropy associated with the internal variable A. Like the latter, $S_{\text{p}}(E_{\text{tot}} \,|\, A)$ depends, in principle, on the microstate of the system. If A is fixed and becomes an external parameter, then $S_{\text{p}}(E_{\text{tot}} \,|\, A)$ is the system's thermodynamic entropy.

When the internal variable A is *macroscopic*, its relative fluctuations are negligible (of the order of 10^{-12}). The probability distribution $P(A)$, described by a Gaussian distribution (2.32), has a very sharp peak centred on its most probable value $A_{\text{eq}} = \langle A \rangle$, interpreted as its equilibrium value. Expression (3.30) for $P(A)$ clearly shows that the equilibrium value of a macroscopic internal variable, and thus the equilibrium macrostate, corresponds to the maximum of S_{p}.

This result, which follows from the fundamental postulate, is expressed in the form of the *principle of maximum entropy*: *the equilibrium macrostate of an isolated system is the one with largest (partial) entropy*[33]. Thus, a formulation of the *second principle of thermodynamics* is recovered with in addition a probabilistic point of view highlighted by Boltzmann (see Section 2.4.3): the macro-equilibrium state of an isolated system is the one that is (by far) the most probable. Thermodynamic entropy, which, in principle, only makes sense for a system in equilibrium, is then (almost) equal to the maximum partial entropy:

$$S(E_{\text{tot}}) = \underbrace{k_{\text{B}} \ln \Omega(E_{\text{tot}}) \simeq k_{\text{B}} \ln \Omega_{\text{p}}(E_{\text{tot}} \,|\, A_{\text{eq}})}_{\text{according to (2.28)}} = S_{\text{p}}(E_{\text{tot}} \,|\, A_{\text{eq}}). \tag{3.31}$$

3.3 PRESSURE AND CHEMICAL POTENTIAL

The example studied in Section 3.1 will be generalised to the case where the two macroscopic systems exchange not only heat, but also work and particles. Using the principle of maximum entropy, the definitions of pressure and chemical potential will be expressed as partial derivatives of the thermodynamic entropy.

---

[32]see W. Grimus, *100th anniversary of the Sackur-Tetrode equation*, Annalen der Physik **525**, A32-A35 (2013).

[33]Within an approach to statistical physics based on information theory, this principle can be generalised to non-isolated systems (see Section 3.4).

3.3.1 THERMAL, MECHANICAL AND CHEMICAL EQUILIBRIUM BETWEEN MACROSCOPIC SYSTEMS

Two macroscopic subsystems $\mathcal{A}$ and $\mathcal{A}'$, forming together an isolated system $\mathcal{A}_{tot}$, are in thermal, mechanical and chemical contact (see Figure 3.4). In other words, the wall between them is diathermic (heat exchange), mobile (work exchange) and permeable (particle exchange). In practice, this fictitious wall can be the interface between two phases (gas and liquid, for example).

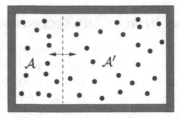

FIGURE 3.4 An isolated system $\mathcal{A}_{tot}$ is composed of two subsystems $\mathcal{A}$ and $\mathcal{A}'$, separated by a diathermic, mobile and permeable wall. Thus, $\mathcal{A}$ and $\mathcal{A}'$ exchange heat, work and particles.

Under these conditions, one gets:

$$
\begin{aligned}
E_{tot} &\simeq E + E' \quad \text{in the weak coupling approximation} \\
V_{tot} &= V + V' \\
N_{tot} &= N + N',
\end{aligned}
$$

where the energies E and E', volumes V and V' and numbers of particles N and N' of $\mathcal{A}$ and $\mathcal{A}'$, respectively, are macroscopic internal variables of the isolated system $\mathcal{A}_{tot}$. By contrast, N_{tot}, V_{tot} and E_{tot} are external parameters of $\mathcal{A}_{tot}$. According to Section 2.3, the probability distribution of the internal variables N, V and E are extremely sharp Gaussian distribution centred on $N_{eq} = \langle N \rangle$, $V_{eq} = \langle V \rangle$ and $E_{eq} = \langle E \rangle$, respectively. In the weak coupling approximation, the partial entropy of $\mathcal{A}_{tot}$ is[34] (see Section 3.2.1):

$$
S_{\text{P}tot}(N_{tot}, V_{tot}, E_{tot} \mid N, V, E) \simeq S_{\text{p}}(N, V, E) + S'_{\text{p}}(N_{tot} - N, V_{tot} - V, E_{tot} - E),
$$

where S_{p} and S'_{p} are the partial entropies of $\mathcal{A}$ and $\mathcal{A}'$, respectively. According to the principle of maximum entropy applied to the isolated system $\mathcal{A}_{tot}$, the equilibrium macrostate corresponds to the maximum of its partial entropy $S_{\text{P}tot}$. Since the three external parameters N, V and E are independent, one gets:

$$
\left. \frac{\partial S_{\text{P}tot}}{\partial E} \right|_{E=E_{eq}} = \left. \frac{\partial S_{\text{p}}}{\partial E} \right|_{E=E_{eq}} - \left. \frac{\partial S'_{\text{p}}}{\partial E'} \right|_{E'=E'_{eq}} = 0
$$

$$
\left. \frac{\partial S_{\text{P}tot}}{\partial V} \right|_{V=V_{eq}} = \left. \frac{\partial S_{\text{p}}}{\partial V} \right|_{V=V_{eq}} - \left. \frac{\partial S'_{\text{p}}}{\partial V'} \right|_{V'=V'_{eq}} = 0
$$

$$
\left. \frac{\partial S_{\text{P}tot}}{\partial N} \right|_{N=N_{eq}} = \left. \frac{\partial S_{\text{p}}}{\partial N} \right|_{N=N_{eq}} - \left. \frac{\partial S'_{\text{p}}}{\partial N'} \right|_{N'=N'_{eq}} = 0.
$$

---

[34]The two subsystems are considered (statistically) quasi-independent, so

$$
\Omega_{\text{P}tot}(N_{tot}, V_{tot}, E_{tot} \mid N, V, E) \simeq \Omega(N, V, E)\,\Omega'(N', V', E') .
$$

According to Equation (3.31), by identifying the partial entropies, S_p and S'_p, estimated at the equilibrium values of their internal variables, with the thermodynamic entropies S and S' of subsystems $\mathcal{A}$ and $\mathcal{A}'$, the thermodynamics equilibrium conditions can be found:

$$T = T', \tag{3.32}$$

$$\frac{P}{T} = \frac{P'}{T'}, \tag{3.33}$$

$$\frac{\mu}{T} = \frac{\mu'}{T'}. \tag{3.34}$$

In Equations (3.32)–(3.34), *temperature* T, *pressure* P and *chemical potential* μ of a macroscopic system in equilibrium are defined as[35]:

$$\frac{1}{T} \equiv \left.\frac{\partial S}{\partial E}\right|_{N,V}, \tag{3.35}$$

$$\frac{P}{T} \equiv \left.\frac{\partial S}{\partial V}\right|_{E,N}, \tag{3.36}$$

$$\frac{\mu}{T} \equiv -\left.\frac{\partial S}{\partial N}\right|_{E,V}. \tag{3.37}$$

The total differential of the entropy $S(N,V,E)$ is then:

$$dS = \frac{\partial S}{\partial E}dE + \frac{\partial S}{\partial V}dV + \frac{\partial S}{\partial N}dN = \frac{1}{T}dE + \frac{P}{T}dV - \frac{\mu}{T}dN.$$

The equivalent expression for the internal energy total differential is easier to remember:

$$dE = TdS - PdV + \mu dN. \tag{3.38}$$

By definitions, T, P and μ are intensive thermodynamic quantities, by contrast to their conjugate variables, S, V and N, which are extensive. Remember that the product of two conjugate variables is homogeneous to an energy. Temperature and pressure are familiar physical quantities. The chemical potential, which is more difficult to grasp, plays a fundamental role in the study of chemical reactions (see Exercise 4.9), osmosis (see Exercise 5.6), solubility of a solute in a solvent (see Exercise 5.5), but also in quantum statistical physics in general (see Chapter 6). In a macroscopic atomic system, the number of microstates, and thus the entropy, increase with energy and volume: according to Equations (3.35) and (3.36), temperature and pressure are therefore always positive[36]. On the other hand, the chemical potential can be positive or negative depending on the interactions between particles and the thermodynamic conditions[37].

Definitions (3.36) and (3.37) of pressure and chemical potential, related to a variation of entropy (at constant energy), are not very intuitive. They can be rewritten as derivatives of

---

[35]During the transformation of a macroscopic system, the variations of the number of particles are large compared to 1. Therefore, N can be considered as a continuous variable with respect to which the entropy can be derived (Equation (3.37)). In 1876, Gibbs introduced the notion of chemical potential in thermodynamics, by analogy with electric and gravitational potentials.

[36]Except for the special systems discussed on page 77.

[37]Adding a particle with no kinetic energy to an ideal gas composed of N particles conserves the energy of the system, since it does not interact with other particles. In this case, the gas entropy increases, because there are more accessible microstates to $N + 1$ particles than to N particles (at fixed E). Thus, according to Equation (3.37), $\mu < 0$. In the general case where interactions are taken into account, the addition of a particle changes the energy of the system and μ can be positive or negative.

the energy (at constant entropy) using the total differential of E given by Equation (3.38):

$$P = -\left.\frac{\partial E}{\partial V}\right|_{S,N} \tag{3.39}$$

$$\mu = \left.\frac{\partial E}{\partial N}\right|_{S,V}. \tag{3.40}$$

Thus, pressure corresponds to the negative of the variation of energy due to the change of volume during an isentropic transformation (at constant S) of a closed system (constant N). Similarly, chemical potential corresponds to a variation of energy due to the addition or removal of particles during an isentropic transformation at constant volume. The negative sign in Definition (3.37) associates a positive chemical potential with an increase of energy resulting from the addition of particles.

More generally, depending on the thermodynamic state and the properties of the studied system, other external parameters and their associated conjugate variables can be introduced. A term of "work" corresponding to the total differential of the internal energy can then be added. For example:

$$dE = T dS \begin{cases} -P dV & \text{pressure force} \\ +\mu dN & \text{particles exchange} \\ +B d\mathcal{M} & \text{magnetic properties} \\ +\mathcal{E} d\mathcal{P} & \text{electrical properties} \\ +\gamma d\mathcal{A} & \text{surface phenomena} \\ +F dL & \text{elasticity} \\ +\dots \end{cases} \tag{3.41}$$

where B and $\mathcal{M}$ are the magnetic field and moment, $\mathcal{E}$ and $\mathcal{P}$, the electric field and dipole moment, γ and $\mathcal{A}$, the surface tension and surface area of the system under consideration, and F the force exerted on a one-dimensional system of length L.

> In summary, for isolated systems described by a Hamiltonian, the calculation of Ω allows, in principle, the determination of the thermodynamic entropy using Boltzmann's formula. Thermodynamic quantities are then deduced from entropy by deriving its expression with respect to external parameters N, V, E...: thermodynamics is recovered.

Example 1: Ideal paramagnetic crystal and Schottky anomaly

In Section 3.2.1, the temperature of the independent spin system was calculated from its entropy. Let us now determine its magnetic properties. In general, the total magnetic moment of a system is defined as:

$$\mathcal{M} \equiv \sum_{j=1}^{N} \mu_M s_j = \mu_M(N_+ - N_-) = \mu_M(2N_+ - N),$$

where N_+ and $N_- = N - N_+$ are the numbers of spins parallel and anti-parallel to the magnetic field, respectively. For an ideal paramagnetic crystal, they are given by Equations (3.25). Put more simply, according to Expression (1.10) of this independent spins system's Hamiltonian, $\mathcal{M} = -E/B$. In this case, the magnetic moment is not an internal variable, its value is fixed by the energy. The expression of the (average) magnetic moment[38] can be deduced directly from Equation (3.24):

$$M = N\mu_M \tanh\left(\frac{\mu_M B}{k_B T}\right). \tag{3.42}$$

At low temperature, when $\mu_M B \gg k_B T$, all spins tend to align with the magnetic field and $\mathcal{M} \simeq N\mu_M$ (indeed, when $y \to \infty$, $\tanh y \simeq 1$). On the contrary, when $\mu_M B \ll k_B T$, interactions with the magnetic field being negligible in front of the thermal energy, the spins' orientations are randomly with probability $p_\pm = 1/2$ ($N_+ \simeq N_- \simeq N/2$) and $\mathcal{M}/N$ tends to zero. As can be seen in Figure 3.5, the magnetic moment per spin given by Equation (3.42) is in very good agreement with measurements performed on paramagnetic salts. In weak magnetic field (or at high temperature), the magnetic moment varies linearly with B: $\mathcal{M} \simeq (N\mu_M^2/k_B T)B$, because $y \ll 1$, $\tanh y \simeq y$. The magnetisation, $M \equiv \mathcal{M}/V$, where V is the volume of the system, is then:

$$M \equiv \frac{\mathcal{M}}{V} \simeq \frac{\chi}{\mu_0}B, \quad \text{with} \quad \chi = \frac{C_{\text{Curie}}}{T} \quad \text{(Curie's law)}, \tag{3.43}$$

where $C_{\text{Curie}} = (N\mu_0\mu_M^2)/(Vk_B)$ is the *Curie constant* (μ_0 is the vacuum permeability). The magnetic susceptibility χ is a dimensionless quantity characterising the capacity of a material to acquire a magnetisation when exposed to a low intensity magnetic field. The law demonstrated experimentally by Pierre Curie in 1895 is recovered.

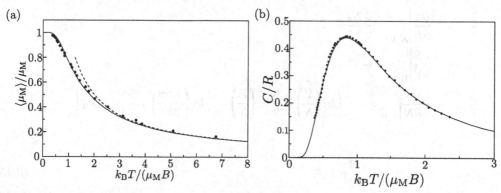

FIGURE 3.5 (a) Average magnetic moment per spin $\mathcal{M}/(N\mu_M)$ and, (b) molar heat capacity C/R as a function of $k_B T/(\mu_M B)$. The solid line curves are given by Equations (3.42) and (3.44), respectively. The experimental values (black dots) are: (a) measurements on a chromium potassium sulfate salt (W.E. Henry, Phys. Rev. **88**, 559 (1952)), and (b) measurements on a cerium chloride crystal (D.P. Landau *et al.*, Phys. Rev. B **7**, 4961 (1973)). The dashed curve represents Curie's law: $\mathcal{M} \propto (T/B)^{-1}$, for $k_B T \gg \mu_M B$.

The heat capacity defined by Equation (3.13) (here the crystal's volume does not intervene) is deduced from the energy expression (3.24):

$$C = \frac{\partial E}{\partial T}\bigg|_N = Nk_B\left(\frac{\frac{\mu_M B}{k_B T}}{\cosh\left(\frac{\mu_M B}{k_B T}\right)}\right)^2 = Nk_B\left(\frac{\frac{\Delta}{2k_B T}}{\cosh\left(\frac{\Delta}{2k_B T}\right)}\right)^2, \tag{3.44}$$

---

[38]The expression of N_+, and thus of $\mathcal{M}$, can be derived from Equation (3.41):

$$\frac{B}{T} = -\frac{\partial S}{\partial \mathcal{M}} = -\frac{1}{2\mu_M}\frac{\partial S}{\partial N_+} = -\frac{k_B}{2\mu_M}\ln\left(\frac{N}{N_+} - 1\right),$$

according to Expression (3.21) of the ideal paramagnetic crystal entropy (note that in this case one cannot fix E independently of $\mathcal{M}$, since $E = -\mathcal{M}B$).

where $\Delta = 2\mu_M B$ is the gap between the two energy levels of a spin, $-\mu_M B$ and $+\mu_M B$, depending on whether it is parallel or anti-parallel to the magnetic field. While in general the heat capacity C tends to increase or remain constant with temperature, C shows here a maximum for $\tanh(\Delta/2k_B T) = 2k_B T/\Delta$, i.e. for $k_B T \simeq 0.42\Delta$ (or $0.84\mu_M B$, see Figure 3.5). This behaviour, peculiar to systems in which particles have two (or a few) energy levels, is called *Schottky anomaly*, named after the German physicist who gave an explanation in 1922. The origin of this maximum can be understood qualitatively: for $k_B T \ll \Delta$, thermal energy is too small for particles to move from the ground level to the excited level ($N_+ \simeq N$ and $N_- \simeq 0$) and a temperature change δT does not change the energy ($\delta E \simeq 0$), so $C \simeq \delta E/\delta T \simeq 0$. On the other hand, when $k_B T \gg \Delta$, particles easily move from one level to the other such that $N_+ \simeq N_-$. The energy of the system thus varies scarcely with T and again $C \simeq 0$. When $k_B T \simeq \Delta$, the populations N_+ and N_- and thus the energy, change abruptly with temperature (see Figure 3.3), leading to a maximum value[39] of C. This quantum-driven macroscopic effect is observed in low-temperature systems, such as cerium chloride (see Figure 3.5), especially when the magnetic field is high enough to lift a degeneracy and separate two energy levels by a gap Δ of the order of $k_B T$.

Example 2: Ideal classical gas

The entropy of the monoatomic ideal gas is given by the Sackur-Tetrode formula (Equation (3.26)) from which all thermodynamic properties can be derived:

$$\frac{1}{T} = \left.\frac{\partial S}{\partial E}\right|_{N,V} = \frac{3}{2}\frac{Nk_B}{E}$$

$$\frac{P}{T} = \left.\frac{\partial S}{\partial V}\right|_{E,N} = \frac{Nk_B}{V}$$

$$\frac{\mu}{T} = -\left.\frac{\partial S}{\partial N}\right|_{E,V} = -k_B\left(\ln\left(\frac{V}{N}\right) + \frac{3}{2}\ln\left(\frac{E}{N}\right) + \frac{3}{2}\ln\left(\frac{4\pi m}{3h^2}\right) + \frac{5}{2}\right) - Nk_B\left(-\frac{5}{2N}\right).$$

So:

$$E = \frac{3}{2}Nk_B T \tag{3.45}$$

$$PV = Nk_B T \quad \text{or} \quad P = \rho k_B T \tag{3.46}$$

$$\mu = -k_B T\left(\ln\left(\frac{V}{N}\right) + \frac{3}{2}\ln\left(\frac{4\pi m}{3h^2}\frac{E}{N}\right)\right) = k_B T \ln \rho\lambda_T^3, \tag{3.47}$$

where $\rho = N/V$ is the particle number density and where, in expression[40] of μ, the energy is replaced by Equation (3.45) and Expression (1.24) of de Broglie's thermal length is used:

$$\lambda_T = \frac{h}{\sqrt{2\pi m k_B T}}. \tag{3.48}$$

Expression (3.46) of the pressure is, of course, the ideal gas equation of state, with k_B equal to R/N_A (as in Section 3.1.2). Starting from the microscopic description of an ideal

---

[39]This maximum value ($C \simeq 0.44R$ J $\cdot$ mol^{-1} $\cdot$ K^{-1}) can be very high compared to the heat capacity associated with other degrees of freedom of the crystal (in particular vibrations), typically of the order of $10^{-5}R$ J $\cdot$ mol^{-1} $\cdot$ K^{-1} at low temperatures. Systems with Schottky anomalies therefore make very good thermostats in cryogenics (if C is large, the temperature hardly varies with energy).

[40]Under the conditions of validity of the classical framework ($\rho\lambda_T^3 \ll 1$), the ideal gas chemical potential is negative, as discussed in Note 37.

classical gas of independent particles, a phenomenological law of thermodynamics has been recovered. Finally, the heat capacity at constant volume defined by Equation (3.13) is:

$$C_V = \left.\frac{\partial E}{\partial T}\right|_{N,V} = \frac{3}{2}Nk_{\mathrm{B}}.$$

As it should be, C_V is extensive, while P and μ are indeed intensive[41]. The value $C_V = 1.5R \simeq 12.5$ J·mol^{-1}·K^{-1}, calculated for one mole, is in very good agreement with experimental data for monoatomic gases at room temperature, such as argon or helium. Using Equations (3.45) and (3.46), the ideal gas entropy can be expressed as a function of pressure and temperature:

$$S(N,P,T) = k_{\mathrm{B}}N\left(\ln\left(\frac{k_{\mathrm{B}}T}{P\lambda_T^3}\right) + \frac{5}{2}\right).$$

The properties of ideal and real gases will be discussed again in Chapter 5.

3.3.2   GENERALISED FORCES AND THERMODYNAMIC TRANSFORMATIONS

In this rather technical section[42], it will be shown that pressure and chemical potential, defined as partial derivatives of the entropy in Section 3.3.1, can be expressed as average values of internal variables. To this end, consider an isolated macroscopic system whose external parameters are its energy E and, for simplicity, a single quantity noted x, which can be the volume, the number of particles, an external field (magnetic, gravitational...), etc.

Generalised forces

The energy $E_m(x)$ of a given microstate m depends on the value of the external parameter x through the Hamiltonian[43]. When x changes by an infinitesimal amount dx, the energy of microstate m changes by:

$$dE_m = \frac{\partial E_m}{\partial x}dx = -X_m dx, \quad \text{where by definition} \quad X_m \equiv -\frac{\partial E_m}{\partial x}$$

is the *generalised force conjugate to the external parameter* x in microstate m. In the particular case where x has the dimension of a length, X has the dimension of a force. If the external parameter varies slowly enough with respect to the relaxation time of the system, the latter passes through a succession of equilibrium states: the transformation is called *quasi-static*. The generalised force then behaves like an internal variable whose average value in the microcanonical ensemble can be calculated. This average is called the *mean generalised force* conjugate to the external parameter x:

$$\langle X \rangle \equiv -\langle \frac{\partial E_m}{\partial x}\rangle = -\frac{\partial E}{\partial x}, \tag{3.49}$$

where E is the system's internal (or mean) energy. Since energy is an extensive quantity, it is obvious that if the external parameter x is intensive (extensive), then $\langle X \rangle$ is extensive

[41]It is the factor $1/N!$, which takes into account the particle indistinguishability, that allows to obtain an intensive chemical potential and an extensive entropy (see Equation (3.26)). The other thermodynamic quantities $(P, T...)$ remain unchanged in the case of discernible particles since they are calculated by deriving the entropy at constant N.

[42]The reading of this section is not essential to the understanding of the rest of the book.

[43]For example, the energy levels of a confined quantum particle depend on the volume via the length L of the container (see Hamiltonian (1.1)).

(intensive). Like every mean value of an internal variable, the mean generalised force depends on the set of external parameters[44] $(E, x \dots)$.

Let us now discuss how the number of accessible microstates $\Omega(E, x)$ varies as a function of x. Formally, the number of microstates having energies between E and $E + \delta E$ can be written as in Equation (2.11). Thus:

$$
\begin{aligned}
\left.\frac{\partial \Omega(E, x)}{\partial x}\right|_E &= \frac{\partial}{\partial x}\left(\int_E^{E+\delta E} dE' \sum_m \delta(E' - E_m(x))\right) \\
&= \int_E^{E+\delta E} dE' \sum_m \frac{\partial}{\partial x}\delta(E' - E_m(x)) \\
&= \int_E^{E+\delta E} dE' \sum_m \left(-\frac{\partial E_m}{\partial x}\right)\frac{\partial}{\partial E'}\delta(E' - E_m(x)).
\end{aligned}
$$

However, for any function $f(E')$:

$$
\int_E^{E+\delta E} dE' \frac{\partial f(E')}{\partial E'} = \frac{\partial}{\partial E}\left(\int_E^{E+\delta E} dE' f(E')\right),
$$

therefore

$$
\left.\frac{\partial \Omega(E, x)}{\partial x}\right|_E = \frac{\partial}{\partial E}\left(\int_E^{E+\delta E} dE' \sum_m \left(-\frac{\partial E_m}{\partial x}\right)\delta(E' - E_m(x))\right). \tag{3.50}
$$

Using the expression of the mean value given by Equation (2.12), one gets:

$$
\langle X \rangle = \langle -\frac{\partial E_m}{\partial x}\rangle = \frac{1}{\Omega(E, x)}\int_E^{E+\delta E} dE' \sum_m \left(-\frac{\partial E_m}{\partial x}\right)\delta(E' - E_m(x)).
$$

Equation (3.50) thus becomes:

$$
\left.\frac{\partial \Omega(E, x)}{\partial x}\right|_E = \frac{\partial}{\partial E}\left(\Omega(E, x)\langle X \rangle\right) = \Omega(E, x)\frac{\partial \langle X \rangle}{\partial E} + \frac{\partial \Omega(E, x)}{\partial E}\langle X \rangle.
$$

Dividing both sides of the equation by $\Omega(E, x)$ and multiplying them by k_B one gets:

$$
\left.\frac{\partial k_B \ln \Omega(E, x)}{\partial x}\right|_E = k_B\frac{\partial \langle X \rangle}{\partial E} + \underbrace{\frac{\partial k_B \ln \Omega(E, x)}{\partial E}}_{1/T}\langle X \rangle.
$$

The definition of temperature (Equation (3.10)) can be recognised in the second member of the equation. Since energy is extensive and temperature is intensive, the first term (which

---

[44]Consider for example a gas contained in a piston with adiabatic walls ($\delta Q = 0$). An operator exerts a force on the piston of section $\mathcal{A}$. The piston's displacement $\vec{dl}$ is infinitesimal such that equilibrium is maintained (reversible transformation, see page 87). The gas, whose volume increases by $dV = \mathcal{A}dl > 0$, exerts on the piston a mean force $\langle \vec{F} \rangle$ of same magnitude but opposite direction to the force exerted by the operator. According to the first law of thermodynamics, the variation of the internal energy (or mean energy) E of the gas is given by Equation (3.53):

$$
dE = \delta Q + \delta W = 0 - \langle \vec{F} \rangle.\vec{dl} = -\frac{\langle F \rangle}{\mathcal{A}}dV = -PdV,
$$

where P is by definition the gas pressure exerted on the piston. Therefore $P = -\frac{\partial E}{\partial V}$, where the partial derivative is such that the transition is reversible and $\delta Q = 0$. It will be shown latter that it is then a isentropic transformation (at constant entropy) and Expression (3.39) of pressure is recovered (the system being closed, N is also constant).

behaves as $\sim k_B \langle X \rangle / f$) is negligible in front of the second (which behaves as $\sim \langle X \rangle$) for $f \gg 1$. Finally,

$$\left.\frac{\partial S}{\partial x}\right|_E \simeq \frac{\langle X \rangle}{T}. \tag{3.51}$$

In this general framework, for $x = V$, one recovers Definition (3.36) of pressure, which appears as the mean generalised force conjugate to the volume. Similarly, for $x = N$, the chemical potential (3.37) is the mean generalised force conjugate to the number of particles. Thus, pressure and chemical potential are indeed statistical mean values of internal variables, which are written without $\langle \ldots \rangle$ for simplicity of notations. If other external parameters x are imposed on the system (see Equation (3.41)), Equation (3.51) allows the calculation of their conjugate mean generalised force $\langle X \rangle$.

Thermodynamic transformations

During an infinitesimal transformation, entropy variation is given by (only one external parameter x is considered in addition to E):

$$dS = \left.\frac{\partial S}{\partial E}\right|_x dE + \left.\frac{\partial S}{\partial x}\right|_E dx,$$

thus from Equations (3.35) and (3.51)

$$dS = \frac{1}{T} dE + \frac{\langle X \rangle}{T} dx. \tag{3.52}$$

Using the first law of thermodynamics[45]:

$$dE = \delta Q + \delta W, \tag{3.53}$$

where δQ are δW are the exchanged heat and work during the infinitesimal transformation, respectively. Thus, dS can be written as:

$$dS = \frac{\delta Q}{T} + \frac{\delta W}{T} + \frac{\langle X \rangle}{T} dx.$$

Moreover, the work received by the system during the infinitesimal transformation is $\delta W = -X_{ext} dx$, where X_{ext} is the generalised force (pressure, chemical potential...) exerted by the external environment on the system. The term work is to be taken here in a broad sense. Thus:

$$dS = \frac{\delta Q}{T} + \frac{(\langle X \rangle - X_{ext})}{T} dx.$$

When a pressure P_{ext} and a chemical potential μ_{ext} are imposed on the system:

$$dS = \frac{\delta Q}{T} + \frac{(P - P_{ext})}{T} dV - \frac{(\mu - \mu_{ext})}{T} dN.$$

By definition, a transformation is *reversible*, if it is quasi-static *and* if the system remains in equilibrium with the external environment, that is with the systems with which it is in

---

[45]Remember that during an infinitesimal transformation, because internal energy and entropy are functions of state, their variations are expressed as exact total differential (written dE and dS, respectively), contrary to the exchanged heat and work (written δQ and δW, respectively), which depend on the path followed during the transformation.

thermal, mechanical and chemical contact[46]. In this idealised situation, $\langle X \rangle = X_{\text{ext}}$ (for example $P = P_{\text{ext}}$ and $\mu = \mu_{\text{ext}}$) and:

$$dS = \frac{\delta Q}{T} \quad \text{for a reversible transformation.}$$

A reversible *and* adiabatic transformation ($\delta Q = 0$) is therefore isentropic ($dS = 0$).

Statistical physics can now be used to interpret work and heat at the microscopic level. Since energy remains equal to the average energy during a quasi-static transformation, according to Equation (1.30), one gets:

$$E \simeq \langle E \rangle = \sum_m E_m \, p_m,$$

so

$$dE = \sum_m E_m \, dp_m + \sum_m p_m \, dE_m. \tag{3.54}$$

At the microscopic level, the system's internal energy variation is thus due either to a variation dp_m of the microstates' probabilities of occurrence, or to a variation dE_m of the energy levels. Moreover, according to the first law of thermodynamics (Equation (3.53)):

$$dE = \delta Q + \delta W = \delta Q - X_{\text{ext}}dx,$$

where X_{ext} is the generalised force exerted by the external environment on the system (e.g., the external pressure). A question then arises: can each of the two terms in the second term of Equation (3.54) be identified with δQ and δW, respectively? From Equations (3.52) and (3.54):

$$TdS = dE + \langle X \rangle dx = \sum_m (E_m \, dp_m + p_m \, dE_m) + \langle X \rangle dx,$$

and since by definition $dE_m = -X_m dx$, then:

$$\sum_m p_m \, dE_m = -\sum_m p_m X_m dx = -\langle X \rangle dx.$$

One deduces:

$$TdS = \sum_m E_m \, dp_m.$$

Finally, for any infinitesimal transformation:

$$dE = \delta Q + \delta W = \underbrace{\sum_m E_m \, dp_m}_{=TdS} + \underbrace{\sum_m p_m \, dE_m}_{=-\langle X \rangle dx}.$$

In the special case of a *reversible* transformation, δQ can be identified as TdS and δW as $-\langle X \rangle dx$, so:

$$\delta Q = TdS = \sum_m E_m \, dp_m \tag{3.55}$$

$$\delta W = -\langle X \rangle dx = \sum_m p_m \, dE_m.$$

[46]To transform a system, external parameters must evolved with time. In the ideal case of a *reversible* transformation, these changes are small enough to be reversible and maintain an equilibrium with the external environment.

Thus, during a reversible transformation, the exchanged heat is associated with a variation of the probabilities of the accessible microstates ($\mathrm{d}p_m \neq 0$), whereas the exchanged work corresponds to a modification of the energy levels[47] ($\mathrm{d}E_m \neq 0$). The subtle distinction between work and heat, so mysterious in the nineteenth century, has therefore a microscopic interpretation.

3.4 GIBBS ENTROPY AND INFORMATION THEORY

The two postulates of statistical physics (see Section 2.1) and the definition of statistical entropy of an isolated system in equilibrium, given by Boltzmann's formula (3.16), render possible the recovery of thermodynamics from a microscopic description of matter. However, there is an alternative approach to statistical physics which is part of the more general framework of information theory[48]. Its object of study is mainly coding, transmission and compression of data, thus covering information and communication technologies, computer science or, for example, molecular and genetic biology. Statistical entropy, a physical quantity at the interface between microscopic and macroscopic, is then defined as a measure of the lack of knowledge about the system under consideration or a lack of information.

In all generality, consider an experiment with an uncertain outcome consisting of a discrete and finite set of Ω events with probabilities p_m where $m = 1, 2, \ldots, \Omega$. For example, consider the roll of a $\Omega = 6$-sided die. In principle, if the initial and experimental conditions of the roll are perfectly known, the laws of solid mechanics allows the determination of the outcome with certainty. In practice, a simple die roll is too complicated to study experimentally and one must resort to probability to predict an outcome. Thus, by moving from a deterministic to a probabilistic approach, information about the state of the system (the expected outcome) is lost. The analogy with statistical physics is obvious: one replaces full (total) information about the system, which would be given formally by the knowledge of its microstate m, by a probability distribution p_m.

It is then a question of evaluation of the missing information associated with a probability distribution p_m, in other words the quantity of information that one would obtain by learning the result of the experiment (the number obtained by rolling a die or the microstate of a system). This quantity, called the *Shannon*[49] *entropy* in information theory, (or *Gibbs entropy* in statistical physics), is written here as $S_{\mathrm{G}}(\{p_m\})$. With $0 \leq p_m \leq 1$ and $\sum_{m=1}^{\Omega} p_m = 1$, it must respect the following obvious criteria:

1. $S_{\mathrm{G}} \geq 0$, the missing information being zero when an event is certain. If for example $p_1 = 1$ and $p_m = 0$ for all $m \neq 1$, then $S_{\mathrm{G}}(1, 0, \ldots, 0) = 0$. The roll of a die which six sides are marked with the number 6 gives a certain result.

---

[47]During Joule expansion (see Section 1.4.1), a gas initially in equilibrium expands in a container with adiabatic walls ($\delta Q = 0$) without exchanging any work with the external environment ($\delta W = 0$). There is therefore no variation of internal energy ($\mathrm{d}E = 0$). Moreover, the transformation is clearly not quasi-static, nor *a fortiori* reversible. As the volume accessible to the gas increases, the energy levels get closer (according to Equation (1.1), the gap varies in $1/L^2$), so $\mathrm{d}E_m \neq 0$. To ensure conservation of energy, the probabilities must therefore also change during the transformation (i.e. $\mathrm{d}p_m \neq 0$), although there is no heat exchange. This is not a contradiction of Equation (3.55), since the transformation is not reversible.

[48]This approach, scarcely developed in classic statistical physics introductory textbooks (K. Huang, R.K. Pathria and P.D. Beale, F. Reif, R.H. Swendsen...), is on the contrary favoured by the less-known (non-anglophone) authors (R. Balian, B. Diu *et al.*, C. and H. Ngô). References of some of these works are given in the foreword.

[49]Claude Elwood Shannon (1916–2001), American mathematician and founder of information theory in 1948 from his work in cryptography within the secret services of the American army during the Second World War. It was the mathematician and physicist John von Neumann who advised him to call "entropy" the quantity S_{G} quantifying the missing information for two reasons: first because its expression, given by Equation (3.57), was already in use in statistical physics (see Note 55) and especially because *"no one really knows what entropy is, so in a discussion you will always have the advantage"* (after M. Tribus and E.C. McIrvine, *Energy and information*, Scientific American **225**, 179 (1971)).

2. Impossible events do not contribute to S_G. If for example $p_1 = 0$ then $S_G(0, p_2, \ldots, p_\Omega) = S_G(p_2, \ldots, p_\Omega)$. If no side of a die has the number 6 on it, then there is no need to consider this event when evaluating the missing information.

3. S_G is maximal for equiprobable events, that is for a probability distribution function $p_m = 1/\Omega$, for any m. Indeed, when the only known information is the number Ω of possible events (minimal information, maximal missing information), there is no reason to favour any of them. This is the case for the roll of a (unpiped) die. Moreover, S_G must increase with Ω. In other words, known information decreases when the number of possible events increases.

4. S_G is an additive quantity: the missing information associated with two *independent* experiments is the sum of the missing information associated with each of them[50]. Let two independent experiments be characterised respectively by the probability distribution functions p_m and p'_n, where $m = 1, 2, \ldots, \Omega$ and $n = 1, 2, \ldots, \Omega'$. For example, rolling a die and tossing a coin simultaneously. The probability of observing the events m and n is $p_m \times p'_n$ and the missing information associated with this double experiment must be written:

$$S_G(\{p_m \times p'_n\}) = S_G(\{p_m\}) + S_G(\{p'_n\}). \tag{3.56}$$

There exists a single function S_G that respects these four criteria. Without presenting a rigorous demonstration[51], the shape of S_G can be guessed: first, Equation (3.56) suggests a behaviour of S_G in $\ln p_m$ and since missing information is a number that characterises a probability distribution as a whole, it must be written as a mean value. Shannon (or Gibbs) entropy is then defined as:

$$S_G(\{p_m\}) = -\lambda \langle \ln p_m \rangle = -\lambda \sum_{m=1}^{\Omega} p_m \ln p_m, \tag{3.57}$$

where λ is a constant that fixes the dimension of S_G. It is easy to show that S_G verifies[52] criteria 1, 2 and 4. To demonstrate that S_G does satisfy criterion 3, the method of Lagrange multipliers can be used (see Appendix A.5): To find the probability distribution $\{p_m\}$ which maximises S_G under the constraint expressed by the normalisation condition ($\sum p_m = 1$), is equivalent to finding the maximum of:

$$L(\{p_m\}, \mu_1) = S_G(\{p_m\}) - \mu_1 \underbrace{\left(\sum_{m=1}^{\Omega} p_m - 1 \right)}_{=0} = -\lambda \sum_{m=1}^{\Omega} p_m \ln p_m - \mu_1 \left(\sum_{m=1}^{\Omega} p_m - 1 \right),$$

---

[50] In addition, the missing information associated with two *non-independent* experiments must verify an equality involving conditional probabilities (see supplement I.G. of B. Diu *et al.*, *Physique statistique* (Hermann, 1996, in French)).

[51] See Fazlollah M. Reza, *An Introduction to Information Theory* (Dover Publications, New York, 1961).

[52] Criteria 1 and 2 are guaranteed by the negative sign in Equation (3.57) (with $0 \le p_m \le 1$) and by the limit $\lim_{p_m \to 0} p_m \ln p_m = 0$. The additivity of S_G (criterion 4) results from the logarithm properties and the normalisation of the probability distribution functions:

$$S_G(\{p_m \times p'_n\}) = -\lambda \sum_{m=1}^{\Omega} \sum_{n=1}^{\Omega'} p_m p'_n \ln(p_m p'_n) = -\lambda \sum_{m=1}^{\Omega} \underbrace{\sum_{n=1}^{\Omega'} p'_n}_{=1} p_m \ln p_m - \lambda \sum_{n=1}^{\Omega'} \underbrace{\sum_{m=1}^{\Omega} p_m}_{=1} p'_n \ln p'_n$$

$$= S_G(\{p_m\}) + S_G(\{p'_n\}).$$

where μ_1 is a parameter, called *Lagrange multiplier* (not to be confused with the chemical potential), which is calculated *a posteriori*. Since the constraint is taken into account in the expression of $L = S_G$, the maximum is obtained by deriving L with respect to each of its $\Omega + 1$ independent variables $p_1, p_2, \ldots, p_\Omega$ and μ_1:

$$\frac{\partial L}{\partial p_m} = -\lambda \ln p_m - \lambda - \mu_1 = 0, \quad \text{for} \quad m = 1, 2, \ldots, \Omega$$

$$\frac{\partial L}{\partial \mu_1} = \sum_{m=1}^{\Omega} p_m - 1 = 0.$$

It can be then deduce that $p_m = e^{-(1+\frac{\mu_1}{\lambda})}$ has the same value for $m = 1, 2, \ldots, \Omega$. The normalisation condition implies $p_m = 1/\Omega$. The missing information S_G is indeed maximal for a discrete and equiprobable probability distribution function.

By convention, in information theory, the constant λ is chosen equal to $1/\ln 2$. Thus, the missing information associated with the probability distribution of two equiprobable events, $p_1 = p_2 = 1/2$, is equal to $S_G = 1$ bit ("binary digit"), the unit of measurement of information in computer science. Consider a simple example, the game of heads or tails. A tossed coin can fall either on the heads side with probability p, or on the tails side with probability $q = 1 - p$. According to Equation (3.57), the Shannon entropy associated with this probability distribution is:

$$S_G(p) = -\frac{1}{\ln 2}\left(p \ln p + (1-p)\ln(1-p)\right).$$

Obviously, if $p = 0$ or $p = 1$, then $S_G = 0$: the information about the coin is total, so the toss brings no information, since the result is certainly heads in the first case and tails in the second. On the other hand, if $p = 1/2$ (equiprobability), the function $S_G(p)$, which can be plotted easily, is maximal and is equal to 1, by definition of the bit: since the missing information about the (unrigged) coin is maximal, the toss result offers maximal information. If for example, $p = 9/10$, the toss result provides less information ($S_G \simeq 0.47$ bit), since it is known that the heads side occurs more frequently (90% of the tosses on average). The binary outcome of a coin toss, also encountered for a spin-1/2 ($s_i = -1$ or $s_i = +1$), is the elementary brick of information theory. Indeed, any information (text, sound, image, video...) can be digitised, that is converted into a succession of 0's and 1's. For instance, a series of eight bits (called *an octet*, for example 10010111) corresponds to $2^8 = 256$ different combinations, each one corresponding to a given character (number, letter or symbol depending on the encoding)[53]. From the concept of entropy to spin systems, information theory thus presents many similarities with statistical physics[54].

The connection between Shannon's mathematical entropy and thermodynamic entropy will now be discussed. According to the third criterion, S_G is maximal when the probability distribution is uniform ($p_m = 1/\Omega$) and it is then written as:

$$S_G(\{p_m = \frac{1}{\Omega}\}) = -\lambda \sum_{m=1}^{\Omega} \frac{1}{\Omega} \ln \frac{1}{\Omega} = \lambda \ln \Omega. \tag{3.58}$$

---

[53]For example, data formats (mp3, jpeg, mpeg...) are based on compression algorithms. Intuitively, a message consisting only of 1 (a blank page) contains little information. It can thus be easily compressed by replacing the succession of N 1's, by a shorter sequence of bits giving the same information: "there are N 1's". On the contrary, a random sequence of N digits, 0 or 1, each with the same probability $p = 1/2$, is incompressible.

[54]For more information, see M. Mézard and A. Montanari, *Information, Physics, and Computation* (Oxford University Press, USA, 2009).

By choosing the constant λ equal to the Boltzmann constant k_B, Boltzmann formula, $S = k_B \ln \Omega$ giving the thermodynamic entropy of an isolated system in equilibrium, is recovered. This result suggests an alternative approach to statistical physics, considered then as a branch of information theory. In this context, the fundamental postulate becomes a principle of maximum entropy: the probability distribution of the microstates of an isolated system in equilibrium is the probability distribution $\{p_m\}$ which maximises the entropy $S_G = -k_B \sum_m p_m \ln p_m$ under the sole constraint of the normalisation condition $\sum p_m = 1$. Shannon entropy S_G, given by Equation (3.57), is then called *Gibbs entropy*[55]. In the framework of classical physics, S_G is associated with a probability density function $\rho(\boldsymbol{q},\boldsymbol{p})$ to find the system in the microstate $(\boldsymbol{q},\boldsymbol{p})$ within $d^f q$ and $d^f p$. It is written by analogy with Equation (3.57):

$$S_G = -k_B \int \rho(\boldsymbol{q},\boldsymbol{p}) \ln \left(\rho(\boldsymbol{q},\boldsymbol{p}) h^f \right) \, d^f q d^f p, \tag{3.59}$$

where f is the number of degrees of freedom of the system. The factor h^f is essential, firstly to make the logarithm argument dimensionless and, secondly to obtain the expression of the statistical entropy, $S = k_B \ln \Omega$, with the microcanonical probability density, $\rho(\boldsymbol{q},\boldsymbol{p}) = 1/(\Omega h^f)$, given by Equation (2.14). Phase space discretisation is needed once again as discussed in Section 2.2.2: the elementary phase space cell volume h^f corresponds to the maximum theoretical precision on the determination of a microstate imposed by quantum mechanics. Thus, the maximum information that one can obtain on a system is the exact elementary cell of phase space containing the system's representative point. In this case[56], $S_G = 0$ (minimal missing information).

As seen in Chapter 1, statistical physics relies first on the possibility of calculating statistical mean values to determine the thermodynamics quantities of a system in equilibrium (first postulate, expressed by Equation (1.27)). It is then necessary to determine i) the

---

[55] In his book on statistical mechanics published in 1902 (see Note 44 in Chapter 1), Gibbs uses the quantity S_G and calls it the "average probability index". It is in the case of a system in contact with a thermostat that he identifies S_G with thermodynamic entropy (see Chapter 4). The use of this statistical entropy formula was then generalised mainly by the American physicist Edwin Thompson Jaynes (1922–1998). Once again, it is Boltzmann who is at the origin of this form of entropy. In his quest for a demonstration of the second law of thermodynamics, he defined several quantities, interpreted as the entropy of a system in equilibrium, in the form of Equation (3.57), but relating to probabilities associated with each *individual* particle and not on the system as a whole, i.e. the system's microstate (probability p_m). In 1872, considering a dilute gas, he introduced the quantity $H(t) = \int f(\vec{v},t) \ln f(\vec{v},t) d^3 v$, where $f(\vec{v},t) d^3 v$ is the probability that a particle has the velocity $\vec{v}$ at time t. He then demonstrates the "H theorem": $-H(t)$ increases with time and becomes constant (in equilibrium) if and only if $f(\vec{v},t)$ takes the form of the Maxwell velocity distribution (see Exercise 1.3). Defining the entropy of a dilute (quasi-ideal) gas as $-Nk_B H$, Boltzmann thus claimed to have demonstrated the second law of thermodynamics in this particular case. In 1877, he found a similar expression using a combinatorial argument applied to the following model. An isolated ideal gas at energy E is comprised of N particles distributed according to their individual states: N_i particles are in a state i of energy e_i, where $i = 1, 2, \ldots, k$. Thus, $N = \sum_{i=1}^{k} N_i$ and $E = \sum_{i=1}^{k} N_i e_i$. The number of microstates compatible with a given distribution $\{N_i\}$ is:

$$\Omega = \frac{N!}{N_1! N_2! \ldots N_k!} \simeq \frac{N^N}{N_1^{N_1} N_2^{N_2} \ldots N_k^{N_k}},$$

where Stirling's formula (A.21) was used ($N_i! \simeq N_i^{N_i} e^{-N_i}$ for $N_i \gg 1$). According to Equation (3.16), the statistical entropy is:

$$S = k_B \ln \Omega \simeq -N k_B \sum_{i=1}^{k} p_i \ln p_i,$$

where $p_i = N_i/N$ is the probability to find *any particle* in state i, while the entropy S_G is expressed as a function of the probability p_m to find the *system* in microstate m.

[56] Indeed, suppose that the system's microstate m is known (with a precision h^f). The normalisation condition (1.33) imposes $\rho(\boldsymbol{q},\boldsymbol{p}) = 1/h^f$ if the system's representative point is located in the elementary cell (of volume h^f) corresponding to m and $\rho(\boldsymbol{q},\boldsymbol{p}) = 0$ otherwise. Then, according to Equation (3.59), $S_G = 0$.

microstates probability distribution $\{p_m\}$ and, ii) the expression of the statistical entropy S which allows the calculation of thermodynamic quantities. Depending on whether the form of p_m or the form of S is postulated, two approaches to equilibrium statistical physics are possible:

- The first, favoured in this book and developed in previous sections, is discussed for the case of an isolated system and postulates the equiprobability of accessible microstates, $p_m = 1/\Omega$ (fundamental postulate). Equilibrium being defined as the most probable macrostate, the form of S can be deduced and is given by Boltzmann's formula, $S = k_B \ln \Omega$.

- The second one, based on the expression of Gibbs entropy and on the principle of maximum (Gibbs) entropy, can be used with any external parameters imposed on the system (and not only for isolated systems). The statistical entropy being given by:

$$S_G(\{p_m\}) = -k_B \sum_m p_m \ln p_m, \tag{3.60}$$

the equilibrium probability distribution $\{p_m\}$ is the one which maximises the entropy $S_G(\{p_m\})$ for the *given constraints*. Indeed, a constraint imposed on a system brings partial information, which must be taken into account in the determination of p_m. For an isolated system, the only constraint is the normalisation of the distribution $\{p_m\}$. In this particular case Boltzmann's formula[57] is recovered. When the system is in contact with the external environment, additional constraints must be taken into

---

[57]Boltzmann entropy is a special case of Gibbs entropy (for $p_m = 1/\Omega$). The latter is expressed as a weighted arithmetic average (see Equation (3.60)). Extending this definition of average, usual in physics, to a more general expression (which includes in particular geometric average), Renyi entropy of order α can be defined (1960). It also verifies the four criteria of Shannon entropy:

$$S_\alpha(\{p_m\}) = \frac{k_B}{1-\alpha} \ln \sum_{m=1}^{\Omega} (p_m)^\alpha,$$

where α is a positive or zero real parameter. For $\alpha = 0$, Boltzmann entropy ($S_0 = k_B \ln \Omega$) is recovered. For $\alpha \to 1$, it can shown that $S_\alpha \to S_G$. Because of its properties, Renyi entropy is used both in quantum information theory, a generalisation of Shannon's information theory relying on quantum systems, and in statistics to estimate the diversity of a population (ecology, biodiversity...). Many other expressions of statistical entropy have been proposed to generalise Gibbs entropy, one of the most famous, for its many applications, was introduced by the Brazilian physicist Constantino Tsallis in 1988:

$$S_q(\{p_m\}) = \frac{k_B}{q-1} \left(1 - \sum_{m=1}^{\Omega} (p_m)^q \right),$$

where q is a real parameter. By writing $p_i^q = p_i e^{(q-1)\ln p_i} \approx p_i(1 + (q-1)\ln p_i)$, it can be easily shown that in the limit $q \to 1$, S_q tends towards Gibbs entropy. The main property of Tsallis entropy is that for $q \neq 1$, it is not additive, unlike Gibbs entropy by construction (criterion 4 of S_G): $S_q(\mathcal{A} + \mathcal{A}') = S_q(\mathcal{A}) + S_q(\mathcal{A}') + (1-q)S_q(\mathcal{A})S_q(\mathcal{A}')$, where $\mathcal{A}$ are $\mathcal{A}'$ are two independent systems (the probability that system $\mathcal{A} + \mathcal{A}'$ is in the (m, n) microstate is thus $p_m p_n'$). For systems with weakly correlated particles or even independent particles, an additive entropy is extensive. On the other hand, when correlations are strong, an additive entropy may not be extensive but a non-additive entropy (S_q with $q \neq 1$) can be. Long controversial, Tsallis approach renders possible the description of the properties of many systems outside of the scope of basic statistical thermodynamics (see Note 5): non-equilibrium states, systems with few particles, long-range correlations or interactions... The applications of non-additive entropy cover a great diversity of systems and physical phenomena in quantum mechanics (cold atoms, quantum chaos), in hydrodynamics (turbulence), in astrophysics (plasma, self-gravitating systems), in geophysics (earthquakes), in computer science (algorithms optimisation). For more information see C. Tsallis, *Introduction ton Nonextensive Statistical Mechanics: Approaching a complex world* (Springer, 2009) and C Tsallis *et al.*, *Introduction to Nonextensive Statistical Mechanics and Thermodynamics*, Proceedings of the 1953–2003 Jubilee "Enrico Fermi" International Summer Schools of Physics.

account and a uniform probabilty distribution function is no longer the equilibrium probability distribution maximising S_G, as it will be seen in Section 4.3.2.

TABLE 3.2 The different approaches to the study of a system in equilibrium.

Approach	Fundamental postulate	Maximum entropy postulate
System	Isolated	Every external parameter
Postulate at equilibrium	Equiprobability of microstates $p_m = 1/\Omega$	Maximum Gibbs entropy $S_G = -k_B \sum_m p_m \ln p_m$
Consequences	Maximum Entropy $S = k_B \ln \Omega$	Expression of p_m depending on constraints

These two approaches, summarised in Table 3.2, are equivalent and lead to the same theoretical predictions of the thermodynamic properties of macroscopic systems in equilibrium, isolated or not. This matter was illustrated by the determination of the entropy expression in the case of an isolated system in equilibrium (Equation (3.58)).

The second approach is attractive, firstly because Gibbs entropy can be used with any constraints imposed on the system by the external parameters, whereas Boltzmann's formula is valid only for isolated systems. Secondly and most importantly, because it fits into the general framework of information theory and thus offers an interpretation of entropy in terms of missing information: thermodynamic entropy measures the lack of knowledge (ignorance) about a system in equilibrium, of which only the macrostate is known. Ignorance is to be taken in the sense of maximal missing information on the system's microstate while taking into account external parameters[58]. A system in equilibrium can, in principle, be in any accessible microstate m with probability p_m. However, the interpretation of entropy in terms of information can be confusing for several reasons. First of all, entropy seems to depend on the observer and on *their* – subjective – ignorance of the system's state. However, entropy is a thermodynamic quantity specific to the studied system, which can be measured objectively with an adapted experimental protocol. Consider a homogeneous fluid in equilibrium. The value of its entropy $S(P,T)$, associated with the macrostate at temperature T and pressure P, is the quantity of information (measured in k_B units) that would be obtained if the exact determination of the system's microstate at a given instant was possible. It is indeed an objective quantity in the sense that entropy is completely determined by the system's macrostate. Note that entropy does not measure how interesting (necessarily subjectively) the information is: the knowledge of the microstate of a gram of water is far much information than the information contained in all books written during the history of humanity[59]. Then, the principle of maximum (Gibbs) entropy must not be confused with the second law of thermodynamics. The latter states that, in equilibrium, the thermodynamic entropy of an isolated system is maximum (see Section 3.5.1). If the system is in contact with the external environment, equilibrium is not defined by the entropy maximum, but by the extremum of another thermodynamic potential.

[58]In particular, all information about the system's initial conditions – i.e. the initial macrostate and *a fortiori* the initial microstate – is lost once equilibrium is reached (only the conserved quantities remain known). In the example of the drop of ink in a glass of water (see Section 1.3.1), once the system is homogeneous, the initial shape of the drop is unknown.

[59]The Bibliothèque Nationale de France (BNF), for example, contains about 10 million books, of 500 pages on average, each containing about 1000 characters (the 26 letters of the alphabet, without taking into account punctuation). There are thus $10^7 \times 500 \times 10^3 = 5 \times 10^{12}$ characters in total. If one supposes, for simplicity, that the occurrences of each characters are equiprobable (which is false, the letter "e" is for example twice as probable as "s" or "a" in French), the content of all the BNF books corresponds to a single "microstate" among $26^{5 \times 10^{12}}$, or an entropy of 2×10^{-10} J·K^{-1}. It is perfectly negligible with respect to the 4 J·K^{-1} corresponding to the entropy of one gram of water at room temperature.

The determination of the microstate probability distribution for systems in contact with a thermostat and a particle reservoir will be discussed in Chapter 4 using these two statistical physics approaches. On the one hand, Gibbs entropy allows for an easy calculation of the probability distribution $\{p_m\}$, but the physical meaning of the additional constraints associated with external parameters is not very intuitive (see Section 4.3.2). On the other hand, reasoning based on the fundamental postulate leads to the same result, while being more rigorous, controlling the necessary approximations and giving a clear meaning to the conditions of applications of the canonical and grand canonical statistical ensembles. Consequently, this book prefers the clarity and rigour of the approach based on the fundamental postulate to the abstract elegance of information theory applied to statistical physics.

3.5 IRREVERSIBILITY AND ENTROPY

Now that Boltzmann (and Gibbs) statistical entropy has been introduced, new light on the second law of thermodynamics and the notion of irreversibility can be shed, giving them a statistical interpretation on the basis of a microscopic description. Like other fundamental aspects of statistical physics, this subject is still a matter of much debate and discussion[60].

The phenomena observed at a macroscopic scale are irreversible. The time evolution of a drop of ink deposited in a glass of water, the fragmentation of a plate, life...occurs in a single time direction, called the *arrow of time*. The film of such a process viewed in reverse seems absurd: something is not going right, or rather, is evolving in the "wrong" direction[61]. It is to take into account the privileged – indeed, unique – direction of time evolution of macroscopic phenomena that entropy and the second law of thermodynamics were introduced in the nineteenth century.

3.5.1 SECOND LAW OF THERMODYNAMICS

In thermodynamics, the existence of the entropy state function and the second law are postulated on a phenomenological basis. There are several formulations of the second law of thermodynamics[62]. Clausius formulation (1854) has the interest of expressing clearly the irreversibility of heat exchange: heat does not spontaneously flow from a cold body to a hot body. The following year, the German physicist gave a more formal expression, using a state function that he named *entropy* in 1865 (see Note 14): *during a transformation in which an adiabatically isolated system passes from a macrostate i to a macrostate f, its thermodynamic entropy increases.* Let:

$$\Delta S = S^{(f)} - S^{(i)} \geq 0.$$

The equality ($\Delta S = 0$) is obtained when the transformation is *reversible*, that is quasi-static (the system passes continuously through a succession of equilibrium states) and such that the system is permanently in equilibrium with the external environment (see Section 3.3.2). Otherwise, the transformation is *irreversible* and $\Delta S > 0$.

---

[60]See Bricmont (1995), Goldstein (2001), Lebowitz (1993), Note 35 in Chapter 2.

[61]On the Internet, some websites show movies in reverse (enter "reverse movie" in a search engine). Some are quite spectacular...

[62]Working on the theory of steam engines, French scientist Nicolas Léonard Sadi Carnot (1796–1832) gave a first formulation in 1824 on an erroneous basis: heat was then considered as a fluid called *caloric*. His theory, first taken over in 1834 by his compatriot Émile Clapeyron (1799–1864), was formalised in the years 1850–1860 by the founders of thermodynamics: the German Rudolf Clausius (1822–1888) and the British William Thomson (1824–1907), known as Lord Kelvin, after having been knighted in 1866 for his work on a transatlantic telegraph project. Note that the second law is followed by a third law (the entropy of a perfect crystal is equal to zero at $T = 0$ K). The latter refers to a limit behaviour ($T \to 0$) and its implications are much more limited than those of the first two laws of thermodynamics (see Section 3.2.1).

Direction of out-of-equilibrium exchanges

The second law of thermodynamics has already been recovered, in the form of the maximum entropy principle for isolated systems (Section 3.2.2). It leads to the definition of temperature, pressure and chemical potential (see Section 3.3.1). The direction of energy, volume and particle exchanges between two macroscopic systems in contact will now be deduced. For that purpose, consider again the case of a macroscopic and isolated system $\mathcal{A}_{\text{tot}}$, divided into two equally macroscopic subsystems, $\mathcal{A}$ and $\mathcal{A}'$, initially separated by an adiabatic, fixed and impermeable wall. The two subsystems are thus isolated from each other and in an initial equilibrium macrostate, characterised by *a priori* different temperatures, T and T', pressures P and P' and chemical potentials μ and μ', respectively. The internal constraints are released, without changing the energy of system $\mathcal{A}_{\text{tot}}$, by making the wall diathermic, mobile and permeable. The two subsystems are then in thermal, mechanical and chemical contact. Their energies E and E', their volumes V and V', and their numbers of particles N and N', initially fixed become internal variables, whose equilibrium average values are determined by the equality of temperatures, pressures and chemical potentials.

If initially, $T \simeq T'$, $P \simeq P'$ and $\mu \simeq \mu'$, the final system macrostate is close to the initial macrostate and the partial entropy variation is given at first order by:

$$\begin{aligned}
\delta S_{\text{Ptot}} &= S_{\text{Ptot}}(N_{\text{tot}}, V_{\text{tot}}, E_{\text{tot}} \mid N + \delta N, V + \delta V, E + \delta E) \\
&\quad - S_{\text{Ptot}}(N_{\text{tot}}, V_{\text{tot}}, E_{\text{tot}} \mid N, V, E) \\
&\simeq \left. \frac{\partial S_{\text{Ptot}}}{\partial E} \right|_{(i)} \delta E + \left. \frac{\partial S_{\text{Ptot}}}{\partial V} \right|_{(i)} \delta V + \left. \frac{\partial S_{\text{Ptot}}}{\partial N} \right|_{(i)} \delta N,
\end{aligned}$$

where $N + \delta N$, $V + \delta V$ and $E + \delta E$ characterise the final macrostate of system $\mathcal{A}$. In the weak coupling approximation, the partial entropy of system $\mathcal{A}_{\text{tot}}$ is the sum of the partial entropies of $\mathcal{A}$ and $\mathcal{A}'$:

$$S_{\text{Ptot}}(N_{\text{tot}}, V_{\text{tot}}, E_{\text{tot}} \mid N, V, E) \simeq S_{\text{p}}(N, V, E) + S_{\text{p}}'(N_{\text{tot}} - N, V_{\text{tot}} - V, E_{\text{tot}} - E).$$

Using Definitions (3.35), (3.36) and (3.37) of temperature, pressure and chemical potential, respectively, one gets:

$$\begin{aligned}
\delta S_{\text{Ptot}} &\simeq \left[\frac{\partial S}{\partial E} - \frac{\partial S'}{\partial E'} \right]_{(i)} \delta E + \left[\frac{\partial S}{\partial V} - \frac{\partial S'}{\partial V'} \right]_{(i)} \delta V + \left[\frac{\partial S}{\partial N} - \frac{\partial S'}{\partial N'} \right]_{(i)} \delta N \\
&\simeq \left[\frac{1}{T} - \frac{1}{T'} \right] \delta E + \left[\frac{P}{T} - \frac{P'}{T'} \right] \delta V - \left[\frac{\mu}{T} - \frac{\mu'}{T'} \right] \delta N > 0, \qquad (3.61)
\end{aligned}$$

because, according to the second law, the entropy change δS_{Ptot} of the isolated system $\mathcal{A}_{\text{tot}}$ is positive. To simplify the discussion, assume that the contact between the two subsystems is[63]:

- only thermal (fixed and impermeable diathermic wall, i.e. $\delta V = \delta N = 0$): if $T > T'$, then $\delta E < 0$, subsystem $\mathcal{A}$ losses energy and its temperature decreases.

- mechanical with an already established thermal equilibrium ($T = T'$, the wall is mobile and impermeable, i.e. $\delta N = 0$): if $P > P'$, then $\delta V > 0$, the volume of subsystem $\mathcal{A}$ increases and its pressure decreases.

- chemical with already established mechanical and thermal equilibria ($T = T'$ and $P = P'$, the wall is permeable): if $\mu > \mu'$, then $\delta N < 0$, subsystem $\mathcal{A}$ looses particles and its chemical potential decreases.

---

[63] If the subsystems exchange energy, volume and particles at the same time, each of the three terms in Inequality (3.61) does not need to be positive, but their sum must be.

Thus, by giving a privileged direction to thermodynamic exchanges, the second law imposes a direction to the arrow of time: an out-of-equilibrium isolated system evolves in a way such that its entropy increases.

3.5.2 STATISTICAL INTERPRETATION OF IRREVERSIBILITY

In statistical physics, the formulation of the second law presented in the previous section can be found very easily using the Boltzmann formula. Consider a system that is kept isolated. Being initially in an equilibrium macrostate $M^{(i)}$, its thermodynamic entropy is $S^{(i)} = k_B \ln \Omega^{(i)}$. If releasing a constraint (without exchanging energy) provides the system with new microstates to explore, the system evolves to a new equilibrium macrostate $M^{(f)}$, of thermodynamic entropy $S^{(f)} = k_B \ln \Omega^{(f)}$, such that $\Omega^{(f)} > \Omega^{(i)}$. Let:

$$\Delta S = S^{(f)} - S^{(i)} = k_B \ln \frac{\Omega^{(f)}}{\Omega^{(i)}} > 0, \tag{3.62}$$

as postulated by the second law of thermodynamics. In the case where constraint release does not change the number of accessible microstates ($\Omega^{(f)} = \Omega^{(i)}$), $\Delta S = 0$, the transformation is reversible. These two types of transformations will be illustrated when mixing entropy is discussed in Section 3.5.3.

Note that there is no demonstration of the second law, in this form or in the form of the principle of maximum entropy, that is based solely on the laws of mechanics, classical or quantum. Boltzmann searched for it for a long time before admitting, as it is done today, that the theory of probabilities is essential to understand the second law and irreversibility in the framework of a microscopic description.

An irreversible process: Joule expansion

Let us illustrate an irreversible process by a concrete example, the Joule expansion[64]: an isolated container is divided into two compartments, of equal volume $V^{(i)} = V/2$, separated by a closed valve (see Figure 3.6). The left compartment contains one mole of ideal gas in equilibrium, in the macrostate $M^{(i)}$, and the right compartment is empty. The valve is then opened (without adding energy to the system). With this constraint released, the gas evolves at constant energy towards a new equilibrium macrostate $M^{(f)}$ by gradually filling the entire available volume ($V^{(f)} = V$). The transformation is clearly irreversible, the gas will not spontaneously return to its initial macrostate[65]. According to Equation (2.20), for an ideal gas in equilibrium, $\Omega(N, V, E) \sim V^N$. Since the energy and the number of particles are conserved during Joule expansion, the variation of the gas entropy between the equilibrium macrostates $M^{(i)}$ and $M^{(f)}$ is

$$\Delta S = k_B \ln \frac{\Omega^{(f)}(N, V^{(f)}, E)}{\Omega^{(i)}(N, V^{(i)}, E)} = k_B \ln \left(\frac{V}{V/2} \right)^N = k_B N \ln 2. \tag{3.63}$$

ΔS is positive, as it should be for an irreversible transformation, in agreement with the thermodynamics result[66].

---

[64] The Joule expansion protocol is described in Section 1.4.1.

[65] To obtain the initial macrostate, the gas must be compressed with a piston. The system is then no longer isolated.

[66] Indeed, since entropy is a function of state, a reversible process can be thought of: starting from the same initial state, a piston is gradually released until the same final state is reached. Since the system is isolated, its internal energy is conserved during the transformation: $dE = TdS - PdV = 0$. Thus,

$$\Delta S = \int_{S^{(i)}}^{S^{(f)}} dS = \int_{V^{(i)}}^{V^{(f)}} \frac{P}{T} dV = \int_{V^{(i)}}^{V^{(f)}} N k_B \frac{dV}{V} = k_B N \ln \frac{V^{(f)}}{V^{(i)}} = k_B N \ln 2,$$

where the ideal gas equation of state was used.

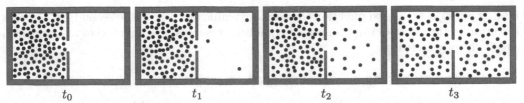

$$t_0 \qquad\qquad t_1 \qquad\qquad t_2 \qquad\qquad t_3$$

FIGURE 3.6 Illustration of Joule expansion: at time t_0, the valve is open and the gas gradually spreads into the right compartment. Four microstates are shown at times $t_0 < t_1 < t_2 < t_3$ (particles' momentum vectors are not drawn).

Beyond a simple reformulation of the second law, statistical physics offers a luminous interpretation of irreversibility in probabilistic terms, essentially thanks to the work of Boltzmann. Why is Joule expansion irreversible? The equations of motion of the N molecules are consistent with a spontaneous return of the gas into the left compartment (see the discussion on the irreversibility paradox, Section 3.5.3). In other words, the $\Omega^{(i)}$ microstates associated with the initial macrostate $M^{(i)}$ remain accessible microstates, once the constraint is released (valve opening). In equilibrium, the system is in the macrostate $M^{(f)}$ and $\Omega^{(f)}$ microstates are equiprobable (fundamental postulate). The probability $p_{(f)\to(i)}$ that the system returns to its initial macrostate $M^{(i)}$ (now an out-of-equilibrium macrostate) is equal to the probability to find the system in one of the $\Omega^{(i)}$ microstates corresponding to $M^{(i)}$:

$$p_{(f)\to(i)} = \frac{\Omega^{(i)}}{\Omega^{(f)}} = \frac{1}{2^N} \simeq 10^{-0.3N} \simeq \underbrace{0,0000000000\ldots000000000000}_{3\times10^{22}\text{ zeros for }N=10^{23}\text{particles !}}1, \qquad (3.64)$$

that is a zero probability in practice for a macroscopic system[67]. Generalising to all non-equilibrium macrostates, as in Section 2.4.3, it can be shown that almost all accessible microstates are typical equilibrium microstates (see Equation (2.41)).

Statistical origin of irreversibility

Two important conclusions can be drawn from the previous result. First, irreversibility is in statistical physics a probabilistic concept by construction: in theory, it is possible to observe a return to the initial macrostate (or any other out-of-equilibrium macrostate), but with a probability $p_{(f)\to(i)}$ so small that this event never occurs in practice. The second law of thermodynamics is therefore not strictly true, but its violation is infinitely unlikely. Second and more importantly, irreversibility, measured by the entropy variation $\Delta S = -k_B \ln\left(p_{(f)\to(i)}\right)$, is clearly a direct consequence[68] of the very large value of N. Thus, a Joule expansion with $N = 2$ particles is not irreversible, since a return to the initial macrostate (2 particles on the left) is then very likely: $p_{(f)\to(i)} = 1/4$. Reversing the time by watching the sequence[69] from right to left in Figure 3.7 is not surprising, unlike the case $N \gg 1$, illustrated in Figure 3.6.

---

[67]The waiting time to observe such an event is of the order of $2^N t_{\text{micro}} \simeq 10^{0.3N} \times 10^{-12} \sim 10^{10^{22}}$ s, where $t_{\text{micro}} \simeq 10^{-12}$ s is the atomic dynamics characteristic time (see Section 1.4.2). This is much longer than the age of the Universe since the Big Bang, estimated at 10^{17} s.

[68]Properties of dynamical systems (chaos, ergodicity, mixing property...), which were encountered in Section 2.4, although sometimes invoked, are not necessary to understand the appearance of irreversibility at the macroscopic scale (see Bricmont (1995), Goldstein (2001), Lebowitz (1993), Note 35 in Chapter 2).

[69]It is even more convincing to simulate Joule expansion for different values of N using molecular dynamics simulations (see Note 43 in Chapter 1).

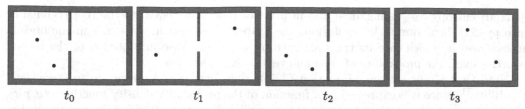

FIGURE 3.7 Joule expansion for $N = 2$ particles: at time t_0, the valve is open and the particles access the right compartment. The sequence of the four microstates, at times $t_0 < t_1 < t_2 < t_3$, does not appear to be temporally ordered, unlike the $N \gg 1$ case (see Figure 3.6).

The irreversibility observed at human scale originates from the gigantic separation between the microscopic and macroscopic scales, that is $N \gg 1$.

Out-of-equilibrium entropy

In principle, thermodynamic quantities, and in particular entropy, are only defined in thermodynamic equilibrium[70]. Thus, as it was shown in Section 3.2, Boltzmann formula is a direct consequence of the fundamental postulate, valid only in equilibrium. Similarly, the entropy variation involved in the second law (Equation (3.62)) is calculated between two equilibrium macrostates $M^{(i)}$ and $M^{(f)}$, as in the Joule expansion example. Entropy, only defined in equilibrium, is therefore sufficient to deal with irreversibility.

Nevertheless, it is possible to extend the definition of entropy to out-of-equilibrium macrostates, as shown in Section 3.2.2: to any macrostate M, a partial entropy $S_p(E_{tot} \mid A) = k_B \ln \Omega_p(E_{tot} \mid A)$ can be defined where A is the value of the internal variable in the macrostate M and is written here as:

$$S_M = k_B \ln \Omega_M, \tag{3.65}$$

where Ω_M is the number of microstates compatible with macrostate M. The *typical* equilibrium Ω_{eq} microstates, that is associated with the equilibrium macrostate M_{eq}, represent almost all accessible microstates Ω, whether they are typical equilibrium microstates or associated with out-of-equilibrium macrostates (see Equation (2.40)). The equilibrium macrostate's partial entropy is therefore practically equal to the thermodynamic entropy defined by Boltzmann's formula (see Equation (3.31)): $S \simeq S_{M_{eq}} = k_B \ln \Omega_{eq}$.

A dynamic interpretation of the second law, which describes the system's evolution *between equilibrium macrostates*, is then obvious: once a constraint is released in an isolated system initially in equilibrium, the system founds itself in an out-of-equilibrium macrostate of low (partial) entropy $S_{M^{(i)}}$. It then evolves according to the laws of mechanics (classical or quantum). Typical equilibrium microstates being incredibly more numerous than atypical microstates (associated with out-of-equilibrium macrostates), the system progressively evolves towards the equilibrium macrostate, during which its (partial) entropy S_M increases. For an isolated system, the arrow of time between two instants t_1 and t_2 is then specified by the growth of S_M:

$$\text{if} \quad S_M(t_1) < S_M(t_2), \quad \text{then} \quad t_1 < t_2.$$

In other words, according to Boltzmann, the arrow of time is oriented from the least likely macrostates to the most likely macrostates ($\Omega_M(t_1) < \Omega_M(t_2)$). Once in equilibrium, the

---

[70]Thermodynamics of out-of-equilibrium systems is a rapidly developing branch of physics that seeks in particular to define entropy and other usual macroscopic quantities (temperature, pressure...) under out-of-equilibrium conditions (see Note 39 in Chapter 1).

(partial) entropy $S_{M_{eq}}$ is maximal and in practice practically equal to the thermodynamic entropy S. Then, during its evolution, the system may find itself in a non-equilibrium macrostate, in which case its (partial) entropy would decrease in violation of the second law. However, the probability of such an event is extremely low.

Note that Gibbs entropy (Equation (3.59)) does not seem appropriate to study irreversibility[71], since it is expressed as a function of the probability density function $\rho(q, p, t)$, which is conserved over time according to Liouville's Equation (2.36). Gibbs entropy thus remains constant during an irreversible process, in contrast to Boltzmann (partial) entropy, which increases until equilibrium is reached[72]. Boltzmann entropy characterises, at a given time, the macrostate of a *single* system, whose evolution is monitored, whereas Gibbs entropy is associated with the probability density $\rho(q, p, t)$ of a *statistical ensemble* of systems that scatter into different phase space macrostates, as seen in Figure 2.7. To study irreversibility, it is indeed the behaviour of a *single* system that is of interest[73].

Initial conditions and the arrow of time

To observe an irreversible phenomenon, a system must necessarily be in an initial low(er) entropy macrostate, but how can it be in such an unlikely macrostate at a given time?

In the case of a Joule expansion, the initial macrostate $M^{(i)}$ does not result from the spontaneous evolution of the isolated system. Instead, the gas was compressed in the left compartment by a piston which in turn was pushed by an experimentalist. During the compression, the gas entropy decreased. This is not surprising because the gas was not isolated, but rather was in contact with the external environment. On the contrary, the entropy of the total system "gas+experimentalist" (which is considered isolated during the compression process) must increase according to the second law. Since the entropy of the gas decreases, the entropy of the experimentalist increases[74]. In turn, how does the experimentalist decrease his own entropy? By consuming air, water and food. Without these inputs, they would rapidly evolve towards their (future) state of maximum equilibrium entropy: death. The set of involved transformations (called *metabolism*), which is a matter of chemistry and biology, is, of course, much more complex than a simple Joule expansion. Nevertheless, food provides living beings energy (calories) which is returned to the external environment in the form of work (literally) and mainly heat. Moreover, food allows living beings to reduce their entropies and keep them (alive) in an out-of-equilibrium state[75]. Food (made of organic matter) is produced by photosynthesis and thus by the light emitted by the Sun, the true source of low entropy.

---

[71] See Bricmont (1995), Goldstein (2001), Lebowitz (1993), Note 35 in Chapter 2.

[72] In the initial and final equilibrium macrostates, S_G is, of course, equal to the Boltzmann entropy.

[73] It is nevertheless possible to force the Gibbs entropy to increase during an irreversible process by replacing $\rho(q, p, t)$ in Equation (3.59) with a probability density function $\bar{\rho}$, averaged in phase space over "large" cells containing a large number of elementary cells, as Gibbs himself proposed in 1902. This operation, known as "coarse-graining", leads by construction to an entropy increase, $S_G(\bar{\rho}) > S_G(\rho)$, since information (or precision) is lost at the microscopic scale. The density ρ, defined as the fraction of occupied elementary cells, does not evolve (Liouville equation), whereas as the statistical ensemble systems evolve, an increasing number of large cells are occupied, indicating an increase of $\bar{\rho}$ and thus of $S_G(\bar{\rho})$ (see Penrose (1979), Note 35 in Chapter 2). It can then be shown by invoking the so-called *mixing property* of dynamical systems (see Note 52 in Section 2.4.2) that $S_G(\bar{\rho})$ tends towards the equilibrium thermodynamic entropy. However, as pointed out by Bricmont, this approach gives the (false) impression that irreversibility is subjective and depends on the degree of precision of the knowledge of the microstates (see Bricmont (1995), Note 35 in Chapter 2).

[74] Before proceeding further, the reader may note that the discussion becomes tricky: how is the entropy of a living being defined? As Jaynes notes (in *Gibbs vs Boltzmann Entropies*, Am. J. Phys. **33**, 391 (1965)), one would have to begin by identifying the thermodynamic variables that characterise the macrostate of such a "system", which is very far from equilibrium and very complex...

[75] See E. Schrödinger, *What is life? The physical aspect of the living cell*, (Cambridge University Press, reprint edition 2012).

Let us continue the reasoning by following in the footsteps of Roger Penrose[76]. The radiative balance of Earth (and the living organisms) is globally zero: the energy it receives from the Sun is entirely re-emitted towards space, keeping its surface temperature constant ($T \simeq 15$ °C when taking into account the greenhouse effect). However, the photons emitted by the Sun are mostly high energy photons (high frequencies, visible spectrum), while the photons emitted by Earth are low energy photons (low frequencies, mostly infrared). Assume, for simplicity, that the incident and emitted radiations by Earth are monochromatic, with frequencies ν_{in} and ν_{em} ($\nu_{in} > \nu_{em}$) and entropies S_{in} and S_{em}, respectively. Earth's radiation balance being zero, the energy E brought by N_{in} incident photons is equal to the energy of N_{em} photons emitted by Earth, so

$$E = N_{in}\, h\nu_{in} = N_{em}\, h\nu_{em} \quad \text{with} \quad \nu_{in} > \nu_{em}, \quad \text{so} \quad N_{in} < N_{em}.$$

Entropy increasing with the number of particles, $S_{in} < S_{em}$. It can be concluded that the Sun provides Earth with a source of low entropy energy which generates the irreversible phenomena observed on the planet[77].

The question of the origin of low entropy states then moves into the domain of cosmology. From the Sun, created in a state of low entropy from a cloud of gas – the solar nebula, itself of lower entropy – compressed under the effect of gravity[78], the entire universe and its origin should be considered as a last resort. Since the Universe can be considered isolated, its entropy must steadily increase until it reaches an equilibrium state, labelled "thermal death of the universe" by the German physicist Hermann von Helmholtz in 1854 (who generalised the dark prophecy of the Solar system's end, formulated by Kelvin two years earlier). In this permanent state, energy would be uniformly distributed in the Universe, suppressing all sources of heat (hot or cold) and therefore all exchange of work: no irreversible process could take place. If the future of the Universe is under discussion, astronomical observations show that it is now expanding. Since its origin, it is in a strongly out-of-equilibrium state, described in the framework of the Big Bang cosmological model[79]. To explain this irreversible evolution, the Universe's initial macrostate, if it can be characterised at all, must be of very low entropy, as Boltzmann already thought. Why the Universe was originally in such an unlikely state remains an open question today[80]. Its rigorous treatment relies on quantum gravitation, a branch of physics currently under development, whose aims at

---

[76]See R. Penrose, *The road to reality: A complete guide to the laws of the universe* (Vintage Books, London, 2007).

[77]More precisely, it will be seen in Chapter 7 that the entropy of a photons gas of energy E emitted by a black body at temperature T is $S = 4E/3T$. Since the incident and emitted radiation energies are equal and the Sun's surface temperature ($T \simeq 5800$ K) is higher than Earth's surface temperature ($T \simeq 288$ K), $S_{in} < S_{em}$.

[78]Long-range gravitational interactions are outside the scope of this book (see Note 5). Simply note that it generates some surprising effects. For example, the entropy of an atomic system decreases when it goes from a diluted state (gas) to a more condensed state (or even a solid), as can be seen in the example of a Joule expansion. On the contrary, the entropy of a system of gravitationally interacting bodies tends to increase during accretion processes transforming a uniform distribution of matter into compact objects (stars, planets...) spread out in space. Entropy is maximal when all matter is condensed in a small region of space, called a *black hole* in astrophysics (see Figure 27.10 of *The road to reality: A complete guide to the laws of the universe* (Vintage Books, London, 2007)).

[79]The Universe's expansion was demonstrated by the American astronomer Edwin Hubble in 1929. He observed the red shift (long wavelength) of the visible spectrum emitted by stars and then deduced the law that bears his name: the speed at which two galaxies move away from each other is (almost) proportional to their separation distance. This observation was then confirmed by the fortuitous discovery of the "cosmological background" (or "2.7 K radiation background") in 1964, reflecting a homogeneous Universe at thermal equilibrium at $T = 3000$ K, 380 000 years after the Big Bang.

[80]Roger Penrose estimated the number Ω_{Univ} of accessible microstates of the Universe, modelled by a system of $N \simeq 10^{80}$ particles called *baryons* (essentially protons and neutrons) which form the observable matter of the Universe. Assuming that in the Universe's final equilibrium state, N particles are condensed in a black hole (Big Crunch hypothesis), whose entropy is given by the Bekenstein-Hawking formula (see

unifying general relativity and quantum mechanics. For the moment, let us settle with a concluding remark from Feynman: "*Therefore I think it necessary to add to the physical laws the hypothesis that in the past the universe was more ordered, in the technical sense* [of lower entropy], *than it is today – I think this is the additional statement that is needed to make sense, and to make an understanding of the irreversibility.*" (R. Feynman, *The character of physical law* (Cambridge, MIT Press, 2001), page 116).

3.5.3 APPARENT PARADOXES OF STATISTICAL PHYSICS

The introduction of probabilities in physics, necessary to explain the transition from microscopic to macroscopic, has raised very instructive paradoxes stated by Boltzmann's contemporaries. As it will be seen, their origin is essentially due to an incorrect definition of the studied system (Gibbs' paradox and Maxwell's demon), or due to the attempt to explain macroscopic phenomena only by means of particles' deterministic dynamics, without taking into account the scales separation imposed by $N \gg 1$ (Loschmidt and Poincaré-Zermelo paradoxes).

Mixing entropy, Gibbs' paradox and Maxwell's demon

Consider again an isolated container separated into two compartments of volume V and V', connected by a closed valve (see Figure 3.8). The left compartment contains a gas $\mathcal{A}$ of $N \gg 1$ *indistinguishable* atoms of type A (for example argon) characterised by their mass m. The right compartment contains a gas $\mathcal{A}'$ of $N' \gg 1$ *indistinguishable* atoms of another type A' (for example helium) of mass $m' \neq m$. In the initial macrostate $M^{(i)}$, the two gases, considered ideal and isolated, are both in equilibrium, $\mathcal{A}$ at energy E and $\mathcal{A}'$ at energy E', at equal temperature T and pressure P.

The valve is opened and the two gases mix. In equilibrium (macrostate $M^{(f)}$), the atoms are distributed uniformly throughout the container. Pressure and temperature of the mixture remaining equal to P and T. The gas mixture is ideal (no interactions between atoms whether different or similar). The final macrostate $M^{(f)}$ can thus be seen as the superposition of two different ideal gases (the atoms of $\mathcal{A}$ are discernible from the atoms of $\mathcal{A}'$). Each gas is in the thermodynamic state defined by P, T and $V + V'$. Both gases have retained their initial energy, E and E', respectively.

By writing the pressure and temperature of gases $\mathcal{A}$ and $\mathcal{A}'$ before the opening of the valve, and of the gas mixture $\mathcal{A} + \mathcal{A}'$ once in equilibrium, one obtains:

$$\frac{P}{k_{\mathrm{B}}T} = \frac{N}{V} = \frac{N'}{V'} = \frac{N+N'}{V+V'} \tag{3.66}$$

$$\frac{3}{2}k_{\mathrm{B}}T = \frac{E}{N} = \frac{E'}{N'} = \frac{E+E'}{N+N'}. \tag{3.67}$$

The mixing of two gases is irreversible: it is very unlikely to observe every atoms of $\mathcal{A}$ and every atoms of $\mathcal{A}'$ returning in their original compartments (probability: $1/2^{N+N'} \ll 1$). To verify this, the entropy variation of the isolated system $\mathcal{A}_{\mathrm{tot}}$ (formed by the two gases)

Exercise 3.7), the fraction of the phase space volume corresponding to the Universe's initial macrostate would be of the order of $1/\Omega_{\mathrm{Univ}} \simeq 1/10^{10^{123}}$. The Universe would thus be born in a very particular microstate (see Chapter 27 of *The road to reality: A complete guide to the laws of the universe* (Vintage Books, London, 2007)). Other models are nevertheless proposed to explain the Universe's entropy increase. Thus, the entropy evolution and the arrow of time could depend on the physical quantities chosen to describe the macrostate of a system (C. Rovelli, *Why do we remember the past and not the future? The 'time oriented coarse graining' hypothesis* arXiv:1407.3384 (2014)) or on the dynamics specific to gravitational systems (S. Carlip, *Arrow of Time Emerges in a Gravitational System*, Physics **7**, 111 (2014)).

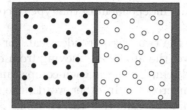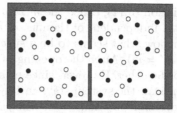

FIGURE 3.8 Mixing of ideal gases: in the initial macrostate (left), the gases $\mathcal{A}$ (black) and $\mathcal{A}'$ (white) are isolated at identical pressure P and temperature T in the left and right compartments, respectively. After the valve opening (right), both gases are distributed uniformly in the two compartments: the transformation is irreversible. If the atoms of $\mathcal{A}$ and $\mathcal{A}'$ were identical (same colour), the transformation would be reversible.

between the initial and the final states can be calculated. Since two ideal gases are considered, they each constitute an *independent subsystem*. In the initial macrostate $M^{(i)}$, they are isolated from each other, that is

$$S_{\text{tot}}^{(i)} = S(N,V,E) + S'(N',V',E').\tag{3.68}$$

In the final macrostate $M^{(f)}$, they are mixed *without interaction*, so

$$S_{\text{tot}}^{(f)} = S(N,V+V',E) + S'(N',V+V',E').\tag{3.69}$$

The atoms from a given type of gas are indistinguishable. The entropies $S(N,V,E)$ and $S'(N',V',E')$ are thus given by the Sackur-Tetrode Formula (3.26):

$$S(N,V,E) = k_{\text{B}}N\left(\ln\left(\frac{V}{N}\right) + \frac{3}{2}\ln\left(\frac{E}{N}\right) + \frac{3}{2}\ln\left(\frac{4\pi m}{3h^2}\right) + \frac{5}{2}\right).\tag{3.70}$$

The entropy variation of $\mathcal{A}_{\text{tot}}$ is thus:

$$
\begin{aligned}
\Delta S_{\text{tot}} &= S_{\text{tot}}^{(f)} - S_{\text{tot}}^{(i)} \\
&= \underbrace{S(N,V+V',E) - S(N,V,E)}_{=\Delta S} + \underbrace{S'(N',V+V',E') - S'(N',V',E')}_{=\Delta S'},
\end{aligned}
$$

where ΔS (and $\Delta S'$) is the entropy variation, Equation (3.63), calculated for the Joule expansion of an ideal gas from a volume $V^{(i)} = V$ (and $V^{(i)} = V'$) to a volume $V^{(f)} = V + V'$. Using Equation (3.66), one obtains:

$$
\begin{aligned}
\Delta S_{\text{tot}} &= k_{\text{B}}N\ln\left(\frac{V+V'}{V}\right) + k_{\text{B}}N'\ln\left(\frac{V+V'}{V'}\right) \\
&= k_{\text{B}}N\ln\left(\frac{N+N'}{N}\right) + k_{\text{B}}N'\ln\left(\frac{N+N'}{N'}\right),
\end{aligned}\tag{3.71}
$$

which is naturally strictly positive, as it should be for an irreversible transformation. This result is, of course, recovered by calculating ΔS_{tot} with Equation (3.70)[81]. Note that the

---

[81] Expression (3.71) can be also obtained by assuming that the atoms of each gas are *distinguishable*. Indeed, in this erroneous case, the (non-extensive) entropy of a gas is given by Equation (3.28), $S_{\text{Distinguishable}}(N,V,E) = S(N,V,E) + k_{\text{B}}\ln(N!)$, which leads to the same ΔS (and $\Delta S'$), since N (and N') is constant.

entropy variation (3.71), called the *mixing entropy*, is independent of the type of considered atoms, that is independent of the masses m and m'.

Consider now what seems to be a special case: the two gases are identical (for example type A, i.e. $m' = m$). After the opening of the valve, the atoms are moving from one compartment to the other. However, since pressure and temperature in each compartment were initially identical, the system stays in the initial macrostate. In other words, the transformation is reversible, so $\Delta S_{\text{tot}} = 0$, whereas according to Equation (3.71), $\Delta S_{\text{tot}} > 0$. This contradictory result is commonly referred to as the *Gibbs paradox*[82]. Where is the mistake? When the two gases are of different natures, the atoms of $\mathcal{A}$ are *distinguishable* from the atoms of $\mathcal{A}'$. This was taken into account in the derivation of $S_{\text{tot}}^{(f)}$ (see Equation (3.69)) and thus ΔS_{tot}. If the atoms of $\mathcal{A}$ and $\mathcal{A}'$ are identical, Expression (3.68) of $S_{\text{tot}}^{(i)}$ remains correct, since in the initial macrostate, the atoms of $\mathcal{A}$ and $\mathcal{A}'$ are located in separate compartments and are therefore distinguishable from each other. On the other hand, once mixed, they must be treated as a unique gas of $N + N'$ *indistinguishable* particles in a volume $V + V'$, at energy $E + E'$. Thus, $S_{\text{tot}}^{(f)}$ is no longer given by Equation (3.69), but by:

$$S_{\text{tot}}^{(f)} = S(N + N', V + V', E + E') = S(N, V, E) + S'(N', V', E') = S_{\text{tot}}^{(i)},$$

since entropy $S(N, V, E)$ is additive. Thus, $\Delta S_{\text{tot}} = 0$, which can be easily verified using Sackur-Tetrode formula (3.70) with $m = m'$, and Relations (3.66) and (3.67). As can be seen, the apparent Gibbs paradox is solved by treating identical gas particles as indistinguishable (see Section 2.2.3)[83].

Another apparent paradox (a thought experiment proposed by Maxwell in 1867) seems to violate the second law. A mixture of two different gases, $\mathcal{A}$ and $\mathcal{A}'$, is in equilibrium macrostate $M^{(f)}$ and the valve is closed. Thermal agitation leads some atoms to move towards the valve, on one side or the other. A small intelligent being, named *Maxwell demon* by Kelvin (in 1874), could open the valve as soon as an atom of $\mathcal{A}$ arrives from the right or an atom of $\mathcal{A}'$ arrives from the left. After a while, all the atoms of $\mathcal{A}$ (of $\mathcal{A}'$) would

---

[82] Actually, Gibbs thought that by reducing progressively the difference between the two types of atoms A and A' (by choosing gases with closer and closer atomic masses, i.e. $m' \to m$), the mixing entropy would decrease continuously ($\Delta S_{\text{tot}} \to 0$), while it abruptly goes from a constant value independent of m and m' (given by Equation (3.71)) to zero (see Uffink (2006), Note 35 in Chapter 2). Conclusion: the particles of a fluid are either identical (and thus indistinguishable), or different (and thus distinguishable), there is no intermediate situation.

[83] This result can be recovered by reasoning directly on the number of accessible microstates. The gases being ideal, temperature and particle number density being equal in the two compartments before and after the opening of the valve, the ensemble of individual microstates (positions and momenta) accessible to all atoms of $\mathcal{A}_{\text{tot}}$ is (almost) the same in the initial $M^{(i)}$ and final $M^{(f)}$ equilibrium macrostates. Indeed, in almost all microstates of $M^{(f)}$, there are (approximately) N atoms on the left and N' on the right. For a given configuration of $N + N'$ individual microstates (e.g. the one represented in Figure 3.8), what eventually distinguishes an initial microstate from a final microstate is the manner of assigning position (and momentum) of each of the $N + N'$ atoms:

- If the atoms of the two gases are identical, the $N + N'$ atoms are indistinguishable and there is only one way to attribute a position to each of them in $M^{(i)}$ and $M^{(f)}$, i.e. $\Omega_{\text{tot}}^{(f)} \simeq \Omega_{\text{tot}}^{(i)}$, so $\Delta S_{\text{tot}} = 0$.

- If the two gases are different, there is one way to assign position to each atom in $M^{(i)}$, because N indistinguishable atoms of $\mathcal{A}$ must be on the left compartment and N' indistinguishable atoms of $\mathcal{A}'$ must be on the right compartment. However, in $M^{(f)}$, the N indistinguishable atoms of $\mathcal{A}$ can be anywhere (the N' indistinguishable atoms of $\mathcal{A}'$ are then in the remaining positions): there are thus $\binom{N+N'}{N}$ ways to assign to each of the N atoms of $\mathcal{A}$ one of the $N + N'$ positions, that is:

$$\Omega^{(f)} \simeq \binom{N + N'}{N} \Omega^{(i)} = \frac{(N + N')!}{N! N'!} \Omega^{(i)} \quad \text{with} \quad N, N' \gg 1.$$

Using Stirling's formula (A.21), Expression (3.71) of $\Delta S_{\text{tot}} = k_{\text{B}} \ln\left(\Omega^{(f)}/\Omega^{(i)}\right)$ is recovered.

be on the left (on the right). As the demon does not exchange energy with the atoms of the gas, the still isolated system would be back in macrostate $M^{(i)}$ and its entropy would have decreased[84].

Although it does not challenge the second law, this little demon has been the subject of much discussion. The key to the paradox lies in the link between entropy and information (see Section 3.4). Indeed, mixing entropy (3.71) can be interpreted as the loss of information on the relative atoms position (left or right) in macrostate $M^{(f)}$ with respect to macrostate $M^{(i)}$: each of the $N + N'$ atoms has a probability $p = V/(V + V') = N/(N + N')$, according to Equation (3.66), to be in the left compartment and a probability $1 - p$ to be in the right compartment. Thus, since Gibbs entropy is additive, Equation (3.60) leads to:

$$\begin{aligned}
\Delta S_{G_{\text{tot}}} &= (N + N')(-k_B p \ln p - k_B(1 - p)\ln(1 - p)) \\
&= (N + N')\left(- k_B \frac{N}{N + N'} \ln\left(\frac{N}{N + N'} \right) - k_B \frac{N'}{N + N'} \ln\left(\frac{N'}{N + N'} \right) \right) \\
&= k_B N \ln\left(\frac{N + N'}{N} \right) + k_B N' \ln\left(\frac{N + N'}{N'} \right).
\end{aligned}$$

In order to make a selection and recover macrostate $M^{(i)}$, the demon must gather information about the position (left or right) and the type of atoms (A ou A') that he is dealing with. It can then be shown that the resulting demon's entropy increase compensates for the gas entropy decrease $-\Delta S_{\text{tot}} < 0$ and that consequently the entropy of the isolated system "gas+demon" increases in agreement with the second law (R. Feynman, Feynman Lectures on Computation (Addison-Wesley, 1996)).

The Hungarian-American physicist Leó Szilárd (in 1929) and the French physicist Léon Brillouin (in 1950) thought that the entropy increase was due to the energy dissipated during the measurements that the demon has to make in order to identify and locate atoms, for example by illuminating them with a light beam. In fact, the German-American physicist, Rolf Landauer, showed in 1961 that a measurement could, in principle, not involve an entropy increase. On the other hand, during atoms selection, the demon accumulates a large amount of information (on their positions), stored in a memory of limited capacity (its brain or a computer). He must therefore regularly erase data and this operation is, of course, irreversible, so energy is dissipated in the form of heat, increasing the system's entropy[85]. The demon increases entropy, not by acquiring information, but by erasing it. Experimental realisations of Maxwell's demon are currently being used to convert information into work on the scale of a particle of a tenth of a micron[86] and to cool gases to temperatures near absolute zero[87] on the order of $T \simeq 10^{-6}$ K.

---

[84]Anxious to contradict the dire prediction of the thermal death of the Universe (see page 101), Maxwell sought to show that the laws of mechanics could be used to violate the second law of thermodynamics and restoring human free will in the face of thermodynamics determinism (P. Ball, *A demon-haunted theory*, Physics World, 36 (April 2013)). The demon proposed by Maxwell actually selects fast particles on one side and slow ones on the other, leading to a temperature difference between the two compartments and thus spontaneously transferring heat from a cold body to a warm one.

[85]A bit of information can be in *two* states (0 or 1) with an entropy $k_B \ln 2$. Its deletion amounts to resetting it to a single predefined state (for example 0) of entropy $k_B \ln 1 = 0$. Its entropy then decreasing by $k_B \ln 2$ and a heat of at least $k_B T \ln 2$ must be released into the environment. This so-called *Landauer's principle* has been verified experimentally by measuring the energy dissipated when a small sphere is moved through a double optical trap (A. Bérut *et al.*, *Experimental verification of Landauer's principle linking information and thermodynamics*, Nature **483**, 187 (2012)).

[86]See S. Toyabe *et al.*, *Experimental demonstration of information-to-energy conversion and validation of the generalized Jarzynski equality*, Nat. Phys. **6**, 988 (2010).

[87]See Mark G. Raizen, *Demons, Entropy, and the Quest for Absolute Zero*, Scientific American **304**(3), 5459 (March 2011).

Microscopic reversibility/Macroscopic irreversibility

At the microscopic scale, the dynamics of atoms is reversible, because the laws of classical (and quantum) mechanics are symmetrical by time reversal. In other words, they are invariant under the transformation $t \rightarrow -t$. Indeed, by performing this change of variable the velocity vector is changed into its opposite, $\vec{v} = d\vec{r}/dt \rightarrow d\vec{r}/d(-t) = -\vec{v}$, but the fundamental principle of dynamics, Equation (1.4), remains unchanged when the forces are independent of time and the particle velocity[88]:

$$m\frac{d}{dt}\vec{v}_j = m\frac{d}{d(-t)}(-\vec{v}_j) = \vec{F}_j \quad \text{for} \quad j = 1, 2, \ldots, N. \tag{3.72}$$

The motion of planets is a good example of the reversible dynamics of (very macroscopic!) "particles". Their direction of rotation around the Sun could be reversed without violating Law (3.72) and without offending common sense. Similarly, at microscopic scale, a movie (necessarily fictitious) representing atomic collisions, viewed in reverse, is not surprising.

Thus, the evolution of *each* atom of a drop of ink deposited in a glass of water is, in principle, reversible, but not the evolution of the drop as a whole (see Figure 1.4). But then, how can atoms reversible dynamics at the microscopic scale generate irreversible behaviours of macroscopic systems? Moreover, most of the equations that describe phenomena at human scale (diffusion equation, Navier-Stokes equation in fluid mechanics...) are not invariant by time reversal.

In order to overcome this paradox, known as the "irreversibility paradox", or Loschmidt's paradox[89], consider again the example of a Joule expansion. Starting from an initial microstate such that all the particles are in the left compartment (low entropy macrostate $M^{(i)}$), the valve is opened at $t = 0$. Once equilibrium has been reached (high entropy macrostate $M^{(f)}$), if the velocity vectors of all particles were reversed at time $t_1 > 0$, each would travel the path it had followed between $t = 0$ and t_1, but in the opposite direction, as if the system were going back in time (see Figure 3.9).

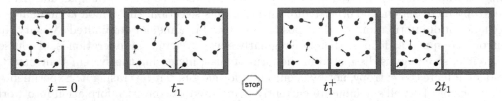

$$t = 0 \qquad\qquad t_1^- \qquad \text{(STOP)} \qquad t_1^+ \qquad\qquad 2t_1$$

FIGURE 3.9 Velocity reversal: at $t = 0$, the valve is opened and the gas gradually spreads into the right compartment. At time t_1, the velocity vectors of all the particles are reversed. The particles then retrace their steps and return to their initial positions at time $t = 2t_1$, with reversed velocity vectors.

At the time instant $2t_1$, particles would have returned to their initial positions in the left compartment (with momenta opposite to those of the initial microstate). The system would

---

[88]Hamilton's Equations (1.11) invariance under the transformation $t \rightarrow -t$ is also easily shown. The dynamics is, of course, no longer reversible in the presence of a magnetic field (Lorentz force depending on velocity) or friction. In quantum mechanics, Schrödinger equation (1.5) is invariant under the transformation $t \rightarrow -t$ and $\Psi(q, t) \rightarrow \Psi^*(q, -t)$, where Ψ^* is the complex conjugate of Ψ. Since the probability density function of the presence of a particle at a point in space at a given time is given by the squared modulus of the wave function $|\Psi|^2 = \Psi\Psi^*$, it is unchanged by time reversal.

[89]Named after the Austrian physicist (see Note 38 in Chapter 1) who called (his friend) Boltzmann in 1876, criticising the irreversibility expressed by the "H theorem" (see Note 47). However, this paradox had already been raised by Maxwell in 1873 and by Kelvin in 1874. The latter rightly attributed irreversibility to the large number of particles constituting a macroscopic system (see C. Cercignani, *Ludwig Boltzmann: The man who trusted atoms* (Oxford University Press, 2006)).

then be back in its initial macrostate $M^{(i)}$, leading to a decrease in entropy in violation of the second law. Confronted with this thought experiment, Boltzmann would have answered "*Well, go ahead, turn them over!*" (according to M. Kac, *Probability and related topics in physical sciences* (American Mathematical Society, 1957)), because how in practice can the velocity vectors of $N \gg 1$ atoms simultaneously be flipped over[90]?

Assume, however, that this reversal operation occurs at time t_1 and that the system returns to its initial macrostate $M^{(i)}$. Since $M^{(i)}$ is a strongly out-of-equilibrium macrostate, particles would quickly redistribute themselves throughout the volume and the system would finally return to its equilibrium macrostate $M^{(f)}$, without any variation of its *thermodynamic entropy* and thus without violating the second law. In any case, the slightest imprecision on the flipped velocities (new initial conditions) and weak interactions with the external environment, necessarily present although the system is considered isolated, would disturb particles trajectories between the time instants t_1 and $2t_1$: the chaotic dynamics would prevent the system from returning to its original macrostate $M^{(i)}$, keeping it in the equilibrium macrostate[91] $M^{(f)}$.

Hence there is no paradox: the underlying microscopic dynamics is reversible and irreversibility emerges at the macroscopic scale with a probability (practically) equal to one, as discussed in Section 3.5.2. It is indeed the separation of microscopic and macroscopic scales that intervenes[92]. To be convinced of this, one can go back in time by reversing the order of microstates: for two particles (Figure 3.7), there is nothing shocking about this, the return to the initial macrostate (two particles on the left) is certain, even without reversing the velocities, contrary to the case with $N \gg 1$ particles (Figure 3.6). More broadly, any evolution equation obtained solely from the laws of atomic physics (classical or quantum) is necessarily symmetric by time reversal. To break symmetry and obtain an irreversible behaviour, an additional element is indispensable, whether it is probabilistic, dissipative (a frictional force) or based on a coarse-graining operation (see Note 73).

---

[90] On the other hand, returning the $N \gg 1$ nuclear spins of a magnetic material at a given time is a laboratory daily practice, in particular in Magnetic Resonance Imaging (MRI), since the discovery of the spin echo phenomenon in 1950 (E. L. Hahn, *Spin echoes*, Phys. Rev. **80**, 580 (1950)). Here is the principle: in the initial macrostate $M^{(i)}$, independent spins are all aligned with an external magnetic field applied to the system, the magnetisation is then maximal and the entropy minimal (ordered state). The external magnetic field is suppressed at $t = 0$ and each spin s_i rotates independently of the others, at a speed that depends only on the local field B_i created by the material heterogeneities. These differences between spins' speeds lead to a macrostate $M^{(f)}$, in which spins are disoriented, each pointing in directions that seem random. The magnetisation is then zero and the entropy maximum (disordered state). At time t_1, an external magnetic field is briefly re-established in a direction chosen to reverse all spins, so that the angular deviations between two spins are reversed: a lagging spin is now leading. The local field B_i being identical, each spin rotates at the same speed as before, but in the opposite direction. At time $t = 2t_1$, each spin recovers at its initial relative orientation, like an echo: the spins are aligned again, the magnetisation is maximal and the entropy minimal. (see animations of this phenomenon on the internet at https://en.wikipedia.org/wiki/Spin_echo and R. Balian, *Information in statistical physics*, Studies in History and Philosophy of Modern Physics **36**, pp. 323353 (2005)).

[91] Once again, molecular dynamics is very useful, on the one hand to simulate numerically the reversal of velocities during a Joule expansion and, on the other hand to observe the role of chaos, which intervenes rapidly because of the time accumulation of numerical rounding errors (see Note 27 in Chapter 1). It is thus observed that when t_1 is too large, the simulated system cannot recover its initial macrostate. In a numerical experiment, it is, however, possible to suppress rounding errors and make the dynamics perfectly reversible (D. Levesque and L. Verlet, *Molecular dynamics and time reversibility*, J. Stat. Phys. **72**, 519 (1993)).

[92] An original and stimulating way to resolve this apparent paradox is to make physics time-independent, by eliminating the variable t from the fundamental classical and quantum theories. Our perception of time and irreversibility would then be due to our ignorance of the microstate of macroscopic systems (C. Rovelli, *Forget time*, Foundations of Physics **41**, pp. 14751490 (2011)).

Poincaré recurrence time

The "recurrence theorem", proved by Henri Poincaré[93] in 1890, has a very general scope, since it applies to any (classical) dynamical system of N isolated and confined particles: in the course of its evolution, the system will return to a microstate arbitrarily close to its initial microstate[94]. Thus, starting from an out-of-equilibrium macrostate, as in a Joule expansion, the system will reach equilibrium, and then, after a period of time called the *recurrence time*, will certainly return to this initial macrostate, leading to a decrease of entropy (and the great surprise of possible observers). Figure 3.10 shows such a phase space trajectory.

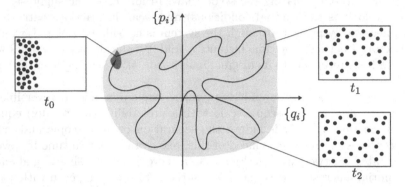

FIGURE 3.10 There and back again: starting at time t_0 from a microstate associated with a out-of-equilibrium macrostate (island), the system reaches equilibrium, when its representative point evolves in the Boltzmann sea (see Section 2.4.3), by following a trajectory in phase space which, according to Poincaré's recurrence theorem, brings it back to its initial macrostate.

The contradiction with the second law of thermodynamics, known as *Zermelo's paradox*[95], is obvious: the equilibrium state and the phenomenon of irreversibility would only be transitory, or as Poincaré wrote in 1893, "*to see the heat pass from a cold body to a hot one [...], a little patience would be enough*" (H. Poincaré, *Le mécanisme et l'expérience* in Revue de Métaphysique et de Morale **1**, 534 (1893, in French)). These assertions are correct, but they do not question in practice the second law of thermodynamics and statistical physics. Indeed, the recurrence time for a macroscopic system of $N \gg 1$ particles, already estimated by Boltzmann, is gigantic, of the order of $10^N \sim 10^{10^{23}}$ years or any other unit of time, and in any case much larger than the age of the Universe ($\sim 10^{17}$ s). Until then, the reassured reader can continue to use the second law of thermodynamics. Boltzmann would have added to his answer to his opponents: "*I wish you to live as long!*" (according to M. Kac, *Probability and related topics in physical sciences*, page 61 (American Mathematical Society, 1957)).

---

[93] French mathematician and physicist (1854–1912) known for his work on the theory of dynamical systems, in particular the three-body problem (see Note 25 in Chapter 1) and founder of the chaos theory.

[94] This theorem, "*easy to prove*" according to its author, is based on three properties verified by the systems studied in statistical physics: reversibility of particle dynamics, the finite volume of the accessible phase space and Liouville's theorem. It does not apply to very particular initial microstates (see page 51), which represent an almost zero fraction of the accessible microstates. Finally, let us note that the recurrence theorem does not prove the (false) ergodic hypothesis (see Section 2.4.2), since a dynamical system can very well return to its initial microstate without having visited the entire accessible phase space, as the ergodic hypothesis stipulates.

[95] According to the German mathematician Ernst Zermelo (1871–1953) who confronted Boltzmann on this subject, in a series of articles, in 1896. The exchanged arguments are very well exposed in C. Cercignani, *Ludwig Boltzmann: The man who trusted atoms* (Oxford University Press, 2006) and in Uffink (2006), Note 35 in Chapter 2.

3.6 EXERCISES

Exercise 3.1. The entropy of water ice

In 1935, Linus Pauling[96] proposed a model to evaluate the entropy of water ice in its usual crystalline form (written I_h), the one observed at atmospheric pressure below 273 K: in a hexagonal lattice (symmetry found in snowflakes), each water molecule is located in the centre of a tetrahedron, as shown in Figure 3.11. The individual state of a molecule is then specified only by its orientation.

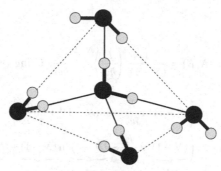

FIGURE 3.11 In I_h water ice, each oxygen atom (black) is located at the centre of a tetrahedron, whose four vertices are occupied by another oxygen atom. A water molecule orients its two hydrogen atoms (gray), only in the direction of two of the four neighbouring oxygen atoms so that each pair of neighbouring oxygen atoms shares a *single* hydrogen atom, thus forming a hydrogen bond.

1. Regardless of the orientation of the four neighbouring water molecules, what is the number k of orientations that the central molecule has to point its two O–H covalent bonds? The two hydrogen atoms of a water molecule are identical.

2. Between two neighbouring oxygen atoms, there can be only one hydrogen atom. What is the probability p_H that a given neighbouring water molecule has directed one of its two hydrogen atoms towards the central molecule? Deduce the probability p_{or} that one of the central molecule k orientations is compatible with the orientations of the neighbouring water molecules.

3. What then is the average number Ω_{or} of acceptable orientations of the central molecule? Neglecting the crystal surface effects, deduce the total number Ω_{tot} of orientations of N_A water molecules and the expression of the residual molar entropy S_0 of the ice. Compare S_0 with its experimental value, $S_{exp} = 3.41$ J·mol^{-1}·K^{-1}.

Solution:

1. *Two directions among four must be chosen, so $k = \binom{4}{2} = 6$.*

2. *$p_H = 1/2$, because the O–H covalent bonds point towards 2 directions among four possible. For each hydrogen atom of the central molecule, there is a fifty-fifty chance that the neighbouring molecule has not already positioned one of its own hydrogen atoms. The central molecule orientation is thus accepted with the probability $p_{or} = (1/2)^2 = 1/4$.*

3. *$\Omega_{or} = k \times p_{or} = 3/2$. Thus, treating molecules as independent particles, with correlations being taken into account (on average) in the expression of Ω_{or}, one obtains $\Omega_{tot} \simeq \Omega_{or}^{N_A} = (3/2)^{N_A}$ and $S_0 = k_B \ln \Omega_{tot} = R \ln(3/2)$, that is $S_0 \simeq 3.37$ J·mol^{-1}·K^{-1}, in very good agreement with the experimental value.*

---

[96] American chemist and physicist (1901–1994), Nobel Prize in Chemistry in 1954 for his work on chemical bonds and Nobel Peace Prize in 1962 for his political commitment against nuclear testing and weapons.

Exercise 3.2. System of independent classical harmonic oscillators

An isolated system consists of $N \gg 1$ distinguishable and independent one-dimensional classical harmonic oscillators of angular frequency ω. The system's energy lies between E and $E + \delta E$.

1. Using the results of Exercise 2.3, give the expression for the system's entropy S. Is it extensive?

2. Deduce the temperature T, the expression of energy as a function of T and the system's heat capacity C.

Solution:

1. From Equation (2.43), $\Omega(N,E) = \dfrac{1}{(N-1)!} \left(\dfrac{E}{\hbar\omega} \right)^{N-1} \dfrac{\delta E}{\hbar\omega}$. Using Stirling's formula (A.21), for $N \gg 1$:

$$
\begin{aligned}
S(N,E) &= k_B \left((N-1) \ln \frac{E}{\hbar\omega} - \ln(N-1)! + \ln \frac{\delta E}{\hbar\omega} \right) \\
&\simeq k_B \Big(\underbrace{(N-1) \ln \frac{E}{\hbar\omega}}_{\simeq N} - \underbrace{(N-1) \ln (N-1)}_{\simeq N} + \underbrace{(N-1)}_{\simeq N} + \underbrace{\ln \frac{\delta E}{\hbar\omega}}_{\ll N} \Big) \\
&\simeq k_B N \left(\ln \frac{E}{N\hbar\omega} + 1 \right),
\end{aligned}
$$

which is naturally extensive.

2. $\dfrac{1}{T} = \dfrac{\partial S}{\partial E} = N k_B \dfrac{1}{E}$, so $E = N k_B T$ and $C = \dfrac{dE}{dT} = N k_B$.

Exercise 3.3. Schottky defects

When thermal agitation is sufficient, one of the N atoms of a crystal can leave its lattice site and diffuse towards the surface, leaving a gap in the crystal lattice, called a *Schottky defect*. It is assumed that the energy $\epsilon > 0$ of a Schottky defect is independent of the site.

1. Let E_0 be the energy of an ideal crystal. What is the energy E of a crystal containing n Schottky defects?

2. What is the number of accessible microstates $\Omega(E)$. Deduce the crystal entropy $S(E)$ as a function of n and N when $n, N \gg 1$ and $(N - n) \gg 1$?

3. Deduce the crystal's temperature T and the expression of the number of defects $n(T)$ as a function of T when $n \ll N$.

4. Knowing that $\epsilon \simeq 1$ eV, is the approximation $N \gg n \gg 1$ justified at room temperature?

Solution:

1. $E = E_0 + n\epsilon$.

2. Ω is the number of way to choose n (distinguishable) gaps among N sites, so $\Omega = \binom{N}{n}$. Thus, using Stirling's formula (A.21):

$$
\begin{aligned}
S &= k_B \ln \frac{N!}{n!(N-n)!} \\
&\simeq k_B \left(N \ln N - N - n \ln n + n - (N-n) \ln(N-n) + (N-n) \right) \\
&\simeq k_B \left(N \ln N - n \ln n - (N-n) \ln(N-n) \right).
\end{aligned}
$$

3. By definition, $\dfrac{1}{T} = \dfrac{\partial S}{\partial E} = \dfrac{1}{\epsilon}\dfrac{\partial S}{\partial n} = \dfrac{k_B}{\epsilon}\left(-\ln n - 1 + \ln(N-n) + 1\right) = \dfrac{k_B}{\epsilon}\ln\dfrac{N-n}{n}$.

Thus $n(T) = N/(1 + e^{\frac{\epsilon}{k_B T}}) \simeq N e^{-\frac{\epsilon}{k_B T}}$, when $n \ll N$.

4. $\dfrac{n}{N} \simeq e^{-\frac{1.6\times 10^{-19}}{1.38\times 10^{-23}\,T}} \simeq e^{-\frac{10^4}{T}}$, where T is in kelvin.

At $T = 300$ K, $\dfrac{n}{N} \simeq e^{-33} \simeq 10^{-15}$. For $N \simeq 10^{23}$, it naturally leads to $N \gg n \gg 1$.

Exercise 3.4. Hydrogen crystal

A hydrogen crystal is made of N H_2 molecules. It is isolated in an adiabatic box. Depending on the orientation of the spins of its two atomic nuclei, a H_2 molecule can be found in one of the following four states:

- State (1), parahydrogen state (singlet state, two proton spins antiparallel) of energy $\varepsilon_1 = 0$,

- States (2), (3), (4), the three orthohydrogen states (triplet states, two proton spins parallel) of energies $\varepsilon_2 = \varepsilon_3 = \varepsilon_4 = \Delta$.

Let n_i be the number of molecules in state (i). The kinetic energies of the molecules and the inter-molecule interactions are ignored.

1. Give the expression for the energy E of the crystal as a function of n_1.

2. Calculate the number of accessible microstates $\Omega(E)$ and the corresponding entropy $S(E)$. In the limit $N \gg 1$, $n_1 \gg 1$, $(N - n_1) \gg 1$, demonstrate that:

$$S = k_B\left[N\ln N + (N-n_1)(\ln 3 - \ln(N-n_1)) - n_1\ln n_1\right].$$

3. Find the system's temperature T and the expression of the energy E as a function of T. Comment on the behaviour of E in the high temperature limit and in the low temperature limit.

4. Calculate the heat capacity C.

Solution:

1. *Since* $N = n_1 + n_2 + n_3 + n_4$, *then* $E = (n_2 + n_3 + n_4)\Delta = (N - n_1)\Delta$.

2. *Since energy is a constant, n_1 is also constant. There are $\binom{N}{n_1}$ ways to position n_1 (distinguishable) parahydrogen molecules among N crystal sites. Each of the $N - n_1$ remaining sites is occupied by an orthohydrogen molecule that can be in 3 possible states, 3^{N-n_1} possible states.*

Therefore $\Omega = \binom{N}{n_1} 3^{N-n_1} = \dfrac{N!}{n_1!(N-n_1)!} 3^{N-n_1}$. *Thus, using Stirling's formula (A.21):*
$S = k_B \ln \Omega \simeq k_B\left(N\ln N + (N-n_1)\left(\ln 3 - \ln(N-n_1)\right) - n_1\ln n_1\right)$.

3. *By definition,* $\dfrac{1}{T} = \dfrac{\partial S}{\partial E} = -\dfrac{1}{\Delta}\dfrac{\partial S}{\partial n_1} = -\dfrac{k_B}{\Delta}\left(-\left(\ln 3 - \ln(N - n_1)\right) + 1 - \ln n_1 - 1\right)$,

thus $\dfrac{\Delta}{k_B T} = \beta\Delta = \ln\dfrac{3n_1}{N - n_1}$, *so* $E = (N - n_1)\Delta = \dfrac{3N\Delta}{3 + e^{\beta\Delta}}$. *When* $T \to 0$ ($\beta \to \infty$), $E \to 0$ *(all molecules are parahydrogen). When* $T \to \infty$ ($\beta \to 0$), $E \to 3N\Delta/4$, *the 4 states are equiprobable (1/4 of molecules are parahydrogen with zero energy and 3/4 of molecules are orthohydrogen with energy Δ).*

4. $C = \dfrac{\partial E}{\partial T} = 3Nk_B(\beta\Delta)^2 \dfrac{e^{\beta\Delta}}{(3 + e^{\beta\Delta})^2}$, *which has a maximum. The H_2 molecules having only two energy levels, the typical behaviour of the Schottky anomaly is recovered (see Section 3.3.1).*

Exercise 3.5. Electron trapping on crystal sites

An isolated system consists of N electrons trapped on N crystal lattice sites. Each site can accommodate at most two electrons and therefore three states are possible:

- no electron is trapped: zero energy state,

- one electron is trapped: energy state $-\epsilon < 0$,

- a pair of electrons is trapped: energy state $-2\epsilon + g$, where $g > 0$ is a constant representing the interaction between the two electrons.

The interactions between electrons located on different sites are ignored.

1. Express the total number of sites N as a function of the numbers of sites occupied by 0, 1 or 2 electrons, written n_0, n_1 and n_2, respectively. Similarly, give the total number of electrons N as a function of n_1 and n_2.

2. Give the expression of the system energy E and show that it is fixed by the values of n_1 and n_2. Deduce the expressions of n_0, n_1 and n_2 as a function of E, N, ϵ and g. Between which limits can the value of E be chosen?

3. Consider the particular case $N = 3$, $n_1 = 1$ and $n_2 = 1$: count the number of accessible microstates.

4. Generalisation: calculate the number $\Omega(N, E)$ of accessible microstates of energy E for any N.

5. Calculate the system entropy $S(N, E)$ when n_0, n_1, n_2 and $N \gg 1$.

6. Express the system's temperature T as a function of the ratio n_1/n_2. Give the expression of E as a function of T and specify the limits of E at low and high temperatures.

7. Give the expressions of n_0, n_1 and n_2 as a function of T. Comment on the limits at low and high temperatures.

8. If electrons spin are taken into account, a single electron on a lattice site can be in two states ($\uparrow$ or $\downarrow$) of equal energy $-\epsilon$. On the other hand, since the electrons of an electron pair occupying the same lattice site are indistinguishable, they must carry opposite spins, according to Pauli exclusion principle, and thus in a single state of energy $-2\epsilon + g$. How do $\Omega(N, E)$ and the other calculated thermodynamic quantities then change?

Solution:

1. *Each (distinguishable) site is in one of the three possible states, so $N = n_0 + n_1 + n_2$. N (indistinguishable) electrons are either alone, or form a pair on a given lattice site. The number of electrons is therefore $N = n_1 + 2n_2$.*

2. *The energy is $E = -\epsilon n_1 + (g - 2\epsilon)n_2$, $n_0 = n_2 = (E + \epsilon N)/g$ and $n_1 = (N(g - 2\epsilon) - 2E)/g$. By definition, $0 \le n_1 \le N$ and $0 \le n_2 \le N/2$, so $-\epsilon N \le E \le (-\epsilon + g/2)N$.*

3. *There are 3 possibilities to choose a site that contains a single electron. And for each of these 3 configurations, there are two possible sites to place the electron pair. So $3 \times 2 = 6$ microstates.*

4. *Ω is the number of ways to choose n_1 sites occupied by one electrons among N sites (i.e. $\binom{N}{n_1}$), multiplied by the number of ways to choose n_2 sites occupied by an electron pair among $N - n_1$ remaining sites (i.e. $\binom{N-n_1}{n_2}$).*

   Thus $\Omega = \binom{N}{n_1}\binom{N - n_1}{n_2} = \dfrac{N!}{n_0!n_1!n_2!} = \dfrac{N!}{n_1!(n_2!)^2}$, because $n_0 = n_2$ (one recovers $\Omega = 6$ for $N = 3$, $n_1 = 1$ and $n_2 = 1$).

5. $S = k_B \ln \Omega \simeq k_B (N \ln N - n_1 \ln n_1 - 2n_2 \ln n_2)$, using Stirling's formula (A.21).

6. $\dfrac{1}{T} = \dfrac{\partial S}{\partial E} = k_B \left(-\dfrac{\partial n_1}{\partial E} \ln n_1 - \dfrac{\partial n_1}{\partial E} - 2\dfrac{\partial n_2}{\partial E} \ln n_2 - 2\dfrac{\partial n_2}{\partial E} \right)$, but $\dfrac{\partial n_1}{\partial E} = -\dfrac{2}{g}$ and $\dfrac{\partial n_2}{\partial E} = \dfrac{1}{g}$,

so $\dfrac{1}{T} = \dfrac{k_B}{g} (2 \ln n_1 + 2 - 2 \ln n_2 - 2) = \dfrac{2k_B}{g} \ln \left(\dfrac{n_1}{n_2} \right)$

then $\dfrac{g}{2k_B T} = \ln \left(\dfrac{n_1}{n_2} \right) = \ln \left(\dfrac{N(g - 2\epsilon) - 2E}{E + \epsilon N} \right) = \ln \left(\dfrac{Ng}{E + \epsilon N} - 2 \right)$.

It can be deduced that $E = N \left(-\epsilon + \dfrac{g}{2 + e^{\frac{g}{2k_B T}}} \right)$, which tends to $-N\epsilon$ when $T \to 0$ and to $(-\epsilon + g/3)N$ when $T \to \infty$.

7. $\dfrac{n_1}{n_2} = e^{\frac{g}{2k_B T}}$ and $N = n_1 + 2n_2$ so $n_1 = N\dfrac{e^{\frac{g}{2k_B T}}}{2 + e^{\frac{g}{2k_B T}}}$ and $n_2 = n_0 = N\dfrac{1}{2 + e^{\frac{g}{2k_B T}}}$. When $T \to 0$, $n_1 \to N$ and $n_2 = n_0 \to 0$ (every site has trapped one electron) and when $T \to \infty$, $n_1 \to N/3$ and $n_2 = n_0 \to N/3$ (the three lattice site states are equiprobable).

8. There are 2^{n_1} ways to choose the spin of single electrons.

So $\Omega(N, E) = 2^{n_1} \dbinom{N}{n_1} \dbinom{N - n_1}{n_2} = 2^{n_1} \dfrac{N!}{n_1! (n_2!)^2}$.

Then $S = k_B (n_1 \ln 2 + N \ln N - n_1 \ln n_1 - 2n_2 \ln n_2)$ and $\dfrac{1}{T} = \dfrac{2k_B}{g} \ln \left(\dfrac{n_1}{2n_2} \right)$.

So $\dfrac{g}{2k_B T} = \ln \left(\dfrac{N(g - 2\epsilon) - 2E}{2(E + \epsilon N)} \right) = \ln \left(\dfrac{Ng}{2(E + \epsilon N)} - 1 \right)$.

It can be deduced that $E = N \left(-\epsilon + \dfrac{g}{2(1 + e^{\frac{g}{2k_B T}})} \right)$, which tends to $-N\epsilon$ (i.e. $n_1 \to N$) when $T \to 0$ and to $(-\epsilon + g/4)N$ when $T \to \infty$ (there are in this case four equiprobable lattice site states).

Exercise 3.6. Maxwell velocity distribution function in the microcanonical ensemble

An ideal gas in equilibrium, isolated at energy E (to within δE), is composed of N *distinguishable* particles of mass m contained in a box of volume V.

1. Show that the probability that a given particle (say number 1) has a momentum between $\vec{p}$ and $\vec{p} + \mathrm{d}\vec{p}$ can be written as:

$$P_1(\vec{p})\mathrm{d}^3 p = \frac{V}{h^3} \frac{\Omega \left(N - 1, V, E - \frac{\vec{p}^2}{2m} \right)}{\Omega(N, V, E)} \mathrm{d}^3 p. \tag{3.73}$$

2. Deduce that in the limit $N \gg 1$, $P_1(\vec{p}) = \left(\dfrac{1}{2\pi m k_B T} \right)^{3/2} e^{-\left(\frac{\vec{p}^2}{2m k_B T} \right)}$. Recover then Maxwell velocity distribution function, $P(\vec{v})$, already encounter in Exercise 1.3.

Solution:

1. The probability $P_1(\vec{p})\mathrm{d}^3 p$ that a particle (say number 1) has a momentum lying between $\vec{p}$ and $\vec{p} + \mathrm{d}\vec{p}$ corresponds to the following two conditions:

   a. This particle has an energy $E_1 = \dfrac{\vec{p}^2}{2m}$ and can be located anywhere in V.

   b. The $N - 1$ other particles can be located anywhere in V and have random momenta under the condition that the sum of the energies of the $N - 1$ particles lies between $E - E_1$ and $E - E_1 + \delta E$.

From Equation (2.14), the probability $\mathrm{d}P$ *to find the system in a given microstate is*

$$\mathrm{d}P = \rho(\boldsymbol{q},\boldsymbol{p})\mathrm{d}^{3N}q\,\mathrm{d}^{3N}p = \frac{1}{h^{3N}\,\Omega(N,V,E)}\prod_{j=1}^{N}\mathrm{d}^3 r_j \prod_{j=1}^{N}\mathrm{d}^3 p_j.$$

The probability $P_1(\vec{p})$ *is therefore equal to* $\int \mathrm{d}P$, *where the integral bounds respect conditions (a) and (b), that is*

$$P_1(\vec{p})\mathrm{d}^3 p = \frac{\mathrm{d}^3 p}{h^{3N}\,\Omega(N,V,E)}\int_V \prod_{j=1}^{N}\mathrm{d}^3 r_j \int_{E-\frac{\vec{p}^2}{2m}\le\sum_{j=2}^{N}\frac{\vec{p}_j^2}{2m}\le E-\frac{\vec{p}^2}{2m}+\delta E}\prod_{j=2}^{N}\mathrm{d}^3 p_j.$$

Indeed $\int P_1(\vec{p})\mathrm{d}^3 p = 1$. *Integrating over the position of particle 1 gives:*

$$P_1(\vec{p}) = \frac{V}{h^3\,\Omega(N,V,E)}\ \underbrace{\frac{1}{h^{3(N-1)}}\int_V \prod_{j=2}^{N}\mathrm{d}^3 r_j \int_{E-\frac{\vec{p}^2}{2m}\le\sum_{j=2}^{N}\frac{\vec{p}_j^2}{2m}\le E+\delta E-\frac{\vec{p}^2}{2m}}\prod_{j=2}^{N}\mathrm{d}^3 p_j}_{=I},$$

where by definition $I = \Omega\left(N-1,V,E-\frac{\vec{p}^2}{2m}\right)$ *is the number of accessible microstates of a system of* $N-1$ *distinguishable particles of total energy* $E-\frac{\vec{p}^2}{2m}$, *in a volume* V. *It therefore proves Equation (3.73).*

2. *From Equation (2.19), without the* $1/N!$ *factor since the particles are here distinguishable, one obtains:* $\Omega(N,V,E) = \dfrac{3N}{2}\dfrac{V^N}{h^{3N}(\frac{3N}{2})!}(2\pi mE)^{\frac{3N}{2}}\dfrac{\delta E}{E}$. *Thus:*

$$
\begin{aligned}
P_1(\vec{p}) &= \frac{V}{h^3}\frac{\Omega\left(N-1,V,E-\frac{\vec{p}^2}{2m}\right)}{\Omega(N,V,E)} \\[2mm]
&= \frac{1}{(2\pi m)^{\frac{3}{2}}}\frac{\left(\frac{3N}{2}-1\right)!}{\left(\frac{3(N-1)}{2}-1\right)!}\frac{\left(E-\frac{\vec{p}^2}{2m}\right)^{\frac{3(N-1)}{2}-1}}{E^{\frac{3N}{2}-1}} \\[2mm]
&= \frac{1}{(2\pi mE)^{\frac{3}{2}}}\frac{\left(\frac{3N}{2}-1\right)!}{\left(\frac{3(N-1)}{2}-1\right)!}\left(1-\frac{\vec{p}^2}{2mE}\right)^{\frac{3(N-1)}{2}-1}.
\end{aligned}
$$

At the limit $N \gg 1$, *Stirling's formula (A.21) implies*[97]:

$$P_1(\vec{p}) \simeq \frac{1}{(2\pi mE)^{\frac{3}{2}}}\left(\frac{3N}{2}\right)^{\frac{3}{2}}\left(1-\frac{\vec{p}^2}{2mE}\right)^{\frac{3N}{2}} = \frac{1}{(2\pi m\frac{2E}{3N})^{\frac{3}{2}}}\,e^{\frac{3N}{2}\ln\left(1-\frac{\vec{p}^2}{2mE}\right)}.$$

---

[97] *Indeed,*

$$
\begin{aligned}
\ln\frac{\left(\frac{3N}{2}-1\right)!}{\left(\frac{3(N-1)}{2}-1\right)!} &\simeq \left(\frac{3N}{2}-1\right)\ln\left(\frac{3N}{2}-1\right)-\frac{3N}{2}+1-\left(\frac{3N}{2}-\frac{5}{2}\right)\ln\left(\frac{3N}{2}-\frac{5}{2}\right)+\frac{3N}{2}-\frac{5}{2} \\[2mm]
&\simeq \frac{3}{2}\ln\left(\frac{3N}{2}-1\right)-\left(\frac{3N}{2}-\frac{5}{2}\right)\underbrace{\ln\left(1-\frac{3}{3N-2}\right)}_{\simeq -3/(3N-2)}-\frac{3}{2} \\[2mm]
&\simeq \frac{3}{2}\ln\left(\underbrace{\frac{3N}{2}-1}_{\simeq 3N/2}\right)+\frac{3}{2}\underbrace{\frac{3N-5}{3N-2}}_{\to 1}-\frac{3}{2} \simeq \frac{3}{2}\ln\left(\frac{3N}{2}\right).
\end{aligned}
$$

Now $E = \frac{3}{2}Nk_BT$, so $\frac{\vec{p}^2}{2mE} \ll 1$ and $\frac{3N}{2}\ln\left(1 - \frac{\vec{p}^2}{2mE}\right) \simeq -\frac{3N}{2}\frac{\vec{p}^2}{2mE} = -\frac{\vec{p}^2}{2mk_BT}$, so

$$P_1(\vec{p}) \simeq \frac{1}{(2\pi mk_BT)^{\frac{3}{2}}}\, e^{-\frac{\vec{p}^2}{2mk_BT}}.$$

By changing momentum into velocity for each component, $p_\alpha = mv_\alpha$, with $\alpha = x, y, z$, one gets

$$P(\vec{v})\mathrm{d}^3v = P_1(\vec{p})\mathrm{d}^3p = P_1(\vec{p} = m\vec{v})\ \underbrace{\mathrm{d}p_x\mathrm{d}p_y\mathrm{d}p_z}_{=m^3\mathrm{d}v_x\mathrm{d}v_y\mathrm{d}v_z} = \left(\frac{m}{2\pi k_BT}\right)^{\frac{3}{2}} e^{-\frac{m\vec{v}^2}{2k_BT}}\,\mathrm{d}v_x\mathrm{d}v_y\mathrm{d}v_z.$$

The Maxwell velocity distribution function is recovered (see Equation (1.37)).

Exercise 3.7. Entropy of a black hole

The physics of black holes is, in principle, a matter of quantum mechanics and general relativity. Their existence was nevertheless predicted at the end of the eighteenth century by the scientists John Michell and Pierre-Simon de Laplace by imagining that light could not escape from a sufficiently massive celestial object[98].

1. Give the expression of the escape velocity v_1, that is the minimum velocity that a particle must have at the surface of a spherical object of mass M and radius R to escape its gravitational attraction. What is the relation between M and R for a black hole, from which a photon cannot escape? Calculate the radius of such an object whose mass would be equal to that of the Earth, $M_T \simeq 6 \times 10^{24}$ kg, then to that of the Sun, $M_\odot \simeq 2 \times 10^{30}$ kg.

2. Because a black hole absorbs matter and radiation without rejecting them, its surface area $\mathcal{A}$ can only increase, as does the entropy of an isolated system. In that respect, Jacob Bekenstein and Stephen Hawking showed in the 1970s that the entropy of a black hole is given by

$$S_{BH} = \frac{k_B}{4}\frac{\mathcal{A}}{l_P^2}, \tag{3.74}$$

where l_P is the Planck length, defined only with the fundamental constants of relativity (c), quantum mechanics ($\hbar$) and gravitation (G). Using dimensional analysis, express and evaluate l_P. Deduce the expression of S_{BH} as a function of these constants and the mass M.

3. In a possible collapse of the universe, called the *Big Crunch*, all the universe visible matter – that is $N \simeq 10^{80}$ baryons (protons, neutrons...) with $m \simeq 1.67 \times 10^{-27}$ kg – would be absorbed into a single black hole. Estimate the entropy of this black hole and deduce an order of magnitude of the total number of accessible microstates of the universe, Ω_{univ}.

4. The black hole's energy is $E = Mc^2$. Calculate its temperature T as a function of its mass M. Evaluate the temperature for a stellar black hole of mass $M = 10M_\odot$.

5. Deduce the heat capacity C of a black hole. How is this result surprising?

---

[98]More rigorously, a black hole is a region of space (of space-time in general relativity) from which matter and radiation cannot escape. It is characterised by its mass M and delimited by a sphere of radius R called the *event horizon* (see for instance J.D. Bekenstein, *Black holes and information theory*, Contemporary Physics **45**, 31 (2004)).

6. One can recover Expression (3.74) of S_{BH} as a function of $\mathcal{A}$ by reasoning on the absorption of a photon by a black hole. This event provides information corresponding to the smallest possible entropy variation δS, equal to one bit: like flipping a coin, the photon is absorbed (or not). Express δS as a function of k_B. The mass of the black hole then increases by $\delta M = \delta E/c^2$, where $\delta E = hc/\lambda$ is the energy of the photon of wavelength λ. No information about the precise location of the absorption of the photon on the surface of the black hole is known, it is only known that it is absorbed (or not). Its wavelength is therefore of the order of the size of the black hole, that is $\lambda \simeq R$. Deduce the increase $\delta\mathcal{A}$ of the black hole's surface as a function of δM, and then as a function of the fundamental constants. What is the ratio $\delta S/\delta\mathcal{A}$?

Solution:

1. _Reasoning on a particle of mass m, the conservation of mechanical energy implies $\frac{1}{2}mv_1^2 - GmM/R = 0$ (at an infinite distance from the object of mass M, the particle has zero velocity), that is $v_1 = \sqrt{2GM/R}$, which is independent of m. In the case of a photon, of velocity c, the radius of a black hole, called Schwarzschild radius, is such that $R = 2GM/c^2$. With $G = 6.7 \times 10^{-11}$ $m^3 \cdot kg^{-1} \cdot s^{-2}$ and $c = 3 \times 10^8$ $m \cdot s^{-1}$, a black hole having the mass of the Earth (of the Sun) would have a radius ~ 1 cm (~ 3 km)!_

2. _By dimensional analysis, $[c] = LT^{-1}$, $[\hbar] = ML^2T^{-1}$ and $[G] = L^3M^{-1}T^{-2}$, one finds $l_P = \sqrt{\hbar G/c^3} \simeq 1.6 \times 10^{-35}$ m, the characteristic length of a theory (still under construction) unifying gravitation and quantum mechanics (see Note 8). At this scale, quantum effects must play an important role in gravitation. Therefore, with $=\mathcal{A} = 4\pi R^2$ and $R = 2GM/c^2$:_

$$S_{BH} = \frac{k_B c^3}{4G\hbar}\mathcal{A} = \frac{\pi k_B c^3}{G\hbar}R^2 = \frac{4\pi k_B G}{\hbar c}M^2. \tag{3.75}$$

The black hole's entropy is correctly additive, but it is not extensive (because S_{BH} is not proportional to M).

3. _With $M = Nm \simeq 1.67 \times 10^{53}$ kg, $S_{BH} \simeq 3.7 \times 10^{-7}$ $M^2 \simeq 10^{100}$ J.K^{-1}._

So $\Omega_{univ} \simeq e^{S/k_B} \sim 10^{10^{23}}$ (see Note 80).

4. _By definition:_
$$\frac{1}{T} = \frac{\partial S_{BH}}{\partial E} = \frac{1}{c^2}\frac{\partial S_{BH}}{\partial M} = \frac{8\pi k_B G}{\hbar c^3}M,$$

or $T = \hbar c^3/(8\pi k_B GM)$, called Hawking temperature. Quantum physics allows to explain that a black hole has a temperature and thus emits a (very weak) radiation coming from vacuum fluctuations, in contradiction with its classical definition. For $M = 10M_\odot \simeq 20 \times 10^{30}$ kg, one finds an extremely low temperature, $T \simeq 6 \times 10^{-9}$ K.

5.
$$C = \frac{\partial E}{\partial T} = c^2\frac{\partial M}{\partial T} = -\frac{\hbar c^5}{8\pi G k_B T^2},$$

which is negative! Indeed, when a black hole absorbs energy (or mass), its temperature decreases ($T \propto 1/M$).

6. _A bit is the entropy associated with the distribution of two equiprobable events, so $\delta S = -k_B(\frac{1}{2}\ln\frac{1}{2} + \frac{1}{2}\ln\frac{1}{2}) = k_B\ln 2$ (see Section 3.4). In addition, $\delta\mathcal{A} = 8\pi R\delta R$ and $\delta R = 2G\delta M/c^2$ thus $\delta\mathcal{A} = 8\pi R(2G/c^2)\delta M$. According to the question, $\delta M = \delta E/c^2 = hc/(\lambda c^2)$ with $\lambda \simeq R$, thus $\delta M = h/(Rc)$. It can be deduced that $\delta\mathcal{A} = 16\pi Gh/c^3$ and $\delta S/\delta\mathcal{A} = \frac{\ln 2}{16\pi}\frac{k_B c^3}{Gh}$, so $S = \alpha\frac{k_B c^3}{G\hbar}\mathcal{A}$, as stated by Formula (3.75), but with the wrong prefactor, $\alpha \neq 1/4$._

Exercise 3.8. Gibbs entropy of a double experiment

During an experiment two events, (a) and (b), are observed simultaneously and are *a priori not independent* of each other. Let P_{ij} be the joint probability that the first event is a_i and that the second is b_j. $P_i^{(a)}$ be the probability that the first event is a_i, whatever is the outcome of the second event and let $P_j^{(b)}$ be the probability that the second event is b_j, whatever the outcome of the first event. For example, by looking out the window, one sees:

- event (a): it rains (a_1) or not (a_2),

- event (b): the neighbour went out with (b_1) or without (b_2) their umbrella.

1. Give the expression of $P_i^{(a)}$ and $P_j^{(b)}$ as a function of P_{ij}. Check the normalisation of the probability distribution functions $P_i^{(a)}$ and $P_j^{(b)}$.

2. Give the Gibbs entropy $S_G^{(ab)}$ associated with the joint experiment (ab) and the entropies $S_G^{(a)}$ and $S_G^{(b)}$ associated with experiments (a) and (b).

3. To compare $S_G^{(ab)}$ and $S_G^{(a)} + S_G^{(b)}$, show that:

$$S_G^{(ab)} - (S_G^{(a)} + S_G^{(b)}) = k_B \sum_{i,j} P_{ij} \ln \frac{P_i^{(a)} P_j^{(b)}}{P_{ij}}. \tag{3.76}$$

4. Prove that for all positive and real number x, $\ln x \leq x - 1$. Deduce the inequality $S_G^{(ab)} \leq S_G^{(a)} + S_G^{(b)}$. Comment.

Solution:

1. $P_i^{(a)} = \sum_j P_{ij}$ and $P_j^{(b)} = \sum_i P_{ij}$. Therefore, $\sum_i P_i^{(a)} = \sum_j P_j^{(b)} = \sum_{ij} P_{ij} = 1$.

2. $S_G^{(ab)} = -k_B \sum_{ij} P_{ij} \ln P_{ij}$, $S_G^{(a)} = -k_B \sum_i P_i^{(a)} \ln P_i^{(a)}$ and $S_G^{(b)} = -k_B \sum_j P_j^{(b)} \ln P_j^{(b)}$.

3. Then

$$S_G^{(ab)} - S_G^{(a)} - S_G^{(b)} = -k_B \left(\sum_{ij} P_{ij} \ln P_{ij} - \sum_i P_i^{(a)} \ln P_i^{(a)} - \sum_j P_j^{(b)} \ln P_j^{(b)} \right)$$

$$= -k_B \left(\sum_{ij} P_{ij} \ln P_{ij} - \sum_i \sum_j P_{ij} \ln P_i^{(a)} - \sum_j \sum_i P_{ij} \ln P_j^{(b)} \right)$$

$$= k_B \sum_{ij} P_{ij} \ln \frac{P_i^{(a)} P_j^{(b)}}{P_{ij}}.$$

In the case of independent events $(P_{ij} = P_i^{(a)} P_j^{(b)})$, entropy is naturally additive:
$S_G^{(ab)} = S_G^{(a)} + S_G^{(b)}$.

4. Let $f(x) = x - 1 - \ln x$, so $f'(x) = 1 - \frac{1}{x}$. $f(x)$ is minimal at $x = 1$ and $f(1) = 0$, thus $f(x) \geq 0$. For $x = P_i^{(a)} P_j^{(b)} / P_{ij}$:

$$S_G^{(ab)} - S_G^{(a)} - S_G^{(b)} \leq k_B \sum_{ij} P_{ij} \left(\frac{P_i^{(a)} P_j^{(b)}}{P_{ij}} - 1 \right)$$

$$\leq k_B \sum_{ij} P_i^{(a)} P_j^{(b)} - k_B \sum_{ij} P_{ij}$$

$$\leq k_B \underbrace{\sum_i P_i^{(a)}}_{=1} \underbrace{\sum_j P_j^{(b)}}_{=1} - k_B \underbrace{\sum_{ij} P_{ij}}_{=1} = 0.$$

Therefore $S_G^{(ab)} \leq S_G^{(a)} + S_G^{(b)}$. When the events are not independent, the result of the joint experiment (ab) offers less information (less missing information), than the results of the two experiments, (a) and (b), observed separately. Indeed, the correlations between (a) and (b) are taken into account in the double experiment, accordingly reducing the missing information. For example, if it is raining, it is likely that the neighbour has taken their umbrella.

4 Canonical and Grand Canonical Statistical Ensembles

"There is no more powerful method for introducing knowledge into the mind than that of presenting it in as many different ways as we can. When the ideas, after entering through different gateways, effect a junction in the citadel of the mind, the position they occupy becomes impregnable."

J. C. Maxwell[1], 1871.

The microcanonical ensemble verifies the conditions of application of the fundamental postulate, valid for isolated systems. As such, it has allowed the development of a statistical approach (Chapter 2) and the recovery of thermodynamics by offering a microscopic interpretation of entropy in terms of number of accessible microstates (Chapter 3). Conceptually simple, microcanonical conditions are ideal conditions that are difficult to implement in practice: a system cannot be perfectly isolated and energy is a delicate quantity to control experimentally. In general, natural or laboratory systems are in contact with an external medium imposing thermodynamic conditions (temperature, pressure...), such as the Earth's atmosphere or a thermostat whose temperature is chosen by an experimentalist.

In this chapter, starting from the microcanonical ensemble (N, V, E), new statistical ensembles corresponding to different sets of external parameters will be constructed: first the canonical ensemble (N, V, T), which describes a closed system of N particles, of constant volume V, in equilibrium with a thermostat at temperature T (Section 4.1), then the grand canonical ensemble (μ, V, T), associated with an open system of fixed volume V exchanging particles and energy with a reservoir imposing the temperature T and the chemical potential μ (Section 4.2). In each of these statistical ensembles, the probability distribution function of the system microstates will be determined and the expressions of thermodynamic quantities will be deduced. As it will be seen, the different procedures and calculations are much simpler than in the microcanonical ensemble. Finally, it will be shown that for any external parameters, statistical ensembles are in general equivalent for macroscopic systems and can be studied in the unifying framework of information theory (Section 4.3).

4.1 CANONICAL ENSEMBLE

As in Section 3.1.1, consider two subsystems $\mathcal{A}$ and $\mathcal{A}'$ in thermal contact, forming together an isolated system $\mathcal{A}_{\text{tot}}$. The two subsystems exchange energy only. Their volumes V and V', and their numbers of particles N and N', respectively, are fixed. Using the weak coupling approximation, the total energy is $E_{\text{tot}} = E + E'$. In the framework of quantum mechanics, according to Equation (3.5), when $\mathcal{A}_{\text{tot}}$ is in equilibrium, the probability p_m that the subsystem $\mathcal{A}$ is in microstate m of energy $E_m = E$ is given by:

$$p_m = \frac{\Omega'(E_{\text{tot}} - E_m)}{\Omega_{\text{tot}}(E_{\text{tot}})} = C\, e^{\frac{S'_p(E_{\text{tot}} - E_m)}{k_B}}, \tag{4.1}$$

where $C = 1/\Omega_{\text{tot}}(E_{\text{tot}})$ is a constant and $S'_p = k_B \ln \Omega'(E_{\text{tot}} - E_m)$ is the partial entropy of subsystem $\mathcal{A}'$ (to ease notations, the external parameters N, N', V and V' are not mentioned in the argument of S'_p).

---

[1] Cited in *Introductory Lecture on Experimental Physics*, 1871.

DOI: 10.1201/9781003272427-4

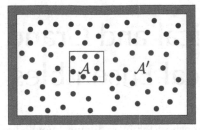

FIGURE 4.1 A closed subsystem $\mathcal{A}$ with rigid walls (fixed N and V) is in thermal contact with a much larger subsystem, the thermostat $\mathcal{A}'$. The two subsystems together form an isolated system $\mathcal{A}_{\text{tot}}$.

Assume now that system $\mathcal{A}$ is much smaller than system $\mathcal{A}'$ (see Figure 4.1). In other words, their respective numbers of degrees of freedom are such that

$$f \ll f' \quad (\text{that is } N \ll N').$$

Energy being an extensive quantity, then $E_m \ll E' = E_{\text{tot}} - E_m$, whatever is the microstate m of $\mathcal{A}$. If the initial temperatures T and T' of $\mathcal{A}$ and $\mathcal{A}'$, respectively, are different, the temperature of the large system remains practically constant and independent of the energy E_m of $\mathcal{A}$ since the exchanged energy (at most of the order of E_m) is very small compared to E'. $\mathcal{A}'$ plays the role of an energy reservoir, called a *thermostat*, for $\mathcal{A}$ and imposes the equilibrium temperature: $T = T'$.

In the rest of this section the properties of a system ($\mathcal{A}$) in contact with a thermostat ($\mathcal{A}'$) will be of interest. The latter must necessarily be macroscopic ($f' \gg 1$). On the other hand, system $\mathcal{A}$, in weak coupling with the thermostat, can be either microscopic (for example a molecule in interaction with the $N-1$ other molecules of an isolated system which then act as a thermostat)[2], or macroscopic, such that $1 \ll f \ll f'$. Note that, by definition, the role of the thermostat is relative to a given smaller system. For example, a large bucket of fresh water acts like a thermostat for a bottle of white wine and Earth's atmosphere ends up fixing the temperature of the bucket's content.

4.1.1 CANONICAL PROBABILITY DISTRIBUTION FUNCTION AND PARTITION FUNCTION

Under these conditions the probability p_m, generally given by Equation (4.1), takes a very simple form. Indeed, since $f \ll f'$ and $E_m \ll E' = E_{\text{tot}} - E_m$, the entropy of the thermostat

---

[2]According to Equation (3.2), the total energy is $E_{\text{tot}} = E + E' + E_{\text{int}}$. If $E + E_{\text{int}} \ll E'$ the large system $\mathcal{A}'$ acts as a thermostat for $\mathcal{A}$. Thus, for an atom (say atom number 1) in a simple fluid, Hamiltonian (1.7) can be decomposed as:

$$H(\{\vec{r}_j, \vec{p}_j\}) = \underbrace{\frac{\vec{p}_1^{\,2}}{2m}}_{E} + \underbrace{\sum_{j=2}^{N} \frac{\vec{p}_j^{\,2}}{2m} + \sum_{2 \le i < j} u(r_{ij})}_{E'} + \underbrace{\sum_{j=2}^{N} u(r_{1j})}_{E_{\text{int}}}. \tag{4.2}$$

For $N \gg 1$ and short range interactions, the weak coupling conditions are naturally met.

$S'_p(E_{tot} - E_m)$ can by expressed as a power series of E_m:

$$S'_p(E_{tot} - E_m) = \underbrace{S'_p(E_{tot})}_{\sim k_B f'} - \underbrace{E_m}_{\sim k_B f} \underbrace{\frac{\partial S'_p}{\partial E'}(E_{tot})}_{=1/T'} + \underbrace{\frac{E_m^2}{2} \frac{\partial^2 S'_p}{\partial E'^2}(E_{tot})}_{\sim k_B \frac{f^2}{f'}} + \ldots$$

$$\simeq S'_p(E_{tot}) - \frac{E_m}{T'},$$

where Definition (3.15) of temperature in the microcanonical ensemble and the extensive nature of energy and entropy were used to estimate the orders of magnitude of the different terms of the series. As can be seen, the terms of order greater than one are negligible[3]. According to Equation (4.1), a system in equilibrium with a thermostat at temperature T, can be found in a given microstate m with the probability

$$p_m = \underbrace{C\, e^{\frac{S'_p(E_{tot})}{k_B}}}_{1/Z} e^{-\frac{E_m}{k_B T'}},$$

where the prefactor, written as $1/Z$, is independent of the microstate m of system $\mathcal{A}$.

The *canonical ensemble*[4] is then defined as a statistical ensemble of systems, each of them closed and consisting of N particles contained in a fixed volume V and in thermal equilibrium with a thermostat at constant temperature $T' = T$. The canonical ensemble external parameters are therefore N, V and T while those of the microcanonical ensemble are N, V and E. The probability to find a system in equilibrium in a given microstate m of energy E_m is given by the canonical probability distribution:

$$p_m = \frac{1}{Z} e^{-\beta E_m}, \tag{4.3}$$

where $\beta = 1/k_B T$. The quantity Z, called the *partition function*[5] of the system, follows from the normalisation condition ($\sum_m p_m = 1$):

$$Z(N, V, T \ldots) \equiv \sum_{\text{microstates } m} e^{-\beta E_m}, \tag{4.4}$$

where the sum relates to *all* microstates m of the N particles confined in volume V, whatever is their energy[6] E_m. By definition, the partition function of a system is a dimensionless quantity which depends only on N, V, T and possibly on other external parameters (for example an external field such as gravitational, magnetic or electric fields...). In some

---

[3]The second order term (and the following ones) is negligible in front of the first two since it behaves as $k_B f^2/f' \ll k_B f \ll k_B f'$, because $E_m^2 \sim (k_B f)^2$ and $\dfrac{\partial^2 S'_p}{\partial E'^2} \sim k_B f'/(k_B f')^2$.

[4]While this statistical ensemble was born under the Boltzmann's pen (see Note 47 in Chapter 1), the term "Canonical" was introduced by Gibbs in 1902 in the sense of "according to a rule or law", the case of a closed system in equilibrium with a thermostat being extremely common. The microcanonical ensemble is considered by Gibbs as a limiting case of the canonical ensemble, in which the energy of the system varies only in a small interval δE (hence the prefix micro).

[5]The partition function describes how the probabilities are distributed among microstates. The notation Z comes from German *Zustandssumme* meaning "sum over states". The term $e^{-\beta E_m}$ is often referred to as *Boltzmann weight factor*.

[6]In Sum (4.4), E_m takes any arbitrarily large value, while by assumption $E_m \ll E_{tot}$. Indeed, since p_m decreases exponentially with E_m, microstates of very high energies with respect to $k_B T$ are very unlikely and have a negligible contribution to the partition function.

situations, it can be useful to write the partition function as a sum over energies rather than a sum over microstates:

$$Z(N,V,T) = \sum_{\text{energies } E} \Omega(N,V,E)\, e^{-\beta E}, \tag{4.5}$$

where the sum is performed over *all* system's energies greater than or equal to the energy E_0 of the fundamental level and where $\Omega(N,V,E)$ is the number of microstates of energy E. It will be shown in Section 4.1.3 that the partition function Z plays the same fundamental role as the number of accessible microstates Ω in the microcanonical ensemble. In principle, it allows the calculation of all thermodynamic quantities of the considered system. The operative advantage of the canonical ensemble over the microcanonical ensemble is that Z is expressed as a sum over all microstates (or all energies) without distinction, whereas $\Omega = \sum_m 1$ relates only to the accessible microstates m at energy E, such that $E_m = E$. Removing this constraint makes canonical ensemble calculations easier, as it will be illustrated by the examples and exercises of this chapter.

In the framework of classical physics, the discrete summation over the microstates is replaced by an integral over the generalised coordinates q and generalised momenta p. For a system of N particles described by f degrees of freedom, the partition function[7] is:

$$Z(N,V,T) = \underbrace{\left(\frac{1}{N!}\right)}_{\text{if indistinguishable}} \int \frac{d^f q\, d^f p}{h^f}\, e^{-\beta H(q,p)}, \tag{4.6}$$

where the factor $1/N!$ is omitted if the particles are (supposedly) distinguishable. The integration domain is such that particles coordinates are bounded by the walls of the container of volume V. Since the energy is not bounded, in particular the kinetic energy, the generalised momenta are not bounded: $-\infty < p_i < +\infty$, where $i = 1, 2, \ldots, f$. Recall that if particles are atoms without internal structure that evolve in a $d = 1, 2$ or 3 dimensional space, then $f = Nd$. By analogy with Equation (4.3), in the canonical ensemble, the probability dP to find the system in the microstate (q,p) to within $d^f q$ and $d^f p$ is:

$$dP = \rho(q,p)d^f q\, d^f p = \underbrace{\left(\frac{1}{N!}\right)}_{\text{if indistinguishable}} \frac{e^{-\beta H(q,p)}}{Z(N,V,T)} \frac{d^f q\, d^f p}{h^f}. \tag{4.7}$$

The canonical probability density function $\rho(q,p)$ does verify the normalisation condition (1.33).

Set of independent systems

The partition function of a system composed of independent (or weakly coupled) subsystems is the product of each subsystem partition functions. To demonstrate this, consider the case

---

[7]It is interesting to recover this result in the following way. The energy of a macroscopic system can be treated as a continuous quantity and Expression (4.5) of Z becomes:

$$Z(N,V,T) = \frac{1}{\delta E} \int dE\ \Omega(N,V,E)\, e^{-\beta E}.$$

Using Expression (2.13) of Ω, one obtains:

$$\begin{aligned}
Z(N,V,T) &= \frac{1}{\delta E} \int dE\ e^{-\beta E} \frac{\delta E}{N!} \int \frac{d^f q\, d^f p}{h^f}\ \delta(E - H(q,p)) \\
&= \frac{1}{N!} \int \frac{d^f q\, d^f p}{h^f} \int dE\ e^{-\beta E}\ \delta(E - H(q,p)) = \frac{1}{N!} \int \frac{d^f q\, d^f p}{h^f}\ e^{-\beta H(q,p)}.
\end{aligned}$$

of two independent subsystems $\mathcal{A}$ and $\mathcal{A}'$, such that $E_{\text{tot}} = E_m + E'_{m'}$, where E_m is the energy of a microstate m of $\mathcal{A}$ and $E'_{m'}$ the energy of a microstate m' of $\mathcal{A}'$. Summing over all the microstates m and m', the partition function of the total system is:

$$Z_{\text{tot}} = \sum_{m,m'} e^{-\beta E_{\text{tot}}} = \sum_{m,m'} e^{-\beta(E_m + E'_{m'})} = \sum_{m,m'} e^{-\beta E_m}\, e^{-\beta E'_{m'}}.$$

It is because the two subsystems are independent that the sum can be separated into a sum over the microstates of $\mathcal{A}$ and a sum over the microstates of $\mathcal{A}'$:

$$Z_{\text{tot}} = \sum_m e^{-\beta E_m} \sum_{m'} e^{-\beta E'_m} = Z\, Z', \tag{4.8}$$

where Z and Z' are the partition functions of subsystems $\mathcal{A}$ and $\mathcal{A}'$, respectively. This result can be generalised immediately to any number of independent subsystems, for example, a sytem of N *identical and independent* particles (see the coming paramagnetic crystal and ideal gas examples). The Hamiltonian is then written in Form (1.8) and the partition function is:

$$Z(N,V,T) = \underbrace{\left(\frac{1}{N!}\right)}_{\text{if indistinguishable}} \sum_{m_1}\sum_{m_2}\cdots\sum_{m_N} e^{-\beta\left(h_1 + h_2 + \cdots + h_N\right)}$$

$$= \underbrace{\left(\frac{1}{N!}\right)}_{\text{if indistinguishable}} \left(\sum_{m_1} e^{-\beta h_1}\right)\left(\sum_{m_2} e^{-\beta h_2}\right)\cdots\left(\sum_{m_N} e^{-\beta h_N}\right)$$

$$= \underbrace{\left(\frac{1}{N!}\right)}_{\text{if indistinguishable}} z^N, \quad \text{where} \quad z = \sum_{m_1} e^{-\beta h_1}, \tag{4.9}$$

where m_j is the microstates of particle j and z is the partition function of a *single particle*. This very useful expression, since in practice z is easier to calculate than Z, shows explicitly that the properties of a system of independent particles reduce to those of a single particle.

The decomposition of a system into independent subsystems is more frequent than it seems, provided that a different meaning to the word "system" is given. So far, we have considered physical objects: independent ensembles of particles (Equation (4.8)), or sets of independent particles (Equation (4.9)). More generally, if the *Hamiltonian of a system can be written as a sum of independent terms*, each term being expressed in terms of different microscopic variables (coordinates, generalised momenta, spins...), the system's partition function is then the product of the partition functions associated to each of these terms. Thus, in the very general case of a *real* fluid, in which the particles interact with each other, Hamiltonian (1.6) can be decomposed into two independent terms (in the absence of a magnetic field): the kinetic energy (which depends only on momenta) and the potential energy (which depends only on positions). The partition function can then be factorised in the Form (4.8), as it will be seen in Chapter 5. Similarly, when internal degrees of freedom (translational, vibrational and rotational) of a polyatomic molecule are assumed independent, the partition function z of the molecule is the product of the partition functions characterising each of these degrees of freedom (see Section 4.1.5 and Exercise 4.8).

Example 1: Ideal paramagnetic crystal

Consider N spins–1/2 carried by atoms located at the nodes of a crystal lattice and immersed in a magnetic field **B**. The interactions between spins are ignored. The system, whose

Hamiltonian H_{IS} is given by Equation (1.10), is in contact with a thermostat at temperature T. The N spins are treated as *independent* and *distinguishable* particles and the system's partition function is (Equation (4.9)):

$$Z(N,T) = \sum_{\{s_j=\pm 1\}} e^{-\beta H_{SI}} = \sum_{s_1=\pm 1}\sum_{s_2=\pm 1}\cdots\sum_{s_N=\pm 1} e^{\beta\mu_M B\left(s_1+s_2+\cdots+s_N\right)} = z^N,$$

where $\{s_j = \pm 1\}$ corresponds to the 2^N system's microstates. The partition function of a single spin can be easily calculated:

$$z = \sum_{s_1=\pm 1} e^{\beta\mu_M B s_1} = e^{\beta\mu_M B} + e^{-\beta\mu_M B} = 2\cosh(\beta\mu_M B),$$

and

$$Z(N,T) = \Big(2\cosh(\beta\mu_M B)\Big)^N. \tag{4.10}$$

Example 2: Ideal classical gas

An ideal classical gas composed of N identical and *indistinguishable* particles of mass m and without internal structure, is enclosed in a cubic container of volume $V = L^3$ in contact with a thermostat at temperature T. The gas Hamiltonian H_{IG} is given by Equation (2.16) and its partition function is (Equation (4.6) with $f = 3N$):

$$
\begin{aligned}
Z(N,V,T) &= \frac{1}{N!}\int \frac{d^{3N}q\,d^{3N}p}{h^{3N}}\, e^{-\beta H_{IG}(q,p)}\\[2mm]
&= \frac{1}{N!}\int \frac{dq_1 dq_2\ldots dq_{3N}dp_1 dp_2\ldots dp_{3N}}{h^{3N}}\, e^{-\beta\left(\frac{p_1^2}{2m}+\frac{p_2^2}{2m}+\cdots+\frac{p_{3N}^2}{2m}\right)}\\[2mm]
&= \frac{1}{N!}\left(\frac{1}{h}\underbrace{\int_0^L dq_1}_{=L}\underbrace{\int_{-\infty}^{+\infty} dp_1\, e^{-\beta\frac{p_1^2}{2m}}}_{=\sqrt{2\pi mk_B T}}\right)^{3N}\\[2mm]
&= \frac{1}{N!}\left(\frac{L\sqrt{2\pi mk_B T}}{h}\right)^{3N},
\end{aligned}
$$

because the $6N$ microscopic variables are independents and $0 \le q_i \le L$ and $-\infty \le p_i \le +\infty$ for $i = 1,2,\ldots,3N$. Since the N particles of the ideal gas are independent, $Z(N,V,T)$ can be written as in Equation (4.9):

$$Z(N,V,T) = \frac{z^N}{N!}, \quad \text{with} \quad z(V,T) = V\left(\frac{\sqrt{2\pi mk_B T}}{h}\right)^3 = \frac{V}{\lambda_T^3}, \tag{4.11}$$

where z is the partition function of a single particle moving freely in the container of volume V and $\lambda_T = h/\sqrt{2\pi mk_B T}$ is the de Broglie thermal wavelength (see Equation (1.24)).

4.1.2   FREE ENERGY

In the canonical ensemble, the energy of a system is a fluctuating internal variable since energy exchanges with a thermostat are possible. The energy probability distribution $P(E)$ can be written formally using Equation (1.32). Since the probability p_m, given by Equation (4.3), depends only on the energy E_m of microstate m, $P(E)$ is simply expressed as a function of the number of accessible microstates $\Omega(N,V,E)$ at energy E:

$$
\begin{aligned}
P(E) &= \sum_{m \text{ such as } E_m=E} p_m = \frac{1}{Z(N,V,T)} \sum_{m \text{ such as } E_m=E} e^{-\beta E_m} \\
&= \frac{1}{Z(N,V,T)} e^{-\beta E} \sum_{m \text{ such as } E_m=E} 1 \\
&= \frac{1}{Z(N,V,T)} e^{-\beta E} \Omega(N,V,E) \tag{4.12} \\
&= \frac{1}{Z(N,V,T)} e^{-\beta\left(E-TS(N,V,E)\right)}, \tag{4.13}
\end{aligned}
$$

where $S(N,V,E) = k_B \ln \Omega(N,V,E)$ is the system's entropy when it is isolated and its energy is equal to E. Note that the probability distribution $P(E)$ is naturally normalised, thanks to the definition of the partition function given by Equation (4.5).

Moreover, it was shown in Section 3.1.1 that the energy distribution $P(E)$ of a *macroscopic* system $\mathcal{A}$, in thermal contact with another macroscopic system $\mathcal{A}'$ (here the thermostat), behaves like an extremely sharp[8] Gaussian distribution (3.6) centred in $E = \langle E \rangle$. This most probable value is interpreted as the equilibrium value of E, the internal energy of the system. Remember that the relative fluctuations are of the order of $\Delta_E/\langle E \rangle \simeq 10^{-12}$ for a mole of particles. The probability $P(E \simeq \langle E \rangle)$ that energy E is close to its equilibrium value is therefore practically equal[9] to 1, that is $\ln P(\langle E \rangle) \simeq 0$. Taking the logarithm of Equation (4.13) for $E = \langle E \rangle$, one thus obtains:

$$
\ln P(\langle E \rangle) = -\ln Z(N,V,T) - \beta\left(\langle E \rangle - TS(N,V,\langle E \rangle)\right) \simeq 0,
$$

where $S(N,V,\langle E \rangle)$ is the thermodynamic entropy of the system in equilibrium, that is for $E = \langle E \rangle$. The function $F(N,V,T)$ is then defined as:

$$
F(N,V,T) \equiv -k_B T \ln Z(N,V,T) \simeq \langle E \rangle - TS(N,V,\langle E \rangle), \tag{4.14}
$$

which is identified with the *free energy* (also called *Helmholtz free energy*), defined in thermodynamics by a Legendre transformation of the internal energy $\langle E \rangle$, that is $F = \langle E \rangle - TS$, allowing the substitution of the external parameters (N,V,E) for (N,V,T). Free energy is an extensive thermodynamic quantity which characterises macroscopic systems in equilibrium. Like the partition function, $F(N,V,T \ldots)$ depends only on the external parameters. According to Equation (4.8), $F(N,V,T)$ is an additive quantity in the case of independent or weakly coupled systems[10]. In the microcanonical ensemble, entropy allows the transition from microscopic to macroscopic, through the number of accessible microstates Ω. In the canonical ensemble, it is the free energy that plays this fundamental role through the partition function Z.

4.1.3 AVERAGE VALUES AND THERMODYNAMIC QUANTITIES

Since the microstates' probability distribution is given by Equation (4.3), the average value of any internal variable A can be calculated. According to Equation (1.30), in the framework

---

[8]Expression (4.12) highlights this behaviour: $P(E)$ is the product of p_m, which decreases exponentially with E, and $\Omega \sim E^f$ (from Equation (2.24)), which grows very rapidly with E for $f \gg 1$. The resulting distribution $P(E)$ is thus very peaked since it is the result of a compromise between the probability p_m which favours low-energy microstates and the number of states Ω which increases the contribution of high-energy microstates.

[9]For a Gaussian distribution, the probability of deviations from the mean value greater than $5\Delta_E$, that is $(|E - \langle E \rangle|/\langle E \rangle > 5\Delta_E/\langle E \rangle \simeq 5 \times 10^{-12})$, is of the order of 3×10^{-7}.

[10]Indeed, $F_{tot} = -k_B T \ln Z_{tot} = -k_B T \ln ZZ' = -k_B T \ln Z - k_B T \ln Z' = F + F'$, where F and F' are the free energies of systems $\mathcal{A}$ and $\mathcal{A}'$, respectively.

of quantum mechanics:

$$\langle A \rangle = \sum_m A_m p_m = \frac{1}{Z} \sum_m A_m \, e^{-\beta E_m} = \frac{\sum_m A_m \, e^{-\beta E_m}}{\sum_m e^{-\beta E_m}}, \tag{4.15}$$

where A_m is the value of the internal variable in microstate m. In the framework of the classical mechanics, the average value of an internal variable $A(q, p)$ is given by Equation (1.34):

$$\begin{aligned} \langle A \rangle &= \int A(q,p)\rho(q,p) \, \mathrm{d}^f q \mathrm{d}^f p = \frac{1}{Z} \int A(q,p) e^{-\beta H(q,p)} \, \frac{\mathrm{d}^f q \mathrm{d}^f q}{h^f} \\ &= \frac{\int A(q,p) e^{-\beta H(q,p)} \, \mathrm{d}^f q \mathrm{d}^f p}{\int e^{-\beta H(q,p)} \, \mathrm{d}^f q \mathrm{d}^f p}. \end{aligned} \tag{4.16}$$

Expressions (4.15) and (4.16) concern any internal variable, whether microscopic (the velocity of a particle for example) or macroscopic (a thermodynamic quantity like energy). It remains now to express thermodynamic quantities, which are by definition equal to the average values of the usual *macroscopic* internal variables.

Internal energy

The average value of the energy, identified with internal energy, is written according to Equation (4.15):

$$\begin{aligned} \langle E \rangle &= \sum_m E_m \, p_m = \frac{1}{Z} \sum_m E_m \, e^{-\beta E_m} \\ &= -\frac{1}{Z} \sum_m \frac{\partial}{\partial \beta}\left(e^{-\beta E_m}\right) = -\frac{1}{Z} \frac{\partial}{\partial \beta}\bigg(\underbrace{\sum_m e^{-\beta E_m}}_{=Z}\bigg) = -\frac{1}{Z} \frac{\partial Z}{\partial \beta}\bigg|_{N,V}, \end{aligned}$$

where the derivative with respect to β and the sum over microstates were swapped. Thus:

$$\langle E \rangle = -\frac{\partial \ln Z}{\partial \beta}\bigg|_{N,V} = \frac{\partial(\beta F)}{\partial \beta}\bigg|_{N,V} = F - T\frac{\partial F}{\partial T}\bigg|_{N,V}. \tag{4.17}$$

Average generalised forces: pressure, chemical potential, magnetisation

In Section 3.3, the average generalised force conjugate to an external parameter x was defined by Equation (3.49):

$$\langle X \rangle = -\left\langle \frac{\partial E_m}{\partial x} \right\rangle,$$

where $\langle \ldots \rangle$ is an average in the considered statistical ensemble. In the canonical ensemble, the ensemble average is:

$$\langle X \rangle = -\frac{1}{Z} \sum_m \left(\frac{\partial E_m}{\partial x}\right) e^{-\beta E_m} = \frac{1}{\beta Z} \frac{\partial}{\partial x}\bigg(\underbrace{\sum_m e^{-\beta E_m}}_{Z}\bigg) = \frac{1}{\beta} \frac{\partial \ln Z}{\partial x}\bigg|_{T \ldots}, \tag{4.18}$$

where the partial derivative is taken by fixing the value of the temperature and the external parameters other than x. Thus, pressure being by definition the average force conjugate to volume ($x = V$), one obtains:

$$P \equiv \frac{1}{\beta} \frac{\partial \ln Z}{\partial V} \bigg|_{N,T} = -\frac{\partial F}{\partial V} \bigg|_{N,T}. \tag{4.19}$$

Similarly, the chemical potential is the average force conjugate to the number of particles ($x = N$):

$$\mu \equiv -\frac{1}{\beta} \frac{\partial \ln Z}{\partial N} \bigg|_{V,T} = \frac{\partial F}{\partial N} \bigg|_{V,T}. \tag{4.20}$$

Remember that pressure and chemical potential are average quantities, which are noted without $\langle \dots \rangle$ by convention. Finally, for a magnetic system, the total magnetic moment is defined as the average force conjugate to the magnetic field ($x = B$):

$$\mathcal{M} \equiv \frac{1}{\beta} \frac{\partial \ln Z}{\partial B} \bigg|_{N,V,T} = -\frac{\partial F}{\partial B} \bigg|_{N,V,T}. \tag{4.21}$$

This result can be obtained directly by writing the Hamiltonian of a spin system in the general form $H = H_{\text{SI}} + H_{\text{int}}$, where H_{SI} is the independent spin term given by Equation (1.10), which describes the interaction of each spin with the magnetic field, and H_{int} is the interaction term between spins, which does not depend on B (an expression of H_{int} in the framework of the Ising model will be derived in Section 9.1.1). By definition the average magnetic moment of a system is given by:

$$\mathcal{M} = \left\langle \sum_{j=1}^{N} \mu_M s_j \right\rangle = \frac{1}{Z} \sum_{\{s_j = \pm 1\}} \left(\sum_{j=1}^{N} \mu_M s_j \right) e^{\beta B \left(\sum_{j=1}^{N} \mu_M s_j \right) - \beta H_{\text{int}}} = \frac{1}{\beta Z} \frac{\partial Z}{\partial B}. \tag{4.22}$$

Depending on the system and the properties being studied, other external parameters associated with their conjugate variables can be introduced (see Equation (3.41)).

Entropy

According to Equations (4.14) and (4.17), entropy is also expressed as a derivative of the free energy:

$$S = \frac{1}{T} \left(\langle E \rangle - F \right) = -\frac{\partial F}{\partial T} \bigg|_{N,V}. \tag{4.23}$$

According to Equations (4.19), (4.20), (4.21) and (4.23), the total differential of F is[11]:

$$dF = -SdT - PdV + \mu dN - \mathcal{M}dB. \tag{4.24}$$

It allows to easily recover the definitions of thermodynamic quantities expressed above.

Heat capacity and energy fluctuation

By deriving the internal energy $\langle E \rangle$ with respect to T, the heat capacity at constant volume[12] (defined by Equation (3.13)) can be obtained:

$$C_V = \frac{\partial \langle E \rangle}{\partial T} \bigg|_{N,V} = \frac{\partial \beta}{\partial T} \frac{\partial \langle E \rangle}{\partial \beta} = -k_B \beta^2 \frac{\partial \langle E \rangle}{\partial \beta} = k_B \beta^2 \frac{\partial^2 \ln Z}{\partial \beta^2} \bigg|_{N,V} = -T \frac{\partial^2 F}{\partial T^2} \bigg|_{N,V}. \tag{4.25}$$

---

[11]In the case of magnetic systems, free energy is written as the Legendre transformation $F = E - TS - B\mathcal{M}$, which allows the following substitution of external parameters: $N, V, E, \mathcal{M} \to N, V, T, B$.

[12]Note that heat capacity can also be written as: $C_V = -T \frac{\partial^2 F}{\partial T^2} \bigg|_{N,V} = T \frac{\partial S}{\partial T} \bigg|_{N,V}$.

Moreover, it is known from Equation (3.14) that the variance Δ_E^2 of the energy is a function of the heat capacities C_V and C_V' of system $\mathcal{A}$ and thermostat $\mathcal{A}'$, respectively. Since $f' \gg f$ and since heat capacity is an extensive quantity, $C_V' \gg C_V$ therefore:

$$\Delta_E^2 = \langle E^2 \rangle - \langle E \rangle^2 = k_{\mathrm{B}} T^2 \frac{C_V C_V'}{C_V + C_V'} \simeq k_{\mathrm{B}} T^2 C_V. \tag{4.26}$$

This result can be obtained directly using the partition function. Writing $\langle E^2 \rangle$:

$$
\begin{aligned}
\langle E^2 \rangle &= \sum_m E_m^2 \, p_m = \frac{1}{Z} \sum_m E_m^2 \, \mathrm{e}^{-\beta E_m} \\
&= \frac{1}{Z} \sum_m \frac{\partial^2}{\partial \beta^2} \left(\mathrm{e}^{-\beta E_m} \right) = \frac{1}{Z} \frac{\partial^2 Z}{\partial \beta^2} \bigg|_{N,V}.
\end{aligned}
$$

The energy variance is therefore, according to Equations (4.17) and (4.25):

$$\langle E^2 \rangle - \langle E \rangle^2 = \frac{1}{Z} \frac{\partial^2 Z}{\partial \beta^2} - \left(-\frac{1}{Z} \frac{\partial Z}{\partial \beta} \right)^2 = \frac{\partial}{\partial \beta} \left(\frac{1}{Z} \frac{\partial Z}{\partial \beta} \right) = \frac{\partial^2 \ln Z}{\partial \beta^2} \bigg|_{N,V} = k_{\mathrm{B}} T^2 C_V. \tag{4.27}$$

Relation (4.26) is, of course, recovered[13]. Since heat capacity is extensive, one again finds that the relative fluctuations, $\Delta_E / \langle E \rangle$, behave as[14] $1/\sqrt{f}$, or of the order of 10^{-12} for $f \simeq 10^{24}$.

Heat capacity, a thermodynamic quantity measured by calorimetry experiments, has a statistical interpretation since it is proportional to the variance of an internal variable (E) treated as a random variable. Moreover, like any variance, C_V is necessarily positive in the canonical ensemble, which is not the case in the microcanonical ensemble under specific conditions (see Note 12 in Chapter 3). Equation (4.27) is a special case of the *fluctuation-dissipation* theorem[15] presented in Chapter 11, that is a relation between the fluctuations of an internal variable in equilibrium (here E) and the response of the system to a small external perturbation placing it in a non-equilibrium state (here a small deviation δT from the equilibrium temperature imposes on the system an energy variation $\delta E = C_V \delta T$).

Similarly, for a magnetic system, the magnetic susceptibility χ characterises the response of the system to a small change δB of the magnetic field B: the resulting variation δM of magnetic moment is proportional to δB. According to Equation (4.21), χ is by definition:

$$\chi \equiv \frac{\mu_0}{V} \frac{\partial \mathcal{M}}{\partial B} \bigg|_{N,V,T} = \mu_0 \frac{\partial M}{\partial B} \bigg|_{N,V,T} = -\frac{\mu_0}{V} \frac{\partial^2 F}{\partial B^2} \bigg|_{N,V,T}, \tag{4.28}$$

where $M = \mathcal{M}/V$ is the magnetisation. By analogy with Equation (4.22), it can be shown that: $\langle \mathcal{M}^2 \rangle = \langle \left(\sum_{j=1}^N \mu_M s_j \right)^2 \rangle = \frac{1}{\beta^2 Z} \frac{\partial^2 Z}{\partial B^2}$, thus

$$\langle \mathcal{M}^2 \rangle - \langle \mathcal{M} \rangle^2 = \frac{1}{\beta^2} \frac{\partial^2 \ln Z}{\partial B^2} = -\frac{1}{\beta} \frac{\partial^2 F}{\partial B^2} \bigg|_{N,V,T} = \frac{k_{\mathrm{B}} T V}{\mu_0} \chi, \tag{4.29}$$

which, as Equation (4.27), is an expression of the fluctuation-dissipation theorem.

---

[13]Note that Equation (4.26) was obtained in the limit $f' \gg f$, as it should be. However in Equation (4.27), this limit is already taken into account by the formalism of the canonical ensemble, in particular by the definition of Z.

[14]Note that $\langle E^2 \rangle$ and $\langle E \rangle^2$ behave as f^2, but their difference is proportional to C_V and thus to f.

[15]The fluctuation-dissipation theorem is part of the linear response theory, developed in the 1950's by Green, Callen and Kubo, to describe in a very general way the behaviour of a system in equilibrium, subjected to a small perturbation. This theory allows in practice to calculate, in the framework of equilibrium statistical physics, response functions such as susceptibilities or transport coefficients (viscosity, diffusion, thermal or electrical conductivity...). See U. Marini Bettolo Marconi *et al.*, *Fluctuationdissipation: Response theory in statistical physics*, Physics Reports **461** (4–6), pp 111–195 (2008) and R. Livi and P. Politi, *Nonequilibrium Statistical Physics: A Modern Perspective* (Cambridge University Press, 2017).

Example 1: Ideal paramagnetic crystal

According to Expression (4.10) of the partition function, the free energy (4.14) of a system of N independent spins immersed in a magnetic field B at temperature T, is:

$$F(N,T) = -k_B T \ln Z(N,T) = -N k_B T \ln \left(2 \cosh(\beta \mu_M B) \right). \qquad (4.30)$$

The system's internal energy (4.17) is:

$$\langle E \rangle = -\left. \frac{\partial \ln Z}{\partial \beta} \right|_N = -N \mu_M B \tanh(\beta \mu_M B), \qquad (4.31)$$

which increases with temperature: $\langle E \rangle = -N \mu_M B$ at $T = 0$ and tends to 0 when $k_B T \gg \mu_M B$. The heat capacity (4.25) can be deduced:

$$C = \left. \frac{\partial \langle E \rangle}{\partial T} \right|_N = k_B \beta^2 \left. \frac{\partial^2 \ln Z}{\partial \beta^2} \right|_N = N k_B \left(\frac{\beta \mu_M B}{\cosh(\beta \mu_M B)} \right)^2. \qquad (4.32)$$

The average magnetic moment $\mathcal{M}$ is given by Equation (4.21), but in the case of independent spins one directly obtains $\mathcal{M} = -\langle E \rangle / B$, from Form (1.10) of the Hamiltonian, so:

$$\mathcal{M} \equiv \frac{1}{\beta} \left. \frac{\partial \ln Z}{\partial B} \right|_T = N \mu_M \tanh(\beta \mu_M B). \qquad (4.33)$$

Expressions (3.24), (3.44) and (3.42) of these thermodynamics quantities, obtained in the microcanonical ensemble[16], are easily recovered. The equivalence of statistical ensembles will be discuss again in Section 4.3. Note, however, that calculations are much simpler in the canonical ensemble. Finally, the magnetic susceptibility defined by Equation (4.28) is:

$$\chi = \frac{\mu_0}{V} \frac{\partial \mathcal{M}}{\partial B} = N \frac{\mu_0}{V} \frac{\beta \mu_M^2}{\cosh^2(\beta \mu_M B)}.$$

At low field (or high temperature), $\beta \mu_M B \ll 1$ and $\cosh(\beta \mu_M B) \simeq 1$, Curie's law (3.43) is recovered: $\chi = C_{\text{Curie}}/T$, where C_{Curie} is the *Curie constant*.

Example 2: Ideal classical gas

According to the ideal classical gas partition function (Equation (4.11)), the free energy is:

$$
\begin{aligned}
F(N,V,T) &= -k_B T \ln Z(N,V,T) = -k_B T \ln \frac{z^N}{N!} \quad \text{with} \quad z = \frac{V}{\lambda_T^3} \\
&\simeq -k_B T \left(N \ln z - N \ln N + N \right) \\
&\simeq -N k_B T \left(\ln \frac{V}{N \lambda_T^3} + 1 \right),
\end{aligned}
\qquad (4.34)
$$

using Stirling's formula (A.21) (F is indeed extensive for $N \gg 1$). Internal energy can be written as:

$$\langle E \rangle = -\left. \frac{\partial \ln Z}{\partial \beta} \right|_{N,V} = -N \left. \frac{\partial \ln z}{\partial \beta} \right|_{N,V} = -N \frac{\partial \ln(1/\beta)^{\frac{3}{2}}}{\partial \beta} = \frac{3}{2} N k_B T, \qquad (4.35)$$

---

[16]The entropy (4.23) can also be calculated and compared to its microcanonical expression (3.22) (see Exercise 4.2).

from which the heat capacity at constant volume is deduced: $C_V = \frac{\partial \langle E \rangle}{\partial T}\big|_{N,V} = \frac{3}{2}Nk_B$. Similarly, the ideal gas pressure (Equation (4.19)) is:

$$P = \frac{1}{\beta}\frac{\partial \ln Z}{\partial V}\bigg|_{N,T} = \frac{N}{\beta}\frac{\partial \ln z}{\partial V}\bigg|_{N,T} = \frac{N}{\beta}\frac{\partial \ln V}{\partial V} = \frac{Nk_BT}{V}, \tag{4.36}$$

which is naturally the ideal gas equation of state[17]. Finally, the chemical potential is given by Equation (4.20). Using Stirling's formula (A.21):

$$\begin{aligned}
\mu &= -\frac{1}{\beta}\frac{\partial \ln Z}{\partial N}\bigg|_{V,T} = -\frac{1}{\beta}\frac{\partial\left(N\ln z - \ln(N!)\right)}{\partial N}\bigg|_{V,T} \\
&\simeq -\frac{1}{\beta}\frac{\partial\left(N\ln(V/\lambda_T^3) - N\ln N + N\right)}{\partial N}\bigg|_{V,T} \\
&\simeq -\frac{1}{\beta}\left(\ln\left(V/\lambda_T^3\right) - \ln N\right) \\
&\simeq k_BT\ln\rho\lambda_T^3, \tag{4.37}
\end{aligned}$$

where $\rho = N/V$. Expressions (3.45), (3.46) and (3.47) of these thermodynamic quantities[18] calculated in the microcanonical ensemble are recovered.

4.1.4   THERMODYNAMIC EQUILIBRIUM AND MINIMUM FREE ENERGY

Like energy, any internal variable A is characterised by its probability distribution function $P(A)$ in equilibrium which, according to Equation (1.32), takes the general form[19]:

$$P(A) = \sum_{\{m|A_m=A\}} p_m = \frac{1}{Z(N,V,T)} \sum_{\{m|A_m=A\}} e^{-\beta E_m}, \tag{4.38}$$

that can be written as:

$$P(A) = \frac{1}{Z(N,V,T)}\,e^{-\frac{F_p(N,V,T\,|\,A)}{k_BT}}, \tag{4.39}$$

with $F_p(N,V,T\,|\,A) \equiv -k_BT\ln\left(\sum_{\{m|A_m=A\}} e^{-\beta E_m}\right)$, where the sum is performed on the microstates m such that $A_m = A$. The function $F_p(N,V,T\,|\,A)$ is by definition the partial free energy associated with the internal variable A. If A is fixed and becomes an external parameter, then $F_p(N,V,T\,|\,A)$ is the free energy of the system. When the internal variable A is macroscopic, its relative fluctuations are negligible (of the order of 10^{-12}). The distribution $P(A)$ then behaves like a Gaussian distribution (2.32), extremely peaked around its most probable value $A_{eq} = \langle A \rangle$, that is its equilibrium value. As Expression (4.39) of $P(A)$ shows,

---

[17]From Equation (4.35), $Nm\langle v^2 \rangle/2 = 3Nk_BT/2$. With Equation (4.36), Relation (1.19), obtained using the very simple model described in Section 1.4.2, is recovered by changing $\bar{v}^2$ to $\langle v^2 \rangle$.

[18]Similarly, the reader can calculate the entropy using Equation (4.23) and easily recover the Sackur-Tetrode formula (3.29).

[19]In the framework of classical physics, the probability density $f(A)$ to find the (continuous) internal variable value lying between A and $A + dA$ is, according to Equation (1.35):

$$P(A) = f(A)dA = \int_{[A,A+dA]} \rho(\boldsymbol{q},\boldsymbol{p})\,d^f q\,d^f p = \frac{1}{Z(N,V,T)}\int_{[A,A+dA]} e^{-\beta H(\boldsymbol{q},\boldsymbol{p})}\,\frac{d^f q\,d^f p}{h^f},$$

where the integral covers all microstates such that the value of the internal variable lies between A and $A + dA$. Remember that a probability being dimensionless, the probability density $f(A)$ has the inverse dimension of the quantity A.

the equilibrium value of a macroscopic internal variable, and more generally the system's equilibrium macrostate correspond to the minimum of F_p.

From the principle of maximum entropy for isolated systems, the following principle follows: *the equilibrium macrostate of a closed system of fixed volume in contact with a thermostat is the one that has the minimum (partial) free energy*. The thermodynamic free energy, which, in principle, has meaning only in equilibrium, is equal to the minimum partial free energy: $F(N,V,T) = F_p(N,V,T \mid A_{eq})$. Since $F = E - TS$, the equilibrium of a system in contact with a thermostat results from a compromise between a low internal energy and a high entropy.

Out-of-equilibrium exchange direction

To illustrate the principle of minimum free energy, consider a macroscopic system $\mathcal{A}_{tot}$ in contact with a thermostat at temperature T. This system is composed of two subsystems $\mathcal{A}$ and $\mathcal{A}'$, also macroscopic, and separated by a wall. Initially, $\mathcal{A}$ and $\mathcal{A}'$ are isolated from each other, they have the same temperature T and their pressures and chemical potentials are nearly equals ($P \simeq P'$ and $\mu \simeq \mu'$). Internal constraints are removed (the wall becomes mobile and permeable to particles) and the two subsystems are now in mechanical and chemical contact. Free energy being additive (in the weak coupling approximation):

$$F_{p_{tot}}(N_{tot}, V_{tot}, T \mid N, V) \simeq F_p(N, V, T) + F_p'(N_{tot} - N, V_{tot} - V, T).$$

Using Definitions (4.19) and (4.20) of pressure and chemical potential, the first order variation of $F_{p_{tot}}$ is given by:

$$\delta F_{p_{tot}} \simeq \left[\frac{\partial F}{\partial V} - \frac{\partial F'}{\partial V'} \right] \delta V + \left[\frac{\partial F}{\partial N} - \frac{\partial F'}{\partial N'} \right] \delta N$$
$$\simeq -\left[P - P' \right] \delta V + \left[\mu - \mu' \right] \delta N \ < 0,$$

because the free energy variation $\delta F_{p_{tot}}$ of system $\mathcal{A}_{tot}$ must be negative (mechanical and chemical equilibrium conditions are obtained for $\delta F_{p_{tot}} = 0$, i.e. $P = P'$ and $\mu = \mu'$). If the contact between the two subsystems is:

- Mechanical such that thermal equilibrium is already established (mobile and impermeable wall, such as $\delta N = 0$): if $P > P'$, then $\delta V > 0$, the volume of subsystem $\mathcal{A}$ increases and its pressure decreases.

- Chemical such that thermal equilibrium is already established (fixed and permeable wall, such as $\delta V = 0$): if $\mu > \mu'$, then $\delta N < 0$, subsystem $\mathcal{A}$ loses particles and its chemical potential decreases.

Of course, the irreversible behaviours evidenced in the microcanonical ensemble (Section 3.5.1) are recovered.

4.1.5 EQUIPARTITION OF ENERGY THEOREM

In the framework of classical physics, there is a property which, because of its general and historical character, is called the *equipartition of energy theorem*[20]: *the average value of a classical Hamiltonian term, independent and of quadratic form, is equal to $k_B T/2$.*

---

[20] The history of this theorem is typical of that which led from the kinetic theory of gases to statistical physics: it originated in 1843 in the long-ignored work of the Scottish physicist John James Waterston. In 1860, Maxwell formulated it in the particular case of the mean kinetic energy of a gas. Finally, in 1868, Boltzmann proved the theorem in a general and more rigorous framework (see C. Cercignani, *Ludwig Boltzmann: the man who trusted atoms* (Oxford University Press, 1998)).

Demonstration: consider a classical system described with f degrees of freedom whose Hamiltonian has the following general form:

$$H(\{q_i, p_i\}) = aq_1^2 + b,$$

where $a(q_2, q_3, \ldots, p_f) > 0$ and $b(q_2, q_3, \ldots, p_f)$ are two functions of the microscopic variables independent of the generalised coordinate q_1. In other words, variable q_1 intervenes in the Hamiltonian *only* as a quadratic term in q_1^2. The demonstration is identical for a generalised momentum p_1 present in an independent quadratic term, which is, of course, the case of the kinetic energy, $p_1^2/2m$, associated to the momentum p_1 (when in general the potential energy is independent of the momenta). According to Equation (4.16), the average value of the independent term aq_1^2 is given by:

$$\langle aq_1^2 \rangle = \frac{\int dq_1 dq_2 \ldots dp_f\, (aq_1^2) e^{-\beta\left(aq_1^2 + b\right)}}{\int dq_1 dq_2 \ldots dp_f\, e^{-\beta\left(aq_1^2 + b\right)}}$$

$$= \frac{\int dq_2 \ldots dp_f\, e^{-\beta b} \overbrace{\left(\int dq_1\, aq_1^2 e^{-\beta aq_1^2}\right)}^{=I_2}}{\int dq_2 \ldots dp_f\, e^{-\beta b} \underbrace{\left(\int dq_1\, e^{-\beta aq_1^2}\right)}_{=I_0}},$$

where, thanks to the change of variable $x = q_1 \sqrt{\beta}$,

$$I_0(q_2, \ldots, p_f) = \int dq_1\, e^{-\beta aq_1^2} = \frac{1}{\sqrt{\beta}} \int dx\, e^{-ax^2}$$

$$I_2(q_2, \ldots, p_f) = \int dq_1\, aq_1^2 e^{-\beta aq_1^2} = -\frac{\partial I_0}{\partial \beta} = \frac{I_0}{2\beta},$$

the function $a(q_2, \ldots, p_f)$ being independent of q_1 and positive, to ensure the convergence of integrals I_0 and I_2. Thus,

$$\langle aq_1^2 \rangle = \frac{\int dq_2 \ldots dp_f\, e^{-\beta b}\, \dfrac{I_0(q_2, \ldots, p_f)}{2\beta}}{\int dq_2 \ldots dp_f\, e^{-\beta b}\, I_0(q_2, \ldots, p_f)} = \frac{1}{2\beta} = \frac{k_B T}{2}.$$

Therefore, in all generality, the average value $\langle aq_1^2 \rangle$ (and $\langle ap_1^2 \rangle$) depends *only* on temperature and not on the form of a, which can be a constant or more generally a function of the other microscopic variables q_2, q_3, $\ldots$, p_f. Let us insist on the fact that this theorem is only true within the framework of *classical physics*, as the demonstration clearly shows. In the following some applications are given.

Kinetic energy of a simple monoatomic fluid

The main application of the equipartition of energy theorem is the determination – without any calculation! – of the mean value of the kinetic energy $\langle E_{kin} \rangle$ of a real fluid of N interacting[21] particles with *no internal structure*. In this very general case, the kinetic energy

---

[21]and in the absence of a magnetic field so that the potential energy is independent of momenta.

is written as the sum of $3N$ independent quadratic terms, $p_i^2/2m$ with $i = 1, 2, \ldots, 3N$ (in three dimensions). Immediately:

$$\langle E_{\text{kin}} \rangle = 3N \frac{k_{\text{B}}T}{2}. \tag{4.40}$$

Let us insist on the generality of this result, valid for any fluid and any form of the interaction potential between the particles[22]. For an *ideal* monoatomic gas, $\langle E_{\text{kin}} \rangle$ is equal to the *total* average energy of the system: each degree of freedom contributes equally to the average energy, hence the name of the equipartition of energy theorem. From Equation (4.40) the molar heat capacity of an *ideal* monoatomic gas can be deduced: $C_V = 3R/2$, which is in very good agreement with experimental measurements on monoatomic gases at room temperature, such as noble gases (He, Ne, Ar...).

Heat capacity of polyatomic gases

Since there are three degrees of freedom per atom, the microscopic state of a diatomic (O_2, HCl...), triatomic (H_2O, CO_2...), or more generally n-atomic molecules is characterised by $3n$ degrees of freedom: 3 so-called *translational* degrees of freedom, associated with the position of the centre of mass and $3n-3$ internal degrees of freedom[23] describing the rotations of the molecule and its vibration modes.

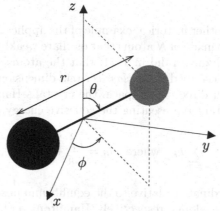

FIGURE 4.2 A diatomic molecule (such as CO) has two degrees of freedom for rotation (described by the angles θ and ϕ) and one for vibration (along r).

Thus, the six degrees of freedom of a diatomic molecule can be broken down as follows (see Figure 4.2): three for translation (position of the centre of mass: x, y and z), two for rotation (orientation of the axis of the molecule: angles θ and ϕ), and one for the bond vibration (distance between the two atoms: r). Assuming that the oscillation amplitudes of the relative motion of the two atoms, of mass m_1 and m_2, are small, the chemical bond can be modelled by a *harmonic* oscillator. The diatomic molecule Hamiltonian is then in the

---

[22]For a single particle, $m\langle v^2 \rangle/2 = 3k_{\text{B}}T/2$. Relation (1.20), obtained thanks to a very simple model described in Section 1.4.2, is recovered by substituting $\tilde{v}^2$ for $\langle v^2 \rangle$.

[23]For a linear molecule (e.g. CO_2), there are 2 rotational degrees of freedom and $3n-5$ vibrational degrees of freedom. For a non-linear molecule (H_2O for example), there are 3 rotational degrees of freedom and $3n-6$ vibrational degrees of freedom.

framework of classical physics[24]:

$$h_j = \frac{p_x^2}{2m} + \frac{p_y^2}{2m} + \frac{p_z^2}{2m} + \underbrace{\frac{p_\theta^2}{2I} + \frac{p_\phi^2}{2I \sin^2 \theta}}_{} + \underbrace{\frac{p_r^2}{2\mu} + \frac{1}{2}\mu\omega^2(r - r_0)^2}_{}, \qquad (4.41)$$

$$\underbrace{\phantom{\frac{p_x^2}{2m} + \frac{p_y^2}{2m} + \frac{p_z^2}{2m}}}_{\text{translation}} \qquad \underbrace{}_{\text{rotation}} \qquad \underbrace{}_{\text{harmonic vibration}}$$

where $m = m_1 + m_2$ is the total mass of the diatomic molecule, I is its moment of inertia, $\mu = m_1 m_2/(m_1 + m_2)$ is its reduced mass, ω is the angular frequency of the harmonic oscillator and r_0 is the equilibrium distance between the two atoms. Note that since the degrees of freedom are independent, the partition function of a molecule can be written as $z = z_{\text{trans}} \, z_{\text{rot}} \, z_{\text{vib}}$, where z_{trans}, z_{rot} and z_{vib} are the translational, rotational and vibrational partition functions, respectively.

Moreover, Hamiltonian (4.41) is written as the sum of seven independent quadratic terms[25]. Be careful not to confuse the number of degrees of freedom (here equal to 6) and the number of independent quadratic terms (here equal to 7). According to the equipartition of energy theorem and neglecting the interactions between molecules, the average total energy of an ideal gas of N diatomic molecules is $\langle E \rangle = 7N \, k_\mathrm{B}T/2$, that is a molar heat capacity, $C_V = 7R/2$. However, this estimate is in disagreement with the experimental value, $C_V \simeq 5R/2$, measured at room temperature as early as the nineteenth century.

Heat capacity of solids

Heat capacity of solids is another historical example of the application of the equipartition of energy theorem. A crystal is made of N atoms that oscillate weakly around their equilibrium positions. Assuming, as Boltzmann did in 1871, that the atoms' motions are independent of each other and can be described by a *classical* one-dimensional harmonic oscillator of angular frequency ω in each direction of space, the crystal's Hamiltonian has Form (1.8), to within one additive constant representing the cohesive energy of the system:

$$H = \sum_{i=1}^{3N} h_i \quad \text{where} \quad h_i = \frac{p_i^2}{2m} + \frac{1}{2}m\omega^2 q_i^2, \qquad (4.42)$$

where q_i and p_i are the coordinate (relative to the equilibrium position) and the momentum associated to the degree of freedom i, respectively. Hamiltonian (4.42) is therefore written as the sum of $6N$ independent quadratic terms (in p_i^2 and q_i^2). According to the equipartition of energy theorem, the crystal average energy is thus $\langle E \rangle = 6Nk_\mathrm{B}T/2$. Its molar heat capacity follows the Dulong and Petit law, $C_V = 3R$, experimentally demonstrated in 1819 by these two French physicists[26]. This success of statistical physics, demonstrating an empirical macroscopic law from the atomic scale, was however qualified by Boltzmann himself. Some solids, such as silicon or carbon (diamond) crystals, have a molar heat capacity well below $3R \simeq 25$ J·mol^{-1}·K^{-1} at room temperature (19.8 J·mol^{-1}·K^{-1} for Si and 8.5 J·mol^{-1}·K^{-1} for C). Moreover, experimental data at low temperature show that C_V increases with T,

---

[24]In reality, the vibrations modify the distance r between the two atoms and thus the moment of inertia I of the molecule, but this coupling effect between vibrational and rotational motions is very weak and can usually be neglected. The rotational kinetic energy describes a rigid rotator of moment of inertia I. Using Hamilton's Equations (1.11), $p_\theta = I\dot{\theta}$ and $p_\phi = I\dot{\phi} \sin^2 \theta$ (kinetic moments), that is the kinetic energy $E_{\text{kin}} = \frac{I}{2}(\dot{\theta}^2 + (\dot{\phi} \sin \theta)^2)$.

[25]Since the angle θ is only involved in the fifth term (in p_ϕ^2), this quadratic term is indeed independent.

[26]This empirical law allowed the first estimates of the molar mass M of solid substances, deduced from the measurement at room temperature of the mass heat capacity C_m (in J·g^{-1}·K^{-1}), since $MC_m \simeq 25$ J·mol^{-1}·K^{-1} (i.e. $3R$), for a large number of chemical elements (Mg, Al, Fe, Cu...).

in contradiction with the Dulong and Petit law and thus with the equipartition of energy theorem[27] (see exercise 4.3).

Quantum explanations

The flagrant disagreement between classical statistical theory and experimental measurements of heat capacities of diatomic gases and some specific solids was widely commented by the scientists by the end of the nineteenth century. This failure of statistical physics seemed to be irremediable. As a sign of the crisis that physics was going through at the turn of the century, before the advent of special relativity and quantum mechanics, Boltzmann tried in vain to explain this contradiction by invoking the interactions between atoms and the "ether", a fictitious substance supposed to transmit electromagnetic forces and waves (L. Boltzmann, *On certain questions of the theory of gases*, Nature **51**, 413 (1895)).

It is Einstein, in 1907, who brought the *quantum* solution to this problem of (classical) statistical physics: the vibrational degrees of freedom, of a diatomic molecule or of an atom in a crystal, must be treated in the framework of quantum physics. More generally, when the thermal energy $k_B T$ is small compared to the typical gap between energy levels (which depends on the considered system) the associated degrees of freedom are frozen. In other words, they play no role in the characterisation of the microscopic state of the particles. Thus, it will be shown in Section 6.5, that at room temperature, thermal energy is large enough to activate the rotational motions of the diatomic molecules, correctly described by classical physics, while the molecules do not (practically) vibrate. Since the degrees of freedom associated with vibrations are frozen, only five independent quadratic terms remain in the classical Hamiltonian (4.41), which, according to the equipartition of energy theorem, gives a molar heat capacity equal to $5R/2$, in very good agreement with experiments. Similarly, the quantum treatment of vibrational degrees of freedom explains the temperature dependence of the heat capacity of solids, as it will be illustrated by the Einstein model (Exercise 4.3) and the Debye model (Section 7.1).

Generalised theorem and virial theorem

The equipartition of energy theorem was generalised in 1918 by Richard Tolman[28]. Let w_l and w_n be two microscopic variables that can be generalised coordinates or momenta, so:

$$\left\langle w_l \frac{\partial H}{\partial w_n} \right\rangle = \frac{1}{Z} \int \left(w_l \frac{\partial H}{\partial w_n} \right) e^{-\beta H} \, d^f q d^f p.$$

Since $\frac{\partial \left(w_l e^{-\beta H} \right)}{\partial w_n} = \delta_{ln} e^{-\beta H} - \beta w_l \frac{\partial H}{\partial w_n} e^{-\beta H}$, with $\delta_{ln} = 1$ if $l = n$ and 0 otherwise, then:

$$
\begin{aligned}
\left\langle w_l \frac{\partial H}{\partial w_n} \right\rangle &= \frac{\delta_{ln}}{\beta} \underbrace{\frac{1}{Z} \int e^{-\beta H} \, d^f q d^f p}_{Z} - \frac{1}{\beta Z} \int \frac{\partial \left(w_l e^{-\beta H} \right)}{\partial w_n} \, d^f q d^f p \\
&= \delta_{ln} k_B T - \frac{1}{\beta Z} \int d^f q' d^f p' \left[w_l e^{-\beta H} \right]_{w_n^{\min}}^{w_n^{\max}},
\end{aligned}
$$

---

[27]Note that Dulong and Petit Law violates the third Law of thermodynamics: when $T \to 0$, the entropy $S(T) = \int_0^T \frac{C_V(T)}{T} dT$ tends to zero (or a constant).

[28]American physicist (1881–1948), Einstein's collaborator and author of a textbook on statistical mechanics, *The principles of statistical mechanics* (Oxford University Press, 1938).

where the integral covers all microscopic variables except w_n. In the very common case[29] where $H = +\infty$, for $w_n = w_n^{\mathrm{min}}$ and for $w_n = w_n^{\mathrm{max}}$, the second term is zero and it only remains:

$$\left\langle w_l \frac{\partial H}{\partial w_n} \right\rangle = \delta_{ln}\, k_{\mathrm{B}} T. \tag{4.43}$$

For a quadratic term independent of the Hamiltonian (of the form aw_n^2), the equipartition of energy theorem is recovered: $\langle aw_n^2 \rangle = k_{\mathrm{B}} T/2$. More generally, using Hamilton's Equations (1.11), one obtains for each degree of freedom $i = 1, 2, \ldots, f$:

$$k_{\mathrm{B}} T = \left\langle p_i \frac{\partial H}{\partial p_i} \right\rangle = \langle p_i \dot{q}_i \rangle \quad \text{and} \quad k_{\mathrm{B}} T = \langle q_i \frac{\partial H}{\partial q_i} \rangle = -\langle q_i \dot{p}_i \rangle. \tag{4.44}$$

Moreover, for a system of particles without internal structure and in the absence of a magnetic field: $\dot{q}_i = p_i/m$ and $\dot{p}_i = -\frac{\partial U}{\partial q_i} = F_i$, where F_i is the force associated with coordinate i which derives from the potential U (fundamental principle of dynamics), thus:

$$k_{\mathrm{B}} T = \left\langle \frac{p_i^2}{m} \right\rangle = -\langle q_i F_i \rangle, \quad \text{for} \quad i = 1, 2, \ldots, f.$$

Summing over the $3N$ degrees of freedom, the virial theorem[30] for a macroscopic system in equilibrium at temperature T is recovered:

$$\langle E_{\mathrm{kin}} \rangle = \left\langle \sum_{i=1}^{3N} \frac{p_i^2}{2m} \right\rangle = -\frac{1}{2} \left\langle \sum_{i=1}^{3N} q_i F_i \right\rangle = \frac{3}{2} N k_{\mathrm{B}} T. \tag{4.45}$$

Example 1: Kappler's experiment

A small mirror suspended by a torsion wire is immersed in an box containing air maintained at the temperature T. In equilibrium, molecules impact on the surface of the mirror constantly changing its orientation, given by an angle θ. In 1931, Eugen Kappler measured the very small fluctuations of this angle using a light beam reflected by the mirror. He finds $\langle \theta^2 \rangle = 4.178 \times 10^{-6}$ rad^2 and deduced very easily an estimate of the Boltzmann's constant. Indeed, the elastic torsion potential energy is written $\frac{1}{2} C\theta^2$, where $C = 9.428 \times 10^{-16}$ kg.m$^2 \cdot$s^{-2} is the torsion constant of the wire. It is a quadratic and independent term of the Hamiltonian, so according to the equipartition of energy theorem:

$$\left\langle \frac{1}{2} C\theta^2 \right\rangle = \frac{1}{2} k_{\mathrm{B}} T,$$

that is $k_{\mathrm{B}} = C\langle \theta^2 \rangle / T \simeq 1.372 \times 10^{-23}$ J$\cdot$K^{-1} at temperature $T = 287.1$ K, in very good agreement with the actual value given by Equation (3.11) (the relative error is 0.6%). This is how the microscopic motion of atoms manifests itself at the macroscopic scale[31].

---

[29]When w_n is a momentum the kinetic energy diverges for $w_n^{\mathrm{min}} = -\infty$ and $w_n^{\mathrm{max}} = +\infty$. Likewise, particles contained in a box have a divergent potential energy when in contact with the repulsive walls (see Note 22 in Chapter 1).

[30]The term "virial" is derived from the Latin word *vires* which means forces. In the more general framework of analytical mechanics, as in the original form derived by Clausius in 1870, the virial theorem is expressed in terms of time averages and not ensemble averages. In Equation (4.45), the equality $\langle E_{\mathrm{cin}} \rangle = 3Nk_{\mathrm{B}}T/2$ is valid only for a system in thermodynamic equilibrium. The virial theorem has many applications both in astrophysics and in quantum mechanics (see G. W. Collins, *The virial theorem in stellar astrophysics* (Pachart Pub. House, 1978) and C. Cohen-Tannoudji, B. Diu, F. Laloë, *Quantum Mechanics, Volume 2: Angular Momentum, Spin, and Approximation Methods* (Wiley-VCH, 2019)).

[31]The equipartition theorem is at the heart of Einstein's analysis of Brownian motion in 1905, and then Langevin's in 1908. A mesoscopic particle (of the order of a micron), suspended in a liquid, is subjected

Example 2: Ideal classical gas in a gravity field

The Hamiltonian of an ideal classical gas in a gravity field is:

$$H_{\mathrm{IG}} = \sum_{j=1}^{N} h_j \quad \text{with} \quad h_j = \frac{p_{x_j}^2}{2m} + \frac{p_{y_j}^2}{2m} + \frac{p_{z_j}^2}{2m} + mgz_j,$$

where $z_j \geq 0$ is the altitude of particle j and g the acceleration of gravity. According to the generalised equipartition theorem (4.43), when the gas is in equilibrium at temperature T, one gets for $j = 1, 2, \ldots, N$:

$$\left\langle z_j \frac{\partial H_{\mathrm{IG}}}{\partial z_j} \right\rangle = \langle mgz_j \rangle = k_{\mathrm{B}} T.$$

The average altitude of a particle is therefore $\langle z_j \rangle = k_{\mathrm{B}} T / mg$.

To go beyond the average values, the density of probability $f(w)$ of a microscopic variable w (position or momentum) can be determined. The probability (4.7) that the system is in a given microstate $x_1, y_1, z_1, \ldots, p_{x_N}, p_{y_N}, p_{z_N}$ to within $\mathrm{d}x_1, \mathrm{d}y_1, \mathrm{d}z_1, \ldots, \mathrm{d}p_{x_N}, \mathrm{d}p_{y_N}, \mathrm{d}p_{z_N}$ is:

$$\begin{aligned}
\mathrm{d}P &= \frac{1}{Zh^f} e^{-\beta H_{\mathrm{IG}}(q,p)} \, \mathrm{d}^f q \mathrm{d}^f p \\
&= \frac{1}{Zh^f} e^{-\beta \sum_{j=1}^{N} \left(\frac{p_{x_j}^2}{2m} + \frac{p_{y_j}^2}{2m} + \frac{p_{z_j}^2}{2m} + mgz_j \right)} \mathrm{d}x_1 \mathrm{d}y_1 \mathrm{d}z_1 \ldots \mathrm{d}p_{x_N} \mathrm{d}p_{y_N} \mathrm{d}p_{z_N}.
\end{aligned}$$

The probability density $f(w)$ is obtained by summing the contributions of all accessible microstates such that the value of w is fixed, that is by integrating $\mathrm{d}P$ over all microscopic variables $x_1, y_1, z_1, \ldots, p_{x_N}, p_{y_N}$ and p_{z_N}, except w:

$$f(w)\mathrm{d}w = \int_{\text{except } w} \mathrm{d}P.$$

Thus, the probability that the variable $w = p_{x_1}$ lies between p_x and $p_x + \mathrm{d}p_x$ is:

$$\begin{aligned}
P_x(p_x)\mathrm{d}p_x &= \int_{\text{except } p_{x_1}} \mathrm{d}P = \left(\int \frac{e^{-\beta H_{\mathrm{IG}}}}{Zh^f} \, \mathrm{d}x_1 \ldots \mathrm{d}z_N \mathrm{d}p_{y_1} \mathrm{d}p_{z_1} \ldots \mathrm{d}p_{z_N} \right) \mathrm{d}p_x \\
&= e^{-\beta \frac{p_x^2}{2m}} \, \mathrm{d}p_x \underbrace{\int \frac{e^{-\beta \left(\frac{p_{y_1}^2}{2m} + \cdots + mgz_N \right)}}{Zh^f} \, \mathrm{d}x_1 \ldots \mathrm{d}z_N \mathrm{d}p_{y_1} \mathrm{d}p_{z_1} \ldots \mathrm{d}p_{z_N}}_{=C_1}.
\end{aligned}$$

There is no need to compute integral C_1, since it is independent of variable p_x and is obtained directly with the normalisation condition[32]. Thus:

$$P_x(p_x)\mathrm{d}p_x = \frac{1}{\sqrt{2\pi m k_{\mathrm{B}} T}} \, e^{-\frac{p_x^2}{2m k_{\mathrm{B}} T}} \, \mathrm{d}p_x.$$

In the canonical ensemble the Maxwell velocity distribution function is easily recovered (expressed here as a function of momentum). It was obtained in a much more laborious

---

to the incessant collision with the fluid molecules. In equilibrium, the particle's velocity fluctuation (in one dimension) is given by $\langle v_x^2 \rangle = k_{\mathrm{B}} T / m$, but its erratic position is described by $\langle x^2 \rangle = 2Dt$, where D is the diffusion coefficient (see Chapter 15 of R. K. Pathria and P. D. Beale, *Statistical Mechanics*, 3rd edition, (Academic Press, Elsevier, 2012)).

[32] Since $\int P_x(p_x) \, \mathrm{d}p_x = 1$, $\frac{1}{C_1} = \int_{-\infty}^{+\infty} e^{-\beta \frac{p_x^2}{2m}} \, \mathrm{d}p_x = \sqrt{\frac{2m\pi}{\beta}}$ (See Appendix A.2).

way in the microcanonical ensemble (see Exercise 3.6). Finally, by integrating dP over all microscopic variables, except $w = z_1 = z$, the density of probability $B(z)$ of the altitude of a particle is immediatelly obtained:

$$B(z)dz = \int_{\text{except } z_1} dP = C_2\, e^{-\beta mgz}\, dz = \frac{1}{z_c}\, e^{-\frac{z}{z_c}}\, dz,$$

where the constant $C_2 = 1/z_c$ is calculated using the normalisation condition of $B(z)$ and where $z_c = RT/(Mg)$ is a characteristic height. It can be deduced that the gas number density $\rho(z)$ and thus the pressure of the supposedly ideal gas $P(z) = \rho(z)k_B T$ follow the barometric law[33]: at constant temperature, the pressure decreases exponentially with the altitude z. For the Earth's atmosphere, assumed isothermal at $T = 15\ °C$, z_c is about 8 km (for air $M \simeq 29$ g $\cdot$ mol^{-1}). The effects of gravity are not observable at low altitudes with respect to z_c, as is the case of a container of a few litres filled with an atomic gas. On the other hand, for heavier particles, $z_c \propto 1/M$ is smaller. Thus, in 1908, Jean Perrin showed that small grains of resin (of the order of one micron with an effective molar mass $M \simeq 10^{10}$ g $\cdot$ mol^{-1}) suspended in water are distributed in altitude according to the barometric law. The measurement of $z_c \simeq 10\ \mu m$ allowed him to estimate the Boltzmann constant (see M. Horne, *An Experiment to Measure Boltzmann's Constant*, Am. J. Phys. **41**, 344 (1973)).

4.2   GRAND CANONICAL ENSEMBLE

Consider now an open system of fixed volume. The system exchanges not only energy (as in the canonical situation), but also particles with a much larger system. Note that the exchange of particles implies a variation of energy (kinetic and potential). The large system then acts as a reservoir of particles and energy (thermostat) for the small system. This is typically the situation of a small volume V, delimited by an imaginary wall, within a macroscopic system (see Figure 4.3). The same method as in Section 4.1 will be used to determine the probability distribution p_m to find the system in a given microstate m and derive the thermodynamic quantities characterising its macroscopic state.

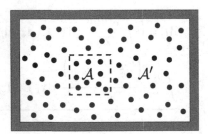

FIGURE 4.3 A subsystem $\mathcal{A}$, of fixed volume V surrounded by fixed diathermic and permeable walls, exchanges energy and particles with a much larger subsystem, the reservoir $\mathcal{A}'$, the whole forming an isolated system.

As in Section 3.1.1, an isolated system $\mathcal{A}_{\text{tot}}$ is composed of two subsystems $\mathcal{A}$ and $\mathcal{A}'$. The latter can exchange particles and energy in the form of heat through a diathermic,

---

[33]This law is easily recovered using the fundamental principle of hydrostatics: $dP = -\rho_m g dz$, where $\rho_m = \rho \times m$ is the mass density of the fluid. Since the gas is ideal, $dP = -(P/k_B T)(M/N_A)g dz$, or by integrating $P(z) = P(0)e^{-z/z_c}$. For the Earth's atmosphere, it is more realistic to take into account the decrease of the temperature with altitude in the troposphere ($0 \le z \le 8 - 15$ km): $T(z) = T_0 - az$, where $a > 0$. Thus:

$$\frac{dP}{P} = -\frac{Mg}{R}\frac{dz}{T_0 - az} \quad \text{so} \quad P(z) = P_0\left(1 - \frac{az}{T_0}\right)^\alpha \quad \text{with} \quad \alpha = \frac{Mg}{aR} \simeq 5.3,$$

where $a \simeq 6.5$ K$\cdot$km^{-1}, $P_0 = 1$ atm and $T_0 = 288$ K, the mean temperature at Earth's surface.

permeable and fixed wall (no work exchange). The volume V is thus an external parameter. In the weak coupling approximation, the energy E and the number of particles N of $\mathcal{A}$ behave as internal variables, free to fluctuate, such that:

$$
\begin{aligned}
E_{\text{tot}} &= E + E' \\
N_{\text{tot}} &= N + N'.
\end{aligned}
$$

In equilibrium, the probability p_m to find subsystem $\mathcal{A}$ in a given microstate m of energy $E_m = E$ with $N_m = N$ particles is given by:

$$
p_m = \frac{\Omega'(E_{\text{tot}} - E_m, N_{\text{tot}} - N_m)}{\Omega_{\text{tot}}(E_{\text{tot}}, N_{\text{tot}})} = C \, e^{\frac{S'_p(E_{\text{tot}} - E_m, N_{\text{tot}} - N_m)}{k_B}}, \tag{4.46}
$$

where $C = 1/\Omega_{\text{tot}}(E_{\text{tot}}, N_{\text{tot}})$ is a constant and $S'_p = k_B \ln \Omega(E_{\text{tot}} - E_m, N_{\text{tot}} - N_m)$ is the partial entropy of subsystem $\mathcal{A}'$. As in the canonical situation, subsystem $\mathcal{A}$ is assumed to be much smaller than subsystem $\mathcal{A}'$, that is $f \ll f'$. Since the energy and the number of particles are extensive quantities, $E_m \ll E' = E_{\text{tot}} - E_m$ and $N_m \ll N' = N_{\text{tot}} - N_m$, for all microstates m of $\mathcal{A}$. Subsystem $\mathcal{A}'$ then behaves as a reservoir of energy and particles for $\mathcal{A}$, which can be microscopic or macroscopic (as long as $f \ll f'$).

4.2.1   GRAND CANONICAL PROBABILITY DISTRIBUTION AND GRAND PARTITION FUNCTION

Since $f \ll f'$, the entropy of the reservoir $S'_p(E_{\text{tot}} - E_m, N_{\text{tot}} - N_m)$ can be expressed as a power series of $E_m \ll E_{\text{tot}}$ and $N_m \ll N_{\text{tot}}$:

$$
\begin{aligned}
S'_p(E_{\text{tot}} - E_m, N_{\text{tot}} - N_m) &= \underbrace{S'_p(E_{\text{tot}}, N_{\text{tot}})}_{\sim k_B f'} - \underbrace{E_m \underbrace{\frac{\partial S'_p}{\partial E'}}_{=1/T'}}_{\sim k_B f} - \underbrace{N_m \underbrace{\frac{\partial S'_p}{\partial N'}}_{=-\mu'/T'}}_{\sim f} \\
&\quad + \underbrace{\frac{E_m^2}{2} \frac{\partial^2 S'_p}{\partial E'^2}}_{\sim k_B \frac{f^2}{f'}} + \underbrace{\frac{N_m^2}{2} \frac{\partial^2 S'_p}{\partial N'^2}}_{\sim k_B \frac{f^2}{f'}} + \underbrace{E_m N_m \frac{\partial^2 S'_p}{\partial N' \partial E'}}_{\sim k_B \frac{f^2}{f'}} + \cdots \\
&\simeq S'_p(E_{\text{tot}}, N_{\text{tot}}) - \frac{E_m}{T'} + \frac{\mu'}{T'} N_m,
\end{aligned}
$$

according to Definitions (3.15) and (3.37) of temperature and chemical potential in the microcanonical ensemble. The extensive nature of energy, particle number and entropy were used to estimate the orders of magnitude of the various terms in the series expansion and neglect those of order greater than one[34]. Therefore, according to Equation (4.46):

$$
p_m = \underbrace{C \, e^{\frac{S'_p(E_{\text{tot}}, N_{\text{tot}})}{k_B}}}_{=1/\Xi} \, e^{-\frac{E_m}{k_B T'} + \frac{N_m \mu'}{k_B T'}},
$$

where the prefactor, written $1/\Xi$, is independent of the microstate m of system $\mathcal{A}$.

By definition, the grand canonical ensemble[35] is a statistical ensemble of open systems of volume V in thermal and chemical equilibrium with an energy and particles reservoir at temperature $T' = T$ and chemical potential $\mu' = \mu$. The external parameters of the grand

---

[34]See Note 3.

[35]Again, the terminology is due to Gibbs who named (in French in the text) "grands ensembles", statistical ensembles for which the number of particles is an internal variable, in contrast with microcanonical and canonical ("petits (small) ensembles"). Indeed a grand ensemble is made of many small ensembles as shown by Equation (4.49) (see J. W. Gibbs, *Elementary principles in statistical mechanics* (Dover, 2015)).

canonical ensemble are thus μ, V and T, while the energy and the number of particles are here internal variables, in contrast to the microcanonical situation. The probability to find a system in equilibrium in a given microstate m of energy E_m and containing N_m particles is given by the grand canonical probability distribution:

$$p_m = \frac{1}{\Xi} \, e^{-\beta\left(E_m - \mu N_m\right)}, \tag{4.47}$$

where $\beta = 1/k_B T$. The quantity Ξ is the *grand partition function* (or grand canonical partition function) of the system and is by definition (according to the normalisation condition $\sum_m p_m = 1$):

$$\Xi(\mu, V, T \ldots) \equiv \sum_{\text{microstates } m} e^{-\beta\left(E_m - \mu N_m\right)}, \tag{4.48}$$

where the sum covers all microstates m, for any energy E_m and any particle number of particles N_m compatible with the studied model. In some cases, the system can only accept one particle at most, so $N_m = 0$ or 1 (see Example 1, as follows). In other situations, such as the ideal gas (see Example 2, as follows), N_m is unbounded[36]. The grand partition function $\Xi(\mu, V, T \ldots)$ is a dimensionless quantity that depends only on μ, V, T and possibly other external parameters. Moreover, $\Xi(\mu, V, T \ldots)$ can be expressed using the canonical partition function by summing over all possible values of N:

$$\Xi(\mu, V, T) = \sum_{N \geq 0} \left(\sum_{\{m \mid N_m = N\}} e^{-\beta E_m} \right) e^{\beta \mu N} = \sum_{N \geq 0} Z(N, V, T) \, e^{\beta \mu N}. \tag{4.49}$$

Note that the first term of the sum (i.e. for $N = 0$ and thus $E = 0$ and $Z = 1$) is equal to 1. In practice, it will be seen that it is sometimes easier to work with the grand partition function $\Xi(\mu, V, T)$ than with the partition function $Z(N, V, T)$, although according to Equation (4.49), the latter seems essential to calculate the former. The technical advantage of the grand canonical ensemble is that the number of particles is no longer fixed, which relaxes a sometimes embarrassing constraint in the calculations. The use of the grand canonical ensemble is obvious when the system is open and exchanges particles with a reservoir. The phenomenon of adsorption illustrates this situation very clearly (see Exercises 4.11 and 4.12): the molecules of a gas, which acts as a reservoir, are trapped on the surface of a solid in a reversible way. In equilibrium, the number of molecules adsorbed on the substrate fluctuates over time. The grand canonical statistical ensemble also proves to be very effective for studying interacting particle systems, in particular when two thermodynamic phases exchange particles, as is the case with the coexistence of a liquid and a gas in equilibrium (see Section 5.3.2). Finally, the grand canonical ensemble is essential when particles obey quantum statistics, as it will be seen in Chapter 6.

In the framework of classical physics, identical particles are necessarily considered *indistinguishable*, since the label (see Section 2.2.3) of particles exchanged with the reservoir has no bearing on the microstate of the system. The grand partition function is then expressed using Equation (4.49) using the classical expression (4.6) for $Z(N, V, T)$ for N indistinguishable particles (see Example 2, as follows).

It is important to understand that in the grand canonical ensemble, a set of particles, or even a single particle, cannot be chosen as the studied system, since in this statistical ensemble the system's number of particles is not fixed. A system studied using the grand canonical ensemble is by definition associated with a region of space, a set of sites or a given

---

[36]In reality, the contact repulsion between molecules limits *a priori* the value of N_m in a finite volume V. However, in a dense medium the probability p_m decreases rapidly with N_m, because the energy E_m becomes strongly positive, even infinite. In order to facilitate the calculations, it is therefore generally assumed that the number of particles contained in a macroscopic volume is not bounded, even if the gas is not ideal.

volume, which can accommodate a fluctuating number of indistinguishable particles, as the following two examples show. Let us insist in the grand canonical ensemble, systems can be distinguishable or not, however, the particles are always indistinguishable.

Set of independent systems

When a system consists of independent (or loosely coupled) subsystems, the grand partition function of the system is the product of the grand partition functions associated with each of the subsystems. For simplicity, consider two independent subsystems $\mathcal{A}$ and $\mathcal{A}'$. Since $E_{\text{tot}} = E_m + E'_{m'}$ and $N_{\text{tot}} = N_m + N'_{m'}$, the grand partition function of the total system is:

$$
\begin{aligned}
\Xi_{\text{tot}} &= \sum_{m,m'} e^{-\beta\left(E_{\text{tot}}-\mu N_{\text{tot}}\right)} = \sum_{m,m'} e^{-\beta\left((E_m+E'_{m'})-\mu(N_m+N'_{m'})\right)} \\
&= \sum_m e^{-\beta\left(E_m-\mu N_m\right)} \sum_{m'} e^{-\beta\left(E'_{m'}-\mu N'_{m'}\right)} = \Xi\,\Xi',
\end{aligned}
\tag{4.50}
$$

where Ξ and Ξ' are the grand partition functions of subsystems $\mathcal{A}$ and $\mathcal{A}'$, respectively. The result is easily generalised to an arbitrary number of independent subsystems.

Exemple 1: Lattice-gas model

In the lattice-gas model, an atomic gas is contained in a macroscopic volume V, which is thoughtfully divided into small cubic cells of volume v_0 (of the order of magnitude of an atomic volume). Each cell can contain at most[37] one atom (see Figure 4.4)). This model takes therefore into account the contact repulsion between atoms: it is a hard sphere model (and not an ideal gas model).

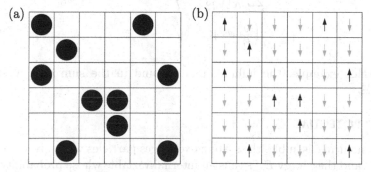

FIGURE 4.4 (a) Gas on a two-dimensional lattice. Each cell i is occupied by $n_i = 0$ or 1 particle. This system is equivalent to the Ising model (b) for which a site i is occupied by a spin $s_i = 2n_i - 1$ equal to -1 or $+1$ (see Section 9.1.1).

If attractive interactions between atoms occupying two neighbour cells are neglected, the system behaves as a set of $N_0 = V/v_0$ distinguishable and independent cells in contact with a reservoir at temperature T and chemical potential μ. According to Equation (4.50), the system's grand partition function is:

$$
\Xi(\mu, V, T) = \xi^{N_0},
$$

---

[37]In contrast to the microcanonical and canonical cases, the ideal paramagnetic crystal is not taken as the first example of application. Indeed, in a crystal, the number of atoms (and consequently of spins) is fixed.

where ξ is the grand partition function of a single cell:

$$\xi = \sum_{n_1=0,1} e^{-\beta\left(e_1-\mu n_1\right)} = 1 + e^{\beta\mu},$$

where n_1 is the occupancy rate of the cell (i.e. 0 or 1) depending on whether it contains an atom or not. The energy e_1 of a cell is zero in both cases, since this model does not explicitly take into account the atoms' kinetic energy. The state of a cell is therefore described by a single microscopic variable that takes two values ($n_1 = 0$ or 1), just like a spin-$\frac{1}{2}$ (see Section 9.1.1) or like any system in which particles can only be in two different states. Finally,

$$\Xi(\mu,V,T) = \left(1 + e^{\beta\mu}\right)^{N_0}. \tag{4.51}$$

Note that the number of cells $N_0 = V/v_0$ is *rigorously fixed*, while the number of particles fluctuates according to whether a cell contains an atom or not.

Example 2: Ideal classical gas

According to Equation (4.49), the grand partition function of an ideal classical gas can be derived from the canonical partition function (Equation (4.11)) for indistinguishable atoms. In the absence of interactions, the system of volume V can contain an infinite number of atoms, so[38]:

$$\begin{aligned}
\Xi(\mu,V,T) &= \sum_{N=0}^{\infty} Z(N,V,T)\, e^{\beta\mu N} = \sum_{N=0}^{\infty} \frac{1}{N!}\left(\frac{V}{\lambda_T^3}\right)^N e^{\beta\mu N} \\
&= \sum_{N=0}^{\infty} \frac{1}{N!}\left(\frac{V}{\lambda_T^3}\, e^{\beta\mu}\right)^N \\
&= e^{\left(\frac{V}{\lambda_T^3}\, e^{\beta\mu}\right)}. \tag{4.52}
\end{aligned}$$

Note how, in this example, the infinite upper bound in the sum over N simplifies the calculation of Ξ.

4.2.2 GRAND POTENTIAL

In the grand canonical ensemble, the system exchanges particles with a reservoir. The particles number N (and the energy E) is thus an internal variable whose probability distribution $P(N)$ is (according to Equation (4.47)):

$$\begin{aligned}
P(N) &= \sum_{\{m|N_m=N\}} p_m = \frac{1}{\Xi(\mu,V,T)} \sum_{\{m|N_m=N\}} e^{-\beta\left(E_m-\mu N_m\right)} \\
&= \frac{1}{\Xi(\mu,V,T)}\, e^{\beta\mu N} \sum_{\{m|N_m=N\}} e^{-\beta E_m} \\
&= \frac{1}{\Xi(\mu,V,T)}\, e^{\beta\mu N}\, Z(N,V,T) \\
&= \frac{1}{\Xi(\mu,V,T)}\, e^{-\beta\left(F(N,V,T)-\mu N\right)}, \tag{4.53}
\end{aligned}$$

[38]Using power series expansion of the exponential function, $\forall\, x \in \mathbb{C},\ \sum_{N=0}^{\infty} \frac{x^N}{N!} = e^x.$

where $F(N,V,T) = -k_B T \ln Z(N,V,T)$ is the system's free energy when it is closed and it contains N particles at temperature T. According to Equation (4.49), $P(N)$ is obviously normalised.

In the case of a macroscopic system, N is distributed (as any macroscopic internal variable) according to an extremely sharp Gaussian distribution centred in $N = \langle N \rangle$, the most probable value, interpreted as the particles number equilibrium value. So $P(N \simeq \langle N \rangle)$ is practically equal to 1 and the logarithm of Equation (4.53) leads to:

$$\ln P(\langle N \rangle) = -\ln \Xi(\mu, V, T) - \beta \Big(F(\langle N \rangle, V, T) - \mu \langle N \rangle \Big) \simeq 0,$$

where $F(\langle N \rangle, V, T)$ is the free energy of the system in equilibrium, that is for $N = \langle N \rangle$. The *grand potential* $J(\mu, V, T)$ can be introduced. It is defined as:

$$J(\mu, V, T) \equiv -k_B T \ln \Xi(\mu, V, T) \simeq F(\langle N \rangle, V, T) - \mu \langle N \rangle, \qquad (4.54)$$

which can also be derived from the thermodynamic free energy by a Legendre transformation, $J = F - \mu \langle N \rangle = \langle E \rangle - TS - \mu \langle N \rangle$, allowing the substitution of the external parameters (N, V, T) for (μ, V, T). The grand potential is an extensive thermodynamic quantity, homogeneous to an energy, which characterises a macroscopic system in equilibrium. Like the grand partition function, $J(\mu, V, T \dots)$ depends only on external parameters. According to Equation (4.50), $J(\mu, V, T)$ is an additive quantity in the case of independent or weakly coupled systems. Like entropy for the microcanonical ensemble and free energy for the canonical ensemble, the grand potential is the thermodynamic potential adapted to the external parameters of the grand canonical ensemble: it allows the determination of every thermodynamic property of the system.

4.2.3 AVERAGE VALUES AND THERMODYNAMIC QUANTITIES

The grand canonical probability distribution (4.47) can be used to express the average value $\langle A \rangle$ of any internal variable A, microscopic or macroscopic. In the quantum mechanics framework:

$$\langle A \rangle = \sum_m A_m p_m = \frac{1}{\Xi} \sum_m A_m \, e^{-\beta(E_m - \mu N_m)} = \frac{\sum_m A_m \, e^{-\beta(E_m - \mu N_m)}}{\sum_m e^{-\beta(E_m - \mu N_m)}},$$

where A_m is the value of the internal variable when the system is in microstate m. Let us now express thermodynamic quantities as the average values of usual *macroscopic* internal variables.

Average number of particles

The average value of N is given by:

$$\begin{aligned}
\langle N \rangle &= \frac{1}{\Xi} \sum_m N_m \, e^{-\beta(E_m - \mu N_m)} = \frac{1}{\beta \Xi} \sum_m \frac{\partial}{\partial \mu} \Big(e^{-\beta(E_m - \mu N_m)} \Big) \Big|_{V,T} \\
&= \frac{1}{\beta \Xi} \frac{\partial}{\partial \mu} \Big(\underbrace{\sum_m e^{-\beta(E_m - \mu N_m)}}_{=\Xi} \Big) \Big|_{V,T} = \frac{1}{\beta} \frac{\partial \ln \Xi}{\partial \mu} \Big|_{V,T} = -\frac{\partial J}{\partial \mu} \Big|_{V,T}, \qquad (4.55)
\end{aligned}$$

where the derivative with respect to μ and the sum over microstates were swapped.

Internal energy

The system's internal energy $\langle E \rangle$ is calculated from the following relationship:

$$\langle E - \mu N \rangle = \frac{1}{\Xi} \sum_m \left(E_m - \mu N_m \right) e^{-\beta \left(E_m - \mu N_m \right)} = -\left. \frac{\partial \ln \Xi}{\partial \beta} \right|_{\mu,V} = \left. \frac{\partial \beta J}{\partial \beta} \right|_{\mu,V}, \tag{4.56}$$

so according to Equation (4.55)

$$\langle E \rangle = \langle E - \mu N \rangle + \mu \langle N \rangle = -\left. \frac{\partial \ln \Xi}{\partial \beta} \right|_{\mu,V} + \frac{\mu}{\beta} \left. \frac{\partial \ln \Xi}{\partial \mu} \right|_{V,T} = J + \beta \left. \frac{\partial J}{\partial \beta} \right|_{\mu,V} - \mu \left. \frac{\partial J}{\partial \mu} \right|_{V,T}.$$

The heat capacity at constant volume can, of course, be deduced by deriving $\langle E \rangle$ with respect to T.

Generalised average forces: pressure

In the grand canonical ensemble, the generalised average force $\langle X \rangle$ conjugate to the external parameter x is:

$$
\begin{aligned}
\langle X \rangle &= -\langle \frac{\partial E_m}{\partial x} \rangle = -\frac{1}{\Xi} \sum_m \left(\left. \frac{\partial E_m}{\partial x} \right|_{\mu,T} \right) e^{-\beta(E_m - \mu N_m)} \\
&= \frac{1}{\beta \Xi} \frac{\partial}{\partial x} \left(\sum_m e^{-\beta(E_m - \mu N_m)} \right)\Bigg|_{\mu,T} \\
&= \frac{1}{\beta} \left. \frac{\partial \ln \Xi}{\partial x} \right|_{\mu,T} = -\left. \frac{\partial J}{\partial x} \right|_{\mu,T}.
\end{aligned}
$$

Pressure being by definition the average force conjugate to volume ($x = V$), then:

$$P \equiv \frac{1}{\beta} \left. \frac{\partial \ln \Xi}{\partial V} \right|_{\mu,T} = -\left. \frac{\partial J}{\partial V} \right|_{\mu,T}. \tag{4.57}$$

When the only external parameters are μ, V and T, which is the case for simple fluids (see Section 5.1), pressure, which is an intensive quantity, depends only[39] on μ and T. On the other hand, the grand potential is extensive so for any real α, $J(\mu, \alpha V, T) = \alpha J(\mu, V, T)$. Deriving this relation with respect to α, one obtains:

$$\frac{\partial J}{\partial \alpha}(\mu, \alpha V, T) = V \frac{\partial J}{\partial V} = J(\mu, V, T),$$

or, according to Equation (4.57),

$$J = -PV. \tag{4.58}$$

Entropy

Since $J = \langle E - \mu N \rangle - TS$, according Relation (4.56), the entropy is:

$$S = \frac{1}{T} \left(\langle E - \mu N \rangle - J \right) = \frac{1}{T} \left(\left. \frac{\partial \beta J}{\partial \beta} \right|_{\mu,V} - J \right) = \frac{1}{k_B T^2} \left. \frac{\partial J}{\partial \beta} \right|_{\mu,V} = -\left. \frac{\partial J}{\partial T} \right|_{\mu,V}. \tag{4.59}$$

---

[39]This can be shown as follows: pressure is intensive, so $P(\mu, \alpha V, T) = P(\mu, V, T)$ for any real α, in particular for $\alpha = 0$. Since the chemical potential μ is a function of ρ and T, pressure can be expressed only in terms of these two intensive variables, ρ and T.

According to Equations (4.55), (4.57) and (4.59), the total differential of J is:

$$dJ = -SdT - PdV - \langle N \rangle d\mu,$$

allowing to easily find the definitions of the thermodynamic quantities expressed above. Moreover, by differentiating Expression (4.58) of J,

$$dJ = -PdV - VdP = -SdT - PdV - \langle N \rangle d\mu,$$

the Gibbs-Duhem thermodynamic relation (which will be useful later on) can be proven:

$$\langle N \rangle d\mu = VdP - SdT. \tag{4.60}$$

Compressibility and particle number fluctuations

Beyond average values of internal variables, it is interesting to calculate their fluctuations, as it was done for energy in the canonical ensemble (see Section 4.1.3). To determine the variance Δ_N^2 of the number of particles, one can start by writing[40]:

$$\langle N^2 \rangle = \frac{1}{\Xi} \sum_m N_m^2 \, e^{-\beta \left(E_m - \mu N_m \right)} = \frac{1}{\beta^2 \Xi} \frac{\partial^2 \Xi}{\partial \mu^2} \bigg|_{V,T}.$$

So according to Equation (4.55)

$$\Delta_N^2 = \langle N^2 \rangle - \langle N \rangle^2 = \frac{1}{\beta^2 \Xi} \frac{\partial^2 \Xi}{\partial \mu^2} - \left(\frac{1}{\beta \Xi} \frac{\partial \Xi}{\partial \mu} \right)^2 = \frac{1}{\beta^2} \frac{\partial}{\partial \mu} \bigg(\underbrace{\frac{1}{\Xi} \frac{\partial \Xi}{\partial \mu} \bigg|_{V,T}}_{\beta \langle N \rangle} \bigg),$$

so

$$\Delta_N^2 = \langle N^2 \rangle - \langle N \rangle^2 = \frac{1}{\beta} \frac{\partial \langle N \rangle}{\partial \mu} \bigg|_{V,T}.$$

In a macroscopic system, the number of particles is not directly measured. Instead the particle number density $\rho = N/V$ is measured and its fluctuations are:

$$\langle \rho^2 \rangle - \langle \rho \rangle^2 = \frac{1}{\beta V} \frac{\partial \langle \rho \rangle}{\partial \mu} \bigg|_{V,T},$$

where $\langle \rho \rangle = \langle N \rangle / V$. From the Gibbs-Duhem Relation (4.60), at constant temperature (that is $dT = 0$), $\langle N \rangle d\mu = VdP$, so $d\mu = dP/\langle \rho \rangle$, thus:

$$\langle \rho^2 \rangle - \langle \rho \rangle^2 = \frac{\langle \rho \rangle}{\beta V} \frac{\partial \langle \rho \rangle}{\partial P} \bigg|_T.$$

The system's *isothermal compressibility* can be defined as:

$$\kappa_T \equiv \frac{1}{\rho_{eq}} \frac{\partial \rho_{eq}}{\partial P} \bigg|_T = -\frac{1}{V} \frac{\partial V}{\partial P} \bigg|_T, \tag{4.61}$$

where $\rho_{eq} = \langle \rho \rangle$ is the density of the system in equilibrium. This intensive thermodynamic quantity characterises the relative change in density (or volume) of a system subjected to

---

[40] One can also derive Δ_E^2 in the grand canonical ensemble (see Exercise 4.10).

a pressure change. Under normal conditions of temperature and pressure, the compressibility of gases is large ($\sim 10^{-5}$ Pa^{-1}) and the compressibility of liquids and solids is small ($\sim 10^{-10}$ Pa^{-1} and $\sim 10^{-12}$ Pa^{-1}, respectively). Finally,

$$\frac{\langle \rho^2 \rangle - \langle \rho \rangle^2}{\langle \rho \rangle^2} = \frac{k_B T}{V} \kappa_T, \tag{4.62}$$

like energy fluctuations are proportional to heat capacity C_V (Equation (4.27)) and magnetic moment fluctuations are proportional to magnetic susceptibility χ (Equation (4.29)), as it was shown in the canonical ensemble. An expression of the fluctuation-dissipation theorem linking the fluctuations of an internal variable (here N or ρ) to the response of the system to a small perturbation (here a pressure change) is recovered[41]. Since ρ and κ_T are intensive quantities and volume is extensive, the relative fluctuations in density (or particles number) behave as:

$$\frac{\sqrt{\langle \rho^2 \rangle - \langle \rho \rangle^2}}{\langle \rho \rangle} \simeq \frac{\sqrt{\langle N^2 \rangle - \langle N \rangle^2}}{\langle N \rangle} \sim \frac{1}{\sqrt{f}} \simeq 10^{-12},$$

for a macroscopic system[42].

Example 1: Lattice-gas model

According to the lattice-gas grand partition function given by Equation (4.51) (with no attraction between atoms), the grand potential is:

$$J(\mu, V, T) = -k_B T \ln \Xi = -N_0 k_B T \ln \left(1 + e^{\beta \mu} \right), \tag{4.63}$$

where $N_0 = V/v_0$ is the number of cells in the system of volume V. The pressure (4.58) and the average atom number (4.55) can be deduced:

$$P = -\frac{J}{V} = \frac{k_B T}{v_0} \ln \left(1 + e^{\beta \mu} \right)$$

$$\langle N \rangle = -\left. \frac{\partial J}{\partial \mu} \right|_{V,T} = \frac{e^{\beta \mu}}{1 + e^{\beta \mu}} N_0.$$

Expressing $e^{\beta \mu}$ as a function of the ratio $\langle N \rangle / N_0$ with the help of the second equation, the equation of state for the lattice-gas model can be obtained:

$$P = -\frac{k_B T}{v_0} \ln \left(1 - v_0 \frac{\langle N \rangle}{V} \right) = -\frac{k_B T}{v_0} \ln \left(1 - v_0 \langle \rho \rangle \right).$$

In the low density limit ($v_0 \langle \rho \rangle \ll 1$), $\ln \left(1 - v_0 \langle \rho \rangle \right) \simeq -v_0 \langle \rho \rangle$, the ideal gas equation of state $P \simeq \langle \rho \rangle k_B T$ is recovered. Remember that the lattice-gas model is different from the ideal gas model since repulsive contact interactions are accounted for by the occupancy rate of a cell which is at most 1.

---

[41] Expression (4.62) shows that compressibility, like a variance, must be positive. It can be understood easily for reasons of mechanical stability: when P increases, V must decrease, so $\frac{\partial V}{\partial P} < 0$. Some systems (foams, neutron stars...) nevertheless exhibit negative compressibilities or heat capacities (see Note 12 in Chapter 3).

[42] On the other hand, it will be seen in Chapter 5 that in the vicinity of the thermodynamic critical point, the compressibility of a fluid diverges. Density fluctuations thus become important and extend over sizes of the order of a micron, leading to an intense light scattering which gives a milky appearance to a fluid observed under these thermodynamic conditions. This is the phenomenon of critical opalescence.

Example 2: Ideal classical gas

According to Equation (4.52), the ideal classical gas grand potential is:

$$J(\mu, V, T) = -k_{\rm B}T \ln \Xi = -k_{\rm B}T \frac{V}{\lambda_T^3}\, e^{\beta\mu}. \tag{4.64}$$

Pressure (4.58) and the average number of atoms (4.55) are thus:

$$
\begin{aligned}
P &= -\frac{J}{V} = \frac{k_{\rm B}T}{\lambda_T^3}\, e^{\beta\mu} \\
\langle N \rangle &= -\left.\frac{\partial J}{\partial \mu}\right|_{V,T} = \frac{V}{\lambda_T^3}\, e^{\beta\mu}.
\end{aligned}
\tag{4.65}
$$

The ideal gas equation of state, $P = \langle N \rangle k_{\rm B}T/V = \langle \rho \rangle k_{\rm B}T$ is recovered. In the same way, the internal energy $\langle E \rangle = \frac{3}{2}\langle N \rangle k_{\rm B}T$ and the gas entropy S can be calculated.

4.2.4 THERMODYNAMIC EQUILIBRIUM AND MINIMUM GRAND POTENTIAL

Like particle number (and energy), any internal variable, microscopic or macroscopic, has a probability distribution in equilibrium given by the general form:

$$P(A) = \sum_{\{m|A_m=A\}} p_m = \frac{1}{\Xi(\mu, V, T)} \sum_{\{m|A_m=A\}} e^{-\beta\left(E_m - \mu N_m\right)},$$

that can be written as:

$$P(A) = \frac{1}{\Xi(\mu, V, T)}\, e^{-\frac{J_{\rm p}(\mu, V, T\,|\,A)}{k_{\rm B}T}}, \tag{4.66}$$

with $J_{\rm p}(\mu, V, T\,|\,A) \equiv -k_{\rm B}T \ln\left(\sum_{\{m|A_m=A\}} e^{-\beta\left(E_m - \mu N_m\right)}\right)$, where the sum is performed over the microstates m such as $A_m = A$. The function $J_{\rm p}(\mu, V, T\,|\,A)$ is by definition the partial grand potential associated to the internal variable A. If A is fixed and becomes an external parameter, $J_{\rm p}(\mu, V, T\,|\,A)$ is then the system's grand potential.

When the internal variable A is macroscopic, its relative fluctuations are negligible (of the order of 10^{-12}). The probability distribution $P(A)$ is a sharp Gaussian distribution (2.32), centred on the most probable value $A_{\rm eq} = \langle A \rangle$, interpreted as the equilibrium value. As shown by Expression (4.66), the equilibrium value of a macroscopic internal variable, and more generally the system's equilibrium macrostate correspond to the minimum of $J_{\rm p}$.

The following principle can be deduced: *the equilibrium macrostate of a system in contact with a reservoir of particles and a thermostat is the one having the minimum (partial) grand potential*. The thermodynamic grand potential, which, in principle, only makes sense in equilibrium, is equal to the minimal partial grand potential: $J(\mu, V, T) = J_{\rm p}(\mu, V, T\,|\,A_{\rm eq})$. The study of the direction of out-of-equilibrium exchanges would lead to the same conclusions as those obtained in the microcanonical (Section 3.5.1) and canonical (Section 4.1.4) ensembles.

4.3 GENERALISATION AND EQUIVALENCE OF THE STATISTICAL ENSEMBLES

To conclude this chapter and the first part of the book devoted to the formalism and foundations of classical equilibrium statistical physics, the main characteristics of the statistical ensembles are reviewed.

4.3.1   STATISTICAL ENSEMBLES

A statistical ensemble is defined by a set of external parameters fixed and imposed on the system (Chapter 1). All other physical quantities are internal variables (written A), which fluctuate with the microstate of the system. To each statistical ensemble is associated a probability distribution of the microstates (p_m), which allows to express the probability distribution of the internal variables $(P(A))$ and their average values $(\langle A \rangle)$. In the microcanonical ensemble (N, V, E), an isolated system in equilibrium is considered. In this case, all accessible microstates are equiprobable $(p_m$ is uniform), as stated in the fundamental postulate (Chapter 2). The relative fluctuations of the macroscopic internal variables are extremely small (of the order of 10^{-12}) and their mean values are interpreted as the thermodynamic quantities of the system (Chapter 3). Using the microcanonical ensemble formalism, systems that are no longer isolated but in contact with much larger macroscopic objects playing the role of energy or particles reservoirs can be studied. The large system imposes new external parameters on the studied system. The canonical (N, V, T) and grand canonical (μ, V, T) ensembles were thus introduced, each characterised by its own probability distribution of the microstates (Chapter 4).

In practice, these three statistical ensembles, named and studied by Gibbs in 1902, are the most commonly used in scientific publications. Their main characteristics are summarised in table 4.1. However, nothing prevents the introduction of other ensembles, by choosing other sets of external parameters.

TABLE 4.1   Characteristics of the three main statistical ensembles.

Statistical Ensemble	Microcanonical	Canonical	grand canonical
External parameters	N, V, E	N, V, T	μ, V, T
Fundamental function	Number of accessible microstate m' $$\Omega(N,V,E) = \sum_{m'} 1$$	Partition function $$Z(N,V,T) = \sum_m e^{-\beta E_m}$$	Grand partition function $$\Xi(\mu,V,T) = \sum_m e^{-\beta(E_m - \mu N_m)}$$
Probability of microstate m	$p_m = \dfrac{1}{\Omega}$	$p_m = \dfrac{1}{Z} e^{-\beta E_m}$	$p_m = \dfrac{1}{\Xi} e^{-\beta(E_m - \mu N_m)}$
Thermodynamic potential	Entropy $$S = k_B \ln \Omega$$ $$dS = \overbrace{\frac{1}{T}dE + \frac{P}{T}dV - \frac{\mu}{T}dN}$$	Free energy $$F = -k_B T \ln Z$$ $$F = \langle E \rangle - TS$$ $$dF = \overbrace{-SdT - PdV + \mu dN}$$	Grand potential $$J = -k_B T \ln \Xi$$ $$J = \langle E \rangle - TS - \langle N \rangle \mu$$ $$dJ = \overbrace{-SdT - PdV + \langle N \rangle d\mu}$$

An additional example of statistical ensemble: the NPT ensemble

The choice of the external parameters N, P and T is natural since they are usually experimentally controllable. The associated statistical ensemble, called *isobaric–isotherm*, is, however, rarely used. Following the method described in Sections 4.1 and 4.2, consider a closed system (containing N particles), in contact with a reservoir of energy and volume which imposes its temperature T and its pressure P to the system (this is for example the case of a balloon filled with air surrounded by Earth's atmosphere). The entropy of the reservoir is:

$$S'_p(E_{tot} - E_m, V_{tot} - V_m) = S'_p(E_{tot}, V_{tot}) - E_m \underbrace{\frac{\partial S'_p}{\partial E'}}_{1/T'} - V_m \underbrace{\frac{\partial S'_p}{\partial V'}}_{P'/T'} + \dots$$

$$\simeq S'_p(E_{tot}, V_{tot}) - \frac{E_m}{T'} - \frac{P'}{T'} V_m.$$

Since in equilibrium $T = T'$ and $P = P'$, the probability $p_m(V) \propto e^{\frac{S'_p}{k_B}}$ to find the system in a given microstate m of volume V_m lying between V and $V + dV$ (V must be treated as a continuous variable) is:

$$p_m(V) = \frac{1}{\Upsilon} e^{-\beta(E_m + PV)} \frac{dV}{v_0},$$

with the isothermal–isobaric partition function Υ defined as:

$$\Upsilon(N, P, T) = \int_{V_{min}}^{\infty} \frac{dV}{v_0} \sum_m e^{-\beta(E_m + PV)} = \int_{V_{min}}^{\infty} \frac{dV}{v_0} Z(N, V, T) e^{-\beta PV}, \tag{4.67}$$

where V_{min} is the minimum volume that can hold N particles and v_0 is a constant having the units of a volume, introduced to make the probability $p_m(V)$ and Υ dimensionless (v_0 does practically not appear in thermodynamic quantities). Note that Υ can be expressed as a function of Z. The Gibbs free energy (or free enthalpy) $G(N, P, T)$ is defined as:

$$G(N, P, T) \equiv -k_B T \ln \Upsilon(N, P, T) \simeq F(N, \langle V \rangle, T) + P\langle V \rangle. \tag{4.68}$$

Its total differential is, according to Equation (4.24), $dG = -SdT + \langle V \rangle dP + \mu dN$, which allows to express entropy, the mean volume and the chemical potential as derivatives of the Gibbs free energy[43]. For a simple fluid, N, P and T are the only external parameters, and it can be then shown that[44] $G = N\mu$.

Equivalence of statistical ensembles

Depending on the considered statistical ensemble, a physical quantity can be an external parameter or an internal variable. However, as it was seen in the case of the ideal gas, the same thermodynamic properties are recovered in the microcanonical, canonical and grand canonical ensembles when $N \gg 1$. The equivalence between statistical ensembles for macroscopic systems in equilibrium is due to the extremely small relative fluctuations of

---

[43] For example, for an ideal classical gas of indistinguishable particles, the reader can show, using Equations (4.67) and (4.11) with $V_{min} = 0$, that

$$\Upsilon(N, P, T) = \frac{1}{v_0 N!} \int_0^{\infty} dV \left(\frac{V}{\lambda_T^3}\right)^N e^{-\beta PV} = \frac{1}{v_0 \lambda_T^{3N}} \left(\frac{k_B T}{P}\right)^{N+1}. \tag{4.69}$$

For $N \gg 1$, the thermodynamic properties of an ideal gas are (once again!) recovered. We then have, $\langle V \rangle = \frac{\partial G}{\partial P} = -k_B T \frac{\partial \ln \Upsilon}{\partial P} = (N+1)k_B T/P \simeq Nk_B T/P$. All thermodynamic quantities are expressed as partial derivatives of G and are independent of v_0. Only G and S depend on v_0. Equation (4.68) allows the deduction of the value of v_0 by using Expression (4.34) of the ideal gas free energy. One finds here $v_0 = k_B T/P$ and according to Equation (4.69), G is naturally extensive.

[44] The intensive quantity μ depends only on P and T. For any real α, $G(\alpha N, P, T) = \alpha G(N, P, T)$, since G is extensive. Taking the derivative of this relation with respect to α, one gets:

$$\frac{\partial G}{\partial \alpha}(\alpha N, P, T) = N\frac{\partial G}{\partial N} = N\mu = G(N, P, T), \quad \text{since} \quad \mu = \frac{\partial G}{\partial N}.$$

any macroscopic internal variable[45] A (of the order of 10^{-12} for $N \simeq 10^{24}$, see Section 2.3). In other words, A remains practically equal to its mean value $\langle A \rangle$. This equivalence is even exact in the *thermodynamic limit*, for which $N \to \infty$, $V \to \infty$ and $\rho = N/V$ is a constant. In this case, the relative fluctuations, typically $\propto 1/\sqrt{N}$, tend to 0: the internal variables values do not fluctuate anymore and are imposed by the external parameters. Nevertheless, it should be remembered that the equivalence between statistical ensembles is valid only when relative fluctuations of the internal variables are negligible, that is i) for a macroscopic system ($N \gg 1$) in which the interactions between particles are short-range[46] and ii) under thermodynamic conditions such that the system does not undergo a phase transition (see Note 42).

The equivalence of statistical ensembles is quite intuitive since the macroscopic properties of a system *in equilibrium* do not depend *a priori* on the conditions under which it has been prepared, whether by isolating it or by putting it in contact with a reservoir of energy and/or particles. Thus, if the quantity A is an external parameter equal to A_0 in a first statistical ensemble and it is an internal variable in a second ensemble, it is sufficient to adjust the conjugate variable imposed by the reservoir to fix its mean value to $\langle A \rangle = A_0$ and obtain the same macrostate in these two ensembles and thus the same thermodynamic properties. For example, suppose that a system is isolated in an equilibrium macrostate characterised by the external parameters N_0, V_0 and E_0 (microcanonical ensemble) and by the mean values of its internal variables, such that its temperature is equal to T. By putting this system in contact with a thermostat at temperature $T_0 = T$ (canonical ensemble N_0, V_0 and T_0), its energy remains practically equal to $\langle E \rangle = E_0$ and it will stay in the same equilibrium macrostate. Figure 4.5 illustrates the equivalence between the three main statistical ensembles.

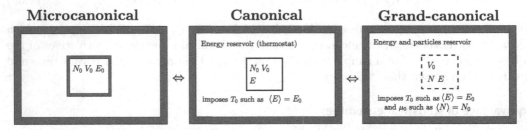

FIGURE 4.5 Equivalence of statistical ensembles for a macroscopic system in equilibrium. For each ensemble, the external parameters are noted with a zero subscript to distinguish them from internal variables associated with exchanges with the reservoir (energy or energy and particle).

In practice, the thermodynamic quantities of a macroscopic system in equilibrium can be calculated using the statistical ensemble which seems the most appropriate to the studied situation or which facilitates calculations. The use of the canonical ensemble or possibly the grand canonical ensemble is therefore generally preferred. Exercises presented in Section 4.4 illustrate the operational power of these two statistical ensembles.

4.3.2 GIBBS ENTROPY AND INFORMATION THEORY

In Section 3.4, an alternative approach to statistical physics based on Gibbs entropy and information theory was presented: the probability distribution p_m of a statistical

---

[45] This property is verified whether the system is isolated or not. If it is not, its internal variable is also an internal variable of the total isolated system $\mathcal{A}_{tot}$ consisting of the studied system and the reservoir.

[46] This is not the case for the systems mentioned in Note 5 of Chapter 3, for which the particle interaction range is comparable to the size of the system (nuclei and atomic aggregates, galaxies...).

ensemble is such that the Gibbs entropy $S_G(\{p_m\}) = -k_B \sum_m p_m \ln p_m$ is maximal under given constraints. Remember that in the microcanonical ensemble, the only constraint is the normalisation condition of p_m, so: $p_m = 1/\Omega$ and $S_G = k_B \ln \Omega$.

In the framework of information theory, the grand canonical probability distribution[47] can also be recovered: energy and the number of particles fluctuate, but their equilibrium average values are fixed at $\langle E \rangle$ and $\langle N \rangle$, respectively. With the normalisation condition, the Gibbs entropy must therefore be maximal under the three following constraints:

$$\sum_m p_m = 1 \tag{4.70}$$

$$\sum_m E_m \, p_m = \langle E \rangle \tag{4.71}$$

$$\sum_m N_m \, p_m = \langle N \rangle. \tag{4.72}$$

Using Lagrange multipliers method (see Appendix A.5), the grand canonical probability distribution p_m is such that

$$\mathcal{F}(\{p_m\}) = \underbrace{S_G(\{p_m\})}_{-k_B \sum_m p_m \ln p_m} - \mu_1 \left(\sum_m p_m - 1 \right) - \mu_2 \left(\sum_m E_m p_m - \langle E \rangle \right) - \mu_3 \left(\sum_m N_m p_m - \langle N \rangle \right)$$

is maximal, where the Lagrange multipliers μ_1, μ_2 and μ_3 are constants to be determined *a posteriori* using the three constraints given by Equations (4.70), (4.71) and (4.72). Thus:

$$\frac{\partial \mathcal{F}(\{p_m\})}{\partial p_m} = -k_B \ln p_m - k_B - \mu_1 - \mu_2 E_m - \mu_3 N_m = 0,$$

so

$$p_m = e^{-(1+\frac{\mu_1}{k_B})} e^{-\frac{\mu_2}{k_B} E_m - \frac{\mu_3}{k_B} N_m}.$$

Parameter μ_1 is determined with the normalisation condition (4.70). The distribution function p_m can then be written as:

$$p_m = \frac{1}{\Xi} \, e^{-\frac{\mu_2}{k_B} E_m - \frac{\mu_3}{k_B} N_m}, \tag{4.73}$$

where the constant $\Xi = e^{(1+\frac{\mu_1}{k_B})} = \sum_m e^{-\frac{\mu_2}{k_B} E_m - \frac{\mu_3}{k_B} N_m}$ was introduced. Moreover, using Equation (4.73) and the constraints (4.71) and (4.72), Gibbs entropy can be written as:

$$S_G = -k_B \sum_m p_m \ln p_m = -k_B \langle \ln p_m \rangle = \mu_2 \langle E \rangle + \mu_3 \langle N \rangle + k_B \ln \Xi. \tag{4.74}$$

Gibbs entropy can be identified with the thermodynamic entropy of an isolated macroscopic system of $\langle N \rangle$ particles at fixed energy $\langle E \rangle$. The system's temperature and chemical potential are given by:

$$\frac{1}{T} = \frac{\partial S_G}{\partial \langle E \rangle} = \mu_2 \quad \text{and} \quad -\frac{\mu}{T} = \frac{\partial S_G}{\partial \langle N \rangle} = \mu_3.$$

The expression for the grand canonical distribution function (4.47) and the grand partition function (4.48) are recovered. Relation (4.74) then takes the form of the Legendre transform defining the grand potential $J = -k_B T \ln \Xi$ as $J = \langle E \rangle - T S_G - \mu \langle N \rangle$. Note that the equilibrium probability distribution p_m is determined by the maximum of S_G, which is then identified

---

[47]The canonical framework is a special case of the grand canonical one in which N is fixed.

with the thermodynamic entropy, under the three constraints (4.70), (4.71) and (4.72), or equivalently by the minimum of J under the single constraint (4.70)[48].

As mentioned in Section 3.4, this approach has the merit of being simple to implement. It is also general and does not privilege any particular statistical ensemble, but it is rather abstract and not very intuitive: what does it mean physically to impose an average energy on a system? The approach based on the fundamental postulate developed in Section 4.2 allows to clearly specify the conditions of application and the approximations made in a concrete framework.

---

[48] Indeed, one easily shows that the grand canonical distribution function (4.47) minimises the grand potential $J = \langle E \rangle - T S_{\mathrm{G}} - \mu \langle N \rangle = \sum_m \left(E_m p_m + k_{\mathrm{B}} T p_m \ln p_m - \mu N_m p_m \right)$ with the constraint $\sum_m p_m = 1$.

4.4 EXERCICES

Exercise 4.1. Hydrogen crystal

In Exercise 3.4, a hydrogen crystal modelled as N independent H_2 molecules was studied in the framework of the microcanonical ensemble. The same crystal is now in contact with a thermostat at temperature T.

1. Calculate the system's partition function Z.

2. Deduce the average energy $\langle E \rangle$, the heat capacity C and the average values $\langle n_i \rangle$ of the number of molecules in state (i) for $i = 1, 2, 3, 4$.

3. Express the entropy S as a function of T. What is the limit of $S(T)$ when $k_B T \gg \Delta$?

Solution:

1. *The N molecules are distinguishable and independent, thus $Z(N,T) = z^N$, where $z(T)$ is the partition function of a single molecule*[49]:

$$z(T) = \sum_{i=1}^{4} e^{-\beta\epsilon_i} = 1 + 3e^{-\beta\Delta} \quad thus \quad Z(N,T) = \left(1 + 3e^{-\beta\Delta}\right)^N.$$

2. $\langle E \rangle = -\dfrac{\partial \ln Z}{\partial \beta} = N\Delta \dfrac{3e^{-\beta\Delta}}{1 + 3e^{-\beta\Delta}}$, *which, of course, tends to 0 (parahydrogen state) for $\beta\Delta \gg 1$ and to $3N\Delta/4$ for $\beta\Delta \ll 1$ (At high temperature the 4 states are equiprobable). Then,* $C = \dfrac{\partial \langle E \rangle}{\partial T} = 3Nk_B(\beta\Delta)^2 \dfrac{e^{\beta\Delta}}{(3 + e^{\beta\Delta})^2}$.

   The probability to find a molecule in state (i) is $p_i = e^{-\beta\epsilon_i}/z$ and $\langle n_i \rangle = Np_i$:

$$\langle n_1 \rangle = \frac{N}{z} = \frac{N}{1 + 3e^{-\beta\Delta}} \quad and \ for \ i = 2, 3, 4 \quad \langle n_i \rangle = \frac{Ne^{-\beta\Delta}}{z} = \frac{Ne^{-\beta\Delta}}{1 + 3e^{-\beta\Delta}}.$$

3. $S = \dfrac{1}{T}(\langle E \rangle - F)$, *where $F = -k_B T \ln Z$. Thus $S = k_B N\left(\dfrac{3\beta\Delta}{3 + e^{\beta\Delta}} + \ln\left(1 + 3e^{-\beta\Delta}\right)\right)$, which tends to $k_B N \ln 4$ when $\beta\Delta \ll 1$ (at high temperature the 4^N microstates are equiprobable).*

Exercise 4.2. Adiabatic demagnetisation refrigeration

A paramagnetic salt crystal consists of N atoms, each carrying a spin-$1/2$ having a magnetic moment μ_M. The interactions between spins are ignored. The crystal is in contact with a thermostat at temperature T and is immersed in a uniform and constant external magnetic field $\vec{B}$. The interaction energy of an atom with the magnetic field is therefore $-\mu_M B$ if the spin is parallel to the field and $+\mu_M B$ if it is antiparallel.

1. Calculate the crystal's partition function Z.

2. Deduce the free energy F, the average energy $\langle E \rangle$, the heat capacity C at constant magnetic field and the average magnetic moment M. Study the behaviour of C as a function of T. What is the value of C in a zero magnetic field?

---

[49] Equation (4.5) and the expression of $\Omega(N, E)$ obtained in Exercise 3.4 can also be used:

$$Z = \sum_E \Omega(N,E)e^{-\beta E} = 3^N e^{-\beta N\Delta} \sum_{n_1} \binom{N}{n_1}\left(\frac{e^{\beta\Delta}}{3}\right)^{n_1} = 3^N e^{-\beta N\Delta}\left(1 + \frac{e^{\beta\Delta}}{3}\right)^N = \left(1 + 3e^{-\beta\Delta}\right)^N,$$

according to Newton binomial formula.

3. Calculate the entropy S and specify its limit values at low and high temperatures. Plot on the same graph $S(T)$ for two different values of the magnetic field, B_1 and $B_2 > B_1$.

4. This type of paramagnetic salt is used in cryogenics to reach very low temperatures. A crystal is prepared at an initial temperature T_i and placed in a strong magnetic field, B_2. The magnetic field is then decreased slowly enough for the system to remain in equilibrium down to a value $B_1 < B_2$. What happens to the system's total entropy during this transformation? Give the expression of the final temperature T_f as a function of T_i. The system is then returned, at constant temperature, to the initial magnetic field B_2 and the previous operation is repeated. What is, in principle, the temperature T_n obtained after n cycle?

Solution:

1. *The ideal paramagnetic crystal partition function is given by Equation (4.10).*

2. *These thermodynamic quantities are given by Equations (4.30), (4.31), (4.32) and (4.33). C exhibits a Schottky anomaly (see Section 3.3.1). For $B = 0$, $C = 0$.*

3. *Entropy can be calculated with Equation (4.23):*

$$S = N k_B \Big(\ln \big(2 \cosh(\beta \mu_M B)\big) - \beta \mu_M B \tanh(\beta \mu_M B)\Big).$$

For $\beta \mu_M B \gg 1$, $S \to 0$ (only one microstate). For $\beta \mu_M B \ll 1$, $S \to N k_B \ln 2$ (2^N equiprobable microstates). The entropy $S/N k_B$ is given by the function $f(x) = \ln(2 \cosh x) - x \tanh x$, where $x = \beta \mu_M B$. Its derivative $f'(x) = -x/\cosh^2(x)$ is negative for all x. Entropy is thus an increasing function of temperature. Moreover the entropy decreases with the magnetic field, as can be seen in Figure 4.6.

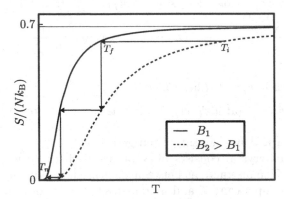

FIGURE 4.6 Entropy $S/N k_B$ as a function of temperature T for two magnetic fields B_1 and $B_2 > B_1$. Principle of adiabatic demagnetisation refrigeration (S = Constant, T = Constant, S = Constant...).

4. *Since the transformation is sufficiently slow, it can be considered reversible. The entropy is therefore constant and since it is expressed only in terms of $x = \beta \mu_M B$, the ratio B/T is also constant during the transformation. It can immediately be deduced $T_f = (B_1/B_2)T_i < T_i$. Performing the operation n times, the obtained temperature is $T_n = (B_1/B_2)^n T_i \to 0$ when $n \to \infty$. However, in this very simple model, the interactions between spins and the degrees of freedom of the crystal lattice are ignored. In practice, this technique applied to nuclear spins allows to reach temperatures of the order of $10 \; \mu K$.*

Exercise 4.3. Heat capacity of solids: Einstein's model

A solid consists of N atoms, located at the nodes of a crystal lattice. They can oscillate around their equilibrium positions in the three directions of space. The atomic oscillations are modelled by a set of $3N$ harmonic oscillators assumed independent of each other. The crystal's temperature is kept equal to T. The cohesive energy of the crystal is not taken into account.

1. Consider *classical* harmonic oscillators. Give without any calculation the heat capacity of the solid. Compare this result to the following experimental values at $T = 300$ K: 24.2 J·mol^{-1}·K^{-1} for silver and 6.1 J·mol^{-1}·K^{-1} for diamond.

2. At low temperatures, the heat capacity of solids is not constant, but increases with temperature (see Section 4.1.5 and Figure 4.7). To understand this behaviour, Einstein proposed in 1907 a very simple model in which every *quantum* oscillator vibrates *independently* at a *unique* angular frequency ω_0. The Hamiltonian of each oscillator is given by Equation (1.2). Calculate the system's partition function Z.

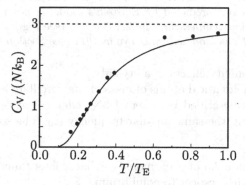

FIGURE 4.7 Diamond molar heat capacity C/R (for $N = N_A$) as a function of temperature T/T_E, with $T_E = 1320$ K. Experimental points are compared to the Einstein's model curve (Equation (4.75). The dashed line shows the Dulong and Petit law ($C = 3R \simeq 24.9$ J · mol^{-1} · K^{-1}). From A. Einstein, Annalen der Physik **22**, 180 (1907).

3. Derive the free energy F, the mean energy $\langle E \rangle$, the entropy S, and the heat capacity C of the solid as a function of the Einstein temperature, $T_E \equiv \hbar\omega_0/k_B$.

4. What is the behaviour of C at high and low temperatures? Plot the curve $C(T)$. What is the value of C for $T = T_E$? Determine the value of T_E for silver and for diamond using the experimental data given in Question 1.

Solution:

1. *According to the equipartition of energy theorem (Section 4.1.5) and the form of Hamiltonian (4.42), the system's average energy is $\langle E \rangle = 3Nk_BT$. For one mole, $C = 3R \simeq 24.9$ J·mol^{-1}.K^{-1} (Dulong and Petit law), in very good agreement with the value for silver, but not for diamond.*

2. *Since the $3N$ harmonic oscillators are independent and distinguishable, $Z = z^{3N}$, where z is the partition function of a single quantum harmonic oscillator:*

$$z = \sum_{n=0}^{\infty} e^{-\beta(n+\frac{1}{2})\hbar\omega_0} = e^{-\frac{\beta}{2}\hbar\omega_0} \sum_{n=0}^{\infty} e^{-n\beta\hbar\omega_0} = \frac{e^{-\frac{\beta}{2}\hbar\omega_0}}{1 - e^{-\beta\hbar\omega_0}} = \frac{1}{2\sinh(\beta\hbar\omega_0/2)}.$$

3. With $k_B T_E = \hbar\omega_0$:

$$F = -k_B T \ln Z = 3N k_B T \ln\left(2\sinh(T_E/2T)\right)$$

$$\langle E \rangle = -\frac{\partial \ln Z}{\partial \beta} = 3N \frac{k_B T_E/2}{\tanh(T_E/2T)}$$

$$S = -\frac{\partial F}{\partial T} = 3N k_B \left(-\ln\left(2\sinh(T_E/2T)\right) + \frac{T_E/2T}{\tanh(T_E/2T)}\right)$$

$$C = \frac{\partial <E>}{\partial T} = 3N k_B \left(\frac{T_E/2T}{\sinh(T_E/2T)}\right)^2. \tag{4.75}$$

4. The heat capacity is, of course, an increasing function of T (see Figure 4.7). For $T \ll T_E$, $C \simeq 3N k_B (T_E/T)^2 e^{-T_E/T} \to 0$, when $T \to 0$. For $T \gg T_E$, ($\sinh x \sim x$ when $x \to 0$), $C \simeq 3N k_B$, i.e. $3R$ for one mole, the Dulong and Petit law is recovered.

For $T = T_E$, $C/3R \simeq 0.92$. For silver, $T_E = 220$ K. Silver heat capacity at room temperature ($T = 300$ K $> T_E$) is therefore in agreement with Dulong and Petit law. On the other hand, for diamond $T_E = 1320$ K, so $T = 300$ K $< T_E$ and $C < 3R$. The growth of C with T is of quantum origin, but the behaviour predicted by Einstein's model is not in quantitative agreement with experiments at low temperature (in fact $C \propto T^3$) (see Figure 4.7). To understand this behaviour, the model was modified by Debye in 1912 (see Section 7.1).

Exercise 4.4. Anharmonicity effects in a crystal

A crystal consists of N identical atoms of mass m that oscillate around their equilibrium position. Their motion is described by a set of 3N *independent* one-dimensional classical harmonic oscillators having the same angular frequency ω_0. The solid is in contact with a thermostat at temperature T.

1. Write the Hamiltonian H_0 of a harmonic oscillator as a function of its momentum p and its position q with respect to equilibrium.

2. According to Lindemann's criterion (1910), a crystal begins to melt when the average oscillation amplitude of the atoms reaches a given fraction α_0 of the average inter-atomic distance d. What is the value of $\langle q \rangle$? Calculate $\langle q^2 \rangle$ and deduce the atoms average oscillation amplitude as a function of temperature.

3. Deduce the expression of α_0 as a function of the melting temperature T_f. Estimate and comment on the value of α_0. The following values are given for copper: $d = 2.5$ Å, $M\omega_0^2 \simeq 10^{26}$ J·m^{-2}·mol^{-1}, where $M = 63.5$ g·mol^{-1} is the molar mass and $T_f \simeq 1400$ K at atmospheric pressure.

4. At high temperature, the heat capacity of solids grows linearly with temperature. For example, for copper at temperatures of about 400–600 °C:

$$C = 23.2 + 2.9 \times 10^{-3} T \quad \text{in} \quad \text{J} \cdot \text{K}^{-1}\text{mol}^{-1}, \tag{4.76}$$

where the temperature is expressed in K. As it will be seen, this effect can be attributed to the anharmonicity of the oscillations. The motion of an oscillator is now described by the Hamiltonian $H = H_0 - fq^4$, where H_0 is the Hamiltonian of the harmonic oscillator (see Question 1) and where f is a positive constant, $-fq^4$ is an anharmonic *correction* term considered small with respect to H_0. Assuming $\beta fq^4 \ll 1$, express the solid's partition function Z as a function of the temperature $T_a = m^2\omega_0^4/(3fk_B)$. Give a physical interpretation to T_a.

5. Deduce the crystal's internal energy $\langle E \rangle$ and its heat capacity C for temperatures $T \ll T_a$. Compare the obtained result with the experimental Expression (4.76) and evaluate T_a. Is $T \ll T_a$?

6. Calculate $\langle q^2 \rangle$ (this average can be written as a derivative of z). Taking now into account the anharmonicity effects, deduce a new value of the fraction α_0 giving the average oscillation amplitude during crystal fusion. Comment on the result.

Solution:

1. $H_0 = \frac{p^2}{2m} + \frac{1}{2}m\omega_0^2 q^2$.

2. By symmetry, $\langle q \rangle = 0$ and the equipartition of energy theorem implies $\langle \frac{1}{2}m\omega_0^2 q^2 \rangle = \frac{k_B T}{2}$. Thus average oscillation amplitude is $\sqrt{\langle q^2 \rangle} = \sqrt{\frac{k_B T}{m\omega_0^2}}$.

3. By definition, for $T = T_f$, $\alpha_0 = \frac{\sqrt{\langle q^2 \rangle}}{d} = \frac{1}{d}\sqrt{\frac{RT_f}{M\omega_0^2}} \simeq 0.04$, which is only 4% of the interatomic distance.

4. For $3N$ distinguishable and independent oscillators, $Z = z^{3N}$, where z is the partition function of one oscillator. Thus, by integrating over momentum:

$$z = \frac{1}{h}\int dp dq \, e^{-\beta\left(\frac{p^2}{2m}+\frac{1}{2}m\omega_0^2 q^2 - f q^4\right)} = \frac{1}{\lambda_T}\int dq \, e^{-\beta\left(\frac{1}{2}m\omega_0^2 q^2 - f q^4\right)},$$

where $\lambda_T = h/\sqrt{2\pi m k_B T}$ and where the integration domain is limited to small values of q. Since $\beta f q^4 \ll 1$, $e^{\beta f q^4} \simeq 1 + \beta f q^4$. Since the exponential $e^{-\frac{\beta}{2}m\omega_0^2 q^2}$ tends quickly to 0, the integration bounds can be extended from $-\infty$ to $+\infty$. Using the Gaussian integrals given in Appendix A.2, one gets:

$$\begin{aligned} z &= \frac{1}{\lambda_T}\int_{-\infty}^{+\infty} dq \, e^{-\frac{\beta}{2}m\omega_0^2 q^2}\left(1 + \beta f q^4 + \dots\right) \\ &\simeq \frac{1}{\lambda_T}\left[\int_{-\infty}^{+\infty} dq \, e^{-\frac{\beta}{2}m\omega_0^2 q^2} + \beta f \int_{-\infty}^{+\infty} dq \, q^4 \, e^{-\frac{\beta}{2}m\omega_0^2 q^2}\right] \\ &\simeq \frac{1}{\lambda_T}\sqrt{\frac{2\pi k_B T}{m\omega_0^2}}\left[1 + \beta f \frac{3}{4}\left(\frac{2k_B T}{m\omega_0^2}\right)^2\right] = \frac{k_B T}{\hbar\omega_0}\left[1 + \frac{3f}{m^2\omega_0^4}k_B T\right]. \end{aligned}$$

Therefore $Z(N,T) = z^{3N} \simeq \left(\frac{k_B T}{\hbar\omega_0}\left[1 + \frac{T}{T_a}\right]\right)^{3N}$. T_a is the temperature above which anharmonicity effects become significant.

5. From Equation (4.17), for $T \ll T_a$:

$$\begin{aligned} \langle E \rangle &= -\frac{\partial \ln Z}{\partial \beta} = -3N\frac{\partial}{\partial \beta}\left(-\ln\hbar\omega_0\beta + \ln\left[1 + \frac{1}{\beta k_B T_a}\right]\right) \\ &\simeq -3N\frac{\partial}{\partial \beta}\left(-\ln\hbar\omega_0\beta + \frac{1}{\beta k_B T_a}\right) = 3N k_B T\left(1 + \frac{T}{T_a}\right). \end{aligned}$$

Thus $C = \frac{\partial\langle E \rangle}{\partial T} = 3N k_B\left(1 + 2\frac{T}{T_a}\right)$ in agreement with Equation (4.76).

It can be deduced that $6R/T_a \simeq 2.9 \times 10^{-3}$, so $T_a \simeq 17000K$ and indeed $T \ll T_a$.

6. $\langle q^2 \rangle = \frac{1}{zh}\int dp dq \, q^2 e^{-\beta\left(\frac{p^2}{2m}+\frac{1}{2}m\omega_0^2 q^2 - f q^4\right)} = -\frac{2}{m\beta}\frac{\partial \ln z}{\partial\omega_0^2}$.

Let $y = \omega_0^2$, $\langle q^2 \rangle = -\frac{2}{m\beta}\frac{\partial}{\partial y}\ln\left(\frac{k_B T}{\hbar}y^{-\frac{1}{2}}(1 + \frac{3f}{m^2}k_B T y^{-2})\right)$, so for $T \ll T_a$:

$$\langle q^2 \rangle \simeq -\frac{2}{m\beta}\left(-\frac{1}{2y} - \frac{6f k_B T}{m^2 y^3}\right) = \frac{k_B T}{m\omega_0^2}\left(1 + 4\frac{T}{T_a}\right).$$

With $T_f/T_a \sim 0.1$, we have $\alpha_0 = \dfrac{1}{d}\sqrt{\dfrac{k_B T_f}{m\omega_0^2}}(1 + 4\dfrac{T_f}{T_a}) \simeq \underbrace{\dfrac{1}{d}\sqrt{\dfrac{k_B T_f}{m\omega_0^2}}}_{harmonic}\underbrace{(1 + 2\dfrac{T_f}{T_a})}_{\sim 1.2} \simeq 5\%,$

that is 20% increase of the average oscillation amplitude during the fusion. The stabilising term of anharmonicity naturally increases the cohesion of the crystal.

Exercise 4.5. Trapped ideal gas

A monoatomic gas, considered ideal, is maintained at temperature T. The gas is placed in a potential field $v(r) = ar$, where $a > 0$ is a constant and r is the distance of an atom to the centre of the device.

1. What is the effect of this potential on the gas atoms? Write the Hamiltonian of the system.

2. Express the system's partition function Z. Use recurrence to calculate the integral $J_n = \int_0^\infty dr\, r^n e^{-\beta a r}$, where n is a positive integer.

3. Deduce the average total energy $\langle E \rangle$ and the average potential and kinetic energies, $\langle E_{kin} \rangle$ and $\langle U_{pot} \rangle$.

4. Calculate $\langle r \rangle$ and its variance Δ_r^2. What is the average volume $\langle V \rangle$ occupied by the gas?

5. Express the gas entropy S as a function of a and then $\langle V \rangle$. By putting the device in an adiabatic enclosure and slowly decreasing the value of parameter a, from a_i to $a_f < a_i$, the gas can be cooled down. Express the final temperature T_f as a function of a_i, a_f and the initial temperature, T_i.

<u>Solution:</u>

1. *The atoms are attracted towards the centre of the device. The Hamiltonian is: $H_{IG} = \sum_{j=1}^N h_j$ with $h_j = \dfrac{\vec{p}^{\,2}}{2m} + ar$, where $\vec{p}$ and r, are the momentum vector and the radial coordinate of atom j, respectively.*

2. *The atoms are independent and indistinguishable. Therefore, $Z = z^N/N!$, where*

$$z = \frac{1}{h^3}\int_{-\infty}^{+\infty} d^3 p\, e^{-\beta \frac{\vec{p}^{\,2}}{2m}} \int_0^\infty dr \int_0^\pi d\theta \int_0^{2\pi} d\phi\, r^2 \sin\theta\, e^{-\beta a r}$$

$$= \frac{4\pi}{\lambda_T^3}\underbrace{\int_0^\infty dr\, r^2 e^{-\beta a r}}_{=J_2 = 2/(\beta a)^3} = 8\pi\left(\frac{k_B T}{\lambda_T a}\right)^3,$$

where the integration over momenta gives $(\sqrt{2\pi m k_B T})^3$ and the integration over angles θ and ϕ is 4π (spherical symmetry). An integration by parts and the use of recurrence lead to: $J_n = \int_0^\infty dr\, r^n e^{-\beta a r} = n!/(\beta a)^{n+1}$, hence the expression of J_2.

3. *Since $z \sim \beta^{-9/2}$, one obtains $\langle E \rangle = -\dfrac{\partial \ln Z}{\partial \beta} = 9N k_B T/2$. According to the equipartition of energy theorem, $\langle E_{kin} \rangle = 3N k_B T/2$, so $\langle U_{pot} \rangle = \langle E \rangle - \langle E_{kin} \rangle = 3N k_B T$.*

4. *By using the integral J_n:*

$$\langle r \rangle = \frac{\int_0^\infty dr\, r^3 e^{-\beta a r}}{\int_0^\infty dr\, r^2 e^{-\beta a r}} = 3\frac{k_B T}{a} \quad \text{and} \quad \langle r^2 \rangle = \frac{\int_0^\infty dr\, r^4 e^{-\beta a r}}{\int_0^\infty dr\, r^2 e^{-\beta a r}} = 12\left(\frac{k_B T}{a}\right)^2.$$

So $\Delta_r^2 = \langle r^2 \rangle - \langle r \rangle^2 = 3(k_B T/a)^2$. The relative fluctuations are $\Delta_r/\langle r \rangle = 1/\sqrt{3}$. And one recovers $\langle U_{pot} \rangle = Na\langle r \rangle = 3N k_B T$. In addition, $\langle V \rangle = \frac{4}{3}\pi\langle r \rangle^2 = 36\pi(k_B T/a)^3$.

5. From Equation (4.23), $S = -\frac{\partial F}{\partial T}\big|_N = Nk_B\left(\ln\left(\frac{8\pi}{N}\left(\frac{k_B T}{\lambda_T a}\right)^3\right) + \frac{11}{2}\right) = Nk_B\left(\ln\left(\frac{2}{9}\left(\frac{\langle V\rangle}{N\lambda_T^3}\right)\right) + \frac{11}{2}\right)$. During a reversible and adiabatic transformation, entropy is conserved (isentropic transformation). Since $S = 3Nk_B\left(\ln\left(T^{\frac{3}{2}}/a\right) + Cte\right)$, then $T_f = (a_f/a_i)^{\frac{2}{3}}T_i < T_i$.

Exercise 4.6. Maxwell velocity distribution function in the canonical ensemble

A gas consists of $N \gg 1$ particles of mass m, without internal structure, contained in a vessel of volume V maintained at temperature T.

1. Initially, the interactions between particles are neglected. Write the gas Hamiltonian and express its partition function Z as a function of the Debye thermal wavelength $\lambda_T = h/\sqrt{2\pi m k_B T}$.

2. Deduce the gas free energy F, the gas average energy $\langle E\rangle$, the gas pressure P, the gas chemical potential μ and the gas entropy S.

3. Now consider that the particles interact through a two-body central potential $u(r)$, where r is the distance between two particles. Write the Hamiltonian of this real gas.

4. Express the probability that a particle has a momentum $\vec{p}$ within $d\vec{p}$ and recover the Maxwell velocity distribution function, $P(\vec{v})$. What is the expression of the distribution function $P_x(v_x)$? Calculate $\langle v_x\rangle$, $\langle v_y\rangle$ and $\langle v_x v_y\rangle$.

5. Deduce that the probability that the *magnitude* of the velocity of a particle lies between v and $v + dv$ is given by the following distribution (shown in Figure 4.8):

$$W(v)dv = 4\pi\left(\frac{m}{2\pi k_B T}\right)^{\frac{3}{2}} v^2\, e^{-\frac{mv^2}{2k_B T}}\, dv. \tag{4.77}$$

Calculate the average velocity $\langle v\rangle$, the most probable velocity v^* and the mean square velocity $\langle v^2\rangle$ (Use the Gaussian integrals given in Appendix A.2). Give an estimate of these velocities for oxygen (O_2) under standard conditions of temperature and pressure.

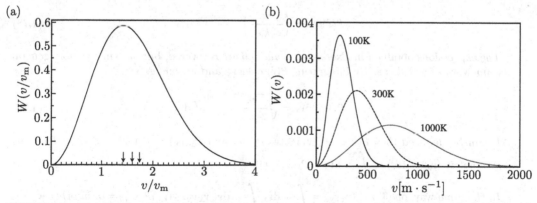

FIGURE 4.8 Distribution W of the velocity magnitude as given by Equation (4.77). (a) W as a function of the scaled velocity v/v_m where $v_m = \sqrt{k_B T/m}$ is a characteristic velocity. The three arrows show in ascending order the three velocities $v^* = \sqrt{2}v_m$, $\langle v\rangle = \sqrt{8/\pi}v_m$ and $\sqrt{\langle v^2\rangle} = \sqrt{3}v_m$. (b) distribution $W(v)$ for gaseous dioxygen (O_2, $M = 32$ g·mol^{-1}) at three different temperatures.

Solution:

1. The ideal gas Hamiltonian and the associated partition function are given by Equations (1.9) and (4.11), respectively.

2. These thermodynamic quantities are given by Equations (4.34), (4.35), (4.36), (4.37) and (3.29). The entropy S can be calculated using Equation (4.23).

3. The Hamiltonian is given by Equation (1.7): $H(\{\vec{r}_j, \vec{p}_j\}) = \sum_{j=1}^{N} \dfrac{\vec{p}_j^{\,2}}{2m} + \sum_{i<j} u(r_{ij})$.

4. The probability to find the gas in a given microstate is given by Equation (4.7) with $f = 3N$:

$$dP = \frac{e^{-\beta H(q,p)}}{Z(N,V,T)} \frac{d^{3N}q\, d^{3N}p}{h^{3N}} = \frac{1}{Zh^{3N}}\, e^{-\beta\left(\sum_{j=1}^{N}\frac{\vec{p}_j^{\,2}}{2m}+\sum_{i<j} u(r_{ij})\right)} \prod_{j=1}^{N} d^3 r_j \prod_{j=1}^{N} d^3 p_j.$$

The probability $P_1(\vec{p})d^3 p$ that a particle (say number 1) has a momentum $\vec{p}$ within $d\vec{p}$ is obtained by integrating dP over all microscopic variables except $\vec{p}_1$, which is fixed and equal to $\vec{p}$:

$$P_1(\vec{p})d^3 p = e^{-\beta \frac{\vec{p}^{\,2}}{2m}} \left(\frac{1}{Zh^{3N}} \int e^{-\beta\left(\sum_{j=2}^{N}\frac{\vec{p}_j^{\,2}}{2m}+\sum_{i<j} u(r_{ij})\right)} \prod_{j=1}^{N} d^3 r_j \prod_{j=2}^{N} d^3 p_j \right) d^3 p,$$

where the term in brackets is a constant independent of $\vec{p}$ that does not need to be calculated. By performing the change of variables $\vec{p} = m\vec{v}$, one obtains:

$$P(\vec{v})d^3 v = C\, e^{-\beta \frac{m\vec{v}^{\,2}}{2}} d^3 v,$$

where the constant C is calculated to ensure the normalisation of the distribution, so

$$\int_{-\infty}^{+\infty} dv_x \int_{-\infty}^{+\infty} dv_y \int_{-\infty}^{+\infty} dv_z\, P(\vec{v}) = C\left(\int_{-\infty}^{+\infty} dv_x\, e^{-\beta \frac{mv_x^2}{2}} \right)^3 = 1,$$

since the variables v_x, v_y and v_z are equivalent. A Gaussian integral equal to $\sqrt{2\pi k_B T/m}$ can be immediately recognised (see Appendix A.2), therefore:

$$P(\vec{v})d^3 v = \left(\frac{m}{2\pi k_B T} \right)^{\frac{3}{2}} e^{-\frac{m\vec{v}^{\,2}}{2k_B T}} d^3 v. \tag{4.78}$$

The expressions obtained in Exercises 1.3 and 3.6 are recovered, but here in the more general framework of a real gas. By integrating $P(\vec{v})$ over v_y and v_z, one gets:

$$P_x(v_x)dv_x = \sqrt{\frac{m}{2\pi k_B T}}\, e^{-\frac{mv_x^2}{2k_B T}} dv_x. \tag{4.79}$$

It can be deduced that $P(\vec{v}) = P_x(v_x)P_y(v_y)P_z(v_z)$ and $P_x(v_x) = P_y(v_y) = P_z(v_z)$. Thus, by symmetry, $\langle v_x \rangle = \langle v_y \rangle = \displaystyle\int_{-\infty}^{+\infty} dv_x\, v_x P_x(v_x) = 0$.

In the same way, one finds $\langle v_x v_y \rangle = \displaystyle\int_{-\infty}^{+\infty} dv_x \int_{-\infty}^{+\infty} dv_y\, v_x v_y P_x(v_x)P_y(v_y) = 0$, in other words, v_x, v_y (and v_z) behave as independent random variables.

5. $W(v)dv = \int P(\vec{v})d^3 v$, where the integral is performed by considering all velocity vectors $\vec{v}$ with magnitudes between v and $v+dv$. The number of such vectors is equal to the volume between the two spheres of radius v and $v + dv$ in velocity space, that is $4\pi v^2 dv$. Since $P(\vec{v})$ depends

only on the velocity magnitude v, one gets[50] $W(v)dv = 4\pi v^2 P(\vec{v})dv$, that is Equation (4.77). The average velocity can then be deduced:

$$\langle v \rangle = \int_0^\infty dv\, v W(v) = 4\pi \left(\frac{m}{2\pi k_B T}\right)^{\frac{3}{2}} \underbrace{\int_0^\infty dv\, v^3 e^{-\frac{mv^2}{2k_B T}}}_{=I_3(m/2k_B T)} = \sqrt{\frac{8k_B T}{\pi m}}, \qquad (4.80)$$

where $I_3(\alpha) = 1/(2\alpha^2)$, according to Appendix A.2. In addition, the most probable velocity is such that $\frac{dW(v)}{dv} = 2v^* e^{-\frac{mv^{*2}}{2k_B T}} + v^{*2}\left(-\frac{mv^*}{k_B T}\right) e^{-\frac{mv^{*2}}{2k_B T}} = 0$, so $v^* = \sqrt{\frac{2k_B T}{m}}$.

The mean square velocity is given by:

$$\langle v^2 \rangle = \int_0^\infty dv\, v^2 W(v) = 4\pi \left(\frac{m}{2\pi k_B T}\right)^{\frac{3}{2}} \underbrace{\int_0^\infty dv\, v^4 e^{-\frac{mv^2}{2k_B T}}}_{=I_4(m/2k_B T)} = \frac{3k_B T}{m}, \qquad (4.81)$$

where $I_4(\alpha) = \frac{3}{8\alpha^2}\sqrt{\frac{\pi}{\alpha}}$. Note that the equipartition of energy theorem gives directly these results since: $\langle mv^2/2\rangle = \langle mv_x^2/2\rangle + \langle mv_y^2/2\rangle + \langle mv_z^2/2\rangle = 3\langle mv_x^2/2\rangle = 3k_B T/2$. The expression of the typical velocity $\tilde{v}$, introduced in the kinetic theory of gases (see Section 1.4.2), is recovered. Finally, all of these velocities are of the order of $\sqrt{k_B T/m}$, i.e. for dioxygen at room temperature $\sim 500\ m\cdot s^{-1}$.

Exercise 4.7. Molecular jet

A vessel of volume V initially contains $N_0 \gg 1$ gas molecules of mass m maintained at temperature T. A small hole of area $\mathcal{A} = (10\ \mu m)^2$ is drilled in the wall. Molecules can thus escape to the outside (vacuum). The dimensions of the hole are assumed smaller than the molecules' mean free path (see Section 1.4.2), which allows to neglect the influence of collisions between molecules when passing through the hole (gas effusion). Moreover, the evolution of the system is considered quasi-static and isothermal, at fixed temperature T.

1. What is the average number of molecules $-dN$ exiting the vessel during a time dt?

2. Deduce the evolution of the average number of molecules $N(t)$ remaining in the container at time t. What is the process characteristic time τ? The vessel contains 1 L of helium at $T = 300$ K and $P = 10^{-3}$ atm. According to Equation (1.23), are these conditions favourable for the observation of gas effusion? Evaluate τ.

3. To create a molecular jet, a diaphragm is built by placing a wall with a hole of identical area $\mathcal{A}$ aligned with the first orifice at a distance $d = 1$ m from the container. What is the solid angle $d\Omega$ of the jet as a function of $\mathcal{A}$ and d? Calculate the molecules flux Φ in the jet as a function of the average velocity of the molecules in the container, $\langle v \rangle$. What is Φ for the helium gas considered in the previous question?

4. Express the jet molecules velocity magnitude probability distribution $g(v_j)$. Calculate the average velocity $\langle v_j \rangle$, the most probable velocity v_j^* and the mean square velocity $\langle v_j^2 \rangle$ of the jet molecules and compare these expressions to those of the container molecules derived in Exercise 4.6.

---

[50]This result can be recovered using spherical coordinates. With $d^3v = dv(vd\theta)(v\sin\theta d\phi)$, one has: $\int d^3v = \int dv\, v^2 \int_0^{2\pi} d\phi \int_0^\pi d\theta\, \sin\theta = \int dv\, 4\pi v^2$.

5. The pressure exerted by a gas on a container is the force per unit area that results from the numerous collisions (assumed elastic) between the molecules and the walls. What is the momentum transferred to the wall by a molecule whose incident velocity is $\vec{v}$? What is the total momentum $\mathrm{d}\vec{\mathcal{P}}$ transferred to the wall due to molecules collisions on a surface element of area $\mathrm{d}\mathcal{A}$ during time $\mathrm{d}t$? Deduce the expression of the kinetic pressure[51] P_{kin} as a function of $\langle v^2 \rangle$.

Solution:

1. _To a first approximation, about $\rho/6$ molecules per unit volume have a velocity vector directed towards the hole, but only the molecules contained in the cylinder of base $\mathcal{A}$ and length $\langle v \rangle \mathrm{d}t$, perpendicular to the hole are close enough to exit the vessel during a time $\mathrm{d}t$. Thus $-\mathrm{d}N \simeq \rho\mathcal{A}\langle v \rangle \mathrm{d}t/6$._

   More precisely, there are on average $\rho P(\vec{v})\mathrm{d}^3 v$ molecules per unit volume with a velocity vector lying between $\vec{v}$ and $\vec{v} + \mathrm{d}\vec{v}$, i.e. with a modulus lying between v and $v + \mathrm{d}v$ and an angular orientation between θ and $\theta + \mathrm{d}\theta$ and between ϕ and $\phi + \mathrm{d}\phi$. Of these, only the molecules at a distance from the wall less than $v\mathrm{d}t$, in the oblique cylinder of base $\mathcal{A}$ and volume $\mathcal{A}v\mathrm{d}t\cos\theta$, can escape during $\mathrm{d}t$ (see Figure 4.9), so

$$-\mathrm{d}N_{\vec{v}} = \mathcal{A}\underbrace{v\cos\theta}_{=v_x}\mathrm{d}t\,\rho P(\vec{v})\mathrm{d}^3 v. \qquad (4.82)$$

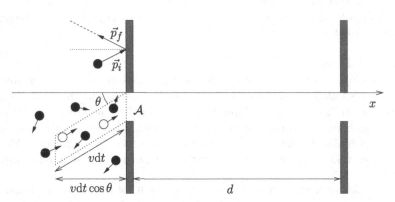

FIGURE 4.9 Only white molecules leave the container through the hole of area $\mathcal{A}$ during $\mathrm{d}t$ with a velocity vector $\vec{v}$ that makes an angle θ with the x-axis. They are contained in the oblique cylinder of volume $\mathcal{A}v\mathrm{d}t\cos\theta$. At the top, a molecule hits the wall with an incident momentum $\vec{p}_i$ and bounces back with momentum $\vec{p}_f$. To create a molecular jet, a perforated wall is placed at a distance d from the container.

   The total average number of molecules escaping during the time $\mathrm{d}t$ is thus obtained by integrating $-\mathrm{d}N_{\vec{v}}$ over all velocities such as $0 \leq v_x \leq \infty$ (velocity towards the wall)[52].

---

[51] That is the component of pressure due to molecules collisions on the wall, without taking into account the molecular interactions present in a real gas (see Chapter 5).

[52] In spherical coordinates, $\mathrm{d}^3 v = v^2 \sin\theta \mathrm{d}v\mathrm{d}\theta\mathrm{d}\phi$, the sum is made over all velocities such as $0 \leq \theta \leq \pi/2$ (velocity directed towards the wall):

$$\begin{aligned}
-\mathrm{d}N &= \mathcal{A}\rho\mathrm{d}t \int_0^\infty \mathrm{d}v\, v^3 P(\vec{v}) \int_0^{2\pi} \mathrm{d}\phi \int_0^{\pi/2} \mathrm{d}\theta\,\cos\theta\sin\theta = \pi\rho\mathcal{A}\mathrm{d}t \int_0^\infty \mathrm{d}v\, v^3 P(\vec{v}) \\
&= \pi\rho\mathcal{A}\mathrm{d}t \left(\frac{m}{2\pi k_{\mathrm{B}}T}\right)^{\frac{3}{2}} \underbrace{\int_0^\infty \mathrm{d}v\, v^3 e^{-\frac{mv^2}{2k_{\mathrm{B}}T}}}_{=I_3(m/2k_{\mathrm{B}}T)} = \rho\mathcal{A}\sqrt{\frac{k_{\mathrm{B}}T}{2\pi m}}\mathrm{d}t,
\end{aligned}$$

where $I_3(\alpha) = \dfrac{1}{2\alpha^2}$, according to Appendix A.2.

Since $P(\vec{v})\mathrm{d}^3v = P_x(v_x)\mathrm{d}v_x\, P_y(v_y)\mathrm{d}v_y\, P_z(v_z)\mathrm{d}v_z$, then according to Equation (4.79):

$$-\mathrm{d}N = \int_{v_x \geq 0} -\mathrm{d}N_{\vec{v}} = \mathcal{A}\rho\mathrm{d}t \int_0^{\infty} \mathrm{d}v_x\, v_x P_x(v_x) \underbrace{\int_{-\infty}^{\infty} \mathrm{d}v_z\, P_z(v_z)}_{=1} \underbrace{\int_{-\infty}^{\infty} \mathrm{d}v_y\, P_y(v_y)}_{=1}$$

$$= \mathcal{A}\rho\mathrm{d}t \sqrt{\frac{m}{2\pi k_B T}} \underbrace{\int_0^{\infty} \mathrm{d}v_x\, v_x e^{-\frac{mv_x^2}{2k_B T}}}_{=I_1(m/2k_B T)} = \rho\mathcal{A}\sqrt{\frac{k_B T}{2\pi m}}\mathrm{d}t = \frac{\rho}{4}\mathcal{A}\langle v\rangle \mathrm{d}t, \qquad (4.83)$$

where $I_1(\alpha) = 1/(2\alpha)$ (see Appendix A.2) and $\langle v\rangle = \sqrt{8k_B T/\pi m}$, according to Equation (4.80). This result is very close to the first estimate. In 1909, the Danish physicist Martin Knudsen measured molecular fluxes, $-\frac{1}{\mathcal{A}}\frac{\mathrm{d}N}{\mathrm{d}t}$, in very good agreement with Equation (4.83), validating the prediction from the kinetic theory of gases. The expression of the phenomenological law discovered in the nineteenth century by the Scottish chemist Thomas Graham is also demonstrated: $-\frac{\mathrm{d}N}{\mathrm{d}t} \sim \frac{1}{\sqrt{M}}$. In other words, the effusion rate of a gas is inversely proportional to the square root of its molar mass. Effusion allows the enrichment of a mixture of gases: the lighter elements leave faster the container and the proportion of heavier element increases. It is a method used for the isotopic separation of a chemical element, for example to enrich uranium in its fissile isotope ^{235}U.

2. *Assuming an ideal gas, $\rho = P/k_B T \simeq 2.4 \times 10^{22}$ m^{-3}. According to Equation (1.23), with $\sigma = 2.6$ Å, the mean free path $l \simeq 200$ μm is larger than the size of the hole $\sqrt{\mathcal{A}} = 10$ μm. It is indeed a gas effusion. According to Equation (4.83) and with $\rho = N/V$, one obtains*

$$N(t) = N_0 e^{-\frac{t}{\tau}}, \text{ where } \tau = \frac{V}{\mathcal{A}}\sqrt{\frac{2\pi M}{RT}} \simeq 9 \text{ hours.}$$

3. $\mathrm{d}\Omega = \sin\theta\mathrm{d}\theta\mathrm{d}\phi = \mathcal{A}/d^2$. *According to Equation (4.82), since $\mathrm{d}^3v = v^2\mathrm{d}v\mathrm{d}\Omega$ and $\theta \simeq 0$:*

$$-\mathrm{d}N_{\vec{v}} = \rho\mathcal{A}v \underbrace{\cos\theta}_{=1} \mathrm{d}t\, P(\vec{v})v^2\mathrm{d}v\mathrm{d}\Omega = \rho\frac{\mathcal{A}^2}{d^2}\mathrm{d}t\, v^3 P(\vec{v})\mathrm{d}v.$$

Integrating over velocities and using Expression (4.78), the flux can be deduced:

$$\Phi = -\int \frac{\mathrm{d}N_{\vec{v}}}{\mathrm{d}t} = \rho\frac{\mathcal{A}^2}{d^2}\left(\frac{m}{2\pi k_B T}\right)^{\frac{3}{2}} \int_0^{\infty} \mathrm{d}v\, v^3 e^{-\frac{mv^2}{2k_B T}} = \frac{\rho\mathcal{A}^2}{\pi d^2}\sqrt{\frac{k_B T}{2\pi m}} = \frac{\rho\mathcal{A}^2}{4\pi d^2}\langle v\rangle.$$

For helium at $T = 300$ K, $\langle v\rangle = \sqrt{8RT/\pi M} \simeq 1260$ m.s^{-1}, so $\Phi \simeq 2.4 \times 10^4$ s^{-1}. Molecular jets are often used in atomic physics such as in the famous Stern and Gerlach experiment (1921) which demonstrated the existence of spin (see for example Claude Cohen-Tannoudji, Bernard Diu and Franck Laloë, Quantum Mechanics Volume I, Basic Concepts, Tools and Applications (Wiley-VCH, 2019)).

4. *Similarly, the distribution $g(v_j)$ can be deduced. It behaves as $v_j^3 P(\vec{v}_j)$, while the distribution of the magnitude of the velocity in the container, $W(v)$, behaves as $v^2 P(\vec{v})$ (see Exercise 4.6). Thus, using the distribution normalisation:*

$$g(v_j)\mathrm{d}v_j = \frac{1}{2}\left(\frac{m}{k_B T}\right)^2 v_j^3 e^{-\frac{mv_j^2}{2k_B T}} \mathrm{d}v_j. \qquad (4.84)$$

Using Appendix A.2, one obtains

$$\langle v_j\rangle = \int_0^{\infty} \mathrm{d}v_j\, v_j g(v_j) = \sqrt{\frac{9\pi k_B T}{8m}} = \frac{3\pi}{8}\langle v\rangle$$

$$\langle v_j^2\rangle = \int_0^{\infty} \mathrm{d}v_j\, v_j^2 g(v_j) = \frac{4k_B T}{m} = \frac{4}{3}\langle v^2\rangle$$

and the most probable velocity is such that $\dfrac{dg(v_{\mathrm{j}})}{dv_{\mathrm{j}}} = 0$, *so* $v_{\mathrm{j}}^{*} = \sqrt{\dfrac{3k_{\mathrm{B}}T}{m}} = \sqrt{\dfrac{3}{2}}v^{*}$. *The velocities are larger than those in the container, since the fastest molecules are the most numerous to escape. The distribution* $g(v_{\mathrm{j}})$ *can be measured using a device that collects atoms in a molecular jet as a function of their velocities. The agreement with Equation (4.84) is excellent (see R.C. Miller and P. Kush, Velocity Distributions in Potassium and Thallium Atomic Beams, Phys. Rev.* **99**, *1314 (1955)).*

5. *During an elastic collision, the momentum component* $m\vec{v}$ *parallel to the wall is conserved, while its normal component,* mv_x, *changes sign (see Figure 4.9). The molecule momentum change is thus:* $\delta\vec{p} = \vec{p}_f - \vec{p}_i = -mv_x\vec{u}_x - (mv_x\vec{u}_x) = -2mv_x\vec{u}_x$, *where* $\vec{u}_x$ *is a unit vector perpendicular to the wall pointing to the outside of the container. The variation of wall momentum is accordingly* $+2mv_x\vec{u}_x$. *According to Equation (4.83):*

$$d\mathcal{P}_x = \int_{v_x \geq 0} dN_{\vec{v}}\, 2mv_x = 2\rho m d\mathcal{A} dt \underbrace{\int_0^{\infty} dv_x\, v_x^2 P_x(v_x)}_{=\langle v_x^2\rangle/2} = \frac{1}{3}\rho m d\mathcal{A} dt \langle v^2\rangle,$$

because $\langle v^2\rangle = 3\langle v_x^2\rangle$ *(see Exercise 4.6). According to the fundamental principle of dynamics, the force per unit area exerted by the gas on the wall, that is the pressure, is:*

$$P_{\mathrm{kin}} = \frac{1}{d\mathcal{A}}\frac{d\mathcal{P}_x}{dt} = \frac{1}{3}\rho m \langle v^2\rangle = \rho k_{\mathrm{B}}T, \tag{4.85}$$

since according to Equation (4.81), $\langle v^2\rangle = 3k_{\mathrm{B}}T/m$ *for all real classical fluid. For such system, the kinetic pressure is therefore always equal the ideal gas pressure*[53]. *In general, the total pressure of a real fluid, which leads to its equation of state, is written as* $P = P_{\mathrm{kin}} + P_{\mathrm{dyn}}$, *where* P_{dyn} *is the dynamic pressure, which takes into account the interactions between molecules (see Chapter 5).*

Exercise 4.8. Hydrochloric gas in an electric field

A vessel of volume V contains gaseous hydrogen chloride maintained at temperature T. Each of the N diatomic HCl molecules of mass m is modelled as a rigid rotator[54] of moment of inertia I. The difference in electronegativity of the two atoms H and Cl creates a permanent electric dipole moment $\vec{d}$ directed along the axis of the molecule and of magnitude $d = 1.08$ D (the Debye (D) is a dipole moment unit, 1 D$\approx 3.335 \times 10^{-30}$ C.m). The gas is immersed in a constant and uniform electric field, $\vec{\mathcal{E}} = \mathcal{E}\vec{u}_z$ directed along the z–axis. The interaction energy of a dipole with the electric field is given by $\epsilon = -\vec{d}.\vec{\mathcal{E}}$. The gas is sufficiently diluted to be considered as ideal (in particular the dipole-dipole interactions are ignored).

1. Let $\vec{p}$ be the centre-of-mass momentum of a given molecule, and the angles θ and ϕ and the associated generalised momenta p_θ and p_ϕ be the variables describing the molecule's rotational motion (see Figure 4.10). Write the gas Hamiltonian H.

2. What is the expression for the gas partition function Z? Express the partition function of a single molecule z in terms of z_{trans} and z_{rot}, the partition functions associated with the translational and rotational degrees of freedom, respectively. Recall the expression of z_{trans}.

3. Express the rotational partition function z_{rot} of a molecule and integrate first on the variables ϕ, p_θ, p_ϕ, and then over θ.

---

[53]Note that Expression (1.19) of the pressure, obtained naively in Section 1.4.2 assuming that all molecules had the same velocity $\tilde{v}$, is rigorously recovered provided that $\tilde{v}^2$ is replaced by $\langle v^2\rangle$.

[54]The (internal) vibrational degrees of freedom associated with oscillations of the H and Cl atoms are ignored.

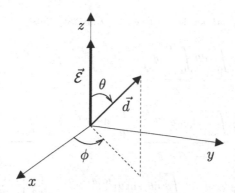

FIGURE 4.10 Dipole moment $\vec{d}$ in an electric field $\vec{\mathcal{E}}$ directed along the z–axis.

4. Calculate the average total energy per particle $\langle e \rangle$. What is the average contribution of the kinetic energy $\langle e_{\text{kin}} \rangle$? Deduce the average value $\langle \epsilon \rangle$ of the interaction energy with the electric field.

5. Express the average values of the dipole moment components $\langle d_x \rangle$, $\langle d_y \rangle$ and $\langle d_z \rangle$. Plot $\langle d_z \rangle$ as a function of $x = \beta \mathcal{E} d$ and comment on its behaviour.

6. Show that in a weak electric field, the gas dipole moment is $\vec{\mathcal{P}} = \chi \epsilon_0 (V/v)\vec{\mathcal{E}}$, where $\epsilon_0 = 8.85 \times 10^{-12}$ m^{-3}.kg^{-1}.s^4.A^2 is the permittivity of vacuum and χ is the electric susceptibility of the gas relative to a molar volume $v = (\mathcal{N}_A/N)V$. Express χ as a function of T and compare to the experimental value: $\chi_{\text{exp}} = (2.01 \times 10^{-2})/T$ m^3.mol^{-1}.

7. In 2005, researchers at the University of Seoul succeeded in forming water ice at $T = 293$ K, by applying a high electric field ($\mathcal{E} = 10^6$ V.m^{-1}) on a liquid water film of nanometric thickness[55]. The dipole moment of H_2O is $d = 1.85$ D. In the framework of the model studied in this exercise, for which values of $\mathcal{E}$ do the water molecules align with each other at room temperature?

Solution:

1. *According to Equation (4.41) and taking into account the energy of the interaction with the electric field: $H = \sum_{i=1}^{N} h_j$, where h_j is the Hamiltonian of molecule j:*

$$h_j = \underbrace{\frac{\vec{p}^{\,2}}{2m}}_{\text{translation}} + \underbrace{\frac{p_\theta^2}{2I} + \frac{p_\phi^2}{2I \sin^2 \theta} - \mathcal{E}d \cos\theta}_{\text{rotation}}.$$

2. *The molecules are independent and indistinguishable, so $Z = z^N/N!$ with $z = z_{\text{trans}} \times z_{\text{rot}}$, since the translational and rotational degrees of freedom are independent. Therefore:*

$$z_{\text{trans}} = \frac{1}{h^3} \int d^3r d^3p\, e^{-\beta \frac{\vec{p}^{\,2}}{2m}} = \frac{1}{h^3} \underbrace{\int d^3r}_{=V} \underbrace{\int d^3p\, e^{-\beta \frac{\vec{p}^{\,2}}{2m}}}_{=(2\pi m k_B T)^{3/2}} = \frac{V}{\lambda_T^3} \sim \beta^{-\frac{3}{2}}.$$

---

[55]See E.-M. Choi *et al.*, *Freezing transition of interfacial water at room temperature under electric fields*, Phys. Rev. Lett. **95**, 085701 (2005).

3. There are $f = 2$ rotational degrees of freedom, θ and ϕ, thus:

$$
\begin{aligned}
z_{\text{rot}} &= \frac{1}{h^2} \int_0^\pi d\theta \underbrace{\int_0^{2\pi} d\phi}_{2\pi} \int_{-\infty}^{+\infty} dp_\theta \int_{-\infty}^{+\infty} dp_\phi \; e^{-\beta\left(\frac{p_\theta^2}{2I} + \frac{p_\phi^2}{2I\sin^2\theta} - \mathcal{E}d\cos\theta\right)} \\
&= \frac{2\pi}{h^2} \underbrace{\int_{-\infty}^{+\infty} dp_\theta \; e^{-\frac{p_\theta^2}{2Ik_BT}}}_{=\sqrt{2I\pi k_BT}} \int_0^\pi d\theta \; e^{\beta\mathcal{E}d\cos\theta} \underbrace{\int_{-\infty}^{+\infty} dp_\phi \; e^{-\frac{p_\phi^2}{2Ik_BT\sin^2\theta}}}_{=\sqrt{2I\pi k_BT\sin^2\theta}} \\
&= \frac{2\pi}{h^2} \, 2I\pi k_BT \int_0^\pi d\theta \; \sin\theta \, e^{\beta\mathcal{E}d\cos\theta} = \frac{I}{\hbar^2\beta}\left[-\frac{e^{\beta\mathcal{E}d\cos\theta}}{\beta\mathcal{E}d}\right]_0^\pi \\
&= \frac{2I}{\hbar^2\mathcal{E}d}\frac{\sinh\beta\mathcal{E}d}{\beta^2}, \quad \text{where} \quad \hbar = h/2\pi.
\end{aligned}
$$

4. The average energy per particle is given by:

$$
\begin{aligned}
\langle e \rangle &= -\frac{\partial \ln z}{\partial \beta} = -\frac{\partial \ln z_{\text{trans}}}{\partial \beta} - \frac{\partial \ln z_{\text{rot}}}{\partial \beta} \\
&= \frac{3}{2\beta} + \frac{2}{\beta} - \mathcal{E}d\frac{\cosh\beta\mathcal{E}d}{\sinh\beta\mathcal{E}d} = \frac{7}{2\beta} - \frac{\mathcal{E}d}{\tanh\beta\mathcal{E}d}.
\end{aligned}
$$

Since the kinetic energy can be written as five independent quadratic terms (3 for translation in p_x^2, p_y^2 and p_z^2, and 2 for rotation in p_θ^2 and p_ϕ^2), the equipartition of energy theorem implies $\langle e_{\text{kin}} \rangle = 5\, k_BT/2 = 5/(2\beta)$, so

$$
\langle \epsilon \rangle = \langle e \rangle - \langle e_{\text{kin}} \rangle = \frac{1}{\beta} - \frac{\mathcal{E}d}{\tanh\beta\mathcal{E}d}.
$$

5. By symmetry, $\langle d_x \rangle = \langle d_y \rangle = 0$. Since $\langle \epsilon \rangle = -\mathcal{E}\langle d_z \rangle$, where $d_z = d\cos\theta$, then

$$
\langle d_z \rangle = -\frac{\langle \epsilon \rangle}{\mathcal{E}} = d\left(\frac{1}{\tanh\beta\mathcal{E}d} - \frac{1}{\beta\mathcal{E}d}\right) = L(x)d \quad \text{wit} \quad x = \beta\mathcal{E}d, \tag{4.86}
$$

where $L(x) = \dfrac{1}{\tanh x} - \dfrac{1}{x}$ is the Langevin function shown in Figure 4.11. At low temperature or high field ($x \gg 1$, $L \to 1$), $\langle d_z \rangle \simeq d$. Each dipole moment is aligned with the electric field along the z–axis. On the other hand, at high temperature or low field, for $x \ll 1$:

$$
L(x) \simeq \left(x - \frac{x^3}{3}\right)^{-1} - \frac{1}{x} \simeq \frac{1}{x}\left(1 + \frac{x^2}{3}\right) - \frac{1}{x} = \frac{x}{3} \to 0,
$$

and $\langle d_z \rangle$ tends to 0, since the dipole moments are randomly and uniformly orientated.

6. For $x = \beta\mathcal{E}d \ll 1$, $\mathcal{P}_z = N\langle d_z \rangle \simeq Nd\dfrac{\beta\mathcal{E}d}{3} = \chi\epsilon_0\dfrac{V}{v}\mathcal{E} = \chi\epsilon_0\dfrac{N}{N_A}\mathcal{E}$, therefore $\chi = \dfrac{N_A d^2}{3\epsilon_0 k_BT}$, or $\chi \simeq (2.13\times10^{-2})/T$, where T is in kelvin, in good agreement with experimental data (relative error of 6%).

7. One must have $\langle d_z \rangle \simeq d$, so $L(x) \simeq 1$, condition obtained for $x = \beta\mathcal{E}d \gg 1$. At $T = 300K$, one finds $\mathcal{E} \gg k_BT/d \simeq 7\times10^8$ V·m^{-1}, values much larger than the experimental ones. The studied model relies on too simple hypotheses (ideal gas) to explain crystal formation. It is necessary to take into account the interactions between molecules and other mechanisms specific to water molecules.

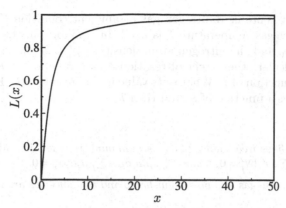

FIGURE 4.11 Langevin function $L(x) = \langle d_z \rangle / d$ (projection of the dipole moment on the z-axis) as a function of $x = \mathcal{E}d/(k_B T)$.

Exercise 4.9. Law of mass action and dissociation of the dinitrogen molecule

A chemical reaction is usually written as $\sum_i a_i A_i \rightleftharpoons \sum_j a'_j A'_j$, where A_i and A'_j are the chemical formulas of the reactants and products, respectively, and where the numbers a_i and a'_j are the stoichiometric coefficients. For example, water dissociation can be written as $H_3O^+ + OH^- \rightleftharpoons 2\,H_2O$. The variations of the numbers of molecules $N_i \gg 1$ and $N'_j \gg 1$ are linked together by:

$$-\frac{dN_i}{a_i} = \frac{dN'_j}{a'_j} = d\xi,$$

where ξ is the extent of reaction.

1. Show that chemical equilibrium at constant temperature and pressure imposes $\sum_i a_i \mu_i = \sum_j a'_j \mu'_j$, where μ_i and μ'_j are the chemical potentials of the reactants and products, respectively.

2. The substances involved in the reaction are (assumed) ideal gases. Write the partition function Z of the gas mixture as a function of z_i and z'_j, the single molecule partition functions. Express μ_i and μ'_j.

3. Deduce the law of mass action:

$$\frac{\prod_j (\rho'_j)^{a'_j}}{\prod_i (\rho_i)^{a_i}} = K(T),$$

where $K(T)$ is the equilibrium constant which depends only on temperature and where $\rho_i = N_i/V$ and $\rho'_j = N'_j/V$ are the particle densities of reactants and products, respectively.

4. Prove the van't Hoff relation[56] $\dfrac{d\ln K}{dT} = \dfrac{Q_V}{k_B T^2}$, where Q_V is the heat of reaction at constant volume. Discuss the chemical equilibrium shift as a function of temperature.

5. In the case of the dinitrogen molecule dissociation, $N_2 \rightleftharpoons N+N$, the internal structure of the N atoms is ignored and the N_2 molecule binding energy[57] is $-\epsilon = -9.8$ eV < 0. Express $K(T)$ and describe its behaviour as a function of T.

---

[56] Jacobus van't Hoff (1852–1911), Dutch chemist who was the first recipient of the Nobel Prize in Chemistry in 1901 for his discoveries of the laws of chemical kinetics and osmotic pressure (see Exercise 5.6).

[57] The rotational and vibrational degrees of freedom of N_2 are ignored.

6. A vessel initially contains dinitrogen gas at ambient condition (i.e. $P_0 = 1$ atm and $T_0 = 300$ K). The gas temperature T is fixed. In equilibrium, the N_2 density is given by $\rho_1 = (1 - \xi)\rho_0$ and the nitrogen atom density is $\rho_1' = 2\xi\rho_0$, where ρ_0 is the initial gas density. Calculate the extent of reaction ξ as a function of $K(T)$ and discuss its behaviour as a function of T. What is its value in the initial state? Express the pressure of the mixture as a function of ξ, and then T.

Solution:

1. _In equilibrium, Gibbs free energy (4.68) is minimum, thus for $dT = dP = 0$, we then have $dG = \sum_i \mu_i dN_i + \sum_j \mu_j' dN_j' = 0$, hence $- \sum_i \mu_i a_i d\xi + \sum_j \mu_j' a_j' d\xi = 0$._

2. _Molecules of the same gas are indistinguishable and all molecules are independent, so:_

$$Z = \prod_{i,j} \frac{(z_i)^{N_i}}{N_i!} \frac{(z_j')^{N_j'}}{N_j'!}. \tag{4.87}$$

According to Equation (4.20), $\mu_i = \left.\frac{\partial F}{\partial N_i}\right|_{V,T} = -k_B T \frac{\partial}{\partial N_i}(N_i \ln z_i - N_i!) \simeq -k_B T \ln \frac{z_i}{N_i}$ (using Stirling's formula (A.21)). Similarly, $\mu_j' = -k_B T \ln \frac{z_j'}{N_j'}$.

3. _So $\sum_i a_i \mu_i = \sum_j a_j' \mu_j'$ becomes $\prod_i \left(\frac{z_i}{N_i}\right)^{a_i} = \prod_j \left(\frac{z_j'}{N_j'}\right)^{a_j'}$, or by dividing N_i and N_j' by V:_

$$\frac{\prod_j (\rho_j')^{a_j'}}{\prod_i (\rho_i)^{a_i}} = V^{-\left(\sum_j a_j' - \sum_i a_i\right)} \frac{\prod_j (z_j')^{a_j'}}{\prod_i (z_i)^{a_i}} \equiv K(T). \tag{4.88}$$

The equilibrium constant $K(T)$ depends only on temperature, because for an ideal gas[58], $z(V,T) = V f(T)$, where the function $f(T)$ is the integral over momenta (which gives the factor $1/\lambda_T^3$) and over the internal degrees of freedom of the molecule and their conjugate momenta.

4. _$\frac{d\ln K}{dT} = -\frac{1}{k_B T^2} \frac{\partial}{\partial \beta}\left(\sum_j a_j' \ln z_j' - \sum_i a_i \ln z_i\right) = \frac{Q_v}{k_B T^2}$, where the heat of reaction is given by $Q_v = \left(\sum_j a_j' \langle e' j \rangle - \sum_i a_i \langle e_i \rangle\right)$. If the reaction is endothermic[59] ($Q_v > 0$), when T increases $K(T)$ increases and more reactant A_i are transformed into products A_j'. If the reaction is exothermic ($Q_v < 0$), when T increases $K(T)$ decreases and less reaction products are obtained. This is an example of the application of the principle of Le Chatelier (1884): "When external modifications to a physical-chemical system in equilibrium cause an evolution towards a new state of equilibrium, the evolution opposes the disturbances which have generated it and moderates their effect"._

5. _In this case, $a_1 = 1$ and $a_1' = 2$. The partition functions of N_2 and N are, respectively:_

$$z_1 = \frac{1}{h^3} \int d^3 r d^3 p \ e^{-\beta\left(\frac{\vec{p}^{\,2}}{2(2m)} - \epsilon\right)} = \frac{V}{(\lambda_T/\sqrt{2})^3} e^{\beta\epsilon} \quad \text{and} \quad z_1' = \frac{V}{\lambda_T^3},$$

where m and λ_T are the mass and the thermal wavelength of the N atom (for N_2, $2m$ and $\lambda_T/\sqrt{2}$, respectively). So according to Equation (4.88),

$$\frac{(\rho_1')^2}{\rho_1} = K(T) = \frac{1}{V} \frac{(z_1')^2}{z_1} = \frac{1}{2\sqrt{2}\lambda_T^3} e^{-\beta\epsilon}. \tag{4.89}$$

Thus $K(T) \sim T^{\frac{3}{2}} e^{-\epsilon/k_B T}$ is an increasing function of T.

---

[58]The integral over the centre of mass position of a molecule gives a factor V (see Equation (4.11)).
[59]Heat is absorbed by the mixture during the reaction, so $Q_v > 0$.

6. $\dfrac{(\rho_1')^2}{\rho_1} = \dfrac{4\rho_0\xi^2}{1-\xi} = K$. *Solving this second degree equation leads to* $\xi = \dfrac{K}{8\rho_0}\left(\sqrt{1 + \dfrac{16\rho_0}{K}} - 1\right) > 0,$

with $\dfrac{K}{\rho_0} = K\dfrac{k_{\rm B}T_0}{P_0} \simeq 142\ T^{\frac{3}{2}}\ e^{-113700/T}$, *where* T *is in kelvin, according to Equation* (4.89).
A low temperature, $K/\rho_0 \to 0$, *so* $\xi \simeq \frac{1}{2}\sqrt{\dfrac{K}{\rho_0}} \simeq 0$, N_2 *molecules are not dissociated. At very*
high temperature, $\rho_0/K \to 0$, *so* $\xi \simeq \dfrac{K}{8\rho_0}\left(\left(1 + \dfrac{16\rho_0}{2K}\right) - 1\right) \simeq 1$, *all dinitrogen molecules are*
dissociated. The behaviour of ξ *as a function of* T *is shown in Figure 4.12. Under standard*
(room) conditions $(P = P_0$ *and* $T = T_0)$, $\xi \simeq 0$.

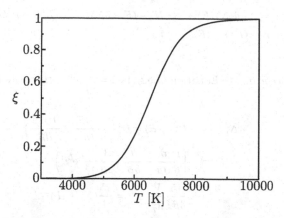

FIGURE 4.12 Extent of reaction ξ for N_2 dissociation as a function of temperature T.

The pressure of a mixture of ideal gases is the sum of the partial pressures of the different gases[60], *thus* $P = \rho_1 k_{\rm B}T + \rho_1' k_{\rm B}T = (1-\xi)\rho_0 k_{\rm B}T + 2\xi\rho_0 k_{\rm B}T = (1+\xi)\rho_0 k_{\rm B}T$. *Since* ξ *increases from 0 to 1 with* T, *the pressure changes continuously from* $\rho_0 k_{\rm B}T$ *to* $2\rho_0 k_{\rm B}T$ *when* T *increases.*

Exercise 4.10. Energy fluctuations

In this exercise, energy fluctuations in the microcanonical, canonical and grand canonical statistical ensembles are compared

1. Recall the value of Δ_E^2 in the microcanonical ensemble.

2. Write Δ_E^2 in the canonical ensemble as a function of the heat capacity C_V.

3. Let the differential operator $D \equiv \left[\dfrac{\mu}{\beta}\dfrac{\partial}{\partial\mu} - \dfrac{\partial}{\partial\beta}\right]$. Show that in the grand canonical ensemble $\Delta_E^2 = D^2(\ln\Xi)$, where Ξ is the grand partition function.

4. Compare relative fluctuations of energy in these three statistical ensembles for an ideal gas. Comment.

Solution:

1. *In the microcanonical ensemble,* E *is fixed, so* $\Delta_E^2 = 0$.

2. *In the canonical ensemble according to Equation* (4.27), $\Delta_E^2 = \langle E^2\rangle - \langle E\rangle^2 = k_{\rm B}T^2 C_V$.

---

[60]This results can be recovered by deriving $\ln Z$ with respect to V, the partition function (4.87) of the mixture being equal to the product of the partition functions of each gas.

3. In the grand canonical ensemble, according to Equation (4.48), $\Xi = \sum_m e^{-\beta(E_m - \mu N_m)}$, thus[61]:

$$
\begin{aligned}
D^2(\ln \Xi) &= D\left(\frac{\mu}{\beta}\frac{\partial \ln \Xi}{\partial \mu} - \frac{\partial \ln \Xi}{\partial \beta}\right) \\
&= D\left(\frac{\mu}{\Xi}\sum_m N_m e^{-\beta(E_m - \mu N_m)} + \frac{1}{\Xi}\sum_m (E_m - \mu N_m)e^{-\beta(E_m - \mu N_m)}\right) \\
&= \left[\frac{\mu}{\beta}\frac{\partial}{\partial \mu} - \frac{\partial}{\partial \beta}\right]\left(\frac{1}{\Xi}\sum_m E_m e^{-\beta(E_m - \mu N_m)}\right) \\
&= -\mu\langle N\rangle\langle E\rangle + \mu\langle EN\rangle - \langle E - \mu N\rangle\langle E\rangle + \langle E(E - \mu N)\rangle \\
&= \langle E^2\rangle - \langle E\rangle^2 = \Delta_E^2.
\end{aligned}
$$

4. For an ideal gas, according to Equation (4.52), $\ln \Xi = \frac{V}{\lambda_T^3} e^{\beta\mu}$, therefore in the grand canonical ensemble:

$$
\begin{aligned}
\Delta_E^2 &= D^2(\ln \Xi) = D\left(\frac{\mu}{\beta}\frac{\partial \ln \Xi}{\partial \mu} - \frac{\partial \ln \Xi}{\partial \beta}\right) \\
&= \left[\frac{\mu}{\beta}\frac{\partial}{\partial \mu} - \frac{\partial}{\partial \beta}\right]\left(\frac{3}{2\beta}\frac{V}{\lambda_T^3} e^{\beta\mu}\right) \\
&= \frac{15}{4\beta^2}\frac{V}{\lambda_T^3} e^{\beta\mu} = \frac{15}{4}\langle N\rangle(k_B T)^2,
\end{aligned}
$$

because according to Equation (4.65), $\langle N\rangle = \frac{V}{\lambda_T^3} e^{\beta\mu}$. In addition $\langle E\rangle = \frac{3}{2}\langle N\rangle k_B T$ in the grand canonical ensemble. In the canonical ensemble, $\langle E\rangle = 3N k_B T/2$ and $C_V = 3N k_B/2$. So

$$
\frac{\Delta_E}{\langle E\rangle} = \begin{cases} 0 & \text{in the microcanonical ensemble,} \\ \sqrt{\dfrac{2}{3N}} & \text{in the canonical ensemble,} \\ \sqrt{\dfrac{5}{3\langle N\rangle}} & \text{in the grand canonical ensemble.} \end{cases}
$$

The expressions of the fluctuations of the energy (and any other internal variable) depends on the considered ensemble, but in the limit $N \gg 1$, the relative fluctuations tend, of course, towards zero, illustrating the equivalence of statistical ensembles in the thermodynamic limit.

Exercise 4.11. Adsorption of a gas on a solid: Langmuir[62] model

The adsorption phenomenon describes the trapping of gas molecules on the surface of a solid called *substrate*. In thermodynamic equilibrium, molecules reversibly move from the gas phase to the adsorbed phase. Since the number of molecules in a given phase is not constant, it is natural to use the grand canonical formalism. The gas, supposed ideal, is confined in a container of volume V at temperature T and acts as a reservoir for the adsorbed phase on one of the walls. The chemical potential of the gas is μ_g.

1. Calculate the gas grand partition function $\Xi_g(\mu_g, V, T)$ and the gas grand potential $J_g(\mu_g, V, T)$.

2. Deduce the average number of molecules $\langle N_g\rangle$ in the gas phase and the gas equation of state. Express the chemical potential as a function of the gas pressure P in the form $\mu_g = k_B T \ln\left(P/P_0(T)\right)$, where $P_0(T)$ is a function of temperature.

---

[61] By using once the operator D, the average value is recovered: $D(\ln \Xi) = \langle E\rangle$.

[62] Irving Langmuir (1881–1957), American chemist and physicist, winner of the Nobel Prize in Chemistry in 1932 for his discoveries in surface chemistry.

3. In the Langmuir model (1916), adsorbed molecules can bind to one of the substrate's N_0 reaction sites with a bond energy $-\epsilon_1 < 0$. Each site can accommodate at most one molecule. Let n_i be the occupation number of site i: $n_i = 1$ if it is occupied by a molecule of the gas and $n_i = 0$, otherwise. The sites are far enough apart so that adsorbed molecules do not interact with each other. The chemical potential of the adsorbed phase is μ_a and its equilibrium temperature is imposed by the gas, i.e. $T_a = T$. Express the number of adsorbed molecules N_a and the energy of the adsorbed phase E_a as a function of n_i.

4. Calculate the grand partition function $\Xi_a(\mu_a, N_0, T)$ and the grand potential $J_a(\mu_a, N_0, T)$ of the adsorbed phase.

5. Deduce the average number of adsorbed molecules $\langle N_a \rangle$, the adsorption rate θ (i.e. the probability of occupation of a site) as a function of μ_a and T, as well as the average energy of the adsorbed phase $\langle E_a \rangle$.

6. Using the condition of chemical equilibrium between the gas phase and the adsorbed phase, establish Langmuir's law:

$$\theta(T,P) = \left(1 + \frac{P_0(T)}{P} e^{-\frac{\epsilon_1}{k_B T}}\right)^{-1}. \tag{4.90}$$

Plot the Langmuir isotherms, $\theta(P)$, as a function of pressure for different temperatures.

7. For a dinitrogen gas adsorbed on charcoal, the experimental adsorption rate is known: $\theta \simeq 3 \times 10^{-5}$ at $T = 20\,°C$ and $P = 85$ atm. Calculate P_0 and deduce the value of ϵ_1.

Solution:

1. According to Equation (4.52), $J_g(\mu_g, V, T) = -k_B T \ln \Xi_g = -k_B T \ln e^{\left(\frac{V}{\lambda_T^3} e^{\beta \mu_g}\right)} = -k_B T \frac{V}{\lambda_T^3} e^{\beta \mu_g}$.

2. $\langle N_g \rangle = -\frac{\partial J_g}{\partial \mu_g} = \frac{V}{\lambda_T^3} e^{\beta \mu_g}$ and $P = -\frac{J_g}{V} = \frac{k_B T}{\lambda_T^3} e^{\beta \mu_g} = \frac{\langle N_g \rangle}{V} k_B T$. By inverting the first relation, one gets $\mu_g = k_B T \ln \frac{\langle N_g \rangle \lambda_T^3}{V} = k_B T \ln \frac{P \lambda_T^3}{k_B T} = k_B T \ln \frac{P}{P_0(T)}$, with $P_0(T) = k_B T / \lambda_T^3$ (so P_0 behaves as $T^{\frac{5}{2}}$).

3. $N_a = \sum_{i=1}^{N_0} n_i$ and $E_a = -\sum_{i=1}^{N_0} \epsilon_1 n_i = -\epsilon_1 N_a$.

4. The N_0 sites are distinguishable and independent, so $\Xi_a(\mu_a, N_0, T) = \xi^{N_0}$, where ξ is the grand partition function of a single site:

$$\xi = \sum_{n_1=0,1} e^{\beta(\epsilon_1 + \mu_a) n_1} = 1 + e^{\beta(\epsilon_1 + \mu_a)}.$$

So $J_a(\mu_a, N_0, T) = -k_B T N_0 \ln \xi = -k_B T N_0 \ln \left(1 + e^{\beta(\epsilon_1 + \mu_a)}\right)$.

5. It can be deduced that:

$$\langle N_a \rangle = -\frac{\partial J_a}{\partial \mu_a} = \frac{N_0}{1 + e^{-\beta(\epsilon_1 + \mu_a)}} \quad \text{and} \quad \theta = \frac{\langle N_a \rangle}{N_0}. \tag{4.91}$$

Moreover, $\langle E_a - \mu_a N_a \rangle = -\frac{\partial}{\partial \beta} \ln \Xi_a(\mu_a, N_0, T) = -\frac{N_0 (\epsilon_1 + \mu_a)}{1 + e^{-\beta(\epsilon_1 + \mu_a)}}$. Therefore, by inverting Equation (4.91), $\langle E_a \rangle = -\langle N_a \rangle \epsilon_1$. The result is expected since each occupied site brings a contribution $-\epsilon_1$ to the energy of the adsorbed phase.

6. In chemical equilibrium $\mu_a = \mu_g = k_B T \ln (P/P_0(T))$, so $e^{\beta \mu_a} = P/P_0(T)$, which reported into Equation (4.91) gives Langmuir's law (4.90) (see Figure 4.13). Thus, θ increases with P and tends to 1 at high pressure. In addition, at fixed pressure θ decreases when the temperature increases (because $P_0(T) e^{-\epsilon_1/k_B T}$ increases with T): substrate outgassing increases with thermal motion.

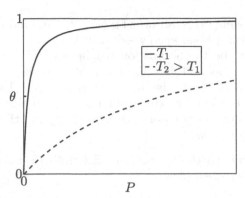

FIGURE 4.13 Plot of Langmuir isotherms, $\theta(P)$, as a function of P for two temperatures T_1 (plain line) and $T_2 > T_1$ (dashed line).

7. For N_2, $M = 28$ g·mol^{-1}, so $P_0 = k_B T/\lambda_T^3 \simeq 5.6 \times 10^6$ atm at $T = 293$ K. By inverting Equation (4.90), $N_A \epsilon_1 = -RT \ln \left(\left(\frac{1}{\theta} - 1 \right) \frac{P}{P_0} \right) \simeq 1.7$ kJ·mol^{-1}, so $\epsilon_1 \simeq 0.02$ eV.

Exercise 4.12. Adsorption of a gas on a solid: BET model

In many situations, the adsorption isotherms have the shape shown in Figure 4.14, in disagreement with the Langmuir model studied in Exercise 4.11. To explain this phenomenon, Brunauer, Emmett and Teller proposed in 1938 the following model[63]:

- Each of the independent and distinct N_0 adsorbing sites of the substrate can trap an unlimited number of molecules;

- the first adsorbed molecule on a given site has an energy $-\varepsilon_1 < 0$;

- every other molecule adsorbed on the same given site is less bonded to the substrate than the first molecule and has an energy $-\varepsilon_2$ such that $-\varepsilon_1 < -\varepsilon_2 < 0$.

The adsorbed phase is in contact with a gas, considered ideal, at temperature T (see exercise 4.11).

1. Write the adsorbed phase grand partition function $\Xi_a(\mu_a, N_0, T)$ as a function of $\lambda = e^{\beta \mu_a}$, $\lambda_1 = e^{\beta \epsilon_1}$ and $\lambda_2 = e^{\beta \epsilon_2}$.

2. Deduce the adsorption rate $\theta = \langle N_a \rangle / N_0$ as a function of λ, λ_1 and λ_2.

3. Using the chemical equilibrium conditions between the gas phase and the adsorbed phase write θ as a function of $x(P,T) = P/P_e$, where $P_e = P_0/\lambda_2 = P_0 e^{-\beta \epsilon_2}$, and $c(T) = e^{\beta(\epsilon_1 - \epsilon_2)} = \lambda_1/\lambda_2$. What is the physical meaning of the pressure P_e?

4. Show that the volume V_a of adsorbed gas obeys the following equation:

$$V_a = \frac{x}{(1-x)(A+Bx)} \quad \text{with} \quad x = \frac{P}{P_e}. \tag{4.92}$$

Write A and B as functions of c and V_m, the volume of adsorbed gas when the substrate is covered by a single layer of molecules. At low pressure, Equation (4.92) fits experimental data perfectly (see figure 4.14). For a dinitrogen gas adsorbed on an iron and aluminum oxide substrate at $T = 77.3$ K, the experimental values are the following $A = 0.3 \times 10^{-4}$ cm^{-3} and $B = 76 \times 10^{-4}$ cm^{-3} for $P_e = 1$ atm. Deduce c, V_m and $\epsilon_1 - \epsilon_2$. Describe the behaviour of the isotherm for $V_a < V_m$ and for $V_a > V_m$.

---

[63] *Adsorption of gases in multimolecular layers*, J. Am. Chem. Soc. **60**, 309 (1938).

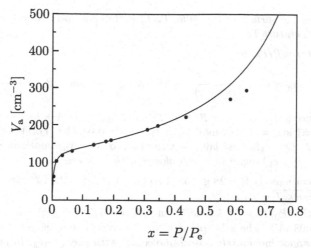

FIGURE 4.14 Adsorbed N_2 volume V_a by an iron and aluminium oxide surface as a function of the relative pressure $x = P/P_e$ at $T = 77.3$ K (V_a is given in standard condition of pressure and temperature. Experimental points (black dots) were extracted from J. Am. Chem. Soc **60**, 309 (1938). The plain curve follows Equation (4.92) with $A = 0.3 \times 10^{-4}$ cm^{-3} and $B = 76 \times 10^{-4}$ cm^{-3}.

5. Evaluate P_0 at $T = 77.3$ K and deduce ϵ_2 and ϵ_1. Compare to the latent heat of vaporisation, $L_{vap} = 5.6$ kJ · mol^{-1} at $P_e = 1$ atm. Why does the BET model overestimate the adsorption rate at high pressure?

Solution:

1. *The N_0 sites are distinguishable and independent, so $\Xi_a(\mu_a, N_0, T) = \xi^{N_0}$, where ξ is the grand partition function of a single site. Let $e_n = -\epsilon_1 - (n-1)\epsilon_2$ be the energy of a site occupied by $n \geq 1$ molecules ($e_0 = 0$), then:*

$$\xi = \sum_{n=0}^{\infty} e^{-\beta(e_n - n\mu_a)} = 1 + \sum_{n=1}^{\infty} \lambda^n \lambda_1 \lambda_2^{n-1} = 1 + \lambda\lambda_1 \sum_{n=0}^{\infty} (\lambda\lambda_2)^n$$

$$= 1 + \frac{\lambda\lambda_1}{1 - \lambda\lambda_2} = \frac{1 + \lambda(\lambda_1 - \lambda_2)}{1 - \lambda\lambda_2},$$

provided that $\lambda\lambda_2 < 1$, i.e. $\mu_a < -\epsilon_2$.

2. *Therefore,*

$$\theta = \frac{\langle N_a \rangle}{N_0} = -\frac{1}{N_0}\frac{\partial J_a}{\partial \mu_a} = k_B T \frac{\partial}{\partial \mu_a} \ln \xi = k_B T \left(\frac{\partial \lambda}{\partial \mu_a}\right)\frac{\partial}{\partial \lambda} \ln \xi$$

$$= \lambda \frac{\partial}{\partial \lambda} \ln\left(\frac{1 + \lambda(\lambda_1 - \lambda_2)}{1 - \lambda\lambda_2}\right)$$

$$= \frac{\lambda\lambda_1}{\left(1 + \lambda(\lambda_1 - \lambda_2)\right)\left(1 - \lambda\lambda_2\right)}.$$

3. *In chemical equilibrium $\mu_a = \mu_g = k_B T \ln(P/P_0(T))$ (see Exercise 4.11), $\lambda = P/P_0(T)$ and $x = \lambda\lambda_2 < 1$. Thus $\lambda\lambda_1 = cx$, with $c > 1$ and*

$$\theta = \frac{cx}{\left(1 + (c-1)x\right)\left(1 - x\right)}.$$

The curve $\theta(x)$ is increasing and diverges at $x = 1$, for $P = P_e(T)$. When $\theta \gg 1$, a very large number of molecular layers cover the substrate. In other words, the gas has liquefied, forming a liquid layer on the surface of the solid. $P_e(T)$ is thus interpreted as the saturating vapour pressure at temperature T.

4. $\theta = V_a/V_m$, with $x = P/P_e$, so

$$V_a = \frac{x}{(A + Bx)(1 - x)} \quad \text{with} \quad A = \frac{1}{V_m c} \quad \text{and} \quad B = \frac{(c - 1)}{V_m c}.$$

It can be deduced that $c = 1 + B/A \simeq 254$ and $V_m = 1/(A + B) \simeq 131$ cm^3. Moreover, $N_A(\epsilon_1 - \epsilon_2) = RT \ln c \simeq 3.6$ kJ·mol^{-1} (so $\epsilon_1 - \epsilon_2 \simeq 0.04$ eV). For $V_a < V_m$, the isotherm increases with P. Once the first layer is saturated, for $V_a \simeq V_m$, molecules are adsorbed with a lower energy ($\epsilon_2 < \epsilon_1$) and θ increases slower with pressure.

5. For N_2, the molar mass is $M = 28$ g·mol^{-1}, so $P_0 = k_B T/\lambda_T^3 \simeq 2 \times 10^5$ atm at $T = 77.3$ K (note that $\lambda_T = h/\sqrt{2\pi m k_B T} \simeq 0.4$ Å).

Then $N_A \epsilon_2 = RT \ln P_0/P_e \simeq 7.8$ kJ·mol^{-1} and $N_A \epsilon_1 \simeq 3.6 + 7.8 = 11.4$ kJ·mol^{-1} (i.e. $\epsilon_1 \simeq 0.12$ eV and $\epsilon_2 \simeq 0.08$ eV). The adsorption rate is overestimated by the BET model, because the substrate cannot indefinitely adsorb molecules with energy $-\epsilon_2$. In fact molecules of the upper layers are less and less bound (thus $N_A \epsilon_1 > N_A \epsilon_2 > N_A \epsilon_3 > \cdots > L_{vap}$), the last layer corresponds to the surface of the liquid in contact with the gas.

5 Simple Classical Fluids

> "The existence and properties of solids and gases are relatively easy to understand on the basis of characteristic atomic or molecular properties and interactions that follow from basic quantum mechanics. Liquids are harder to understand. Why should there be a state with density nearly equal to that of solids, but with such relatively low viscosity that the shape can vary readily?"
>
> H.J. Bernstein and V.F. Weisskopf[1], 1987

The properties of physical systems are determined by the interactions between their constituting particles: from the expression of the Hamiltonian and in particular the potential energy term, follows the calculation of the (grand) partition function and finally the estimation of the thermodynamic quantities (see Chapter 4). The ideal gas approximation is quite justified in the case of dilute gases, but to understand the behaviour of a real fluid, gas or liquid, interactions must be taken into account. First, to obtain a more realistic gas model, whose predictions are in better agreement with the experimental data. Secondly and more importantly, because the interactions between molecules generate collective behaviours that manifest at human scale by the emergence of the different states of matter and phase transitions. Thus, the existence of liquid, an intermediate phase between solid and gas, is not obvious, as it will be discussed in this chapter.

The simple classical fluid model (Section 5.1) falls within a sufficiently general framework to study interaction effects and the thermodynamic properties of real fluids (at least to a first approximation). Despite its simplicity, the (grand) partition function cannot be calculated exactly and very common approximations in physics must be used: series expansion, called *virial expansion* (Section 5.2) and the mean field approximation, the simplest method to handle a system of interacting particles (Section 5.3). The van der Waals equation of state, the first model proposed to study phase transition, will be then derived (Section 5.3.2) and the universal behaviour of the thermodynamic quantities at the critical point, that is independent of the microscopic details specific to a fluid, will be highlighted (Section 5.3.3). In Section 5.4, it will be shown how to describe the microscopic structure of a fluid in terms of the *pair correlation function*, $g(r)$, which gives the probability to find a particle at a distance r of a given particle. The main thermodynamic quantities will be expressed as functions of $g(r)$.

5.1 SIMPLE CLASSICAL FLUID MODEL

A simple fluid is a set of $N \gg 1$ identical and indistinguishable particles, of mass m, without internal structure, which interact through a two-body central potential $u(r)$, depending only on the distance r between two particles. These assumptions are fully justified to describe monoatomic fluids such as noble gases (He, Ne, Ar...) and to understand the general properties of real gases and liquids to a first approximation. However, they are insufficient to correctly account for all observed behaviours of molecular fluids (H_2O, CO_2...) and complex fluids (polymers in solution, emulsions, colloidal suspensions...)[2]. In the framework of

---

[1] Quoted from H.J. Bernstein and V.F. Weisskopf, *About liquids*, Am. J. Phys. **55**, 974 (1987).

[2] In the case of a molecular fluid, internal degrees of freedom (rotation, vibration, even molecule dipole moment) must be taken into account. Two or three body (or more) interactions then depend not only on the distance between molecules, but also on their relative orientation. Thus, hydrogen bonds due to the polarity of the H_2O molecule favour the alignment of the oxygen and hydrogen atoms of a water molecule with the oxygen atom of a neighbour molecule. The specific properties of water (e.g. ice floats because liquid

DOI: 10.1201/9781003272427-5

classical physics, valid at sufficiently high temperature (see Section 1.4.2), the Hamiltonian of a simple fluid is given by Equation (1.7):

$$H(\{\vec{r}_i, \vec{p}_i\}) = \sum_{i=1}^{N} \frac{\vec{p}_i^2}{2m} + U_{\text{pot}}(\{\vec{r}_i\}) \quad \text{with} \quad U_{\text{pot}}(\{\vec{r}_i\}) = \sum_{i<j} u(r_{ij}),$$

where $\vec{r}_i$ and $\vec{p}_i$ are the position and momentum vectors of particle i and where r_{ij} is the distance between particles i and j. The sum over i and j, such as $i < j$, runs over $N(N-1)/2$ interacting particle pairs.

Every thermodynamic property of a simple fluid then relies on the choice of the pair interaction potential $u(r)$. The simplest general form is $u(r) = \epsilon \, f(r/\sigma)$, where f is a dimensionless function and the two parameters σ and ϵ characterise the size of the particles and the energy scale of the interactions, respectively[3]. The main features of atomic interactions are: (i) a strong contact repulsion between the electron clouds of atoms basically originating from the Pauli exclusion principle, and (ii) a short-range attraction in $-1/r^6$, the so-called *van der Waals attraction*, which is due to interactions between the electric dipoles (induced or permanent) of atoms and molecules. To take into account these two effects, Lennard-Jones proposed in 1924 a potential named after him:

$$u_{\text{LJ}}(r) = 4\epsilon \left[\left(\frac{\sigma}{r}\right)^{12} - \left(\frac{\sigma}{r}\right)^{6} \right], \tag{5.1}$$

where the term in $1/r^{12}$ is arbitrarily chosen to achieve a strong contact repulsion[4]. As seen in Figure 5.1, the length σ is the contact distance between two particles (effective "diameter") and the energy ϵ is the depth of the potential well in $r = 2^{1/6}\sigma$. The values of parameters $\sigma > 0$ and $\epsilon > 0$ depend on the fluid under consideration and are determined by quantum calculations or fitted using experimental data (see Section 5.3.3). Values of these parameters are given in Table 5.1 for various fluids.

TABLE 5.1 Values of parameters ϵ and σ of the Lennard-Jones potential deduced from experimental data ($100 \text{ K} \simeq 8.6 \text{ meV}$).

	ϵ/k_{B} (K)	σ (Å)
He	10.9	2.6
Ar	119.8	3.4
O_2	117.5	3.6
CO_2	189	4.5

To simplify the expression of $u(r)$, while retaining the main features of atomic interactions, a square potential well can be used. It has an infinite contact repulsion in $r = \sigma$, a depth $\epsilon > 0$ and a range $\lambda\sigma$, where $\lambda \geq 1$ is a dimensionless number (see Figure 5.1):

$$u_{\text{well}}(r) = \begin{cases} \infty & \text{if} \quad r \leq \sigma \\ -\epsilon & \text{if} \quad \sigma < r \leq \lambda\sigma \\ 0 & \text{if} \quad r > \lambda\sigma. \end{cases} \tag{5.2}$$

---

water is denser than ice at ambient pressure) result from this peculiarity of H_2O molecules which escapes the simple fluid model.

[3] When different size and energy scales are present, $u(r)$ depends on a larger number of parameters.

[4] The exponential form of the Buckingham potential (1938), $u(r) = Ae^{-Br} - C/r^6$, for example, describes more realistically the interatomic repulsion.

For this potential, the values $\sigma = 3.07$ Å, $\lambda = 1.7$ and $\epsilon/k_B = 93.3$ K reproduce very well the thermodynamic properties of argon (see Section 5.2.1). λ is typically between 1 and 2. The case $\epsilon = 0$ describes a system of hard spheres of diameter σ, i.e. $u(r) = \infty$, if $r \leq \sigma$ and $u(r) = 0$, if $r > \sigma$ (see Figure 5.1). In other words, there is no attraction between the particles. This model without characteristic energy is the simplest way to study the properties of a fluid beyond the ideal gas approximation.

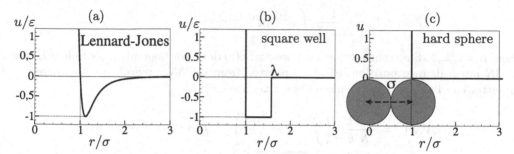

FIGURE 5.1 Examples of interaction potentials $u(r)/\epsilon = f(r/\sigma)$, where ϵ and σ (particle diameter) set the energy and length scales. (a) The Lennard-Jones potential and (b) the square potential well (of range $\lambda\sigma$) are given by Equations (5.1) and (5.2), respectively. (c) The hard sphere potential has no characteristic energy ϵ.

5.1.1 VALIDITY OF THE IDEAL GAS APPROXIMATION

Taking into account particle interactions allows to estimate the conditions under which the potential energy of the system is negligible in front of the kinetic energy (ideal gas approximation), that is $H = E_{\text{kin}} + U_{\text{pot}} \simeq E_{\text{kin}}$. For simplicity, assume that the interaction potential is the square potential well (5.2). Two particles have an interaction energy $u = -\epsilon$ if the second is at a distance less than $\lambda\sigma$ of the former, in the sphere of volume $v_{\text{at}} = \frac{4}{3}\pi(\lambda\sigma)^3$. Neglecting correlations between particles, yet due to attraction, this event has a probability of about v_{at}/V, where V is the volume of the container. The average interaction energy between two particles is therefore $\langle u \rangle \simeq -\epsilon v_{\text{at}}/V$. Since there are $N(N-1)/2 \simeq N^2$ particle pairs, the average potential energy of the gas is about:

$$\langle U_{\text{pot}} \rangle \simeq N^2 \langle u \rangle = -\frac{4\pi(\lambda\sigma)^3}{3V}\epsilon N^2,$$

which is naturally extensive. Moreover, the average kinetic energy of the gas is given by the equipartition of energy theorem (see Section 4.1.5): $\langle E_{\text{kin}} \rangle = \frac{3}{2}N k_B T$. Thus the gas can be considered ideal if:

$$\frac{|\langle U_{\text{pot}} \rangle|}{\langle E_{\text{kin}} \rangle} \simeq \frac{8}{9}\pi\rho(\lambda\sigma)^3 \frac{\epsilon}{k_B T} \sim \rho\sigma^3 \frac{\epsilon}{k_B T} \ll 1,$$

i.e. at high temperature ($k_B T \gg \epsilon$) and/or at low density ($\rho\sigma^3 \ll 1$). In standard conditions of temperature and pressure ($T = 300$ K and $P = 1$ atm), the gas particle number density is $\rho \simeq 2 \times 10^{25}$ m^{-3} (see Section 1.4.2). For simple fluids, $\epsilon/k_B \sim 100$ K and $\sigma \simeq 3$ Å, the ratio $|\langle U_{\text{pot}} \rangle|/\langle E_{\text{kin}} \rangle$ is of the order of 10^{-3}. The ideal gas approximation is thus justified under these thermodynamic conditions. In contrast, for a liquid $|\langle U_{\text{pot}} \rangle| \sim \langle E_{\text{kin}} \rangle$. For example, for water in ambient conditions ($\rho_m = 1$ g·cm^{-3}), $\rho \simeq 3 \times 10^{28}$ m^{-3}. With $\sigma = 2.5$ Å, $\lambda = 1.2$ and $\epsilon/k_B = 1330$ K (see Note 12), $|\langle U_{\text{pot}} \rangle|/\langle E_{\text{kin}} \rangle \simeq 10$.

5.1.2 CONFIGURATION INTEGRAL

In the canonical ensemble, every thermodynamic property of a system can be deduced from the partition function: the equation of state $P(\rho, T)$, the existence and characteristics of phase transitions, heat capacity... For a simple fluid made of $N \gg 1$ indistinguishable particles enclosed in a container of volume V and maintained at temperature T:

$$Z(N,V,T) = \frac{1}{N!h^{3N}} \int d^{3N}r_i d^{3N}p_i \ e^{-\beta\left(\sum_{i=1}^{N} \frac{\vec{p}_i^2}{2m} + \sum_{i<j} u(r_{ij})\right)},$$

where $\beta = 1/k_B T$. Since the interaction potential $u(r)$ does not depend on particle velocities, the $3N$ integrals over positions can be separated from the $3N$ integrals over momenta and the latter can be explicitly calculated (see Exercise 4.6):

$$
\begin{aligned}
Z(N,V,T) &= \frac{1}{N!h^{3N}} \underbrace{\left(\int_{-\infty}^{+\infty} dp\, e^{-\beta \frac{p^2}{2m}}\right)^{3N}}_{=\sqrt{2\pi m k_B T}} \int_V d^{3N}r_i \ e^{-\beta \sum_{i<j} u(r_{ij})} \\
&= \frac{1}{N!\lambda_T^{3N}} Q_N(V,T) \quad \text{where} \quad Q_N \equiv \int_V d^{3N}r_i \ e^{-\beta \sum_{i<j} u(r_{ij})}
\end{aligned}
\tag{5.3}
$$

is called the *configuration integral* and where $\lambda_T = h/\sqrt{2\pi m k_B T}$ is the de Broglie thermal wavelength. The determination of the partition function and the thermodynamic quantities is therefore reduced to the calculation of Q_N, whose expression depends on the two-body potential[5] $u(r)$. Thus, the internal energy of the fluid is given by Equation (4.17):

$$
\begin{aligned}
\langle E \rangle &= \left. -\frac{\partial \ln Z}{\partial \beta}\right|_{N,V} = \frac{3}{2} N k_B T - \left.\frac{\partial \ln Q_N}{\partial \beta}\right|_{N,V} \\
&= \underbrace{\frac{3}{2} N k_B T}_{=\langle E_{kin}\rangle} + \underbrace{\frac{1}{Q_N} \int_V d^{3N}r_i \left(\sum_{i<j} u(r_{ij})\right) e^{-\beta \sum_{i<j} u(r_{ij})}}_{=\langle U_{pot}\rangle},
\end{aligned}
\tag{5.4}
$$

where $\langle E_{kin}\rangle$ and $\langle U_{pot}\rangle$ are the average kinetic and potential energies, respectively. Regardless of particles interactions, the average translational kinetic energy[6] of a simple fluid is therefore equal to $3N k_B T/2$, as stated by the equipartition of energy theorem (see Section 4.1.5). The pressure of a simple fluid depends only on the intensive variables T and $\rho = N/V$ (see Note 39 in Chapter 4). According to Equation (4.19), the equation of state $P(\rho, T)$ is formally written as:

$$P(\rho, T) = k_B T \left.\frac{\partial \ln Z}{\partial V}\right|_{N,T} = k_B T \left.\frac{\partial \ln Q_N}{\partial V}\right|_{N,T}.
\tag{5.5}
$$

On the need for approximations

When interactions between particles are taken into account, the configuration integral cannot be calculated exactly, even for the simplest form of $u(r)$: the hard sphere model[7]. To

---

[5] In the ideal gas case $u(r) = 0$, and $Q_N(V,T) = \left(\int_V d^3 r_i\right)^N = V^N$, partition function (4.11) is recovered.

[6] When internal rotational or vibrational degrees of freedom are considered, the corresponding terms must be added to kinetic energy.

[7] In this case, $Q_N(V)$ is obviously independent of temperature, because $e^{-\beta u(r)}$ is either equal to 0 ($u(r) = \infty$ if $r \leq \sigma$), or equal to 1 ($u(r) = 0$ if $r > \sigma$). It can be deduced that the hard sphere fluid average

determine the system's properties, it is therefore necessary to resort to approximations. The first one studied in this book is the virial expansion (Section 5.2). It consists in expressing a physical quantity (here pressure) as a power series of a small parameter (here density). The second approach is the mean field approximation (Section 5.3), which allows to approximate a system of interacting particles by a system of independent particles immersed in an effective field. These two types of approximation are frequently used in theoretical physics whatever the system under consideration.

With the development of computer science, numerical simulations have become today a widespread alternative technique (see D. Frenkel and B. Smit, *Understanding Molecular Simulation: From Algorithms to Applications* (Academic Press Inc., 2001)). Molecular dynamics (see Section 1.3.2) and Monte Carlo simulations[8] then allow the study of realistic fluid models, such as the Lennard-Jones fluid, without making any approximations to the shape of the interaction potential $u(r)$. The price to pay is that physical quantities are not expressed in analytical forms, but are evaluated by calculating averages over a limited number of configurations. Moreover, the biases inherent to numerical approaches must be taken into account, in particular the finite size effects due to the small numbers of considered particles ($N \simeq 10^3$–10^6). Once these reservations are made, numerical simulations are a very powerful method, not only to test the validity of the implemented approximations in a theoretical approach (for example the mean field approximation), but also to model complex systems (with realistic interactions) and compare the predicted results with experimental data.

5.2 VIRIAL EXPANSION

The principle of the virial expansion is to express pressure as a power series of the number density $\rho = N/V$. The expansion terms of order greater than one provide a correction to the ideal gas equation of state:

$$\beta P = \underbrace{\rho}_{\text{ideal gas}} + B_2(T)\rho^2 + B_3(T)\rho^3 + \ldots, \qquad (5.6)$$

where the function $B_n(T)$ is called n^{th} *virial coefficient*. Since pressure is intensive, it is expressed as a function of quantities that are also intensive, T and ρ. The dependence in ρ takes the form of the series expansion (5.6) and the coefficients $B_n(T)$ depend only on temperature. By construction, the expansion is more reliable at low density and when the number of terms is large. It is thus perfectly adapted for the study of a real gas, but much less for a liquid.

5.2.1 EXPRESSIONS OF THE VIRIAL COEFFICIENTS

In order to eliminate the constraint of a fixed particle number, the virial coefficients will now be expressed in the grand canonical ensemble. According to Equation (5.3), the grand

---

energy given by Equation (5.4) is equal to $3Nk_{\text{B}}T/2$. Moreover, according to Equation (5.5), the ratio $P/k_{\text{B}}T$ is a function of ρ only, but its exact form cannot be calculated in three dimensions (for the one-dimensional gas, see F. Gürsey, *Classical statistical mechanics of a rectilinear assembly*, Mathematical Proceedings of the Cambridge Philosophical Society **1** (Cambridge University Press, 1950)).

[8]In Monte Carlo simulations, the system evolves randomly using the probability distribution of the considered statistical ensemble. The resulting phase space sampling makes the evaluation of ensemble averages of physical quantities possible.

partition function (4.49) is:

$$\Xi(\mu,V,T) = 1 + \sum_{N=1}^{\infty} \underbrace{Z(N,V,T)}_{=\frac{1}{N!\lambda_T^{3N}}Q_N(V,T)} e^{\beta\mu N} = 1 + \sum_{N=1}^{\infty} Q_N(V,T)\frac{x^N}{N!},$$

where

$$x(\mu,T) = \frac{e^{\beta\mu}}{\lambda_T^3}. \qquad (5.7)$$

The grand potential can be deduced:

$$J(\mu,V,T) = -k_BT \ln \Xi(\mu,V,T) = -k_BT \ln\left(1 + \sum_{N=1}^{\infty} Q_N(V,T)\frac{x^N}{N!}\right).$$

Pressure, given by Equation (4.58), is:

$$P = -\frac{J(\mu,V,T)}{V} = \frac{k_BT}{V} \ln\left(1 + \sum_{N=1}^{\infty} Q_N(V,T)\frac{x^N}{N!}\right). \qquad (5.8)$$

To express $B_n(T)$, a Taylor expansion of Equation (5.8) to the order n must be written. Restricting here the expansion to order three[9], one gets:

$$
\begin{aligned}
\beta PV &= \left(Q_1 x + \frac{Q_2}{2}x^2 + \frac{Q_3}{6}x^3 + \dots\right) - \frac{1}{2}\left(Q_1 x + \frac{Q_2}{2}x^2 + \frac{Q_3}{6}x^3 + \dots\right)^2 \\
&\quad + \frac{1}{3}\left(Q_1 x + \frac{Q_2}{2}x^2 + \frac{Q_3}{6}x^3 + \dots\right)^3 + \dots \\
&= Q_1 x + \frac{1}{2}\left(Q_2 - Q_1^2\right)x^2 + \frac{1}{6}\left(Q_3 - 3Q_2Q_1 + 2Q_1^3\right)x^3 + \dots,
\end{aligned}
$$

where the same order terms were grouped together. The following expansion is thus obtained:

$$\beta PV = \sum_{n=1}^{\infty} a_n x^n, \qquad (5.9)$$

of which the first three coefficients are:

$$a_1 = Q_1, \quad a_2 = \frac{1}{2}\left(Q_2 - Q_1^2\right) \quad \text{and} \quad a_3 = \frac{1}{6}\left(Q_3 - 3Q_2Q_1 + 2Q_1^3\right). \qquad (5.10)$$

In addition, the average particle number N (written without $\langle\dots\rangle$ for $N \gg 1$) is given by Equation (4.55) and since $J = -PV$, according to Equation (5.9):

$$N = -\frac{\partial J}{\partial \mu}\bigg|_{V,T} = k_BT\frac{\partial}{\partial\mu}\left(\sum_{n=1}^{\infty} a_n x^n\right) = k_BT\underbrace{\frac{\partial x}{\partial\mu}}_{=\beta x}\frac{\partial}{\partial x}\left(\sum_{n=1}^{\infty} a_n x^n\right) = \sum_{n=1}^{\infty} n a_n x^n, \qquad (5.11)$$

where Equation (5.7) was used. To express P as a function of $\rho = N/V$, x must be eliminated in Expansions (5.9) and (5.11). Firstly, Expansion (5.11) can be inverted by posing:

$$x = \sum_{m=1}^{\infty} b_m N^m, \qquad (5.12)$$

---

[9]With $\ln(1 + \epsilon) = \epsilon - \frac{\epsilon^2}{2} + \frac{\epsilon^3}{3} + \dots$, for all $|\epsilon| < 1$.

where the coefficients b_m are obtained by transferring Expansion (5.11) into Equation (5.12) and identifying term by term the coefficients of the series expansion in power of x. For all x:

$$x = b_1 \left(a_1 x + 2a_2 x^2 + 3a_3 x^3 + \ldots \right) + b_2 \left(a_1 x + 2a_2 x^2 + 3a_3 x^3 + \ldots \right)^2$$
$$+ b_3 \left(a_1 x + 2a_2 x^2 + 3a_3 x^3 + \ldots \right)^3 + \ldots$$

$$x = \underbrace{b_1 a_1}_{=1} x + \underbrace{\left(2b_1 a_2 + b_2 a_1^2 \right)}_{=0} x^2 + \underbrace{\left(3b_1 a_3 + 4b_2 a_1 a_2 + b_3 a_1^3 \right)}_{=0} x^3 + \ldots .$$

The first three coefficients b_m are thus:

$$b_1 = \frac{1}{a_1}, \quad b_2 = -2\frac{a_2}{a_1^3} \quad \text{and} \quad b_3 = \frac{1}{a_1^5} \left(8a_2^2 - 3a_1 a_3 \right). \tag{5.13}$$

Finally, pressure is obtained by substituting x, given by Expression (5.12), in Equation (5.9). Then by grouping the terms with the same power of N:

$$\beta PV = \sum_{n=1}^{\infty} a_n x^n = \sum_{n=1}^{\infty} a_n \left(b_1 N + b_2 N^2 + b_3 N^3 + \ldots \right)^n$$
$$= a_1 \left(b_1 N + b_2 N^2 + b_3 N^3 + \ldots \right) + a_2 \left(b_1^2 N^2 + 2b_1 b_2 N^3 + \ldots \right)$$
$$+ a_3 \left(b_1^3 N^3 + \ldots \right) + \ldots$$
$$= (a_1 b_1) N + \left(a_1 b_2 + a_2 b_1^2 \right) N^2 + \left(a_1 b_3 + 2a_2 b_1 b_2 + a_3 b_1^3 \right) N^3 + \ldots ,$$

therefore, dividing by V and replacing the coefficients b_m by Expressions (5.13) as functions of the coefficients a_n:

$$\beta P = \frac{N}{V} + V \left(-\frac{a_2}{a_1^2} \right) \left(\frac{N}{V} \right)^2 + V^2 \left(4\frac{a_2^2}{a_1^4} - 2\frac{a_3}{a_1^3} \right) \left(\frac{N}{V} \right)^3 + \ldots$$
$$= \rho + B_2(T)\rho^2 + B_3(T)\rho^3 + \ldots .$$

The virial expansion (5.6) is obtained with ($B_1 = 1$)

$$B_2 = -V\frac{a_2}{a_1^2} = \frac{V}{2Q_1^2}(Q_1^2 - Q_2) \tag{5.14}$$

$$B_3 = \frac{2V^2}{a_1^4} \left(2a_2^2 - a_1 a_3 \right) = \frac{V^2}{3Q_1^4} \left(Q_1^4 + 3Q_2^2 - 3Q_1^2 Q_2 - Q_1 Q_3 \right)$$
$$\ldots$$

where Expressions (5.10) of the coefficients a_n in terms of the configuration integrals were used. In principle, this systematic method allows to express B_n in terms of the configuration integrals Q_N with $N \leq n$. However, the difficulty of the calculations increases rapidly with n and only the lowest order virial coefficients are calculated in practice. In the following, the correction made to the equation of state by the second virial coefficient, B_2, will be considered as accurate enough.

5.2.2 SECOND VIRIAL COEFFICIENT

From the configuration integral definition (5.3):

$$Q_1 = \int_V d^3r_1 = V \quad \text{and} \quad Q_2 = \int_V d^3r_1 d^3r_2 \, e^{-\beta u(r_{12})} = \underbrace{\int_V d^3r_1}_{V} \int_V d^3r \, e^{-\beta u(r)},$$

where $\vec{r} = \vec{r}_2 - \vec{r}_1$. According to Equation (5.14),

$$B_2(T) = \frac{V}{2V^2}\left(\underbrace{\int d^3r \int_V d^3r - V \int_V d^3r \, e^{-\beta u(r)}}_{=V} \right) = \frac{1}{2}\int_V d^3r \left(1 - e^{-\beta u(r)}\right).$$

In spherical coordinates $d^3r = r^2 \sin\theta dr d\theta d\phi$, which integrated over the angles gives $4\pi r^2 dr$ (the central potential $u(r)$ depends only on r). The distance r is limited by the macroscopic dimensions of the container of volume V, but given the small interaction range, the integrated function tends quickly to 0 and the integral over r can be extended from 0 to infinity, that is:

$$B_2(T) = 2\pi \int_0^\infty dr \, r^2 \left(1 - e^{-\beta u(r)}\right). \tag{5.15}$$

Hard sphere model

In the most simple case, one immediately finds:

$$B_2 = 2\pi \left(\int_0^\sigma dr \, r^2(1 - e^{-\beta \times \infty}) + \int_\sigma^\infty dr \, r^2(1 - e^{-\beta \times 0}) \right) = \frac{2\pi}{3}\sigma^3,$$

or four times the volume $v_0 = \pi\sigma^3/6$ of a particle of diameter σ. To improve the accuracy of the equation of state and extend its range of validity to higher densities and pressures, higher order virial coefficients must be taken into account. For the hard sphere gas, the first twelve virial coefficients have been calculated: B_3 is easily calculated, Boltzmann expressed B_4 in 1899 and the following eight coefficients have been evaluated numerically[10]. Although rudimentary, the hard sphere model reproduces very well the properties of homogeneous liquids, suggesting that at high density it is essentially repulsion, and not attraction, that controls the behaviour of a liquid.

---

[10]Introducing the volume fraction $\eta = Nv_0/V = \pi\sigma^3\rho/6$, that is the fraction of the container volume V occupied by the N particles, each taking up a volume $v_0 = \pi\sigma^3/6$, one obtains (see R.K. Pathria and P.D. Beale, *Statistical Mechanics* (Academic Press, 2011) and R.J. Wheatley, *Calculation of high-order virial coefficients with applications to hard and soft spheres*, Phys. Rev. Lett. **110**, 200601 (2013)):

$$\frac{\beta P}{\rho} \simeq 1 + 4\eta + 10\eta^2 + 18.4\eta^3 + 28.2\eta^4 + 39.8\eta^5 + 53.3\eta^6 + 68.5\eta^7 + 85.8\eta^8 + 105.8\eta^9 \ldots.$$

This expression is in very good agreement with the hard sphere fluid equation of state determined using numerical simulations. Note that the fluid is stable for $\eta \leq 0.49$, or $v \geq v_s \simeq 1.06\sigma^3$ (see Figure 5.6). To estimate the series convergence, the reader can plot $\beta P/\rho$ by gradually adding terms.

Square potential well model

To go beyond the hard sphere model, consider now the square potential well given by Equation (5.2) and plotted in Figure 5.1. The second virial coefficient (5.15) is easily calculated:

$$
\begin{aligned}
B_2(T) &= 2\pi \left(\int_0^\sigma dr\, r^2(1-0) + \int_\sigma^{\lambda\sigma} dr\, r^2(1 - e^{\beta\epsilon}) + \int_{\lambda\sigma}^\infty dr\, r^2(1-1) \right) \\
&= \underbrace{\frac{2\pi}{3}\sigma^3}_{\text{hard spheres}} \left(1 + (1 - e^{\beta\epsilon})(\lambda^3 - 1) \right).
\end{aligned}
\tag{5.16}
$$

As seen in Figure 5.2, at high temperature ($\beta\epsilon \ll 1$), B_2 is positive and tends to $2\pi\sigma^3/3$, so the pressure $P \simeq k_B T(\rho + B_2\rho^2)$ is larger than the pressure of an ideal gas under the same conditions (ρ and T). It is the contact repulsion between atoms that then dominates the interactions. In contrast, at low temperature ($\beta\epsilon \gg 1$), B_2 is negative: attraction between atoms decreases the frequency and intensity of the collisions with the walls, thus reducing the fluid pressure compared to the ideal gas case (or the hard spheres fluid).

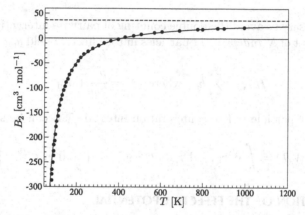

FIGURE 5.2 Second virial coefficient B_2 of argon as a function of temperature T. The black dots are the experimental values (from R.B. Stewart and R.T. Jacobsen, J. Phys. Chem. Ref. Data. **18**, 639 (1989)). The plain curves is a plot of Equation (5.16), for a square potential well with $\sigma = 3.07$ Å, $\lambda = 1.7$ and $\epsilon/k_B = 93.3$ K. The dashed line shows the constant value B_2 of a hard spheres fluid.

The second virial coefficient can be extracted from the experimental equation of state $P_{\exp}(\rho, T)$ of a given fluid: $B_2(T) \simeq (\beta P_{\exp} - \rho)/\rho^2$. A second order virial expansion (5.6) is justified if $B_2(T)\rho \ll 1$, where $\rho = \beta P$ to a first approximation (ideal gas). This is indeed the case for argon at room pressure ($P = 1$ atm), since[11] $-4 \times 10^{-2} < B_2\beta P < 3 \times 10^{-4}$ in the temperature range shown in Figure 5.2. Furthermore, by fitting the experimental curve with the theoretical expression of $B_2(T)$ (5.15), it is possible to evaluate the microscopic parameters of the interaction potential $u(r)$ characterising a fluid (e.g. ϵ and σ for the Lennard-Jones potential). In the case of the square potential well, a very good agreement with argon experimental data[12] is found over the entire temperature range shown in Figure 5.2.

---

[11] In contrast, at $P = 100$ atm, $-4 < B_2\beta P < 3 \times 10^{-2}$. To be reliable, especially at low temperatures, the virial expansion must include higher order terms.

[12] The full range of thermodynamic properties of a polar liquid such as H_2O cannot be understood within the framework of a simple fluid model. Nevertheless, the $B_2(T)$ curve for water vapour is very well fitted using a square potential well with $\sigma \simeq 2.5$ Å, $\lambda \simeq 1.2$ and $\epsilon/k_B \simeq 1330$ K, or $\epsilon \simeq 0.11$ eV, the order of magnitude of a hydrogen bond between two water molecules (see Note 2).

5.3   MEAN FIELD APPROXIMATION AND VAN DER WAALS EQUATION OF STATE

The "mean field" approximation is a very common method in theoretical physics. Its principle is to reduce the study of a system of *mutually interacting particles* to the study of a system of independent particles subject to an external mean field that effectively replaces the interactions with other particles. The advantage of this approach is that the properties of a system of independent particles (such as the ideal gas) can easily be calculated. On the other hand, particles correlations, without being fully suppressed, are poorly taken into account. As for any approximation, the predictions of a model made within the framework of a mean field theory must be compared to experimental data (or numerical simulations) or confronted to a more rigorous theory to be validated.

For a simple fluid, the mean field approximation amounts to replacing two-body interactions, $u(r_{ij})$, by the (one-body) interactions of each particle with an effective mean field u_{eff}, that is:

$$\sum_{i<j} u(r_{ij}) = \frac{1}{2} \sum_{i=1}^{N} \sum_{j\neq i} u(r_{ij}) \simeq \sum_{i=1}^{N} \frac{1}{2} \underbrace{\langle \sum_{j\neq i} u(r_{ij}) \rangle}_{=u_{\text{eff}}} . \tag{5.17}$$

The factor $\frac{1}{2}$ is necessary to avoid the double counting of pairwise interactions. The real fluid is then treated as a set of N *independent* particles in the effective field u_{eff}. The Hamiltonian becomes:

$$H_{\text{mf}} = \sum_{i=1}^{N} h_i \quad \text{where} \quad h_i = \frac{\vec{p}_i^2}{2m} + u_{\text{eff}}$$

is the Hamiltonian of particle i. The configuration integral (5.3) can be simplified to:

$$Q_{N_{\text{mf}}}(V,T) = \int_V \mathrm{d}^3 r_1 \ldots \mathrm{d}^3 r_N \; \mathrm{e}^{-\beta \sum_{i=1}^{N} u_{\text{eff}}} = \left(\int \mathrm{d}^3 r_i \; \mathrm{e}^{-\beta u_{\text{eff}}} \right)^N .$$

5.3.1   DETERMINATION OF THE EFFECTIVE POTENTIAL

Considering any particle i interacting with the $N-1$ other particles, the effective potential must take into account the main features of the pairwise interaction potential $u(r_{ij})$: contact repulsion assumed infinite ($u(r_{ij}) = \infty$ for $r_{ij} \leq \sigma$), and short range attraction. The $N-1$ particles occupy a region of space, of average volume V_e called *excluded volume*, into which particle i cannot enter, i.e. $u_{\text{eff}} = +\infty$. In the accessible space of volume $V - V_e$, particle i evolves in a mean attractive potential, assumed homogeneous, due to the remaining $N-1$ particles, so $u_{\text{eff}} = u_0$, where u_0 is a negative constant. Therefore:

$$Q_{N_{\text{mf}}}(V,T) = \left(\int_{V_e} \mathrm{d}^3 r_i \; \mathrm{e}^{-\beta \times \infty} + \int_{V-V_e} \mathrm{d}^3 r_i \; \mathrm{e}^{-\beta u_0} \right)^N = \left((V - V_e)\mathrm{e}^{-\beta u_0} \right)^N . \tag{5.18}$$

It remains to express the quantities V_e and u_0. The excluded volume increases with the number of particles. To obtain an estimate, the particles can be added one by one in the initially empty container: the first particle has access to the whole volume (its excluded volume $v_e^{(1)}$ is zero). The second particle cannot approach the first at a distance less than σ, so its excluded volume is the exclusion sphere of radius σ, so $v_e^{(2)} = 4\pi\sigma^3/3$ as seen in Figure 5.3. The exclusion spheres of the first two particles create an excluded volume for the third particle equal to $v_e^{(3)} = 2 \times 4\pi\sigma^3/3$, assuming a low enough density to neglect exclusion sphere overlaps (when the distance between two particles is between σ and 2σ)[13].

---

[13]The overlap of exclusion spheres causes a surprising effect in colloidal systems discussed in Exercise 5.7.

Similarly for the ith particle, $v_e^{(i)} = (i-1) \times 4\pi\sigma^3/3$. The *average* excluded volume for any particle among $N \gg 1$ is therefore:

$$V_e = \frac{1}{N}\sum_{i=1}^{N} v_e^{(i)} = \frac{4\pi\sigma^3}{3N}\sum_{i=1}^{N}(i-1) = \frac{4\pi\sigma^3}{3N}\frac{N(N-1)}{2} \simeq bN, \quad \text{where} \quad b = \frac{2}{3}\pi\sigma^3. \quad (5.19)$$

The parameter $b > 0$ is thus four times the volume of a particle, like the second virial coefficient of the hard spheres model.

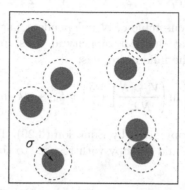

FIGURE 5.3 Each particle of radius $\sigma/2$ is surrounded by a spherical excluded volume of radius σ (dashed circle) in which other particles cannot enter. However, exclusion spheres can overlap each other when the distance between two particles is between σ and 2σ.

To express the *mean* attractive potential u_0 experienced by particle i, particles are supposed uniformly distributed with a constant number density $\rho = N/V$ (the spatial correlations between neighbour particles precisely due to interactions are neglected). At distance $r > \sigma$ from particle i, in the volume $4\pi r^2 dr$ between the spheres of radius r and $r + dr$, there are on average $4\pi r^2 dr\rho$ particles, each having an interaction energy $u(r)$ with particle i. According to Equation (5.17):

$$u_0 = \frac{1}{2}\langle\sum_{j\neq i} u(r_{ij})\rangle \simeq \frac{1}{2}\int_{\sigma}^{\infty} dr\, 4\pi\rho r^2 u(r) = -a\frac{N}{V} \quad \text{where} \quad a = -2\pi\int_{\sigma}^{\infty} dr\, r^2 u(r) > 0.$$
$$(5.20)$$

The upper bound on the integral is extended to infinity given the small range of the interatomic interactions relative to the size of the container. Thus, the average attractive field $u_{\text{eff}} = u_0$ is proportional to particle density $\rho = N/V$. Regardless of the form of the pair interaction potential $u(r)$, parameter a is finite[14], positive ($u(r)$ is essentially negative for $r > \sigma$) and of the order of $\epsilon\sigma^3$.

5.3.2 VAN DER WAALS FLUID

According to Equations (5.18), (5.19) and (5.20), when using the mean field approximation, the partition function of a simple fluid can therefore be written as:

$$Z_{\text{mf}} = \frac{1}{N!\lambda_T^{3N}} Q_{N_{\text{mf}}}(V,T) = \frac{1}{N!}\left(\frac{1}{\lambda_T^3}(V - bN)\, e^{\beta a\frac{N}{V}}\right)^N \quad \text{and} \quad V \geq V_e = bN.$$

---

[14]Short rang interactions, such as $u(r) \sim 1/r^\alpha$ where $\alpha > 3$, ensure the convergence of the integral defining a, as well as the extensivity of potential energy (see Note 5 in Chapter 3). Trivially, $a = \frac{16\pi}{9}\epsilon\sigma^3$ for Lennard-Jones potential and $a = \frac{2\pi}{3}(\lambda^3 - 1)\epsilon\sigma^3$ for the square well potential (and, of course, $a = 0$ for hard spheres).

Using Stirling's formula (A.21), the free energy is then:

$$F_{\mathrm{mf}} = -k_{\mathrm{B}}T \ln Z_{\mathrm{mf}} = -Nk_{\mathrm{B}}T\left(\ln\left[\left(\frac{2\pi m k_{\mathrm{B}}T}{h^2}\right)^{3/2} \left(\frac{V}{N} - b\right)\right] + 1 + \frac{aN}{k_{\mathrm{B}}TV}\right).$$

The average energy, given by

$$\langle E \rangle = -\frac{\partial \ln Z_{\mathrm{mf}}}{\partial \beta} = \frac{3}{2}Nk_{\mathrm{B}}T - a\frac{N^2}{V} = \frac{3}{2}Nk_{\mathrm{B}}T + u_0 N, \tag{5.21}$$

has the expected form for an ensemble of N independent particles, each subjected to the effective potential u_0. Note that the heat capacity at constant volume is equal to that of the ideal gas, $C_V = 3Nk_{\mathrm{B}}/2$. The fluid entropy is:

$$S = -\frac{\partial F_{\mathrm{mf}}}{\partial T} = k_{\mathrm{B}}N\left(\ln\left(\frac{V - bN}{N\lambda_T^3}\right) + \frac{5}{2}\right) = S_{\mathrm{IG}} + Nk_{\mathrm{B}} \ln\left(1 - b\frac{N}{V}\right) < S_{\mathrm{IG}}, \tag{5.22}$$

where S_{IG} is the ideal gas entropy given by Equation (3.29). The inclusion of the excluded volume, $V_{\mathrm{e}} = bN$, reduces the fluid entropy with respect to the ideal gas case. Finally the chemical potential is:

$$\mu = \frac{\partial F_{\mathrm{mf}}}{\partial N} = \mu_{\mathrm{IG}} + kT \ln\left(\frac{V}{V - bN}\right) + \frac{bNk_{\mathrm{B}}T}{V - bN} - 2a\frac{N}{V}, \tag{5.23}$$

where $\mu_{\mathrm{IG}} = k_{\mathrm{B}}T \ln \rho\lambda_T^3$ is the ideal gas chemical potential.

Van der Waals equation of state

The fluid pressure is given by Equation (4.19):

$$P = -\frac{\partial F_{\mathrm{mf}}}{\partial V} = \frac{Nk_{\mathrm{B}}T}{V - bN} - a\frac{N^2}{V^2} = \frac{k_{\mathrm{B}}T}{v - b} - \frac{a}{v^2} \quad \text{and} \quad v \geq b, \tag{5.24}$$

where the intensive quantity $v = V/N = 1/\rho$ is the volume per particle. One immediately recognises the van der Waals[15] equation of state proposed in 1873 to correct the ideal gas equation of state phenomenologically, based on the atomic assumption. In the spirit of van der Waals, the parameter $a > 0$ reflects the presence of attractive interactions between atoms, which decrease the pressure exerted by the fluid on the container walls compared to the ideal gas case[16]. As for parameter b, it takes into account the specific volume of the atoms which reduces their accessible space and increases pressure compared to the ideal gas case (Note that for $a = b = 0$, the ideal gas equation of state is recovered). The values of a and b, which depend on the considered fluid, are estimated by comparing Equation (5.24) with experimental data.

The important point is that statistical physics has allowed the deduction of the van der Waals equation of state in a more rigorous way. Starting from a microscopic description of matter in terms of atomic interactions, treated in the mean field approximation, the

---

[15] Johannes Diderik van der Waals (1837–1923), Dutch physicist who was awarded the Nobel Prize in Physics in 1910 for his work on the equation of state of gases and liquids.

[16] Consider the particles contained in a cubic volume d^3 in contact with a wall (the interactions with the wall itself are neglected), where d is the atomic interactions range. Each of the $\sim \rho d^3$ particles is attracted to the interior of the container by the $\sim\rho d^3$ particles in the adjacent cube of identical volume. So there are roughly $(\rho d^3)^2$ interactions reducing the force (per unit area) exerted by the fluid on the wall, that is a pressure decrease proportional to ρ^2, as shown by the term in a/v^2 in Equation (5.24).

two parameters a and b are expressed in terms of the microscopic characteristics of the interaction potential (ϵ, σ...) specific to a given fluid:

$$a = -2\pi \int_{\sigma}^{\infty} dr\, r^2 u(r) > 0 \quad \text{and} \quad b = \frac{2}{3}\pi\sigma^3. \tag{5.25}$$

Note that the van der Waals equation of state can also be recovered using a second order virial expansion, which is justified only at low density and in the limit of high temperatures (see Exercise 5.1). The mean field approximation, although rudimentary (quasi-independent particles), is not limited by such constraints on thermodynamic conditions[17]. The van der Waals equation of state (5.24) is considered a valid first approximation to describe a classical fluid, gas or liquid, at all density and temperature[18].

Van der Waals isotherms

To study van der Waals fluids, the pressure $P(v)$ can be plotted along an isotherm as a function of the volume per particle v. According to Equation (5.24):

$$\left.\frac{\partial P}{\partial v}\right|_{T} = -\frac{k_{\rm B}T}{(v-b)^2} + 2\frac{a}{v^3} = -\frac{1}{v\kappa_T} < 0, \tag{5.26}$$

where the isothermal compressibility κ_T, defined by Equation (4.61), must be positive to ensure the mechanical stability of the homogeneous fluid in equilibrium. In other words when the volume increases, the pressure decreases (the isotherm $P(v)$ must be a decreasing function of v). The stability condition (5.26) is verified at a given temperature T, if:

$$f(v) = 2a\frac{(v-b)^2}{v^3} < k_{\rm B}T,$$

where the function $f(v)$, defined for $v > b$, shows a maximum in $v_{\rm C} = 3b$. Three behaviours of the isotherms $P(v)$ can be identified depending on the system's temperature (see Figure 5.4):

- If $T > T_{\rm C}$, where $k_{\rm B}T_{\rm C} = f(v_{\rm C}) = 8a/(27b)$, the stability condition is verified for any volume v. The isotherm $P(v)$ is naturally decreasing (and $\kappa_T > 0$).

- If $T = T_{\rm C}$, the isotherm is decreasing and has an inflexion point[19] located in $v = v_{\rm C}$, where κ_T diverges. It is the fluid *critical point* defined as:

$$v_{\rm C} = 3b, \quad k_{\rm B}T_{\rm C} = \frac{8a}{27b} \quad \text{and} \quad P_{\rm C} = \frac{a}{27b^2}, \tag{5.27}$$

where the critical pressure $P_{\rm C}$ was calculated using the van der Waals equation of state (5.24).

- if $T < T_{\rm C}$, there exist two volumes $v_{\rm s_l}$ and $v_{\rm s_g}$ such as $f(v_{\rm s_l}) = k_{\rm B}T = f(v_{\rm s_g})$, for which κ_T diverges. The stability condition $f(v) < k_{\rm B}T$ is then respected only:

  - for $v < v_{\rm s_l}(T)$: the volume per particle is small, the fluid is therefore relatively dense and weakly compressible (isotherms are rapidly decreasing, i.e. κ_T is small). The fluid is a liquid.

---

[17]Only Expression (5.19) of parameter b is based on the assumption of a sufficiently low density in order to neglect overlaps of exclusion spheres.

[18]The van der Waals equation of state does not describe the crystalline phase in which a system of interacting particles is at low temperature and/or high density. Nevertheless, the mean field approach can be extended to the qualitative study of crystals (see A. Clark and M. de Llano, *A van der Waals theory of the crystalline state*, Am. J. Phys. **45**, 247 (1977)).

[19]It can be easily shown that for $T = T_{\rm C}$ and $v = v_{\rm C}$, $\left.\frac{\partial P}{\partial v}\right|_{T_{\rm C}} = 0$ and $\left.\frac{\partial^2 P}{\partial v^2}\right|_{T_{\rm C}} = 0$.

- for $v > v_{s_g}(T)$: the volume per particle is large, the fluid is thus diluted and highly compressible (isotherms are slowly decreasing, i.e. κ_T is large). The fluid is a gas.

In contrast, for $v_{s_l}(T) < v < v_{s_g}(T)$, the stability condition (5.26) is broken: the van der Waals isotherms are increasing ($\kappa_T < 0$) along the s_l–s_g branch, as seen in Figure 5.4, and they cannot describe an equilibrium thermodynamic state. This erroneous behaviour is a consequence of the mean field approximation[20]. The set of points s_l and s_g in the phase diagram such that $f(v) = k_B T$ (i.e. $\kappa_T^{-1} = 0$) defines the *spinodal curve* shown in Figure 5.4.

It remains to understand how the van der Waals fluid can result in the coexistence of gas and liquid. This is the object of Maxwell's construction.

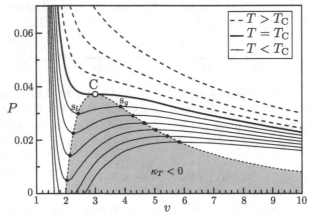

FIGURE 5.4 Van der Waals isotherms $P(v)$ as a function of the volume per particle v for temperatures T that decrease from top to bottom (for $a = 1$ and $b = 1$). The critical point C is in $v_C = 3$ and $P_C \simeq 0.037$. The shadowed region, in which the isotherms are increasing ($\kappa_T < 0$, unstable system), is bounded by the spinodal curve along which the compressibility diverges (locus of points s_l and s_g).

Maxwell's construction and liquid-gas coexistence

The van der Waals equation of state is that of a single homogeneous fluid. To show the coexistence of two fluids of different densities, gas and liquid, phase separation must be introduced in an *ad hoc* manner. In equilibrium, the temperatures, pressures and chemical potentials of the gas (g) and the liquid (l) are equal:

$$T_g = T_l = T, \quad P_g = P_l = P_{eq}(T) \quad \text{and} \quad \mu_g = \mu_l,$$

---

[20] In addition, note that pressure (5.24) can be negative at low temperature if $k_B T < a(v-b)/v^2$ (see Figure 5.4), which is impossible for a homogeneous fluid in equilibrium. On the other hand, negative pressure can exist in metastable (non-equilibrium) states, as it is the case in tree sap. Indeed, with a pressure $P_0 = 1$ atm at its base, a water column in static equilibrium cannot rise above $P_0/(\rho_m g) \simeq 10.3$ m (where ρ_m is the mass density of water). It is primarily plant transpiration that allows trees to exceed this height (the tallest trees are about 100 m tall!). Evaporation of water through leaves induces a strong depression and a traction effect on the sap which is stretched while keeping its cohesion thanks to the attractive interactions between water molecules. The negative pressures measured in tree sap can thus reach values of the order of −100 atm.

where $P_{eq}(T)$ is *the* equilibrium pressure at temperature T. However, the two phases have different volumes per particle (and densities), $v_g(T)$ and $v_l(T)$. For $T \leq T_C$, the pressure $P(v,T)$ given by the van der Waals equation of state corresponds to two values of v (the third value belonging to the unstable branch is forbidden). However, the shape in ∞ of the van der Waals isotherm must be substituted "manually" for a plateau at pressure $P = P_{eq}(T)$ between $v = v_l(T)$ and $v = v_g(T)$. The equilibrium pressure $P_{eq}(T)$ and the volumes per particle $v_l(T)$ and $v_g(T)$ are then determined by the equality of chemical potential which, according to the Gibbs-Duhem relation (Equation (4.60)), is (along an isotherm, $dT = 0$ so $d\mu = vdP$):

$$0 = \mu_g(T) - \mu_l(T) = \int_{\mu_l}^{\mu_g} d\mu = \int_{P_l}^{P_g} v \, dP = [Pv]_l^g - \int_{v_l}^{v_g} P(v,T)dv,$$

where an integration by parts was performed. The term in square brackets can be written as $\left(P_{eq}(T)v_g - P_{eq}(T)v_l\right)$, since $P_g = P_l = P_{eq}(T)$, thus:

$$0 = \int_{v_l(T)}^{v_g(T)} \left(P_{eq}(T) - P(v,T)\right) dv. \tag{5.28}$$

Since pressure $P(v,T)$ is given by the van der Waals equation of state, Equation (5.28) has a simple geometrical interpretation proposed by Maxwell in 1875: at a temperature T, the plateau at pressure $P_{eq}(T)$ is positioned in the phase diagram $v - P$ in such a way that the areas bounded by the van der Waals isotherm $P(v,T)$ are equal on either side of the line $P = P_{eq}(T)$ (see Figure 5.5). $v_g(T)$ and $v_l(T)$ can then be deduced graphically.

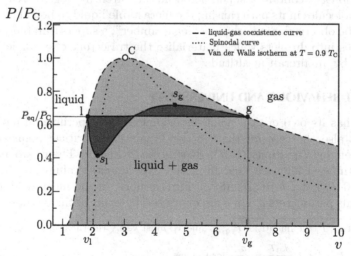

FIGURE 5.5 Liquid-gas coexistence curve (dashed) and spinodal curve (dotted) in the $P - v$ plane for a van der Waals fluid (with $a = 1$ and $b = 1$, so $v_C = 3$). The loop (l-s_l-s_g-g) of the van der Waals isotherm is replaced by the l-g plateau at equilibrium pressure P_{eq}, using Maxwell's construction, which imposes equality of the two darker areas.

By repeating Maxwell's construction for different temperatures, the coexistence curve[21] between the gas and liquid phases can be plotted. It gives the equilibrium pressure P_{eq} as a function of the volumes per particle, v_g and v_l (see Figure 5.5). The liquid-gas change of phase is a first-order phase transition, accompanied by an exchange of energy with the outside environment called *latent heat*. Thus, to transform one mole of liquid into a gas (vaporisation) at constant temperature T, it is necessary to provide the latent heat of vaporisation L_{vap} to the system, that is the energy necessary for molecules to escape neighbour molecules attraction in the condensed phase (see Exercise 5.3). Using Expression (5.22) of the van der Waals fluid entropy:

$$L_{vap} = T\Delta S(T) = T\left(S_g(T) - S_l(T)\right) = RT \ln \frac{v_g(T) - b}{v_l(T) - b} \geq 0, \quad \text{because} \quad v_g(T) \geq v_l(T).$$

Thus $L_{vap}(T)$ decreases with T and cancels at the critical point, where $v_g(T_C) = v_l(T_C) = v_C$.

The behaviour of the fluid *in equilibrium* along a subcritical isotherm ($T < T_C$) is as follows: as one mole of gas is gradually compressed (v decreases), a first liquid drop appears when $v = v_g(T)$ (point g in Figure 5.5). The liquid fraction increases and in $v = v_l(T)$ (at point l), the last gas bubble disappears, leaving a dense fluid with low compressibility. The extension of the van der Waals isotherms, uncorrected by Maxwell's construction (l-s_l-s_g-g branch in Figure 5.5), into the region between the coexistence curve and the spinodal curve describes metastable states. Instead of observing a liquid-gas phase separation, the fluid remains homogeneous and forms a supersaturated vapour on the s_g-g branch, or a superheated liquid on the l-s_l branch (that is raised to a temperature above its vaporisation temperature). These metastable states are observed in fluids pure enough to delay the dynamic nucleation processes leading to the formation of a gas bubble in a superheated liquid or a droplet in a supersaturated vapour. For example, subatomic particle detectors called *bubble chambers* contain a superheated liquid (often hydrogen). A charged particle locally ionises molecules in its path causing the metastable liquid to boil along its trajectory, which can then be observed. Similarly, in a "fog chamber", a supersaturated water or alcohol vapour condenses into droplets which materialise the trajectory of a particle, just like the white trails left by an aircraft at altitude[22].

5.3.3 CRITICAL BEHAVIOUR AND UNIVERSALITY

Each real fluid has its own critical point, as evidenced by experimental data given in Table 5.2. This is also the case for the van der Waals fluid: critical point coordinates v_C, T_C and P_C, given by Equations (5.27), depend on a and b. These two parameters can therefore be estimated for each fluid from the experimental values of T_C and P_C. Thus, $a = 27(k_B T_C)^2/(64 P_C)$ and $b = k_B T_C/(8 P_C)$. Furthermore, in the mean field approximation, the van der Waals parameters a and b are expressed in terms of the parameters (ϵ, σ...) of

---

[21] At a temperature T, the quantities P_{eq}, v_g and v_l, must verify the three following equations:

$$P_{eq} = \frac{k_B T}{v_g - b} - \frac{a}{v_g^2} \quad \text{for the gas,}$$

$$P_{eq} = \frac{k_B T}{v_l - b} - \frac{a}{v_l^2} \quad \text{for the liquid,}$$

$$P_{eq}(v_g - v_l) = \int_l^g P(T, v)\mathrm{d}v = k_B T \ln \frac{v_g - b}{v_l - b} + a\left(\frac{1}{v_g} - \frac{1}{v_l}\right) \quad \text{according to (5.28).}$$

By eliminating the pressure, the temperature or volumes per particle, the coexistence curve can be drawn numerically, as shown in Figure 5.5.

[22] In this case, small ice crystals spontaneously form around the impurities emitted by jet engines in a sufficiently cold and humid air.

the chosen interaction potential, according to Equations (5.25): for the Lennard-Jones potential, $b = 2\pi\sigma^3/3$ and $a = (16\pi/9)\epsilon\sigma^3$. Inverting these relations, the values of $\epsilon = 3a/(8b)$ and $\sigma = (3b/(2\pi))^{1/3}$ are obtained in the framework of the mean field theory (see Table 5.2).

TABLE 5.2 Experimental values of temperature T_C, pressure P_C, molar volume V_C and compressibility factor $\mathcal{Z}_C = P_C v_C/(k_B T_C)$ at the critical point for different fluids. Parameters a and b of the van der Waals equation of state and parameters ϵ and σ of the Lennard-Jones potential are accordingly deduced in the framework of the mean field approximation.

	Experimental values				Mean field theory values			
	T_C (K)	P_C (bar)	V_C (cm$^3 \cdot$ mol^{-1})	$\mathcal{Z}_C$	a (eV.Å^3)	b (Å^3)	ϵ/k_B (K)	σ (Å)
He	5.2	2.27	57	0.300	0.06	39.5	6.6	2.7
Ar	150.7	48.63	75	0.291	2.34	53.5	190	2.9
O$_2$	154.6	50.43	73	0.286	2.37	52.9	195	2.9
CO$_2$	304.1	73.75	94	0.274	6.29	71.1	385	3.2
H$_2$O	647.1	220.6	56	0.230	9.52	50.6	819	2.9

However, these values must be taken with caution. First, while the use of a simple fluid model and the Lennard-Jones potential is justified for noble gases (He and Ar), it is not realistic for molecular fluids (O$_2$ and CO$_2$) and *a fortiori* for polar liquids (H$_2$O). Secondly, the mean field approximation is rather rudimentary and its predictions are in disagreement with experiments[23]. Numerical simulations allow to test the validity of the Lennard-Jones potential, without using approximation, and to evaluate ϵ and σ. A very good agreement with experimental data is found when choosing the values presented in Table 5.1, which can be compared with those obtained using mean field theory (Table 5.2).

Existence of the critical point and of the liquid-gas transition

Despite quantitative disagreements with experiments, the mean field approximation provides a very good theoretical framework for understanding phase transitions at the microscopic level. Expressions (5.27) of v_C, T_C and P_C clearly show that the existence of the critical point and the liquid-gas phase transition are due to attractive interactions between molecules: if $a = 0$, then $T_C = 0$ and there is no phase separation between liquid and gas, as is the case for an ideal gas ($a = b = 0$) and for hard spheres ($a = 0$ and $b \neq 0$).

This point may seem purely theoretical, yet the properties of gases (dilute and disordered phase) and the properties of solids (dense and ordered phase) follow intuitively from the atomic hypothesis and from quantum mechanics, whereas the existence of liquids – an intermediate phase that is both dense and disordered – is not at all obvious[24]. In addition, the relationship between the states of matter and microscopic interactions in colloidal suspensions can be studied experimentally. These systems, such as milk, blood or clays, are made of mesoscopic particles, of the order of a micrometer or less, homogeneously dispersed in a molecular fluid (water, oil. . .)[25]. When the influence of the carrier fluid is taken into account, the effective interactions between colloidal particles show a great similarity with interatomic interactions (contact repulsion and short-range attraction), despite the scale

---

[23] For example, after an estimation of a and b using experimental values of T_C and P_C, the molar volume at the critical point calculated using mean field theory, $V_C = 3b N_A$, is higher by ~30% with respect to the measured values (see Table 5.2).

[24] See the epigraph to this chapter.

[25] For example, milk is formed from fat globules dispersed in water, blood from globules (especially red ones) in plasma and clays from silicate minerals in water. Many industrial products, such as paints or cosmetic creams, are colloidal suspensions.

differences. Nevertheless, unlike interatomic interactions, the range and intensity of the attraction between colloidal particles can be experimentally controlled or even suppressed to create a physical hard sphere systems (see Exercise 5.7).

Under specific conditions, a type of liquid-gas phase transition is observed. It is evidenced by the coexistence of two fluid phases, one poor in colloidal particles (dilute phase) and the other rich in colloidal particles (dense phase)[26]. This phenomenon disappears when the range or intensity of the attraction between the particles is decreased. This observation can be understood by expressing T_C as a function of the square potential well parameters (5.2) in the mean field approximation. According to Equations (5.27) and Note 14:

$$k_B T_C = \frac{8a}{27b} = \frac{8}{27}\epsilon(\lambda^3 - 1) \to 0 \quad \text{when } \epsilon \to 0 \text{ or } \lambda \to 1.$$

When T_C is sufficiently low (typically for $\lambda < 1.3$), the critical point becomes metastable. A fluid-solid transition still remains, but there is no longer coexistence of two colloidal fluids (dense phase and dilute phase, see Figure 5.6).

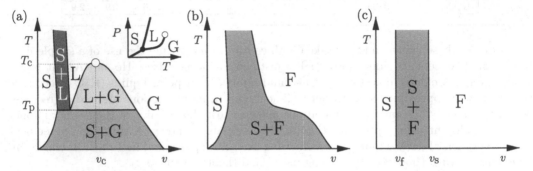

FIGURE 5.6 Phase diagrams of a simple fluid in the $T-v$ plane, where T is the temperature and v is the volume per particle, for three interaction potentials. (a) When the range (or intensity) of attraction is large enough, as in the case of the Lennard-Jones potential (or square potential well with $\lambda > 1.3$), typical of atomic systems. There are then three phases, solid (S), liquid (L), and gas (G), which can coexist two by two (shaded areas in the phase diagram). The critical point (v_C, T_C) is indicated by a white circle. The three phases coexist at the triple point at $T = T_P$ (black circle in the $P-T$ plane, where P is the pressure, shown at the top right). (b) When the range (or intensity) of attraction is too small, as for a square potential well with $\lambda < 1.3$, there is no longer a critical point: only one fluid phase (F) can coexist with the solid phase. This is the case, for example, of biological protein systems (see the references given in Note 26). (c) For a system of hard spheres ($\lambda = 0$, no attraction), the phase diagram is independent of temperature. Solid and fluid coexist when $v_f \leq v \leq v_s$, where $v_f \simeq 0.96\,\sigma^3$ and $v_s \simeq 1.06\,\sigma^3$ are the volumes per particle of melting and crystallisation, respectively (σ is the particle diameter). After V.J. Anderson and H.N.W. Lekkerkerker, Nature **416**, 811 (2002).

It is indeed the contact repulsion ($b \neq 0$) that induces a phase transition between a fluid and a crystalline solid: solidification/melting and condensation/sublimation observed in atomic systems. Even a fluid of hard spheres ($a = 0$) can crystallise[27]. This non-intuitive result was predicted theoretically in 1939, evidenced in numerical simulations in 1957 and

[26]See D. Frenkel, *Entropy-driven phase transition*, Physica A **263**, 26 (1999) and V.J. Anderson and H.N.W. Lekkerkerker, *Insights into phase transition kinetics from colloid science*, Nature **416**, 811 (2002).

[27]The fluid-crystal phase transition illustrates the ambiguity of entropy as a measure of disorder (see Section 3.2). Thus, for volumes per particle lower than $v_f \simeq 0.96\,\sigma^3$, the crystal phase of a hard sphere system is more stable than the fluid phase for any temperature. Therefore $\Delta F = F_{crystal} - F_{fluid} < 0$. Since

finally observed experimentally in a colloidal suspension in 1986 (see A.P. Gast and W.B. Russell, *Simple ordering in complex fluid*, Physics Today **51**, 24 (1998)).

Law of corresponding states

If each fluid has its own critical point (v_C, T_C, P_C) and its own liquid-gas coexistence curve, it is possible to identify generic behaviours, shared by systems of very different nature or chemical composition.

A first illustrative example is the compressibility factor at critical point, defined as $\mathscr{Z}_C = P_C v_C/(k_B T_C)$. For a van der Waals fluid, $\mathscr{Z}_C$ is exactly $3/8 = 0.375$ according to Equations (5.27). It is a universal value, independent of parameters a and b, and thus of the considered system. This universal value is certainly about 25% higher than the experimental data presented in Table 5.2. It can be noticed that the latter depend relatively weakly on the studied fluid, as suggested by the mean field approximation.

Furthermore, by introducing the dimensionless variables $v^* = v/v_C$, $T^* = T/T_C$ and $P^* = P/P_C$, where the critical point coordinates are given by Equations (5.27), the van der Waals equation of state (5.24) can be written in a form independent of parameters a and b, and consequently of the considered fluid:

$$P^*(v^*, T^*) = \frac{8T^*}{3v^* - 1} - \frac{3}{v^{*2}}. \tag{5.29}$$

This type of generic behaviour, called the *law of corresponding states*, is observed for very different real fluids, even molecular ones[28] (O_2, CH_4...), as shown by the liquid-gas coexistence curve plotted using scaled variables in Figure 5.7. Based on this similarity between fluids, the liquefaction of dihydrogen (at $T < T_C = 33$ K) and helium (at $T < T_C = 5.2$ K) were predicted, before being observed in 1898 and 1908, respectively. However, the theoretical curve calculated using the mean field approximation is not in agreement with experimental data[29].

Critical exponents in the mean field approximation

Consider now the behaviour of the thermodynamic quantities of a van der Waals fluid in the vicinity of the critical point. With the help of new dimensionless variables,

$$\phi = v^* - 1 = \frac{v - v_C}{v_C}, \quad t = T^* - 1 = \frac{T - T_C}{T_C} \quad \text{and} \quad p = P^* - 1 = \frac{P - P_C}{P_C},$$

---

the internal energy of a hard spheres system depends only on T (see Note 7), the two phase in equilibrium (at identical temperature) have the same internal energy ($\Delta E = 0$), so:

$$\Delta F = F_{crystal} - F_{fluid} = \Delta E - T\Delta S = -T(S_{crystal} - S_{fluid}) < 0.$$

In other words, the so-called *ordered phase* (crystal) has a higher entropy than the disorder phase (fluid). This phase transition is referred to as *entropy controlled* (see Note 21 in Chapter 3).

[28] As seen in Figure 5.7, the liquid phase of water (for $\rho > \rho_C$) cannot be approximated by a simple fluid: it is necessary to take into account the H_2O molecule polarity.

[29] Without resorting to the mean field approximation, when the interaction potential is written in general terms $u(r) = \epsilon f(r/\sigma)$, the dimensionless variables $\tilde{T} = k_B T/\epsilon$, $\tilde{v} = v/\sigma^3$ and $\tilde{P} = \sigma^3 P/\epsilon$ can be introduced. The equation of state can then be written into the form $\tilde{P}(\tilde{v}, \tilde{T})$ which depends only on the function f and not on the parameters ϵ and σ, specific to the fluid under consideration. According to Figure 5.7, many fluids (but not water) have therefore a behaviour very well described by the Lennard-Jones potential (for which Function f is given by Equation (5.1)).

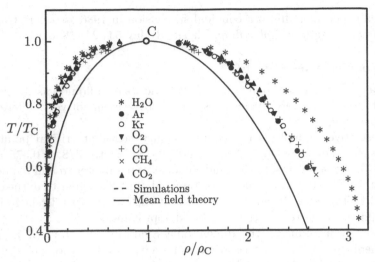

FIGURE 5.7 Liquid-gas coexistence curve in the normalised temperature-density plane (dimensionless variables), measured for different fluids (after E.A. Guggenheim, J. Chem. Phys. **13**, 253 (1945)), calculated numerically for a van der Waals fluid (mean field theory) and estimated numerically for a Lennard-Jones fluid (after H. Watanabe *et al.*, J. Chem. Phys. **136**, 204102 (2012)). C indicates the critical point in $T/T_C = \rho/\rho_C = 1$.

which cancel at the critical point, the equation of state (5.29) can be written as a power expansion of ϕ, for ϕ, $t \ll 1$:

$$
\begin{aligned}
p &= -1 + 4(1+t)\left(1 + \frac{3}{2}\phi\right)^{-1} - 3(1+\phi)^{-2} \\
&= -1 + 4(1+t)\left(1 - \frac{3}{2}\phi + \frac{9}{4}\phi^2 - \frac{27}{8}\phi^3 + \ldots\right) - 3\left(1 - 2\phi + 3\phi^2 - 4\phi^3 + \ldots\right) \\
&= 4t - 6t\phi + 9t\phi^2 - \underbrace{\frac{3}{2}(1+9t)}_{\simeq 1}\phi^3 + \cdots \simeq 4t - 6t\phi + 9t\phi^2 - \frac{3}{2}\phi^3 + \ldots.
\end{aligned} \tag{5.30}
$$

Along the critical isotherm ($t = 0$), $p \simeq -\frac{3}{2}\phi^3$, thus:

$$
P - P_C \simeq \frac{3P_C}{2v_C^3}(v_C - v)^3 \sim (v_C - v)^{\delta_{mf}} \quad \text{with} \quad \delta_{mf} = 3.
$$

Furthermore, the coexistence curve, defined for $t \leq 0$, $p \leq 0$ and $\phi_l \leq 0 \leq \phi_g$, has a maximum at the critical point in $\phi = 0$ (see figure 5.7). In a first approximation, in the vicinity of the critical point, $\phi_l \simeq -\phi_g$. In addition, the equality of pressures, expressed by Equation (5.30), imposes:

$$
p_g = 4t - 6t\phi_g + 9t\phi_g^2 - \frac{3}{2}\phi_g^3 = p_l = 4t - 6t(-\phi_g) + 9t(-\phi_g)^2 - \frac{3}{2}(-\phi_g)^3 \quad \text{with} \quad t \leq 0.
$$

It is easily found that $\phi_g = -\phi_l = 2\sqrt{-t} > 0$. Along the coexistence curve, in the vicinity of the critical point, one thus obtains:

$$
v_g - v_l \simeq \frac{4v_C}{\sqrt{T_C}}\sqrt{T_C - T} \sim (T_C - T)^{\beta_{mf}} \quad \text{with} \quad \beta_{mf} = \frac{1}{2}.
$$

Finally, the isothermal compressibility diverges at the critical point (as $\frac{\partial P}{\partial v} = 0$) with the following behaviour[30], for $\phi \simeq 0$ and $t \geq 0$:

$$\kappa_T = -\frac{1}{v}\frac{\partial v}{\partial P} = -\frac{1}{P_C(\phi + 1)}\underbrace{\left(\frac{\partial p}{\partial \phi}\right)^{-1}}_{\simeq -6t} \simeq \frac{1}{P_C 6t} \sim |T - T_C|^{-\gamma_{mf}} \quad \text{with} \quad \gamma_{mf} = 1.$$

According to Equation (4.62), the density fluctuations $\langle \rho^2 \rangle - \langle \rho \rangle^2$, proportional to κ_T, diverge as a power law at the critical point giving rise to the phenomenon of critical opalescence (see Note 42 in Chapter 4).

Critical behaviour and universality

These power law behaviours in the vicinity of the critical point are confirmed by experimental data, whatever the real fluid under consideration. For the following thermodynamic variables, one gets:

$$C_V \sim |T - T_C|^{-\alpha}$$
$$\rho_l - \rho_g \sim (T_C - T)^{\beta}$$
$$\kappa_T \sim |T - T_C|^{-\gamma}$$
$$P - P_C \sim (\rho - \rho_C)^{\delta}$$

where α, β, γ and δ are positive exponents: the so-called *critical exponents*. Critical exponents are *universal* since experimental values are (almost) independent from the considered fluid[31]:

$$\alpha \simeq 0.11, \quad \beta \simeq 0.34, \quad \gamma \simeq 1.2 \quad \text{and} \quad \delta \simeq 4.8.$$

As discussed earlier, the mean field approximation predicts critical behaviours with exponents independent of the fluid characteristics (i.e. van der Waals parameters a and b). The van der Waals critical exponents values[32] are:

$$\alpha_{mf} = 0, \quad \beta_{mf} = \frac{1}{2}, \quad \gamma_{mf} = 1 \quad \text{and} \quad \delta_{mf} = 3.$$

They are not in agreement with experimental data. Mean field theory and more generally phase transitions will be discussed again in Chapter 9 and the following chapters. It will be shown that the universality characterising critical behaviour extends well beyond fluids.

5.4 MICROSCOPIC STRUCTURE OF A FLUID AND PAIR CORRELATION FUNCTION

The N particles of an ideal gas in equilibrium are uniformly distributed in a container of volume V. The probability of finding a particle at a given distance r from a reference particle, chosen within the fluid, is uniform and simply proportional to the particle density, $\rho = N/V$. In contrast, in a real fluid, the interactions generate correlations between the particles' positions: contact repulsion prevents particles from interpenetrating at distances smaller than the particle diameter. In addition, short-range attraction tends to bring them closer to each other, thus increasing the probability of a particle to be found in the vicinity

---

[30] It can be shown that κ_T diverges with the same exponent along the coexistence curve, for $t \leq 0$.

[31] Voir R.K. Pathria and P.D. Beale, *Statistical Mechanics* (Academic Press, 2011).

[32] The heat capacity of a van der Waals fluid is constant and equal to the ideal gas one, $C_V = 3Nk_B/2$ and therefore does not diverge at the critical point, that is $\alpha = 0$.

of the reference particle. As the attractive potential rapidly approaches zero, one can expect the correlations to decay with the distance r. To describe the resulting spatial distribution of particles, the *pair correlation function*, giving the probability of finding a particle at a distance r from the reference particle, will be defined. Then, the manner in which it can be used to express the main thermodynamic quantities will be explained.

5.4.1 PAIR CORRELATION FUNCTION

To begin, let the particles be treated as distinguishable. In the canonical ensemble, the probability $P_N(\vec{r}_1, \vec{r}_2, \ldots, \vec{r}_N) d^3 r_1 d^3 r_2 \ldots d^3 r_N$ that particle "1" is located in $\vec{r}_1$ within $d^3 r_1$, particle "2" in $\vec{r}_2$ within $d^3 r_2$, ... and particle "N" in $\vec{r}_N$ within $d^3 r_N$, is derived from the probability (4.7) to find the system in a given microstate by integrating over all momenta:

$$
\begin{aligned}
P_N(\vec{r}_1, \vec{r}_2, \ldots, \vec{r}_N) d^3 r_1 d^3 r_2 \ldots d^3 r_N &= \int_{\text{momenta}} dP \\
&= \left(\frac{1}{h^{3N} Z} \int d^{3N} p_i \, e^{-\beta H(\{\vec{r}_i, \vec{p}_i\})} \right) d^{3N} r_i \\
&= \frac{e^{-\beta U_{\text{pot}}(\{\vec{r}_i\})}}{Q_N(V,T)} d^{3N} r_i,
\end{aligned}
$$

where the contribution of the integral over momenta is the same as in the partition function Z (note that Z does not include the factor $1/N!$ since the particles are distinguishable). This leaves only the configuration integral $Q_N(V,T)$ in the denominator. The probability $P_2(\vec{r}_1, \vec{r}_2) d^3 r_1 d^3 r_2$, that particle "1" is in $\vec{r}_1$ within $d^3 r_1$ and that particle "2" is in $\vec{r}_2$ within $d^3 r_2$, is then obtained by integrating P_N over the positions of the other $N - 2$ particles, so

$$
\begin{aligned}
P_2(\vec{r}_1, \vec{r}_2) d^3 r_1 d^3 r_2 &= \left(\int_V d^3 r_3 \ldots d^3 r_N \, P_N(\vec{r}_1, \vec{r}_2, \ldots, \vec{r}_N) \right) d^3 r_1 d^3 r_2 \\
&= \left(\frac{1}{Q_N} \int_V d^3 r_3 \ldots d^3 r_N \, e^{-\beta U_{\text{pot}}(\{\vec{r}_i\})} \right) d^3 r_1 d^3 r_2.
\end{aligned}
$$

The probability that one of the N particles, chosen as the reference particle, is in $\vec{r}_1$ within $d^3 r_1$ and that one of the $N - 1$ other particles is in $\vec{r}_2$ within $d^3 r_2$ for any two particles labels, is therefore $N(N - 1) P_2(\vec{r}_1, \vec{r}_2) d^3 r_1 d^3 r_2$. A dimensionless quantity characterising the local deviation from the fluid mean density, $\rho = N/V$, in $\vec{r}_2$ when a reference particle is located in $\vec{r}_1$ is needed. For that purpose, let the *pair correlation function* be defined as:

$$
g(\vec{r}_1, \vec{r}_2) = \frac{N(N - 1)}{\rho^2} P_2(\vec{r}_1, \vec{r}_2) \simeq \frac{V^2}{Q_N} \int_V d^3 r_3 \ldots d^3 r_N \, e^{-\beta U_{\text{pot}}(\{\vec{r}_i\})}, \quad \text{for } N \gg 1. \qquad (5.31)
$$

In the case of a two-body central potential, $u(r)$, the isotropic fluid pair correlation function, $g(\vec{r}_1, \vec{r}_2) = g(r)$, depends only on the distance $r = r_{12}$ between two particles and is called the *radial distribution function*. Thus, $\rho g(r)$ is the fluid mean density at a distance r within dr from the reference particle and $dN = \rho g(r) d^3 r$ is the average number of particles contained in the volume element $d^3 r$ located at a distance r from the reference particle[33]. In the ideal gas case, $U_{\text{pot}} = 0$ and $Q_N = V^N$, so Equation (5.31) implies $g(r) = (1 - 1/N) \simeq 1$ for all r, reflecting a perfectly uniform average spatial distribution of the particles.

---

[33] By integrating $dN = \rho g(r) d^3 r$ over the entire system's volume:

$$
\int_V d^3 r \, \rho g(r) = \rho \int_V \frac{d^3 r_1}{V} \int_V d^3 r_2 \, g(\vec{r}_1, \vec{r}_2) = \frac{\rho}{V} \frac{N(N-1)}{\rho^2} \underbrace{\int_V d^3 r_1 \int_V d^3 r_2 \, P_2(\vec{r}_1, \vec{r}_2)}_{=1} = N - 1.
$$

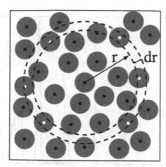

FIGURE 5.8 Principle of calculation of $g(r)$ in a numerical simulation. Here, six particles (white centres) are at a distance r within dR from the chosen reference particle at the centre of the figure.

In general, the exact form of the radial distribution function, which depends on the studied fluid (and thus on the shape of the potential $u(r)$) and on the thermodynamic conditions, P and T, cannot be determined. However, $g(r)$ can be easily calculated in numerical simulations or measured in neutron or X-ray diffraction experiments. In colloidal suspensions or in complex plasma crystals[34], the pair correlation function can be calculated directly from the measured particle positions. The calculation principle of $g(r)$ in a numerical simulation is illustrated in Figure 5.8: for each configuration and from each particle, the number of neighbour particles whose centre lies between r and $r + dr$ is counted. By averaging over all simulated configurations, an estimate of $g(r)$ is then obtained.

In Figure 5.9(a), the typical shape of $g(r)$ in a liquid is shown. Here, $g(r)$ was obtained by molecular dynamics simulation of a Lennard-Jones fluid: at short range, for $r \leq 0.9\,\sigma$, $g(r) \simeq 0$, contact repulsion prevents the particles from interpenetrating. The short-range attraction tends to create a first layer of particles around the reference particle, which manifests itself as the peak at $r \simeq \sigma$. Subsequent peaks, of decreasing heights, correspond to the successive layers exhibiting less and less structure. Indeed, liquid only have short range order. At long range, for $r \gg \sigma$, correlations tend to zero and $g(r) \to 1$, as in an ideal gas. In contrast, in a crystal, $g(r)$ exhibits a succession of peaks separated by minimas which reflects the symmetries of the long-range crystal order (see Figure 5.9(b)).

5.4.2 THERMODYNAMIC QUANTITIES

The radial distribution function is all the more important as it allows the calculation of the thermodynamic quantities of a fluid, demonstrating once again how the microscopic description (here the atomic scale structure) determines the macroscopic characteristics of

---

[34]Complex plasmas are partially ionised gases containing electrically charged microparticles. Due to the electrostactic interactions, the microparticles can under specific conditions arrange themselves into ordered structures: the so-called *complex plasma crystals*. Complex plasma crystals are often used as model systems to study generic phenomena in classical condensed matter (see A. Ivlev, G. E. Morfill and H. Lowen, *Complex Plasmas and Colloidal Dispersions: Particle-Resolved Studies of Classical Liquids and Solids* (World Scientific Publishing Company, 2012)).

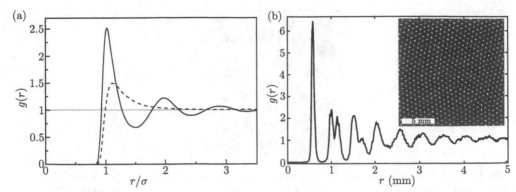

FIGURE 5.9 (a) Radial distribution function $g(r)$ as a function of distance r/σ, calculated from molecular dynamics simulation data. The interaction potential is a Lennard-Jones $u_{\mathrm{LJ}}(r)$, for argon (with $\sigma = 3.4$ Å and $\epsilon/k_{\mathrm{B}} = 119.8$ K) at $T = 300$ K and $P = 5$ kbar (bold line). The dashed line corresponds to the low density approximation (for a gas), at the same temperature, $g(r) = e^{-\beta u_{\mathrm{LJ}}(r)}$ (see Section 5.4.3). (b) Experimental radial distribution function $g(r)$ of a two-dimensional plasma crystal (inset) created in a parallel-plate radio-frequency (rf) argon discharge with a background argon pressure $p_{\mathrm{Ar}} = 1$ Pa at a rf power $P_{\mathrm{W}} = 20$ W. 9.2 μm diameter melamine-formaldehyde spherical particles were used to create the crystal (L. Couëdel (Private communication, 2022)). Note the sharp peaks of $g(r)$ at small r.

a system. Thus, according to Equation (5.4), the average potential energy is:

$$
\begin{aligned}
\langle U_{\mathrm{pot}}\rangle &= \frac{1}{Q_N}\int_V \mathrm{d}^{3N}r_i\, U_{\mathrm{pot}}\, e^{-\beta U_{\mathrm{pot}}} = \frac{1}{Q_N}\int_V \mathrm{d}^{3N}r_i \left(\sum_{i<j} u(r_{ij})\right) e^{-\beta U_{\mathrm{pot}}(\{\vec{r}_i\})} \\
&= \frac{N(N-1)}{2}\int_V \mathrm{d}^3 r_1 \mathrm{d}^3 r_2\, u(r_{12}) \underbrace{\left(\frac{1}{Q_N}\int_V \mathrm{d}^3 r_3 \dots \mathrm{d}^3 r_N\, e^{-\beta U_{\mathrm{pot}}(\{\vec{r}_i\})}\right)}_{=g(r_{12})/V^2} \\
&\simeq \frac{N^2}{2V^2}\underbrace{\int_V \mathrm{d}^3 r_1}_{=V}\int_V \mathrm{d}^3 r\, u(r)g(r) = \frac{1}{2}N\rho \int_0^\infty \mathrm{d}r\, 4\pi r^2 u(r)g(r), \qquad (5.32)
\end{aligned}
$$

for $N \gg 1$, where the contributions of the $N(N-1)/2$ pairs of particles to the potential energy have been taken into account and the change of variable $\vec{r} = \vec{r}_2 - \vec{r}_1$ was performed. The integral over r is extended to infinity, because the interaction range $u(r)$ is very short compared to the macroscopic size of the system. The expression of $\langle U_{\mathrm{pot}}\rangle$ is easily interpreted: starting from a reference particle, there are on average $\mathrm{d}N = 4\pi r^2 \rho g(r)\mathrm{d}r$ particles at a distance r within $\mathrm{d}r$, which make a contribution $u(r)\mathrm{d}N$ to the fluid potential energy. The integral over r then sums up the potential energies of the interactions of the $N-1$ particles with the reference particle. The operation is then repeated by choosing successively each of the N particles as the reference particle. The results is divided by 2 in order to avoid double counting of the interaction pairs (prefactor $N/2$ in front of the integral).

For a macroscopic system, pressure is independent of the shape of the container. For simplicity, a cubic container with side of length L is considered in the following. Pressure is given by Equation (5.5), that is

$$
P(\rho, T) = k_{\mathrm{B}}T \left.\frac{\partial \ln Q_N}{\partial V}\right|_{N,T} \quad \text{with } Q_N = \int_0^L \mathrm{d}^{3N}r_i\, e^{-\beta U_{\mathrm{pot}}(\{\vec{r}_i\})} = V^N \int_0^1 \mathrm{d}^{3N}\tilde{r}_i\, e^{-\beta U_{\mathrm{pot}}},
$$

with the change of variable $\tilde{r}_i = r_i/L$. Thus:

$$\frac{\partial Q_N}{\partial V}\bigg|_{N,T} = NV^{N-1}\int_0^1 d^{3N}\tilde{r}_i\, e^{-\beta U_{\text{pot}}} - \beta V^N \int_0^1 d^{3N}\tilde{r}_i\, \frac{\partial U_{\text{pot}}}{\partial V}\, e^{-\beta U_{\text{pot}}}$$

$$= \rho\underbrace{\int_0^L d^{3N}r_i\, e^{-\beta U_{\text{pot}}}}_{=Q_N} -\beta \int_0^L d^{3N}r_i\, \frac{\partial U_{\text{pot}}}{\partial V}\, e^{-\beta U_{\text{pot}}},$$

with the reverse change of variable, $r_i = L\tilde{r}_i$. Additionally,

$$\frac{\partial U_{\text{pot}}}{\partial V} = \sum_{i<j}\frac{du(r_{ij})}{dr_{ij}}\frac{dr_{ij}}{dL}\frac{dL}{dV} = \sum_{i<j}\frac{du(r_{ij})}{dr_{ij}}\frac{r_{ij}}{L}\frac{1}{3L^2}$$

so

$$P = \frac{k_B T}{Q_N}\frac{\partial Q_N}{\partial V}\bigg|_{N,T} = \rho k_B T - \frac{1}{3VQ_N}\int_V d^{3N}r_i\left(\sum_{i<j}\frac{du(r_{ij})}{dr_{ij}}r_{ij}\right)e^{-\beta U_{\text{pot}}}$$

$$= \rho k_B T - \frac{1}{3V}\frac{N(N-1)}{2}\int_V d^3r_1 d^3r_2\, \frac{du(r_{12})}{dr_{12}}r_{12}\underbrace{\left(\frac{1}{Q_N}\int_V d^3r_3\ldots d^3r_N\, e^{-\beta U_{\text{pot}}}\right)}_{=g(r)/V^2}$$

$$\simeq \rho k_B T - \frac{N^2}{6V^3}\underbrace{\int_V d^3r_1}_{=V}\int_V d^3r\, r\frac{du(r)}{dr}g(r) \quad \text{for } N \gg 1$$

$$\simeq \rho k_B T - \frac{2\pi\rho^2}{3}\int_0^\infty dr\, r^3\frac{du(r)}{dr}g(r), \tag{5.33}$$

where the contributions of the $N(N-1)/2$ pairs of particles have been taken into account and the change of variable $\vec{r} = \vec{r}_2 - \vec{r}_1$ was performed. Note that this expression for the equation of state can be found using the Virial theorem[35]. Of course, the ideal gas equation of state is recovered when $u(r) = 0$.

5.4.3 LOW DENSITY APPROXIMATION

For a given interaction potential $u(r)$, the expression of $g(r)$ cannot be calculated exactly. On the other hand, taking the low density approach, the interactions between particles can be neglected in a first approximation, that is $u(r_{ij}) \simeq 0$, so $Q_N \simeq V^N$, and only the contribution $u(r)$ between the two particles "1" and "2" is kept in the numerator of $g(r)$. Thus, for $N \gg 1$:

$$g(r) = \frac{V^2}{Q_N}e^{-\beta u(r)}\int d^3r_3\ldots d^3r_N\, e^{-\beta \sum'_{i<j}u(r_{ij})} \simeq \frac{V^2}{V^N}e^{-\beta u(r)}V^{N-2} = e^{-\beta u(r)}, \tag{5.34}$$

where the sum in the argument of the exponential runs over all pairs of particles except pair $1-2$. Expression (5.34), shown in Figure 5.9(a), has the expected shape for $g(r)$ in a gas: a value of zero at small distances where the potential is strongly repulsive, a peak corresponding to the minimum of $u(r)$, the higher the lower the temperature, followed by a

---

[35]The second term in the expression of P is equal to the virial mean value, $\frac{1}{3V}\left\langle\sum_{i<j}(\vec{r}_j - \vec{r}_i).\vec{f}_{ij}\right\rangle$, where $\vec{f}_{ij}$ is the force exerted by particle j on particle i (see for example J.-P. Hansen and I.R. McDonald, *Theory of Simple Liquids: with Applications to Soft Matter* (Academic Press, 4th edition, 2013)).

decay to 1 (for $r \to \infty$, $u(r) \to 0$ so $g(r) \to 1$). In this approximation, the potential energy (5.32) is then:

$$\langle U_{\text{pot}} \rangle \simeq 2\pi N \rho \int_0^\infty dr \, r^2 u(r) \, e^{-\beta u(r)} \simeq N \rho \frac{dB_2}{d\beta},$$

where $B_2(T) = 2\pi \int_0^\infty dr \, r^2 \left(1 - e^{-\beta u(r)}\right)$ is the second virial coefficient defined by Equation (5.15). Similarly, pressure (5.33) can be written as:

$$
\begin{aligned}
P \quad & \simeq \quad \rho k_B T - \frac{2\pi \rho^2}{3} \int_0^\infty dr \, r^3 \frac{du(r)}{dr} \, e^{-\beta u(r)} \simeq \rho k_B T - \frac{2\pi \rho^2}{3\beta} \int_0^\infty dr \, r^3 \frac{d}{dr} \left(1 - e^{-\beta u(r)}\right) \\
& \simeq \quad \rho k_B T - \frac{2\pi \rho^2}{3} k_B T \left(\underbrace{\left[r^3 \left(1 - e^{-\beta u(r)}\right) \right]_0^\infty}_{=0} - \int_0^\infty dr \, 3r^2 \left(1 - e^{-\beta u(r)}\right) \right) \\
& \simeq \quad \rho k_B T + 2\pi \rho^2 k_B T \int_0^\infty dr \, r^2 (1 - e^{-\beta u(r)}) = \rho k_B T + \rho^2 k_B T B_2(T),
\end{aligned}
$$

where an integration by parts was done. Thus, the first two terms of the virial expansion given by Equation (5.6) can be recovered consistently in the low density approximation. More sophisticated approximations have been proposed in physical theories of liquids in order to determine more realistic forms of the radial distribution function[36].

---

[36]See J.-P. Hansen and I.R. McDonald, *Theory of Simple Liquids: with Applications to Soft Matter* (Academic Press, 4th edition, 2013).

5.5 EXERCISES

Exercise 5.1. Joule expansion of a van der Waals fluid

Consider a fluid of particles that repel each other as hard spheres ($u(r) = \infty$ for $r \leq \sigma$) and attract each other through a short range pairwise interaction potential of the general form $u(r) = \epsilon f(r/\sigma) < 0$ for $r > \sigma$, where f is a function that tends rapidly to 0.

1. Express the fluid second virial coefficient $B_2(T)$ in the high-temperature limit in terms of the parameters $a = -2\pi \int_\sigma^\infty dr \, r^2 \, u(r) > 0$ and $b = 2\pi\sigma^3/3$.

2. For a helium gas, the following experimental behaviour is observed: $B_2(T) = B - A/T$, with $A = 421.77 \text{ cm}^3 \cdot \text{K} \cdot \text{mol}^{-1}$ and $B = 23.05 \text{ cm}^3 \cdot \text{mol}^{-1}$. Deduce the values of a in $\text{meV} \cdot \text{Å}^3$, b in Å^3 and σ in Å^3.

3. Write the pressure using a second order virial expansion and the expression of $B_2(T)$ found in Question 1. Show that at low density, the van der Waals equation is recovered.

4. The van der Waals fluid undergoes a Joule expansion (described in Section 3.5.2): from an initial volume $V^{(i)} = V/2$, it expends into a final volume $V^{(f)} = V$. Calculate the entropy change ΔS and the temperature change ΔT of the fluid during this expansion. Comment on the sign of ΔT and ΔS. Estimate ΔT for one mole of helium under standard conditions of temperature and pressure at the end of the expansion.

5. When $f(r/\sigma) = -(\sigma/r)^\alpha$, with $\alpha > 0$ (generalised Sutherland potential), show that a is defined only for specific values of α. Estimate ϵ for $\alpha = 6$.

6. Show that for the generalised Sutherland potential, $B_2(T)$ can be written at any temperature as:

$$B_2(T) = \frac{2}{3}\pi\sigma^3 \sum_{p=0}^{\infty} \frac{-3(\beta\epsilon)^p}{p!(\alpha p - 3)}.$$

7. Assuming $\alpha = 6$ and using the inequality $2(p+1) > 2p - 1$, show that $B_2(T) < \frac{\pi\sigma^3}{3\beta\epsilon}\left(1 + 3\beta\epsilon - e^{\beta\epsilon}\right)$.

Solution:

1. *At high temperature:* $0 \leq -\beta u(r) \ll 1$ *for* $r > \sigma$. *According to Equation (5.15):*

$$B_2(T) = 2\pi\left(\int_0^\sigma dr \, r^2(1 - e^{-\beta \times \infty}) + \int_\sigma^\infty dr \, r^2 \underbrace{\left(1 - e^{-\beta u(r)}\right)}_{\approx \beta u(r)}\right)$$

$$= \frac{2\pi}{3}\sigma^3 - \frac{1}{k_B T}\left(-2\pi \int_\sigma^\infty dr \, r^2 \, u(r)\right) = b - \frac{a}{k_B T}.$$

2. $a = k_B A/N_A \simeq 60.3 \text{ meV.Å}^3$, $b = B/N_A \simeq 38.3 \text{ Å}^3$ *and* $\sigma = (3b/(2\pi))^{1/3} \simeq 2.6 \text{ Å}$ *(comparable to the values reported in Table 5.2).*

3. *According to Equation (5.6),* $P = k_B T\left(\rho + B_2(T)\rho^2 + \ldots\right) \simeq \rho k_B T(1 + b\rho) - a\rho^2$. *At low density $b\rho \ll 1$, thus $(1 + b\rho) \simeq 1/(1 - b\rho)$ and the van der Waals equation of state is recovered:* $P \simeq \rho k_B T/(1 - b\rho) - a\rho^2$.

4. According to Equation (5.22), $\Delta S = k_B N \left(\ln 2 + \ln \left[\frac{V - bN}{V - 2bN} \right] \right) > k_B N \ln 2 = \Delta S_{IG} > 0$ (irreversible process). During the (isolated) gas expansion, the gas internal energy (5.21) is conserved, $E = E_{kin} + U_{pot} = \text{Const}$, its density decreases so $|\langle U_{pot}\rangle|$ decreases ($\langle U_{pot}\rangle < 0$ increases) and $\langle E_{kin}\rangle$ decreases: the gas cools down[37]. Therefore:

$$E = \frac{3}{2} N k_B T^{(i)} - a\frac{N^2}{V^{(i)}} = \frac{3}{2} N k_B T^{(f)} - a\frac{N^2}{V^{(f)}}.$$

For a mole of helium ($N = N_A$) in a volume $V^{(f)} = V = 22.4$ L, we have the temperature variation $\Delta T = -2aN_A/(3k_B V) = -2A/(3V) < 0$, so $\Delta T \simeq -0.01$ K.

5. In this case, $a = 2\pi\epsilon\sigma^\alpha \int_\sigma^\infty dr\, r^{2-\alpha} = \frac{2\pi\epsilon\sigma^3}{\alpha - 3}$, which is defined only for $\alpha > 3$ (short-enough attraction range in $1/r^\alpha$). For $\alpha = 6$ (van der Waals force), $\epsilon = 3a/(2\pi\sigma^3) = a/b \simeq 1.6$ meV (i.e. $\epsilon/k_B \simeq 18$ K).

6. Using the series expansion of the exponential function:

$$\begin{aligned}
B_2(T) &= \frac{2}{3}\pi\sigma^3 + 2\pi \int_\sigma^\infty dr\, r^2 \left(1 - e^{\beta\epsilon(\frac{\sigma}{r})^\alpha}\right) \\
&= \frac{2}{3}\pi\sigma^3 - 2\pi \int_\sigma^\infty dr\, r^2 \sum_{p=1}^\infty \frac{1}{p!}\left(\beta\epsilon\left(\frac{\sigma}{r}\right)^\alpha\right)^p \\
&= \frac{2}{3}\pi\sigma^3 - 2\pi \sum_{p=1}^\infty \frac{(\beta\epsilon\sigma^\alpha)^p}{p!} \int_\sigma^\infty dr\, r^{2-\alpha p} \\
&= \frac{2}{3}\pi\sigma^3 + \frac{2}{3}\pi\sigma^3 \sum_{p=1}^\infty \frac{-3(\beta\epsilon)^p}{p!(\alpha p - 3)} \quad \text{with} \quad \alpha > 3.
\end{aligned}$$

7. For $\alpha = 6$:

$$\begin{aligned}
B_2(T) &= \frac{2}{3}\pi\sigma^3 \left(1 - \sum_{p=1}^\infty \frac{(\beta\epsilon)^p}{p!(2p-1)}\right) \\
&< \frac{2}{3}\pi\sigma^3 \left(1 - \frac{1}{2}\sum_{p=1}^\infty \frac{(\beta\epsilon)^p}{p!(p+1)}\right) = \frac{2}{3}\pi\sigma^3 \left(1 - \frac{1}{2\beta\epsilon}\sum_{p=1}^\infty \frac{(\beta\epsilon)^{p+1}}{(p+1)!}\right) \\
&< \frac{2}{3}\pi\sigma^3 \left(1 - \frac{1}{2\beta\epsilon}\sum_{p=2}^\infty \frac{(\beta\epsilon)^p}{p!}\right) = \frac{2}{3}\pi\sigma^3 \left(1 - \frac{1}{2\beta\epsilon}(e^{\beta\epsilon} - 1 - \beta\epsilon)\right) \\
&< \frac{2}{3}\pi\sigma^3 \frac{1}{2\beta\epsilon}\left(1 + 3\beta\epsilon - e^{\beta\epsilon}\right).
\end{aligned}$$

Exercise 5.2. Virial expansion and Dieterici fluid

A simple fluid is made of $N \gg 1$ particles interacting with each other through a two-body central potential $u(r)$. The potential considered in this exercise is the Lennard-Jones potential, with a well of depth $-\epsilon$.

---

[37] During a Joule expansion, a temperature increase is sometimes measured under specific conditions (such as for a helium gas at room temperature). Indeed, at high temperature atoms can have enough kinetic energy to interpenetrate at distances such that $u(r) > 0$ (i.e. $r < \sigma$ for the Lennard-Jones potential), which is impossible with the hard sphere system studied in this exercise. These positive contributions to the mean potential energy, although limited in number, may be sufficient to dominate the negative contributions due to the attractive part of $u(r)$. As the volume increases, the density decreases, these collisions are rarer and $\langle U_{pot}\rangle$ then decreases from a positive to a possible negative value. Consequently $\langle E_{kin}\rangle$ and therefore T increase (see J.-O. Goussard and B. Roulet, *Free expansion for real gases*, Am. J. Phys. **61**, 845 (1993)).

1. Show that the configuration integral of the fluid is:

$$Q_N(V,T) = \int_V d^{3N}r_i \prod_{i<j}(1+f_{ij}) \quad \text{with} \quad f_{ij} = e^{-\beta u(r_{ij})} - 1.$$

Plot f_{ij} as a function of r_{ij}. What are the limits of f_{ij}?

2. Expand the product in the configuration integral and show that, in the high temperature limit, the terms that contain products of f_{ij} functions can be neglected and $Q_N(V,T)$ approximated by:

$$Q_N(V,T) \simeq V^N\left(1 - \frac{N^2}{V}B_2(T)\right),$$

where $B_2(T)$ is the second virial coefficient. Deduce the expression of the partition function $Z(N,V,T)$.

3. Calculate the fluid free energy F in the low density limit. Deduce the pressure P and the chemical potential μ.

4. The particles interact now through a square potential well (Equation (5.2)). Calculate $B_2(T)$ in the high temperature limit. What happens to the expression of F derived in the previous question? Deduce the fluid internal energy $\langle E \rangle$.

5. In 1899, Dieterici proposed the following phenomenological equation of state:

$$P = \frac{\rho k_B T}{1 - b\rho} e^{-\frac{a\rho}{k_B T}} \quad \text{with} \quad \rho = \frac{N}{V},$$

where $a > 0$ and $b > 0$ are fluid-dependant coefficients. Expand this equation of state in the low-density limit, keeping only the terms in ρ and ρ^2. Compare this expansion to the expression of P obtained in Question 3 and derive the coefficients a and b as functions of the parameters σ, ϵ and λ.

6. Determine the coordinates $v_C = 1/\rho_C$, T_C and P_C of the critical point of the Dieterici fluid. Deduce the value of $\mathscr{Z}_C = P_C v_C/(k_B T_C)$ (critical compressibility factor). Does it depend on the considered fluid? For argon, $v_C = 124.6$ Å^3, $T_C = 150.7$ K and $P_C = 48.6$ bar. Evaluate a and b, then σ and $\epsilon(\lambda^3 - 1)$.

Solution:

1. *According to Equation (5.3):* $Q_N = \int_V d^{3N}r_i\, e^{-\beta \sum_{i<j} u(r_{ij})} = \int_V d^{3N}r_i \prod_{i<j}\underbrace{e^{-\beta u(r_{ij})}}_{=1+f_{ij}}.$

   When $r_{ij} \to 0$, $u(r_{ij}) \to \infty$, thus $f_{ij} = e^{-\beta u(r_{ij})} - 1 \to -1$ and when $r_{ij} \to \infty$, $u(r_{ij}) \to 0$, thus $f_{ij} \to 0$. Moreover $u(r_{ij})$ has a minimum, for which f_{ij} has a maximum equal to $e^{\beta\epsilon} - 1$, so $-1 \le f_{ij} \le e^{\beta\epsilon} - 1$. f_{ij} is plotted in Figure 5.10.

2. *At high temperature $\beta\epsilon \ll 1$, thus $f_{ij} \lesssim \beta\epsilon \ll 1$. f_{ij} is therefore a small parameters that can be used for the high temperature expansion of $Q_N(V,T)$:*

$$\begin{aligned}
Q_N &= \int_V d^{3N}r_i\, (1+f_{12})(1+f_{13})(1+f_{14})\ldots(1+f_{(N-1)N}) \\
&= \int_V d^{3N}r_i \left(1 + \sum_{i<j} f_{ij} + \sum_{i<j}\sum_{k<l} f_{ij}f_{kl} + \sum_{i<j}\sum_{k<l}\sum_{m<n} f_{ij}f_{kl}f_{mn} + \ldots\right) \\
&\simeq \int_V d^{3N}r_i + \sum_{i<j}\int_V d^{3N}r_i\, f_{ij},
\end{aligned}$$

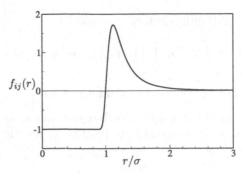

FIGURE 5.10 $f_{ij}(r) = e^{-\beta u(r)} - 1$ as a function of distance r/σ for the Lennard-Jones potential with $\beta\epsilon = 1$.

where the terms in $f_{ij}f_{kl}$, $f_{ij}f_{kl}f_{mn}\ldots$ were ignored. The sum $\sum_{i<j}$ contains $N(N-1)/2$ integrals of the form

$$\int_V d^3r_1 d^3r_2 \ldots d^3r_N \; f_{12} = V^{N-2}\int_V d^3r_1 d^3r_2 \; f_{12} = V^{N-1}\int_0^\infty dr \; 4\pi r^2 f(r),$$

where r is the distance between particles 1 and 2, with $f(r) = e^{-\beta u(r)} - 1$. Thus:

$$\begin{aligned}
Q_N &\simeq V^N + \frac{N(N-1)}{2} V^{N-1}\int_0^\infty dr \; 4\pi r^2 f(r) \quad \text{with } N \gg 1 \\
&\simeq V^N - 2\pi N^2 V^{N-1}\int_0^\infty dr \; r^2\left(1 - e^{-\beta u(r)}\right) \\
&\simeq V^N\left(1 - \frac{N^2}{V}B_2(T)\right) \quad \text{with} \quad B_2(T) = 2\pi\int_0^\infty dr \; r^2\left(1 - e^{-\beta u(r)}\right).
\end{aligned}$$

According to Equation (5.3), $Z(N,V,T) = \dfrac{1}{N!\lambda_T^{3N}}Q_N(V,T) \simeq \dfrac{1}{N!}\left(\dfrac{V}{\lambda_T^3}\right)^N\left[1 - \dfrac{N^2}{V}B_2(T)\right]$.

3. By definition, $F = -k_B T \ln Z \simeq -Nk_B T\left(\ln\dfrac{V}{N\lambda_T^3} + 1 + \dfrac{1}{N}\ln\left[1 - \dfrac{N^2}{V}B_2(T)\right]\right)$ (using Stirling's formula A.21). At low density, $N^2 B_2/V \ll 1$ (since B_2 is of the order of a microscopic volume), so $F \simeq -Nk_B T\left(\ln\dfrac{V}{N\lambda_T^3} + 1 - \dfrac{N}{V}B_2(T)\right)$, which is naturally extensive. It can be deduced:

$$P = -\frac{\partial F}{\partial V} = \frac{Nk_B T}{V} + k_B T\frac{N^2}{V^2}B_2(T) = \rho k_B T + k_B T B_2(T)\rho^2. \tag{5.35}$$

The second order virial expansion is recovered. By taking into account additional terms containing products of f_{ij} in the expansion of Q_N, higher order virial coefficients could be expressed.

Furthermore, $\mu = \dfrac{\partial F}{\partial N} = -k_B T\left(\ln\dfrac{V}{N\lambda_T^3} - 2\dfrac{N}{V}B_2(T)\right)$, which is naturally intensive.

4. According to Equation (5.16), $B_2 = \dfrac{2\pi}{3}\sigma^3\left(1 + (1 - e^{\beta\epsilon})(\lambda^3 - 1)\right) \simeq \dfrac{2\pi}{3}\sigma^3\left(1 - \beta\epsilon(\lambda^3 - 1)\right)$ for $\beta\epsilon \ll 1$, therefore at low density and high temperature:

$$F = -Nk_B T\left(\ln\frac{V}{N\lambda_T^3} + 1 - \frac{2\pi}{3}\sigma^3\frac{N}{V}\left(1 - \beta\epsilon(\lambda^3 - 1)\right)\right).$$

Therefore $\langle E\rangle = -\dfrac{\partial \ln Z}{\partial\beta} = \dfrac{\partial(F/k_B T)}{\partial\beta} \simeq \dfrac{3}{2}Nk_B T - \dfrac{2\pi}{3}\sigma^3\epsilon(\lambda^3 - 1)\dfrac{N^2}{V}$, which is, of course, extensive. The second term can be written as $N\langle u_{\text{pot}}\rangle$, where the average potential energy per particle *is given by (see Equation (5.20)):*

$$\langle u_{pot} \rangle = \frac{1}{2} \int_\sigma^\infty dr \; 4\pi r^2 \rho u(r) = 2\pi \frac{N}{V} \int_\sigma^{\lambda\sigma} dr \; r^2 (-\epsilon) \simeq -\frac{2}{3}\pi\sigma^3 \epsilon (\lambda^3 - 1)\frac{N}{V}.$$

5. *For* $b\rho \ll 1$: $P = \rho k_B T (1 - b\rho)^{-1} e^{-\beta a\rho} \simeq \rho k_B T (1 + b\rho)(1 - \beta a\rho) \simeq \rho k_B T (1 + (b - \beta a)\rho).$ *Comparing with Equation (5.35), one gets* $B_2 = \frac{2\pi}{3}\sigma^3 \left(1 - \beta\epsilon(\lambda^3 - 1)\right) = b - \beta a$, *so*

$$a = 2\pi\frac{\sigma^3}{3}\epsilon(\lambda^3 - 1) \quad \text{and} \quad b = 2\pi\frac{\sigma^3}{3}. \tag{5.36}$$

6. *The equation can be written as a function of* $v = 1/\rho$: $P = \frac{k_B T}{v - b} e^{-\frac{a}{k_B T v}}$. *Since the critical isotherm has an inflexion point at the critical point* ($v = v_C$ *and* $T = T_C$), *one gets:*

$$\frac{\partial P}{\partial v} = \left(-\frac{k_B T_C}{(v_C - b)^2} + \frac{a}{v_C^2}\frac{1}{v_C - b}\right)e^{-\beta_C \frac{a}{v_C}} = 0$$

$$\frac{\partial^2 P}{\partial v^2} = \left(\frac{2k_B T_C}{(v_C - b)^3} - \frac{2a}{v_C^3(v_C - b)} - \frac{2a}{v_C^2(v_C - b)^2} + \frac{\beta_C a^2}{v_C^4(v_C - b)}\right)e^{-\beta_C \frac{a}{v_C}} = 0.$$

The relation $2v_C k_B T_C = a$ *is recovered and*

$$v_C = 2b, \quad k_B T_C = \frac{a}{4b} \quad \text{and} \quad P_C = \frac{a}{4(eb)^2}.$$

Therefore $\mathscr{Z}_C = P_C v_C/(k_B T_C) = 2/e^2 \simeq 0.27$, *a universal value which does not depend on the fluid and which is very close to the experimental data (see Table 5.2). It can be deduced* $b = v_C/2 \simeq 62.3$ Å^3 *and* $a = 2v_C k_B T_C \simeq 3.2$ eV·Å^3. *Note that with these values of* a *and* b, *one finds* $P_C = a/(4(eb)^2) \simeq 45$ *bar, in good agreement with experiments. Finally, with Equations (5.36), one gets* $\sigma = (3b/(2\pi))^{1/3} \simeq 3.1$ Å *and* $\epsilon(\lambda^3 - 1) = a/b \simeq 0.05$ eV *(or* $(\epsilon/k_B)(\lambda^3 - 1) \simeq 580$ *K)*. ϵ *and* λ *cannot be determined, but the values obtained thanks to numerical simulations give* $\epsilon/k_B \simeq 93.3$ K *and* $\lambda \simeq 1.7$, *so* $(\epsilon/k_B)(\lambda^3 - 1) \simeq 365$ K.

Exercise 5.3. The liquid-gas coexistence curve of water
   Liquid water is in equilibrium with its vapour at constant temperature T. The liquid and gas phases are made of $N_l \gg 1$ and $N_g \gg 1$ molecules and have volumes V_l and V_g, respectively. The two fluids are treated as ideal gases, assuming that in the liquid the potential energy per particle is equal to a constant, $-\epsilon_0 < 0$. The temperature at the triple point (where the three phases solid, liquid and gas, coexist) and the temperature at the critical point are $T_t = 273$ K and $T_C = 647$ K, respectively (see the phase diagram[38] in Figure 5.6(a)).

1. Calculate the partition gas function Z_g, its free energy F_g, its entropy per particle s_g and its chemical potential μ_g which will be expressed as functions of the pressure P_g of the gas.

2. Express the liquid partition function Z_l as a function of V_l. To take into account the compressibility of the liquid, the volume of the liquid is assumed to obey this phenomenological law: $V_l = N_l v_t \; e^{[\frac{T}{T_t}]^2}$, where v_t is the volume per particle in the liquid at $T = T_t$. How is the expression of Z_l modified? Deduce the liquid free energy F_l, the liquid average energy $\langle E_l \rangle$ and the liquid heat capacity C_{V_l}. Experimentally, C_{V_l} increases with temperature along the coexistence curve. Is this the case in this model? Calculate the entropy *per particle* of the liquid s_l and the liquid chemical potential μ_l.

---

[38]However, note the water fusion-solidification curve is decreasing in the $P - T$ plane.

3. What are the thermodynamic equilibrium conditions between the two phases? Derive the expression for the *gas* equilibrium pressure $P_{eq}(T)$ as a function of temperature (the pressure of the liquid will not be calculated). Show that $P_{eq}(T)$ is increasing for $T < T_{max}$ (the expression of T_{max} will not be determined).

4. Calculate the difference of entropy *per particle*, $\Delta s = s_g - s_l$, between the two phases in equilibrium and derive the latent heat of vaporisation *per particle*, $l \equiv T\Delta s$. Show that for $T > T_{min}$, l decreases with temperature. Give the expression of T_{min}. Is $T_{min} < T_t$? Under standard conditions of temperature and pressure, the molar latent heat is $L = N_A l \simeq 45$ kJ·mol^{-1}. Deduce an estimate of ϵ_0/k_B. Compare $k_B T$ to ϵ_0 in the existence domain of the liquid.

5. At low temperature, for $k_B T_t \le k_B T \ll \epsilon_0$, show that $v_g \simeq v_t e^{\beta \epsilon_0} \gg v_l$, where v_g and v_l are the volume *par particule* in the gas and in the liquid, respectively. Does Clapeyron's formula[39] (1834), $l = T(v_g - v_l)\frac{dP_{eq}}{dT}$, hold true?

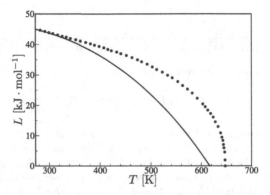

FIGURE 5.11 Latent heat of vaporisation of water L as a function of temperature T. The experimental points are compared to the curve given by Equation (5.37) with $\epsilon_0/k_B = 5688$ K.

6. What is the value of $l(T_{max})$? How to interpret the temperature T_{max}? Plot $l(T)$ for $T_t < T < T_{max}$, with the value of ϵ_0/k_B found in Question 4. Graphically, one finds $T_{max} \simeq 617$ K (see Figure 5.11). Comment on this value and the behaviour of $l(T)$ in the vicinity of T_{max}.

Solution:

1. $Z_g = \frac{z_g^{N_g}}{N_g!}$ with $z_g = V_g/\lambda_T^3$. Then one obtains $F_g = -k_B T N_g\left(\ln \frac{V_g}{N_g \lambda_T^3} + 1\right)$, $P_g = \frac{N_g k_B T}{V_g}$,

   $s_g = k_B\left(\ln \frac{V_g}{N_g \lambda_T^3} + \frac{5}{2}\right) = k_B\left(\ln \frac{k_B T}{P_g \lambda_T^3} + \frac{5}{2}\right)$ and $\mu_g = k_B T \ln \frac{N_g \lambda_T^3}{V_g} = k_B T \ln \frac{P_g \lambda_T^3}{k_B T}$.

2. Likewise, $Z_l = \frac{z_l^{N_l}}{N_l!}$ with $z_l = \frac{1}{h^3}\int e^{-\beta(\frac{\vec{p}^2}{2m}-\epsilon_0)}d^3q d^3p = \frac{V_l}{\lambda_T^3}e^{\beta\epsilon_0} = \frac{N_l v_t}{\lambda_T^3}e^{\beta\epsilon_0 + [\frac{T}{T_t}]^2}$. Then one

   gets $F_l = -N_l\epsilon_0 - k_B T N_l\left(\ln \frac{v_t}{\lambda_T^3} + 1 + [\frac{T}{T_t}]^2\right)$, $\langle E_l \rangle = N_l\left(\frac{3}{2}k_B T - \epsilon_0 + 2k_B \frac{T^3}{T_P^2}\right)$

   and $C_{V_l} = d\langle E \rangle/dT = k_B N_l\left(\frac{3}{2} + 6\left[\frac{T}{T_P}\right]^2\right)$, which indeed increases with temperature. Finally,

   $s_l = k_B\left(\ln \frac{v_t}{\lambda_T^3} + \frac{5}{2} + 3\left[\frac{T}{T_t}\right]^2\right)$ and $\mu_l = -\epsilon_0 - k_B T\left(\ln \frac{v_t}{\lambda_T^3} + 1 + \left[\frac{T}{T_t}\right]^2\right)$.

[39]see Note 62 in Chapter 3.

3. In equilibrium, temperatures, pressures and chemical potentials of the two phases are equal, so $\mu_g = k_B T \ln \frac{P_{eq} \lambda_T^3}{k_B T} = \mu_l = -\epsilon_0 - k_B T \left(\ln \frac{v_l}{\lambda_T^3} + 1 + \left[\frac{T}{T_t} \right]^2 \right)$, that is

$$P_{eq}(T) = \frac{k_B T}{v_t} e^{-\left(1 + \beta \epsilon_0 + \left[\frac{T}{T_t} \right]^2 \right)}.$$

Therefore $\frac{dP_{eq}}{dT} = \frac{k_B}{v_t} e^{-\left(1 + \beta \epsilon_0 + \left[\frac{T}{T_t} \right]^2 \right)} \left(1 + \frac{\epsilon_0}{k_B T} - 2 \left[\frac{T}{T_t} \right]^2 \right)$, which is positive if the function

$f(T) = 1 + \frac{\epsilon_0}{k_B T} - 2 \left[\frac{T}{T_t} \right]^2 > 0$, however, $f(T)$ decreases such that $f(T) \to +\infty$ for $T \to 0$ and $f(T) \to -\infty$ for $T \to +\infty$. There is therefore a temperature T_{max} such that $f(T) > f(T_{max}) = 0$ for $T < T_{max}$.

4. $\Delta s = s_g - s_l = k_B \left(\ln \frac{k_B T}{v_t P_{eq}} - 3 \left[\frac{T}{T_t} \right]^2 \right)$. Using the expression of $P_{eq}(T)$, we have:

$\Delta s = k_B \left(1 + \frac{\epsilon_0}{k_B T} - 2 \left[\frac{T}{T_t} \right]^2 \right)$. So:

$$l = T \Delta s = k_B T \left(1 + \frac{\epsilon_0}{k_B T} - 2 \left[\frac{T}{T_t} \right]^2 \right), \tag{5.37}$$

which is decreasing for $T \geq T_{min} = \frac{T_t}{\sqrt{6}}$. And indeed $T_{min} < T_t$, where T_t is the temperature (at triple point) above which liquid-gas coexistence is observed.

For $T \simeq T_t$, $l/k_B \simeq \epsilon_0/k_B - T_t = L/R \simeq 5415$ K, so $\epsilon_0/k_B \simeq 5688$ K, a temperature well above the critical temperature $T_C = 647$ K, above which there is no liquid-gas coexistence. Thus, $\epsilon_0 \gg k_B T$ in the liquid existence domain[40].

5. $v_g = k_B T / P_{eq} = v_t e^{1 + \beta \epsilon_0 + \left[\frac{T}{T_t} \right]^2} \simeq v_p e^{\beta \epsilon_0} \gg v_l = v_t e^{\left[\frac{T}{T_t} \right]^2}$. According to the expressions of l and $\frac{dP_{eq}}{dT}$, $l = T v_t e^{1 + \beta \epsilon_0 + \left[\frac{T}{T_t} \right]^2} \frac{dP_{eq}}{dT} \simeq T v_t e^{\beta \epsilon_0} \frac{dP_{eq}}{dT}$ for $\beta \epsilon_0 \gg 1$, so $l \simeq T(v_g - v_l) \frac{dP_{eq}}{dT}$, and Clapeyron's relation hold true at low temperature.

6. Since $f(T_{max}) = 0$, $l(T_{max}) = 0$. T_{max} can therefore be interpreted as the critical temperature of the model, that is $T_{max} = T_C$. The curve of $l(T)$, plotted in Figure 5.11, is comparable to the experimental one, especially at low temperature. However, this very rudimentary model predicts a critical temperature $T_{max} \simeq 617$ K, lower than the experimental value ($T_C = 647$ K) and an incorrect critical behaviour, since the slope of $l(T)$ given by Equation (5.37) is not infinite in T_{max}.

Exercise 5.4. Lattice-gas model: mean field theory

A fluid contained in a vessel of volume V is treated as an ensemble of $N_0 = V/v_0$ discernible cells of volume v_0, in contact with a reservoir at temperature T and chemical potential μ (see page 146). Each cell i contains at most one atom: its occupancy number is either $n_i = 0$ or 1 (see Figure 4.4). The lattice coordination number q is the number of nearest neighbour cells ($q = 4$ for a square lattice and $q = 6$ for a cubic lattice). When two nearest neighbour cells are occupied, their two atoms have an interaction energy $-\epsilon$. The Hamiltonian of the lattice-gas model is consequently:

$$H(\{n_i\}) = -\epsilon \sum_{\langle i,j \rangle} n_i n_j,$$

---

[40] Since ϵ_0 is the potential energy per particle in the liquid and each molecule is surrounded by approximately 12 neighbours, the interaction energy between two molecules is of the order of $\epsilon_0/(6k_B) \simeq 950$ K, in agreement with the value found using mean field theory (see Table 5.2). In reality, water molecules form hydrogen bonds with typically four neighbours (see Note 2).

where the sum is over all pairs of nearest neighbour cells i and j. Since the partition function of this system can generally not be exactly calculated, mean field theory will be used.

1. Using the mean field approximation, the Hamiltonian can be written as for independent particles: $H_{\mathrm{mf}}(\{n_i\}) = \sum_{i=1}^{N_0} h_i$, where h_i is the Hamiltonian of cell i. Assuming that the occupancy rate of every other cell is equal to the average value $n = \langle n_i \rangle$, write h_i within one additive constant c that will be later determined.

2. Count the number $\sum_{\langle i,j \rangle} 1$ of nearest-neighbour cell pairs. Periodic boundary conditions are assumed in order to ignore edge effects.

3. Calculate the constant c to obtain the equality on average of Hamiltonians H and H_{mf} (i.e. $\langle H \rangle \simeq \langle H_{\mathrm{mf}} \rangle$), assuming that $\langle n_i n_j \rangle \simeq \langle n_i \rangle \langle n_j \rangle = n^2$.

4. Express the system's grand partition function $\Xi(\mu, V, T)$ and the system's grand potential $J(\mu, V, T)$.

5. Show that $n = \langle N \rangle / N_0$, where $\langle N \rangle$ is the average number of atoms in the system, must verify a self-consistent equation. Derive the fluid's equation of state $P(\rho, T)$, where $\rho = \langle N \rangle / V$ is the density. What happens at low density?

6. Give the general shape of the isotherm $P(\rho)$ and show that if the temperature is sufficiently low, the fluid's isothermal compressibility κ_T can be negative. Determine the critical point coordinates ρ_C, T_C and P_C.

Solution:

1. *Cell i has an interaction energy $-\epsilon n n_i$ with each of its q nearest neighbours that all have an occupation rate n, so $h_i = -\epsilon q n n_i + c$.*

2. *Each of the N_0 cells has q neighbours. There are therefore $q N_0 / 2$ nearest neighbour cells pairs (the factor $1/2$ avoids double counting of the pairs).*

3. *On one hand $\langle H_{\mathrm{mf}} \rangle = \sum_{i=1}^{N_0} \langle -\epsilon q n n_i + c \rangle = (-\epsilon q n \langle n_i \rangle + c) N_0 = (-\epsilon q n^2 + c) N_0$ and on the other hand $\langle H \rangle = -\epsilon \sum_{<i,j>} \langle n_i n_j \rangle \simeq -\epsilon n^2 \sum_{<i,j>} 1 = -\epsilon n^2 (q N_0 / 2)$. Thus $\langle H_{\mathrm{mf}} \rangle \simeq \langle H \rangle$ implies $c = \epsilon q n^2 / 2$. So, $h_i = -\epsilon q n n_i + \epsilon q n^2 / 2$.*

4. *In mean field theory, the cells are treated as independent particles. Therefore, we have $\Xi(\mu, V, T) = \xi^{N_0}$, where ξ is the grand partition function of a single cell:*

$$\xi = \sum_{n_1 = 0,1} e^{-\beta(h_1 - \mu n_1)} = \sum_{n_1 = 0,1} e^{-\beta(-(\epsilon q n + \mu) n_1 + \frac{q}{2} \epsilon n^2)}$$

$$= e^{-\frac{q}{2} \beta \epsilon n^2} \left(1 + e^{\beta(\epsilon q n + \mu)} \right).$$

Thus $J = -k_B T N_0 \ln \xi = \frac{q}{2} \epsilon n^2 N_0 - k_B T N_0 \ln \left(1 + e^{\beta(\epsilon q n + \mu)} \right).$

5. *$\langle N \rangle = -\frac{\partial J}{\partial \mu} = N_0 \frac{e^{\beta(\epsilon q n + \mu)}}{1 + e^{\beta(\epsilon q n + \mu)}}$ and $n = \langle N \rangle / N_0$, hence the self-consistent equation verified by n:*

$$n = \frac{e^{\beta(\epsilon q n + \mu)}}{1 + e^{\beta(\epsilon q n + \mu)}}.$$

Pressure is given by $P = -\frac{J}{V} = -\frac{N_0}{V} \left(\frac{q}{2} \epsilon n^2 - k_B T \ln \left(1 + e^{\beta(\epsilon q n + \mu)} \right) \right)$. By inverting the self-consistent equation, one gets: $1 + e^{\beta(\epsilon q n + \mu)} = 1/(1 - n)$, and thus with $v_0 = V/N_0$:

$$P = -\frac{1}{v_0} \left(\frac{q}{2} \epsilon n^2 + k_B T \ln(1 - n) \right) = -\frac{q}{2} \epsilon v_0 \rho^2 - \frac{k_B T}{v_0} \ln(1 - v_0 \rho), \qquad (5.38)$$

where $\rho = \langle N \rangle / V = n/v_0$ is the fluid density. For $n = v_0\rho \ll 1$,

$$P = -\frac{q}{2}\epsilon v_0 \rho^2 - \frac{k_B T}{v_0}\left(-v_0\rho - \frac{(v_0\rho)^2}{2} + \dots\right) \simeq -\frac{q}{2}\epsilon v_0 \rho^2 + \rho k_B T\left(1 + \frac{v_0\rho}{2}\right),$$

so $P = -a\rho^2 + \rho k_B T/(1 - b\rho)$, with $a = qev_0/2$ and $b = v_0/2$. Once again, the van der Waals equation of state is recovered

6. The isotherm $P(\rho)$ given by Equation (5.38) resembles the van der Waals fluid isotherms shown in Figure 5.4. According to Equation (4.61),

$$\kappa_T = \frac{1}{\rho}\left(\frac{\partial P}{\partial \rho}\right)^{-1} = \frac{1}{\rho}\left(-q\epsilon v_0\rho + \frac{k_B T}{1 - v_0\rho}\right)^{-1},$$

so $\kappa_T > 0$, if $k_B T > f(\rho) = q\epsilon v_0\rho(1 - v_0\rho)$. The function $f(\rho)$ has a maximum in $\rho_C = 1/(2v_0)$, which implies $k_B T_C = f(\rho_C) = q\epsilon/4$. For $T > T_C$, $\kappa_T > 0$ for any density. It can be deduced that $P_C(\rho_C, T_C) = \frac{q\epsilon}{4v_0}(\ln 2 - \frac{1}{2})$. Note the critical compressibility factor of the lattice-gas model studied with mean field theory has, of course, a universal value: $P_C/(\rho_C k_B T_C) = 2\ln 2 - 1 \simeq 0.39$.

Exercise 5.5. Widom insertion method and Henry's law

Consider a simple fluid of $N + 1$ identical particles contained in a vessel of volume V at temperature T. In the following, particle $i = 0$ will be distinct from the $N \gg 1$ other particles[41].

1. Show that the system's Hamiltonian H_{N+1} can be decomposed as: $H_{N+1} = H_N + \frac{\vec{p}_0^2}{2m} + \Psi_{0N}$, where H_N is the Hamiltonian of N particles and Ψ_{0N} is the interaction energy of particle "0", with the N other particles.

2. Express the system's partition function Z_{N+1} as a function of Z_N, the partition function of a fluid composed of N indistinguishable particles, and show that

$$\langle e^{-\beta\Psi_{0N}}\rangle_u = (N + 1)\frac{Z_{N+1}}{Z_N}\frac{\lambda_T^3}{V},$$

where λ_T is the de Broglie thermal wavelength and $\langle \dots \rangle_u$ is an average quantity whose meaning must be explained.

3. Deduce the expression of the chemical potential μ as a function of $\langle e^{-\beta\Psi_{0N}}\rangle_u$ and the ideal gas chemical potential, $\mu_{IG} = k_B T \ln \rho\lambda_T^3$.

4. When a gas composed of molecules A (CO_2 for example) is in contact with a liquid composed of molecules B (H_2O) at pressure P and temperature T, the gas molecules A can dissolve in the liquid. The density of gas A, considered ideal, is ρ_g^A, the density of molecules B in the liquid is ρ_l^B and the density of molecules A in the liquid is ρ_l^A. The chemical potential of a single molecule A dissolved in the liquid is given by the expression derived in the previous question, by substituting Ψ_{0N} for Ψ_{AB}, the interaction energy between a molecule A and the N_B molecules B in the liquid. Show that in equilibrium:

$$\rho_l^A = \frac{P}{k_B T}\langle e^{-\beta\Psi_{AB}}\rangle_u. \tag{5.39}$$

---

[41]After B. Widom, *Some Topics in the Theory of Fluids*, J. Chem. Phys. **39**, 2808 (1963).

5. Deduce that at low dilution ($\rho_1^A \ll \rho_1^B$), the molar fraction of molecules A dissolved in the liquid, $x_1^A = \rho_1^A/(\rho_1^A + \rho_1^B)$, is given by Henry's law (1803): $x_1^A = P/K_{A,B}$, where $K_{A,B}$ is the Henry's constant of gas A in liquid B. Give the expression of $K_{A,B}$.

6. Show that in the framework of the mean field approximation studied in Section 5.3.1, $\langle e^{-\beta\Psi_{AB}}\rangle_u \simeq (1 - 2b\rho)e^{2a\beta\rho}$, where a and b are defined by Equation (5.25). Deduce the expression of μ and $K_{A,B}$ as functions of T and ρ_1^B. Comment on the effect of global warming on the quantity of CO_2 dissolved in the oceans.

7. Henry's constant K_{CO_2,H_2O} of CO_2 in water is $K_1 = 1.67 \times 10^3$ bar at $T_1 = 25$ °C and $K_2 = 3.52 \times 10^3$ bar at $T_2 = 60$ °C. In this case, deduce the effective values of a and b, assuming that water density is independent of temperature.

Solution:

1. _The Hamiltonian (1.7) of a simple fluid composed of $N + 1$ particles can be written as:_

$$
\begin{aligned}
H_{N+1} &= \sum_{i=0}^{N} \frac{\vec{p}_i^2}{2m} + \sum_{i<j} u(r_{ij}) = \sum_{i=0}^{N} \frac{\vec{p}_i^2}{2m} + \sum_{i=0}^{N-1}\sum_{j=i+1}^{N} u(r_{ij}) \\
&= \frac{\vec{p}_0^2}{2m} + \underbrace{\sum_{i=1}^{N} \frac{\vec{p}_i^2}{2m} + \sum_{i=1}^{N-1}\sum_{j=i+1}^{N} u(r_{ij})}_{=H_N} + \underbrace{\sum_{j=1}^{N} u(r_{0j})}_{=\Psi_{0N}}.
\end{aligned}
$$

2. _By definition,_

$$
\begin{aligned}
Z_{N+1} &= \frac{1}{(N+1)!h^{3(N+1)}} \int \prod_{i=0}^{N} d^3r_i \prod_{i=0}^{N} d^3\vec{p}_i \ e^{-\beta\overbrace{\left(\frac{\vec{p}_0^2}{2m} + H_N + \Psi_{0N}\right)}^{=H_{N+1}}} \\
&= \frac{1}{(N+1)h^3} \frac{1}{N!h^{3N}} \int d^3r_0 d^3p_0 \ e^{-\beta\frac{\vec{p}_0^2}{2m}} \int \prod_{i=1}^{N} d^3r_i \prod_{i=1}^{N} d^3p_i \ e^{-\beta(H_N+\Psi_{0N})} \\
&= \frac{Z_N}{(N+1)\lambda_T^3} \int d^3r_0 \underbrace{\frac{1}{Z_N N! h^{3N}} \int d^{3N}r_i d^{3N}p_i \ e^{-\beta H_N} \left(e^{-\beta\Psi_{0N}}\right)}_{=\langle e^{-\beta\Psi_{0N}}\rangle},
\end{aligned}
$$

where $\langle\ldots\rangle$ is the usual canonical ensemble average. Thus:

$$
\langle e^{-\beta\Psi_{0N}}\rangle_u \equiv \int \frac{d^3r_0}{V} \ \langle e^{-\beta\Psi_{0N}}\rangle = (N+1)\frac{Z_{N+1}}{Z_N} \frac{\lambda_T^3}{V},
$$

where $\langle\ldots\rangle_u$ is by definition an average value calculated as follows: the N particles configurations $\{\vec{r}_i\}$ are created in the canonical ensemble at temperature T. For each of these configurations, particle "0" is inserted among the N frozen particles, with the uniform probability $\frac{d^3r_0}{V}$. At each insertion, the integration energy Ψ_{0N} between particle "0" and the N others is calculated, to finally evaluate $\langle e^{-\beta\Psi_{0N}}\rangle_u$. This method is easy to implement in numerical simulations.

3. _According to Equation (4.20), $\mu = \frac{\partial F}{\partial N} \simeq F_{N+1} - F_N = -k_BT \ln(Z_{N+1}/Z_N)$ for $N \gg 1$, so[42]:_

$$
\mu = -k_BT \ln\left(\frac{V}{N\lambda_T^3}\langle e^{-\beta\Psi_{0N}}\rangle_u\right) = \mu_{IG} - k_BT \ln\langle e^{-\beta\Psi_{0N}}\rangle_u.
$$

---

[42] $\langle E \rangle$ and P can also be expressed as a function of an average value $\langle\ldots\rangle_u$ (see the article cited in Note 41).

4. In equilibrium the chemical potentials of molecules A in the gas and in the liquid are equal: $\mu_g^A = \mu_l^A$. Since the gas is ideal, $\mu_g^A = k_BT \ln \rho_g^A \lambda_T^3$, where λ_T is the de Broglie thermal wavelength of molecules A and $P = \rho_g^A k_BT$. According to the previous question, $\mu_l^A = k_BT \ln \rho_l^A \lambda_T^3 - k_BT \ln \langle e^{-\beta \Psi_{AB}} \rangle_u$, from which Relation (5.39) can be deduced.

5. For $\rho_l^A \ll \rho_l^B$, $x_l^A = \rho_l^A/(\rho_l^A + \rho_l^B) \simeq \rho_l^A/\rho_l^B = \dfrac{P}{\rho_l^B k_BT} \langle e^{-\beta \Psi_{AB}} \rangle_u = P/K_{A,B}$,

   where $K_{A,B} = \dfrac{\rho_l^B k_BT}{\langle e^{-\beta \Psi_{AB}} \rangle_u}$: the solubility (x_l^A) increases with pressure[43].

6. $\langle e^{-\beta \Psi_{AB}} \rangle_u \simeq p_{ins} \times e^{-\beta \langle \Psi_{AB} \rangle}$, where p_{ins} is the insertion probability of molecule A in the liquid, which can be written as $p_{ins} = (V - v_e)/V$, where v_e is the excluded volume of molecule A due to contact repulsion with the N_B molecules B of the liquid: $v_e = N_B \times 4\pi\sigma^3/3 = 2bN_B$, with $b = 2\pi\sigma^3/3$ (here, σ is the contact distance between molecules A and B treated as attractive hard spheres). Once molecule A is inserted, it remains to calculate its average attractive interaction energy with the molecules B of the liquid. According to Equation (5.20),

$$\langle \Psi_{AB} \rangle = \langle \sum_{j=1}^{N_B} u_{AB}(r_{Aj}) \rangle \simeq \int_\sigma^\infty dr \, 4\pi\rho_l^B r^2 u_{AB}(r) = -2a\rho_l^B,$$

where the sum is over the N_B molecules B, $u_{AB}(r)$ is the interaction potential between molecules A and B and $a = -2\pi \int_\sigma^\infty dr \, r^2 u_{AB}(r)$. It can be easily deduced:

$$\langle e^{-\beta \Psi_{AB}} \rangle_u \simeq (1 - v_e/V)e^{2a\beta\rho_l^B} = (1 - 2b\rho_l^B)e^{2a\beta\rho_l^B}, \text{ with } \rho_l^B < 1/(2b). \text{ Therefore}[44]:$$

$$\mu = \mu_{IG} - k_BT \ln \langle e^{-\beta \Psi_{AB}} \rangle_u \simeq \mu_{IG} - k_BT \ln\left(1 - 2b\rho_l^B\right) - 2a\rho_l^B,$$

where $\mu_{IG} = k_BT \ln \rho_l^B \lambda_T^3$ and

$$K_{A,B} = \frac{\rho_l^B k_BT}{\langle e^{-\beta \Psi_{AB}} \rangle_u} \simeq \frac{\rho_l^B k_BT}{1 - 2b\rho_l^B} e^{-2a\beta\rho_l^B}.$$

At constant pressure, the solubility $(x_l^A = P/K_{A,B})$ decreases when the temperature, the liquid density ρ_l^B or the size of the solute molecules A (hence b and σ) increase[45]. Thus, the solubility of CO_2 in seawater decreases with increasing ocean temperature. The amount of carbon dioxide that can no longer be dissolved is then released into the atmosphere, increasing the greenhouse effect and thus the global temperature, and so on.

7. Eliminating b, $(K_1/T_1)e^{-2a\rho_l^B/(k_BT_1)} = (K_2/T_2)e^{-2a\rho_l^B/(k_BT_2)}$. With $\rho_l^B \simeq 3.3 \times 10^{28}$ m^{-3} (or 1 g·cm^{-3}) one finds $a \simeq 2.3$ eV·Å^{3}e and $b \simeq 15$ Å^3.

Exercise 5.6. Dilute solution and osmotic pressure

A solution is a homogeneous mixture of solute molecules of mass m, dissolved in a solvent consisting of molecules of mass m' of another chemical substance (for example sugar dissolved in water). Since there are two types of molecules, the potential energy U of the fluid is broken down into:

$$U_{pot}(\mathbf{q}, \mathbf{q}') = U_0(\mathbf{q}) + U'(\mathbf{q}') + U_{int}(\mathbf{q}, \mathbf{q}'),$$

---

[43] The CO_2 pressure in a soda bottle is about 2 to 4 atm. When it is opened, the pressure drops (the CO_2 partial pressure in ambient air is 3.4×10^{-4} atm), as well as the solubility. The amount of gas that can no longer be dissolved in the liquid then forms bubbles that gradually escape from the bottle. Similarly, in scuba diving, the amount of nitrogen dissolved in the blood increases with depth (pressure). During the ascent, decompression stop must be observed. The decompression time must be respected to give time to nitrogen to be gradually eliminated from the body without forming bubbles that would obstruct blood vessels (embolism).

[44] In the low density limit, the chemical potential of a van der Waals fluid (Equation (5.23)) is recovered. Indeed, for $x = b\rho_l^B \ll 1$, $\ln(1 - 2x) = \ln[(1 - x)(1 - x/(1 - x))] \simeq \ln(1 - x) - x/(1 - x)$.

[45] A cool soda contains more CO_2 and is much more fuzzy.

where $\boldsymbol{q}$ and $\boldsymbol{q}'$ are the set of generalised coordinates of the N solvent molecules and the N' solute molecules, respectively: U_0, U' and U_{int} are the potential energies of the pure solvent, the solute and interaction between the two types of molecules, respectively.

1. The fluid is maintained at temperature T in a container of volume V. Write the partition function $Z(N, N', V, T)$ as a function of the thermal wavelengths λ_T and λ_T' of the solvent molecules and the solute molecules, respectively.

2. In the case of a dilute solution ($N \gg N' \gg 1$), the molar fraction of the solute is very small ($x = N'/(N+N') \simeq N'/N \ll 1$). Therefore, the interactions between the N' solute molecules can be ignored since they are far from each other and each is surrounded by solvent molecules. In a mean field approximation, each solute molecule is subjected to a uniform effective interaction potential $u_{\text{eff}}(\rho)$, which depends only on the solvent molecule density $\rho = N/V$. Derive the expression for Z as a function of Z_0 and z', the partition functions of the pure solvent and a solute molecule immersed in the solvent, respectively.

3. Write the fluid's free energy F as a function of the pure solvent's free energy F_0, N, N' and ρ. Assuming that the solvent density ρ is not changed by the presence of the solute, express the chemical potentials μ and μ' of the solvent and the solute, respectively, as functions of the pure solvent chemical potential μ_0, the solute molar fraction x and ρ.

4. A solution with a small solute molar fraction is in contact with the pure solvent through a semipermeable membrane that allows only solvent molecules to pass through (see Figure 5.12). In equilibrium, an overpressure P_{osmo}, called the *osmotic pressure*, is observed in the solution. A small increase of the solvent density ρ in the solution leads to a small change of the chemical potential $d\mu_0$ with respect to the pure solvent value $\mu_0(\rho, T)$. Deduce van't Hoff's law (1886): $P_{\text{osmo}} \simeq N' k_B T / V$.

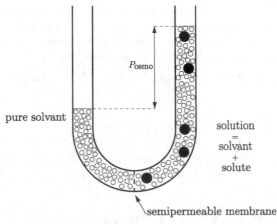

FIGURE 5.12 The osmotic pressure P_{osmo} is the pressure difference between the pure solvent (white molecules) and the solution that contains solute molecules (black molecules). Only the solvent molecules can pass through the semipermeable membrane.

5. Physiological saline contains 0.9% sodium chloride diluted in pure water (that is $m_{\text{salt}} = 9$ g of salt for $m_{\text{water}} = 1$ kg of water). This value corresponds to the concentration of solutes in human cells (especially blood cells). The cells are surrounded by a membrane permeable to water and impermeable to solute molecules. This salt

dissociates into Na^+ and Cl^- ions of molar masses $M_{Na} = 23$ g·mol^{-1} and $M_{Cl} = 35.5$ g·mol^{-1}. Calculate the solute molar fractions x and the osmotic pressure at $T = 37°C$. What happens when cells are immersed in seawater (about 3.5% NaCl)? In pure water?

Solution:

1. *Solvent molecules are indistinguishable (and so are the solute molecules) so:*

$$Z = \frac{1}{N!h^{3N}}\frac{1}{N'!h^{3N'}}\int d^{3N}q\, d^{3N}p \int d^{3N'}q'\, d^{3N'}p'$$
$$e^{-\beta\left(\sum_{i=1}^{N}\frac{\vec{p}_i^2}{2m}+\sum_{j=1}^{N'}\frac{\vec{p}_j'^2}{2m'}+U(q,q')\right)}$$
$$= \frac{1}{N!\lambda_T^{3N}}\frac{1}{N'!\lambda_T'^{3N'}}\int d^{3N}q \int d^{3N'}q'\, e^{-\beta(U_0(q)+U'(q')+U_{int}(q,q'))},$$

where the integrations over momenta p and p' of solvent and solute molecules were performed.

2. *Thus, $U'(q') \ll U_{int}(q,q')$, the solute molecules are independent of each other. In addition, $U_{int}(q,q') \simeq N'u_{eff}(\rho)$, which is independent of positions q and q'. Therefore:*

$$Z = \frac{1}{N!\lambda_T^{3N}}\int d^{3N}q\, e^{-\beta U_0(q)}\frac{1}{N'!\lambda_T'^{3N'}}\int d^{3N'}q'\, e^{-\beta N'u_{eff}(\rho)}$$
$$= Z_0 \times \frac{1}{N'!\lambda_T'^{3N'}}\left(\int d^3q'\, e^{-\beta u_{eff}(\rho)}\right)^{N'} = Z_0\frac{z'^{N'}}{N'!} \quad \text{where } z' = \frac{Ve^{-\beta u_{eff}(\rho)}}{\lambda_T'^3}.$$

3. *Therefore, $F = -k_B T \ln Z = F_0(N,V,T) + N'\left(u_{eff}(\rho) - k_B T[1 + \ln\frac{N}{N'\rho\lambda_T'^3}]\right)$, where $\rho = N/V$. By deriving F (at constant ρ) one gets, with $x \simeq N'/N$,*

$$\mu = \frac{\partial F}{\partial N}\Big|_{\rho,T,N'} = \mu_0(\rho,T) - k_B T\frac{N'}{N} \simeq \mu_0(\rho,T) - xk_B T$$
$$\mu' = \frac{\partial F}{\partial N'}\Big|_{\rho,T,N} = u_{eff}(\rho) + k_B T \ln\left(\rho\lambda_T'^3\right) + k_B T \ln x.$$

4. *In equilibrium, the chemical potentials of the pure solvent ($\mu_0(\rho,T)$) and the solvent in the solution ($\mu \simeq \mu_0(\rho,T) + d\mu_0 - xk_B T$) are equal, so $d\mu_0 = xk_B T$. According to the Gibbs-Duhem relation (Equation (4.60)) at constant temperature, $\rho d\mu_0 = dP_0$, so the excess pressure can be written as $P_{osmo} \simeq dP_0 = \rho xk_B T = N'k_B T/V$: at low dilution, the osmotic pressure is equal to the pressure exerted by an imaginary ideal gas having the density $\rho' = N'/V$ of the solute in the solution. However, the solute is not treated as an ideal gas since the interactions between the solute molecules and the solvent molecules are taken into account[46].*

5. *Since there are two type of ions in the solution, one gets (with $M_{water} = 18$ g·mol^{-1}):*

$x = \frac{N_{salt}}{N_{water}} = 2\left(\frac{m_{salt}}{M_{Na}+M_{Cl}}\right)/\left(\frac{m_{water}}{M_{water}}\right) \simeq 0.0055$, *so $P_{osmo} \simeq 2 \times \frac{m_{salt}}{M_{Na}+M_{Cl}}\frac{RT}{V} \simeq 7.9$ bar, with $V = 1$ L and $T = 310$ K. In seawater, the concentration of solutes is higher outside the cell than inside. Consequently, the cell dehydrates and contracts under the osmotic pressure exerted by seawater. On the other hand, when immersed in pure water, the oscmotic pressure is larger inside the cell than outside; water penetrates to dilute the solutes: the cell swells, or even bursts.*

---

[46]When these interactions are ignored ($u_{eff} = 0$), van't Hoff law is recovered by calculating the solution's pressure: $P = -\frac{\partial F}{\partial V} = \frac{N}{V^2}\frac{\partial F}{\partial \rho} = P_0 + \frac{NN'}{V^2}\left(\frac{du_{eff}}{d\rho} + \frac{k_B T}{\rho}\right) \simeq P_0 + \frac{N'k_B T}{V}$.

Exercise 5.7. Colloidal suspension and attractive depletion interactions

Colloidal silica particles suspended in a liquid behave as a system of hard spheres of radius $R \sim 10$ nm with no attractive interactions between them. Polymers added to the solution form small spherical balls of radius $\Delta < R$ that interpenetrate each other and collide with the colloidal particles. This mixture gives rise to very interesting thermodynamic behaviours, such as the coexistence between a phase rich (dense) in colloidal particles and a phase poor (dilute) in colloidal particles, analogous to the liquid-gas phase separation (see references in Note 26). To explain this phenomenon, S. Asakura and F. Oosawa[47] demonstrated in 1954 the existence of an *effective* attractive interaction between colloidal particles due to the presence of polymer pellets. This interaction, called *depletion interaction* (whose expression will be determined in this exercise), is added to the contact repulsion u_{cc} between two colloidal particles at a distance r:

$$u_{cc}(r) = \begin{cases} +\infty & \text{if } r \leq 2R \\ 0 & \text{otherwise.} \end{cases} \tag{5.40}$$

In this model, the interpenetrable polymer pellets behave among *each other* as an ideal gas ($u_{pp}(r) = 0$, for any r), but have a contact repulsion with the colloidal particles:

$$u_{pc}(r) = \begin{cases} +\infty & \text{if } r \leq R + \Delta \\ 0 & \text{otherwise,} \end{cases} \tag{5.41}$$

where r is the distance between a colloidal particle and a polymer ball. The system is studied in the grand canonical ensemble: the mixture is contained in a vessel of volume V in contact with a thermostat at temperature T and with a reservoir of colloidal particles of chemical potential μ_c and a reservoir of polymers of chemical potential μ_p.

1. Give the mixture grand partition function $\Xi(\mu_c, \mu_p, V, T)$ as a function of $\Xi_p\left(\mu_p, V, T | \{\vec{r}_i\}_{i=1,N_c}\right)$, the grand partition of the polymer ideal gas interacting with N_c colloidal particles. The colloidal particles' positions are $\{\vec{r}_i\}_{i=1,N_c}$. Use λ_{Tc} and λ_{Tp}, the thermal wavelength of the colloidal particles and the polymer pellets, respectively.

2. Show that $k_B T \ln \Xi_p\left(\mu_p, V, T | \{\vec{r}_i\}_{i=1,N_c}\right) = P_p V_p$, where P_p is pressure of the polymer ideal gas and V_p is the volume accessible to the polymer pellets for a given configuration $\{\vec{r}_i\}_{i=1,N_c}$ of colloidal particles.

3. The volume accessible to the polymer balls can be written as:

$$V_p = V - N_c \frac{4\pi}{3}(R + \Delta)^3 + \sum_{i<j} V_{ij}^{(2)} - \sum_{i<j<k} V_{ijk}^{(3)} + \dots, \tag{5.42}$$

where the second term of the right-hand side is the excluded volume due to the N_c colloidal particles, $V_{ij}^{(2)}$ is the overlapping volume of the two spheres of radius $R + \Delta$ centred in $\vec{r}_i$ and $\vec{r}_j$. $V_{ijk}^{(3)}$ is the overlapping volume of three spheres of radius $R + \Delta$ centred in $\vec{r}_i$, $\vec{r}_j$ and $\vec{r}_k$ (see Figure 5.13). What situation maximises the volume $V_{ijk}^{(3)}$? Show that for any colloidal particles configuration, $V_{ijk}^{(3)}$ is zero if the ratio Δ/R is lower than a specific value d_{CO}. Determine d_{CO}.

4. Calculate $V_{ij}^{(2)}$ as a function of the distance r between two colloidal particles.

---

[47] *On interaction between two bodies immersed in a solution of macromolecules*, J. Chem. Phys. **22**, 1255 (1954).

5. Show that if three spheres overlaps are ignored, the system's grand partition function Ξ can be written as the grand partition function of a fluid of colloidal particles interacting with each other through the effective pairwise potential (to within one additive constant):

$$u_{\text{eff}}(r) = \begin{cases} +\infty & \text{if } r \leq 2R \\ f(r) & \text{if } 2R < r \leq 2(R+\Delta) \\ 0 & \text{if } r > 2(R+\Delta), \end{cases} \tag{5.43}$$

where $f(r)$ is a third order polynomial in r to determine. Plot $u_{\text{eff}}(r)$ as a function of r.

6. What are the physical quantities associated with polymers that control the intensity and extent of the effective interaction?

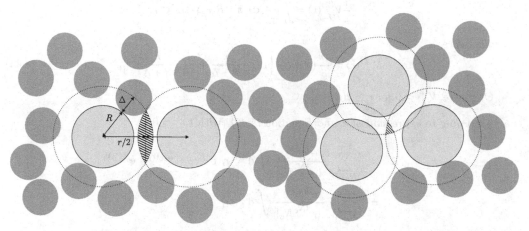

FIGURE 5.13 Mixture of colloidal particles of radius R and polymer pellets of radius $\Delta < R$. The hatched regions correspond to the overlap volumes of the excluded spheres of radius $R + \Delta$ (dashed curve) surrounding each colloidal particle: on the left the overlap volume $V^{(2)}(r)$ for two particles separated by the distance r, on the right the overlap volume $V^{(3)}$ for three colloidal particles.

Solution:

1. *Let $\vec{l}_i$ be the position of the ith polymer pellet. $\Xi(\mu_c, \mu_p, V, T)$ is given by:*

$$\Xi = \sum_{N_c \geq 0} \sum_{N_p \geq 0} \frac{e^{\beta \mu_c N_c}}{\lambda_{T_c}^{3N_c} N_c!} \frac{e^{\beta \mu_p N_p}}{\lambda_{T_p}^{3N_p} N_p!} \int d^{3N_c} r_i \int d^{3N_p} l_j$$
$$e^{-\beta \left(\sum_{i<j} u_{cc}(r_{ij}) + \sum_{i=1}^{N_c} \sum_{j=1}^{N_p} u_{pc}(|\vec{r}_i - \vec{l}_j|) \right)}.$$

Separating the sum over N_c and N_p, one gets:

$$\Xi = \sum_{N_c \geq 0} \frac{e^{\beta \mu_c N_c}}{\lambda_{T_c}^{3N_c} N_c!} \int d^{3N_c} r_i \, e^{-\beta \sum_{i<j} u_{cc}(r_{ij})}$$
$$\underbrace{\left(\sum_{N_p \geq 0} \frac{e^{\beta \mu_p N_p}}{\lambda_{T_p}^{3N_p} N_p!} \int d^{3N_p} l_j \, e^{-\beta \sum_{i=1}^{N_c} \sum_{j=1}^{N_p} u_{pc}(|\vec{r}_i - \vec{l}_j|)} \right)}_{\Xi_p \left(\mu_p, V, T \,\middle|\, \{\vec{r}_i\}_{i=1, N_c} \right)},$$

where the term in parentheses is the grand partition function of the ideal polymer gas interacting with N_c colloidal particles in a given configuration $\{\vec{r}_i\}_{i=1, N_c}$.

2. *The grand potential of a simple fluid, a fortiori an ideal one, is given by Equation(4.58), that is $J = -k_B T \ln \Xi_p = -P_p V_p$, where V_p is the volume accessible to the ideal polymer gas, when the colloidal particles are in the configuration $\{\vec{r}_i\}_{i=1,N_c}$.*

3. *The volume $V_{ijk}^{(3)}$ is maximal when the three colloidal particles of radius R are in contact. In this case, the three centres form an equilateral triangle of side $2R$ whose centre C is at the distance d_{CO} from the centre O of one of the spheres, that is*

$$d_{CO} = \frac{2}{3}\sqrt{(2R)^2 - R^2} = \frac{2}{\sqrt{3}}R.$$

Thus, $V_{ijk}^{(3)} = 0$, if $R + \Delta < d_{CO}$, that is $\Delta/R < 2/\sqrt{3} - 1 \simeq 0.15$.

4. *According to Figure 5.13, for $r \leq 2(R + \Delta)$:*

$$\frac{1}{2}V_{ij}^{(2)}(r) = \int_{r/2}^{R+\Delta} dx \, \pi \left((R + \Delta)^2 - x^2\right).$$

Therefore,

$$V_{ij}^{(2)}(r) = \frac{4\pi}{3}(R + \Delta)^3 \left(1 - \frac{3}{4}\frac{r}{R + \Delta} + \frac{1}{16}\left(\frac{r}{R + \Delta}\right)^3\right). \qquad (5.44)$$

Of course, $V_{ij}^{(2)} = 0$ for $r \geq 2(R + \Delta)$.

5. *According to expression of Ξ and Equation (5.42) (with $V_{ij}^{(3)} = 0$):*

$$\begin{aligned}
\Xi &= \sum_{N_c \geq 0} \frac{e^{\beta \mu_c N_c}}{\lambda_{T_C}^{3N_c} N_c!} \int d^{3N_c} r_i \, e^{-\beta \sum_{i<j} u_{cc}(r_{ij})} \, e^{\beta P_p V_p} \\
&= \sum_{N_c \geq 0} \frac{e^{\beta \mu_c N_c}}{\lambda_{T_C}^{3N_c} N_c!} \int d^{3N_c} r_i \\
&\quad e^{-\beta\left(\sum_{i<j} u_{cc}(r_{ij}) - P_p\left(V - N_c \frac{4\pi}{3}(R+\Delta)^3 + \sum_{i<j} V_{ij}^{(2)}(r_{ij})\right)\right)} \\
&= \sum_{N_c \geq 0} \frac{e^{\beta \mu_c N_c}}{\lambda_{T_C}^{3N_c} N_c!} \int d^{3N_c} r_i \, e^{-\beta \sum_{i<j} u_{\text{eff}}(r_{ij})}
\end{aligned}$$

which is the grand partition function of a fluid of particles interacting through an effective pairwise potential $u_{\text{eff}}(r)$ defined as:

$$u_{\text{eff}}(r_{ij}) = u_{cc}(r_{ij}) - P_p V_{ij}^{(2)}(r_{ij}),$$

within a negative constant $-P_p\left(V - N_c \frac{4\pi}{3}(R + \Delta)^3\right)$. Equation (5.43) is, of course, recovered with, according to Equation (5.44),

$$f(r) = -P_p V_{ij}^{(2)}(r) = -P_p \frac{4\pi}{3}(R + \Delta)^3 \left(1 - \frac{3}{4}\frac{r}{R + \Delta} + \frac{1}{16}\left(\frac{r}{R + \Delta}\right)^3\right).$$

Function f is increasing from $r = 2R$ (with $f(2R) < 0$) and cancels in $r = 2(R + \Delta)$. The effective interaction is thus attractive for $2R < r \leq 2(R + \Delta)$.

6. *The polymer pellet size Δ controls the extent of the effective attraction equal to $2(R + \Delta)$. The intensity of the attraction depends on Δ, but also on the amount of polymer present in the mixture, since $P_p = \rho_p k_B T$ (ideal gas), where ρ_p is the polymer density in the system. At a given temperature, the attraction intensity increases with the polymer concentration. If the range and intensity of the attraction is sufficient, a phase transition between the two fluid phases of the colloidal particles system can be observed (see Section 5.3.3).*

Exercise 5.8. Plasmas and electrolyte solutions: Debye-Hückel theory

Plasmas are ionised gases made of positive ions and electrons. They are making up more than 99% of ordinary matter in the Universe (stars, nebulae, solar wind, etc.). In these systems, as in molten salts or electrolyte solutions (which contain positive and negative ions dissolved in a solvent, typically water), Coulomb interactions are predominant and cannot be treated as short-range interactions as it was down so far (see Note 14 of this chapter). In this exercise[48], a mean field approach, called the *Debye-Hückel theory* (1923), will be described. It allows the study, to a first approximation, of the properties of fully ionised systems, such as strong electrolytes (molten salts, hydrochloric acid...) or "hot" plasmas, in which ions and electrons are in equilibrium at the same very high temperature.

Consider the case of a plasma, an electrically neutral system of $N/2$ ions of charge $+e$ and $N/2$ electrons of charge $-e$, in equilibrium in a chamber of volume V. The plasma is maintained at a sufficiently high temperature T so that the particles kinetic energy is large in comparison with the electrostatic interaction potential energy. The ions and electrons densities are both equal to $\rho = N/(2V)$.

To determine the plasma equation of state $P(\rho,T)$, a single particle (ion or electron) of charge $q_0 = \pm e$ is considered. Its position is chosen as the origin of the coordinate system. The $N - 1$ other particles are distributed around it, according to a spherically symmetric non-uniform distribution (the charge q_0 attracts particles of opposite charge and repels same charge particles). It is described by a continuous average charge density, $\rho_c(r)$ (in C·m^{-3}) where r is the distance from the origin. Let $\rho_+(r)$ and $\rho_-(r)$ be the *mean particle densities* of ions and electrons, respectively. The central charge and the surrounding ion and electron clouds generate a *mean* electrostatic potential $\psi(r)$ in which the plasma ions and electrons evolve. Poisson equation is:

$$\Delta\psi(r) + \frac{\rho_c(r)}{\epsilon_0} = 0, \qquad (5.45)$$

where $\epsilon_0 \simeq 8.85 \times 10^{-12}$ m^{-3}·kg^{-1}·s^4·A^2 is the electric permittivity of vacuum and Δ is the Laplacian operator. In spherical coordinates, for a function $f(r)$: $\Delta f = \dfrac{1}{r}\dfrac{d^2(rf)}{dr^2}$.

1. What is the potential energy of a particle, ion $(+e)$ or electron $(-e)$, located in $\vec{r}$? Deduce the probability $P_\pm(r)dV$ to find a particle in the volume element dV located at a distance r from the origin as a function of $\psi(r)$ and k_BT within a multiplicative constant. Knowing that $\psi(r)$ tends to zero for $r \to \infty$, express $\rho_+(r)$ and $\rho_-(r)$ as functions of ρ, $\psi(r)$ and k_BT.

2. Express $\rho_c(r)$ as a function of $\rho_+(r)$ and $\rho_-(r)$. Show that at high temperature:

$$\rho_c(r) \simeq -2\beta\rho e^2 \psi(r). \qquad (5.46)$$

3. Deduce that

$$\psi(r) = \frac{A}{r}e^{-\frac{r}{\lambda_D}} + \frac{B}{r}e^{\frac{r}{\lambda_D}},$$

where A and B are constants and λ_D is a characteristics length called *Debye length*. What is the value of B?

4. What is the expression of $\psi(r)$ at the limit $r \to 0$? Deduce the expression of A and $\rho_c(r)$. What is the meaning of the Debye length? What is the value of the total charge Q of the ion and electron cloud surrounding the central charge q_0?

---

[48]This exercise is based on complement III.H of *Physique statistique* by B. Diu, C. Guthmann, D. Lederer et B. Roulet (Hermann, 1996, in French).

5. To determine the mean electrostatic potential $\langle\phi_0\rangle$ created at the origin by the $N-1$ other particles, the contribution of the central particle itself must be subtracted from $\psi(r)$, so:

$$\langle\phi_0\rangle = \lim_{r\to 0}\left[\psi(r) - \frac{q_0}{4\pi\epsilon_0 r}\right].$$

Calculate $\langle\phi_0\rangle$. Deduce the mean electrostatic potential $\langle\phi_i\rangle$ at the position of particle i of charge q_i created by the $N-1$ other particles. Give the expression of the plasma mean potential energy $\langle U_{\text{pot}}\rangle = \frac{1}{2}\sum_{i=1}^{N} q_i\langle\phi_i\rangle$, where the factor $1/2$ avoids the double counting of the particle pair interaction energy.

6. Considering monoatomic ions, what is the mean total kinetic energy $\langle E_{\text{kin}}\rangle$ of the plasma particles? Deduce the system's mean energy $\langle E\rangle$.

7. Under which conditions is the Debye-Hückel approximation justified? Deduce a condition using λ_D and the mean interparticle distance d.

8. Express the mean energy $\langle E\rangle$ as a partial derivative of the free energy F. Show that, for a plasma, F can be written as $F = F_{\text{IG}} + F_{\text{int}}$, where F_{IG} is the free energy of an ideal gas and F_{int} is the free energy associated to the electrostatic interactions.

9. Calculate and express the plasma pressure P in the form $P = P_{\text{IG}} + P_{\text{int}}$, where P_{IG} is the ideal gas pressure and P_{int} is the pressure associated with the electrostatic interactions that will be expressed as a function of λ_D.

10. **Application to plasmas**: The ionosphere is the upper layer of the atmosphere located, for Earth, beyond 60 km altitude. Under the action of solar radiation, a fraction of the molecules is ionised. The electron density is $\rho \simeq 10^{12}$ m^{-3} and the electron temperature is $T \simeq 10^3$ K. Ions, much heavier, have a much lower temperature and are much less mobile. The ion contribution is thus negligible and only the electron density and temperature intervene in the expression of the Debye length. Is Debye-Huckel approximation justified? What about solar wind, the plasma ejected by the sun upper atmosphere with a density $\rho \simeq 10^6$ m^{-3} and an electron temperature $T \simeq 10^5$ K? What is then the correction to pressure $P_{\text{int}}/P_{\text{IG}}$?

11. **Application to electrolytes**: Pure water contains 10^{-7} mol·l^{-1} of H$^+$ ions and OH$^-$ ions. Sea water contains around $c = 35$ g·l^{-1} of NaCl dissolved in Na$^+$ and Cl$^-$ ions. Calculate λ_D in pure water and in sea water at room temperature (in that case, vacuum permittivity must be substituted by water permittivity in the previously derived expressions, i.e. $80\epsilon_0$). The molar mass of NaCl is $M \simeq 58$ g·mol^{-1}.

Solution:

1. _The potential energy of an ion ($+e$) or an electron ($-e$) at a distance r is $\pm e\psi(r)$, thus:_

$$P_{\pm}(r)\mathrm{d}V = Ce^{\mp\beta e\psi(r)}\mathrm{d}V,$$

where C is a constant. Therefore, $\rho_{\pm}(r) = \rho\, e^{\mp\beta e\psi(r)}$, because for $r \to \infty$, $\psi(r) \to 0$ and $\rho_{\pm}(r) \to \rho$, the uniform density, unperturbed by the central charge in $r = 0$.

2. _The total density of charge is, for $k_B T \gg e\psi(r)$ (with $\sinh x \simeq x$ for $x \ll 1$):_

$$\rho_c(r) = e\left(\rho_+(r) - \rho_-(r)\right) = e\rho\left(e^{-\beta e\psi(r)} - e^{\beta e\psi(r)}\right) = -2e\rho\sinh(\beta e\psi(r)) \simeq -2\beta\rho e^2\psi(r).$$

3. Using Equation (5.46) in (5.45), $\Delta\psi(r) + \dfrac{\rho_c(r)}{\epsilon_0} = \Delta\psi(r) - \dfrac{2\beta\rho e^2}{\epsilon_0}\psi(r) = \Delta\psi(r) - \dfrac{\psi(r)}{\lambda_D^2} = 0$, where

$\lambda_D = \sqrt{\dfrac{\epsilon_0 k_B T}{2\rho e^2}}$ has, of course, the dimension of a length. So:

$$\Delta\psi(r) = \frac{1}{r}\frac{d^2(r\psi(r))}{dr^2} = \frac{\psi(r)}{\lambda_D^2}.$$

With $u(r) = r\psi(r)$, one gets $\dfrac{d^2 u(r)}{dr^2} = \dfrac{u(r)}{\lambda_D^2}$, thus $u(r) = r\psi(r) = A\,e^{-\frac{r}{\lambda_D}} + B\,e^{\frac{r}{\lambda_D}}$, where A and B are constants. In the limit $r \to \infty$, $\psi(r) \to 0$, therefore $B = 0$.

4. For $r \to 0$, $\psi(r) \to \dfrac{A}{r} = \dfrac{q_0}{4\pi\epsilon_0 r}$, the potential of the unscreened central charge q_0. Thus $A = \dfrac{q_0}{4\pi\epsilon_0}$, so:

$$\psi(r) = \frac{q_0}{4\pi\epsilon_0 r}e^{-\frac{r}{\lambda_D}}.$$

It can be deduced:

$$\rho_c(r) \simeq -2\beta\rho e^2\psi(r) = -\frac{q_0}{4\pi\lambda_D^2 r}e^{-\frac{r}{\lambda_D}}.$$

Thus, λ_D is the screening length: beyond a few λ_D, the central charge q_0 is screened by an ion and electron cloud and $\rho_c(r) \simeq 0$. Charge neutrality imposes $Q = -q_0$, which can be checked with an integration by parts:

$$\begin{aligned} Q &= \int d^3 r\,\rho_c(r) = \int_0^\infty dr\,4\pi r^2 \rho_c(r) = -\frac{q_0}{\lambda_D^2}\int_0^\infty dr\,r\,e^{-\frac{r}{\lambda_D}} \\ &= -\frac{q_0}{\lambda_D^2}\Big(\underbrace{\Big[-\lambda_D r\,e^{-\frac{r}{\lambda_D}}\Big]_0^\infty}_{=0} + \lambda_D\underbrace{\int_0^\infty dr\,e^{-\frac{r}{\lambda_D}}}_{=\lambda_D}\Big) = -q_0. \end{aligned}$$

5. Therefore, $\langle\phi_0\rangle = \lim\limits_{r\to 0}\Big[\psi(r) - \dfrac{q_0}{4\pi\epsilon_0 r}\Big] = \lim\limits_{r\to 0}\Big[\dfrac{q_0}{4\pi\epsilon_0 r}e^{-\frac{r}{\lambda_D}} - \dfrac{q_0}{4\pi\epsilon_0 r}\Big] \simeq -\dfrac{q_0}{4\pi\epsilon_0\lambda_D}$. Since the central particle is arbitrarily chosen, the expression of $\langle\phi_0\rangle$ is thus valid for every particle i of charge $q_i = \pm e$, that is:

$$\langle\phi_i\rangle \simeq -\frac{q_i}{4\pi\epsilon_0\lambda_D}.$$

And $\langle U_{\text{pot}}\rangle = \dfrac{1}{2}\sum\limits_{i=1}^{N} q_i\langle\phi_i\rangle = -\sum\limits_{i=1}^{N}\dfrac{q_i^2}{8\pi\epsilon_0\lambda_D} = -\dfrac{Ne^2}{8\pi\epsilon_0\lambda_D} = -\dfrac{Nk_B T}{16\pi\rho\lambda_D^3} = -\dfrac{Vk_B T}{8\pi\lambda_D^3}$, which is naturally extensive and homogeneous to an energy.

6. According to the equipartition of energy theorem, $\langle E_{\text{kin}}\rangle = \dfrac{3}{2}\left(\dfrac{N}{2} + \dfrac{N}{2}\right)k_B T$, thus

$$\langle E\rangle = \langle E_{\text{kin}}\rangle + \langle U_{\text{pot}}\rangle = \frac{3}{2}Nk_B T - \frac{Vk_B T}{8\pi\lambda_D^3} = \frac{3}{2}N\frac{1}{\beta} - \frac{V}{8\pi}\left(\frac{2\rho e^2}{\epsilon_0}\right)^{\frac{3}{2}}\sqrt{\beta}.$$

7. The approximation is justified for $\langle E_{\text{kin}}\rangle \gg |\langle U_{\text{pot}}\rangle|$, so $\dfrac{3}{2}Nk_B T \gg \dfrac{Vk_B T}{8\pi\lambda_D^3}$ or

$1 \gg \dfrac{1}{12\pi}\left(\dfrac{2e^2}{\epsilon_0}\right)^{\frac{3}{2}}\dfrac{\sqrt{\rho}}{(k_B T)^{\frac{3}{2}}}$ (high temperature or low density approximation), or additionally

$\lambda_D \gg \left(\dfrac{V}{N}\right)^{\frac{1}{3}} = d$, where d is the mean interparticle distance. This condition is consistent with the continuous charge distribution assumption $\rho_c(r)$.

8. Integrating $\langle E \rangle = -\dfrac{\partial \ln Z}{\partial \beta} = \dfrac{\partial \beta F}{\partial \beta}$, one obtains $F = F_{\mathrm{IG}} + F_{\mathrm{int}}$, where F_{IG} is the free energy on an ideal gas made of a mixture of $N/2$ ions and $N/2$ electrons, with a mean energy $\frac{3}{2} N k_B T$ and

$$\beta F_{\mathrm{int}} = -\frac{2}{3} \frac{V}{8\pi} \left(\frac{2\rho e^2}{\epsilon_0} \right)^{\frac{3}{2}} \beta^{\frac{3}{2}} + K(N, V),$$

where $K(N, V)$ is an integration constant equal to zero, because $F_{\mathrm{int}} \to 0$ for $T \to \infty$. Therefore:

$$F(N, V, T) = F_{\mathrm{IG}}(N, V, T) - \frac{V}{12\pi} \left(\frac{2\rho e^2}{\epsilon_0} \right)^{\frac{3}{2}} \beta^{\frac{1}{2}} = F_{\mathrm{IG}}(N, V, T) - \frac{V}{12\pi\sqrt{k_B T}} \left(\frac{N e^2}{V \epsilon_0} \right)^{\frac{3}{2}},$$

which is, of course, extensive and homogeneous to an energy.

9. $P = P_{\mathrm{IG}} - \dfrac{\partial F_{\mathrm{int}}}{\partial V}\bigg|_{N,T} = 2\rho k_B T - \dfrac{1}{24\pi\sqrt{k_B T}} \left(\dfrac{N e^2}{V \epsilon_0} \right)^{\frac{3}{2}} = 2\rho k_B T \left(1 - \dfrac{1}{48\pi\rho\lambda_{\mathrm{D}}^3} \right)$, which is intensive.

   An "ideal" plasma, being a mixture of $N/2$ ions and $N/2$ electrons, the pressure of this mixture of ideal gases is $P = \frac{k_B T}{V} \left(\frac{N}{2} + \frac{N}{2} \right) = 2\rho k_B T$. The correction term $P_{\mathrm{int}} = -k_B T/(24\pi\lambda_{\mathrm{D}}^3)$ is small with respect to $P_{\mathrm{IG}} = 2\rho k_B T$ under the considered approximation (high temperature or low density).

10. Ignoring the ion contribution, only the electron density intervenes in the Debye length expression: $\lambda_{\mathrm{D}} = \sqrt{\epsilon_0 k_B T/(\rho e^2)} \simeq 2$ mm, which is slightly larger than $d = (1/\rho)^{\frac{1}{3}} \simeq 0.1$ mm. The Debye-Hückel approximation is here questionable. On the contrary, the approximation is valid for the solar wind since in this case $\lambda_{\mathrm{D}} \simeq 20$ m $\gg d \simeq 10^{-2}$ m and $P_{\mathrm{int}}/P_{\mathrm{IG}} = -1/(48\pi\rho\lambda_{\mathrm{D}}^3) \simeq 10^{-12}$.

11. For pure water ($\rho \simeq 6 \times 10^{19}$ m^{-3}): $\lambda_{\mathrm{D}} = \sqrt{80\epsilon_0 k_B T/(2\rho e^2)} \simeq 10^{-6}$ m and $d \simeq 0.2 \times 10^{-6}$ m. For see water, with $\rho = \frac{c}{M} N_A \simeq 3.6 \times 10^{26}$ m^{-3}, $\lambda_{\mathrm{D}} \simeq 0.4$ nm, and $d \simeq 1.5$ nm.

6 Quantum Statistical Physics

In the previous chapters, the focus was on systems of particles following the law of classical mechanics. Of course, this was mainly done within the framework of classical mechanics but a quantum description of individual (discretised) states was also used, while assuming *distinguishability* of the particles (for instance, particles located at the nodes of a lattice (spin system, quantum harmonic oscillators, ...))[1]. The present chapter as well as Chapters 7 and 8 are dedicated to the study of statistical physics in a fully quantum framework.

The first obstacle encountered by a quantum approach is the impossibility to define a quantum phase space (in which the ergodic hypothesis would hold for instance). This problem is inherent to quantum mechanics: because of Heisenberg's uncertainty principle, one cannot define both the momentum and the position of a particle.

Besides, statistical physics considers systems made of a large number of particles, *a priori* in mutual interaction. In classical physics, model situations for which the interactions between the constituents of the system are neglected can be studied. In quantum mechanics, the indistinguishability of identical particles makes the study of the problem in the absence of interaction already difficult: even for non-interacting particles, the symmetrisation (or antisymmetrisation) of the wave function of N identical particles induces correlations which have non-trivial thermodynamical effects such as the existence of a "Fermi sea" for fermions (see Section 8.1) or the Bose–Einstein condensation phenomenon for bosons (see Section 7.4).

The present chapter addresses these two points consecutively.

6.1 STATISTICAL DISTRIBUTIONS IN HILBERT SPACE

Classical statistical physics assigns to each microstate of phase space a statistical weight that corresponds to the probability of observing this state. Heisenberg's uncertainty principle forbids the use of the concept of phase space: in quantum mechanics, the position and momentum of a state cannot be precisely defined simultaneously.

The natural generalisation of the concept of phase space to quantum mechanics is the space of all possible quantum states of a given system, called *Hilbert space*. A statistical distribution in Hilbert space can be realised as follows: let $\{|\phi_n\rangle\}_{n\in\mathbb{N}}$ be an orthonormal basis of a given Hilbert space and let p_n be the probability assigned to a state $|\phi_n\rangle$. Among all possible types of probability distributions, it is sufficient for our purpose to consider equilibrium distributions where the probabilities p_n are time independent and the $|\phi_n\rangle$'s are stationary states. Canonical and microcanonical distributions in classical mechanics depend only on the energy of the microstates, and it is therefore appropriate to choose the set of eigenstates of the system's Hamiltonian $\hat{H}$, $\{|\phi_n\rangle\}_{n\in\mathbb{N}}$, as the basis of Hilbert space (the eigenvalue associated to the quantum state $|\phi_n\rangle$ is denoted by E_n). Quantum states $|\phi_n\rangle$ play a similar role to phase space microstates in classical physics.

As any probability distribution, $\{p_n\}$ should satisfy:

$$p_n \in [0,1], \quad \text{with} \quad \sum_{n=0}^{\infty} p_n = 1, \tag{6.1}$$

---

[1] Section 2.2.3 explains how to treat the indistinguishably of the particles of a fluid in the framework of classical physics.

DOI: 10.1201/9781003272427-6

The average over a large number of measurements of the value of an observable $\hat{O}$ is the quantity

$$\langle \hat{O} \rangle = \sum_{n \in \mathbb{N}} p_n \langle \phi_n | \hat{O} | \phi_n \rangle . \tag{6.2}$$

Formula (6.2) is at the heart of quantum statistical physics. It has a simple interpretation: since the system is in a state $|\phi_n\rangle$ with a probability p_n, the average value of the measurement of a quantum observable is the p_n-weighted average of its average value $\langle \phi_n | \hat{O} | \phi_n \rangle$ in each state. The equilibrium statistical distribution is characterised by an operator $\hat{D}$, the so-called *density matrix*, defined as:

$$\hat{D} = \sum_{n \in \mathbb{N}} p_n |\phi_n\rangle\langle\phi_n| . \tag{6.3}$$

In terms of the density matrix, the normalisation (6.1) and average (6.2) read:

$$\mathrm{Tr}\,\hat{D} = 1 , \quad \text{and} \quad \langle \hat{O} \rangle = \mathrm{Tr}\,(\hat{O}\,\hat{D}) . \tag{6.4}$$

Form (6.3) of the density matrix encompasses all the relevant statistical distributions of equilibrium quantum statistical mechanics.

6.1.1   STATE COUNTING

Consider a system described by a quantum Hamiltonian[2] $\hat{H}$ and let $d(E)$ be the associated density of states. $d(E)$ is defined as follows: consider a small energy interval $[E, E + \delta E]$ (with $\delta E \ll E$), $d(E)\delta E$ is the number of eigenstates of $\hat{H}$ with energies lying within this interval ($\Omega(E)$ in Equation (2.25)). It is also convenient to define the cumulative density of states (sometimes called the *spectral staircase*) $\Phi(E)$: $\Phi(E)$ is the number of eigenstates of $\hat{H}$ with energies lower than E. By writing $\{E_n\}_{n \in \mathbb{N}}$ the set of eigenlevels of $\hat{H}$ ordered by increasing values, and Θ the Heaviside[3] function, then:

$$\Phi(E) = \sum_{n=0}^{\infty} \Theta(E - E_n) . \tag{6.5}$$

It should be noted that, in this sum, a degenerate level appears as many times as there are eigenstates corresponding to its energy.

The number of eigenstates of $\hat{H}$ in the interval $[E, E + \delta E]$ can also be written in terms of the spectral staircase as the difference $\Phi(E + \delta E) - \Phi(E) = \delta E \, \mathrm{d}\Phi/\mathrm{d}E$. The density of state $d(E)$ is thus the derivative of $\Phi(E)$ and according to Equation (6.5):

$$d(E) = \frac{\mathrm{d}\Phi}{\mathrm{d}E} = \sum_{n=0}^{\infty} \delta(E - E_n) , \tag{6.6}$$

where δ is the Dirac distribution. Of course, Θ is not differentiable at zero, and $d(E)$ is a singular function, but this is not a real issue: it is enough to "round off" the Heaviside function by convolving it, for example, with a Gaussian of very small width (small with respect to the typical spacing between two energy levels) to calculate a meaningful derivative and regularise Expression (6.6). From the left term of Equality (6.6), the cumulative density of state $\Phi(E)$ can be expressed in terms of $d(E)$ as:

$$\Phi(E) = \int_{-\infty}^{E} d(\epsilon)\,\mathrm{d}\epsilon . \tag{6.7}$$

---

[2]In this section, as in Section A.7, quantum observables are written with a hat to distinguish them from their classical counterpart.

[3]$\Theta(x) = 1$ if $x \geq 0$ and 0 otherwise.

By defining an operator $\hat{\Phi}(E) = \Theta(E - \hat{H})$, it is possible to cast Formula (6.5) under the form[4] $\Phi(E) = \mathrm{Tr}\,\hat{\Phi}(E)$. Then using Expression (A.44), the trace of the operator $\hat{\Phi}$ can be evaluated by computing an integral in phase space:

$$\Phi(E) = \frac{1}{h^f} \int_{\mathbb{R}^{2f}} \mathrm{d}^f q \, \mathrm{d}^f p \, \Theta(E - H(\boldsymbol{q}, \boldsymbol{p})) = \frac{1}{h^f} \int_{H(\boldsymbol{q},\boldsymbol{p})<E} \mathrm{d}^f q \, \mathrm{d}^f p \ . \qquad (6.8)$$

This formula has a natural interpretation: $\Phi(E)$ counts quantum states with eigenenergies smaller than E. Relation (6.8) means that everything happens as if, in the phase space of a system with f degrees of freedom, an eigenstate occupied a "volume" h^f. This picture is consistent with the Heisenberg uncertainty principle which imposes that for each degree of freedom $j \in \{1, ..., f\}$, the uncertainties Δq_j and Δp_j on the quantum measurements of the observables $\hat{q}_j$ and $\hat{p}_j$j verify $\Delta q_j \Delta p_j \geq \hbar/2$: a non-rigorous reasoning on orders of magnitude then suggests that each quantum state is associated to an elementary volume $\Delta q_1 \Delta p_1 \times ... \times \Delta q_f \Delta p_f$ of order h^f. This is indeed what Relation (6.8) demonstrates.

As stated in Appendix A.7, Formula (6.8) is a semi-classical approximation. It is valid when the energy E is large compared to the energy of the ground level of $\hat{H}$, which is quite a sound assumption: it is only at high energy that $\Phi(E)$ takes on large values and that the staircase curve defined by Equation (6.5) can be approximated by the continuous function (6.8).

Formula (6.8) has two main interests. The first one is of fundamental nature: it allows the counting of classical microstates in a region of phase space, whereas these are, in principle, infinite in number. The following analogy will be made: 1 classical microstate $\Longleftrightarrow$ 1 eigenstate of $\hat{H}$ (see the discussion at the beginning of Section 6.1). Since the counting of states is an essential ingredient of statistical physics, classical statistical physics can only be defined consistently as the semi-classical limit $h \to 0$ of quantum statistical mechanics. Secondly, Formula (6.8) is of practical interest: in the N-body problem, the system's Hamiltonian is often written in first approximation as a sum of "one-body Hamiltonians" (this amounts to neglecting the interactions between particles, see Section 6.4.2). In the thermodynamic limit of a large number of particles, since the energy of the system is an extensive quantity (i.e. proportional to the number of particles in the system), it has typically a large value and the corrections to the approximate law (6.8) are negligible. It will be seen on several examples (in Chapters 7 and 8) that the knowledge of the density of states of the one-body Hamiltonian enables the explicit calculation of many thermodynamic quantities.

Let us end this section by writing explicitly the useful semi-classical expression of the density of states that stems from Equation (6.8). Combined with the left equality of Equation (6.6), it yields:

$$d(E) = \frac{\mathrm{d}}{\mathrm{d}E} \left(\frac{1}{h^f} \int_{H(\boldsymbol{q},\boldsymbol{p})<E} \mathrm{d}^f q \, \mathrm{d}^f p \right) \ . \qquad (6.9)$$

6.1.2 MICROCANONICAL ENSEMBLE

In the case of an isolated quantum system with a fixed total energy E, the analogue of the classical microcanonical distribution now becomes a statistical distribution of states for which:

$$p_n = p(E_n) = \begin{cases} C_{\mathrm{qu}}, & \text{if } E < E_n < E + \delta E \ , \\ 0, & \text{otherwise} \ , \end{cases} \qquad (6.10)$$

where C_{qu} is a constant that depends on E. The statistical distribution characterised by Equation (6.10) corresponds indeed to a microcanonical configuration where every states

---

[4]To verify this statement, it is sufficient to calculate the trace in the basis $\{|\phi_n\rangle\}_{n \in \mathbb{N}}$ formed by the eigenstates of $\hat{H}$. One gets $\mathrm{Tr}\,\hat{\Phi}(E) = \sum_n \langle \phi_n | \Theta(E - \hat{H}) | \phi_n \rangle = \sum_n \langle \phi_n | \Theta(E - E_n) | \phi_n \rangle = \Phi(E)$.

located in the energy interval $[E, E + \delta E]$ are equiprobable. A non-strictly zero width δE of the distribution in energy needs to be introduced because in quantum mechanics the system's energy spectrum is discrete. It is then necessary to work in the regime:

$$\begin{array}{l}\text{Typical spacing of the}\\ \text{eigenvalues of the Hamiltonian } \hat{H}\end{array} \sim \frac{1}{d(E)} \ll \delta E \ll E \sim E_{\text{macroscopic}} \,. \qquad (6.11)$$

The inequality on the right is necessary to define the micro-canonical regime: it imposes that the width of Distribution (6.10) (which is a kind of uncertainty in the determination of the energy) be small with respect to the energy of the system (which is supposed to be well determined in the microcanonical ensemble). In practice, Condition (6.11) is not binding: the systems typically considered in statistical physics have a large number f of degrees of freedom, and since the energy is an extensive quantity (proportional to f), the right-hand side inequality of Equation (6.11) is easily satisfied. For the same reason, the spacing of the energy levels of macroscopic systems is very small and the left equality in Equation (6.10) is easily verified[5].

In terms of the notations introduced in Section 6.1.1, the density matrix corresponding to the distribution (6.10) can be written as $\hat{D} = C_{\text{qu}}(\hat{\Phi}(E+\delta E) - \hat{\Phi}(E))$ and the normalisation (6.4) reads

$$1 = \text{Tr}\,\hat{D} = C_{\text{qu}}\left(\Phi(E + \delta E) - \Phi(E)\right) = C_{\text{qu}}d(E)\,\delta E \,, \quad \text{i.e.,} \quad C_{\text{qu}} = \frac{1}{d(E)\,\delta E} \,. \qquad (6.12)$$

Since the energy E is a macroscopic quantity, typically much larger than the energy of the quantum ground state of $\hat{H}$, the approximate formula (6.9) of the density of state $d(E)$ can be used to determine C_{qu} in Equation (6.12). Therefore, although Expression (6.9) results from a semi-classical approximation, its use in the present context is legitimate and does not violate any law of quantum mechanics.

Exercise 6.1. Show that in the limit (6.11) the exact value of the quantity δE has no influence on the value of a statistical average $\langle \hat{O} \rangle$ resulting from a large number of measurements of an observable $\hat{O}$.

Solution : _For the same reason that, in the limit (6.11) it is legitimate to use the approximate Formula (6.9) to determinate the normalisation constant_ C_{qu}, _one can legitimately calculate_ $\langle \hat{O} \rangle = \text{Tr}(\hat{D}\hat{O})$ _in Equation (6.12) using Equation (A.44). Let_ $O(\boldsymbol{q}, \boldsymbol{p})$ _and_ $D(\boldsymbol{q}, \boldsymbol{p})$ _be the classical versions of operators_ $\hat{O}$ _and_ $\hat{D}$, _respectively. Then:_

$$\langle O \rangle = \frac{1}{h^f} \int \mathrm{d}^f q\, \mathrm{d}^f p\, D(\boldsymbol{q},\boldsymbol{p}) O(\boldsymbol{q},\boldsymbol{p}) = \frac{C_{\text{qu}}}{h^f} \int_{E<H(\boldsymbol{q},\boldsymbol{p})<E+\delta E} \mathrm{d}^f q\, \mathrm{d}^f p\, O(\boldsymbol{q},\boldsymbol{p})$$

$$= \frac{1}{h^f d(E)} \frac{\partial}{\partial E} \left\{ \int_{H(\boldsymbol{q},\boldsymbol{p})<E} \mathrm{d}^f q\, \mathrm{d}^f p\, O(\boldsymbol{q},\boldsymbol{p}) \right\} \,,$$

which is indeed independent on δE.

6.1.3 CANONICAL ENSEMBLE

When the system is in contact with a thermostat at temperature T, the quantum distribution in Hilbert space equivalent of the classical canonical distribution is (see Formula (4.3)):

$$p_n = p(E_n) = \frac{\exp\{-\beta E_n\}}{Z(T)} \,, \quad \text{where} \quad \beta = \frac{1}{k_{\text{B}}T} \,. \qquad (6.13)$$

---

[5]See Note 6 of Chapter 2.

In this equation $Z(T)$ is the quantum canonical partition function ensuring the normalisation (Equation (6.1)):

$$Z(T) = \sum_{n=0}^{\infty} \mathrm{e}^{-\beta E_n} . \tag{6.14}$$

Since the system is in contact with a thermostat, its energy varies and according to Equation (6.2), its average value is:

$$
\begin{aligned}
\langle \hat{H} \rangle &= \sum_{n=0}^{\infty} p_n \langle \phi_n | \hat{H} | \phi_n \rangle = \sum_{n=0}^{\infty} p_n E_n = \frac{1}{Z} \sum_{n=0}^{\infty} E_n \mathrm{e}^{-\beta E_n} \\
&= -\frac{1}{Z}\frac{\partial Z}{\partial \beta} = -\frac{\partial \ln Z}{\partial \beta} = k_\mathrm{B} T^2 \frac{\partial \ln Z}{\partial T} .
\end{aligned}
\tag{6.15}
$$

Exercise 6.2. Calculate the average energy of a one-dimensional harmonic oscillator.

<u>Solution</u> : *Let ω be the angular frequency of the oscillator. The eigenlevels are nondegenerate and their corresponding energies are $E_n = \hbar\omega(n + \frac{1}{2})$ with $n \in \mathbb{N}$. Thus Equation (6.14) reads:*

$$Z = \mathrm{e}^{-\beta\hbar\omega/2} \sum_{n=0}^{\infty} \mathrm{e}^{-n\beta\hbar\omega} = \frac{\mathrm{e}^{-\beta\hbar\omega/2}}{1 - \mathrm{e}^{\beta\hbar\omega}} = \frac{1}{2\sinh(\beta\hbar\omega/2)} . \tag{6.16}$$

Then Equation (6.15) gives:

$$\langle \hat{H} \rangle = -\frac{\partial \ln Z}{\partial \beta} = \frac{\hbar\omega/2}{\tanh(\beta\hbar\omega/2)} . \tag{6.17}$$

Figure 6.1 represents the dependence of $\langle \hat{H} \rangle$ on temperature. At low temperature $\beta \to \infty$ and thus $\langle \hat{H} \rangle \to \hbar\omega/2$: this is expected since the statistical distribution tends to a pure state which is simply the system's ground state. At high temperature, Equation (6.17) leads to $\langle \hat{H} \rangle \to \beta^{-1} = k_\mathrm{B}T$: one recovers the classical result which stems form the equipartition theorem (see Section 4.1.5). This is a generic feature: at high temperature a quantum system behaves classically. This point will be justified shortly on a more general basis (Equation (6.21)).

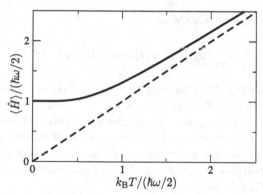

FIGURE 6.1 Average energy (6.17) of the one-dimensional harmonic oscillator plotted as a function of temperature. The dashed curve corresponds to the classical result.

The quantum canonical partition function (6.14) can be expressed as an integral over the exact quantum density of states $d(E)$ (Equation (6.6)):

$$Z = \int_{\mathbb{R}} d(E)\,\mathrm{e}^{-\beta E}\,\mathrm{d}E . \tag{6.18}$$

Writing $d(E)$ as the derivative of the cumulative density of states $\Phi(E)$ [see Equations (6.5) and (6.6)], an integration by parts gives:

$$Z = \left[\Phi(E)\,e^{-\beta E}\right]_{-\infty}^{+\infty} + \beta \int_{\mathbb{R}} \Phi(E)\,e^{-\beta E}\,dE \ . \tag{6.19}$$

The term in square brackets is zero: the exponential cancels in $+\infty$. The exponential diverges in $-\infty$ but this divergence is cancelled by $\Phi(E)$ which is strictly zero for energies below the ground state energy of the Hamiltonian. The same reason allows to substitute the lower bound of the integration domain in the right-hand side member of Equation (6.19) for the energy E_0 of the fundamental (ground) level of $\hat{H}$.

At low temperature β is very large, the $\exp(-\beta E)$ term in the integral of the right-hand side term of Equation (6.19) cancels out very quickly and only the few lowest eigenstates of the quantum Hamiltonian contribute to the integral via $\Phi(E) = \sum_n \Theta(E - E_n)$ (eigenstates with energies $\sim k_B T$ away from E_0). In contrast, at high temperature $\beta \to 0$ and many quantum levels contribute to the integral. The more numerous have a sufficiently high energy to legitimate the use of the semi-classical approximation (Equation (6.8)) over the whole energy domain, leading to:

$$Z \simeq \beta \int_{\mathbb{R}} dE\,e^{-\beta E}\,\frac{1}{h^f}\int_{H(q,p)<E} d^f q\, d^f p = \frac{\beta}{h^f}\int d^f q\, d^f p \int_{H(q,p)}^{+\infty} dE\,e^{-\beta E} \ . \tag{6.20}$$

In the last expression of Equation (6.20), the order of integration has been swapped. The integral over E can be calculated immediately, giving $\beta^{-1}\exp[-\beta H(q,p)]$, and the high-temperature approximation of the quantum partition function reads:

$$Z \simeq \frac{1}{h^f}\int d^f q\, d^f p\, \exp\left[-\beta\,H(q,p)\right] \ . \tag{6.21}$$

The classical expression of the canonical partition function is recovered as the high temperature limit of the quantum partition function.

Exercise 6.3.   Prove Relation (6.21) using the approach of Appendix A.7.

Solution: Definition (6.14) can be written as:

$$Z = \sum_{n=0}^{\infty} \langle \phi_n | \exp\{-\beta\hat{H}\}|\phi_n\rangle = \mathrm{Tr}\left[\exp\left(-\beta\hat{H}\right)\right] , \tag{6.22}$$

using Relation (A.44), Equation (6.21) is immediately recovered as a semi-classical approximation of Equation (6.14). Note that in Exercise 6.1, the use of a semi-classical formula was legitimate because is was restricted to a small domain of energies which were all large compared to the energy of the ground state. Here the semi-classical formula is also used for energies in the vicinity of the ground state. This is an approximation which is only acceptable at high temperature, when the contribution of low energy states to the sum (6.14) – or equivalently of low energy regions of phase space to the integral (6.21) – is small.

Exercise 6.4.   Show that, as is the case in classical mechanics, the relative energy fluctuations around the mean value $\langle \hat{H} \rangle$ are negligible in the thermodynamic limit.

Solution: The fluctuations are measured by the quantity ΔE defined as:

$$(\Delta E)^2 = \left\langle \left(\hat{H} - \langle \hat{H}\rangle\right)^2 \right\rangle = \langle \hat{H}^2\rangle - \langle \hat{H}\rangle^2 \ . \tag{6.23}$$

ΔE is the standard deviation of the energy distribution function and $(\Delta E)^2$ is the variance. Since $\langle \hat{H}^2\rangle = \sum_n p_n E_n^2 = Z^{-1}\partial^2 Z/\partial\beta^2$ and Formula (6.15) allows to express Equation (6.23) as $(\Delta E)^2 =$

$\partial^2 \ln Z / \partial \beta^2$. *This relation shows that* $(\Delta E)^2$ *is an extensive quantity, because* $\ln Z$ *is extensive [this last point follows itself directly from Formula (6.15) linking* $\langle \hat{H} \rangle$ *and* $\ln Z$*]. Therefore*

$$\frac{\Delta E}{\langle \hat{H} \rangle} \propto \frac{1}{\sqrt{f}} , \tag{6.24}$$

where f *is the number of degrees of freedom of the system. This demonstrates the required result. In the thermodynamic limit, the energy fluctuations of a system in equilibrium with a thermostat can be neglected: therefore, the canonical ensemble is equivalent to the micro-canonical ensemble in this limit.*

It was shown in the previous exercise that energy fluctuations can be neglected in the thermodynamic limit and it is thus possible to identify the system's energy E to $\langle \hat{H} \rangle$. Relation (6.15) yields:

$$E(T) = -\frac{\partial \ln Z}{\partial \beta} = k_B T^2 \frac{\partial \ln Z}{\partial T} . \tag{6.25}$$

In practice, in order to obtain every thermodynamic quantity in the canonical ensemble, it is appropriate to calculate the partition function Z and deduce from it the system's free energy[6] $F = E - TS$ using the following method: in Formula (6.18), the density of state can be expressed as a function of the entropy S: $d(E) = \exp\{S(E)/k_B\}$. This formula was demonstrated in the microcanonical case, cf. Equation (3.10), but it can be used here since the two ensembles are equivalent in the thermodynamic limit (demonstrated above). By writing $\epsilon = E/N_f$ and $s(\epsilon) = S(E)/N_f$ where N_f is the number of degrees of freedom of the system[7], Formula (6.18) becomes:

$$Z = N_f \int d\epsilon \, \exp\left\{ -\frac{N_f}{k_B T}\left(\epsilon - T s(\epsilon)\right)\right\} . \tag{6.26}$$

In the thermodynamic limit where N_f is large, this integral is of the form (A.17) and can be evaluate using the saddle-point method presented in Appendix A.3. Writing $f(\epsilon) = \epsilon - T s(\epsilon)$, it gives:

$$Z \simeq \sqrt{\frac{2\pi N_f k_B T}{f''(\epsilon_0)}} \exp\left\{ -\frac{N_f}{k_B T} f(\epsilon_0)\right\} , \tag{6.27}$$

where ϵ_0 is the energy that minimises $f(\epsilon)$ and f'' is the second derivative of f with respect to ϵ. The quantity f is, of course, an intensive quantity: it is the free energy $F = E - TS$ divided by the number of degrees of freedom N_f. The logarithm of Z is therefore of the form[8]:

$$\ln Z = -\frac{1}{k_B T} F(E_0) + O\left(\ln N_f\right) . \tag{6.28}$$

Since the energy fluctuations are negligible in the thermodynamic limit, it is legitimate to identify $F(E_0)$ with the system's free energy and write:

$$F = -k_B T \ln Z . \tag{6.29}$$

This relation determines the free energy which then enables the calculation of every usual thermodynamic quantity.

---

[6]F is the Helmoltz free energy.

[7]From now on, and only until the end of this subsection, we depart from our usual notations and denote as N_f —and not f— the number of degrees of freedom. The quantity f is momentary used for the intensive version of the free energy.

[8]The terms omitted in Equation (6.27) have a contribution to $\ln Z$ smaller than the first correction term – in $\ln N_f$ – to Equation (6.28), hence the sign "=" and not "≃" in Formula (6.28).

Exercise 6.5.   Show that Relation (6.28) is compatible with Equation (6.25).

Solution : _The right-hand term of Equation (6.25) combined with Equation (6.28) gives_

$$E = F - T \frac{\partial F(E_0)}{\partial T}, \tag{6.30}$$

where the logarithmic corrections were neglected. In this expression, the partial derivative indicates that $F(E_0)$ must be differentiated with respect to temperature while keeping the other thermodynamic quantities $(V, N, ...)$ constant. In this configuration, however, one must consider that $F(E_0)$ depends on T for two reasons: explicitly through its definition: $F = N_f \times [\epsilon - Ts(\epsilon)]$, and also through ϵ_0. Thus, $\partial_T F(E_0) = -S + N_f \times f'(\epsilon_0) \times (d\epsilon_0/dT)$. Since by definition of the saddle-point, $df/d\epsilon_0 = 0$, then $\partial_T F(E_0) = -S$ and Formula (6.30) gives $E = F - TS$, in agreement with the thermodynamic definition of these quantities.

6.1.4   GRAND CANONICAL ENSEMBLE

In the grand canonical ensemble, the system is in contact with a reservoir, i.e. another much larger system with which it exchanges energy and particles without neither the reservoir's temperature T, nor the reservoir's chemical potential μ being affected. In this case, the space of quantum states is no longer a Hilbert space, but a Fock space, which is a direct sum of Hilbert spaces for zero, one, two, three, etc. particles. A basis of a Fock space corresponds to a set of eigenstates, each having a given number of particles. The index n that was previously used to label the states must therefore be supplemented by an index N indicating the number of particles in the considered state. The grand canonical distribution corresponds to [see Equation (4.47)]:

$$p_{n,N} = \frac{\exp\{-\beta(E_{n,N} - \mu N)\}}{\Xi}, \tag{6.31}$$

In this expression, $\Xi(\mu, T)$ is the grand canonical partition function (or grand partition function) that enforces normalisation (Equation (6.1)) which reads here $\sum_{n,N} p_{n,N} = 1$, i.e.:

$$\Xi(\mu, T) = \sum_{N=0}^{\infty} e^{\beta\mu N} \sum_{n=0}^{\infty} e^{-\beta E_{n,N}} = \sum_{N=0}^{\infty} e^{\beta\mu N} Z(N, T), \tag{6.32}$$

where $Z(N, T)$ is the canonical partition function [Equation (6.14)] of a system of N particles. In a manner similar to how energy was studied in the canonical case, the average values and variances of the energy and the number of particles can now be studied. For this purpose, it is more appropriate to consider Ξ as a function of $\alpha = \beta\mu$ and μ rather than a function of β and μ. One gets:

$$\langle \hat{H} \rangle = \sum_{N=0}^{\infty} \sum_{n=0}^{\infty} p_{n,N} E_{n,N} = \frac{-1}{\Xi}\left(\frac{\partial \Xi}{\partial \beta}\right) = -\left(\frac{\partial \ln \Xi}{\partial \beta}\right), \tag{6.33}$$

$$\langle N \rangle = \sum_{N=0}^{\infty} \sum_{n=0}^{\infty} p_{n,N} N = \frac{1}{\Xi}\left(\frac{\partial \Xi}{\partial \alpha}\right) = \left(\frac{\partial \ln \Xi}{\partial \alpha}\right). \tag{6.34}$$

The fluctuations of the number of particles are measured by the standard deviation ΔN defined as:

$$(\Delta N)^2 = \langle N^2 \rangle - \langle N \rangle^2 = \frac{1}{\Xi}\left(\frac{\partial^2 \Xi}{\partial \alpha^2}\right) - \frac{1}{\Xi^2}\left(\frac{\partial \Xi}{\partial \alpha}\right)^2 = \frac{\partial}{\partial \alpha}\left(\frac{1}{\Xi}\frac{\partial \Xi}{\partial \alpha}\right) = \left(\frac{\partial^2 \ln \Xi}{\partial \alpha^2}\right). \tag{6.35}$$

In the following, the notation $\bar{N}$ will be used instead of $\langle N \rangle$ for readability. While α and β are intensive, it is clear from Equations (6.33) and (6.34) that $\ln \Xi$ is an extensive quantity, so Equation (6.35) shows that $(\Delta N)^2 \propto \bar{N}$. The relative fluctuations $(\Delta N)/\bar{N}$ are therefore of order $\bar{N}^{-1/2}$, i.e. negligible in the thermodynamic limit. The same analysis can be done for the energy, showing that the grand canonical ensemble is equivalent to the canonical and micro-canonical ensembles in the thermodynamic limit.

In the grand canonical case considered here, the logarithm of Ξ is related to the grand potential $J = F - \mu N = E - TS - \mu N$. This can be seen as follows: the probability of observing N particles in the system is:

$$\sum_{n=0}^{\infty} p_{n,N} = \frac{1}{\Xi} e^{\beta \mu N} Z(N,T) . \tag{6.36}$$

This probability is maximum for $N = \bar{N}$. Thus, the most important contribution to Sum (6.32) is coming from the term $N = \bar{N}$, which however is not the only term to consider: the ΔN neighbours are equally important. Since $\Delta N \propto \bar{N}^{1/2}$, one can write

$$\Xi = \sum_{N=0}^{\infty} e^{\beta \mu N} Z(N,T) = C_{\bar{N}} \bar{N}^{1/2} e^{\beta \mu \bar{N}} Z(\bar{N},T) , \tag{6.37}$$

where $C_{\bar{N}}$ is a term of order $O(\bar{N}^0)$ which depends weakly on $\bar{N}$. Thus, one gets:

$$\ln \Xi = \ln \left\{ e^{\beta \mu \bar{N}} Z(\bar{N},T) \right\} + \ln \left\{ C_{\bar{N}} \bar{N}^{1/2} \right\} = \frac{\mu \bar{N}}{k_{\mathrm{B}} T} - \frac{F(\bar{N},T)}{k_{\mathrm{B}} T} + \ln \left\{ C_{\bar{N}} \bar{N}^{1/2} \right\} . \tag{6.38}$$

In the thermodynamics limit, the logarithmic term can be ignored in this expression and, given the small relative fluctuations of N around its nominal value $\bar{N}$, Formula (6.38) yields

$$J(\mu,T) = -k_{\mathrm{B}} T \ln \Xi . \tag{6.39}$$

6.2 PARAMAGNETISM

In this section, the concepts of quantum statistical mechanics will be exemplified by determining the equation of state of a paramagnetic substance, i.e. the relation between the total magnetic moment of the substance (or its magnetisation $\vec{M}$, which is the magnetic moment per unit volume), its temperature T and the magnetic induction $\vec{B}$. It will give the opportunity to compare in a realistic case the results of classical and quantum statistical mechanics.

6.2.1 CLASSICAL ANALYSIS

Consider N independent atoms, fixed at the nodes of a crystal lattice, each having a magnetic moment $\vec{\mu}_M$ of constant modulus, treated initially as a classical vector.

Since the atoms do not interact with each other, it is enough to consider only one of them. In the absence of magnetic field, its magnetic moment can be oriented in any direction, marked by the polar and azimuthal angles θ and φ. In the presence of a magnetic field $\vec{B}$ directed along the z-axis, the magnetic moment acquires an energy:

$$H_M = -\vec{\mu}_M \cdot \vec{B} = -\mu_M B \cos \theta . \tag{6.40}$$

Assuming that a magnetic moment behaves as a symmetric spinning top of moment of inertia I, its dynamics is governed by the Hamiltonian[9]:

$$H = \frac{1}{2I}\left(p_\theta^2 + \frac{p_\varphi^2}{\sin^2\theta}\right) + H_M \; . \tag{6.41}$$

The canonical partition function z associated with this Hamiltonian is:

$$
\begin{aligned}
z &= \frac{1}{h^2}\int_0^\pi \mathrm{d}\theta \int_0^{2\pi} \mathrm{d}\varphi \int_\mathbb{R} \mathrm{d}p_\theta \int_\mathbb{R} \mathrm{d}p_\varphi \exp\left\{-\frac{\beta}{2I}\left(p_\theta^2 + \frac{p_\varphi^2}{\sin^2\theta}\right)\right\} \mathrm{e}^{-\beta H_M} \\
&= \frac{4I\pi^2}{h^2\beta}\int_0^\pi \mathrm{d}\theta \sin\theta\, \mathrm{e}^{\beta\mu_M B\cos\theta} = z_{\mathrm{rot}}^{\mathrm{clas}} \times z_M \; .
\end{aligned}
\tag{6.42}
$$

In the last expression of the partition function z has been separated in two parts: $z_{\mathrm{rot}}^{\mathrm{clas}}$ corresponding to the classical partition function of a free rotator (for which $H_M \equiv 0$):

$$z_{\mathrm{rot}}^{\mathrm{clas}} = \frac{8\,I\,\pi^2}{h^2\beta} \; , \tag{6.43}$$

and the "paramagnetic" part z_M which reads:

$$z_M = \frac{\sinh(\beta\mu_M B)}{\beta\mu_M B} \; . \tag{6.44}$$

This separation and the terminology used for z_M are justified by the following reasoning: the probability $\mathrm{d}^2 P$ that the magnetic moment points in a direction specified by the polar and azimuthal angles lying in the intervals $[\theta, \theta + \mathrm{d}\theta]$ and $[\varphi, \varphi + \mathrm{d}\varphi]$ is:

$$\mathrm{d}^2 P = \frac{\mathrm{d}\theta\,\mathrm{d}\varphi}{h^2 z}\int_\mathbb{R} \mathrm{d}p_\theta \int_\mathbb{R} \mathrm{d}p_\varphi\, \mathrm{e}^{-\beta H} = \frac{\exp(\beta\mu_M B\cos\theta)}{4\pi z_M}\sin\theta\,\mathrm{d}\theta\,\mathrm{d}\varphi. \tag{6.45}$$

The term z_M in the denominator of this expression ensures normalisation: $\int \mathrm{d}^2 P = 1$. Equation (6.45) can be interpreted physically by noting that in the absence of field B the magnetic moment orientation distribution is isotropic (since then $\mathrm{d}^2 P = \mathrm{d}^2\Omega/(4\pi)$ where $\mathrm{d}^2\Omega = \sin\theta\mathrm{d}\theta\mathrm{d}\varphi$ is the infinitesimal solid angle in $\mathbb{R}^3$), and that when $B \neq 0$, each orientation (i.e. each solid angle) acquires a statistical weight proportional to $\exp(-\beta H_M) = \exp(\beta\mu_M B\cos\theta)$.

From Distribution (6.45) of the magnetic moment, the average value $\langle\vec{\mu}_M\rangle$ can be calculated. One immediately obtains $\langle\mu_{Mx}\rangle = \langle\mu_{My}\rangle = 0$ and then $\langle\vec{\mu}_M\rangle = \langle\mu_{Mz}\rangle\,\vec{e}_z$ with:

$$\langle\mu_{Mz}\rangle = \int \mu_M \cos\theta\,\mathrm{d}^2 P = \frac{1}{z_M}\frac{\partial z_M}{\partial(\beta B)} = \mu_M\, L\!\left(\frac{\mu_M B}{k_\mathrm{B}T}\right) , \tag{6.46}$$

where

$$L(x) = \frac{1}{\tanh x} - \frac{1}{x} \tag{6.47}$$

is the "Langevin function". In the low-temperature limit ($k_\mathrm{B}T \ll \mu_M B$), $\lim_{x\to\infty} L(x) = 1$ and the moments are all aligned with the field $\vec{B}$. At high temperatures, thermal agitation

---

[9]This result is easily recovered considering a pendulum of length l and mass m: in this case $I = ml^2$ and the kinetic energy is: $E_{\mathrm{kin}} = \frac{I}{2}[\dot\theta^2 + \dot\varphi^2\sin^2\theta]$. Then $p_\theta = \partial E_{\mathrm{kin}}/\partial\dot\theta = I\dot\theta$ and $p_\varphi = \partial E_{\mathrm{kin}}/\partial\dot\varphi = I\dot\varphi\sin^2\theta$. Therefore, $E_{\mathrm{kin}} = (p_\theta^2 + p_\varphi^2\sin^{-2}\theta)/(2I)$.

prevails and the magnetisation cancels ($\lim_{x \to 0} L(x) = 0$). In practice, the degree of magnetisation of a substance is evaluated by its magnetisation $\vec{M}$, the density of magnetic moment per unit volume:

$$M = n_V \langle \mu_{Mz} \rangle , \tag{6.48}$$

where n_V is the volume density of magnetic dipoles. Writing the applied magnetic field as $\vec{H} = \vec{B}/\mu_0$, the magnetic susceptibility is defined by:

$$\chi = \left(\frac{\partial M}{\partial H} \right)_{H=0} = \mu_0 \left(\frac{\partial M}{\partial B} \right)_{B=0} . \tag{6.49}$$

χ is a dimensionless quantity that measures the propensity of the substance to magnetise in the presence of an external magnetic field. At high temperature ($k_B T \gg \mu_M B$), a Taylor expansion of the Langevin function near the origin ($L(x) = \frac{1}{3}x + O(x^3)$) leads to "Curie's law" stating that the magnetic susceptibility is inversely proportional to temperature:

$$\chi = C_{\text{Curie}}/T , \tag{6.50}$$

where C_{Curie} is called *Curie constant*. According to the present model:

$$C_{\text{Curie}} = \frac{\mu_0}{3} n_V \mu_M^2 / k_B , \tag{6.51}$$

where n_V is the volume density of magnetic dipoles in a substance in which non-interacting[10] identical moments μ_M are uniformly distributed.

6.2.2 QUANTUM ANALYSIS

In this subsection, the dynamics of a magnetic moment will be described now in the framework of quantum mechanics. Here, it is governed only by the Hamiltonian H_M [Equation (6.40)]. This is made possible by considering the quantum origin of the magnetic moment $\vec{\mu}_M$ of the atoms. Let $\vec{j}$ be the total angular momentum, i.e. the sum of the orbital and spin angular momentum of the electrons of an atom in its ground state, and let j be the associated quantum number. The quantum analysis of the interaction of an atom with a magnetic field[11] shows that the magnetic moment $\vec{\mu}_M$ is proportional to $\vec{j}$:

$$\vec{\mu}_M = -g\mu_B \vec{j}/\hbar , \tag{6.52}$$

where

$$\mu_B = \frac{|q_e|\hbar}{2m_e} \simeq 9.27 \times 10^{-24} \ \text{A.m}^2 \tag{6.53}$$

is the Bohr magneton (q_e and m_e are the electron charge and the electron mass, respectively). The dimensionless constant g in Equation (6.52) is the Landé factor, typically of the order of unity. More precisely, if the angular momentum of the dipole is caused by the electron spin alone, then[12] $g = 2$. If it is caused only by the orbital motion, then $g = 1$. If its origin is mixed, then $g = 3/2 + [s(s + 1) - \ell(\ell + 1)]/[2j(j + 1)]$ where s and ℓ are the intrinsic and orbital quantum numbers and $\vec{j} = \vec{\ell} + \vec{s}$.

---

[10] Interactions will be considered in Chapter 9, Section 9.1.1 and the following sections.

[11] see for example Claude Cohen-Tannoudji, Bernard Diu and Franck Laloë. *Quantum Mechanics Volume I-III* (Wiley-VCH, 2019).

[12] Quantum electrodynamics corrections show that $g = 2.002319304....$

The eigenenergies of Hamiltonian (6.40) are of the form $E_m = g\mu_B B m$ where $m \in \{-j, -j+1, .., j\}$. The associated canonical partition function is[13]:

$$z = \sum_{m=-j}^{j} e^{-\beta E_m} = \sum_{m=-j}^{j} e^{ym/j} = \frac{\sinh\left[(1 + 1/2j)y\right]}{\sinh\left(y/2j\right)}, \quad \text{where} \quad y = \frac{g\mu_B jB}{k_B T}. \tag{6.54}$$

In the particular case $j = 1/2$, it is possible to write $z = \sum_{m=-1/2}^{1/2} e^{2my} = 2\cosh(y)$ which is equivalent to Equation (6.54). Then the projection of the average magnetic moment on the axis of the magnetic field (chosen as the z-axis[14]) is:

$$\langle \mu_{Mz} \rangle = \frac{1}{z} \sum_{m=-j}^{j} (g\mu_B m) e^{-\beta E_m} = \frac{g\mu_B}{z} \sum_{m=-j}^{j} m\, e^{ym/j}, \tag{6.55}$$

which can be written as

$$\langle \mu_{Mz} \rangle = g\mu_B j \mathcal{B}_j(y), \tag{6.56}$$

where $\mathcal{B}_j(y)$ is the Brillouin function:

$$\mathcal{B}_j(y) = \frac{d}{dy} \ln\left\{ \sum_{m=-j}^{j} e^{my/j} \right\} = \frac{1 + \frac{1}{2j}}{\tanh\left[(1 + \frac{1}{2j})y\right]} - \frac{\frac{1}{2j}}{\tanh\left(\frac{y}{2j}\right)}. \tag{6.57}$$

For $j = 1/2$, an equivalent expression is $\mathcal{B}_{1/2}(y) = \tanh(y)$. At high temperature, $y \to 0$ and in the vicinity of the origin $\mathcal{B}_j(y) = \frac{j+1}{3j} y + O(y^3)$: Curie's law is recovered. The high temperature susceptibility is given by Expression (6.50) where the Curie constant is here:

$$C_{\text{Curie}} = \frac{\mu_0}{3} nv(g\mu_B)^2 j(j+1)/k_B. \tag{6.58}$$

This expression is in practice identical to Expression (6.51) since according to Equation (6.52) and the quantum properties of angular momentum $\vec{\mu}_M^2 = (g\mu_B)^2 j(j+1)$.

In the low temperature limit, since $\lim_{y\to\infty} \mathcal{B}_j(y) = 1$, one obtains $\langle \mu_{Mz} \rangle = g\mu_B j$. As in the classical case, the projection of the magnetic moment along the axis of the field $\vec{B}$ is maximal at low temperature in order to minimise the system's energy. However, the non commutativity of the components of the quantum angular momentum does not allow its maximum value to be equal to $|\vec{\mu}_M| = g\mu_B\sqrt{j(j+1)}$. Experiments clearly discriminate these two behaviours: for $j = s = 3/2$ at high field and low temperature $\langle \mu_{Mz} \rangle$ tends towards $3\mu_B$ and not towards $|\vec{\mu}_M| = \sqrt{15}\,\mu_B = 3.87\,\mu_B$.

Experiment not only confirms Curie's law (at high temperature) and the saturation of magnetisation (at low temperature) but also provides a detailed test of the magnetisation law (6.56) as shown in Figure 6.2 which shows measurements of magnetisation performed at a temperature ~ 1 K and for fields up to 5 T. In order to observe a paramagnetic behaviour not disturbed by the interactions between neighbour spins[15] the studied substance are salts (chromium and potassium sulfate, ferric ammonium sulfate and gadolinium sulfate octahydrate) with structures isolating the magnetic centres (Cr^{+++}, Fe^{+++} and Gd^{+++} ions).

---

[13]Note that m takes integer or half-integer values whether j is integer or a half-integer. Formula (6.54) is obtained by noticing that $e^{y/j}z = z + e^{(1+1/j)y} - e^{-y}$.

[14]It is easy to demonstrate that $\langle \mu_{Mx} \rangle = \langle \mu_{My} \rangle = 0$. This is recommended as an exercise.

[15]The interaction between spins gives rise to phenomena of fundamental interest which will be discussed in Chapter 9, Section 9.1.1 and followings.

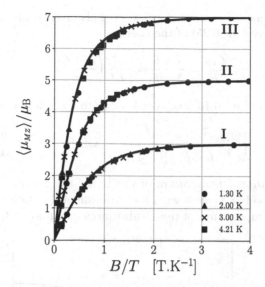

FIGURE 6.2 Average magnetic moment per ion (in units of Bohr magneton) plotted as a function of B/T for a few paramagnetic salts: (I) Cr^{3+} ($j = 3/2$), (II) Fe^{3+} ($j = 5/2$) and (III) Gd^{3+} ($j = 7/2$). In all cases $g = 2$ ($\ell = 0$ and $j = s$). The dots are experimental data and the solid lines correspond to the results obtained using the functions $\mathcal{B}_{3/2}$, $\mathcal{B}_{5/2}$ and $\mathcal{B}_{7/2}$. (adapted from W. E. Henry, Phys. Rev. **88**, 559 (1952))

6.3 VIBRATIONAL CONTRIBUTION TO THE HEAT CAPACITY OF DIATOMIC MOLECULES

In this section, a simple system, consisting of a set of independent harmonic oscillators having all the same angular frequency, will be studied. This system illustrates some fundamental aspects of quantum statistical physics and has also played a major role in the development of the field. Einstein[16] used this model to calculate the heat capacity of solids, with the idea that the excitations of a solid with respect to its ground state corresponded to the vibrations of the crystal lattice and could be described by a quantum harmonic model (see Exercise 4.3). This idea is correct, but the vibrational modes of the lattice do not have all the same frequency. This phenomenon will be studied in details in Section 7.1.

The model of independent harmonic oscillators having all same frequency is suitable to describe the vibrational degrees of freedom of an ideal gas of diatomic molecules. Indeed, a diatomic molecule has different kinds of degrees of freedom (see Figure 4.2): centre of mass translation as well as internal degrees of freedom (overall rotation of the molecule, vibration, and possible excitation of the orbital electrons). For an ideal gas, the theoretical treatment of the translational degree of freedom is rather simple. The internal degrees of freedom (vibrational, rotational and electronic) imply, in the majority of cases, energies with very different orders of magnitude: $\sim 10^{-3}$ eV for rotation, ~ 0.3 eV for vibration and 1–10 eV for electron excitation[17]. Therefore, these different degrees of freedom can be treated separately. In this case, the vibration of the diatomic molecule is described by the Hamiltonian of one dimensional harmonic oscillator of frequency ω. The energy of such system in the canonical ensemble at temperature T was already calculated in Exercise 6.2. The vibrational contribution to the heat capacity of an ensemble of N molecules is therefore

---

[16] A. Einstein, Ann. Physik **22**, 180 (1907).

[17] There are a few noticeable exceptions to this rule when fine structure effects lead to electron excitation energy of the order of $1 - 2 \times 10^{-2}$ eV, as in the case of the NO molecule for instance.

that of N independent one-dimensional oscillators. This gives, using Definition (3.13) of the heat capacity and Expression (6.25) of the energy:

$$\frac{C_{\text{vib}}}{N} = \frac{1}{N}\frac{\partial E}{\partial T} = -\frac{\partial}{\partial T}\left(\frac{\partial \ln Z}{\partial \beta}\right), \tag{6.59}$$

where Z is given by Equation (6.16) and corresponds to a single oscillator. This leads to[18]:

$$\frac{C_{\text{vib}}}{N k_{\text{B}}} = \left(\frac{\Theta_{\text{vib}}}{T}\right)^2 \frac{e^{\Theta_{\text{vib}}/T}}{\left(e^{\Theta_{\text{vib}}/T} - 1\right)^2}, \quad \text{where} \quad \Theta_{\text{vib}} \equiv \frac{\hbar \omega}{k_{\text{B}}}. \tag{6.60}$$

Θ_{vib} is called the *vibrational temperature* of the molecule. As can be seen in Figure 6.3, Equation (6.60) is in good agreement with the experimental data for CO, Cl_2, O_2 and N_2, while the vibrational temperatures of these substances are quite different[19].

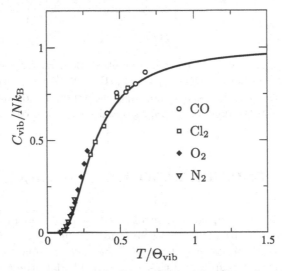

FIGURE 6.3 Vibrational contribution to the heat capacity of a diatomic gas. The solid curve is the theoretical result [Equation (6.60)]. Experimental data (dots) are taken from Sherratt and Griffiths, Proc. Roy. Soc. A **147**, 292 (1934) for CO; from the book "*Specific heats at low temperatures*" by Gopal (Plenum Press, 1966) for Cl_2; and from Henry, Proc. Roy. Soc. A **133**, 492 (1931) for O_2 and N_2.

Note that, as usual, the classical result is recovered at high temperature: $C_{\text{vib}}/(N k_{\text{B}}) \to 1$ when $T \gg \Theta_{\text{vib}}$[20]. A discussion of the low temperature limit is also instructive:

$$\frac{C_{\text{vib}}}{N k_{\text{B}}} \to \left(\frac{\Theta_{\text{vib}}}{T}\right)^2 e^{-\Theta_{\text{vib}}/T}, \quad \text{when} \quad T \ll \Theta_{\text{vib}}, \tag{6.61}$$

which corresponds to the contribution of the first excited state alone, obtained by keeping only the terms $n = 0$ and $n = 1$ in Sum (6.16). This contribution is exponentially small when $T \ll \Theta_{\text{vib}} \iff \beta \hbar \omega \gg 1$: in this temperature regime, the thermal fluctuations correspond to energies of the order of $k_{\text{B}} T$, small compared to the energy difference $\hbar \omega$ between the first

---

[18]In the model proposed by Einstein to describe the vibrations of a solid (a model superseded by Debye's model, see Section 7.1), the oscillators have three degrees of freedom. Formula (6.60) must therefore be multiplied by a factor 3 when used to study the vibrations of a solid.

[19]For CO, $\Theta_{\text{vib}} = 3080$ K; for Cl_2, $\Theta_{\text{vib}} = 813$ K; for O_2, $\Theta_{\text{vib}} = 2251$ K; for N_2, $\Theta_{\text{vib}} = 3374$ K.

[20] This corresponds to the classical equipartition theorem, see Section 4.1.5.

excited state and the ground state. This indicates that quantisation of energy levels leads, at low temperature, to an inhibition of the degree of freedom under consideration. This is a specificity of quantum statistical physics, totally absent in the classical framework.

Two important points should be noted here:

1. C_{vib} is only one of several contributions to the heat capacity of the gas. At the temperatures considered here ($T \approx \Theta_{\text{vib}}$), the translational and rotational degrees of freedom must also be taken into account. Their contribution[21] to the heat capacity C_V are $\frac{3}{2}Nk_B$ and Nk_B, respectively. Hence, the zero of $C_{\text{vib}}/(Nk_B)$ in Figure 6.3 corresponds to the value $\frac{5}{2}$ of the total $C_V/(Nk_B)$. And for $T \gg \Theta_{\text{vib}}$, $C_V \simeq \frac{7}{2}Nk_B$.

2. At high temperatures (i.e. at high excitation energies) the agreement between the result of the linear oscillator model [Equation (6.60)] and the experimental data deteriorates because anharmonic effects come into play[22]. This is proven by the fact that the agreement of experimental data with the heat capacity calculated with the *exact* spectrum of the molecule remains good even at high temperatures[23].

6.4 INDISTINGUISHABLE PARTICLES IN QUANTUM MECHANICS

In quantum mechanics, due to the possible overlap of particle wave functions, identical quantum particles cannot be labelled and are thus indistinguishable[24].

Consider first the case of 2 identical particles. The system's quantum states obtained by exchange of the two particles must be physically equivalent. Let $\psi(\xi_1, \xi_2)$ be the wave function of the system[25] (ξ denoting the set of position and spin coordinates), then one must have:

$$\psi(\xi_2, \xi_1) = e^{i\alpha}\,\psi(\xi_1, \xi_2)\,. \qquad (6.62)$$

It means that the two quantum states can only differ by an unimportant global phase. Performing a new permutation leads to $\exp\{2\,i\,\alpha\} = 1$, i.e. $\exp\{i\,\alpha\} = \pm 1$. Thus, the wave function $\psi(\xi_1, \xi_2)$ is either even or odd under the the exchange of the two particles. This result can be generalised immediately to a system consisting of any number N of identical particles: either the wave function $\psi(\xi_1, \cdots, \xi_N)$ is even under the effect of a transposition of two indices and the particles are called *bosons*, or it is odd and the particles are called *fermions*.

It is of interest to consider not only exchange of two particles, but also permutations possibly involving many or possibly all particles. Let σ be such a permutation $(1 \rightarrow \sigma(1), \cdots, N \rightarrow \sigma(N))$, it corresponds to a quantum operator $\hat{\Sigma}$ which effect on the N-body quantum state is:

$$\hat{\Sigma}\,\psi(\xi_1, \cdots, \xi_N) = \psi(\xi_{\sigma(1)}, \cdots, \xi_{\sigma(N)})\,. \qquad (6.63)$$

---

[21] This point will be proven in Section 6.5.1: at $T \approx \Theta_{\text{vib}}$ these degrees of freedom can be studies in the framework of classical physics.

[22] The electron contribution must also be taken into account, as well as a coupling of vibrational degrees of freedom with the rotational degrees of freedom.

[23] See, for example in the case of hydrogen, H. W. Woolley, R. B. Scott and F. G. Brickwedde, J. Res. Natl. Bur. Stand. (U.S.), **41**, 379 (1948).

[24] Think of the case for a gas or a plasma: when two particles A and B collide, their wave functions overlap, at least partially (the relative particle has a wave function not strictly zero at the origin). It is thus impossible to identify with certainty after the collision which of the two particles moving away was initially particle A and which was initially particle B.

[25] ψ is not necessarily normalised. As a general rule, the question of normalisation will not be discussed in this subsection. It is always possible to normalise the wave function of a system by multiplying it by the appropriate scalar, but this is irrelevant for the present discussion.

Mathematicians call a permutation that exchanges only two particles a transposition. Any permutation can be decomposed into a product of transpositions. If the number of transpositions is even the permutation is also even and, by definition, the sign (sometimes signature or signum) of the permutation is $\varepsilon_\sigma = +1$. If the number of transpositions is odd, the permutation is odd and, by definition, $\varepsilon_\sigma = -1$. Then, it implies:

$$\hat{\Sigma}\psi = \begin{cases} \psi & \text{for bosons}\,, \\ \varepsilon_\sigma\,\psi & \text{for fermions}\,, \end{cases} \tag{6.64}$$

meaning that the wave function of a system of identical bosons is totally symmetric under the effect of a permutation of its particles, whereas a system of identical fermions is totally antisymmetric. This is referred to as *Bose–Einstein* and *Fermi-Dirac* statistics, respectively. A fully symmetric or antisymmetric wave-function can be constructed starting from any N-particle wave function as explained in the following.

Let us begin with the case $N = 2$ and consider a given wave function $\psi(\xi_1,\xi_2)$ having no particular symmetry. Let σ_{12} be the permutation of indices 1 and 2, $\hat{\Sigma}_{12}$ its effect on the wave function ψ and $\hat{I}_d$ the identity operator. Consider the operators $\hat{A}$ and $\hat{S}$ defined as follows:

$$\hat{A} = \frac{1}{2}\left(\hat{I}_d - \hat{\Sigma}_{12}\right) \quad \text{and} \quad \hat{S} = \frac{1}{2}\left(\hat{I}_d + \hat{\Sigma}_{12}\right)\,. \tag{6.65}$$

Since $(\hat{\Sigma}_{12})^2 = \hat{I}_d$, it is easy to check that $\hat{A}$ and $\hat{S}$ are projectors (i.e. $\hat{A}^2 = \hat{A}$ and $\hat{S}^2 = \hat{S}$). It is also easy to check that $\hat{\Sigma}_{12}\hat{A} = -\hat{A}$ and $\hat{\Sigma}_{12}\hat{S} = \hat{S}$ such that $\hat{A}\psi = \frac{1}{2}(\psi(\xi_1,\xi_2) - \psi(\xi_1,\xi_2))$ verifies the fermionic version of Equation (6.64) and $\hat{S}\psi = \frac{1}{2}(\psi(\xi_1,\xi_2) + \psi(\xi_1,\xi_2))$ verifies the bosonic version. Therefore $\hat{A}$ is the operator that projects any wave function onto a totally antisymmetric state, the only acceptable type of state for fermions, and $\hat{S}$ projects on a totally symmetric state (the only one valid for bosons). The antisymmetry properties of the wave function of a system of identical fermions imply that two fermions cannot occupy the same quantum state: this is the "Pauli exclusion principle". Indeed a two-fermion state of the form $\psi(\xi_1,\xi_2) = \phi(\xi_1)\phi(\xi_2)$ is projected by operator $\hat{A}$ onto the null ket. Nothing of the sort exists for bosons which can sometimes all accumulate in the same state (a phenomenon called *Bose–Einstein condensation* studied in Section 7.4).

The Hilbert space of N identical particles is thus reduced either to a set of totally symmetric states (bosons), or to a set of totally anti-symmetric states (fermions) under the effect of particles permutation. To construct such a space, the tensor product of the N one-particle Hilbert spaces must be calculated from which a basis constructed. Then, the basis is symmetrised or antisymmetrised[26]. When the system includes a small number of particles, the symmetrisation or antisymmetrisation (similar to the example given above for a two-particle wave function) is performed in a simple way. As the number of particles grows the task become more involved. The general operators $\hat{A}$ and $\hat{S}$ can be constructed in the following way: let S_N be the set of $N!$ permutations of the N indices, $\hat{A}$ and $\hat{S}$ can be defined as:

$$\hat{A} = \frac{1}{N!}\sum_{\sigma\in S_N}\epsilon_\sigma\hat{\Sigma} \quad \text{and} \quad \hat{S} = \frac{1}{N!}\sum_{\sigma\in S_N}\hat{\Sigma}\,. \tag{6.66}$$

It is relatively easy to be convinced that operators $\hat{A}$ and $\hat{S}$ in Equation (6.66) are the generalisation to the N-particle case of operators (6.65) defined for two particles. For example,

---

[26]Some vectors of the initial basis may, under the effect of symmetrisation or antisymmetrisation, lead to non-independent vectors. Of course, one must keep only a set of linearly independent basis vectors in the final space.

for any permutation[27] σ of S_N: $\hat{\Sigma}\hat{S} = \hat{S}$ and $\hat{\Sigma}\hat{A} = \varepsilon_\sigma \hat{A}$ from which it is easily deduced that $\hat{S}^2 = \hat{S}$ and $\hat{A}^2 = \hat{A}$. Thus $\hat{A}$ and $\hat{S}$ are undeniably projectors on totally symmetric and totally antisymmetric state spaces, respectively. This procedure, which may seem cumbersome, is actually quite simple when the "occupation number" formalism is used, as illustrated for several types of particles (distinguishable particles, bosons, spinless fermions, spin-1/2 fermions) by a schematic example in Section 6.4.1, and then in a more general setting in Section 6.4.2.

The indistinguishability of identical particles, so important at the quantum level, also has repercussions in classical statistical physics. For example, it is imposed that $(\vec{q}_1, \vec{p}_1, \vec{q}_2, \vec{p}_2)$ corresponds to the same microstate as $(\vec{q}_2, \vec{p}_2, \vec{q}_1, \vec{p}_1)$ (see Section 2.2.3). Therefore, for N identical particles, in every integral in the classical phase space of the present (and following) chapters:

$$\int d^f q\, d^f p \quad \text{must be replaced by} \quad \frac{1}{N!} \int d^f q\, d^f p \,. \tag{6.67}$$

The $1/N!$ factor avoids the overcounting of identical configurations obtained by permutation of the N particles ($N!$ is the total number of permutations among N objects). Rule (6.67) which has just been justified intuitively, can be demonstrated by taking the classical limit of the quantum symmetrisation (or antisymmetrisation) process. For example, in the calculation of the number of microstates $\Phi(E)$ with energies less than E, the reasoning leading to Equation (6.8) should be applied but one must calculate the trace over a space of symmetric (or antisymmetric) states. This amounts to calculating $\text{Tr}[\hat{\Phi}(E)\hat{S}]$ (or $\text{Tr}[\hat{\Phi}(E)\hat{A}]$) instead of $\text{Tr}[\hat{\Phi}(E)]$ in the Hilbert space defined by the tensor product of the N unsymmetrised one-particle Hilbert spaces. In the calculation of this trace, it can be shown that the dominant contribution to Sum (6.66) defining $\hat{S}$ (or $\hat{A}$) is the term $(N!)^{-1}\hat{I}_d$. Rule (6.67) naturally follows[28]. Finally note that the factor $(N!)^{-1}$ in Rule (6.67) can be easily justified for the case of non-interacting particles in the Maxwell-Boltzmann approximation, as it will be seen in Section 6.4.3.

Particles with half-integer spin (electrons, protons, neutrons, ...) are fermions whereas particles with integer spin (photons, phonons, mesons, ...) are bosons. The "spin-statistic" theorem, proven in quantum field theory, allows to interpret this rule as a consequence of very general assumptions. However, it is possible to imagine that some of these assumptions are violated, or bypassed by quantum statistics more complex than those of simple bosons and fermions.

A particle can be composite, i.e. formed of several "elementary" particles, but experimental conditions may not allow to probe the internal structure of this particle[29]. If such a particle is composed of bosons or of an even number of fermions, the composite particle behaves like a boson. If it is formed by an odd number of fermions the composite particle behaves like a fermion[30]. If, for instance, the internal structure of helium atoms is not

---

[27]For two permutations σ and π, the product $\sigma\pi$ is a permutation having the signature $\varepsilon_\sigma \varepsilon_\pi$. Consequently:

$$\hat{\Sigma}\hat{A} = \frac{1}{N!}\hat{\Sigma}\sum_{\pi \in S_N} \varepsilon_\pi \hat{\Pi} = \frac{\varepsilon_\sigma}{N!}\sum_{\pi \in S_N} \varepsilon_\sigma \varepsilon_\pi \hat{\Sigma}\hat{\Pi} \,.$$

Since, when σ is fixed and π goes over S_N, the product $\sigma\pi$ goes over S_N, the right hand side term of the above formula can be written as $\varepsilon_\sigma (N!)^{-1} \sum_{\pi \in S_N} \varepsilon_\pi \hat{\Pi} = \varepsilon_\sigma \hat{A}$.

[28]This calculation was done explicitly in a very similar case: the calculation of the density of state of a Hamiltonian with spatial symmetries (see N. Pavloff, J. Phys. A **27**, 4317 (1994) and also B. Lauritzen and N.D. Whelan, Ann. Phys. **244**, 112 (1995)).

[29]This is the case, for example, of the proton, which was initially considered as an elementary particle by Rutherford at the time of its discovery in 1920, whereas it is now known that it is made up of three quarks which are themselves fermions.

[30]This property is consistent with the spin-statistic theorem, as can be easily verified on the example of two identical particles (either bosons or fermions): if two identical spins s_1 and s_2 (either integer of

probed, the isotope ^{3}He (with two electrons orbiting around a nucleus made of two protons and one neutron) being composed of a total of 5 fermions is a fermion, whereas ^{4}He which contains 6 fermions (two neutrons in its nucleus) is a boson. And indeed, at low temperature, in the regime where quantum effects become important, the physical properties of these two isotopes are quite different.

The important and special case of a system made of identical non-interacting particles (fermions or bosons) will be considered in the next sections. In the framework of quantum mechanics, for such a system, it is possible to calculate the grand canonical partition function Ξ while a general method to calculate the canonical partition function Z is unknown. This point is discussed in Section 6.4.1.

6.4.1 CANONICAL TREATMENT

This section illustrates on a simple example how Law (4.9) is not valid for an ideal quantum gas and how it is not possible to obtain a general expression of the canonical partition function of a system of N *indistinguishable* quantum particles even in the absence of interaction.

Consider a schematic quantum Hamiltonian with only three non-degenerate eigenstates of energies 0, ϵ and 2ϵ, respectively. If the system contains only a single spinless particle, then the canonical partition function Z_1 is:

$$Z_1 = 1 + e^{-\beta\epsilon} + e^{-2\beta\epsilon} . \tag{6.68}$$

If the system contains two *distinguishable* spinless particles without mutual interaction, then the canonical partition function Z_2^{dis} is:

$$Z_2^{\text{dis}} = 1 + 2\,e^{-\beta\epsilon} + 3\,e^{-2\beta\epsilon} + 2\,e^{-3\beta\epsilon} + e^{-4\beta\epsilon} . \tag{6.69}$$

This result is obtained by counting the number of configurations at a given energy. There are for example 2 possible configurations of total energy ϵ (see Figure 6.4) corresponding to the term $2\,e^{-\beta\epsilon}$ in Equation (6.69).

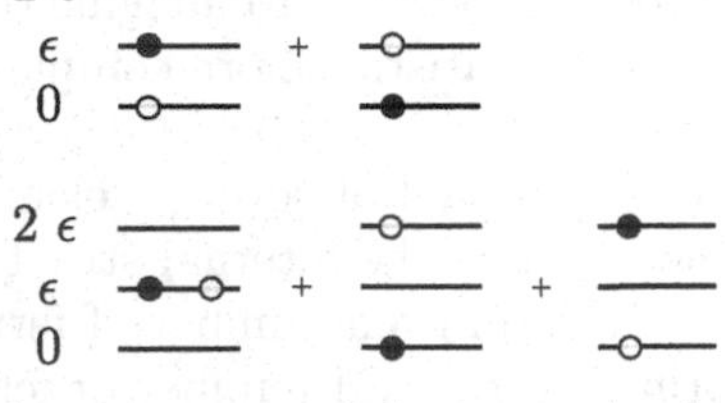

FIGURE 6.4 Representation of the occupancy of the energy levels of a one-body schematic Hamiltonian with three levels by two distinguishable particles (white and black balls). The top figure represents the two configurations of total energy ϵ. The lower figure shows the three configurations of total energy 2ϵ.

Exercise 6.6. It can be easily directly verified using Expressions (6.68) and (6.69) that $Z_2^{\text{dis}} = (Z_1)^2$, in agreement with the classical intuition [see, for example, Equation (4.8)]. Generalise this result to the case of N distinguishable particles (classical or quantum) without interactions, all subjected to the same Hamiltonian[31].

---

half-integer) are coupled together, the only possible values of the projection of the total spin on an axis are $-s_1 - s_2$, $-s_1 - s_2 + 1$, $-s_1 - s_2 + 2,...,s_1 + s_2$, i.e. integer values. The total spin is naturally an integer and the composite particle is a boson.

[31]i.e. the non-realistic case of identical but distinguishable particles.

Solution: _The results is immediate in the classical case: particle distinguishability simply removes the factor_ $1/N!$ _in Equation_ (4.9). _For the quantum case, one uses almost the same reasoning: The Hilbert space of N distinguishable particles is the direct product of N one-particle Hilbert spaces. In other words, the generic contribution to the partition function_ Z_N^{dis} _for which particle "1" is in state_ ϕ_{i_1}, _particle "2" in state_ ϕ_{i_2}, _etc... is one of the terms of the product of the N partition functions (all identical) of each individual particle, and this without omission nor double counting. Therefore:_

$$Z_N^{\mathrm{dis}} = (Z_1)^N \, . \tag{6.70}$$

If two indistinguishable spinless bosons (still without mutual interaction) are now placed in the three-level Hamiltonian, some of the configurations shown in Figure 6.4 disappear: for example, the two arrangements of total energy ϵ become identical if the white and the black particle are indistinguishable (both white for instance). A trivial counting of the microstates now leads to:

$$Z_2^{\mathrm{B}} = 1 + e^{-\beta\epsilon} + 2\,e^{-2\beta\epsilon} + e^{-3\beta\epsilon} + e^{-4\beta\epsilon} \, . \tag{6.71}$$

If, on the other hand, the system contains two identical spinless fermions[32] (still without mutual interaction), taking into account the Pauli exclusion principle, the canonical partition function is:

$$Z_2^{\mathrm{F}} = e^{-\beta\epsilon} + e^{-2\beta\epsilon} + e^{-3\beta\epsilon} \, . \tag{6.72}$$

Two spin-1/2 fermions (still indistinguishable and without mutual interaction) can also be placed in the system (considering that the three-levels schematic Hamiltonian has a purely spatial origin and no spin contribution) and the associated canonical partition function $Z_2^{\frac{1}{2},\mathrm{F}}$ can be calculated. In this case, the two particles can occupy the same level of the Hamiltonian without violating the Pauli exclusion principle and the associated wave function is (for the level of energy ϵ for example):

$$\frac{1}{\sqrt{2}}\Big(|1 : \epsilon \uparrow\rangle|2 : \epsilon \downarrow\rangle - |1 : \epsilon \downarrow\rangle|2 : \epsilon \uparrow\rangle\Big) \, , \tag{6.73}$$

where the notations are simple: the contribution $|1 : \epsilon \uparrow\rangle$ corresponds for example to a one-body wave function where particle "1" is in the eigenstate of energy ϵ with spin up. Function (6.73) is simply the direct product of a function where the two particles sit on the same spatial state with an antisymmetric spin function (a so-called _singlet state_).

On the other hand, the configuration where the two particles are on two distinct levels corresponds to four possible two-particles states: in two of these states, the particles have identical spin and an antisymmetric space wave function [cases (a) and (b) in Figure 6.5] and in the other two states, the antisymmetry is shared by the space and spin states [cases (c) and (d)]. By counting the configurations and their energies for the three-level schematic

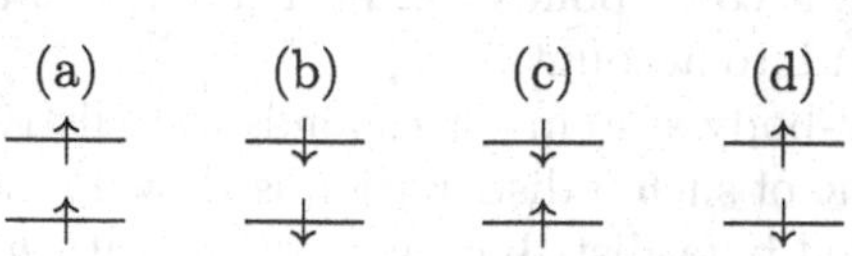

FIGURE 6.5 Schematic representation of the four possible occupation states of two distinct levels by two spin-1/2 fermions. It is advised as an exercise to write the four wave functions corresponding to the states (a)–(d).

---

[32]The spin-statistic theorem is violated in this model system!

Hamiltonian, $Z_2^{\frac{1}{2},F}$ is obtained:

$$Z_2^{\frac{1}{2},F} = 1 + 4\,e^{-\beta\epsilon} + 5\,e^{-2\beta\epsilon} + 4\,e^{-3\beta\epsilon} + e^{-4\beta\epsilon}\,. \tag{6.74}$$

The same partition function is obtained by placing two spinless fermions in a system having the same sequence of levels, but each degenerated twice. This result can be generalised and will be often used in the following.

Exercise 6.7.   Repeat the above study when three particles are placed in the system. Consider the case of distinguishable particles, spinless bosons, spinless fermions and spin-1/2 fermions.

Solution: Writing $\varepsilon = \exp(-\beta\epsilon)$, one gets:

$$Z_3^{\text{dis}} = (1 + \varepsilon + \varepsilon^2)^3 = 1 + 3\,\varepsilon + 6\,\varepsilon^2 + 7\,\varepsilon^3 + 6\,\varepsilon^4 + 3\,\varepsilon^5 + \varepsilon^6\,,$$

$$Z_3^{0,B} = 1 + \varepsilon + 2\,\varepsilon^2 + 2\,\varepsilon^3 + 2\,\varepsilon^4 + \varepsilon^5 + \varepsilon^6\,,$$

$$Z_3^{0,F} = \varepsilon^3\,, \quad Z_3^{\frac{1}{2},F} = 2\,\varepsilon + 4\,\varepsilon^2 + 8\,\varepsilon^3 + 4\,\varepsilon^4 + 2\,\varepsilon^5\,.$$

6.4.2   CALCULATION OF THE GRAND CANONICAL PARTITION FUNCTION

In the previous section, an example illustrated the fact that there is no general method to write the canonical partition function of an ideal quantum gas (i.e. a gas of N identical indistinguishable particles without mutual interaction). It will now be shown that, within the grand canonical approach, it is possible to get around the problem of counting symmetric (for bosons) or anti-symmetric (for fermions) N-body states.

Consider a system governed by a Hamiltonian of the form:

$$\hat{H}(\vec{r}_1,\vec{p}_1,\vec{s}_1,...,\vec{r}_N,\vec{p}_N,\vec{s}_N) = \sum_{i=1}^{N} \hat{h}(\vec{r}_i,\vec{p}_i,\vec{s}_i)\,, \tag{6.75}$$

where $\hat{h}$ is called a _one-body Hamiltonian_. Hamiltonian (6.75) is separable into N independent Hamiltonians and naturally corresponds to a system of non-interacting particles. The cases of bosons and fermions will be treated in parallel. Let

$$\epsilon_0 \leq \epsilon_1 \leq \epsilon_2 \leq ... \tag{6.76}$$

be the eigenlevels of the one-body Hamiltonian $\hat{h}$. The eigenlevels ϵ_λ are labelled with the indices $\lambda \in \mathbb{N}$, each associated with a single state (written $|\varphi_\lambda\rangle$). Several consecutive ϵ_λ are equal in the case of a degenerate level. Note here that, if the particles carry a spin s and if $\hat{h}$ is spin independent, then each level appears at least $2s + 1$ times in Sequence (6.76). Indeed, even if an eigenlevel of $\hat{h}$ is non-degenerate in the Hilbert space of wave functions depending only on position, it corresponds to $2s + 1$ different states in the Hilbert space of $|\varphi_\lambda\rangle$ where the spin is taken into account.

A configuration of the N-body system corresponds to a distribution of particles over the eigenstates of $\hat{h}$. An example of such a distribution is shown schematically in Figure 6.6. A configuration is characterised by a distribution $\{n_\lambda\}_{\lambda\in\mathbb{N}}$, with n_λ indicating the number of particles occupying the state $|\varphi_\lambda\rangle$ of energy ϵ_λ. n_λ is called the _occupation number_ of the state.

A given configuration corresponds to a total number of particles $N\{n_\lambda\}$ and a total energy $E\{n_\lambda\}$ such that:

$$N\{n_\lambda\} = \sum_{\lambda=0}^{\infty} n_\lambda \quad \text{and} \quad E\{n_\lambda\} = \sum_{\lambda=0}^{\infty} n_\lambda\,\epsilon_\lambda\,. \tag{6.77}$$

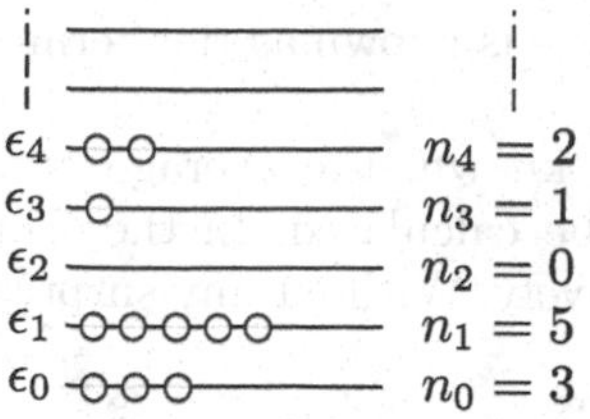

FIGURE 6.6 Schematic representation of the occupation of states of the one-body Hamiltonian. The subscripts n_λ on the right side of the figure indicate the number of particles occupying state $|\varphi_\lambda\rangle$ of energy ϵ_λ. To simplify the schematic, a sequence of equidistant levels is represented, but the method is general. The example shown in this figure corresponds to bosons, since for fermions Pauli's exclusion principle restricts occupation numbers to $n_\lambda = 0$ or 1.

The grand canonical partition function is given by Equation (6.32):

$$\Xi = \sum_{N=0}^{\infty} e^{\beta\mu N} Z_N = \sum_{N=0}^{\infty} e^{\beta\mu N} \sum_{N\{n_\lambda\}=N} e^{-\beta E\{n_\lambda\}} = \sum_{\{n_\lambda\}} \exp\left\{-\beta[E\{n_\lambda\} - \mu N\{n_\lambda\}]\right\}. \quad (6.78)$$

In the central term of Equation (6.78), the sum is restricted to configurations such that $N\{n_\lambda\} = N$. Because N is not fixed in the grand canonical ensemble, this constraint can be relaxed in the right term of Equation (6.78), making the explicit calculation of the partition function Ξ possible and leading to:

$$\Xi = \sum_{\{n_\lambda\}} \exp\left\{-\beta \sum_{\lambda=0}^{\infty} n_\lambda(\epsilon_\lambda - \mu)\right\} = \sum_{\{n_\lambda\}} \prod_{\lambda=0}^{\infty} \left[e^{-\beta(\epsilon_\lambda-\mu)}\right]^{n_\lambda} = \prod_{\lambda=0}^{\infty} \sum_{n_\lambda=0}^{n_{\lambda,\max}} \left[e^{-\beta(\epsilon_\lambda-\mu)}\right]^{n_\lambda}, \quad (6.79)$$

where the term $n_{\lambda,\max}$ is ∞ for bosons and 1 for fermions (this stems from Pauli exclusion principle). It takes a bit of thinking to convince oneself of the correctness of the last equality in Formula (6.79). In these two cases, the summation $\sum_{n_\lambda}$ can be calculated exactly[33] and the final expression of the grand canonical partition function is:

$$\Xi(V, \mu, T) = \begin{cases} \displaystyle\prod_{\lambda=0}^{\infty} \left[1 + e^{-\beta(\epsilon_\lambda-\mu)}\right] & \text{, for fermions,} \\ \displaystyle\prod_{\lambda=0}^{\infty} \frac{1}{1 - e^{-\beta(\epsilon_\lambda-\mu)}} & \text{, for bosons.} \end{cases} \quad (6.80)$$

From this expression, the average particle number $\langle n_\nu\rangle$ occupying a state $|\varphi_\nu\rangle$ of energy ϵ_ν can be calculated. By definition:

$$\langle n_\nu\rangle = \frac{1}{\Xi} \sum_{\{n_\lambda\}} n_\nu \exp\left\{-\beta \sum_{\lambda=0}^{\infty} n_\lambda(\epsilon_\lambda - \mu)\right\} = -\frac{1}{\beta}\frac{\partial}{\partial \epsilon_\nu} \ln \Xi. \quad (6.81)$$

This leads to the (very useful) expression:

$$\langle n_\nu\rangle = \begin{cases} \dfrac{1}{e^{\beta(\epsilon_\nu-\mu)} + 1} \equiv n_F(E) & \text{for fermions,} \\ \dfrac{1}{e^{\beta(\epsilon_\nu-\mu)} - 1} \equiv n_B(E) & \text{for bosons.} \end{cases} \quad (6.82)$$

---

[33] For fermions, the results is trivial: $\sum_{n=0}^{1} x^n = 1 + x$, and for bosons, the following formula is used: $\sum_{n=0}^{\infty} x^n = 1/(1 - x)$ valid for $|x| < 1$.

The expression of $\langle n_\nu \rangle$ for fermions is known as the Fermi-Dirac statistics. For bosons, it is the Bose–Einstein statistics.

Once the expression of Ξ is known, the average number of particles (6.34) and the average energy (6.33) can also be calculated. In the thermodynamics limit, they can be identified with N and E, respectively. Without any surprises, one obtains (to compare with Equation (6.77)):

$$N = \sum_{\lambda=0}^{\infty} \langle n_\lambda \rangle \quad , \qquad E = \sum_{\lambda=0}^{\infty} \epsilon_\lambda \langle n_\lambda \rangle \, . \tag{6.83}$$

These expressions will be often used in Chapters 7 and 8 when dealing with non-interacting bosons and fermions.

6.4.3 MAXWELL-BOLTZMANN APPROXIMATION

When dealing with a gas of non-interacting identical quantum particles (fermions or bosons) in the canonical ensemble, the following approximation is commonly made:

$$Z(N,T) \simeq \frac{1}{N!} [z(T)]^N \, , \tag{6.84}$$

where z is the canonical one-particle partition function, i.e.:

$$z = \sum_{\lambda=0}^{\infty} e^{-\beta \epsilon_\lambda} \, , \tag{6.85}$$

where ϵ_λ are the energy levels of the one-body Hamiltonian [see Equations (6.75) and (6.76)]. In this case, the canonical partition function Z is evaluated using Formula (4.9) valid for an ideal classical gas. It was however shown in Section 6.4.1 using an example that this formula does not apply in the quantum case. Equation (6.84) is thus only an approximation. It corresponds to the so-called *Maxwell-Boltzmann* approximation which is equivalent to discarding quantum statistical effects. In the Maxwell-Boltzmann framework, the grand canonical partition function can be written as:

$$\Xi(T,V,\mu) = \sum_{N=0}^{\infty} e^{\beta \mu N} Z(N,T) \simeq \sum_{N=0}^{\infty} e^{\beta \mu N} \frac{z^N}{N!} = \exp\{z\, e^{\beta \mu}\} \, , \tag{6.86}$$

and then the occupation number reads [using (6.81)]:

$$\langle n_\nu \rangle = -\frac{1}{\beta} \frac{\partial \ln \Xi}{\partial \epsilon_\nu} = \exp\{-\beta(\epsilon_\nu - \mu)\} \, , \tag{6.87}$$

where the ϵ_ν's are the energies of the one-body Hamiltonian. This last result does not distinguish bosons and fermions. It is an approximation of the exact formulas (6.82) which is only valid in the limit:

$$\exp\{\beta(\epsilon_\nu - \mu)\} \gg 1 \, . \tag{6.88}$$

It is not immediately obvious but Limit (6.88) corresponds to a high temperature limit. Indeed, for Limit (6.88) to be valid for all energies ϵ_ν it is necessary and sufficient that it is valid for the lowest one:

$$e^{-\beta \mu} \gg 1 \, , \tag{6.89}$$

where the zero of the energy scale is taken as the ground sate of the one-body Hamiltonian ($\epsilon_0 = 0$) which is always possible. To go further into the physical interpretation of Limit (6.88), consider an ideal classical gas contained in a box of volume V. In this case, the direct application of Equation (6.84) leads to $e^{-\beta\mu} = V/(N\lambda_T^3)$, where $\lambda_T = h\,(2\pi m k_B T)^{-1/2}$ is the thermal wavelength (3.48). Unsurprisingly, the classical result for the chemical potential [Equation (3.47)] is recovered. The condition (6.89) of applicability of the Maxwell-Boltzmann approximation can thus be written as:

$$\left(\frac{V}{N}\right)^{1/3} \gg \lambda_T . \tag{6.90}$$

This corresponds indeed to a high temperature limit, and more precisely, to a limit where the interparticle distance $\sim (V/N)^{1/3}$ is large compared to de Broglie thermal wavelength.

Conversely, in the limit (6.89), the grand canonical partition function (6.80) can be written in an approximate form:

$$\ln \Xi = \pm \sum_{\lambda=0}^{\infty} \ln\left(1 \pm e^{-\beta(\epsilon_\lambda - \mu)}\right) \simeq \sum_{\lambda=0}^{\infty} e^{-\beta(\epsilon_\lambda - \mu)} = z\,e^{\beta\mu} , \tag{6.91}$$

where, in the second term, the "+" corresponds to the fermionic version of Equation (6.80) and the "−" to its bosonic version. Expression (6.86) is recovered which is a direct consequence of the approximate formula (6.84). Approximation (6.84) is therefore strictly equivalent to Limit (6.89).

Let us add a bonus to the discussion of this very common approximation. Rule (6.67) was demonstrated in the particular case of a system without interactions: in the classical limit corresponding to Limit (6.88), the canonical partition function of indistinguishable particles must effectively include the factor $1/N!$ since it corresponds precisely to using the semi-classical Maxwell-Boltzmann approximation of the quantum grand canonical partition function.

Note finally that the opposite of Limit (6.89) is called the *degenerate gas limit*. In this regime, the quantum statistical effects are important, and fermions and bosons can no be longer treated on an equal basis. Each quantum statistic will be studied separately in Chapters 7 and 8.

6.5 ROTATIONAL CONTRIBUTION TO THE HEAT CAPACITY OF DIATOMIC MOLECULES

This section is devoted to the study of an ideal gas composed of N diatomic molecules. The study is performed at low enough temperatures so that the vibrational degrees of freedom (studied in Section 6.3) can be ignored (i.e. $T \ll \Theta_{\text{vib}}$). The Hamiltonian of a molecule is thus of the form $h = h_{\text{tr}} + h_{\text{rot}}$, where $h_{\text{tr}} = \vec{p}^2/(2\,M)$ describes the motion of the centre of mass, and $h_{\text{rot}} = \vec{L}^2/(2\,I)$ describes the rotational degrees of freedom of the molecule where $\vec{L}$ is the kinetic orbital momentum and I is the moment of inertia[34]. The study of these degrees of freedom confirms the importance, firstly, of the quantum approach already illustrated in Section 6.3 for vibrational degrees of freedom, and, secondly, of the correct treatment of the quantum statistics effects and their links with the spin of the particles, as will be explained in Section 6.5.2.

6.5.1 HETERO-NUCLEAR MOLECULES

The case of "hetero-nuclear" molecules (such as HCl) will first be considered. When a molecule is homo-nuclear (like H_2 for example), the study is complicated by symmetry

---

[34]The expression of h_{rot} can be easily recovered by modelling the molecule as a pendulum of length l and mass m: in that case $I = ml^2$ and h_{rot} is the rotation energy which can be written as $h_{\text{rot}} = \frac{I}{2}[\dot\theta^2 + \dot\varphi^2 \sin^2\theta]$. The kinetic orbital momentum $\vec{L} = mr^2(\dot\theta\,\vec{e}_\varphi - \dot\varphi \sin\theta\,\vec{e}_\theta)$ such that $h_{\text{rot}} = \vec{L}^2/(2I)$.

effects which will be discussed in Section 6.5.2. The eigenstates of h_{rot} which form a basis of the Hilbert space are the spherical harmonics $Y_{\ell,m}$. The eigenenergies are $\epsilon_\ell = \hbar^2 \ell(\ell+1)/(2I)$ and are $2\ell + 1$ fold degenerate. Thus, the canonical partition function associated to h_{rot} is:

$$z_{\text{rot}} = \sum_{\ell=0}^{+\infty} (2\ell + 1)e^{-\beta \epsilon_\ell} = \sum_{\ell=0}^{+\infty} f(\ell) , \qquad (6.92)$$

with

$$f(\ell) = (2\ell + 1) \exp\left[-\frac{\Theta_{\text{rot}}}{T} \ell(\ell+1) \right] , \quad \text{where} \quad k_B \Theta_{\text{rot}} \equiv \frac{\hbar^2}{2I} . \qquad (6.93)$$

A simple order-of-magnitude calculation gives for the temperature Θ_{rot} a value of the order[35] of 10 K. In the low temperature limit ($T \ll \Theta_{\text{rot}}$) the argument of the exponential in Equation (6.93) is always large, except for $\ell = 0$. It is therefore justify to consider only the first two terms in Series (6.92) and write $z_{\text{rot}} \simeq 1 + 3\exp(-2\Theta_{\text{rot}}/T)$. The rotational energy of a gas made of N molecules is, according to Equation (6.15), $E_{\text{rot}} = Nk_B T^2 \partial \ln z_{\text{rot}}/\partial T$. The contribution of the rotational degrees of freedom to the heat capacity is therefore:

$$C_{\text{rot}} = \frac{\partial E_{\text{rot}}}{\partial T} \Rightarrow \frac{C_{\text{rot}}}{Nk_B} = 12 \left(\frac{\Theta_{\text{rot}}}{T} \right) \exp\left(-\frac{\Theta_{\text{rot}}}{T} \right) + \cdots$$

Hence, at low temperatures, the discrete nature of the quantum energy spectrum leads to an inhibition of the rotational degree of freedom[36]. The heat capacity vanishes as $\exp(-\Theta_{\text{rot}}/T)$ and becomes extremely small when $T \lesssim 0.2\Theta_{\text{rot}}$. Most substances are liquid or solids at temperatures of the order of Θ_{rot}, and the subtle effects just discussed are generally not observable. The situation is different for hydrogen H_2 and hydrogen deuteride HD (D is the symbol for deuterium: an isotope of hydrogen whose nucleus is composed of one neutron and one proton) whose low mass is associated with a small moment of inertia and a relatively high value of Θ_{rot}[37]: these substances are still gaseous at temperatures $T \lesssim \Theta_{\text{rot}}$.

In the high temperature limit ($T \gg \Theta_{\text{rot}}$), since the function f defined by Equation (6.93) is analytic on $[0, +\infty[$, the Euler-MacLaurin summation formula[38] can be used (valid for a function f tending sufficiently quickly to 0 at infinity and all the better as f varies little with ℓ):

$$\sum_{\ell=0}^{+\infty} f(\ell) = \int_0^{+\infty} f(x)\,\mathrm{d}x + \frac{1}{2}f(0) - \frac{1}{12}f'(0) + \frac{1}{720}f^{(3)}(0) + \cdots \qquad (6.94)$$

This enables to re-calculate Expression (6.92) and obtain the first terms of the high temperature series expansion of z_{rot}:

$$z_{\text{rot}} = \frac{T}{\Theta_{\text{rot}}} + \frac{1}{3} + \frac{1}{15} \frac{\Theta_{\text{rot}}}{T} + \cdots \qquad (6.95)$$

The first term of this expansion (corresponding to the integral in the right-hand side of Equation (6.94)) is the classical result [Equation (6.43)]. Expression (6.95) corresponds to the first terms of the so-called *Mulholland expansion*[39]. Adapting Expression (6.59) of the

---

[35] For instance, in the case of the HCl molecule, $\Theta_{\text{rot}} = 15.2$ K whereas for CO, $\Theta_{\text{rot}} = 2.77$ K.

[36] A phenomenon already discussed in Section 6.3.

[37] For H_2, $\Theta_{\text{rot}} = 85.4$ K; for HD, $\Theta_{\text{rot}} = 65.7$ K.

[38] See for example https://en.wikipedia.org/wiki/Euler-Maclaurin_formula

[39] The expansion is named after H.P. Mulholland, Proc. Camb. Phil. Soc. **24**, 280 (1928). With $a = \Theta_{\text{rot}}/T$, the values of the first and third derivative of the function $f(x) = (2x + 1)\exp[-ax(x + 1)]$ with respect to x in $x = 0$ are:

$$f'(0) = 2 - a \quad \text{and} \quad f^{(3)}(0) = -a(12 - 12a + a^2) .$$

The terms proportional to $a = \Theta_{\text{rot}}/T$ in these two expressions are sufficient to obtain the last term of Expansion (6.95). To reach the next order in $(\Theta_{\text{rot}}/T)^2$, the Euler-MacLaurin expansion must be pushed to a higher order than Equation (6.94) (by adding the next term in $f^{(5)}(0)$).

heat capacity[40], one then obtains:

$$\frac{C_{\text{rot}}}{N k_{\text{B}}} = -\frac{1}{k_{\text{B}}} \frac{\partial}{\partial T} \left(\frac{\partial \ln z_{\text{rot}}}{\partial \beta} \right) = \frac{\partial}{\partial T} \left(T^2 \frac{\partial \ln z_{\text{rot}}}{\partial T} \right) = 1 + \frac{1}{45} \left(\frac{\Theta_{\text{rot}}}{T} \right)^2 + \cdots \qquad (6.96)$$

The constant term in Equation (6.96) is the classical contribution to the heat capacity[20] as can be easily verified from Equation (6.43). The next corrective term is positive. As can be seen in Figure 6.7, it is consistent with the small hump observed around $T \simeq \Theta_{\text{rot}}$ in the curve depicting C_{rot} as a function of temperature. Despite their uncertainties, the experimental data for the HD molecule shown in Figure 6.7 clearly demonstrate the existence of this hump.

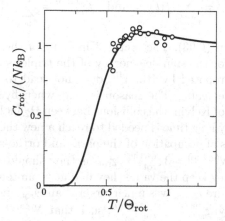

FIGURE 6.7 Heat capacity of a heteronuclear diatomic molecule. Solid lines: value calculated using expression (6.92) of z_{rot}. Dots: experimental values for the HD molecule (from K. Clusius and E. Bartholomé, Zeit. Elektrochem. **40**, 524 (1934)).

In the more common temperature range $\Theta_{\text{rot}} \ll T \ll \Theta_{\text{vib}}$ where $\Theta_{\text{vib}} = \hbar\omega/k_{\text{B}}$ where ω is a characteristic angular frequency of the vibrations of the molecule[41], the vibrational degrees of freedom are inhibited whereas $C_{\text{rot}} = N k_{\text{B}}$. Adding the contribution of the translational degrees of freedom of the centre of mass of the molecule (described by h_{trans} and treated classically in this temperature range), one obtains $C_v = \frac{5}{2} N k_{\text{B}}$ for an ideal diatomic gas. This result remains valid for homo-nuclear molecules: quantum statistical effects (discussed in Section 6.5.2) play no role when $T \gg \Theta_{\text{rot}}$.

6.5.2 HOMO-NUCLEAR MOLECULES

The study of the case of homo-nuclear molecules is particularly interesting because it requires a discussion of quantum statistical effects. This section is devoted to the study of the H_2 molecule. The nucleus of the hydrogen atom consists of a single proton: a fermion of spin 1/2. The Hilbert space is the direct product of the Hilbert space of spatial wave functions with the Hilbert space of two spins 1/2. The basis of the spin states suitable for the discussion of the symmetry effects is the basis of the singlet[42] and triplets[43] states.

---

[40] The heat capacity of a gas of N independent rotators is N times that of a single isolated rotator calculated with the partition function z_{rot}.

[41] See Equation (6.60). For HCl, $\Theta_{\text{vib}} = 4227$ K; for H_2, $\Theta_{\text{vib}} = 6332$ K, for HD, $\Theta_{\text{vib}} = 5300$ K.

[42] The singlet state is $\frac{1}{\sqrt{2}}(|\downarrow\uparrow\rangle - |\uparrow\downarrow\rangle)$ (the notation used here is a condensed version Formula (6.73)). This state is antisymmetric under the exchange of the two spins and has a zero total spin.

[43] The triplet states are the three states of total spin equal to 1: $|\uparrow\uparrow\rangle$, $\frac{1}{\sqrt{2}}(|\downarrow\uparrow\rangle + |\uparrow\downarrow\rangle)$ and $|\downarrow\downarrow\rangle$. They are symmetric under the exchange of the two spins.

The functions forming a basis of the space for the relative position of the two nuclei are the spherical harmonics $Y_{\ell m}(\theta, \varphi)$. The exchange of the two nuclei corresponds to $\theta \to \pi - \theta$ and $\varphi \to -\varphi$ and, in this case, the space function is multiplied by $(-1)^\ell$. The only way to obtain a fully antisymmetric wave function for the two protons is to either associate a symmetric space states (even ℓ) with the singlet (antisymmetric) spin state (this configuration is called *para-hydrogen*) or to associate an odd space states with the triplet (symmetric) spin states (*ortho-hydrogen*). The canonical partition function is thus:

$$z_{\text{rot}} = z_{\text{rot}}^{(\text{ortho})} + 3\, z_{\text{rot}}^{(\text{para})} , \tag{6.97}$$

where

$$z_{\text{rot}}^{(\text{para})} = \sum_{\ell \text{ even}} f(\ell) , \quad \text{and} \quad z_{\text{rot}}^{(\text{ortho})} = \sum_{\ell \text{ odd}} f(\ell) . \tag{6.98}$$

$f(\ell)$ is defined by Equation (6.93). The factor 3 in front of the contribution of $z_{\text{rot}}^{(\text{para})}$ to Equation (6.97) accounts for the spin degeneracy of the triplet state[44].

Formula (6.97), first proposed by Hund[45], does not lead to a heat capacity agreeing with experimental measurements. The reason is that when hydrogen is cooled down to temperatures of a few tens of Kelvin, the collisions between H_2 molecules have little effect on the nuclear spins and a very long time is needed to reach a new thermodynamic equilibrium. It follows that the numbers of occupation of the para- and ortho-states are not respectively $N^{(\text{para})} = z_{\text{rot}}^{(\text{para})}/z_{\text{rot}}$ and $N^{(\text{ortho})} = 3\, z_{\text{rot}}^{(\text{ortho})}/z_{\text{rot}}$ as they should in thermal equilibrium at temperature T. Instead, they keep the value they had at room temperature, before cooling. At that time, the temperature T_{room} was much larger than Θ_{rot} and the rotational degrees of freedom behaved classically: $z_{\text{rot}}^{(\text{para})} \simeq z_{\text{rot}}^{(\text{ortho})}$[46], such that $N^{(\text{para})} = \frac{1}{4}N$ and $N^{(\text{ortho})} = \frac{3}{4}N$. To take this effect into account, one does not use the formula valid in thermal equilibrium and corresponding to the heat capacity of N independent rotators in thermal equilibrium (analogous to Equation (6.59))

$$C_{\text{rot}} = \frac{\partial}{\partial T}\left(-N\frac{\partial \ln z_{\text{rot}}}{\partial \beta}\right) = \frac{\partial}{\partial T}\left(N k_{\text{B}} T^2 \frac{\partial \ln z_{\text{rot}}}{\partial T}\right) , \tag{6.99}$$

where z_{rot} is given by Equation (6.97), but rather:

$$C_{\text{rot}} = \frac{1}{4}\frac{\partial}{\partial T}\left(N k_{\text{B}} T^2 \frac{\partial \ln z_{\text{rot}}^{(\text{para})}}{\partial T}\right) + \frac{3}{4}\frac{\partial}{\partial T}\left(N k_{\text{B}} T^2 \frac{\partial \ln z_{\text{rot}}^{(\text{ortho})}}{\partial T}\right) . \tag{6.100}$$

In other words, one quarter of the heat capacity is due to para-hydrogen molecules and three quarters to ortho-hydrogen molecules. However, the rotational states of each species are determined by the equilibrium with a reservoir at temperature T which is the correct temperature at which the experiment is conducted. This mechanism, first proposed by Dennison[47] leads to a very good agreement with the experimental data, as shown in Figure 6.8. Note that at high temperature ($T \gg \Theta_{\text{rot}}$), quantum statistical effects are no longer relevant and the classical result $C_{\text{rot}} = N k_{\text{B}}$ is recovered.

---

[44]Strictly speaking, the nuclear spin degeneracy factor should also have been put in front of the partition function of heteronuclear molecules [Equation (6.92)], but a global multiplicative factor has, of course, no consequence.

[45]F. Hund, Zeit. Physik **42**, 93 (1927).

[46]Their joint value was $\simeq \frac{1}{2}T_{\text{room}}/\Theta_{\text{rot}}$, cf. Expansion (6.95).

[47]D. Dennison, Proc. Roy. Soc. A **115**, 483 (1927).

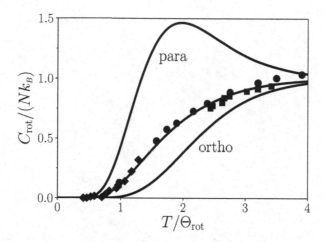

FIGURE 6.8 Heat capacity of H_2. The solid curves are the theoretical results for para-hydrogen, ortho-hydrogen and the combination of both as described by Equation (6.100). The circles, squares and diamonds are the experimental data of F.A. Giacomini, Phil. Mag. **50**, 146 (1925) and Eucker cited in the previous reference and in R. H. Fowler and E. A. Guggenheim, *Statistical thermodynamics* (The Macmillan Company, New York City, 1940).

7 Bosons

The statistical study of a system of bosons has a peculiar status in the history of physics: it is for describing what would nowadays be called a *gas of photons* that Planck introduced the first quantum hypothesis in 1900. The first part of this chapter is devoted to the statistical analysis of a system of bosons of zero chemical potential, corresponding to the elementary excitations of a given system, as those initially studied by Planck (Sections 7.1, 7.2 and 7.3). The canonical and grand canonical point of views are strictly equivalent when dealing with such quanta of excitation (see Section 7.1.2). This is no longer the case when the number of particles is conserved, and is associated to the phenomenon of Bose–Einstein condensation which is studied in Section 7.4.

7.1 HEAT CAPACITY OF SOLIDS: DEBYE MODEL

Debye introduced in 1912 a model that describes the oscillations of a crystal lattice improving on Einstein model[1]. Debye model treats the lattice as an ensemble of interacting harmonic oscillators. The difference between the two approaches is illustrated in Figure 7.1 for a two-dimensional configuration.

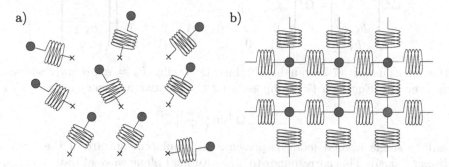

FIGURE 7.1 Illustration of the differences between a) Einstein model and b) Debye model. In Einstein model, each atom is connected to its equilibrium position by a harmonic oscillator which is independent of its neighbours. In Debye model each atom interacts with its neighbours *via* a harmonic potential.

7.1.1 SEPARATION OF VARIABLES

Debye's approach consists in considering the crystal as a lattice of interacting harmonic oscillators, see Fig. 7.1(b). The purpose of the present subsection is to separate the corresponding Hamiltonian into a sum of independent oscillators. To simplify the calculation, a one-dimensional system is considered. According to Debye it should be described as a chain of harmonically interacting particles:

$$H = \frac{m}{2} \sum_{n=1}^{S} \dot{x}_n^2 + \frac{m}{2} \Omega^2 \sum_{n=1}^{S} (x_{n+1} - x_n)^2 , \tag{7.1}$$

---

[1]Einstein model was introduced in Exercise 4.3 and also at the beginning of Section 6.3. It was then used to describe the vibrational contribution to the heat capacity of diatomic gases, see Formula (6.60) and Figure 6.3.

DOI: 10.1201/9781003272427-7

with $x_{S+1} = x_1$ and $x_0 = x_S$ (periodic boundary conditions are used for simplicity). This Hamiltonian describes the harmonic modes of vibration of an atomic lattice containing S sites with a lattice spacing a. x_n is the deviation of the nth atom from its equilibrium position na, see figure 7.2.

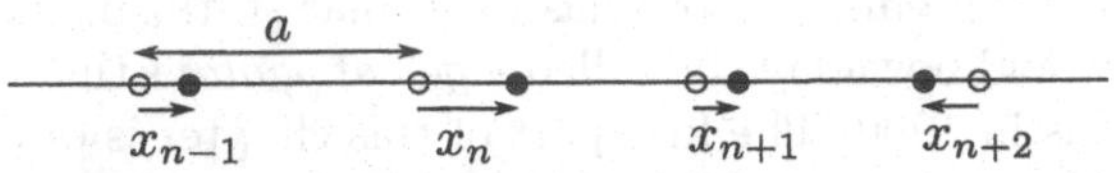

FIGURE 7.2 Representation of a one-dimensional chain of harmonic oscillators. The white circles represent the atoms' equilibrium positions. The black circles are their positions at a given time t. Neighbouring white circles are separated by a distance a called *lattice spacing*. $x_n(t)$ is the departure of the position of the nth atom at time t from its equilibrium position.

The equations of motion are $\ddot{x}_n = \Omega^2(x_{n+1} - 2\,x_n + x_{n-1})$. The normal modes correspond to solutions for which the displacements are of the form $x_n(t) = b_n \exp\{-i\omega t\}$ (they form a basis of the solution space). For such modes, the equations of motion can be written in matrix form as:

$$\omega^2 \begin{pmatrix} b_1 \\ b_2 \\ .. \\ .. \\ b_{S-1} \\ b_S \end{pmatrix} = \Omega^2 \begin{pmatrix} 2 & -1 & 0 & .. & 0 & -1 \\ -1 & 2 & -1 & 0 & .. & 0 \\ & .. & .. & .. & & \\ & & .. & .. & .. & \\ 0 & .. & 0 & -1 & 2 & -1 \\ -1 & 0 & .. & 0 & -1 & 2 \end{pmatrix} \begin{pmatrix} b_1 \\ b_2 \\ .. \\ .. \\ b_{S-1} \\ b_S \end{pmatrix}. \tag{7.2}$$

A solution of Equation (7.2) can be obtained by taking $b_n = \exp\{iqan\}$, where $q \in \mathbb{R}$. Carrying over into Equation (7.2) imposes that the angular frequency ω_q of such a mode reads:

$$\omega_q = 2\Omega \left| \sin\left(\frac{q\,a}{2}\right) \right|. \tag{7.3}$$

When only a single normal mode is present, the displacements are of the form $x_n(t) = A_q \exp\{i(qna - \omega_q t)\}$. This corresponds to a longitudinal plane wave of wave vector $\vec{q} = q\,\vec{e}_x$ and angular frequency ω_q. Relation (7.3) is called the *dispersion relation*. In the limit of small values of q ($q\,a \ll 1$), i.e., for modes which wavelength is large with respect to the lattice spacing ($\lambda = \frac{2\pi}{q} \gg a$), a first order Taylor expansion of the sine function in Equation (7.3) yields:

$$\omega_q = c_s\,|q|, \quad \text{with} \quad c_s = a\,\Omega. \tag{7.4}$$

The parameter c_s above has the dimension of a speed: it is the speed of sound waves in the system.

What are the possible values of q? If $q \to q' = q + 2\pi/a$, the values of b_n and ω_q are not affected and thus q and q' describe the same normal mode. The values of q can therefore be restricted to an interval $q \in [-\pi/a, +\pi/a]$ (or any other equivalent choice). The condition $x_S = x_0$ imposes $q = \frac{n}{S}\frac{2\pi}{a}$ with (if S is odd, which can always be assumed) $n \in \{-\frac{S-1}{2}, \ldots, \frac{S-1}{2}\}$. In the limit $S \gg 1$ (i.e. a long chain which will be considered from now on), the wave vector q takes continuous values in the whole interval $[-\pi/a, +\pi/a]$ which is called the *first Brillouin zone*.

The interest of the analysis of the dynamics in terms of normal modes lies in the possibility it offers to decompose the Hamiltonian into a sum of *independent* oscillators. To this end, let us define the quantity:

$$\xi(q,t) = \sum_{n \in \mathbb{Z}} x_n(t) \exp\{-iqan\}. \tag{7.5}$$

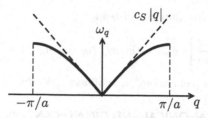

FIGURE 7.3 Dispersion relation of the normal modes of the one-dimensional harmonic oscillator chain [Equation (7.3)]. The dashed lines correspond to the long-wavelength (small wave vector) approximation [Equation (7.4)]: $\omega_q \simeq c_s |q|$ with $c_s = a\,\Omega$.

Multiplying Equation (7.2) by $\exp\{-iqan\}$, summing over n and noting that:

$$\sum_{n\in\mathbb{Z}} x_{n\pm1}\, e^{-inqa} = e^{\pm iqa} \sum_{n\in\mathbb{Z}} x_{n\pm1}\, e^{-i(n\pm1)qa} = e^{\pm iqa} \sum_{n\in\mathbb{Z}} x_n\, e^{-inqa} = e^{\pm iqa}\, \xi(q,t)\,, \qquad (7.6)$$

one gets:

$$\frac{\partial^2 \xi}{\partial t^2} = -\omega_q^2\, \xi\,. \qquad (7.7)$$

The quantity $\xi(q,t)$ describes a coherent oscillatory motion of all the atoms of the chain with given wave vector q and angular frequency ω_q. Equation (7.5) can be inverted (see Appendix A.6.2) leading to:

$$x_n(t) = \frac{a}{2\pi} \int_{-\pi/a}^{\pi/a} dq\, \xi(q,t)\, e^{iqan}\,. \qquad (7.8)$$

This relation combined with Equation (7.7) is enough to demonstrate that the most general oscillations of the lattice can be decomposed into a sum (continuous sum over the variable q) of oscillations of uncoupled harmonic oscillators of angular frequency ω_q.

Exercise 7.1. Show explicitly that Hamiltonian (7.1) can be written as a sum of independent harmonic oscillator.

Solution : _A first step consists in rewriting the interaction term of Hamiltonian (7.1) using the following formula:_

$$\sum_{n=-\infty}^{+\infty} (x_n - x_{n+1})^2 = \frac{a}{2\pi} \int_{-\pi/a}^{\pi/a} \left|\left(1 - e^{iqa}\right)\xi(q,t)\right|^2 dq \quad \text{where} \quad \left|1 - e^{iqa}\right|^2 = \left(\frac{\omega_q}{\Omega}\right)^2\,.$$

To prove this equality, note that, according to the definition of ξ, the coefficients x_n are the Fourier coefficients[2] of $\xi(q)$ and x_{n+1} can be seen as a Fourier coefficient of $e^{iqa}\xi(q)$ [according to Formula (7.6)]. Additionally, for any function $f(q)$ the Parseval-Plancherel equality (A.38), gives[3]:

$$\frac{a}{2\pi} \int_{q_0}^{q_0 + \frac{2\pi}{a}} dq\, |f(q)|^2 = \sum_{n=-\infty}^{+\infty} |f_n|^2 \quad \text{when} \quad f(q) = \sum_{n=-\infty}^{+\infty} f_n\, e^{-inqa}\,.$$

The desired relation follows from the choice $f_n = x_n - x_{n+1}$. The same type of reasoning yields:

$$\sum_{n=-\infty}^{+\infty} \dot{x}_n^2 = \frac{a}{2\pi} \int_{-\pi/a}^{\pi/a} \left|\frac{\partial \xi(q,t)}{\partial t}\right|^2 dq\,.$$

---

[2] See Appendix A.6.2.
[3] The formula is valid for any q_0, in particular for $q_0 = -\pi/a$.

Altogether Hamiltonian (7.1) can thus be written as:

$$H = \frac{a}{2\pi} \int_{-\pi/a}^{\pi/a} \mathcal{H}_q\left[\xi, \frac{\partial \xi}{\partial t}\right] dq, \quad \text{with} \quad \mathcal{H}_q\left[\xi(q,t), \frac{\partial \xi}{\partial t}(q,t)\right] = \frac{m}{2}\left|\frac{\partial \xi}{\partial t}\right|^2 + \frac{m}{2}\omega_q^2|\xi|^2, \quad (7.9)$$

which indeed corresponds to a (continuous) sum of uncoupled harmonic oscillators.

7.1.2   EQUIVALENCE OF CANONICAL AND GRAND CANONICAL ANALYSIS

As demonstrated in the previous subsection, the Hamiltonian describing a Debye solid can be written as a sum of independent harmonic oscillators. Once quantised, the corresponding modes can be described in terms of quanta called *phonons*. Phonons are bosons (they do not obey Pauli exclusion principle) and since their number is not conserved (they can be created or annihilated at will by heating the system, for example) it is natural to set their chemical potential at zero. Furthermore, the average number of phonons in a state of energy ϵ_ν is given by the bosonic version of Equation (6.82) with $\mu = 0$. The purpose of this short subsection is to prove these statements.

As previously shown, lattice vibrations can be decomposed into a sum of M independent harmonic oscillators (in d dimensions, $M = dS$ where S is the number of lattice sites). The quantum energy of the system is then:

$$E = E_0 + \sum_{\lambda=0}^{M} \hbar\omega_\lambda n_\lambda, \quad (7.10)$$

where n_λ is the number of phonons in the vibration mode of energy ω_λ and E_0 is the ground state energy[4]. In this case, the canonical partition function can be calculated precisely because the sum $\sum_\lambda n_\lambda$ is not constrained, in contrast to the situation considered in Section 6.4.2. Here, one gets:

$$Z = e^{-\beta E_0} \sum_{\{n_\lambda\}} \exp\left\{-\beta\hbar\sum_{\lambda=0}^{M}\omega_\lambda n_\lambda\right\} = e^{-\beta E_0} \prod_{\lambda=0}^{M}\sum_{n=0}^{\infty}\exp\{-\beta\hbar\omega_\lambda n\}$$

$$= e^{-\beta E_0} \prod_{\lambda=0}^{M} \frac{1}{1 - \exp[-\beta\hbar\omega_\lambda]}. \quad (7.11)$$

The proof is completed by verifying that Expression (7.11) is equal to the grand canonical partition function of an ideal Bose gas with chemical potential $\mu = 0$ [Equation (6.80)]. The average number of phonons $\langle n_\nu \rangle$ in the vibration mode ω_ν can also be calculated:

$$\langle n_\nu \rangle = \frac{e^{-\beta E_0}}{Z} \sum_{\{n_\lambda\}} n_\nu \exp\left\{-\beta\hbar\sum_{\lambda=0}^{M}\omega_\lambda n_\lambda\right\} = -\frac{1}{\beta\hbar}\frac{\partial \ln Z}{\partial \omega_\nu} = \frac{1}{\exp\{\beta\hbar\omega_\nu\} - 1}. \quad (7.12)$$

This is the expected result, i.e. Equation (6.82) for bosons with $\mu = 0$.

Let us summarise Sections 7.1.1 and 7.1.2 in one sentence: an excited state of a crystal lattice can typically be seen as a superposition of independent quantised excitations (phonons) behaving as non-interacting bosons of zero chemical potential.

7.1.3   PHYSICAL DISCUSSION

In three dimensions, phonons have a dispersion relation that depends on the orientation of the vibrations of the considered mode. Three independent directions are possible and

---

[4] $E_0 = \hbar\sum_{\lambda=0}^{M}\omega_\lambda/2$ with, according to Equation (7.3), $\omega_\lambda = 2\Omega\left|\sin\left[\frac{\pi}{S}(\lambda - \frac{S-1}{2})\right]\right|$.

the phonons are said to have 3 possible polarisations. In a simplified approach, it can be admitted that these three types of waves exhibit the same dispersion relation: $\omega = c_s |\vec{k}|$ where c_s is the speed of sound in the solid[5].

Since a phonon is the quantum of a given harmonic oscillator, it has a characteristic frequency ω and an energy $\hbar\omega$. A microstate of the lattice where a single phonon is present corresponds, in the classical limit, to a displacement of the atoms with respect to their equilibrium position of the form: $\vec{u}(\vec{r},t) = \vec{U} \exp\{i(\vec{k}\cdot\vec{r} - \omega t)\}$ with $k = \omega/c_s$. The corresponding density of states in reciprocal space is $d(k_x, k_y, k_z) = 3V/(2\pi)^3$, where the factor 3 comes from the three possible directions of polarisation in a solid[6]. The distribution is isotropic in reciprocal space, and accordingly, the density of states depends only on the modulus $k = |\vec{k}|$: $d(k) = 4\pi k^2 3V/(2\pi)^3$, or in terms of the angular frequency[7] ω : $d(\omega) = (3V\omega^2)/(2\pi^2 c_s^3)$. The long wavelength approximation leading to this dispersion relation has a drawback: it neglects the upper cut-off frequency which exists for example in the one-dimensional model [Equation (7.3)]. This cut-off is due to the discrete nature of matter: in a system where atoms are separated by a distance a, excitations with a wavelength shorter than $\sim a$, i.e. with a wave vector larger than $\sim 1/a$, cannot be observed. Debye's idea consists in imposing a sharp cutoff by modelling the density of state as follows:

$$d(\omega) = \frac{3V\omega^2}{2\pi^2 c_s^3} \Theta(\omega_D - \omega), \quad \text{with} \quad \omega_D = c_s \left(6\pi^2 \frac{N}{V}\right)^{1/3}, \tag{7.13}$$

where Θ is the Heaviside function and ω_D is the Debye frequency which was determined by imposing $\int_0^{+\infty} d(\omega)\mathrm{d}\omega = 3N$ [Equation (7.13)]. This condition is equivalent to state that the number of excited modes cannot exceed the number of degrees of freedom (there are N atoms and 3 degrees of freedom per atom)[8].

Then, the energy E of the system is obtained using Formulas (7.10) and (7.12), or equivalently using Formula (6.83):

$$E = \int_0^{+\infty} \mathrm{d}\omega \, \frac{\hbar\omega \, d(\omega)}{\exp(\beta\hbar\omega) - 1}. \tag{7.14}$$

Subsequently, the heat capacity $C_V = (\partial E/\partial T)_V$ is:

$$C_V = 3Nk_B \, D\left(\frac{T_D}{T}\right), \tag{7.15}$$

where $T_D = \hbar\omega_D/k_B$ is called the *Debye temperature*, and the function $D(x)$ is defined as:

$$D(x) = \frac{3}{x^3} \int_0^x \mathrm{d}t \, \frac{t^4 e^t}{[\exp(t) - 1]^2} = \begin{cases} 1 - x^2/20 + \dots & (x \ll 1), \\ 4\pi^4/(5x^3) + O(e^{-x}) & (x \gg 1). \end{cases} \tag{7.16}$$

---

[5]There are two simplifications here. The first one consists in making the approximation of long wavelengths. This is the same approximation as in the one-dimensional model [Hamiltonian (7.1)] to go from the exact dispersion relation [Equation (7.3)] to the long wavelength approximation [Equation (7.4)]. The other approximation consists in assuming that the dispersion relation is isotropic. This is not correct: there are differences in the propagation speed of longitudinal and transverse waves (see Pathria and Beale, *Statistical Mechanics* (Academic Press, 2011), or Ashcroft and Mermin, *Solid State Physics* (Thomson Press, 2003) for a justification of this simplifying assumption).

[6]In a liquid, there exit only waves with longitudinal polarisation, because shear and torsion displacements have no associated restoring force.

[7]The number of microstates in an interval $[k, k + \mathrm{d}k]$ is $d(k)\mathrm{d}k$. It is written as $d(\omega)\mathrm{d}\omega$ if the interval is defined in terms of angular frequency, i.e. $[\omega, \omega + \mathrm{d}\omega]$: thus $d(k)\mathrm{d}k = d(\omega)\mathrm{d}\omega$ where $\omega = c_s k$ and $\mathrm{d}\omega = c_s \mathrm{d}k$. If the frequency $\nu = \omega/2\pi$ is used then $d(\nu) = 12\pi V\nu^2/c_s^3$, to be compared with Formula (7.17) obtained with a different method that does not include the factor 3 related to the three possible polarisations of the sound wave.

[8]The same value for ω_D can be easily obtained (up to dimensionless numerical factors) by requiring the minimum wavelength of the sound waves to be the interparticle distance.

The high temperature limit of C_V is simple: when $T \gg T_D$, the upper bound of the integrand in Equation (7.16) tends to 0 and $D(x) \simeq 3x^{-3} \int_0^x t^2 dt = 1$. The series expansion of the integrand can be continued to obtain higher order corrections [cf. Formula (7.16)]. The leading order corresponds to the Dulong-Petit law[9]: $C_V(T \gg T_D) = 3Nk_B$: as expected, the N harmonic oscillators behave classically at high temperature.

If, on the contrary, T is small with respect to T_D, then the upper bound of the integral defining $D(x)$ can be substitute for $+\infty$ leading[10] to Approximation (7.16): the heat capacity behaves as T^3 at low temperature. The departure from the Dulong-Petit law is due to a quantum effect already discussed in Section 6.3: at low temperature the excitations of the high energy modes are inhibited. In this sense one can say that diamond shows quantum effects at room temperature: as it is a very rigid material, with a very high speed of sound, and therefore a very high Debye temperature: $T_D = 1900$ K. Quantum inhibition of the degrees of freedom, which is seen up to about $T \sim T_D/5$ (see Figure 7.4), is therefore, for diamond, always in order at room temperature.

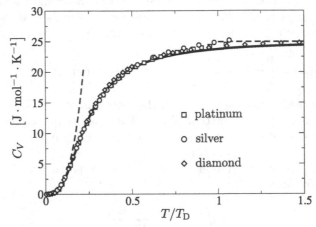

FIGURE 7.4 Experimental heat capacity for some elements having a crystalline structure simple enough for the phonon spectrum to be properly described by the Debye approximation. The solid line is a plot of Equation (7.15) and the dashed curves are the low and high temperature limits: $C_V \propto T^3$ and Dulong-Petit law (the molar heat capacity tends to $3R$), respectively.

The agreement with experimental results is rather good, as can be seen in Figure 7.4. The considered substances have significantly different Debye temperatures (240 K for platinum, 215 K for silver, and 1900 K for diamond), but the molar heat capacities overlap exactly when plotted as a function of the normalised temperature T/T_D. The behaviour in T^3 at low temperature is in good agreement with the experiments, contrary to the behaviour in $\exp(-1/T)$ predicted by Einstein model [see Equation (6.61)]. This is illustrated in more details in Figure 7.5 which shows the measurements performed on solid argon[11]. The values of T_D providing the best agreement at low temperatures between theory and experiment are equal, to within a few percent, to those calculated from the sound velocities. For instance, for argon[12], the results presented in Figure 7.5 lead to $T_D = 92.0 \pm 0.3$ K, whereas speed of sound measurements (see Equations (7.15) and (7.13)) give $T_D = 90.5 \pm 1.5$ K.

Debye's approach leads to results that fit experimental data remarkably well, but the agreement is not always perfect. First, it is the heat capacity at constant pressure C_P that

---

[9] Dulong and Petit Law corresponds to the classical equipartition theorem, see Section 4.1.5.
[10] see Note 28 page 264.
[11] after L. Finegold and N.E. Phillips, Phys. Rev. **177**, 1383 (1969).
[12] G. J. Keeler and D. N. Batchelder, J. Phys. C: Solid State Phys. **3**, 510 (1970).

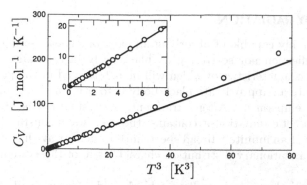

FIGURE 7.5 Heat capacity of solid argon plotted as a function of T^3 for temperature lower than ~4.5 K. The straight line is a fit of the experimental data. The inset is a zoom at low temperature ($T < 2$ K). The observed behaviour corresponds to a Debye temperature $T_D = 92$ K.

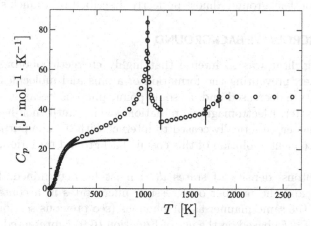

FIGURE 7.6 Heat capacity of iron below 3000 K. The solid curve is the heat capacity C_V calculated using Debye model (Equation (7.15) with $T_D = 460$ K). The points are experimental measurements of C_P (after P. D. Desai, J. Phys. Chem. Ref. Data **15**, 967 (1986)). Vertical lines mark temperatures of phase transitions: $T_C = 1043$ K, $T_{\alpha-\gamma} = 1185$ K, $T_{\gamma-\delta} = 1667$ K and $T_{\text{melt}} = 1811$ K.

is measured rather than the heat capacity capacity at constant volume C_V (the quantity that is calculated with Debye model). For solids, the difference between C_P and C_V is small but it increases with temperature and can be as high as a few percent. Second, although in many cases the use of the schematic density of state (7.13) is appropriate, the use of realistic densities of state cannot always be avoided[13]. Thus, when trying to fit the experimental heat capacity with Equation (7.15) by adjusting each time the Debye temperature T_D to obtain the best possible agreement, it is found that T_D is not absolutely constant over the whole temperature range but can vary by about 10%. Finally, as temperature increases, the considered material can undergo a phase transition which results in a discontinuity in the behaviour of the heat capacity (such a discontinuity is typical during a phase transition, see Section 9.2.3). This is observed, for example, in iron (Fe) which has a Debye temperature of 460 K, undergoes a ferromagnetic/paramagnetic transition at 1043 K, then changes its crystal structure (allotropic transitions[14]) at 1183 K and 1667 K, and finally melts at 1811 K (see Figure 7.6).

---

[13] This issue will be discussed in more details in Section 7.3.4 for the special case of liquid helium.

[14] Below 1185 K one speaks of alpha iron (α-Fe), and above 1667 K, delta iron (δ-Fe) while the crystalline structure is the same for both phases (cubic centred). Between these two temperature, the crystalline structure is face-centred cubic gamma iron (γ-Fe).

7.2   BLACK BODY RADIATION

A black body is a body capable of absorbing all electromagnetic radiation incident on its surface, without reflection nor scattering. A black body in thermal equilibrium can also emit radiation, the characteristics of which will be determined in the present section.

One way to create an approximate black body is to bring a closed oven with absorbing walls to a uniform temperature. After a transient, the system reaches thermal equilibrium with the electromagnetic radiation it contains: the oven walls constantly emit and absorb radiation (ideally over an infinitely broad spectrum) in such a way that the radiation energy is distributed over a stationary distribution characteristic of the system's equilibrium (i.e. its temperature).

Experimental and theoretical studies of black body physics had, at the turn of the twentieth century, a fundamental importance for statistical physics and quantum mechanics, with the description of the emission spectrum by Planck in 1900. In this section, the problem will be considered by studying an excellent black body model: the whole universe which is bathed in a radiation background almost perfectly described by Planck's law.

7.2.1   COSMIC MICROWAVE BACKGROUND

After the Big Bang, heat was so intense that highly energetic photons continuously interacted with matter, preventing the formation of atoms and molecules. As the Universe expanded, temperature decreased and, at some point, photons became no longer energetic enough to ionise matter. Electromagnetic radiations, which until then had been in thermal equilibrium with matter, practically ceased to interact with it. The purpose of this section is to study the subsequent evolution of this cosmic background radiation, sometimes called *fossil radiation*.

Firstly, the photons' density of states $d(E)$ must be determined. Photons are non-interacting bosons, and the number of accessible microstates in a container of volume V can be obtained in the same manner as for phonons (see previous section). Here, an alternative calculation of $d(E)$ based on the use of Equation (6.9) is presented. The Hamiltonian linking the momentum and the energy of a photon is simply $H = cp$, where c is the speed of light. It is customary to work with the frequency ν rather than with energy E ($E = h\nu$). The corresponding version of Equation (6.9) is accordingly:

$$d(\nu) = \frac{d}{d\nu} \left\{ \frac{1}{h^3} \int_V d^3 r \int_{cp \leq h\nu} d^3 p \right\} = 4\pi V \frac{\nu^2}{c^3} . \tag{7.17}$$

Like phonons, photons describe the quantum excitation modes of a Hamiltonian. In the case considered here, it is the Hamiltonian of the electromagnetic field[15]. As it was seen in Section 7.1.2, it implies that the system of photons at temperature T can be described by set of bosons with zero chemical potential. Thus, at temperature T, the number of photons in a cube of side L and with a frequency lying between ν and $\nu + d\nu$ is:

$$dN(L,\nu) = 2\, d(\nu) n_B(\nu) d\nu = \frac{8\pi L^3}{c^3} \frac{\nu^2 \, d\nu}{\exp\{\beta h\nu\} - 1} . \tag{7.18}$$

---

[15]For phonons, the corresponding Hamiltonian is the harmonic oscillator chain Hamiltonian. Its expression in one dimension is given by Equation (7.1), or equivalently (7.9).

In the middle term of Equation (7.18), the factor "2" reflects the two possible polarisation states of the photon[16] and n_B is the Bose–Einstein distribution [Equation (6.82)].

The expansion of the Universe induces a dilation of all distances: lengths follow a scaling law of the type $L \to L' = \alpha L$ ($\alpha > 1$). In particular, wavelengths evolve according to the same law and, since the speed of light remains constant, this results in the frequencies scaling: $\nu \to \nu' = \nu/\alpha$. The dN photons that originally had a frequency lying between ν and $d\nu$ and were contained in volume L^3 have then a frequency lying between ν' and $\nu' + d\nu'$ and are contained in volume L'^3 (dN does not change, because matter is not absorbing nor creating photons anymore). It is then easy to be convinced that $dN(L', \nu')$ still obeys Law (7.18) with a temperature $T' = T/\alpha$. Therefore it can be considered (as it will be done from now on) that radiations, although isolated, evolves while remaining in a state of thermal equilibrium as if they were in contact with a thermostat with a decreasing temperature.

The grand potential $J = -k_B T \ln \Xi$ [where Ξ is given by Equation (6.80)] can be written as:

$$J = k_B T \sum_{\lambda=0}^{\infty} \ln \left(1 - e^{-\beta h \nu_\lambda}\right) = 2\, k_B T \int_0^\infty d\nu\, d(\nu) \ln \left(1 - e^{-\beta h \nu}\right)$$
$$= \frac{k_B T}{\pi^2} \frac{L^3}{c^3} \left(\frac{k_B T}{\hbar}\right)^3 \int_0^\infty dx\, x^2 \ln(1 - e^{-x}) \,. \tag{7.19}$$

In the central term of Equation (7.19), the factor "2" corresponds, as previously stated, to the two possible polarisations of a photon. The last integral of Equation (7.19) can be computed by means of an integration by parts and by using the results of Appendix A.1, in particular Equations (A.6), (A.8) and Table A.1. This yields

$$\int_0^\infty dx\, x^2 \ln(1 - e^{-x}) = -\frac{1}{3} \int_0^\infty dx\, \frac{x^3}{e^x - 1} = -\frac{1}{3}\Gamma(4) g_4(1) = -2\zeta(4) = -\frac{\pi^4}{45} \,. \tag{7.20}$$

From this result, the grand potential and the entropy of the system can be calculated:

$$J = -\frac{\pi^2}{45} \left(\frac{L}{\hbar c}\right)^3 (k_B T)^4 \,, \quad \text{then} \quad S = -\left(\frac{\partial J}{\partial T}\right)_{V,\mu} = \frac{4\pi^2}{45} k_B \left(\frac{k_B T L}{\hbar c}\right)^3 \,. \tag{7.21}$$

Since during the expansion, $T \to T/\alpha$ and $L \to \alpha L$, the entropy of the photon gas remains constant.

Exercise 7.2. Using Hubble's Law which states that two galaxies separated by a distance L move away from each other at a speed $\dot{L}$ proportional to L, show that the distances in the universe vary approximately as $L(t) \propto t^{2/3}$ where t is the time elapsed since the Big Bang (which took place 14 billion years ago). Knowing that the decoupling of matter and radiation occurred when the temperature was $T_0 = 3000$ K and that the Universe's present temperature is $T \simeq 2.7$ K, calculate the age of the Universe at the time of the decoupling of radiation and matter.

Solution: Consider a sphere of radius L. It contains a mass $M = 4\pi L^3 \rho/3$, where ρ is the average mass density of the universe. The gravitational interaction potential energy between M and a galaxy of mass m located at the periphery of the sphere is $E_{\text{pot}} = -GmM/L$, where G is the gravitational constant. The kinetic energy of the galaxy of mass m is $E_{\text{cin}} = \frac{1}{2} m \dot{L}^2 = \frac{1}{2} m H^2 L^2$ where H is the constant of proportionality (called Hubble's constant) between L and $\dot{L}$. The total energy of the

---

[16]An electromagnetic wave is transverse, in contrast to phonons which can have 3 polarisation states (one longitudinal, two transverse).

galaxy at the periphery of the sphere of radius L is therefore:

$$E_{\text{tot}} = mL^2(t) \left[\frac{1}{2} H^2(t) - \frac{4}{3} \pi \rho(t) G \right] . \qquad (7.22)$$

The quantity of matter contained inside the sphere of radius L is constant, i.e. $\frac{4}{3}\pi L^3(t)\rho(t)$ is independent of time and consequently ρ behaves as L^{-3}. For E_{tot} to remain constant, it is necessary that, at the leading order, the bracketed terms in Equation (7.22) have the same dependence in terms of power of L, i.e. $H \propto \rho^{1/2} \propto L^{-3/2}$. This gives $\dot{L} = HL \propto L^{-1/2}$, which after integration leads to $L^{3/2} \propto t$.

L is thus proportional to $t^{2/3}$. Then, the relation linking the evolution of temperatures and distances leads to a temperature evolving as $t^{-2/3}$. So, if t_0 is the age of the universe at the time of decoupling and T_0 the corresponding temperature, then $t_0 = t \cdot (T/T_0)^{3/2}$ giving $t_0 \simeq 400,000$ years.

The black body spectrum is characterised by a density of energy per unit volume and frequency $u_T(\nu)$ of the photon gas. By definition, $u_T(\nu)L^3 d\nu = h\nu \, dN$, or:

$$u_T(\nu) = \frac{8\pi h}{c^3} \frac{\nu^3}{\exp\left[\frac{h\nu}{k_B T}\right] - 1} . \qquad (7.23)$$

At high temperature and low frequency (i.e. $k_B T \gg h\nu$), a series expansion of the exponential leads to the Rayleigh-Jeans Law:

$$u_T(\nu) \simeq \frac{8\pi k_B T}{c^3} \nu^2 = 2k_B T \frac{d(\nu)}{V} \quad \text{when} \quad k_B T \gg h\nu . \qquad (7.24)$$

This law can be obtained by treating electromagnetic waves classically and assigning to each degree of freedom an average energy[17] $k_B T$. It predicts a divergence at high frequency, known as the ultraviolet catastrophe (term introduced by Erhenfest in 1911). The Planck spectrum [Equation (7.23)] does not suffer from this deficiency. In the (opposite) high frequency and low temperature limit (i.e. $k_B T \ll h\nu$), the asymptotic behaviour of Planck's spectrum correspond to Wien's Law:

$$u_T(\nu) \simeq \frac{8\pi h\nu^3}{c^3} \exp\left[-\frac{h\nu}{k_B T}\right] = 2 \, h\,\nu\, e^{-\beta h\nu} \frac{d(\nu)}{V} \quad \text{when} \quad k_B T \ll h\nu . \qquad (7.25)$$

This law can be obtained by treating the system of photons as a classical ideal gas with each particle having an energy $h\nu$.

Exercise 7.3. Measurements of the cosmic background radiation spectrum show that the maximum radiation is reached for a frequency $\nu_m = 1.605 \times 10^{11}$ Hz. What is the corresponding temperature T of the cosmic background?

<u>Solution</u>: With $x = h\nu/k_B T$, $u_T(\nu) = 8\pi h(k_B T/hc)^3 x^3/(e^x - 1)$. $u_T(\nu)$ reaches a maximum when $x/3 = 1 - e^{-x}$, i.e. $x = x_m = 2.82144\ldots$. The frequency at which $u_T(\nu)$ is maximum is thus $\nu_m = x_m k_B T/h$. The measured value of ν_m corresponds to $T = 2.728$ K.

Satellites launched since the late 1980's have measured the spectrum of the cosmic background radiation with increasing precision. Figure 7.7 shows data acquired by the COBE satellite. It indicates that the fossil radiation obeys Planck's law [Equation (7.23)] almost perfectly for a temperature $T = 2.728 \pm 0.004$ K.

---

[17]i.e. one considers that each wave corresponds to two one-dimensional harmonic oscillator (the factor two corresponds to the two possible transverse polarisations) with, for each oscillator, two quadratic terms in the Hamiltonian, see Section 4.1.5.

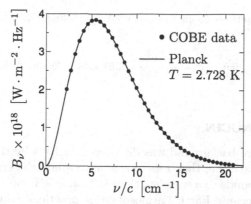

FIGURE 7.7 COBE satellite data for the cosmic background radiation spectrum (black dots). The y-axis represents the irradiance $B_\nu = c\,u_T(\nu)/(4\pi)$, the energy current per unit solid angle and frequency (see Exercise 7.4). The solid line corresponds to the Planck spectrum [Equation (7.23)] for a temperature of $T = 2.728$ K.

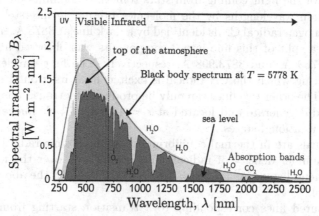

FIGURE 7.8 Solar irradiance as a function of wavelength. Note that, unlike Figure 7.7, high energies are on the left. Global Warming Art, Wikipedia, cc by sa 3.0 (https://en.wikipedia.org/wiki/Sunlight\#Solar_irradiance)

It is interesting to note that the sun also approximately behaves as a black body. Figure 7.8 shows the solar irradiance B_λ as a function of wavelength λ ($B_\lambda d\lambda = B_\nu d\nu$). The light-coloured data represent the spectrum at the top of the atmosphere, with a maximum irradiance in the green ($\lambda = 504$ nm). The irradiance at sea level (dark spectrum) is decreased by atmospheric absorption. In the ultraviolet domain, ozone (O_3) is mainly responsible for this decrease. At other places on the spectrum, the influence of oxygen, water and carbon dioxide can be seen. Note that the redistribution of light by Rayleigh scattering is responsible for the blue colour of the sky.

Exercise 7.4. Calculate the power per unit area radiated by a black body.

Solution: Consider an element of area d^2S of the black body's surface. The energy of the radiation emitted during a time dt in a solid angle $d^2\Omega$ in a direction making an angle θ with the normal vector to the surface and having a frequency lying between ν and $\nu + d\nu$ is $dE = d^2S\,dt\,c\cos\theta\,u_T(\nu)d\nu d^2\Omega/4\pi$. The radiated power per unit area is therefore $R = c\int d^2\Omega\cos\theta\int_0^\infty u_T(\nu)d\nu = \frac{1}{4}c\int_0^\infty u_T(\nu)d\nu$. The integral over the frequencies is proportional to $g_4(1)$

(see the definition of the functions g_ν in Appendix (A.6)). Using Equation (A.8), the so-called Stefan-Boltzmann law *is finally obtained:*

$$R = \sigma T^4, \quad \text{with} \quad \sigma = \frac{\pi^2 k_B^4}{60 c^2 \hbar^3} = 5.67 \times 10^{-8} \ \text{W} \cdot \text{m}^{-2} \cdot \text{K}^{-4}. \tag{7.26}$$

7.2.2 INTERSTELLAR CYANOGEN

The cosmological microwave background was discovered in 1964 by the radio astronomers Penzias and Wilson. At that time, they were not able to study the entire spectrum whose maximum intensity corresponds to a wavelength of 0.2 cm (see Figure 7.7, B_ν is maximum for $\lambda^{-1} = \nu/c = 5 \ \text{cm}^{-1}$) because Earth's atmosphere is practically opaque to wavelengths shorter than 3 cm. In the years around Penzias and Wilson's discovery, a clever technique was used to obtain a fairly accurate estimate of the temperature of the cosmic background radiation. This technique relies on a fine understanding of the thermal occupancy of the rotational lines of a molecule (a problem already studied in Section 6.5).

The spectrum of the light coming from stars sometimes exhibits dark lines due to the absorption of certain wavelengths by the molecules present in interstellar gas clouds. In particular, the cyanogen radical CN is identified by a dark line at 3875 Å. In the light coming from ζ Ophiuchi, a split of this line into 3 components was observed at the wavelengths 3874.608 Å, 3875.763 Å and 3873.998 Å, respectively. The first wavelength corresponds to a transition during which the CN radical is excited from its ground state ($\ell = 0$) to a vibrational state. The other two lines can only be produced by transitions from a rotational state $\ell = 1$ (3 fold-degenerate and located at $\epsilon = 4.70 \times 10^{-4}$ eV above the fundamental state) to various vibrational states.

If the CN radicals are in thermal equilibrium with the cosmic radiations at $T = 2.7$ K, the ratio of the number density of radicals in a level $\ell = 1$ over the number density of radicals in fundamental state is[18]: $n_1/n_0 = 3 \exp[-\epsilon/k_B T]$. With the above value for ϵ, one obtains $n_1/n_0 = 0.4$.

The two measured lines corresponding to a transition starting from the state $\ell = 1$ must have the same intensity (proportional to the absorbed light). This intensity measured relative to the intensity of the transition starting from the state $\ell = 0$ allows an estimate of the ratio n_1/n_0. Early measurements[19] gave $n_1/n_0 \sim 0.5$, compatible with a temperature[20] of $T \simeq 3$ K. More recent measurements[21] using the same technique gave a much more accurate estimate: $T = 2.729 \, (+0.023 \, ; -0.031)$ K.

7.3 LIQUID HELIUM

Helium has the lowest boiling point of all elements (3.19 K for its fermionic isotope ^{3}He and 4.21 K for its bosonic isotope ^{4}He). Under saturating vapour pressure, both isotopes remain liquid until absolute zero (this also makes helium unique) and it is necessary to apply high pressures to see the solid phase appear (at T $=$ 0 K, 25 atm for ^{4}He and 30 atm for ^{3}He). It is the effect of the low inter-atomic potential combined with the low mass of helium (which induces a high zero-point energy) that explains the difficulty to

---

[18]From Equation (6.92), the density n_ℓ of a state characterised by an orbital quantum number ℓ is $n_\ell = z_{\text{rot}}^{-1}(2\ell + 1) \exp[-\epsilon_\ell/k_B T]$.

[19]G. B. Field and J. L. Hitchcock, Phys. Rev. Lett. **16**, 817 (1966).

[20]An excellent discussion of this question and its historical aspects is given in S. Weinberg's great scientific popularisation book, *The First Three Minutes: A Modern View of The Origin of The Universe* (Basic Books, 1993).

[21]K. C. Roth, D. M. Meyer, and I. Hawkins, ApJ, **413**, L67 (1993).

liquefy and solidify. Thanks to these particularities, helium remains liquid at temperatures where quantum effects become important. It is for example the only liquid to which,*mutatis mutandis*, Debye approach can be applied to calculate heat capacity (see Section 7.3.4).

Liquid helium has been the object of sustained experimental interest because it has remained for a long time one of the few quantum fluids on which direct measurements could be made. Moreover, it is quite easy to study under a large variety of configurations and its chemical purity makes it a model system for the study of many physical phenomena (surfaces, thin films, crystal growth and roughening transition...).

7.3.1 SUPERFLUID HELIUM

When cooled down to a temperature lower than 2.17 K, liquid ^{4}He undergoes a phase transition: it goes from liquid helium I to liquid helium II. It is known as the "lambda transition" for reasons which will be explained in Section 11.4.3. The transition temperature is $T_\lambda = 2.17$ K. Helium II has very peculiar characteristics: no boiling at vaporisation, thermomechanical effect, second sound... Its most spectacular property is the phenomenon of superfluidity: helium II can flow without apparent friction through the narrowest capillaries. These properties are interpreted by a two-fluid phenomenological model in which the volumetric mass density of helium is divided into two parts:

$$\rho = \rho_s + \rho_n . \tag{7.27}$$

ρ_n corresponds to the density of normal fluid and ρ_s corresponds to the density of inviscid fluid that carries no entropy: the superfluid component. The ratio between these densities varies with temperature: at $T = T_\lambda$ (the superfluid transition temperature) $\rho = \rho_n$ ($\rho_s = 0$) while at $T = 0$, $\rho = \rho_s$ (and $\rho_n = 0$). This behaviour is illustrated in Figure 7.9.

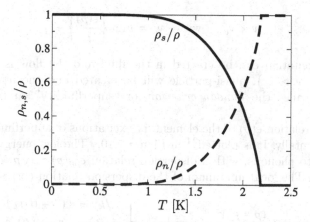

FIGURE 7.9 Schematic evolution of the superfluid fraction ρ_s/ρ (solid lines) and the normal fraction ρ_n/ρ (dashed lines) as a function of temperature. Quantitative data are presented in Fig. 7.11.

7.3.2 ZERO TEMPERATURE. LANDAU CRITERION

The goal of this section is to study the flow of helium II at $T = 0$ K in a capillary and to determine under which condition it is energetically favourable to dissipate energy by emitting an elementary excitation.

Consider a fluid moving in a capillary with a uniform speed $\vec{v}$ and let $\mathcal{R}$ be the frame of reference attached to the fluid. In this frame of reference the fluid is at rest and has an energy E_0. Possible excitations have a momentum $\vec{p}$ and an energy $\epsilon(p)$, where $\epsilon(p)$ is the dispersion relation of the excitations in the fluid at rest. If an excitation is created then the total momentum becomes $\vec{P} = \vec{p}$ and the total energy energy $E = E_0 + \epsilon(p)$. However, it is in the frame of reference $\mathcal{R}'$ attached to the capillary that the reasoning must be carried on, because it is in this frame that stationary obstacles are present (imperfections of the wall of the capillary) and it is also $\mathcal{R}'$ that can be in contact with a thermostat (see Section 7.3.3). The appearance of an excitation of momentum $\vec{p}$ in $\mathcal{R}$ corresponds to a change of energy $\Delta E' = \epsilon(p) + \vec{p}.\vec{v}$ and momentum $\Delta \vec{P}' = \vec{p}$ in $\mathcal{R}'$.

Exercise 7.5. Prove this result.

Solution: Let m_i be the mass of the ith particle and, $\vec{v}_i$ and $\vec{v}_i'$ its velocities measured in $\mathcal{R}$ and $\mathcal{R}'$, respectively. Since $\mathcal{R}'$ has a translation velocity $-\vec{v}$ with respect to $\mathcal{R}$, $\vec{v}_i$ and $\vec{v}_i'$ are related by $\vec{v}_i = \vec{v}_i' - \vec{v}$. The total momentum of the fluid in $\mathcal{R}'$ is thus: $\vec{P}' = \sum_i m_i \vec{v}_i' = \vec{P} + M\vec{v}$ where $\vec{P}$ is the momentum measured in $\mathcal{R}$ and $M = \sum_i m_i$ is the total mass of the fluid.

The energy in $\mathcal{R}'$ can be written as $\sum_i \frac{m_i}{2}(\vec{v}_i')^2 + E_{\text{pot}}$, where $E_{\text{pot}} = \sum_{i<j} U_{\text{int}}(\vec{r}_i - \vec{r}_j)$ is an interaction energy term which is invariant under a Galilean change of frame of reference. Therefore, $E' = E + \frac{1}{2}Mv^2 + \vec{P} \cdot \vec{v}$, where E is the energy measured in $\mathcal{R}$.

Therefore, when a quasi-particle appears in a fluid initially at rest in $\mathcal{R}$, the energy and the momentum which were $E_0 + \frac{1}{2}Mv^2$ and $M\vec{v}$ in $\mathcal{R}'$, respectively, are modified and become $E' = E_0 + \frac{1}{2}Mv^2 + \vec{p} \cdot \vec{v} + \epsilon(p)$ and $\vec{P}' = M\vec{v} + \vec{p}$.

The creation of an excitation of momentum[22] $\vec{p}$ will therefore be energetically favourable in $\mathcal{R}'$ if:

$$\vec{p} \cdot \vec{v} + \epsilon(p) < 0 \ . \tag{7.28}$$

Let θ be the angle between $\vec{p}$ and $\vec{v}$. Equation (7.28) can be written as $\epsilon(p)/p < -v\cos\theta$ and has consequently no solution if:

$$v < v_c \ , \quad \text{with} \quad v_c = \min\left\{\frac{\epsilon(p)}{p}\right\} \ . \tag{7.29}$$

In this case, no excitation can be created in the fluid and the flow is superfluid. In the opposite situation ($v > v_c$), quasi-particle will be created and the flow is dissipative[23]. Relation (7.29) is called the *Landau criterion* for "superfluidity", and v_c is the Landau critical velocity.

The dispersion relation $\epsilon(p)$ of the elementary excitations of superfluid helium has been measured experimentally. It is plotted[24] in Figure 7.10. The low energy part of the spectrum corresponds to phonons, with a dispersion relation $\epsilon_{\text{ph}}(p) = c_s\, p$ ($c_s = 238$ m $\cdot$ s^{-1} in superfluid helium). The local mimimum of the dispersion relation corresponds to "rotons" with:

$$\epsilon_{\text{rot}}(p) = \Delta + \frac{(p - p_0)^2}{2\,m^*} \quad \text{with} \quad \begin{cases} \Delta/k_{\text{B}} = 8.65 \pm 0.04 \text{ K} \ , \\ p_0/\hbar = 1.92 \pm 0.01 \text{ Å}^{-1} \ , \\ m^*/m = 0.16 \pm 0.01 \ . \end{cases} \tag{7.30}$$

From these data, it is easy to determine the value of Landau critical velocity v_c of helium II: let p_c be the momentum for which $\epsilon(p_c)/p_c = \min\{\epsilon(p)/p\}$. Then, $d\epsilon_{\text{rot}}/dp|_{p_c} =$

---

[22] As just demonstrated, the momentum of the excitation is the same in $\mathcal{R}$ and $\mathcal{R}'$.

[23] This is a rather peculiar type of dissipation: the total energy is always conserved but the system irreversibly creates a wake sending quasi-particles towards infinity. This corresponds to a mechanism called *wave resistance* in hydrodynamics, which is for instance the predominant cause of energy dissipation of boats and ships (by creation of a wake).

[24] L. Landau had guessed the shape of the dispersion relation before it was measured, by demonstrating that it could explain the temperature dependence of the heat capacity of helium II described in Section 7.3.4 [see L. Landau, Phys. Rev. **60**, 356 (1941)].

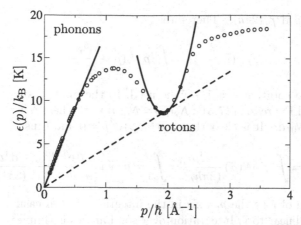

FIGURE 7.10 Dispersion relation of the excitations in liquid helium II. The white circles are data compiled from R. J. Donnelly's website http://pages.uoregon.edu/rjd/vapor1.htm. The "phonons" and "rotons" curves correspond to the schematic models presented in this book. The dashed straight line as a slope slope $v_c = 59$ m·s^{-1} corresponding to Landau critical velocity.

$\epsilon_{\rm rot}(p_c)/p_c = v_c$ and one gets[25] $p_c = (p_0^2 + 2\,m^*\,\Delta)^{1/2} \simeq p_0$ and $v_c = \epsilon_{\rm rot}(p_c)/p_c \simeq \Delta/p_0 = 59$ m·s^{-1}.

7.3.3 FINITE TEMPERATURE. NORMAL FRACTION

At finite temperature, excitations appear in the fluid. They form a gas of quasi-particles, considered non-interacting if it has a sufficiently small number density (i.e. at low enough temperature). These quasi-particles collide with the walls with which they exchange momentum and energy, inducing dissipation. The quasi-particles thus correspond to the viscous part of the flow, they are immobile in the capillary frame of reference $\mathcal{R}'$ and form the "normal fraction". Another part of the flow corresponds to the "superfluid fraction". Let $\vec{v}_n$ and $\vec{v}_s$ be the velocities of the normal and the superfluid fractions, respectively. The equilibrium distribution of the quasi-particles must be evaluated in $\mathcal{R}'$ where the normal fraction is at rest: it is in $\mathcal{R}'$ that thermodynamic equilibrium occurs. The velocity of $\mathcal{R}'$ with respect to $\mathcal{R}$ in which the superfluid is at rest is $\vec{v}_n - \vec{v}_s$. The same reasoning as in Exercise 7.5 shows here that the appearance of a quasi-particle corresponds in $\mathcal{R}'$ to a momentum $\vec{p}$ and an energy $\epsilon(p) + \vec{p} \cdot (\vec{v}_s - \vec{v}_n)$. The equilibrium distribution of the quasi-particles (which are bosons of zero chemical potential) is therefore:

$$N_v(\vec{p}) = \frac{1}{\exp\left(\dfrac{\epsilon(p) + \vec{p} \cdot (\vec{v}_s - \vec{v}_n)}{k_{\rm B}T}\right) - 1}. \tag{7.31}$$

This definition is consistent with the Landau criterion: it is only below the Landau critical velocity – i.e., when $|\vec{v}_s - \vec{v}_n| < v_c$ – that Distribution (7.31) makes sense (since it is positive for all $\vec{p}$). In the frame of reference $\mathcal{R}$ in which the superfluid is at rest, the mass current of quasi-particles is $\vec{J}_{\rm excit} = \rho_n(\vec{v}_n - \vec{v}_s)$. $\vec{J}_{\rm excit}$ is also the quasi-particle momentum density $\vec{J}_{\rm excit} = \frac{1}{V}\sum_i \vec{p}_i$ (where i numbers the quasi-particles) and can thus be written

---

[25] Indeed $2\,m^*\Delta/p_0 = 0.06 \ll 1$.

as[26] $\vec{J}_{\text{excit}} = \int \vec{p} N_\nu(\vec{p})\mathrm{d}^3 p/(2\pi\hbar)^3$. Therefore:

$$\rho_n(\vec{v}_n - \vec{v}_s) = \int_{\mathbb{R}^3} \vec{p} N_\nu(\vec{p}) \frac{\mathrm{d}^3 p}{(2\pi\hbar)^3} . \tag{7.32}$$

In the following, the notation $\vec{v} = \vec{v}_s - \vec{v}_n$ is used. In the limit of small values of $|\vec{v}|$, it is legitimate to expand Expression (7.31): $N_\nu(\vec{p}) \simeq N_0(p) + \delta N$ where $N_0(p) = (\exp(\beta\epsilon(p)) - 1)^{-1}$ and $\delta N(\vec{p}) = \vec{p} \cdot \vec{v} \, \mathrm{d}N_0/\mathrm{d}\epsilon$. It is clear that $\int \vec{p} N_0(p) \, \mathrm{d}^3 p = 0$ and thus:

$$\rho_n \vec{v} = \int_{\mathbb{R}^3} \vec{p} \, \delta N(\vec{p}) \frac{\mathrm{d}^3 p}{(2\pi\hbar)^3} = \int_{\mathbb{R}^3} \vec{p}(\vec{p} \cdot \vec{v}) \frac{\beta \, e^{\beta\epsilon(p)}}{(e^{\beta\epsilon(p)} - 1)^2} \frac{\mathrm{d}^3 p}{(2\pi\hbar)^3} . \tag{7.33}$$

Taking the direction of $\vec{v}$ as the p_z-axis in the integral, one can easily check the left-hand term of (7.33) is collinear to $\vec{v}$. Integration over solid angles is simple[27] and yields:

$$\rho_n = \frac{4\pi \, \beta}{3 \, h^3} \int_0^\infty \frac{p^4 \, e^{\beta\epsilon(p)}}{(e^{\beta\epsilon(p)} - 1)^2} \, \mathrm{d}p . \tag{7.34}$$

In the limit of low temperatures, excitations have a small energy and it is sound to approximate $\epsilon(p) \simeq c_s \, p$ in Expression (7.34). This yields[28]:

$$\rho_n = \frac{16 \, \pi^5}{45 \, h^3 \, c_s^5} (k_{\text{B}}T)^4 . \tag{7.35}$$

Taking for helium $\rho = 0.1455$ g·cm^{-3} and $c_s = 238$ m·s^{-1}, gives $\rho_n/\rho = 1.22 \times 10^{-4} \, T^4$ (where T is in K), in very good agreement with the experimental data in the low-temperature regime ($T < 0.6$ K, see Figure 7.11).

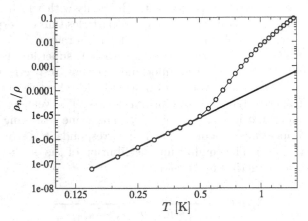

FIGURE 7.11 Normal fluid fraction ρ_n/ρ plotted as a function of temperature in superfluid helium. White dots represent experimental data. At low temperature, the agreement with Formula (7.34) (solid line) is excellent. Experimental data are taken from R. J. Donnelly's website `http://pages.uoregon.edu/rjd/vapor1.htm` [see also D. De Klerk, R.P. Hudson and J.R. Pellan, Phys. Rev. **89**, 662 (1953)].

---

[26]The integral $\int_{\mathbb{R}^3} N_\nu(\vec{p})\mathrm{d}^3 p/(2\pi\hbar)^3$ is the number of elementary excitations per unit volume.
[27]$\int_0^\pi \mathrm{d}\theta \cos^2\theta \sin\theta = 2/3$.
[28]An integration by parts and the result (7.20) show that
$\int_0^\infty \mathrm{d}x \, x^4 \exp(x)[\exp(x) - 1]^{-2} = 4 \int_0^\infty \mathrm{d}x \, x^3 [\exp(x) - 1]^{-1} = 24 \, \zeta(4) = 4\pi^4/15$.

7.3.4 FINITE TEMPERATURE. HEAT CAPACITY

The phonon model used in Section 7.1 to describe solids is also valid for liquid helium. Helium is the only element that remains liquid near absolute zero and it is the only liquid for which the low temperature behaviour in T^3 of $C_V(T \to 0)$ can be tested.

In a liquid, sound waves are only longitudinal. The low temperature formula obtained for solids using Equations (7.15) and (7.16) must therefore be divided by 3. Then, the mass heat capacity[29] is:

$$(c_V)_{\text{ph}} = \frac{k_B}{\rho} \frac{2\pi^2}{15} \left(\frac{k_B T}{\hbar c_s} \right)^3 = 0.0208 \times T^3 \quad \text{J} \cdot \text{g}^{-1} \cdot \text{K}^{-1}, \tag{7.36}$$

where T is in K. This expression is in very good agreement with experimental results[30] for $0 < T < 0.6$ K : $c_V = (0.0204 \pm 0.0004) T^3$ J·g^{-1}·K^{-1}.

In helium, the Debye temperature[31] is $T_D = 19.8$ K. One would thus expect Law (7.36) to be accurate up to temperatures ~5–10 K. However, experiments clearly show that when $T \geq 0.6$ K data deviate from the T^3-law. This is due to the peculiar shape of the excitation spectrum of ^{4}He (Figure 7.10). As previously explained, it has a phononic and a rotonic branch. The energy of the gas of quasi-particles is accordingly divided into two contributions: $E(T,V) = E_{\text{ph}}(T,V) + E_{\text{rot}}(T,V)$. The phonon term has a behaviour in T^4 leading to Formula (7.36) (hence the index "ph" in this formula). The rotonic term is:

$$E_{\text{rot}} = \frac{V}{(2\pi\hbar)^3} \int_0^\infty 4\pi p^2 dp \, \frac{\epsilon_{\text{rot}}(p)}{\exp[\beta\epsilon_{\text{rot}}(p)] - 1}, \tag{7.37}$$

where $\epsilon_{\text{rot}}(p)$ corresponds to Parameterisation (7.30). When $T \ll \Delta/k_B = 8.65$ K, then $\exp[\beta\epsilon_{\text{rot}}(p)] \gg 1$ and the Maxwell-Boltzmann approximation can be used (see Section 6.4.3). This leads to (taking $q = p - p_0$)

$$E_{\text{rot}} = \frac{V}{(2\pi\hbar)^3} \int_{-p_0}^\infty 4\pi (q + p_0)^2 dq \left(\Delta + \frac{q^2}{2m^*} \right) e^{-\beta\Delta} \exp\left[-\beta \frac{q^2}{2m^*} \right]. \tag{7.38}$$

The last term of the integrand in Equation (7.38) is a Gaussian of typical width $q_1 = \sqrt{2m^* k_B T}$. When[32] $p_0 \gg q_1$, the lower bound of integration in Equation (7.38) can be substituted for $-\infty$:

$$E_{\text{rot}}(T) \simeq \frac{V}{2\pi^2\hbar^3} e^{-\Delta/(k_B T)} \int_{-\infty}^{+\infty} dq \, (q + p_0)^2 \left(\Delta + \frac{q^2}{2m^*} \right) e^{-\beta q^2/(2m^*)}. \tag{7.39}$$

The dominant term in this integral is $\int_{\mathbb{R}} dq \, p_0^2 \Delta \exp[-\beta q^2/(2m^*)] = p_0^2 \Delta \sqrt{2\pi m^* k_B T}$, therefore:

$$E_{\text{rot}}(T) \simeq \left(\frac{V p_0^3}{2\pi^{3/2}\hbar^3} \right) \Delta e^{-\Delta/(k_B T)} \sqrt{\frac{2m^* k_B T}{p_0^2}}. \tag{7.40}$$

Then, when calculating the contribution $(C_V)_{\text{rot}} = (\partial E_{\text{rot}}/\partial T)_V$ of the rotonic branch to the heat capacity, the main contribution to the derivative with respect to temperature comes

---

[29] In Equation (7.36), the heat capacity is written with index "ph" for reasons that will become clear in the following.

[30] J. Wiebes, C.G. Niels-Hakkenberg and H.C. Kramers, Physica **23**, 625 (1957).

[31] The formula on the left-hand side of (7.15) still holds here. Thus, $T_D = (\hbar c/k_B)(6\pi^2 N/V)^{1/3}$. For helium, $N/V = 2.184 \times 10^{-2}$ atom/Å^3.

[32] The condition $p_0 \gg q_1$ is equivalent to $T \ll p_0^2/(2m^* k_B) \simeq 3.9$ K.

from the exponential in Equation (7.40). Taking only this term into account yields the following approximate expression for the roton part of the mass heat capacity:

$$(c_V)_{\text{rot}} = \frac{2\,k_{\text{B}}(p_0/\hbar)^3}{\rho\,(2\pi)^{3/2}} \sqrt{\frac{m^*\Delta}{p_0^2}} \left(\frac{\Delta}{k_{\text{B}}T}\right)^{3/2} \exp\left[-\frac{\Delta}{k_{\text{B}}T}\right]. \qquad (7.41)$$

The dominant roton contribution to the heat capacity thus behaves as $T^{-3/2}\exp\{-\Delta/(k_{\text{B}}T)\}$. The sum $(c_V)_{\text{ph}} + (c_V)_{\text{rot}}$ defined by Equations (7.36) and (7.41) reproduces very well the experimental data up to temperatures of about 1.5 K. The phonon contribution is dominant only up to about 0.6 K and contributes to only 10% of the total heat capacity at $T = 1.5$ K (see Figure 7.12).

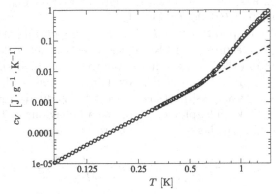

FIGURE 7.12 The "phonons+rotons" model (the solid curve is a plot of Equations (7.36) and (7.41)) explains very well the temperature dependence of the mass heat capacity. White circles are the experimental results compiled on R. J. Donnelly's website http://pages.uoregon.edu/rjd/vapor1.htm. The dashed curve takes only into account the phonon contribution (Equation (7.36)).

7.4 BOSE–EINSTEIN CONDENSATION

In this section, a gas composed of N point-like non-interacting bosonic atoms of mass m and zero spin is considered. Such a system is realised experimentally by ultracold atoms placed in an external potential created thanks to a magneto-optical trap. Contrary to the cases of photons (Section 7.2) and phonons (Section 7.1), which are elementary excitations whose number depends on energy and temperature, the number of atoms in a given experiment is fixed. This simple remark, combined with the boson statistics is at the origin of a very peculiar behaviour of the system at low temperature.

For the present study, the case of an harmonic trap is considered. The Hamiltonian of the system is $H_{\text{tot}} = \sum_{i=1}^{N} H(\vec{r}_i, \vec{p}_i)$, where $H(\vec{r}, \vec{p}) = \vec{p}^2/(2m) + m\,\omega^2\,\vec{r}^2/2 - 3\hbar\,\omega/2$ is the one-body (single particle) Hamiltonian. The constant term, $-3\hbar\,\omega/2$, in H has no influence on the system's dynamics; it is added to place the ground state of H at the origin of the energy scale. The eigenlevel of H are of the form $\varepsilon_{\underline{\nu}} = \hbar\omega(\nu_1 + \nu_2 + \nu_3)$, where $\underline{\nu}$ is a notation for the triplet (ν_1, ν_2, ν_3) of $\mathbb{N}^3$.

The cumulative density of state [Equation (6.5)] of the one-body Hamiltonian H is calculated by means of the semi-classical rule [Equation (6.8)]:

$$\Phi(E) = \int \frac{\mathrm{d}^3p\,\mathrm{d}^3r}{(2\pi\hbar)^3} \Theta\left(E - \frac{p^2}{2m} - \frac{m\omega^2 r^2}{2}\right) = \frac{1}{(\pi\hbar\omega)^3} \int_{\vec{P}^2 + \vec{Q}^2 < E} \mathrm{d}^3P\,\mathrm{d}^3Q\,, \qquad (7.42)$$

where the changes of variables $\vec{P} = \vec{p}\,(2m)^{-1/2}$ and $\vec{Q} = \vec{r}\,(m\omega^2/2)^{1/2}$ have been made. The integral of the right-hand side term of Equation (7.42) is the volume of a sphere of radius $\sqrt{E}$ in $\mathbb{R}^6$. Using Equation (A.25), one obtains $\Phi(E) = \frac{1}{6}(E/\hbar\omega)^3$ and thus according to Equation (6.6)[33], the density of state is:

$$d(E) = \frac{E^2}{2(\hbar\omega)^3}\,. \tag{7.43}$$

The trapped gas corresponds to a physical implementation of a system of bosons in the microcanonical ensemble (fixed energy and fixed particle number). However, by virtue of ensemble equivalence (see Section 6.1.4), the grand canonical formulation can be used in equilibrium at the thermodynamic limit. The average occupancy number of the energy level $\varepsilon_{\underline{\nu}}$ is then $\langle n(\varepsilon_{\underline{\nu}})\rangle = (\exp(\beta(\varepsilon_{\underline{\nu}} - \mu)) - 1)^{-1}$ [see Equation (6.82)]. The ground level is $\varepsilon_{\underline{0}} = 0$, imposing a negative chemical in order to ensure a finite $\langle n(\varepsilon_{\underline{\nu}})\rangle$ for all $\underline{\nu}$.

Let us first consider a system at high enough temperature so that the condition $\mu < 0$ is easily fulfilled (as in the classical case at high temperature, see Note 37 in Chapter 3). In this case, Equation (6.83) reads:

$$N = \sum_{\underline{\nu}} \langle n(\varepsilon_{\underline{\nu}})\rangle = \int_0^\infty dE\, \frac{d(E)}{\exp(\beta(E - \mu)) - 1}\,. \tag{7.44}$$

This equation allows the determination of the value of μ from the knowledge of N and T. A semi-graphical resolution can be achieved by rewriting Equation (7.44) under the form:

$$N\left(\frac{\hbar\omega}{k_BT}\right)^3 = g_3(e^{\beta\mu})\,, \tag{7.45}$$

where Function g_3 belongs to the class of the "Bose functions" introduced in Appendix A.1. For legibility, we recall their definition [Equation (A.6)]:

$$g_\nu(z) = \frac{1}{\Gamma(\nu)}\int_0^\infty \frac{x^{\nu-1}\,dx}{z^{-1}\exp(x) - 1}\,, \tag{7.46}$$

where $\Gamma(\nu)$ is defined by Equation (A.2). The quantity

$$z = \exp(\beta\mu) \tag{7.47}$$

is called *the fugacity*. Function $g_3(z)$ is plotted in Figure 7.13 for $z \in [0,1]$.

The knowledge of g_3 allows the determination of the fugacity by graphically solving Equation (7.45). At high temperatures, the chemical potential is large and negative. In this case, $e^{\beta\mu} \to 0$ and $g_3(e^{\beta\mu})$ can be approximated by keeping only the first term of the corresponding series expansion [Equation (A.7)]. This immediately yields:

$$\mu(T,N) \simeq \mu_{\text{class}} = -3k_BT\ln(T/T^*) \quad \text{where} \quad k_BT^* = \hbar\omega\,N^{1/3}\,. \tag{7.48}$$

It can be checked *a posteriori* that $\beta\mu$ is indeed large and negative. As temperature decreases, the left-hand side of Equation (7.45) becomes larger and eventually a solution can no longer be found. The last acceptable solution is obtained when $e^{\beta\mu} = 1$, i.e. $\mu = 0$ which,

---

[33]Expression (7.43) omits the effect of the $-3\hbar\omega/2$ shift included in the one-body Hamiltonian. Taking it into account would lead to a density of states $\tilde{d}(E) = (E + 3\hbar\omega/2)^2/2(\hbar\omega)^3$. It has the same leading order contribution as Equation (7.43) which does not need to be corrected in the following. $\tilde{d}(E)$ also describes the first correction to Equation (7.43): $\tilde{d}(E) = \frac{1}{2}\frac{E^2}{(\hbar\omega)^3} + \frac{3}{2}\frac{E}{(\hbar\omega)^2} + \cdots$ [see Equation (7.62)], but not the next order corrections, as shown in Exercise 7.8.

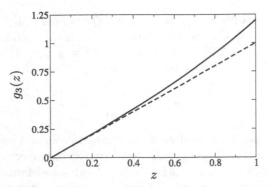

FIGURE 7.13 Solid curve: $g_3(z)$; dashed curve: approximation of $g_3(z)$ using the first term of Equation (A.7): $g_3(z) \simeq z$ which corresponds to a Taylor expansion around the origin. g_3 is strictly increasing on the interval $[0,1]$ and its maximum value is $g_3(1) = \zeta(3) = 1.202\ldots$, see (A.8) and table A.1.

as discussed above, is the maximum possible value of the chemical potential. The value $\mu = 0$ is reached when the temperature T is equal to the Bose–Einstein temperature T_{BE} defined as:

$$k_{\mathrm{B}}T_{\mathrm{BE}} = \hbar\omega \left[\frac{N}{\zeta(3)}\right]^{1/3} , \quad \text{where} \quad \zeta(3) = g_3(1) = 1.202\cdots \qquad (7.49)$$

As temperature decreases and approaches T_{BE}, the chemical potential tends to 0 and the occupation of the ground state starts to increase. Ultimately, the use of a continuous approximation for the discrete summation in Equation (7.44) is no longer justified and a more sophisticated mathematical approach is needed to deal with this issue. Alternatively, the following two *ad hoc* assumptions can be made. When $T \leq T_{\mathrm{BE}}$

(i) $\mu(T, N)$ remains fixed at 0.

(ii) The average number of particles in the ground state $\langle n_0 \rangle$ can no longer be evaluated using the usual expression (6.82) which is now diverging. One writes instead:

$$N = N_0(T) + N_{\mathrm{th}}(T) \quad \text{with} \quad N_{\mathrm{th}}(T) = \sum_{\underline{\nu} \neq (0,0,0)} \frac{1}{e^{\beta \varepsilon_{\underline{\nu}}} - 1}, \qquad (7.50)$$

where $N_0(T)$ is the number of particles accumulated in the ground state (the so-called *condensate*) and $N_{\mathrm{th}}(T)$ is the number of particles distributed over the excited states (the non-condensed fraction or, equivalently, the "thermal cloud").

Assumption (i) fixes the temperature dependence of the chemical potential below T_{BE}. Figure 7.14 shows the behaviour of μ in the whole range of temperature.

The contributions to the sum allowing the calculation of $N'(T)$ in Equation (7.50) are regular and can therefore be evaluated by substituting the sum for an integral:

$$N_{\mathrm{th}}(T) = \int_{\varepsilon_1}^{\infty} \mathrm{d}E\, \frac{d(E)}{e^{\beta E} - 1} \simeq \int_0^{\infty} \mathrm{d}E\, \frac{d(E)}{e^{\beta E} - 1} . \qquad (7.51)$$

In this expression, ε_1 is the Hamiltonian's first excited level, of energy $\hbar\omega$. The right-hand expression of Equation (7.51) trivially gives:

$$N_{\mathrm{th}}(T) = \left(\frac{k_{\mathrm{B}}T}{\hbar\omega}\right)^3 g_3(1) = N(T/T_{\mathrm{BE}})^3 . \qquad (7.52)$$

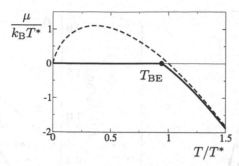

FIGURE 7.14 Chemical potential plotted as a function of the normalised temperature T/T^* for a non-interacting Bose gas trapped in a harmonic potential. The solid curve is the solution of Equation (7.45) when $T > T_{BE}$. The value of μ is fixed at zero when $T < T_{BE}$, in agreement with assumption (i). The dashed curve is the classical value of μ_{class} given by Equation (7.48). Here, the normalised Bose–Einstein temperature is $T_{BE}/T^* \simeq 0.9405$.

Note that it is an approximate expression since the lower integration bound ε_1 of the integral was replaced by 0. An upper estimate of the induced error is[34]: $\int_0^{\varepsilon_1} dE\, d(E)/(\beta E) = \frac{1}{4}[N/\zeta(3)]^{1/3}(T/T_{BE})$. This is typically a small correction when compared to $N'(T)$ which is of order $O(N)$. Nonetheless, in the vicinity of T_{BE}, $N_{th}(T)$ becomes small and the approximation is no longer valid. A simple calculation shows that the approximation is invalid when $N^{-2/3} \gg (T - T_{BE})/T_{BE}$, which corresponds to a tiny temperature interval in the vicinity of T_{BE}.

Now that $N_{th}(T)$ is known, Assumption (ii) can be used to calculate the number $N_0(T)$ of ground state. According to Equation (7.50), $N_0(T) = N - N_{th}(T)$ and the fraction of condensed atoms is (when $T < T_{BE}$)

$$\frac{N_0(T)}{N} = 1 - \left(\frac{T}{T_{BE}}\right)^3 . \tag{7.53}$$

This behaviour is indeed observed in trapped ultra-cold atomic vapours. Typical experimental values are $N = 10^6$ and $\omega = 2\pi \times 100$ Hz, giving $T_{BE} = 450$ nK. This is an extremely low temperature, but it can be reached with modern cooling techniques. Below T_{BE}, atoms accumulate in the ground state according to Equation (7.53), as it has been reported in many experiments. The results of Cornell and Wieman group at JILA are reproduced in Figure 7.15.

In a manner similar to the one that lead to Equation (7.53), it is also possible to determine the system's energy [Equation (6.83)]:

$$E = \sum_\nu \varepsilon_\nu \langle n(\varepsilon_\nu)\rangle = \int_0^\infty d\epsilon\, \frac{\epsilon\, d(\epsilon)}{\exp(\beta(\epsilon - \mu)) - 1} = 3\frac{(k_BT)^4}{(\hbar\omega)^3} g_4(e^{\beta\mu}) . \tag{7.54}$$

At high temperature, it is legitimate to use Maxwell-Boltzmann approximation and the energy is:

$$E = e^{\beta\mu} \int_0^\infty d\epsilon\, \epsilon\, d(\epsilon) e^{-\beta\epsilon} = 3N k_BT . \tag{7.55}$$

In Expression (7.55) of the energy, $\exp(\beta\mu)$ has been calculated by solving Equation (7.45) in the high-temperature limit where $g_3(\exp(\beta\mu)) \simeq \exp(\beta\mu)$. Expression (7.55) is identical to the energy of N non-interacting classical particles trapped in a harmonic well.

---

[34]For all $x \in \mathbb{R}$, $\exp(x) > x + 1$. It allows to calculate an upper limit of the integrand by using βE as a lower limit of the denominator.

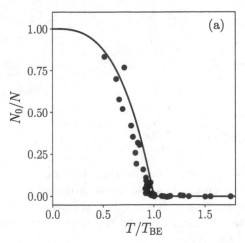
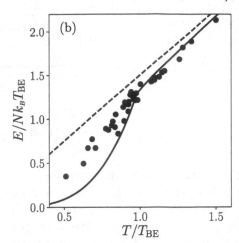

FIGURE 7.15 (a) $N_0(T)/N$ as a function of T/T_{BE} where T_{BE} is given by Equation (7.49). The plain curve is a plot of Equation (7.53), the points are the experimental results of the JILA group for 40000 ^{87}Rb atoms [Phys. Rev. Lett. **77**, 4984 (1996)]. (b) Energy measured after the opening of the trap as a function of T/T_{BE}. The points are the experimental results. The straight dashed line is the result for an ideal classical gas. Note that the potential energy disappears when the trap is opened: only half of the contribution to Equation (7.55) remains (see the discussion in Exercise 14.5 page 408). The solid line corresponds to Formula (7.58) corrected for finite size effects (see Equation (7.67), still with a factor 1/2). A detailed analysis of the experimental data indicates a transition at a temperature $T = 0.94\, T_{BE}$. A cause of this decrease with respect to Equation (7.49) is discussed in Exercise 7.8.

Exercise 7.6. Calculate the next correction term to Equation (7.55) at high temperature.

Solution: The denominator of the integrand of Equation (7.54) can be expended using the formula $(\exp(x) - 1)^{-1} \simeq \exp(-x) + \exp(-2x)$. _This yields_

$$E \simeq e^{\beta\mu} \int_0^\infty d\epsilon\, \epsilon\, d(\epsilon)\, e^{-\beta\epsilon} + e^{2\beta\mu} \int_0^\infty d\epsilon\, \epsilon\, d(\epsilon)\, e^{-2\beta\epsilon}\ , \qquad (7.56)$$

where $z = \exp(\beta\mu)$ is determined by solving Equation (7.45) at the second order z ($z \ll 1$), yielding: $\exp(\beta\mu) = N(\hbar\omega/k_BT)^3 - \frac{1}{8}N^2(\hbar\omega/k_BT)^6$. _Carrying over into Equation (7.56) and keeping only the terms up order N^2 gives:_

$$E = 3Nk_BT\left[1 - \frac{1}{16}N\left(\frac{\hbar\omega}{k_BT}\right)^3\right] = 3Nk_BT\left[1 - \frac{\zeta(3)}{16}\left(\frac{T_{BE}}{T}\right)^3\right]. \qquad (7.57)$$

Below T_{BE}, the chemical potential is zero and from the relation[35] $g_4(1) = \zeta(4)$, one gets

$$\frac{E}{N k_B T_{BE}} = \frac{3\,\zeta(4)}{\zeta(3)}\left(\frac{T}{T_{BE}}\right)^4. \qquad (7.58)$$

These results are compared with experimental data shown in Figure 7.15(b). The good agreement between the predicted values of $N_0(T)$ and $E(T)$ and the experimental results confirm the analysis presented in this section: when $T < T_{BE}$ a macroscopic number of

---

[35]see Equation (A.8).

particles occupy the ground state of the harmonic trap. This phenomenon is called *Bose–Einstein condensation*. It was first observed by the group of Cornell and Wieman at JILA and the group of Ketterle at MIT[36]. This phenomenon is intimately linked to the Bose–Einstein statistics, as shown by the role of the Bose distribution function in the previous calculations. Pauli exclusion principle prevents fermions to behave in a similar manner. The reasoning can be pushed further by considering the case of fictitious identical particles, which would be neither bosons nor fermions. The particles would instead obey the approximate Maxwell-Boltzmann statistics (see Section 6.4.3), with an occupation number given by Equation (6.87). It can be easily shown that these particles would also undergo a form of condensation on the lowest level, but this would only occur when the thermal motion energy k_BT is lower than energy $\varepsilon_1 = \hbar\omega$ of the first excited level of the well (i.e. at a very low temperature of the order of $T_{BE}\, N^{-1/3}$). Hence, it can be considered that, while Fermi statistics opposes condensation (this is Pauli exclusion principle), Bose statistics amplifies the phenomenon instead.

The $N_0(T)$ atoms in the fundamental level of the trap form what is called a *condensate*. Bose–Einstein condensation is a phase transition whose order parameter (see the definition of this concept in Section 9.2.2) is the number of atoms in the condensate[37]. As already discussed, $N = 10^4$ to 10^5 atoms are typically confined in a harmonic trap with an eigenfrequency of the order of a few hundreds of Hertz, leading to a condensation temperature of the order of $T_{BE} \approx 100$ nK.

Exercise 7.7. Give an approximate order of magnitude of the condensation temperature T_{BE} [Equation (7.49)] by identifying it with the temperature (known as the quantum degeneracy temperature) below which the thermal wavelength λ_T [Equation (3.48)] is larger than the interparticle distance.

Solution: A classical ideal gas confined in a harmonic trap occupies a sphere whose radius R can be estimated using the equipartition theorem [Equation 4.1.5]: $\frac{1}{2}m\omega^2 R^2 = \frac{3}{2}k_B T$. The typical interparticle distance is then $d \simeq R N^{-1/3}$ and the condition $\lambda_T > d$ can be approximately written as:

$$k_B T \lesssim \sqrt{\frac{2\pi}{3}}\, T^* , \tag{7.59}$$

where T^* is defined in Equation (7.48). It has been seen in Section 6.4.3 that the opposite regime ($\lambda_T \ll d$) corresponds instead to the limit where a classical description is valid. Therefore, Bose–Einstein condensation, which is an effect of Bose quantum statistics, naturally occurs in the so-called quantum degeneracy regime $\lambda_T > d$, and Estimate (7.59) yields a qualitative agreement with the exact Expression (7.49). It will be seen in the next chapter that, for fermions, it is also in the regime $\lambda_T > d$ that quantum statistical effects are important.

Exercise 7.8. Calculate the correction to the value of the condensation temperature T_{BE} due to the finite number of particles [Ketterle and van Druten, Phys. Rev. A **54**, 656 (1996)].

Solution: Above the condensation temperature, Equation (7.44) can be written as

$$N = \sum_{\underline{\nu}} \frac{1}{\exp\bigl(\beta(\varepsilon_{\underline{\nu}} - \mu)\bigr) - 1} = \sum_{n=0}^{\infty} \frac{d_n}{\exp(\beta(\varepsilon_n - \mu)) - 1} , \tag{7.60}$$

where d_n is the degeneracy of the energy level $\varepsilon_n = n\hbar\omega$. Thus, instead of summing over the states identified by the three indices $(\nu_1, \nu_2, \nu_3) = \underline{\nu}$, the summation is performed, in the right-hand term

---

[36]M. H. Anderson, J. R. Ensher, M. R. Matthews, C. E. Wieman, and E. A. Cornell, Science **269**, 198 (1995); Davis, K. B., M.-O. Mewes, M. R. Andrews, N. J. van Druten, D. S. Durfee, D. M. Kurn, and W. Ketterle, Phys. Rev. Lett. **75**, 3969 (1995).

[37]A finer description uses a complex position-dependent order parameter, see Chapter 14.

of Equation (7.60), over the eigenenergies (written as $n\hbar\omega$) taking into account the degeneracy d_n of the energy levels $n\hbar\omega$. d_n is equal to the number of different ways to write an integer number as the sum of three other integers: $n = \nu_1 + \nu_2 + \nu_3$. This calculation was carried out – in a more general way – for the first time by Planck (see Exercise 2.4): a possible combination can be schematised in a diagram where ν_1 balls are shown, then a bar, then ν_2 balls, then a bar, and then ν_3 balls. Altogether, there are n balls and two bars. Permutations of the balls or the bars do not count. Only the different ways of distributing the n balls, or equivalently, of placing the 2 bars on one of the $n + 2$ sites matter[38]. Hence the degeneracy is: $d_n = \binom{n+2}{2} = (n + 1)(n + 2)/2 \simeq \frac{1}{2}n^2 + \frac{3}{2}n$.

By multiplying the right-hand side term of Equation (7.60) by $\hbar\omega$ (which is the spacing between the two adjacent energy levels), a Riemann sum is obtained. In the formal limit $\hbar\omega \to 0$, it can be identified to an integral:

$$\hbar\omega \sum_{n=0}^{\infty} f(\hbar\omega n) \simeq \int_0^{\infty} f(\epsilon)\,d\epsilon \ . \tag{7.61}$$

This gives here:

$$N = \int_0^{\infty} \left[\frac{1}{2} \frac{\epsilon^2}{(\hbar\omega)^3} + \frac{3}{2} \frac{\epsilon}{(\hbar\omega)^2} \right] \frac{d\epsilon}{\exp(\beta(\epsilon - \mu)) - 1} \ . \tag{7.62}$$

Comparing with Equation (7.44), it can be seen that the above calculation corresponds to the next correction to the dominant term of the density of state [Equation (7.43)]. This expression can be rewritten in terms of Bose functions (A.6)

$$N = \left(\frac{k_B T}{\hbar\omega} \right)^3 g_3(e^{\beta\mu}) + \frac{3}{2} \left(\frac{k_B T}{\hbar\omega} \right)^2 g_2(e^{\beta\mu}) \ . \tag{7.63}$$

Condensation occurs when the chemical potential cancels, i.e. when $\exp(\beta\mu) = 1$. The corresponding transition temperature (written T_{BE}^*) is then a solution of:

$$N = \left(\frac{k_B T_{BE}^*}{\hbar\omega} \right)^3 g_3(1) + \frac{3}{2} \left(\frac{k_B T_{BE}^*}{\hbar\omega} \right)^2 g_2(1) \ . \tag{7.64}$$

where $g_2(1) = \zeta(2)$ and $g_3(1) = \zeta(3)$ (see Equation (A.8) and Table A.1). In this equation, the second term of the right-hand side is a small correction to the main term. If it is ignored, Equation (7.49) is recovered. If it is taken into account perturbatively, the following expression is obtained:

$$\frac{T_{BE}^*}{T_{BE}} = 1 - \frac{1}{2} \frac{\zeta(2)}{\zeta(3)^{2/3} N^{1/3}} = 1 - 0.7275\, N^{-1/3} \ . \tag{7.65}$$

In a trap containing 40000 atoms (corresponding to the number of atoms trapped at the transition in Figure 7.15), Formula (7.65) leads to a decrease of 3% of the condensation temperature compared to Formula (7.49). A larger decrease is measured experimentally (of the order of 6%, see figure 7.15) which is explained by trap anisotropies and by interaction effects between particles which are not taken into account in our model.

Exercise 7.9. Calculate the energy of the system of N bosons trapped in a harmonic well for values of temperature above and below the condensation temperature including the correction to the leading order given by Equation (7.58) [S. Grossmann and M. Holthaus, Phys. Lett. A **208**, 188 (1995)].

Solution: Using the expression of the energy level density obtained in the previous exercise $d(\epsilon) = \frac{1}{2}\epsilon^2/(\hbar\omega)^3 + \frac{3}{2}\epsilon/(\hbar\omega)^2$, it immediately comes:

$$E = \int_0^{\infty} \frac{\epsilon\, d(\epsilon)}{z^{-1}\exp(\beta\epsilon) - 1}\, d\epsilon = 3 \frac{(k_B T)^4}{(\hbar\omega)^3} g_4(z) + 3 \frac{(k_B T)^3}{(\hbar\omega)^2} g_3(z) \ , \tag{7.66}$$

---

[38]It is even possible to place the vertical bars at the beginning or end of the diagram. For example, if the bars are in the first two places, it is equivalent to writing $n = 0 + 0 + n$. If they are one at the beginning and the other at the end, it is equivalent to writing $n = 0 + n + 0$. These two configurations must be accounted for in the calculation of the degeneracy.

where $z = \exp(\beta\mu)$ is solution of Equation (7.63). A fugacity $z = 1$ is obtained below the transition temperature T_{BE}^*. This temperature is not exactly the temperature T_{BE} given by Equation (7.49), as seen in the previous exercise. The above expression can be rewritten as:

$$\frac{E}{Nk_BT_{BE}} = 3\left(\frac{T}{T_{BE}}\right)^4 \frac{g_4(z)}{\zeta(3)} + \frac{3}{N^{1/3}}\left(\frac{T}{T_{BE}}\right)^3 \frac{g_3(z)}{[\zeta(3)]^{2/3}} \,. \tag{7.67}$$

Exercise 7.10. Discuss the occurrence of Bose–Einstein condensation of an ideal gas of N bosons uniformly distributed in a vessel of volume V in the absence of external field.

Solution: For an ideal system without interactions, the density of state $d(E)$ is given by Equation (8.2). Hence, the number of particle is:

$$N = \int_0^\infty d(E)\,n_B(E)\,dE = \frac{2\pi V}{h^3}(2m)^{3/2}\int_0^\infty \frac{\sqrt{E}\,dE}{\exp(\beta(E-\mu))-1} = \frac{V}{\lambda_T^3}g_{3/2}(z)\,, \tag{7.68}$$

where $z = \exp(\beta\mu)$, $\lambda_T = h(2\pi mk_BT)^{-1/2}$ is the de Broglie thermal wavelength (3.48). Definition (A.6) of the Bose functions $g_\nu(z)$ and the relation $\Gamma(3/2) = \frac{1}{2}\sqrt{\pi}$ were used [see Equation (A.5)]. This expression allows the calculation of the chemical potential μ of a gas of density $n = N/V$ at temperature T. At high temperature [in the Maxwell-Boltzmann limit (6.90)], $g_{3/2}(z) \simeq z$ [see Equation (A.7)] and the classical mechanics result for the chemical potential of an ideal gas $(\mu_{IG} = k_BT\ln(n\lambda_T^3) < 0)$ is recovered. As temperature decreases, μ increases and cancels for the temperature[39]:

$$T_{BE} = \frac{h^2}{2\pi mk_B}\left[\frac{N}{\zeta(3/2)V}\right]^{2/3}. \tag{7.69}$$

As discussed earlier, below the condensation temperature T_{BE} the chemical potential remains zero and the number of atoms N_0 occupying the ground state becomes macroscopic. The densities of the thermal fraction and of the condensed fraction are then, respectively:

$$n_{th}(T) = \frac{N_{th}}{V} = \frac{\zeta(3/2)}{\lambda_T^3} = n\left(\frac{T}{T_{BE}}\right)^{3/2}\,, \quad n_0(T) = \frac{N_0}{V} = n\left[1 - \left(\frac{T}{T_{BE}}\right)^{3/2}\right], \tag{7.70}$$

where $n = N/V$ is the total gas number density. The pressure of the system can be calculated from the grand potential $J = -PV = -k_BT\ln\Xi$, see Equations (4.58) and (6.39). Using Expression (6.80) of the grand potential of a system of non-interacting bosons leads to:

$$P = -\frac{k_BT}{V}\int_0^E dE\, d(E)\ln\left(1-\exp[-\beta(E-\mu)]\right) = -\frac{2\,k_BT}{\sqrt{\pi}}\frac{1}{\lambda_T^3}\int_0^\infty dx\sqrt{x}\ln\left(1-ze^x\right)\,, \tag{7.71}$$

where $z = \exp(\beta\mu)$. An integration by parts gives:

$$P = \frac{k_BT}{\lambda_T^3}g_{5/2}(z)\,. \tag{7.72}$$

At high temperature z tends to 0, $g_{5/2}(z) \simeq z \simeq g_{3/2}(z)$ and, combining Equations (7.68) and (7.72), one recovers, as expected, the equation of state of a classical ideal gas: $P = nk_BT$. At temperatures below T_{BE}, $z = 1$ and Equation (7.72) shows that pressure is independent of n. Thus, the system's isothermal compressibility [Equation (4.61)] becomes infinite: the condensate constitutes a reservoir of particles and a density variation modifies the equilibrium between the condensate and the thermal cloud without modifying the pressure.

---

[39] Use is made of Relation (A.8): $g_{3/2}(1) = \zeta(3/2)$. The numerical value of $\zeta(3/2)$ is given in Table A.1.

8 Fermions

The concepts of statistical physics have been used to study fermionic systems rapidly after the birth of the "new" quamtum mechanics in 1925. This proved to be extremely fruitful and led to major advances in atomic and solid state physics. Fermi-Dirac statistics was introduced in 1926 by Enrico Fermi and Paul Dirac. It was used by Ralph Fowler in 1926 to describe the collapsed state of a star (a so-called *white dwarf*) and by Arnold Sommerfeld in 1928 to describe conduction electrons in solids. This chapter first presents some general results and then discusses a choice of specific systems.

8.1 FREE FERMION GAS

Consider a gas of free fermions (i.e. without mutual interaction) contained in a box of volume V (for example, an ideal quantum gas modelling the conduction electrons in a metal). The corresponding one-particle Hamiltonian is a free Hamiltonian $\vec{p}^2/(2m)$ with Dirichlet boundary conditions, i.e. the wave function cancels at the edges of the box (e.g. because a potential barrier prevents electrons from leaving the solid). The corresponding density of states $d(E)$ can be calculated as explained in Section 6.1.1 [Formulas (6.5), (6.6) and (6.8)]. The number $\Phi(E)$ of eigenstates of the Hamiltonian with an energy less than E is:

$$\Phi(E) = \frac{1}{h^3} \int_{H(\vec{r},\vec{p})<E} \mathrm{d}^3 r\, \mathrm{d}^3 p = \frac{V}{h^3} \int_{\frac{p^2}{2m}<E} \mathrm{d}^3 p = \frac{4\pi V}{3h^3}(2mE)^{3/2}\Theta(E), \qquad (8.1)$$

where $\Theta(E)$ is the Heaviside function[1]. The presence of $\Theta(E)$ is consistent with the definition of Φ and Formula (6.8) appearing in the left-hand side term of Equation (8.1): it ensures that the Hamiltonian $\vec{p}^2/(2m)$ has no negative energy state. However, in order to simplify notations, the presence of $\Theta(E)$ will be implicit and it will not be explicitly written in the following formulae. The density of states $d(E)$ is then:

$$d(E) = \frac{\mathrm{d}\Phi}{\mathrm{d}E} = \frac{2\pi V}{h^3}(2m)^{3/2}\sqrt{E}. \qquad (8.2)$$

The grand canonical approach described in Chapter 6 enables to relate the chemical potential μ to the system's average number of particles $\langle N \rangle$ [Formula (6.83)]. In the thermodynamic limit, $\langle N \rangle$ is identified with the exact number of fermions, N, and the distribution of energy level is continuous, as already assumed in Equation(8.2). Then, Formula (6.83) yields:

$$N = d_s \int_0^\infty d(E)\, n_F(E)\, \mathrm{d}E, \qquad (8.3)$$

where $d_s = 2s+1$ is the spin degeneracy ($d_s = 1$ for spinless fermions, or $d_s = 2$ for spin-1/2) and $n_F(E)$ is the Fermi distribution function [Formula (6.82)]: $n_F(E) = (\exp[\beta(E - \mu)] + 1)^{-1}$. The integrand in Equation (8.3) consists in a power law function multiplying a Fermi distribution function. Similar integrands will often be encountered in the following. The corresponding integrals are called *Fermi-Dirac integrals*. They are defined as follows (see Appendix A.1)

$$f_\nu(z) \equiv \frac{1}{\Gamma(\nu)} \int_0^\infty \frac{x^{\nu-1}\, \mathrm{d}x}{z^{-1}\exp(x) + 1}, \qquad (8.4)$$

---

[1] $\Theta(x > 0) = 1$ and $\Theta(x < 0) = 0$.

DOI: 10.1201/9781003272427-8

where $\Gamma(\nu)$ is defined by Equation (A.2). Relation (8.3) can thus be written as:

$$N = \frac{2\pi V}{h^3}(2m)^{3/2}d_s \int_0^\infty \frac{E^{1/2}\,dE}{e^{\beta(E-\mu)}+1} = \frac{d_s V}{\lambda_T^3}\,f_{3/2}(e^{\beta\mu})\,, \tag{8.5}$$

where

$$\lambda_T = \frac{h}{(2\pi m k_B T)^{1/2}}\,, \tag{8.6}$$

is the De Broglie thermal wavelength [Equation (3.48)]. The argument of the Fermi-Dirac function in Equation (8.5) is, as already encountered in Chapter 7, the *fugacity* $\exp(\beta\mu)$ often noted z (hence the notation z in Equation (A.9) for the generic argument of f_ν).

Expression (6.83) of the system's energy also involves a Fermi-Dirac integral:

$$E = d_s \int_0^\infty E\,d(E)\,n_F(E)\,dE = \frac{2\pi V}{h^3}(2m)^{3/2}d_s \int_0^\infty \frac{E^{3/2}\,dE}{e^{\beta(E-\mu)}+1}$$
$$= \frac{3}{2}k_B T\,\frac{d_s V}{\lambda_T^3}\,f_{5/2}(e^{\beta\mu})\,. \tag{8.7}$$

The equation of state can be obtained thanks to the formula $-k_B T \ln \Xi = J = -PV$, (see Equation (4.58)). According to Expression (6.80), it yields:

$$\frac{PV}{k_B T} = d_s \int_0^\infty d\epsilon\,d(\epsilon)\ln\left(1 + e^{-\beta(\epsilon-\mu)}\right) = \frac{d_s V}{\lambda_T^3}\,f_{5/2}(e^{\beta\mu})\,, \tag{8.8}$$

where an integration by parts was performed. Comparing with Equation (8.7), one obtains

$$E = \frac{3}{2}PV\,. \tag{8.9}$$

Expression (8.7) has been derived within the grand canonical formalism and thus naturally connects the energy to the temperature and the chemical potential. In order to confront with physical intuition, it is more convenient to express E as a function of temperature and particle number. This can be achieved by inverting Relation (8.5) and reporting the result in Equation (8.7). In order to achieve this, one must numerically calculate Integrals (8.5) and (8.7). However, there are two particularly interesting limit cases for which analytic results can be obtained: the classical and the degenerate regimes, both studied in the next subsections.

8.1.1 CLASSICAL LIMIT: HIGH TEMPERATURE AND LOW DENSITY

When $|z| \leq 1$, it is appropriate to use the series expansion of the Fermi-Dirac integrals [Equation (A.10)] to evaluate the above integrals. In particular, when $z \to 0$, only the first term of the expansion matters and Equation (8.5) can be approximated by:

$$\exp(\beta\mu) = \frac{N\lambda_T^3}{d_s V}\,. \tag{8.10}$$

This result, reported in Equation (8.7), leads – still in the limit $z \to 0$ – to the classical ideal gas energy [Expression (4.35)]:

$$E = \frac{3}{2}N k_B T\,. \tag{8.11}$$

The approximation leading to this result is valid provided $z \ll 1$. In this regime $f_{3/2}(z)$ is also small [see Equation (A.10)] and therefore, according to Equation (8.5) (and taking into account that $d_s^{1/3} \sim 1$):

$$\left(\frac{V}{N}\right)^{1/3} \gg \lambda_T . \tag{8.12}$$

Conversely, in the Limit (8.12), $f_{3/2}(z)$ must be small, which is only possible when $z \to 0$. Inequality (8.12) imposes a typical inter-particle distance be large with respect to the thermal wavelength λ_T. This condition was already encountered [see Equation (6.90)]: it is the condition of applicability of the Maxwell-Boltzmann approximation. It is verified in a regime of high temperature and/or low density. In this regime, the quantum statistics is irrelevant and the fermion gas behaves classically: indeed the combination of Equation (8.11) and (8.9) leads to the classical ideal gas equation of state.

8.1.2 DEGENERATE LIMIT: LOW TEMPERATURE AND HIGH DENSITY

The degenerate limit corresponds to the regime in which the occupancy of the phase space cells is of the order of the maximum capacity allowed by the Heisenberg uncertainty principle. In this limit, quantum statistics play a prominent role. This is the limit opposite to Limit (8.12) as can be confirmed by calculating orders of magnitude based on an approximate form of the uncertainty relation: the degenerate limit is reached when $\Delta x \cdot \Delta p \sim h$. The order of magnitude of position uncertainty Δx is the inter-particle spacing $(V/N)^{1/3}$. Δp can be roughly evaluated as caused by thermal motion[2]: $\Delta p \simeq \sqrt{2mk_BT} = h/\lambda_T$. This order-of-magnitude calculation confirms that the degenerate regime is reached when $(V/N)^{1/3} \simeq \lambda_T$.

In the classical limit, the first term of Series (A.10) could be obtained by approximating $1/(z^{-1}e^x + 1)$ by $z\,e^{-x}$, which, in integrals of the type of Equation (8.3), is equivalent to the approximation $n_F(\epsilon) \simeq e^{-\beta(\epsilon-\mu)}$: this corresponds to the framework of the Maxwell-Boltzmann approximation (see Section 6.4.3) where the occupation of the levels is far from the maximum value $n_F(\epsilon) = 1$ allowed by the Pauli exclusion principle. For a degenerate gas, the opposite situation occurs. The corresponding behaviour of $n_F(\epsilon)$ is shown in Figure 8.1: the occupation number plotted as a function of energy is constant and equal to unity until it shows a steep decrease to zero. In the limit $T \to 0$, $n_F(\epsilon) \to \Theta(\mu - \epsilon)$ and Equation (8.3) can then be written as:

$$N = d_s \int_0^{\mu(T=0)} d\epsilon\, d(\epsilon) , \tag{8.13}$$

which imposes $\mu(T = 0) > 0$. It remains so at low temperature, as shown in Figure 8.1. Formula (8.13) means that at zero temperature, all states are occupied up to the so-called *Fermi energy* $E_F = \mu(T = 0)$: the system minimises its energy while fulfilling the constraint of the exclusion principle. This is an important characteristic of the ideal Fermi gas[3]. The ensemble of occupied states at $T = 0$ forms the so-called *Fermi sea*. Formula (8.13) is valid for all non-interacting Fermi systems subjected to an external field. In the three-dimensional case of a free Fermi gas contained in an box, for which the density of states is given by Equation (8.2), the Fermi energy is:

$$E_F = \frac{\hbar^2}{2m}\left(\frac{6\pi^2 N}{d_s V}\right)^{2/3} , \tag{8.14}$$

---

[2] The reasoning is done in the classical limit for which $p^2/(2m) = \frac{3}{2}k_BT$. Once at the degenerate limit, Δp is no longer evaluated this way, but through the saturation of the quantum occupancy.

[3] In the degenerate limit, the ideal Bose gas has a totally different behaviour, also due quantum statistical effects (see Section 7.4).

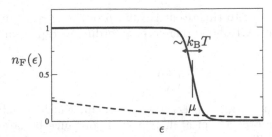

FIGURE 8.1 Fermi-Dirac distribution $n_F(\epsilon)$ in the degenerate limit (continuous curve). The vertical line shows the value of the chemical potential. The dashed curve is $n_F(\epsilon)$ for $T \gg T_F$; in this case the large and negative value of μ cannot be shown on the plot.

which is an explicit form of the Fermi energy stemming from Definition (8.13). At $T = 0$ K all the states are occupied up to the Fermi energy, therefore the number of particle is $N = d_S \, \Phi(E_F)$.

The degeneracy condition $(V/N)^{1/3} \ll \lambda_T$ [reverse of Equation (8.12)] can also be put in the form $T \ll T_F$ where $T_F = E_F/k_B$ is called the *Fermi temperature*. It is interesting to evaluate the order of magnitude of the Fermi temperature in several systems:

- For the conduction electrons in a metal, the electron density N/V is typically of the order of $\sim 5 \times 10^{28}$ m^{-3}. This corresponds to a Fermi energy of the order of a few eV. For potassium for instance, at low temperature ($T \simeq 5$ K) $N/V \simeq 1.4 \times 10^{28}$ m^{-3}, the Fermi energy is $E_F \simeq 2.12$ eV corresponding to a Fermi temperature $T_F \simeq 25000$ K.

- Ultra-cold vapours of fermionic atoms can be trapped with typical densities of the order of $N/V \simeq 10^{21}$ m^{-3}. For example, for ^{6}Li atoms, Fermi temperatures on the order of $T_F \sim 4\ \mu$K are typically obtained.

- In a white dwarf, the celestial object resulting from the "death" of a star whose mass was initially of the order of the mass of the sun (see Section 8.3), the electron density is $N/V \simeq 3 \times 10^{35}$ m^{-3}, the white dwarf temperature is $T \simeq 10^7$ K while its Fermi temperature is $T_F \simeq 2 \times 10^9$ K.

The chemical potential at low, but non zero temperature can be determined by calculating approximation of the Fermi integrals [Equation (A.9)] for large values of z (a result first obtained by Sommerfeld in 1928). Let us consider a general framework and determine, for a function $g(\epsilon)$ regular in the vicinity of μ, an expansion of the following integral:

$$I = \int_0^\infty \frac{g(\epsilon)\,d\epsilon}{e^{\beta(\epsilon - \mu)} + 1} = \frac{1}{\beta} \int_0^\infty \frac{g(x/\beta)\,dx}{e^{x-\alpha} + 1} = \int_{-\alpha}^\infty G\left(\tfrac{y+\alpha}{\beta}\right) \frac{e^y}{(e^y + 1)^2}\,dy . \qquad (8.15)$$

In the last term of Equation (8.15), an integration by part was first performed where G is the antiderivative of g that cancels in $\epsilon = 0$, and then the change of variable $y = x - \alpha$ was used (with $\alpha = \beta\mu$ and $x = \beta\epsilon$). The function $e^y(e^y + 1)^{-2}$ appearing in the integrand of the last term is peaked around zero. If $\beta\mu$ is large and positive (which is the case in the degenerate limit $T \ll T_F$), the lower integration bound can be substituted for $-\infty$ (at the cost of exponentially small errors in $\beta\mu$). A Taylor expansion of $G(\tfrac{y+\alpha}{\beta})$ around 0 can then be done. Only the even terms contribute to the integral because $e^y(e^y + 1)^{-2}$ is an even function and the integration over $\mathbb{R}$ eliminates the odd terms. Therefore, I is given by:

$$I = G(\alpha/\beta) + \frac{G''(\alpha/\beta)}{2\,\beta^2} \int_{\mathbb{R}} \frac{y^2\,e^y\,dy}{(e^y + 1)^2} + \frac{G^{(4)}(\alpha/\beta)}{24\,\beta^4} \int_{\mathbb{R}} \frac{y^4\,e^y\,dy}{(e^y + 1)^2} + \cdots \qquad (8.16)$$

In this expression, the parity of the integrands allows the restriction of the domain of integration to the interval $[0,\infty[$, then an integration by parts [inverse to the one used to obtain the last term of Equation (8.15)] leads to:

$$I = G(\alpha/\beta) + \frac{2g'(\alpha/\beta)}{\beta^2}\Gamma(2)f_2(1) + \frac{g^{(3)}(\alpha/\beta)}{3\beta^4}\Gamma(4)f_4(1) + \cdots \tag{8.17}$$

where the Γ function is defined by Equation (A.2). The values of the Fermi-Dirac integrals $f_2(1)$ and $f_4(1)$ are obtained using Equation (A.11) and Table A.1: $f_2(1) = \frac{1}{12}\pi^2$ and $f_4(1) = \frac{7}{720}\pi^4$. One finally obtains:

$$\int_0^\infty \frac{g(\epsilon)\,d\epsilon}{e^{\beta(\epsilon-\mu)}+1} = \int_0^\mu d\epsilon\, g(\epsilon) + \frac{\pi^2}{6}(k_BT)^2 g'(\mu) + \frac{7\pi^4}{360}(k_BT)^4 g^{(3)}(\mu) + \cdots \tag{8.18}$$

Taking $z = e^\alpha$, the generic relation (8.18) allows to write:

$$f_\nu(e^\alpha) = \frac{\alpha^\nu}{\Gamma(\nu+1)}\left[1 + \nu(\nu-1)\frac{\pi^2}{6}\frac{1}{\alpha^2} + \nu(\nu-1)(\nu-2)(\nu-3)\frac{7\pi^4}{360}\frac{1}{\alpha^4} + \cdots\right]. \tag{8.19}$$

If one wishes, for instance, to determine the chemical potential in the degenerate limit, Formula (8.5) and Expansion (8.19) lead to[4]:

$$1 = \left(\frac{\mu}{E_F}\right)^{3/2}\left[1 + \frac{\pi^2}{8}\left(\frac{k_BT}{\mu}\right)^2 + \frac{7\pi^4}{640}\left(\frac{k_BT}{\mu}\right)^4 + \cdots\right]. \tag{8.20}$$

This equation can be solved iteratively with increasing accuracy as a function of the small parameter T/T_F, as illustrated by a simple example in Exercise 8.1. As expected, the dominant term (the only one to survive at $T = 0$) is $\mu = E_F$. The following contributions read:

$$\mu(T,N) = E_F\left[1 - \frac{\pi^2}{12}\left(\frac{T}{T_F}\right)^2 - \frac{\pi^4}{80}\left(\frac{T}{T_F}\right)^4 + \cdots\right]. \tag{8.21}$$

This approximation is compared to the numerical solution of Equation (8.5) in Figure 8.2. It can be seen that it leads to satisfactory results up to $T \simeq 0.6\,T_F$.

The expression of the energy is obtained by combining Equation (8.21) with Equations (8.7) and (8.19), yielding:

$$E(T,N) = \frac{3}{5}NE_F\left[1 + \frac{5\pi^2}{12}\left(\frac{T}{T_F}\right)^2 - \frac{\pi^4}{16}\left(\frac{T}{T_F}\right)^4 + \cdots\right]. \tag{8.22}$$

Combining this result with Formula (8.9), it can be seen that at $T = 0$ the pressure of a non-interacting fermionic gas is not zero: $P(T = 0)$ is a function of density only. This is a direct consequence of Pauli exclusion principle and the existence of a Fermi sea.

At low temperature, the knowledge of the energy of the system enables to calculate the heat capacity $C_V = (\partial E/\partial T)_V$. The dominant contribution corresponds to the second term of Expansion (8.22) and leads to a molar heat capacity:

$$C_V = \gamma T \quad\text{with}\quad \gamma = N_A\, m\left(\frac{k_B\pi}{\hbar}\right)^2\left(\frac{d_S}{6\pi^2}\frac{V}{N}\right)^{2/3}. \tag{8.23}$$

This gives, for potassium, $\gamma = 1.67$ mJ·mol^{-1}·K^{-2}. Experimental data reported in Figure 8.3 indicate instead a value $\gamma_{exp} = 2.08 \pm 0.03$ mJ·mol^{-1}·K^{-2}. The difference between

---

[4]It is convenient to write $d_S V/\lambda_T^3 = \frac{3}{4}\sqrt{\pi}N(k_BT/E_F)^{3/2}$.

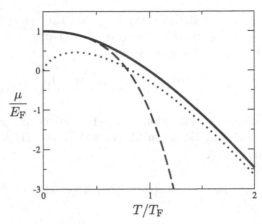

FIGURE 8.2 Chemical potential plotted as a function of temperature for a Fermi gas confined in a box of finite volume. The solid curve is the chemical potential of the system, solution of Equation (8.5). The dotted curve is the classical value of the chemical potential $\mu_{class} = k_B T \ln\left[\frac{4}{3\sqrt{\pi}}(T_F/T)^{3/2}\right]$ [Equation (8.10)]. The dashed curve is a plot of Approximation (8.21), valid in the degenerate regime $T \ll T_F$.

the two values is interpreted as an effective mass effect: everything happens as if electrons in potassium had an effective mass m^* different from their bare mass with, since $m \propto \gamma$ [see Equation (8.23)] $m^*/m = \gamma_{exp}/\gamma = 1.25$. The physical effects leading to this renormalised mass are of two kinds: (i) the electron's Hamiltonian is impacted by the atom lattice and, as a result, the density of states [Equation (8.2)] is modified, (ii) electrons can interact with each other or with the phonons of the atom lattice inducing a drag that results in an effective mass.

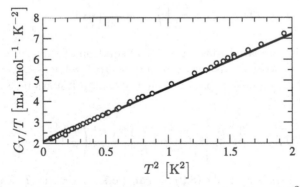

FIGURE 8.3 Heat capacity C_V/T of potassium plotted as a function of T^2 [adapted from W.H. Lien and N.E. Phillips, Phys. Rev. **133**, A1370 (1964)]. Experimental data are well described by the curve $C_V/T = 2.08 + 2.57T^2$. The linear part of the dependence of C_V on T corresponds to an effective electron mass $m^* = 1.25\,m$.

In Figure (8.3), it can be noticed that the heat capacity also has a contribution proportional to T^3. This term is not a correction induced by the term following the dominant contribution (8.23) in Expansion (8.22) in T/T_F (this term is quite negligible) but rather the phonon contribution at low temperature [Equation (7.15)]. According to experimental data, this contribution corresponds to a Debye temperature $T_D = 91.1$ K, in very good agreement with theoretical calculations[5].

---

[5] see W.H. Lien and N.E. Phillips, Phys. Rev. **133**, A1370 (1964).

Exercise 8.1. This exercise illustrates the iterative method used to determine $\mu(T)$ for a degenerate Fermi gas by solving Equation (8.20) perturbatively in successive orders of T/T_F. Consider the following third order polynomial equation:

$$x^3 + 3\,\epsilon\,x - 1 = 0 . \tag{8.24}$$

Determine its real solution perturbatively in successive orders in ϵ in the limit $|\epsilon| \ll 1$. Compare with the exact root (for $\epsilon > -4^{-1/3} = -0.63\ldots$):

$$x = \left(\frac{1}{2} + \sqrt{\frac{1}{4} + \epsilon^3}\right)^{1/3} + \left(\frac{1}{2} - \sqrt{\frac{1}{4} + \epsilon^3}\right)^{1/3} = 1 - \epsilon + \frac{\epsilon^3}{3} + \frac{\epsilon^4}{3} - \frac{4\,\epsilon^6}{9} + \cdots$$

Solution: When $\epsilon = 0$, the real solution is $x = 1$. For $\epsilon \neq 0$, the solution has a regular perturbation expansion of the form $x = 1 + a\,\epsilon + b\,\epsilon^2 + c\,\epsilon^3 + \cdots$. Carrying over in Equation (8.24) and cancelling all successive orders in ϵ, one gets $a = -1$, then $b = 0$, then $c = \frac{1}{3}$, etc.

8.2 RELATIVISTIC FERMI GAS

Consider a relativistic gas of free spin-1/2 fermions (spin degeneracy term $d_S = 2$). The energy ε is related to the momentum $\vec{p}$ by the relation[6]:

$$\varepsilon^2 = p^2 c^2 + m^2 c^4 . \tag{8.25}$$

The completely degenerate limit ($T = 0$ K) is here the main focus.

Since Relation (8.25) between energy and momentum is more complicated than its classical counterpart, the expression of the density of state as a function of energy is not as simple as Formula (8.2). It is easier to carry out calculations using the density of state in momentum space $d(p) = d(\varepsilon)\mathrm{d}\varepsilon/\mathrm{d}p$. One gets:

$$d(p) = \frac{\mathrm{d}}{\mathrm{d}p}\left(\frac{1}{h^3}\int_{p'<p} \mathrm{d}^3 r\,\mathrm{d}^3 p'\right) = \frac{\mathrm{d}}{\mathrm{d}p}\left(\frac{V}{h^3}\frac{4\pi p^3}{3}\right) = \frac{4\pi V}{h^3} p^2 . \tag{8.26}$$

The Fermi momentum p_F is defined by $N = 2\int_0^{p_F} d(p)\mathrm{d}p$ leading to:

$$p_F = h\left(\frac{3\,N}{8\,\pi\,V}\right)^{1/3} . \tag{8.27}$$

As this expression was derived by assuming no specific relation between energy and momentum, it is valid in both the relativistic and non-relativistic limits. The Fermi energy is here $E_F = \sqrt{(p_F c)^2 + m^2 c^4}$. In the non-relativistic limit where $p_F \ll mc$, Expression (8.14) is recovered with the addition of the mass term mc^2, and in the ultra-relativistic limit, the Fermi energy is $E_F = p_F\,c$. The total energy of the gas is:

$$E = \int_0^{p_F} \mathrm{d}p\,d(p)\sqrt{p^2 c^2 + m^2 c^4} = \frac{8\pi}{h^3} V c (mc)^4 F(X_F) = 3\,N\,mc^2\,\frac{F(X_F)}{X_F^3} , \tag{8.28}$$

where $X_F = p_F/(mc)$ and

$$
F(X) = \int_0^X \mathrm{d}x\,x^2 \sqrt{x^2 + 1} = \frac{1}{8}\left\{X(2X^2 + 1)\sqrt{1 + X^2} - \ln\left(X + \sqrt{1 + X^2}\right)\right\}
$$

$$
= \begin{cases} \dfrac{X^3}{3} + \dfrac{X^5}{10} + \ldots & \text{when } X \ll 1, \\[2mm] \dfrac{X^4}{4} + \dfrac{X^2}{4} + \ldots & \text{when } X \gg 1. \end{cases} \tag{8.29}
$$

---

[6]Remember that in special relativity the energy of a free particle is $\varepsilon = m\gamma c^2$ and its momentum $\vec{p} = m\gamma\vec{v}$, where $\gamma = (1 - \vec{v}^2/c^2)^{-1/2}$ is the Lorentz factor. The combination of these two relations gives (8.25).

Thus, one gets:

$$\frac{E}{N} \simeq \begin{cases} mc^2 + \dfrac{3}{5}\dfrac{p_{\mathrm{F}}^2}{2m} & \text{when} \quad p_{\mathrm{F}} \ll mc\,, \\[2ex] \dfrac{3}{4}\,p_{\mathrm{F}}c & \text{when} \quad p_{\mathrm{F}} \gg mc\,. \end{cases} \tag{8.30}$$

At $T = 0$ K, the grand potential $J = E - TS - \mu N$ is equal to $E - E_{\mathrm{F}}N$, leading, according to Relation (4.58) $J = -PV$, to the equation of state:

$$P = \begin{cases} \dfrac{h^2}{5m}\left(\dfrac{3}{8\pi}\right)^{2/3} n^{5/3} & \text{when} \quad n \ll n_{crit}\,, \\[3ex] \dfrac{hc}{4}\left(\dfrac{3}{8\pi}\right)^{1/3} n^{4/3} & \text{when} \quad n \gg n_{crit}\,. \end{cases} \tag{8.31}$$

In these expressions, $n = N/V$ is the fermion number density and

$$n_{\mathrm{crit}} = \frac{8\pi}{3}(mc/h)^3 \tag{8.32}$$

is the value of the density for which $p_{\mathrm{F}} = mc$, meaning that the limit $n \gg n_{\mathrm{crit}}$ is the ultra-relativistic limit.

In both the non-relativistic or the ultra-relativistic case, the pressure of a non-interacting fermion gas at $T = 0$ K is non-zero. This is a consequence of Pauli exclusion principle: fermions cannot all occupy the same quantum state resulting in a finite pressure driven by the quantum statistics of the particles[7]. This result has a consequence of primary importance on the stability of white dwarfs, as discussed in the next section.

8.3 WHITE DWARFS, CHANDRASEKHAR MASS

When a star with a mass of the order of the sun mass has exhausted all its nuclear fuel, it collapses and ejects a planetary nebula[8]. A small celestial object then remains (of size typically comparable to Earth). Such object, called a *white dwarf*, has a very low luminosity (due to its small surface). White dwarfs are very dense (the mass of the sun and the size of a planet) and, as discussed in this section, their collapse under their own weight is prevented by the Fermi pressure (this point was first understood by R. Fowler in 1926).

For concreteness, consider a white dwarf of mass $m = 0.6M_\odot$ ($M_\odot = 1.989 \times 10^{30}$ kg is the solar mass) and mainly consisting in helium atoms. It comprises a number N of atoms of the order of $N \sim 0.6M_\odot/(4m_{\mathrm{H}}) = 0.2 \times 10^{57}$, where m_{H} is the mass of the hydrogen atom. If it has a radius of the order of the Earth's radius $R_\oplus = 6371$ km, then the average distance between the helium atoms is $d \sim R_\oplus N^{-1/3} = 10^{-2}$ Å. Under such conditions, no atoms can be formed: the inter-nuclear distances are too small to accommodate electronic orbitals. Instead, a completely ionised state of matter is formed, with nuclei embedded in a fluid of electrons, the whole system being electrically neutral.

The internal temperature of the white dwarf is $T \sim 10^7$ K. The corresponding thermal wavelength is $\lambda_{T\mathrm{n}} \simeq 10^{-3}$ Å for the nuclei and $\lambda_{T\mathrm{el}} \simeq 10^{-1}$ Å for the electrons. The nuclei can therefore be treated as classical particles while the electrons must be studied in the

---

[7]This can be stated in a different manner by noting that in the non-relativistic (relativistic) limit $PV = 2E/3 \ (= E/3)$: this is the fermionic version of the expression of the pressure in the kinetic theory of gases (Bernoulli relation, see Note 37 on page 16). The existence of a Fermi sea results in an average motion of the fermions which is non-zero even at $T = 0$ K, and thus in collisions with the wall of the container inducing a finite pressure.

[8]According to a scenario presented, for example, in Hugh M. Van Horn, *Unlocking the Secrets of White Dwarf Stars* (Astronomers' Universe, Springer, 2015).

framework of quantum mechanics. If the electrons are non-relativistic, their density is $n_{\text{el}} = N_{\text{el}}/V \simeq 3 \times 10^{35}$ m^{-3} and their Fermi temperature is, according to Equation (8.14), $T_{\text{F}} \simeq 2 \times 10^9$ K $\gg T$: the electron gas is completely degenerate.

The pressure of the classical nuclei is $n_{\text{n}} k_{\text{B}} T$ and the pressure of the quantum electrons is $\frac{2}{5} n_{\text{el}} k_{\text{B}} T_{\text{F}}$ [this last formula is a combination of Equations (8.31) and (8.14) with $E_{\text{F}} = k_{\text{B}} T_{\text{F}}$]: the electron pressure is by a factor of the order of T_{F}/T higher than the nuclei pressure. Therefore, only the electron contribution can be be considered to evaluate the equilibrium profile of the white dwarf. Note, however, that the electron density $n_{\text{el}} \sim 3 \times 10^{35}$ m^{-3} is close to the critical density $n_{\text{crit}} = \frac{8\pi}{3}(m_{\text{e}} c/h)^3 \simeq 5 \times 10^{35}$ m^{-3} [Equation (8.32)]. So it is appropriate to take relativistic effects into account. This was first done by Chandrasekhar in 1931. However, the general relation between pressure and density of a relativistic Fermi gas at $T = 0$ K has no simple analytical expression. Therefore, two limit cases will be considered: the non-relativistic and ultra-relativistic regimes for which pressure can be expressed as a function of density [see formulas (8.31)]. The results will be compared the numerical solution interpolating between these two regimes first obtained by Chandrasekhar.

The hydrostatic equilibrium of the star relates the local pressure $P(\vec{r})$ to the mass density $\rho(\vec{r})$ rather than to the electron number density $n_{\text{el}}(\vec{r})$ as in Formulas (8.31). If the white dwarf is composed only of protons and electrons then $\rho = m_{\text{H}} n_{\text{el}}$, where m_{H} is the hydrogen atom mass. However, this configuration is unlikely: the white dwarf has consumed its hydrogen in the earlier stage of its evolution. If the white dwarf is composed instead helium nuclei (2 neutrons and 2 protons) and electrons (as numerous as protons) then $\rho \simeq 2 m_{\text{H}} n_{\text{el}}$. In the following, the notation $\rho(\vec{r}) = \mu_{\text{e}} m_{\text{H}} n_{\text{el}}(\vec{r})$ will be used where μ_{e} is a dimensionless constant called the *molecular weight per electron*. μ_{e} is certainly close to 2 because ~2 nucleons per electron are expected (as for ^{4}He, ^{12}C and ^{16}O which are the main constituents of most white dwarfs).

Relations (8.31) can be written in the generic form:

$$P = K\rho^\gamma , \qquad (8.33)$$

where $\gamma = \frac{5}{3}$ and $K = \frac{\hbar^2}{5m}(3/8\pi)^{2/3}/(\mu_{\text{e}} m_{\text{H}})^\gamma$ in the non-relativistic case and, $\gamma = \frac{4}{3}$ and $K = \frac{hc}{4}(3/8\pi)^{1/3}/(\mu_{\text{e}} m_{\text{H}})^\gamma$ in the ultra-relativistic case. The hydrostatic equilibrium of the star relates the pressure gradient to the local gravitational field[9]:

$$\vec{\nabla} P = \rho(\vec{r}) \vec{g}(\vec{r}) . \qquad (8.34)$$

Of course, the star has a spherical profile and the equilibrium Equation (8.34) can be written as:

$$\frac{\mathrm{d}P}{\mathrm{d}r} = -\rho(r) \frac{GM(r)}{r^2} , \quad \text{where} \quad M(r) = 4\pi \int_0^r \mathrm{d}r'\, r'^2 \rho(r') \qquad (8.35)$$

is the mass contained in a sphere of radius r, and G is the universal gravitational constant. Using the equation of state [Equation (8.33)], pressure can be eliminated from Equation (8.35) leading to the differential equation:

$$\frac{\gamma K}{4\pi G} \frac{\mathrm{d}}{\mathrm{d}r}\left(r^2 \rho^{\gamma-2} \frac{\mathrm{d}\rho}{\mathrm{d}r}\right) = -r^2 \rho . \qquad (8.36)$$

---

[9]This is obtained by writing the mechanical equilibrium of a volume element $\mathrm{d}^3 v = \mathrm{d}x\mathrm{d}y\mathrm{d}z$ located around point $\vec{r}$. Local coordinates are used such as the z-axis is directed along $-\vec{g}(\vec{r})$. The equilibrium of the pressure forces exerted on the lateral sides implies $\partial_x P = \partial_y P = 0$. Along the z-axis gravity must be taken into account leading to $-\mathrm{d}x\mathrm{d}y P(\vec{r} + \mathrm{d}z\vec{e}_z) + P(\vec{r})\mathrm{d}x\mathrm{d}y - \mathrm{d}^3 v \rho(\vec{r})g(\vec{r}) = 0$, i.e. $\partial_z P = -\rho(\vec{r})g(\vec{r})$. Combining with the component along x, y and z, Equation (8.34) is obtained.

It is convenient to make the change of variable $\theta = (\rho/\rho_0)^{\gamma-1}$ and $\xi = r/r_0$ with $r_0 = (\frac{\rho_0^{\gamma-2}}{\gamma-1}\frac{\gamma K}{4\pi G})^{1/2}$. It leads to the so-called *Lane-Emden equation*:

$$\frac{d}{d\xi}\left(\xi^2 \frac{d\theta}{d\xi}\right) = -\xi^2 \theta^n, \quad \text{where} \quad n = \frac{1}{\gamma-1}. \tag{8.37}$$

The free parameter ρ_0 in the change of variable is chosen to be equal to the density at the centre of the star, such that $\theta(0) = 1$ and additionally[10] $\theta'(0) = 0$. With these initial conditions, the Lane-Emden equation can be numerically integrated for $n = 3/2$ ($\gamma = 5/3$, non-relativistic limit) and $n = 3$ ($\gamma = 4/3$, ultra-relativistic limit). In these two cases, $\theta(\xi)$ cancels for a value ξ_1 of the parameter ξ which defines the radius of the star. The following numerical values are obtained:

$$\text{Non-relativistic limit:} \quad \gamma = \tfrac{5}{3}, \quad \xi_1 = 3.65375, \quad \xi_1^2\, \theta'(\xi_1) = -2.71406,$$

$$\text{Ultra-relativistic limit:} \quad \gamma = \tfrac{4}{3}, \quad \xi_1 = 6.89685, \quad \xi_1^2\, \theta'(\xi_1) = -2.01824. \tag{8.38}$$

The star radius R is then $R = \xi_1 r_0$ or

$$R = \xi_1 \left(\frac{\rho_0^{\gamma-2}}{\gamma-1}\frac{\gamma K}{4\pi G}\right)^{1/2}, \tag{8.39}$$

and its mass is[11]:

$$M = 4\pi \int_0^R dr\, r^2 \rho(r) = 4\pi r_0^3 \rho_0 \int_0^{\xi_1} d\xi\, \xi^2 \theta^n = -4\pi r_0^3 \rho_0\, \xi_1^2\, \theta'(\xi_1). \tag{8.40}$$

Eliminating ρ_0 and r_0 from Expressions (8.39) and (8.40), the relation between the mass of the star and its radius is obtained:

$$M = 4\pi \frac{|\theta'(\xi_1)|}{\xi_1}\left(\frac{\xi_1^2 \gamma}{\gamma-1}\frac{K}{4\pi G}\right)^{\frac{1}{2-\gamma}} \cdot R^{\frac{3\gamma-4}{\gamma-2}}. \tag{8.41}$$

In the non-relativistic limit $M \propto R^{-3}$, whereas in the ultra-relativistic limit, R disappears from Equation (8.41) and the mass is a universal constant. The correct relation interpolates between these two limits and is shown in Figure 8.4. A low mass white dwarf has also a low density and the non-relativistic description of the electron gas can be used in this case (since $n \ll n_{\text{crit}}$). When the mass increases relativistic corrections become more and more important and fix an upper limit to the white dwarf mass: the universal value found in the ultra-relativistic limit. Formula (8.41) with $\gamma = 4/3$ gives the value of this so-called *Chandrasekhar mass*:

$$M_{\text{Ch}} = \left.|\xi_1^2\, \theta'(\xi_1)|\right|_{\gamma=\frac{4}{3}} \frac{\sqrt{3\pi}}{2}\left(\frac{\hbar c}{G}\right)^{3/2}\frac{1}{(\mu_e m_{\text{H}})^2}. \tag{8.42}$$

If $\mu_e = 2$ then $M_{\text{Ch}} = 1.44\, M_\odot$. For a mass higher than M_{Ch}, the pressure of the electron gas is not sufficient to counterbalance gravity and the star at the end of its life becomes a neutron star, or, if it is even more massive, a black hole. An interesting additional read on this subject is the science popularisation book by Kip S. Thorne *Black Holes and Time Warps: Einstein's Outrageous Legacy* (W.W. Norton, New York, 1995), which gives a lively account of the controversy which opposed Chandrasekhar to Eddington on the subject.

---

[10]Indeed, according to Equation (8.35) when $r \to 0$, $M(r) \propto r^3$ and thus $\frac{dP}{dr} \to 0$.

[11]The Lane-Emden Equation (8.37) has been used to obtain the last equality in Equation (8.40).

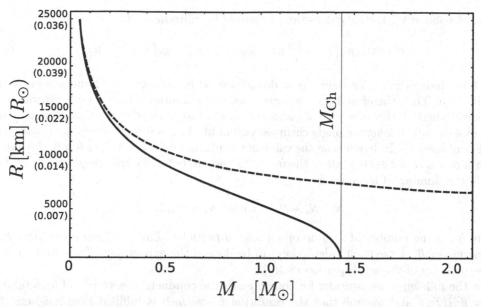

FIGURE 8.4 Radius as a function of mass for a white dwarf. The solid curve was obtained using the general pressure law for an ideal Fermi gas at $T = 0$: $P = (NE_F - E)/V$ where E is given by Equation (8.28). The dashed curve is the non-relativistic limit: Equation (8.41) with $\gamma = 5/3$. The vertical straight line is the Chandrasekhar mass [Equation (8.42)] (adapted from https://en.wikipedia.org/wiki/White_dwarf).

8.4 PAULI PARAMAGNETISM

Consider a gas of non-interacting spin-1/2 fermions governed by a Hamiltonian H_0. H_0 can describe free particles confined in a box of finite volume V ($H_0 = \vec{p}^2/(2\,m)$ with boundary conditions imposing a wave function that cancels at the box boundaries), or particles trapped in a harmonic well. The system is immersed in a uniform and constant magnetic field $\vec{B} = B\,\vec{e}_z$ ($B > 0$). $\vec{\mu}_M = -g\mu_B\,\vec{s}/\hbar$ is the magnetic moment of a particle, where $\vec{s}$ is the spin of a particle, $\mu_B = \frac{|q|\hbar}{2m}$ is the Bohr magneton and g is the dimensionless Landé factor [see Equation (6.52)]. For a free electron, the Landé Factor is $g = 2$ (see the discussion at the beginning of Section 6.2.2). In the following of this section, we suppose that the fermions interact with the magnetic field $\vec{B}$ field only *via* their magnetic moment. So the interaction Hamiltonian is:

$$H = H_0(\vec{r}, \vec{p}) - \vec{\mu}_M \cdot \vec{B}. \tag{8.43}$$

Since $\vec{\mu}_M \cdot \vec{B} \propto B\,s_z$ and since s_z has only two possible values ($\pm\hbar/2$), Hamiltonian (8.43) has two possible forms:

$$H_\pm = H_0(\vec{r}, \vec{p}) \pm \frac{g\mu_B}{2}\,B. \tag{8.44}$$

The symbol $\pm$ in Equation (8.44) refers to the situation where the magnetic moment is parallel ($-$) or antiparallel ($+$) to $\vec{B}$ (i.e. $\pm \equiv \mathrm{sgn}(s_z)$). It is convenient to treat the system as if there were two kinds of particles: the "up" spin particles and the "down" spin particles. Each kind obeys a Hamiltonian H_+ or H_- which is simply H_0 shifted by a constant energy proportional to the intensity of the magnetic field.

Assuming that the density of states $d_0(E)$ of Hamiltonian H_0 and the integrated density of states $\Phi_0(E) = \int_{-\infty}^{E} d_0(\epsilon)\mathrm{d}\epsilon$ are already known [see Equation (8.2) for a gas contained in a box of volume V], then the densities of state of Hamiltonians $H_\pm$ are:

$$\Phi_\pm(E) = \Phi_0 \left(E \mp \frac{g\mu_B}{2} B \right) \quad \text{and} \quad d_\pm(E) = d_0 \left(E \mp \frac{g\mu_B}{2} B \right) . \qquad (8.45)$$

The two kinds of particles (spin up or down) are artificially separated and are in thermal equilibrium. They therefore have the same chemical potential, which sounds obvious since, in reality, there is only one type of particle subjected to a well-defined Hamiltonian [Equation (8.43)] and therefore a single chemical potential. At $T = 0$ K, this chemical potential is the Fermi level E_F. It depends on the value of the magnetic field B. When $B = 0$, the number of fermions is $N = 2\,\Phi_0(E_F)$ where the factor "2" accounts for the spin degeneracy[12]. When $B \neq 0$, the number of fermions is:

$$N = N_+ + N_- \quad \text{with} \quad N_\pm = \Phi_\pm(E_F), \qquad (8.46)$$

where $N_\pm$ is the number of spin-up or spin-down particles. This equation determines E_F as a function of B. An explicit calculation can be done if the expressions of $\Phi_\pm$ are known, i.e the expression of Φ_0 [see Equation (8.45)].

In the following, we consider for concreteness the conduction electrons of a metal [with $H_0 = \vec{p}^2/(2m)$] and assume that the condition $E_F \gg \mu_B B$ is fulfilled (low magnetic field limit). This is the relevant experimental limit since typically $E_F \simeq 1$ eV while for an electron $\mu_B = 5.8 \times 10^{-5}$ eV/T and typical laboratory magnetic fields are lower than 1 Tesla. A first order series expansion of Equation (8.46) in the small parameter $\mu_B B/E_F$ yields:

$$N = 2\,\Phi_0(E_F), \qquad (8.47)$$

which implies that the Fermi energy is independent of the value of B, or more precisely that the corrections are of order $O(|\mu_B B|^2)$. The numbers N_+ and N_- of electrons with spin up and down are therefore:

$$N_\pm = \Phi_0(E_F) \mp \frac{g\mu_B}{2} B\, d_0(E_F) . \qquad (8.48)$$

This formula is often illustrated by a diagram as shown in Figure 8.5.

The magnetisation, i.e. the magnetic moment per unit volume, is $M = \frac{1}{2} g\mu_B (N_- - N_+)/V$. Using the relation $d_0(E) = \frac{3}{2}\Phi_0(E)/E$ [Equations (8.1) and (8.2)], one obtains

$$M = \left(\frac{g\mu_B}{2} \right)^2 \frac{3\,N/V}{2\,E_F} B . \qquad (8.49)$$

and, for the magnetic susceptibility, $\chi = \mu_0(\partial M/\partial B)_{B=0}$ [see Equation (6.49)], with $g = 2$:

$$\chi_{\text{Pauli}} = \frac{3}{2} \frac{N}{V} \frac{\mu_0\,\mu_B^2}{2E_F} . \qquad (8.50)$$

Expressions (8.49) and (8.50) correspond to the degenerate[13] and low magnetic field limit. This calculation was first made by Pauli in 1927 to explain why most metals are weakly paramagnetic and why their magnetic susceptibility does not behave as $1/T$ at low temperature as expected from Curie's Law [Equation (6.50)], also valid for quantum spins [see the discussion in Section 6.2.2 around formula (6.58)]. The physical argument is simple: the alignment of magnetic moments favoured by Hamiltonian (8.44) is inhibited by Pauli exclusion principle: it is indeed impossible to fill a Fermi sea only with polarised spins (i.e.

---

[12]Equivalently, in the absence of a magnetic field there are as many "up" particles as "down" particles, hence the factor 2.

[13]To simplify the presentation, the calculations are done at $T = 0$ K. A study at finite temperature is presented in Section 8.5 in a two-dimensional configuration.

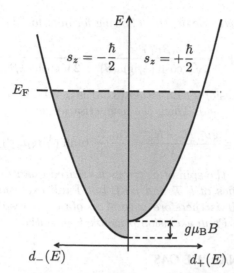

FIGURE 8.5 Density of state d_+ and d_- [Equation (8.45)] associated with the Hamiltonians H_+ and H_-. At $T = 0$ K the states are filled up to the Fermi energy E_F. This atypical representation (rotated by 90 degrees with respect to a usual graph) is chosen because the Fermi sea (shaded area in the diagram) appears as a fluid filling a vessel.

with $N_+ = N$ and $N_- = 0$). This would only be possible for a colossal magnetic field intensity allowing to almost empty the "$s_z = +\hbar/2$" side of the "vessel" shown in Figure 8.5 in favour of the "$s_z = -\hbar/2$" side.

While the study presented above correctly explains the weak paramagnetism of many metals, quantitative comparison with experimental data is made difficult due to the omission of several important effects: (i) there is a diamagnetic contribution, which, for free electrons, decreases the magnetisation given by Equation (8.49) by one third [see Equation (8.91)], (ii) in a real metal, the electrons move more or less freely while the magnetisation of the still ionic nuclei must be taken into account and (iii) electron interactions also significantly affect Result (8.49).

Exercise 8.2. Study the behaviour of the magnetisation of a free Fermi gas at high temperature, i.e. when $T \gg T_F$.

Solution: In the high temperature limit, the Maxwell-Boltzmann approximation can be used (see Sections 6.4.3 and 8.1.1): this amounts to substitute the Fermi distribution function $n_F(E)$ _for Approximation (6.87):_

$$N_\pm = \int_{\mathbb{R}} d_\pm(E) n_F(E)\, dE \simeq \int_{\mathbb{R}} d_\pm(E) \exp\{-\beta(E - \mu)\}\, dE \ . \tag{8.51}$$

The integral of the right-hand side term of Equation (8.51) can be calculated exactly[14]. Writing the density of state as [see Equations (8.2) and (8.14)]:

$$d_0(E) = \alpha\sqrt{E}\ , \quad \text{with} \quad \alpha = \frac{3N}{4 E_F^{3/2}} = \frac{2V \beta^{3/2}}{\sqrt{\pi}\,\lambda_T^3}\ , \tag{8.52}$$

---

[14]$d_\pm(E)$ is a square-root function, some simple changes of variable allow to reduce the integral to an expression proportional to $\int_0^{+\infty} \sqrt{x}\,\exp(-x)\,dx = \frac{1}{2}\sqrt{\pi}$.

and calculating $N = N_- + N_+$, yields, the following formula for the chemical potential:

$$\exp\{\beta\mu\} = \frac{4(\beta E_{\mathrm{F}})^{3/2}}{3\sqrt{\pi}\,\cosh\left(\frac{1}{2}\beta g\mu_{\mathrm{B}}B\right)} = \frac{N\lambda_T^3}{2V\cosh\left(\frac{1}{2}\beta g\mu_{\mathrm{B}}B\right)}\,. \tag{8.53}$$

The right-hand side expression of Equation (8.53) is a version of Equation (8.10) valid in the presence of a magnetic field. Then, the magnetisation is:

$$M = \frac{g\mu_{\mathrm{B}}}{2}\frac{N_- - N_+}{V} = \frac{g\mu_{\mathrm{B}}}{2}\frac{N}{V}\tanh\left(\frac{1}{2}\beta g\mu_{\mathrm{B}}B\right)\,. \tag{8.54}$$

Note that in this exercise, the spin of electrons is treated quantumly (no assumption were made on the relative values of $k_{\mathrm{B}}T$ and μB), but Pauli exclusion principle is not taken into account $(T \gg T_{\mathrm{F}})$. It is therefore natural to obtain a result more standard[15] than Equation (8.49) for which Pauli exclusion principle is essential.

8.5   TWO-DIMENSIONAL ^{3}HE GAS

Helium has two isotopes: ^{3}He which is a fermion (its nucleus contains two protons and one neutron, around which orbit two electrons) and ^{4}He which is a boson (two neutrons in its nucleus). At low temperature these two isotopes are liquid. When the concentration of ^{3}He is low in a ^{3}He/^{4}He mixture, ^{3}He, which has a lower mass and thus a higher zero point energy, "floats" on the free surface of the ^{4}He forming, in a first approximation, an ideal two-dimensional Fermi gas.

This section is devoted to the study of the thermodynamic properties of ^{3}He atoms in this system. The ^{3}He atoms (mass m and spin-1/2) move freely but their motion is restricted to a surface of area $\mathcal{A}$. The macroscopic dimensions of this surface authorises a continuous approximation. By repeating for the two-dimensional case, the calculations made in three dimensions at the beginning of Section 8.1, the density of state can be determined:

$$d(E) = \frac{m\mathcal{A}}{2\pi\hbar^2}\,. \tag{8.55}$$

Exercise 8.3.   Prove Formula (8.55).

Solution: According to Equations (6.6) and (6.8), the number $\Phi(E)$ of eigenstates of the Hamiltonian with energy less than E is [see Equation (6.5)]:

$$\Phi(E) = \frac{1}{h^2}\int_{H(\vec{r},\vec{p})<E}\mathrm{d}^2r\,\mathrm{d}^2p = \frac{\mathcal{A}}{h^2}\int_{\frac{p^2}{2m}<E}\mathrm{d}^2p = \frac{\pi\mathcal{A}}{h^2}(2mE)\,\Theta(E)\,, \tag{8.56}$$

where $\Theta(E)$ is the Heaviside function. For legibility, its implicit presence will not appear in the following formulas. The density of state $d(E)$ is then:

$$d(E) = \frac{\mathrm{d}\Phi}{\mathrm{d}E} = \frac{2\pi m\mathcal{A}}{h^2} = \frac{m\mathcal{A}}{2\pi\hbar^2}\,. \tag{8.57}$$

The Fermi energy E_{F} is defined by $N = d_S\,\Phi(E_{\mathrm{F}})$, where N is the total number of ^{3}He atoms and $d_S = 2$ is the spin degeneracy. The Fermi temperature reads:

$$T_{\mathrm{F}} = E_{\mathrm{F}}/k_{\mathrm{B}} = \frac{\pi\hbar^2}{mk_{\mathrm{B}}}\left(\frac{N}{\mathcal{A}}\right)\,. \tag{8.58}$$

---

[15]Indeed, Expression (8.54) is in agreement with Formula (6.56) for $j = 1/2$.

Typical values of the surface density for which the ^{3}He atoms can be assimilated to a two-dimensional Fermi gas are very low (of the order of one atomic layer[16]) and for ^{3}He, $\hbar^2/(mk_B) = 16.1$ K.Å^2, which gives $T_F = 1.6$ K for a half atomic layer of ^{3}He.

The finite temperature chemical potential $\mu(T)$ can be calculated by writing [see Equation (8.3)]:

$$N = 2 \int_0^\infty dE \, d(E) n_F(E) = \frac{m\mathscr{A}}{\pi\hbar^2} \int_0^\infty \frac{dE}{\exp\{\beta(E - \mu)\} + 1} = N \frac{T}{T_F} \int_0^\infty \frac{dx}{z^{-1}e^x + 1} \,. \tag{8.59}$$

In the right-hand side of Equation (8.59), the change of variable $x = \beta E$ was made and the fugacity $z = \exp(\beta\mu)$ was used. The integral can be calculated exactly[17] yielding:

$$\mu = k_B T \ln\left(e^{T_F/T} - 1\right) \,. \tag{8.60}$$

The energy of the N ^{3}He atoms is:

$$E(T) = 2 \int_0^\infty dE \, E \, d(E) n_F(E) = N \frac{k_B T^2}{T_F} \int_0^\infty \frac{x \, dx}{1 + z^{-1} \exp(x)} = N \frac{k_B T^2}{T_F} f_2(z) \,, \tag{8.61}$$

where f_2 is a Fermi integral [Equation (A.9)]. At $T \ll T_F$, $\mu \simeq T_F$ and $z \to +\infty$. More precisely, $\beta\mu(T) = T_F/T + O(e^{-T_F/T})$. Expansion (8.19) reads here

$$f_2(z) = \frac{1}{2}\left[\ln(z)\right]^2 + \frac{\pi^2}{6} + O\left(\frac{1}{\ln^2(z)}\right) \,, \tag{8.62}$$

and one obtains:

$$E(T) = \frac{1}{2} N E_F \left[1 + \frac{\pi^2}{3}\left(\frac{T}{T_F}\right)^2 + \cdots\right] \,, \tag{8.63}$$

which is the two-dimensional analogue of Expansion (8.22). The heat capacity $C_\mathscr{A} = (\partial E/\partial T)_\mathscr{A}$ has the low temperature behaviour: $C_\mathscr{A}(T) \simeq \frac{\pi^2}{3} N k_B T/T_F = \gamma T$ with $\gamma = \frac{\pi}{3}(k_B/\hbar)^2 m\mathscr{A}$. This linear dependence of the heat capacity to temperature is observed experimentally[18], but with a slope that is not exactly equal to the coefficient γ. This difference is interpreted as an effective mass effect, as for the electronic contribution to the heat capacity of solids (see the discussion around Figure 8.3). For surface densities of ^{3}He less than half an atomic layer, a weakly density-dependent effective mass, ranging for 1.4 to 1.8 m is found. At very low density, the effective mass is $m^*/m \to 1.400 \pm 0.003$ in good agreement with theoretical simulations[19]. This effective mass is mainly caused by the interaction of the ^{3}He atoms with the ^{4}He bath.

Other experiments confirm this "ideal gas with an effective mass" approach: let us apply to the system a magnetic field $\vec{B} = B \, \vec{e}_z$ with $B > 0$. Each atom has a magnetic moment $\vec{\mu}_M = -2 \mu_B \vec{s}/\hbar$, where $\vec{s}$ is the spin and μ_B the Bohr magneton [see Equation (6.52)]. The atomic energies are shifted by a factor $\mu_B B$ for spins parallel to $\vec{B}$ and by $-\mu_B B$ for spins anti-parallel to $\vec{B}$. Thus, there are two densities of states $d_\pm(E)$, one for each spin projection along the z-axis. As in Section 8.4, the two densities of state are (taking here $g = 2$)

---

[16]It is customary to use the atomic layer as the unit of surface density. It is defined as $(n_3)^{2/3} = 6.4$ nm^{-2}, where n_3 is the number density (number of atoms per unit volume) of liquid ^{3}He at $T = 0$ K.

[17]For all positive real number z, $\int_0^{+\infty} dx \, [1 + z^{-1} \exp(x)]^{-1} = \ln(1 + z)$.

[18]M. Dann, J. Nyéki, B. P. Cowan, and J. Saunders, Phys. Rev. Lett. **82**, 4030 (1999).

[19]N. Pavloff and J. Treiner, J. Low Temp. Phys. **83**, 331 (1991); B. E. Clements, E. Krotscheck, and M. Saarela, J. Low Temp. Phys. **100**, 175 (1995).

$d_{\pm}(E) = d(E \mp \mu_B B)$. The only difference with Section 8.4 is that here $d(E)$ is constant (for $E > 0$) while in three dimensions $d(E) \propto \sqrt{E}$ [compare Equations (8.55) and (8.2)].

Let N_+ and N_- be the average numbers of atoms in the spin state $|+\rangle_z$ and the spin state $|-\rangle_z$, respectively, then:

$$N_{\pm} = \int_{\pm\mu_B B}^{\infty} dE\, d_{\pm}(E)\, n_F(E) = \int_0^{+\infty} d\varepsilon\, d(\varepsilon)\, n_F(\varepsilon \pm \mu_B B)$$

$$= \frac{N}{2\, k_B T_F} \int_0^{+\infty} d\varepsilon\, \frac{1}{\exp\{\beta[\varepsilon \pm \mu_B B - \tilde{\mu}]\} + 1}\,, \tag{8.64}$$

where $\tilde{\mu} \equiv \mu(B,T)$ is the chemical potential in the presence of magnetic field.

Exercise 8.4. Determine the expression of $\tilde{\mu} = \mu(T,B)$.

Solution: Integrals of the type of Equation (8.64) have already been calculated (see Note 17) and one obtains:

$$N = N_+ + N_- = \frac{N}{2}\frac{T}{T_F}\left\{\ln\left[1 + e^{\beta(\tilde{\mu}-\mu_B B)}\right] + \ln\left[1 + e^{\beta(\tilde{\mu}+\mu_B B)}\right]\right\}$$

$$= \frac{N}{2}\frac{T}{T_F}\ln\left[1 + 2\,e^{\beta\tilde{\mu}}\cosh(\beta\mu_B B) + e^{2\beta\tilde{\mu}}\right]\,. \tag{8.65}$$

This relation can be written in the form of a second degree equation for the fugacity $z = \exp(\beta\tilde{\mu})$:

$$z^2 + 2\cosh(\beta\mu_B B)z + 1 - e^{2T_F/T} = 0\,. \tag{8.66}$$

The solution of this equation yields

$$\tilde{\mu} = k_B T \ln\left[-\cosh(\beta\mu_B B) + \sqrt{\sinh^2(\beta\mu_B B) + e^{2T_F/T}}\right]\,. \tag{8.67}$$

In the absence of magnetic field ($B = 0$) this result reduces to (8.60) as expected. For a weak field, an expansion of Equation (8.67) yields a correction of order $O(B^2)$:

$$\tilde{\mu} \simeq \mu(T,0) - \frac{1}{2}\frac{(\mu_B B)^2}{k_B T}\exp(-T_F/T) + O(\beta\mu_B B)^4\,. \tag{8.68}$$

In the following, only the regime $\mu_B B \ll k_B T$ is considered, because it is the regime relevant to the experiments discussed below. Expression (8.68) shows that it is then legitimate to make the approximation $\mu(T,B) \simeq \mu(T,0)$. The chemical potential will thus be noted as μ without further specification. In this regime, Expression (8.64) is:

$$N_{\pm} = \frac{N}{2}\frac{T}{T_F}\left[\ln\left(1 + e^{\beta\mu}\right) \mp \beta\mu_B B\,\frac{e^{\beta\mu}}{1 + e^{\beta\mu}} + \cdots\right]\,, \tag{8.69}$$

and the logarithm in this expression[20] is equal to T_F/T (see Equation (8.60)).

---

[20] Note that Expression (8.69) leads to the result $\mu(T, B) = \mu(T, 0) + O(B^2)$ without requiring the detailed calculation of Exercise 8.4. Indeed, if N_+ and N_- are calculated without the use of the results of Exercise 8.4, Expression (8.69) is recovered where μ is substituted for $\tilde{\mu}$. Then:

$$N = N_+ + N_- = N\frac{T}{T_F}\ln\left(1 + e^{\beta\tilde{\mu}}\right) + O(B^2)\,.$$

Comparing with the formula $N = N(T/T_F)\ln(1 + e^{\beta\mu})$ which is an intermediate step in the calculation leading to Expression (8.60) of $\mu(T, 0)$, the desired result is obtained.

The magnetisation $M(T)$ (magnetic moment per unit area along the z-axis) is $M(T) = \mu_B(N_- - N_+)/\mathcal{A}$. Using Equations (8.69) and (8.60), one gets

$$M(T) = M_0 \left[1 - e^{-T_F/T} \right], \quad \text{with} \quad M_0 = (N/\mathcal{A}) \mu_B \frac{\mu_B B}{k_B T_F}. \qquad (8.70)$$

At low temperature $M(T \to 0) \to M_0$ and, in two dimensions, M_0 plays the role of M given by Equation (8.49) in three dimensions. Experimental data[21] reproduced in Figure 8.6 do not exactly show the expected low temperature limit: $M(T)/M_0$ tends indeed towards a constant, but this constant is not unity. This is again interpreted as an effective mass effect: the density of states has a behaviour similar to Equation (8.55) but where the bare mass is replaced by an effective mass m^*. This substitution should be made in all expressions

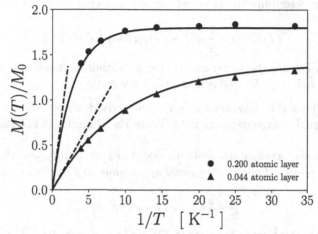

FIGURE 8.6 NMR measurements[21] of the magnetisation of a two-dimensional ^{3}He system plotted as a function of $1/T$. The black circles and triangles correspond to different ^{3}He surface densities $N/\mathcal{A}$. The solid curves represent $M^*(T)/M_0$ calculated using Equation (8.70) with an effective mass chosen to fit the experimental data. The dashed lines are the tangents at the origin.

involving the mass m (including the expression of T_F). From now on, all quantities calculated with the effective mass will be written with an asterisk: $T_F^* = T_F(m^*)$, $M^*(T)$, M_0^*... Note that in Figure 8.6, data are normalised to M_0 calculated with the real mass. It is thus appropriate to compare experimental results with $M^*(T)/M_0$: when $T \to 0$, $M^*(T)/M_0 \to m^*/m$. It can be seen in Figure 8.6 that some experimental points (circles) are consistent with an effective mass $m^* = 1.6\,m$, while for other data (triangles), the lowest temperatures are not yet in the regime where $M^*(T)$ saturates, but the evolution of the magnetisation is compatible with an effective mass $m^* = 1.4\,m$.

It is interesting to note that when $T \to \infty$ the limit $M^*(T)/M_0 \to (T_F^*/T) \times (M_0^*/M_0) = T_F/T$ is independent of the effective mass[22]. The slope at the origin of the curves shown in Figure 8.6 thus enables the determination of the value of T_F for each experimental data set and therefore the surface density $N/\mathcal{A}$ [see Equation (8.58)].

For the data represented by triangles in in Figure 8.6, the slope at the origin leads to $T_F = 0.14$ K and $N/\mathcal{A} = m k_B T_F/(\pi \hbar^2) = 0.28$ nm^{-2}, or 0.044 atomic layer. It is a very low surface density and, in this case, an effective mass ($m^* = 1.4\,m$) identical to the one obtained by the analysis of the behaviour of the heat capacity at low temperature and low ^{3}He surface

[21] R. H. Higley, D. T. Sprague, and R. B. Hallock, Phys. Rev. Lett. **63**, 2570 (1989).

[22] This is the high temperature regime where the Curie law applies. Indeed, a little algebra shows that in this limit $M = (N/\mathcal{A}) \mu_B^2 B/k_B T$.

density is recovered. The data represented by circles in Figure 8.6 lead to $T_F = 0.67$ K and $N/\mathcal{A} = 1.3$ nm^{-2}, or 0.2 atomic layer. The effective mass determined in this case from the magnetisation data ($m^* = 1.8\,m$) is also compatible with the heat capacity experimental data.

8.6 LANDAU DIAMAGNETISM

In 1930, Landau calculated the possible energy states of a charged particle in a uniform magnetic field and demonstrated the phenomenon of orbital diamagnetism: unlike the case of paramagnetism studied in the previous section, an external magnetic field induces in a conductor – via the orbital motion of free electrons – a magnetic field opposite to the applied field.

The Lagrangian describing the classical motion of a charged particle in an electromagnetic field is[23]:

$$L(\vec{r}, \vec{v}, t) = \frac{1}{2}m\vec{v}^2 - qV(\vec{r}, t) + q\vec{A}(\vec{r}, t) \cdot \vec{v}\,, \tag{8.71}$$

where $V(\vec{r}, t)$ and $\vec{A}(\vec{r}, t)$ are the scalar and vector potentials, respectively, related to electromagnetic fields by $\vec{E}(\vec{r}, t) = -\vec{\nabla}V - \partial_t\vec{A}$ and $\vec{B}(\vec{r}, t) = \vec{\nabla} \wedge \vec{A}$.

Exercise 8.5. Check that Lagrangian (8.71) correctly describes the motion of a charged particle in an external electromagnetic field. Write the Hamiltonian of the system.

Solution: Let $\vec{p}$ be the generalised momentum associated with Lagrangian (8.71). By definition, $p_x = \partial L/\partial v_x$ (with similar definitions for p_y and p_z), leading to $\vec{p} = m\vec{v} + q\vec{A}$. The Euler-Lagrange equations are:

$$\frac{d\vec{p}}{dt} = \vec{\nabla}L = -q\vec{\nabla}V + q\vec{\nabla}\left(\vec{A} \cdot \vec{v}\right)\,. \tag{8.72}$$

One can check that the last term in Equation (8.72) can be written as $\vec{\nabla}\left(\vec{A} \cdot \vec{v}\right) = (\vec{v}\cdot\vec{\nabla})\vec{A} + \vec{v}\wedge(\vec{\nabla}\wedge\vec{A})$, where $(\vec{v}\cdot\vec{\nabla}) = v_x\partial_x + v_y\partial_y + v_z\partial_z$. In order to calculate explicitly the left-hand term of Equation (8.72), the derivative $d\vec{A}/dt$ must be calculated, where $\vec{A}(\vec{r}, t)$ depends explicitly on t, but also implicitly via the position $\vec{r}$ of the particle, thus: $d\vec{A}/dt = (\vec{v} \cdot \vec{\nabla})\vec{A} + \partial_t\vec{A}$. As expected, the equations of motion (8.72) are finally:

$$m\frac{d\vec{v}}{dt} = q\left(-\vec{\nabla}V - \partial_t\vec{A}\right) + q\vec{v} \wedge \left(\vec{\nabla} \wedge \vec{A}\right) = q\left(\vec{E} + \vec{v} \wedge \vec{B}\right)\,, \tag{8.73}$$

The Hamiltonian $H = \vec{p} \cdot \vec{v} - L$ is:

$$H(\vec{r}, \vec{p}, t) = \frac{1}{2m}\left(\vec{p} - q\,\vec{A}(\vec{r}, t)\right)^2 + q\,V(\vec{r}, t)\,. \tag{8.74}$$

In the case of interest here, the system is contained in a cube of side L to which a constant magnetic field $\vec{B} = B\,\vec{e}_z$ is applied. This field derives from the vector potential[24] $\vec{A}(\vec{r}) = Bx\vec{e}_y$. The quantum Hamiltonian of an electron (with an electric charge $q = -e$ with $e > 0$) subjected to this magnetic field is obtained using the substitution $\vec{p} \rightarrow -i\hbar\vec{\nabla}$, and reads:

$$H = \frac{1}{2m}\left(-i\hbar\vec{\nabla} + e\vec{A}(\vec{r})\right)^2 = -\frac{\hbar^2}{2m}(\partial_x^2 + \partial_z^2) + \frac{1}{2m}(-i\hbar\partial_y + eBx)^2\,. \tag{8.75}$$

---

[23]see for example, Landau and Lifshitz, _Mechanics, Course of Theoretical Physics, Volume 1_ (Butterworth-Heinemann 1976).

[24]The so-called _Landau gauge_ has been chosen here. Another possible choice is $\vec{A} = -By\vec{e}_x$. A symmetric gauge can also be chosen. In that case, the vector potential is half of the sum of the two previous gauges, i.e. $\vec{A} = \frac{1}{2}\vec{B} \wedge \vec{r}$.

The eigenstates ψ of the system correspond to the solutions of the stationary Schrödinger equation:

$$\frac{1}{2m}\left(-i\hbar\vec{\nabla} + e\vec{A}(\vec{r})\right)^2 \psi(x,y,z) = E\,\psi(x,y,z)\,. \tag{8.76}$$

Since neither y nor z appear in Hamiltonian (8.75), one can look for solutions of the form:

$$\psi(x,y,z) = e^{ip_y \cdot y/\hbar} e^{ip_z \cdot z/\hbar} \phi(x)\,, \tag{8.77}$$

where p_y and p_z are constants. Reporting this expression in Equation (8.76) leads to the following equation for $\phi(x)$:

$$-\frac{\hbar^2}{2m}\frac{d^2\phi}{dx^2} + \frac{e^2B^2}{2m}\left(x + \frac{p_y}{eB}\right)^2 \phi = \left(E - \frac{p_z^2}{2m}\right)\phi\,. \tag{8.78}$$

This is the Schrödinger equation of a one-dimensional harmonic oscillator[25] of angular frequency ω_c such that $\frac{1}{2}m\omega_c^2 = (eB)^2/2m$, i.e. $\omega_c = eB/m$. The cyclotron frequency of the classical motion of an electron subjected to a constant magnetic field can be immediately recognised. The eigenlevels of Equation (8.78), $\epsilon_n = E - p_z^2/2m$, can thus be written as $\hbar\omega_c(n + \frac{1}{2})$, where $n \in \mathbb{N}$ and the system's energy levels read:

$$E_n(p_z) = \frac{p_z^2}{2m} + \frac{e\hbar B}{m}\left(n + \frac{1}{2}\right)\,. \tag{8.79}$$

This expression has the following physical interpretation: the longitudinal and transverse motions to the magnetic field are decoupled as in classical mechanics. The first term of the right-hand side of Equation (8.79) corresponds to the unperturbed motion along the z-axis (the axis of the magnetic field) and the second term to the cyclotron motion in the xOy plane (perpendicular to the magnetic field). The gauge choice puts special emphasis on the wave function along x, but another choice can singularise y instead or make x and y play a symmetric role (see Note 24).

In the xy-plane, the free classical motion in the absence of a magnetic field is replaced, when B is non-zero, by cyclotron orbits and the initially equidistant quantum levels with a density of state $d(E)$ given by Equation (8.55) now have a spacing $\hbar\omega_c = 2\mu_B B$, where μ_B is the Bohr magneton (6.53).

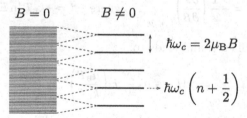

FIGURE 8.7 Representation of the energy spectrum of the Hamiltonian in the xy-plane with and without magnetic field. When $B = 0$, the levels are equidistant, with a density given by Equation (8.55) (two-dimensional free fermions).

In the presence of the magnetic field B, the levels of the transverse Hamiltonian (the so-called *Landau levels*) have high degeneracy since a level $\hbar\omega_c(n + \frac{1}{2})$ gathers states initially lying between $\hbar\omega_c n$ and $\hbar\omega_c(n + 1)$. The degeneracy of a Landau level is therefore:

$$g_L = \int_{\hbar\omega_c n}^{\hbar\omega_c(n+1)} d(E)\,dE = \hbar\omega_c \frac{m\mathcal{A}}{2\pi\hbar^2} = \frac{eB}{h}\mathcal{A}\,, \tag{8.80}$$

---

[25] Not centred at the origin, but in $-p_y/eB$.

where $d(E)$ is the density of states of a two-dimensional free Fermi gas [Equation (8.55)] and $\mathcal{A}$ is the area of the sample in the xy-plane ($\mathcal{A} = L^2$).

The grand canonical partition function Ξ can now be calculated using Expression (6.80):

$$\ln \Xi = d_s \, g_{\rm L} \, \frac{L}{h} \int_{\mathbb{R}} {\rm d}p_z \sum_{n=0}^{\infty} \ln \left(1 + z \, {\rm e}^{-\beta E_n(p_z)} \right) . \tag{8.81}$$

In this formula, d_S is the spin degeneracy factor, and L/h is the density of levels in p_z-space[26].

8.6.1 NON DEGENERATE LIMIT

This is the high temperature and low density limit for which $z \to 0$. It is valid in the regime defined by Equation (8.12). A series expansion in Equation (8.81) leads to:

$$
\begin{aligned}
\ln \Xi &\simeq d_S \, g_{\rm L} \, \frac{zL}{h} {\rm e}^{-\beta \hbar \omega_c /2} \sum_{n=0}^{\infty} {\rm e}^{-n\beta \hbar \omega_c} \int_{-\infty}^{\infty} {\rm d}p_z \, {\rm e}^{-\beta p_z^2/(2m)} \\
&= d_S \, g_{\rm L} \, \frac{zL}{\lambda_T} \frac{{\rm e}^{-\beta \hbar \omega_c /2}}{1 - {\rm e}^{-\beta \hbar \omega_c}} = d_S \, \frac{L^3}{\lambda_T^3} e^{\beta \mu} \frac{\beta \mu_{\rm B} B}{\sinh(\beta \mu_{\rm B} B)} .
\end{aligned}
\tag{8.82}
$$

In this expression, λ_T is the thermal wavelength [Equation (8.6)]. The chemical potential is determined as a function of the number of electrons using Formula (6.34) (where $\alpha = \beta \mu$). This immediately gives (here $V = L^3$)

$$N = d_S \, \frac{V}{\lambda_T^3} e^{\beta \mu} \frac{\beta \mu_{\rm B} B}{\sinh(\beta \mu_{\rm B} B)} . \tag{8.83}$$

Expression (8.83) which relates N to μ has been written in a form which makes apparent that, in the zero field limit, the usual result [Equation (8.10)] is recovered.

The system's magnetisation M (magnetic moment per unit volume) is [see Equation (10.64) for a proof of the formula relating M to J]:

$$M = -\frac{1}{V} \left(\frac{\partial J}{\partial B} \right)_{T,V,\mu} , \quad \text{where} \quad J = -k_{\rm B} T \ln \Xi . \tag{8.84}$$

The calculation is not difficult, and leads to:

$$M = -\mu_{\rm B} \frac{N}{V} L \left(\frac{\mu_{\rm B} B}{k_{\rm B} T} \right) , \tag{8.85}$$

where L is the Langevin function defined by Equation (6.47). It is important to note that the magnetisation is negative, i.e. opposite to the magnetic field inducing it: this is a dia-magnetism response. Note here that Expression (8.85) was obtained discarding the Pauli exclusion principle (which is legitimate in the non-degenerate limit). However, this results remains "quantum" is some sense: the wave nature of the particles and the discreteness of the Landau levels are fully taken into account.

---

[26]This can be found by imposing periodic boundary conditions along the z−direction (which has a length L) and then noting that the p_z are equidistant and separated by $2\pi\hbar/L = h/L$.

8.6.2 LOW FIELD LIMIT

This is the regime $\mu_B B \ll k_B T$. In this limit the variation of the integrand in Equation (8.81) when n is increased by one is small and the Euler-MacLaurin formula [Equation (6.94)] can be used. Here, the sum can be written as[27]:

$$\sum_{n=0}^{\infty} \varphi(n + \tfrac{1}{2}) = \int_0^{\infty} dx\, \varphi(x) + \tfrac{1}{24}\, \varphi'(0) + \cdots \tag{8.86}$$

This leads to [with $J = -k_B T \ln \Xi$, where Ξ is given by: Equation (8.81)]

$$\begin{aligned} J \simeq &-2d_s k_B T g_L \frac{L}{h} \int_0^{\infty} dx \int_0^{\infty} dp_z \ln\left[1 + \exp\left(\frac{\mu - 2\mu_B Bx - p_z^2/2m}{k_B T}\right)\right] \\ &+ \frac{\mu_B B}{6} d_s g_L \frac{L}{h} \int_0^{\infty} \frac{dp_z}{1 + \exp\{\beta(p_z^2/2m - \mu)\}} \,. \end{aligned} \tag{8.87}$$

In the following, let J_0 and J_B be the two right-hand terms of (8.87), respectively. To calculate J_0, the change variable $(x, p_z) \rightarrow (\epsilon, p)$ is made with $p_z = p$ and $x = (\epsilon - p^2/2m)/(2\mu_B B)$. So, $dx\, dp_z = dp\, d\epsilon/(2\mu_B B)$, with $\epsilon \in \mathbb{R}^+$ and $p \in [0, \sqrt{2m\epsilon}\,]$. Thus:

$$J_0 = -2k_B T d_s g_L \frac{L}{h} \frac{1}{2\mu_B B} \int_0^{\infty} d\epsilon \int_0^{\sqrt{2m\epsilon}} dp \ln\left[1 + \exp\left(\frac{\mu - \epsilon}{k_B T}\right)\right] \,. \tag{8.88}$$

The integration over p is trivial and leads to the following expression:

$$J_0 = -k_B T d_s \int_0^{\infty} d\epsilon\, d(\epsilon) \ln\left(1 + e^{-\beta(\epsilon - \mu)}\right) , \tag{8.89}$$

where $d(\epsilon)$ is the three dimensional density of state [Equation (8.2)]. J_0 corresponds to the grand potential in the absence of the external magnetic field as can be seen when comparing Equation (8.89) to Equation (8.8) (in which $PV = -J_0$).

The term J_B is the low field correction to J_0. With the change of variable $x = \beta p_z^2/2m$, it can be written as:

$$J_B = \frac{1}{3} \frac{(\mu_B B)^2}{k_B T} \frac{V}{\lambda_T^3} f_{1/2}(z) \,. \tag{8.90}$$

In this expression, the spin degeneracy is $d_s = 2$. In addition, the electric charge e was written in terms of the Bohr magneton $\mu_B = e\hbar/(2m)$.

It is instructive to study the non-degenerate limit of Expression (8.90). In this limit, $f_{1/2}(z) \simeq z \simeq N\lambda_T^3/(2V)$ [see Equations (A.10) and (8.10)] which yields $J_B = \tfrac{1}{6}(\mu_B B)^2(k_B T)^{-1}N$ then, with Equation (8.84), the magnetisation is $M = -\tfrac{1}{3}\mu_B^2 B(k_B T)^{-1}N/V$. It is, of course, identical to the low field expansion of expression Equation (8.85)[28].

The opposite limit is the degenerate regime. Expansion (8.19) must now be used giving $f_{1/2}(\exp(\beta\mu)) \simeq 2(\beta\mu/\pi)^{1/2}$. In this limit, at the leading order $\mu = E_F$ and using the expression of Note 4 of page 279, one obtains:

$$J_B = \frac{N (\mu_B B)^2}{4 E_F} \,, \quad \text{and} \quad M = -\frac{1}{V}\frac{\partial J}{\partial B} = -\frac{1}{2}\frac{N}{V}\frac{\mu_B^2 B}{E_F} \,. \tag{8.91}$$

---

[27]To recover Equation (8.86) from Equation (6.94), let $f(x) = \varphi(x + 1/2)$ be a function of x. Then in the right-hand term of Equation (6.94), one can write $\int_0^{\infty} f(x)dx = \int_0^{\infty} \varphi(x)dx - \int_0^{1/2} \varphi(x)dx$. After a series expansion of $\varphi(x)$ around zero, this last integral can be written as: $\tfrac{1}{2}\varphi(0) + \tfrac{1}{8}\varphi'(0)$. Again, in the right-hand term of (6.94), one can write $f(0) = \varphi(1/2) \simeq \varphi(0) + \tfrac{1}{2}\varphi'(0)$ and, with the same level of accuracy, $f'(0) = \varphi'(1/2) \simeq \varphi'(0)$. Then, Formula (8.86) is recovered.

[28]This can be checked by using the series expansion of Langevin function when $x \ll 1$: $L(x) = \tfrac{1}{3}x + O(x^3)$.

The magnetic susceptibility is given by Equation (6.49) leading to

$$\chi_{\text{Landau}} = -\frac{1}{2}\frac{N}{V}\frac{\mu_0\mu_B^2}{E_F} . \tag{8.92}$$

This expression of the diamagnetic susceptibility of a system of degenerate free electrons is exactly equal to $-1/3$ of the corresponding paramagnetic susceptibility [Equation (8.50)].

9 Phase Transition – Mean Field Theories

This and the four next chapters are devoted to the study of several aspects of the theory of phase transition. This has been for more than one century, and still remains, a subject of intense experimental and theoretical activity, with important breakthroughs having impacts on different fields in physics but also on other disciplines such as mathematics, economics, sociology, biology. . .

In the first section of this chapter, the transition between the ferromagnetic and paramagnetic phases of the Ising system will be studied in the framework of a mean field approximation. The problem will first be addressed by means of the intuitive Weiss molecular field approximation (Section 9.1.2) and then by means of the more detailed Bragg-Williams approximation (Section 9.1.3) which will serve as a base to develop, in Section 9.2, a general theory of continuous phase transitions: the Landau mean field theory for which the concept of order parameter will first be introduced. Other types of phase transitions will then be studied using the mean field approximation[1]. Phase transitions in liquid crystals will be studied in Section 9.3. These are systems with a tensor order parameter which exhibit a rich phenomenology. Finally, in Section 9.4, a transition of a fairly more elaborate nature than a simple first or second order phase transition will be discussed.

9.1 FERROMAGNETISM

Some solids, such as pure iron, cobalt and nickel, gadolinium, dysprosium, as well as certain alloys, have so-called *ferromagnetic properties*: they tend to acquire a strong magnetisation in the presence of an external magnetic field, and even can spontaneously magnetise (i.e. in the absence of a magnetic field) at low temperature.

The direct interaction between magnetic moments is the first phenomenon that comes to mind when attempting to interpret this behaviour. However, a simple order of magnitude estimate rules out this mechanism. Indeed, the interaction energy of two classical magnetic dipoles, $\vec{m}_1$ and $\vec{m}_2$ located in $\vec{r}_1$ and $\vec{r}_2$, respectively, is:

$$\mathcal{H}_{MD} = \frac{\mu_0}{4\pi r^3} \left[\vec{m}_1 . \vec{m}_2 - 3 \left(\vec{m}_1 . \hat{r} \right) \left(\vec{m}_2 . \hat{r} \right) \right] , \tag{9.1}$$

where $r = |\vec{r}_2 - \vec{r}_1|$ and $\hat{r} = (\vec{r}_2 - \vec{r}_1)/r$. For two moments of the order of the Bohr magneton $m \sim \mu_B = |q_e|\hbar/(2 m_e) \sim 10^{-23}$ J/T, considering (an optimistic assumption) a small distance between dipoles of the order of 1 Å, the interaction energy is of the order of $\mathcal{H}_{MD}|_{\text{typ}}/k_D \sim 1$ K. Thus, the direct magnetic dipole interaction [given by Hamiltonian (9.1)] is completely neutralised by thermal motion for temperatures of the order of $10^2 - 10^3$ K at which ferromagnetism is, however, observed: direct magnetic dipole interactions play no role in the ferromagnetic properties at room temperature.

9.1.1 HEISENBERG AND ISING HAMILTONIANS

The mechanism at the origin of magnetic order is called the *exchange mechanism*. It is based on a quantum statistical effect. For simplicity, consider a two-electron system described by the Hamiltonian $\mathcal{H} = \mathcal{H}_0(\vec{r}_1, \vec{p}_1) + \mathcal{H}_0(\vec{r}_2, \vec{p}_2) + W$. $\mathcal{H}_0 = p^2/(2m) + V(\vec{r} - \vec{R}_a) + V(\vec{r} - \vec{R}_b)$

---

[1]In Section 5.3, the mean field approximation was already used to study the liquid-gas transition.

DOI: 10.1201/9781003272427-9

is a Hamiltonian with two potential wells centred in $\vec{R}_a$ and $\vec{R}_b$, respectively. $W(\vec{r}_1, \vec{r}_2) = q^2/(4\pi\varepsilon_0 r)$ describes the interaction between the two electrons ($r = |\vec{r}_2 - \vec{r}_1|$). The wave function describing the quantum state of the two spin-1/2 electrons must be antisymmetric: Since $\mathcal{H}$ is spin-independent, the system is either (i) in a space-antisymmetric and spin-symmetric $s = 1$ triplet state or (ii) in a space-symmetric and spin-antisymmetric $s = 0$ singlet state[2].

Before taking into account symmetry considerations, the Hilbert space of spatial wave functions of the two particles is reduced to only the two following states: $\Phi_{ab}(\vec{r}_1, \vec{r}_2) = \varphi_a(\vec{r}_1)\varphi_b(\vec{r}_2)$ and $\Phi_{ba}(\vec{r}_1, \vec{r}_2) = \varphi_b(\vec{r}_1)\varphi_a(\vec{r}_2)$, where $\varphi_{a/b}$ is a normalised wave function localised around $\vec{R}_{a/b}$ and an approximate solution[3] of $\mathcal{H}_0\,\varphi_{a/b} = e_0\varphi_{a/b}$. Since $\Phi_{ba}(\vec{r}_1, \vec{r}_2) = \Phi_{ab}(\vec{r}_2, \vec{r}_1)$, the system's symmetrised ($s = 0$) or anti-symmetrised ($s = 1$) spatial wave functions take the form of the so-called *Heitler-London ansatz*:

$$\Phi_s(\vec{r}_1, \vec{r}_2) = \frac{1}{\sqrt{2}(1 + (-1)^s\Sigma^2)}\left[\Phi_{ab}(\vec{r}_1, \vec{r}_2) + (-1)^s\Phi_{ba}(\vec{r}_1, \vec{r}_2)\right], \qquad (9.2)$$

where $s = 0$ or 1, depending on whether a singlet or a triplet spin configuration is considered. The denominator of Expression (9.2) is a normalisation factor chosen to ensure $\langle\Phi_s|\Phi_s\rangle = 1$. It involves the overlap integral $\Sigma = \langle\varphi_a|\varphi_b\rangle$ (Σ^2 is real). When using the variational ansatz (9.2), the system's energy is:

$$E_s = \langle\Phi_s|\mathcal{H}|\Phi_s\rangle = 2e_0 + \frac{C + (-1)^s\mathcal{J}}{1 + (-1)^s\Sigma^2}, \qquad (9.3)$$

where $2e_0$ is the energy of the two electrons when the interactions are ignored and the real constants C and $\mathcal{J}$ are defined by:

$$C = \langle\Phi_{ab}|W|\Phi_{ab}\rangle = \langle\Phi_{ba}|W|\Phi_{ba}\rangle, \quad \mathcal{J} = \langle\Phi_{ab}|W|\Phi_{ba}\rangle = \langle\Phi_{ba}|W|\Phi_{ab}\rangle.$$

C and $\mathcal{J}$ are called the *Coulomb integral* and the *overlap integral*, respectively. C is obviously positive, as is[4] $\mathcal{J}$. Thus, the triplet configuration is energetically favoured in the very standard case $\Sigma^2 \ll 1$ (which will be assumed from now on). The electron density profiles associated with the two wave functions are sketched in Figure 9.1. As can be seen, the electron density profile of the antisymmetric space wave function Φ_1 is more clearly separated into two peaks than that of the symmetric space wave function Φ_0. This explains the lower energy of Φ_1: the repulsive interaction between electrons is weaker in the spatial anti-symmetric configuration.

Note that

$$\vec{s}_1.\vec{s}_2 = \frac{1}{2}\left[(\vec{s}_1 + \vec{s}_2)^2 - s_1^2 - s_2^2\right] = \frac{\hbar^2}{2}\left[s(s+1) - \frac{3}{2}\right] = \begin{cases} \hbar^2/4 & \text{in a triplet state,} \\ -3\hbar^2/4 & \text{in the singlet state,} \end{cases}$$
$$(9.4)$$

where we have used the fact that, for an angular momentum j, the eigenvalue of $\vec{j}^2$ is $\hbar^2 j(j+1)$. The factor $(-1)^s$ can consequently be written in terms of the scalar product $\vec{s}_1.\vec{s}_2$,

---

[2]The triplet state has three possible spin projections along the quantisation axis: $|s = 1, s_z = 1\rangle = |\uparrow\uparrow\rangle$, $|s = 1, s_z = 0\rangle = 2^{-1/2}(|\uparrow\downarrow\rangle + |\downarrow\uparrow\rangle)$ and $|s = 1, s_z = -1\rangle = |\downarrow\downarrow\rangle$. These spin wave functions are all symmetric under the exchange of the two spins. On the other hand, the singlet spin state $|s = 0, s_z = 0\rangle = 2^{-1/2}(|\uparrow\downarrow\rangle - |\downarrow\uparrow\rangle)$ is anti-symmetric.

[3]Wave functions of the type $\varphi_a(\vec{r}_1)\varphi_a(\vec{r}_2)$ where both electrons are centred on the same site are discarded: once the interaction between electrons is taken into account, such wave functions are energetically very unfavourable.

[4]Let us introduce the quantity $f(\vec{r}) = \varphi_a(\vec{r})\varphi_b^*(\vec{r})$. $\hat{f}(\vec{k})$ and $\hat{W}(\vec{k}) = q^2/(\varepsilon_0 k^2) > 0$ are the Fourier transforms of f and W, respectively. It can be shown that $\mathcal{J} = \int d^3k\,\hat{W}(\vec{k})|\hat{f}(\vec{k})|^2$, which implies that $\mathcal{J}$ is positive.

(a) $\rho_0(x)$ (b) $\rho_1(x)$

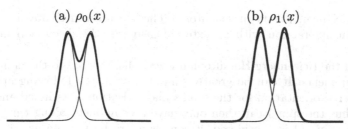

FIGURE 9.1 Electron density profiles $\rho_s(x)$ (bold curves) for (a) the state $s = 0$ (space symmetric) and (b) the state $s = 1$ (space antisymmetric). For simplicity, a one dimension space is considered where $\rho_s(x) = 2 \int |\Phi_s(x,y)|^2 \mathrm{d}y$. The thin line curves are plots of $|\varphi_a(x)|^2$ and $|\varphi_b(x)|^2$ (each represented by a Gaussian function normalised to unity).

and the original interaction term W can be substituted for an effective interection term:

$$\mathcal{H}_{\text{eff}} = C + (-1)^s \mathcal{J} = C - \frac{1}{2}\left(1 + \frac{4}{\hbar^2} \vec{s}_1 . \vec{s}_2\right)\mathcal{J} \ . \tag{9.5}$$

In Equation (9.5), the order of magnitude of the term proportional to spins is: $\mathcal{J} \sim [q^2/(4\pi\varepsilon_0 r)] \times \Sigma^2$. The order of magnitude of the overlap term Σ^2 is certainly small but difficult to estimate. The order of magnitude of the prefactor $q^2/(4\pi\varepsilon_0 r)$ is large ($\sim 3 \times 10^4$ K for a distance $r \simeq 5$ Å) and the exchange mechanism can lead to an overall quite significant (ferro)magnetic interaction.

The mechanism presented above is called *direct exchange*. It is effective in materials for which the electrons involved in the exchange are localised (ground hypothesis of the Heitler-London ansatz) such as chromium dioxide (CrO_2) and $CrBr_3$. For transition metals of the iron series, an overlap of the 3d electron orbitals creates energy bands in which electrons cannot be treated as localised particles (band-ferromagnetism). In other solids, different mechanisms, often more relevant to describe magnetic interactions, can be involved (super-exchange, indirect exchange, itinerant exchange). The direct exchange was presented here because it is the simplest. However, every mechanism relies on the same two ingredients: electrostatic interaction and quantum statistical effects.

The Hamiltonian describing the magnetic interactions between spins in a solid and their coupling to a possible external field $\vec{B}$ is the Heisenberg Hamiltonian:

$$H = -\frac{2}{\hbar^2}\sum_{i=1}^{N}\sum_{j=i+1}^{N} \mathcal{J}_{ij}\, \vec{s}_i \cdot \vec{s}_j - \sum_{i=1}^{N} \vec{\mu}_i \cdot \vec{B} \ . \tag{9.6}$$

In this expression, $\mathcal{J}_{ij}$ is the overlap integral between sites i and j of a regular (crystalline) lattice. The constant terms in Equation (9.5) are not included. The double sum in Equation (9.6) is written in a way which avoids double counting of the interaction between spins i and j. Given the small range of the exchange interaction, it is assumed for simplicity that $J_{ij} \neq 0$ only if the lattice sites i and j are nearest neighbours. The nearest neighbours overlap integral is written as $\mathcal{J}_{ij} = 2J$ (the factor 2 is here for aesthetic reasons). With this simplifying assumption the Hamiltonian of the system reads:

$$H = -\frac{4J}{\hbar^2}\sum_{\langle i,j\rangle} \vec{s}_i \cdot \vec{s}_j - \sum_{i=1}^{N} \vec{\mu}_i \cdot \vec{B} \ , \tag{9.7}$$

where the first sum is over all pairs of nearest neighbour sites of the lattice. If $J > 0$, the material is ferromagnetic and if $J < 0$, it is anti-ferromagnetic (a phenomenon explained by the "super-exchange" theory in insulators). The last term in Expression (9.6) describes the paramagnetic interaction with an external field [see the discussion in Section 6.2.2, Formula (6.52)].

The study of the Heisenberg Hamiltonian is complicated due to the non-commutativity of the spin components. It can be greatly simplified with the following approximation: a (strongly) preferred orientation for the solid's magnetisation is assumed and the magnetic moment (and thus the spins) can then only have a component along this axis (chosen as the z-axis). If a spin-1/2 is considered, it can therefore be written as: $\vec{s}_i = \sigma_i \frac{\hbar}{2} \vec{e}_z$, where $\sigma_i = \pm 1$. Here, instead of the quantum relation (9.4), one simply gets $\vec{s}_i \cdot \vec{s}_j = \frac{\hbar^2}{4} \sigma_i \sigma_j$ and Relation (6.52) leads to $\mu_i|_z = \frac{1}{2} g \mu_B \sigma_i$. Taking $h = \frac{1}{2} g \mu_B B_z$, the Ising[5] Hamiltonian is obtained:

$$H = -J \sum_{\langle i,j \rangle} \sigma_i \sigma_j - h \sum_{i=1}^{N} \sigma_i . \tag{9.8}$$

If the constant J is chosen positive, the first term corresponds to a short range ferromagnetic interaction (only among nearest neighbours). The second sum runs over all spins and corresponds to a paramagnetic term.

9.1.2   WEISS MOLECULAR FIELD

Consider a system described by Hamiltonian (9.8) in contact with a thermostat at temperature T. Let us examine the following mean value:

$$\bar{m} = \frac{1}{N} \left\langle \sum_{i=1}^{N} \sigma_i \right\rangle , \tag{9.9}$$

where the mean is calculated over all spins (hence the sum over i) but also in a statistical way. This statistical average is represented by the notation $\langle \ldots \rangle$ and corresponds to a sum over all possible spin arrangements on the lattice, called here *spin configurations*. A given configuration has an energy E_{conf} and must be weighted by the canonical factor $\exp(-\beta E_{\text{conf}})/Z$. Sometimes $\bar{m}$ is referred to as the *magnetisation* of the system. This abuse of language will be used throughout this chapter: magnetisation has indeed a precise definition, it is the density of magnetic moment per unit volume, noted M [see, e.g., Equation (6.48)]. According to Definition (9.9), the two quantities are related by:

$$M = \mu n_V \bar{m}, \tag{9.10}$$

where $\mu = |\mu_z| = g \mu_B / 2$ is the z-axis component of the magnetic moment carried by each site and n_V is the number of sites per unit volume.

Using the canonical partition function $Z = \sum_{\text{conf}} \exp(-\beta E_{\text{conf}})$ and the free energy $F = -k_B T \ln Z$, the magnetisation is:

$$\bar{m} = \frac{1}{NZ} \sum_{\text{conf}} \left(\sum_{i=1}^{N} \sigma_i \right)_{\text{conf}} \exp(-\beta E_{\text{conf}}) = \frac{1}{N\beta Z} \frac{\partial Z}{\partial h} = -\frac{1}{N} \frac{\partial F}{\partial h} . \tag{9.11}$$

This formula allows the determination of $\bar{m}$ from the knowledge of the canonical partition function[6]. However the spins interactions render the calculation of Z complicated. The problem is greatly simplified by simply eliminating the interactions, i.e. by taking

---

[5]Note that the sign of h has no peculiar relevance: changing h to $-h$ is equivalent to changing the direction of the z-axis or changing σ_i to $-\sigma_i$ for all i.

[6]It is the dimensionless counterpart of Formula (10.64) derived for a general case in Chapter 10.

$J = 0$ in Equation (9.8). The corresponding Hamiltonian is then separable into N independent identical one-body Hamiltonians with the associated canonical partition function (see Equation (6.54) and the following):

$$z = \sum_{\sigma_i \in \{-1,1\}} \exp(\beta h \sigma_i) = 2\cosh(\beta h) . \tag{9.12}$$

Then, the magnetisation is:

$$\bar{m} = \langle \sigma_i \rangle = z^{-1} \sum_{\sigma_i \in \{-1,1\}} \sigma_i \exp(\beta h \sigma_i) = \tanh(\beta h) . \tag{9.13}$$

This formula has a clear physical interpretation: at low temperature ($k_{\mathrm{B}}T \ll h$), the spin aligns with the magnetic field to minimise energy and the magnetisation is then maximum ($\langle \sigma_i \rangle = 1$); in contrast, at high temperature ($k_{\mathrm{B}}T \gg h$), thermal motion prevails and the statistical average $\langle \sigma_i \rangle$ is zero.

In the presence of interaction ($J \neq 0$), every spin sees the external field h in addition to an effective field created by its nearest neighbours on the lattice: $J \sum_{j \in \mathrm{nn}\{i\}} \sigma_j$, where $\mathrm{nn}\{i\}$ denotes the set of nearest neighbours[7] of the lattice site i. In the mean field approximation, this fluctuating effective field is approximated by the Weiss "molecular field", h_{mol}, which is a statistical average calculated as if each spin were independent: $h_{\mathrm{mol}} = qJ\bar{m}$. The system is then reduced to a set of independent spins whose magnetisation in the presence of an external field $h + h_{\mathrm{mol}}$ is calculated using Equation (9.13):

$$\bar{m} = \tanh[\beta(h + qJ\bar{m})] . \tag{9.14}$$

This is a self-consistent equation for $\bar{m}$. It can be solved graphically as illustrated in Figure 9.2 in the absence of an external field ($h = 0$). To do so, it is simpler to make the change of variable $X = \beta qJ\bar{m}$ and to look for the intersection of the curve $\tanh(X)$ with the straight line $k_{\mathrm{B}}TX/(qJ)$ for several values of $k_{\mathrm{B}}T/(qJ)$.

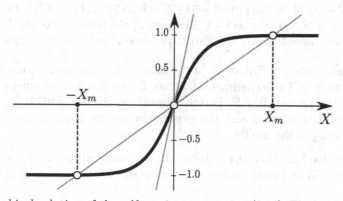

FIGURE 9.2 Graphical solution of the self-consistent equation (9.14). The bold curve is a plot of $\tanh(X)$. The grey strait lines going through the origin are plots of $k_{\mathrm{B}}TX/(qJ)$ for two values of $k_{\mathrm{B}}T/(qJ)$, one larger and the other smaller than unity. In the latter case there are three intersections with the hyperbolic tangent, corresponding to $X = 0$ and $\pm X_m$.

When $k_{\mathrm{B}}T > qJ$, the graph shows that $\bar{m} = 0$ is the only solution of Equation (9.14). The molecular field approach predicts that the magnetisation behaves as if the spins interaction had no role. It corresponds to a paramagnetic phase because, in the presence of an external field, the paramagnetic term of Equation (9.8) has the dominant contribution here.

---

[7]The number q of nearest neighbours is called the *lattice coordination number*.

When $k_B T < qJ$ there are three possible solutions: the non-magnetised solution $\bar{m} = 0$ and two magnetised solutions with opposite values of $\bar{m}$. They can be determined graphically as illustrated by the construction in Figure 9.2 ($\bar{m} = \pm X_m/\beta Jq$). These two solutions reveal the appearance of a spontaneous magnetisation appears in the absence of an external magnetic field (remember that the situation $h = 0$ is for now considered). Equation (9.14) does not indicate which of the three possible solutions is preferred. An unambiguous answer to this question is given in Section 9.1.3. But on physical grounds, it is clear that the magnetised solutions are energetically more favourable than the solution $\bar{m} = 0$ since they correspond to a decrease of the spins interaction energy [the first left-hand term in Equation (9.6)]. It is one of these two solutions that the system will choose. In practice, the degeneracy between these two possible and opposite values of the magnetisation is lifted by small initial fluctuations or by uncontrollable perturbations of the system's magnetic environment: the preferred orientation of the magnetisation cannot be *a priori* predicted, and, in fact, it is unimportant.

Above the critical temperature $T_C = qJ/k_B$, the system looses its permanent magnetisation. This phenomenon corresponds to a transition between a high-temperature phase in which the thermal energy $k_B T$ prevails over the interaction energy qJ with the q neighbours and a low-temperature phase in which, on the contrary, the interactions between neighbour spins (which tends to align them) supersede thermal disorder. Although Hamiltonian (9.8) has both a ferromagnetic component (associated with spins interactions) and paramagnetic component (the interaction with the external field), the high-temperature phase is called the *paramagnetic phase* and the low-temperature magnetic phase is called the *ferromagnetic phase*. The temperature T_C is known as the Curie temperature.

Exercise 9.1. Solve Equation (9.14) in the absence of external magnetic field when $T \leq T_C$: (i) in vicinity of T_C and (ii) when $T \to 0$ K.

Solution: In the vicinity of T_C, when $T \leq T_C$, the magnetisation $\bar{m}$ is non-zero but very weak. In this regime, Equation (9.14) can be solved using a Taylor expansion of the hyperbolic tangent $[\tanh(\epsilon) = \epsilon - \epsilon^3/3 + O(\epsilon^5)]$. It leads to $\bar{m}^2 \simeq 3\,k_B(T_C - T)(k_B T)^2/(qJ)^3 \simeq 3(T_C - T)/T_C$.

When $T \to 0$, it is obvious that $\bar{m} \to 1$. By studying the behaviour of the hyperbolic tangent function when its argument becomes very large, one obtains[8] $\bar{m} \simeq 1 - 2\exp(-2\beta qJ)$.

The numerical solution of Equation (9.14) leads to the magnetisation curve shown as a solid line in Figure 9.3. The experimental points for various ferromagnetic substances are also reported in the figure [after F. Bitter, Phys. Rev. **39**, 337 (1932)][9]. The agreement between the numerical solution and the experimental data is particularly satisfactory in view of the simplicity of the model.

Exercise 9.2. Solve Equation (9.14) above the critical temperature T_C in the presence of a weak magnetic field and then evaluate the magnetic susceptibility in the paramagnetic phase.

Solution: In the regime $\beta h \ll 1$ in the paramagnetic phase ($T > T_C$), the magnetisation is weak and a Taylor expansion can be used in Equation (9.14) ($\tanh x \simeq x$). This yields $\bar{m}/h = 1/(k_B T - k_B T_C)$. The magnetic susceptibility χ is the ratio of the magnetisation over the applied magnetic field: $\chi = M/H$ (where $H = B/\mu_0$). It is thus dimensionless and, according to Equation (9.10), it is

---

[8]$\tanh(x) \simeq 1 - 2e^{-x}$ when $x \to \infty$.

[9]The results were obtained by P. Weiss and R. Forrer, Ann. de Physique **5**, 153 (1926); O. Bloch, PhD thesis, Zurich (1912); F. Hegg, PhD thesis, Zurich (1910) and P. Curie, Ann. de Chim. et de Phys. **5**, 289 (1895).

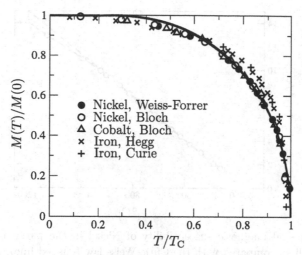

FIGURE 9.3 Magnetisation of several ferromagnetic materials as a function of the normalised temperature T/T_C. The solid line corresponds to the numerical solution of Equation (9.14) when $h = 0$. The points are the experimental results. The Curie temperatures are $T_C = 1043$ K for iron, $T_C = 650$ K for nickel and $T_C = 650$ K for cobalt.

proportional to the ratio $\bar{m}/h$. One obtains the Curie-Weiss law:

$$\chi = \frac{C_{\text{Curie}}}{T - T_C}, \quad \text{where} \quad C_{\text{Curie}} = (\tfrac{1}{2} g \mu_B)^2 n_V \mu_0 / k_B \tag{9.15}$$

is the Curie constant, in agreement with Equation (6.58) for $j = \tfrac{1}{2}$. The existence of a ferromagnetic phase transition modifies the purely paramagnetic law decribed by Equation (6.50). A comparison of Law (9.15) with experimental data is shown in Figure 9.4. The agreement is satisfactory, but, as will be seen in Section 11.6.3, the magnetic susceptibility is in reality no exactly inversely proportional to $(T - T_C)$: the Curie-Weiss law (9.15) must be modified to achieve a better agreement with experiment.

Exercise 9.3.   Study the magnetic susceptibility in the ferromagnetic phase and in the paramagnetic phase near the critical temperature

Solution: To calculate the susceptibility[10] $\chi = \partial \bar{m}/\partial h|_{h=0}$, the two terms of Equation (9.14) must be differentiated with respect to h. This yields

$$\frac{\partial \bar{m}}{\partial h} = \beta \left(1 + q J \frac{\partial \bar{m}}{\partial h}\right)(1 - \bar{m}^2). \tag{9.16}$$

Writing $qJ = k_B T_C$, the suceptibility is:

$$\chi = \left.\frac{\partial \bar{m}}{\partial h}\right|_{h=0} = \frac{1}{k_B} \left.\frac{1 - \bar{m}^2}{T - T_C(1 - \bar{m}^2)}\right|_{h=0}. \tag{9.17}$$

This expression is valid both above and below T_C. At a temperature larger than T_C, the value of the magnetisation is $\bar{m} = 0$. When T is lower but close to T_C, the magnetisation takes the form given in Exercise 9.1: $\bar{m}^2 \simeq 3(T_C - T)/T_C$. Thus, Expression (9.17) reads

$$\chi = \frac{1}{k_B(T - T_C)} \quad \text{when} \quad T > T_C, \quad \text{and} \quad \chi = \frac{1}{2 k_B(T - T_C)} \quad \text{when} \quad T < T_C, T \simeq T_C. \tag{9.18}$$

---

[10]In this exercise, the true dimension of the susceptibility is unimportant. This aspect is studied in Exercise 9.2.

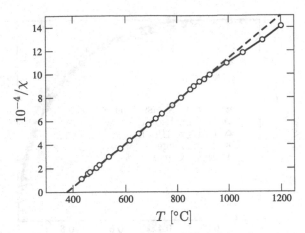

$$T \ [°C]$$

FIGURE 9.4 Experimental magnetic susceptibility of Nickel in the paramagnetic phase (circles connected by a solid line) compared with the Curie-Weiss law (dashed line). A good agreement is obtained for $T_C = 377$ °C and $C_{Curie} = 0.00548$ K [after data published in W. Sucksmith and R. R. Pearce, Proc. R. Soc. Lond. A **167**, 189 (1938).]

This asymmetry in the behaviour of thermodynamic quantities on either side of the critical point, encountered in all mean field theories [see Equation (9.46)], is also observed experimentally, (see Sections 11.5 and 11.6).

Exercise 9.4. Determine the magnetisation curve of a system of spins-j in ferromagnetic interaction.

Solution: The behaviour of interacting spins-j is described by a Hamiltonian similar to Hamiltonian (9.7). Performing the transformation which, for spins-1/2, leads to Equation (9.8), gives here:

$$H = \frac{4J}{\hbar^2} \sum_{\langle i,j \rangle} j_{i,z} j_{j,z} - \frac{g\mu_B}{\hbar} \sum_i j_{i,z} B \ . \tag{9.19}$$

The Weiss molecular field approach is adapted to the present case. Using the results from Section 6.2.2, the molecular field is $B_{mol} = (4J)/(g\mu_B)^2 q \langle \mu_{Mz} \rangle$, where q is the number of nearest neighbours of a lattice site and:

$$\langle \mu_{Mz} \rangle = g\mu_B j \, \mathcal{B}_j \left(\frac{4Jqj}{g\mu_B k_B T} \langle \mu_{Mz} \rangle \right) , \tag{9.20}$$

where $\mathcal{B}_j$ is the Brillouin function [Equation (6.57)]. It is convenient to do the change of variable $X = 4Jq\langle \mu_{Mz} \rangle/(g\mu_B k_B T)$ and then to solve $(k_B T X)/(4Jqj^2) = \mathcal{B}_j(X)$ instead of Equation (9.20). Since $\mathcal{B}_j$ has a slope $(j+1)/(3j)$ at the origin (see Section 6.2.2), the reasoning used to solve Equation (9.14) graphically leads here to a critical temperature $k_B T_C = \frac{4}{3} Jqj(j+1)$. The magnetic susceptibility in the paramagnetic phase can also be calculated: $\chi = C_{Curie}/(T - T_C)$ and the Curie constant is here $C_{Curie} = \frac{\mu_0}{3} n_V (g\mu_B)^2 j(j+1)/k_B$ in agreement with Equation (6.58). Figure 9.5 shows the spontaneous magnetisation of a gadolinium crystal. Solid gadolinium is a ferromagnetic material with a Curie temperature $T_C = 293$ K (20 °C). According to Figure 6.2, gadolinium magnetic centres carry a spin-7/2, and the spontaneous magnetisation is indeed properly described by the mean field theory for spins-7/2.

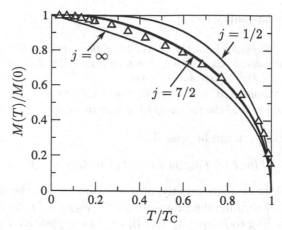

FIGURE 9.5 Magnetisation of Gadolinium as a function of the normalised temperature T/T_C. The triangles are the experimental data. The curve $j = 7/2$ is a plot of the self-consistent solution of Equation (9.20) with $j = 7/2$. The other curves correspond to $j = 1/2$ and the limit $j \to \infty$ [after H. E. Nigh, S. Legvold and F. H. Spedding, Phys. Rev. **132**, 1092 (1963)].

9.1.3 BRAGG-WILLIAMS APPROXIMATION

This section is devoted to an alternative study of the Ising system but still in the framework of a mean field approach. The N spins are located at the nodes of a lattice. Let N^+ be the number of spins "up", and N^- be the number of spins "down". The average magnetisation[11] $m = \frac{1}{N} \sum_i \sigma_i$ can be expressed as $m = (N^+ - N^-)/N$. Conversely, the knowledge of the value of m determines N^+ and N^- [$N^\pm = \frac{N}{2}(1 \pm m)$]. Let us define an incomplete canonical partition function, $Z(m, T)$, corresponding to all spins configurations leading to the same value of m:

$$Z(m, T) = \sum_{\text{conf } m} \exp\{-\beta E_{\text{conf}}\}, \qquad (9.21)$$

where the sum $\sum_{\text{conf } m}$ is restricted the configurations leading to the same, fixed, value of m. The total canonical partition function of the system is the sum of $Z(m, T)$ over all possible values of m:

$$Z(T) = \sum_m Z(m, T). \qquad (9.22)$$

The canonical probability of observing a given magnetisation m of the system is $P(m) = Z(m, T)/Z(T)$. The most probable value of m is therefore the one for which $Z(m, T)$ reaches its maximum, or equivalently, the one for which $F(m, T) = -k_B T \ln Z(m, T)$ reaches its minimum. It is important to understand that the quantities $Z(m, T)$ and $F(m, T)$ are, for the time being, only intermediate tools useful for the study of the magnetisation of the system.

Let $W(m)$ be the number of configurations with a given fixed value m of the system's magnetisation. The following inequality can be obtained:

$$Z(m, T) \geq W(m) \exp\{-\beta \langle E_{\text{conf}} \rangle_m\} \equiv Z_{\text{BW}}(m, T), \qquad (9.23)$$

where the equality on the right-hand side defines the quantity $Z_{\text{BW}}(m, T)$ and $\langle E_{\text{conf}} \rangle_m$ is the average value of E_{conf} over all configurations corresponding to the fixed magnetisation m: $\langle E_{\text{conf}} \rangle_m = \sum_{\text{conf } m} E_{\text{conf}}/W(m)$.

---

[11]Note the difference with Definition (9.9) of $\bar{m}$: $\bar{m} = \langle m \rangle$.

Exercise 9.5.   Prove inequality (9.23).

Solution: *Inequality (9.23) results from the convexity of the exponential function. Indeed, as can be clearly seen on the graphical construction shown in Figure 9.6, for two real numbers a and b and for $\lambda \in [0,1]$, $\lambda e^a + (1 - \lambda)e^b \geq \exp\{\lambda a + (1 - \lambda)b\}$.*

Then, by inductive reasoning, it can be proven that for n terms, the mean of the exponentials is greater than the exponential of the mean, which is exactly equivalent to Inequality (9.23).

Using Inequality (9.23), it can be seen that:

$$F(m,T) \leq F_{\mathrm{BW}}(m,T) = -k_{\mathrm{B}}T \ln Z_{\mathrm{BW}}(m,T) .\tag{9.24}$$

Since $F(m,T)$ is difficult to evaluate exactly, instead of looking for the value of m minimising $F(m,T)$, it is simpler to determine the value minimising $F_{\mathrm{BW}}(m,T)$ whose dependence in m is easy to evaluate. According to Inequality (9.24), the exact quantity $F(m,T)$ is smaller than $F_{\mathrm{BW}}(m,T)$. So F_{BW} certainly has a minimum, but it is not necessarily reached at the true most probable value of m: this is the limitation of the Brag-Williams approximation.

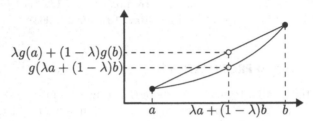

FIGURE 9.6   Illustration of the convexity property. For a convex function g (such as the exponential function), the barycentre of $g(a)$ and $g(b)$ weighted by coefficients λ and $1-\lambda$ is larger than the image by g of the barycentre of a and b weighted by the same coefficients, respectively: $\lambda g(a)+(1-\lambda)g(b) \geq g(\lambda a + (1 - \lambda)b)$.

For a given spin configuration, the energy is $E_{\mathrm{conf}} = -J(N^{++} + N^{--} - N^{+-})$, where N^{++} (N^{--}) is is the number of spin-up (spin-down) pairs of nearest neighbours and N^{+-} is the numbers of nearest neighbour pairs with different spin orientations. The number of spin-up pairs can be written as:

$$N^{++} = \frac{1}{2} \sum_{\substack{i \text{ with } j \in \mathrm{nn}\{i\} \\ \sigma_i = +1}} \begin{bmatrix} 1 & \text{if} & \sigma_j = +1 \\ 0 & \text{if} & \sigma_j = -1 \end{bmatrix} .\tag{9.25}$$

Then, taking the average over the configurations corresponding to the same magnetisation m, one obtains $\langle N^{++}\rangle = \frac{1}{2}N^+ q\, p^+$, where $p^+ = N^+/N$ is the spin-up probability. Defining p^- in a similar manner leads to $\langle N^{--}\rangle = \frac{1}{2}qN^-p^-$. In these formulas, q is the lattice coordination number and the factor $\frac{1}{2}$ is there to avoid double counting. In addition, $\langle N^{+-}\rangle = qN^+p^- = qN^-p^+$, and thus $\langle E_{\mathrm{conf}}\rangle = -\frac{1}{2}qJNm^2$. Evaluating[12] $\ln W(m)$ with the Stirling's formula (A.21) leads to:

$$\frac{F_{\mathrm{BW}}(m,T)}{N} = -\frac{q}{2}Jm^2 + k_{\mathrm{B}}T\left\{\frac{m+1}{2}\ln(m+1) + \frac{1-m}{2}\ln(1-m) - \ln 2\right\} .\tag{9.26}$$

In order to determine how m depends on T within the Bragg-Williams approximation, one must find, at fixed T, the minima of $F_{\mathrm{BW}}(m,T)$ treated as a function of m, or equivalently, the

---

[12]$W(m)$ is the number of ways to distribute N^+ spins-up among N spins: $W(m) = \binom{N}{N^+} = \binom{N}{N^-} = N!/(N^+!\,N^-!)$.

minima of $f_{BW}(m,T) \equiv (F_{BW}(m,T) - F_{BW}(0,T))/N$. In the vicinity of $m = 0$, $(m+1)\ln(m+1) + (1-m)\ln(1-m) = m^2 + m^4/6 + O(m^6)$, so that:

$$f_{BW}(m,T) = \frac{k_B(T-T_C)}{2}m^2 + \frac{k_B T}{12}m^4 + O(m^6), \qquad (9.27)$$

where the critical temperature T_C is defined by $k_B T_C = qJ$. As can be seen in Figure 9.7, when $T > T_C$, the minimum is reached for $m = 0$, while when $T < T_C$, $m = 0$ corresponds to a local maximum (and is therefore not an equilibrium solution). The value of m minimising

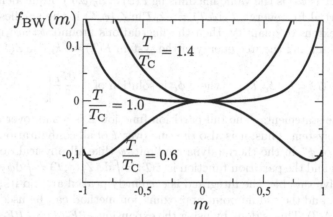

FIGURE 9.7 $f_{BW}(m,T)$ as a function of m for 3 values of T/T_C. The magnetisation corresponds to the value of m minimising $f_{BW}(m,T)$. When $T \geq T_C$ the minimum is reached when $m = 0$, in agreement with the series expansion (9.27). When $T < T_C$ there are two minima obtained for a non-zero value of the mean magnetisation m.

$f_{BW}(m,T)$ is determined by solving $\partial f_{BW}/\partial m = 0$. From Equation (9.26), this reads:

$$-qJm + \frac{k_B T}{2}\ln\frac{1+m}{1-m} = 0. \qquad (9.28)$$

Remembering that $\frac{1}{2}\ln\frac{1+m}{1-m} = \text{artanh}(m)$, it is obvious that this equation is equivalent to the self-consistent equation (9.14) obtained using the Weiss molecular field theory. The advantage of the Bragg-Williams approach is that it presents the calculation of the magnetisation in the framework of the minimisation of an (approximate) free energy.

Note that it is the most probable value of the magnetisation that has been determined here, not its mean value $\bar{m}$. The two concepts are certainly very close, since the equation leading here the most probable magnetisation determines also its mean value when using the Weiss molecular field approach. This point will be discussed further in Section 9.2.1 and finally fully explained in Section 11.2 [Formula (11.15)].

9.2 LANDAU THEORY OF PHASE TRANSITION

In this section, the approach previously used on the Ising model is extended and a general theory of phase transitions is presented.

9.2.1 PARTIAL PARTITION FUNCTION

Consider a system of N interacting particles in thermal equilibrium with a thermostat at temperature T (canonical ensemble). The canonical partition function is $Z(T) = \sum_{\text{conf}} \exp\{-\beta E_{\text{conf}}\}$. Let ϕ be a parameter that depends on the state of the system (energy, magnetisation,...). The probability of observing the system in one of the states corresponding to a specific value of ϕ is:

$$P(\phi) = \frac{Z_L(\phi, T)}{Z(T)} \qquad \text{where} \qquad Z_L(\phi, T) \equiv \sum_{\text{conf } \phi} e^{-\beta E_{\text{conf}}} \,. \tag{9.29}$$

Z_L is a "partial partition function"[13] that only considers the configurations where the parameter takes the value ϕ [hence the notation $\sum_{\text{conf } \phi}$ in Formula (9.29)]. The most probable value of ϕ (written as ϕ_c) is the value maximising $P(\phi) \propto Z_L(\phi, T)$. Equivalently, it is the one for which the partial free energy $F_L(\phi, T) = -k_B T \ln Z_L(\phi, T)$ is the smallest. Furthermore, if $F_L(\phi, T)$ is an extensive quantity, then the fluctuations around ϕ_c are negligible in the thermodynamic limit and the free energy of the system $F(T) = -k_B T \ln Z(T)$ can be written as:

$$F(T) \simeq F_L(\phi_c, T) \,, \quad \text{where } \phi_c \text{ is solution of} \quad \left.\frac{\partial F_L}{\partial \phi}\right|_{\phi_c} = 0 \,. \tag{9.30}$$

Let us prove these statements. The full partition function Z is a sum over all possible configurations of the system. Thus, it is also the sum over ϕ of all configurations corresponding to a "fixed value of ϕ". In the thermodynamic limit, ϕ (like all physical observables) takes continuous values and the partition function is $Z(T) = \int d\phi \, Z_L(\phi, T) = \int d\phi \exp\{-\beta F_L(\phi, T)\}$. If F_L is an extensive variable, the integrand is extremely peaked around its maximum value $\exp\{-\beta F_L(\phi_c, T)\}$, and the saddle point approximation method can be used to evaluate the integral (see Sec. A.3). This is done by using the expansion $-\beta F_L(\phi) \simeq -\beta F_L(\phi_c) - \kappa(\phi - \phi_c)^2$, with $\kappa = \frac{\beta}{2} F_L''(\phi_c)$, where F_L'' is the second derivative of F_L with respect to ϕ. Then, the partition function is approximately [see Equation (A.18)]:

$$Z(T, N) \simeq \sqrt{\frac{\pi}{\kappa}} e^{-\beta F_L(\phi_c, N, T)} \,. \tag{9.31}$$

$P(\phi)$ is therefore a Gaussian distribution of the form: $P(\phi) \simeq \sqrt{\kappa/\pi} \exp\{-\kappa(\phi - \phi_c)^2\}$. Knowing the probability distribution function of the random variable ϕ, its typical fluctuations around the most probable value ϕ_c can be determined by calculating the standard deviation $\sigma_\phi = \langle(\phi - \phi_c)^2\rangle^{1/2}$ which immediately gives $\sigma_\phi = (2\kappa)^{-1/2}$. Now, let us assume that ϕ and its most probable value ϕ_c take values typically behaving like N^γ (γ being any real). Since the partial free energy is considered extensive ($F_L \propto N$), one gets $\kappa \propto F_L''(\phi_c) \propto N^{1-2\gamma}$ so $\sigma_\phi \propto N^{\gamma - 1/2}$ and:

$$\frac{\sigma_\phi}{\phi_c} \propto \frac{1}{\sqrt{N}} \,. \tag{9.32}$$

As already discussed, the ratio σ_ϕ/ϕ_c is a measure of the relative fluctuations which, according to Equation (9.32), are negligible in the thermodynamic limit. Moreover Equation (9.31) yields:

$$F(T) = -k_B T \ln Z(T) = F_L(\phi_c, T) - k_B T \ln \sqrt{\frac{\pi}{\kappa}} \,. \tag{9.33}$$

The first term of the right-hand side of Equation (9.33) is an extensive term, of order $O(N)$. Since the next term is of order $O(1)$, it is negligible in the thermodynamic limit: this justifies that indeed $F(T) \simeq F_L(\phi_c, T)$. However, Relation (9.33) shows that the identification of these two energies is problematic when $\kappa \to 0$; it will be seen later that this happens near a second order phase transition (see Section 9.2.3: at the transition, the second derivative

---

[13]The index L in the notation of Z_L (and later F_L) refers to Landau, as will become obvious in Section 9.2.3.

of the Landau free energy cancels). This is a serious issue because it is precisely in the vicinity of a phase transition that the partial partition function method is used. The use of the method (known as the Landau theory of phase transition) is nevertheless pushed quite far in the following. The aporia will only be solved in Chapter 11 (Section 11.3).

Far from a phase transition – i.e. in the non-pathological case where $\kappa \neq 0$ – the relative fluctuations of ϕ around its most probable value are small and $Z_L(\phi, T)$ can be seen as the system's partition function. However, ϕ has then has a special status: its value (ϕ_c) is not externally fixed (as is the case for T, V or N) but is determined by the equilibrium condition $\partial F_L / \partial \phi = 0$.

9.2.2 ORDER PARAMETER AND SECOND ORDER PHASE TRANSITION

The preceding discussion is theorised in this section to establish a mean field approach to phase transitions generalising the study presented in Section 9.1.3 for the special case of the Ising system[14]. Let us assume that any transition can be characterised by an "order parameter", ϕ, that is zero above a critical temperature, T_C, and takes a finite value below T_C. For the Ising model, $\phi = \bar{m} = \langle \sigma_i \rangle$, for the lambda transition in liquid helium, ϕ is the superfluid fraction (see Figure 7.9), but in some cases the order parameter may be difficult to identify. ϕ may be real, as is the case for the Ising system, or complex as is the case in a superconductor (see Chapter 10), it may also not be a scalar but a vector or a tensor (see Section 9.3 dealing with liquid crystals). The order parameter characterises the transition in the sense illustrated in Figure 9.8.

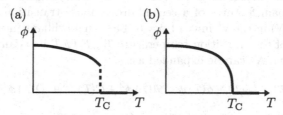

FIGURE 9.8 Order parameter, ϕ, as a function of temperature, T, for (a) a first order phase transition and, (b) a second order phase transition

Phase transitions are classified as first or second order transitions, depending on whether ϕ is continuous or not in T_C. Both behaviours are illustrated in Figure 9.8. First order transitions involve a latent heat, like the solid-liquid transition or the liquid-gas transition away from the critical point. Second order transitions such as the ferromagnetic or the superfluid transition are sometimes called *continuous transitions* and their characteristics will be studied at length in this chapter and the following ones. Landau theory outlined in Section 9.2.3 is designed for the study of second-order phase transitions, but it can be extended to first-order transitions when the discontinuity of ϕ at T_C is small. This is what happens for a second order phase transition in the presence of an external field (see Section 9.2.4). Each phase is characterised by the properties summarised in Table 9.1.

The high temperature phase is symmetric: for example, for the Ising system above the Curie temperature, there are as many spins pointing up as spins pointing down. In this case, the mean magnetisation is zero and invariant by reversal of all spins: the high temperature state verifies the same symmetry property as Hamiltonian (9.8) which is invariant by spins reversal (when $h = 0$). In the low temperature phase, the symmetry is clearly broken: the mean magnetisation points, for example, upwards which breaks the symmetry. Of course,

---

[14]See also the example of a gas on lattice (Exercise 5.4) and the van der Waals theory of the liquid-gas transition discussed in Section 5.3.

TABLE 9.1 General properties of the order parameter during a phase transition.

Low temperature	High temperature
Ordered phase	Disordered phase
Broken symmetry	Symmetric
$\phi \neq 0$	$\phi = 0$

there is also a solution with a magnetisation pointing downwards, which has the same free energy, but the system, driven by tiny and uncontrollable perturbations, chooses one of these two states, which is separated from its "inverted twin" by a potential barrier (as in the case "$\mathcal{A} < 0$" in Figure 9.7). The system's equilibrium configuration does not preserve the symmetry of the Hamiltonian. This is called a *broken symmetry*.

9.2.3 LANDAU FREE ENERGY

The study of the Ising model presented in Section 9.1.1 suggests that, in a mean field approximation, the main features of a second-order phase transition depend only on the behaviour of $F_L(\phi,T,N)$ in the vicinity of $\phi = 0$. Landau's brilliant intuition was to assume that, in the vicinity of the transition temperature T_C, $F_L(\phi,T,V)$ is an analytic function of ϕ which (at fixed T and N) can be expanded as:

$$F_L(\phi,T) = F_0(T) + \alpha(T)\,\phi + \mathcal{A}(T)\,\phi^2 + C(T)\,\phi^3 + \mathcal{D}(T)\,\phi^4 + \ldots \qquad (9.34)$$

Systems at fixed (N,P,T) are often considered (see page 147). In this case, it is the Gibbs free energy (or free enthalpy) $G_L(T,P,\phi) = F_L + PV$ that is expanded in a manner similar to Equation (9.34). In the remainder of this chapter, a general reasoning at fixed (N,V,T) is presented and the Helmoltz free energy $F_L(\phi,T)$ is employed, but the method applies equally well at fixed pressure.

It will be seen later (when studying the influence of an external field, Section 9.2.4) that the term linear in ϕ in Expansion (9.34) introduces a discontinuity of the function $\phi(T)$ when $T = T_C$. It is therefore not appropriate for the description of a second-order transition: henceforth, $\alpha = 0$ is imposed in Expansion (9.34). This *ad hoc* choice sometimes directly results from the symmetries of the problem: for instance, for the Ising system in which $\phi = m$, it is obvious that F_L must have the same symmetry properties as the Hamiltonian H. Thus, every term of its series expansion must be invariant under the transformation $\phi \leftrightarrow -\phi$. In this case odd terms of expansion (9.34) must be zero: $\alpha(T)$ and $C(T)$ are identically zero. These symmetry properties depend on the Hamiltonian and are not universally valid. The choice of the value of the parameter $C(T)$ must thus be justified by other considerations.

According to the example of expansion (9.27) used for the Ising model, it can be expected that coefficient $\mathcal{A}(T)$ in Equation (9.34) changes sign at the transition so that $\mathcal{A}(T \lessgtr T_C) \lessgtr 0$. This behaviour – illustrated in Figure 9.9 – is indeed generic: it corresponds to a situation in which the order parameter [i.e. the value of ϕ minimising $F_L(\phi,T)$] is zero for $T > T_C$ and non-zero $T < T_C$.

In order to obtain a stable transition point when $T = T_C$ (when $\mathcal{A}(T_C) = 0$), it is necessary that $F_L(\phi,T_C)$ has an absolute minimum in $\phi = 0$ so that the transition is indeed a second

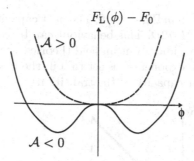

FIGURE 9.9 Behaviour of the Landau free energy (9.34) for two values of $\mathcal{A}$. The case $\mathcal{A} \gtrless 0$ corresponds to $T \gtrless T_C$.

order transition (see Exercise 9.7 at the end of this section). This imposes:

$$\mathcal{D}(T_C) > 0 \quad \text{and} \quad C(T_C) = 0 . \tag{9.35}$$

In the following, the transition is described considering the simplest case: C and $\mathcal{D}$ keep their values in $T = T_C$ for all temperatures, i.e. $C(T) \equiv 0$ and $\mathcal{D}(T) \equiv D = \text{Const} > 0$. The change of sign of $\mathcal{A}(T)$ in $T = T_C$ is simply described by a linear function: $\mathcal{A}(T) = A \times (T - T_C)$ with $A > 0$ [compare to the identical behaviour of term quadratic term in m in Equation (9.27)], i.e.:

$$F_L(\phi, T) = F_0(T) + A \times (T - T_C)\,\phi^2 + D\,\phi^4 + \ldots \tag{9.36}$$

This expression is expected to be only valid in the vicinity of T_C. The minimisation $\partial F_L / \partial \phi = 0$ is obtained for a value ϕ_c given by:

$$\phi_c = \begin{cases} \pm\sqrt{\frac{A}{2D}}(T_C - T)^{1/2} & \text{when } T < T_C , \\ 0 & \text{when } T > T_C . \end{cases} \tag{9.37}$$

Therefore, according to Landau theory, in the vicinity of the critical temperature when $T < T_C$, the order parameter behaves as a power law $(T_C - T)^\beta$ with $\beta = 1/2$. The exponent β is a "critical exponent" such as those already encountered for the liquid-gas transition in Section 5.3.3. These exponents play an important role in the theory of second order phase transitions: mean field approaches predict universal values (for instance, the same mean field value $\beta = 1/2$ was already obtained for the van der Waals theory described in Section 5.3.3). This universality can be tested experimentally (see Section 11.4).

From Equations (9.36) and (9.37), one gets:

$$F(T) = F_L(\phi_c, T) = \begin{cases} F_0(T) - \frac{A^2}{4D}(T_C - T)^2 & \text{when } T < T_C , \\ F_0(T) & \text{when } T > T_C . \end{cases} \tag{9.38}$$

This relation yields for the heat capacity

$$C_V = \left(\frac{\partial E}{\partial T}\right)_V = -T\left(\frac{\partial^2 F}{\partial T^2}\right)_V , \tag{9.39}$$

the following expression:

$$C_V = -T\left(\frac{\partial^2 F_0}{\partial T^2}\right)_V + \begin{cases} A^2 T/2D & \text{when } T < T_C , \\ 0 & \text{when } T > T_C . \end{cases} \tag{9.40}$$

The typical shape of C_V is shown in Figure 9.10. The heat capacity is discontinuous at $T = T_C$, with a discontinuity $\Delta C_V = A^2 T_C/2D$. This behaviour can be compared with experiment. It provides a precise test of the validity of mean field theories, see Section 11.4. In the vicinity of T_C, $C_V \propto |T - T_C|^{-\alpha}$ and the exponent α is set to 0 in the mean field theory ($\alpha > 0$ would mean a divergence of the heat capacity at the transition).

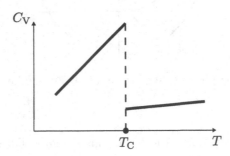

FIGURE 9.10 Schematic shape of the heat capacity in the vicinity of T_C for a second order phase transition. C_V has a discontinuity $\Delta C_V/V = a^2 T_C/2d$ in $T = T_C$, where the parameters a and d are defined by Formula (9.41).

Exercise 9.6. Prove the right-hand side formula of Equation (9.39) used for the calculation of C_V [Equation (9.40)] starting from Definition (3.13): $C_V = (\partial E/\partial T)_V$.

Solutions:
○ *Thermodynamic method: the definition $F = E - TS$ of the Helmoltz free energy allows to rewrite the thermodynamic identity (3.38) as $dF = -SdT - PdV$. Thus $S = -(\partial F/\partial T)_V$, from which it can be deduced $C_V = (\partial [F + TS]/\partial T)_V = (\partial F/\partial T)_V + S + T(\partial S/\partial T)_V = T(\partial S/\partial T)_V = -T(\partial^2 F/\partial T^2)_V$.*
○ *Statistical method: In the thermodynamic limit, the system's energy is $E = \langle E \rangle = Z^{-1} \sum_{\text{conf}} E_{\text{conf}} \exp\{-\beta E_{\text{conf}}\} = -Z^{-1} \partial Z/\partial \beta$, so $E = -\partial \ln Z/\partial \beta$, see Equation (6.15). The relation $\partial(\ldots)/\partial T = \partial(\ldots)/\partial \beta \times \partial \beta/\partial T = -k_B \beta^2 \partial(\ldots)/\partial \beta$ between the derivatives with respect to T and β allows the writing of the heat capacity (3.13) in the form $C_V = k_B \beta^2 \partial^2 \ln Z/\partial \beta^2 = -k_B \beta^2 \partial^2(\beta F)/\partial \beta^2$ where $F = -\beta^{-1} \ln Z$. It is then easy to substitute for $\partial/\partial T$ and obtain the desired expression.*

The free energy is an extensive quantity and it is more convenient to work with quantities per unit volume: $f_L = (F_L - F_0)/V$, $f_0 = F_0/V$, $a = A/V$ and $d = D/V$, such that:

$$\frac{F_L(\phi, T)}{V} = f_0(T) + f_L(\phi, T) = f_0(T) + a \times (T - T_C)\phi^2 + d\phi^4 \, . \qquad (9.41)$$

Exercise 9.7. Consider the case of a system having the following free energy $f_L(T, \phi) = a \times (T - T_0)\phi^2 + c\phi^3 + d\phi^4$, where the parameters a, c and d are positive. Show that the system undergoes a first-order phase transition at the critical temperature $T_C = T_0 + c^2/(4ad)$.

Solution: The solutions of Equation $\partial f_L/\partial \phi = 0$ are $\phi = 0$ and the possible solutions of the equation

$$4d\phi^2 + 3c\phi + 2a(T - T_0) = 0 \, , \qquad (9.42)$$

whose discriminant is $\Delta = 9c^2 - 32ad(T - T_0)$. Δ is positive when $T < T_0 + 9c^2/(32ad)$. In this case, f_L has two extrema other than $\phi = 0$. When $T > T_0 + 9c^2/(32ad)$, f_L has only one extremum (in $\phi = 0$); it is thus a minimum since f_L tends toward infinity for $\phi \to \pm\infty$. This case corresponds to the disordered high temperature phase. When $T < T_0$, $f_L''(0) = 2a(T - T_0) < 0$: $\phi = 0$ is a local maximum, and the two other extrema are local minima: the system is already in the ordered phase where the order parameter is non-zero. The phase transition occurs thus at a temperature between

T_0 and $T_0 + 9c^2/(32ad)$. *For such a temperature, the two roots $\phi^{(\pm)}$ of Equation (9.42) are negative and the behaviour of f_L is represented in Figure 9.11.*

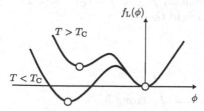

FIGURE 9.11 The two possible behaviours of $f_L(\phi)$ as a function of the order parameter ϕ for temperatures between T_0 and $T_0 + 9c^2/(32ad)$. The minima are shown with white dots. The situation for which the absolute minimum is not reached in $\phi = 0$ is encountered when $T < T_C$.

It is clear that the system is in the ordered phase when $f_L(\phi^{(-)}) < 0$, where $\phi^{(-)}$ is the most negative root of Equation (9.42). $\phi^{(-)} = (-3c - \sqrt{\Delta})/(8d)$ and

$$f_L(\phi^{(-)}) = (\phi^{(-)})^2 \left[a(T - T_0) + c\,\phi^{(-)} + d\,(\phi^{(-)})^2 \right] = (\phi^{(-)})^3 (-d\,\phi^{(-)} - c/2)\,.$$

Thus, the condition $f_L(\phi^{(-)}) \leq 0$ is equivalent to $(-d\,\phi^{(-)} - c/2) \geq 0$. It is easy to check that this last condition is verified when $T < T_C \equiv T_0 + c^2/(4ad)$, indicating that T_C is the system's critical temperature. It is a first order transition since the order parameter ϕ jumps discontinuously from 0 to $\phi^{(-)}(T_C) = -c/(2d)$ in $T = T_C$.

9.2.4 EFFECT OF AN EXTERNAL FIELD. PHASE DIAGRAM

Let us consider the effect of an external field h. The coupling of h with the system is described by the modified free energy[15]

$$f_L(\phi, T, h) = a(T - T_C)\phi^2 + d\phi^4 - h\phi\,. \tag{9.43}$$

The extremisation condition is:

$$\frac{\partial f_L}{\partial \psi} = 0\,, \quad \text{which reads} \quad 2a(T - T_C)\phi + 4d\phi^3 = h\,. \tag{9.44}$$

A quantity of interest is the susceptibility which describes how the order parameter responds to a weak field:

$$\chi \equiv \lim_{h \to 0} \frac{\partial \phi}{\partial h} = \left. \frac{\partial^2 f_L(T, \phi, h)}{\partial h^2} \right|_{h=0}\,. \tag{9.45}$$

By differentiating Equation (9.44) with respect to h, one finds:

$$\chi = \begin{cases} \dfrac{1}{2\,a(T - T_C)} & \text{when } T > T_C\,, \\[2mm] \dfrac{1}{4\,a(T_C - T)} & \text{when } T < T_C\,. \end{cases} \tag{9.46}$$

The susceptibility diverges at the critical temperature. This can be easily understood for $T \to T_C^+$: the system is evolving in the vicinity of a region of temperature at which it spontaneously (i.e. without external forcing) acquires a non-zero value of the order parameter ϕ

---

[15]The coupling term "$-h\phi$" with the external field is similar to the paramagnetic term of the Ising Hamiltonian (9.8). It can be justified that it corresponds to a very general case, but more complex situations are sometimes encountered [see, e.g., the coupling of a superconductor with a magnetic field described by Equation (10.34)].

(for $T \leq T_{\mathrm{C}}$). It is natural that, under such conditions, it is extremely sensitive to an external field, even a weak one, which tends to increase ϕ. The divergence of χ at the critical point is associated with a critical exponent, γ: $\chi \propto |T - T_{\mathrm{C}}|^{-\gamma}$ when $T \to T_{\mathrm{C}}$. Equation (9.46) shows that $\gamma = 1$ according to mean field theory.

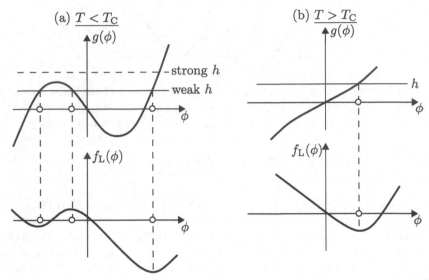

FIGURE 9.12 Behaviour of $f_L(T, \phi, h)$ and $g(\phi) = 2a(T - T_{\mathrm{C}})\phi + 4d\phi^3$ as a function of ϕ at fixed T and h. The extrema of f_L correspond to the solutions of $g(\phi) = h$, [see Equation (9.44)] and are graphically determined with the help of the curves at the top of the figure. (a): case $T < T_{\mathrm{C}}$, (b): case $T > T_{\mathrm{C}}$.

To get more detailed information on the behaviour of ϕ as a function of the external field h, Equation (9.44) can be solved graphically as shown in Figure 9.12 [in which $g(\phi) = 2a(T - T_{\mathrm{C}})\phi + 4d\phi^3$]. In this figure, when $T < T_{\mathrm{C}}$, the "weak h" case corresponds to the situation in which h is smaller than the local maximum $h^*(T)$ of $g(\phi)$: $|h| < h^*(T) = \frac{4}{3}[a(T_{\mathrm{C}} - T)]^{3/2}/(6d)^{1/2}$. The curve $f_L(\phi)$ is plotted in that case.

A graphical analysis of Figure 9.12 can provide a plot of the equation of state $\phi = \phi(T, h)$. To do so, one must, among other things, discuss the behaviour of ϕ as a function of h at fixed T. If $T \geq T_{\mathrm{C}}$, it is obvious from the right part of Figure 9.12 that ϕ is a monotonic and odd function of h because g is. A critical exponent, δ, characterises the dependence of ϕ on h at $T = T_{\mathrm{C}}$:

$$\phi \propto \mathrm{sgn}(h) \times h^{1/\delta}, \quad \text{when} \quad T = T_{\mathrm{C}}. \tag{9.47}$$

Equation (9.44) shows that Landau mean field theory predicts $\delta = 3$ (the equation of the critical isotherm is $\phi = [h/(4d)]^{1/3}$).

The dependence of ϕ on h is more complicated when $T < T_{\mathrm{C}}$, as shown in Figure 9.13: Equation (9.44) has 3 roots if $|h| < h^*(T)$. The section MM' of the curve is unstable (it corresponds to a local maximum of f_L). Similarly, the PM and $P'M'$ sections are metastable: they do correspond to minima of f_L, but these minima are higher than those along the $N'P'$ and PN sections. This can be proven by a direct calculation, but it is clear on physical ground: a solution where ϕ does not have the same sign as h is not energetically favourable. In a situation where the external field varies rapidly, the dependence illustrated in Figure 9.13 induces a hysteresis in the behaviour of $\phi(h)$ as a function of h: if h varies rapidly from a value lower than $-h^*$ to a value larger than h^*, ϕ will follow the path $N' \to P' \to M' \to N$ etc... and will instead follow the path $N \to P \to M \to N'$ if the external field is reversed.

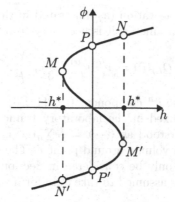

FIGURE 9.13 ϕ as a function of h when $T < T_C$. The section MM' of the curve is unstable. The sections MP and $M'P'$ are metastable.

By contrast, for slow variation of h, the behaviour of $\phi(h)$ shows no hysteresis but simply a jump at $h = 0$ from P to P' (or vice versa): this corresponds to a first order phase transition driven by the external field h.

In Figure 9.14(a), the equation of state $\phi = \phi(h,T)$ is represented along with three isotherms: the critical isotherm ($T = T_C$), a subcritical isotherm ($T < T_C$) and a supercritical isotherm ($T > T_C$). Figure 9.14(b) is the projection of Figure 9.14(a) in the (h,T) plane and Figure 9.14(c) its projection in the (ϕ,h) plane (along with the 3 isotherms). In the (T,h) plane, a first-order phase transition line is thus observed at $h = 0$ for $0 < T < T_C$. The line ends in $h = 0$ and $T = T_C$ on a critical point where the transition becomes a second order transition, see Figure 9.14(b). It is exactly the same phenomenology as for the liquid-gas transition for example.

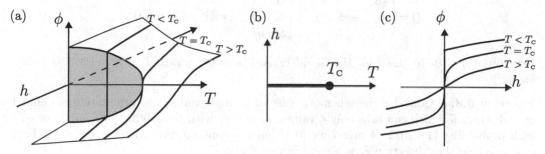

FIGURE 9.14 (a) Phase diagram in the (h,T,ϕ) space. Three isotherms are represented: the critical isotherm ($T = T_C$), a subcritical isotherm ($T < T_C$) and a supercritical isotherm ($T > T_C$). (b) projection in (T,h) plane, (c) projection in the (h,ϕ) plane.

9.3 LIQUID CRYSTALS

This section is devoted to the study of the rich phenomenology of phase transitions in liquid crystals. It gives the opportunity to consider order parameters with a more complex structure than the simple scalar parameter ϕ studied so far.

A liquid crystal is formed of elongated molecules often modelled as rods (see Figure 3.2). The orientation of a rod α is given by a unit vector $\vec{\nu}^{(\alpha)}$ called the *director*. Since a rod has no "head" nor "tail", $\vec{\nu}^{(\alpha)}$ or $-\vec{\nu}^{(\alpha)}$ can be chosen indifferently to indicate the rod's orientation. As a result, the average $\langle \vec{\nu}^{(\alpha)} \rangle$ taken over a small volume is zero, even if the rods have a preferred spatial orientation (as illustrated in the right plot of Figure 3.2). To define an orientational order parameter, it is necessary to introduce a tensor Q, quadratic in $\vec{\nu}^{(\alpha)}$:

$$Q_{i,j}(\vec{r}) = \left\langle \nu_i^{(\alpha)} \nu_j^{(\alpha)} - \frac{1}{3} \delta_{i,j} \right\rangle . \tag{9.48}$$

In Expression (9.48), $\nu_i^{(\alpha)}$ is the ith component ($i \in \{1,2,3\}$) of the unit vector $\vec{\nu}^{(\alpha)}$ on an orthonormal basis $\{\vec{e}_1, \vec{e}_2, \vec{e}_3\}$ fixed in the laboratory frame. Additionally, the average in Equation (9.48) is to be understood as $\langle \cdots \rangle = \frac{1}{N} \sum_\alpha (\cdots)$, where the sum is over the N molecules contained in a small volume around point $\vec{r}$. The spatial ($\vec{r}$) dependence of the tensor order parameter Q will only be considered in Section 10.4 and in the remaining of the present section, Tensor Q is assumed to take a uniform value over the entire considered sample.

9.3.1 PROPERTY OF THE ORDER PARAMETER Q

The aim of this section is to obtain some acquaintance with the tensor order parameter Q by studying its elementary properties.

If the orientation of the directors $\vec{\nu}^{(\alpha)}$ is isotropically distributed as is the case in the left plot of Figure 3.2, then $Q_{ij} = 0$. Rather than averaging over several molecules as in Equation (9.48), this result can be obtained by considering only one director $\vec{\nu}$ of random and isotropic orientation and averaging over its orientation. The result is straightforward: for an isotropic distribution $\langle \nu_i \nu_j \rangle = \frac{1}{3} \delta_{i,j}$ and $Q \equiv 0$.

According to Definition (9.48), it is clear that for all configurations $\text{Tr}(Q) = 0$. This implies that the 3 eigenvalues of Q can be expressed in terms of only 2 real parameters (noted S and η in the following). In a coordinate system diagonalising Q, the following choice can be made for Q:

$$Q = \begin{pmatrix} \frac{2}{3}S & 0 & 0 \\ 0 & -\frac{1}{3}S + \eta & 0 \\ 0 & 0 & -\frac{1}{3}S - \eta \end{pmatrix}, \quad \text{with} \quad S \geq \eta \geq 0 . \tag{9.49}$$

Very often $\eta = 0$. In this case the liquid crystal is called *uniaxial* (otherwise it is called *biaxial*).

Exercise 9.8.   Consider the schematic case of a single rod whose orientation is defined by a director $\vec{\nu}$ which can take only 4 values: $\vec{e}_1$ or $-\vec{e}_1$ with probability $p/2$ and $\vec{e}_2$ or $-\vec{e}_2$ with probability $(1-p)/2$. *A priori* $p \in [0,1]$ but the configuration where $\frac{1}{2} < p \leq 1$ is here considered so that direction $\vec{e}_1$ is always favoured over $\vec{e}_2$.

Show that in this case Q is diagonal in the reference frame $\{\vec{e}_1, \vec{e}_2, \vec{e}_3\}$ and determine the values of S and η. Discuss the notions of uniaxiality and biaxiality on this example.

Solution: Here, Q can be written as:

$$Q = \begin{pmatrix} p - \frac{1}{3} & 0 & 0 \\ 0 & \frac{2}{3} - p & 0 \\ 0 & 0 & -\frac{1}{3} \end{pmatrix} \equiv \text{diag}\left(p - \frac{1}{3}, \frac{2}{3} - p, -\frac{1}{3} \right) ,$$

Comparison with Equation (9.49) gives $0 \leq \eta = \frac{1}{2} - \frac{p}{2} \leq \frac{1}{4} \leq S = \frac{3}{2}p - \frac{1}{2} \leq 1$. For $p = 1$, $S = 1$ and $\eta = 0$, and for $p = \frac{1}{2}$ and $\eta = S = \frac{1}{4}$. When $p = 1$, $S = 1$ and $\vec{\nu}$ is necessarily parallel to $\vec{e}_1$ and then $Q = \text{diag}\left(\frac{2}{3}, -\frac{1}{3}, -\frac{1}{3} \right)$. Then, when p decreases, the system becomes biaxial and η increases until it *becomes equal to S when $p = \frac{1}{2}$. In this case, the configurations $\vec{\nu} \parallel \vec{e}_1$ and $\vec{\nu} \parallel \vec{e}_2$ are equiprobable and $Q = \text{diag} \left(\frac{1}{6}, \frac{1}{6}, -\frac{1}{3} \right)$.*

In the case of a uniaxial crystal (i.e. $\eta = 0$), the preferred orientation $\vec{n}$ is the eigenvector associated with the highest eigenvalue of Q. The expression, valid in all frames of reference, is then:

$$Q_{ij} = S(n_i n_j - \tfrac{1}{3}\delta_{i,j}) . \tag{9.50}$$

To prove this property, it is sufficient to use a specific basis set since the tensor notation [Equation (9.50)] ensures that the two terms of the equality are transformed in the same way and will remain equal in all basis if they are equal in a given one. The equality is immediate in the basis diagonalising Q and in which $\vec{n}$ is the first vector; hence the result. In an isotropic phase $S = 0$ and, in a so-called *nematic phase*, $0 \leq S \leq 1$ measures the degree of alignment of the molecules.

In the presence of a magnetic field $\vec{H}(\vec{r})$, the liquid crystal acquires a magnetisation $\vec{M}(\vec{r})$ with $M_i = \sum_{j=1}^{3} \chi_{ij} H_j$, where χ_{ij} is the magnetic susceptibility tensor written as $\chi_{ij} = \chi_s \delta_{ij} + \chi_a Q_{ij}$, where χ_s and χ_a are the isotropic and anisotropic components of the magnetic susceptibility, respectively. In the case of a uniaxial liquid crystal, the corresponding contribution to the free energy is then:

$$F_{\text{mag}} \equiv -\frac{\mu_0}{2} \int d^3 r \, \vec{M} \cdot \vec{H} = -\frac{\mu_0}{2} \int d^3 r \left(\chi_s H^2 + \chi_a S \left[\left(\vec{n} \cdot \vec{H} \right)^2 - \frac{1}{3} H^2 \right] \right) . \tag{9.51}$$

9.3.2 LANDAU'S THEORY OF THE NEMATIC-ISOTROPIC TRANSITION

In this section, the Landau free energy $F_L(Q)$ is formulated in the most general form for a uniaxial crystal. Then, it is used to describe the phenomenology of the system. F_L is a scalar which must be invariant under a global rotation of the system, i.e., under a transformation in which the orientation of all rods is changed globally without changing their relative orientations. The only scalar invariants constructed from Q having this property are the traces of the powers of Q: $\text{Tr}(Q^2)$, $\text{Tr}(Q^3)$... In the absence of an external field. the free energy density is thus:

$$\frac{F_L(Q,T)}{V} = f_L(Q,T) = \frac{3a}{4} \times (T - T^*) \, \text{Tr}\,(Q^2) - \frac{3b}{2} \, \text{Tr}\,(Q^3) + \frac{9c}{16} \left(\text{Tr}\,(Q^2) \right)^2 + \cdots \tag{9.52}$$

where a, b and c are real positive constants. The numerical prefactors are chosen for future convenience. $\text{Tr}(Q)$ does not appear since it is equal to zero. In the following, the notation $\tilde{a} = a \times (T - T^*)$ shall be used. For a uniaxial liquid crystal, Parametrisation (9.49) leads to:

$$f_L(S,T) = \frac{\tilde{a}}{2} S^2 - \frac{b}{3} S^3 + \frac{c}{4} S^4 . \tag{9.53}$$

Expression (9.53) shows that the legitimate inclusion of a term proportional to $\text{Tr}(Q^4)$ in (9.52) is not necessary for a uniaxial liquid crystal since it would simply add another contribution in S^4, as $\left(\text{Tr}\,(Q^2) \right)^2$ already does.

As will be seen shortly, the free energy given by Equation (9.52) describes a first-order transition between an isotropic and a nematic phase. $f_L(S,T)$ reaches an extremum for the values of S solutions of $(\tilde{a} - bS + cS^2)S = 0$. $S = 0$ is, of course, always a solution, but when $b^2 > 4\tilde{a}c$ there are two other solutions (see Figure 9.15). This condition corresponds to $T < T^{**} = T^* + b^2/(4ac)$.

The nematic phase becomes stable at a temperature T_C which corresponds to the case illustrated in Figure 9.15(c): the function $f_L(S,T_C)$ has two minima (one of which being

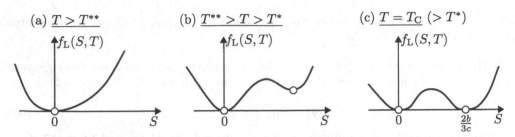

(a) $\underline{T > T^{**}}$ (b) $\underline{T^{**} > T > T^*}$ (c) $\underline{T = T_C \ (> T^*)}$

FIGURE 9.15 Shape of $f_L(S,T)$ as a function of S for different temperatures, decreasing from left to right.

$S = 0$), solutions of $f_L(S,T_C) = 0$ and $\partial f_L(S,T_C)/\partial S = 0$. These two equations read:

$$\begin{cases} \dfrac{\tilde{a}}{2} - \dfrac{b}{3}S + \dfrac{c}{4}S^2 = 0\,, \\[2mm] \tilde{a} - bS + cS^2 = 0\,. \end{cases} \qquad (9.54)$$

The nontrivial solution of this system of equations is $S = 2b/(3c)$ giving $\tilde{a} = S(b - cS) = 2b^2/(9c)$ and so $T_C = T^* + 2b^2/(9ac)$ (in agreement with the result of Exercise 9.7).

Note that the isotropic phase $(S = 0)$ is stable as long as $T > T_C$ since it corresponds to the absolute minimum of the free energy, and when $T^* < T < T_C$ it remains metastable. It is only below T^* that the isotropic case becomes unstable. As for the nematic phase, it is metastable when $T_C < T < T^{**}$ and it becomes stable below T_C.

The transition is a first order transition since when $T > T_C$ the minimisation of the free energy leads to $S = 0$ and, at $T = T_C$, the order parameter jumps by $2b/(3c)$. The transition is clearly visible in the experimental results shown in Figure 9.16 where the value of the magnetic anisotropy $\Delta\chi$ of the 5CB liquid crystal is plotted as a function of temperature. 5CB is an organic compound with the following organic structure: $C_5H_{11} -\!\!\bigcirc\!\!-\!\!\bigcirc\!\!- CN$. The magnetic anisotropy $\Delta\chi$ is the difference between the two eigenvalues of the magnetic susceptibility tensor χ_{ij}. According to the results of Section 9.3.1 this tensor reads:

$$\chi_{ij} = \chi_s \begin{pmatrix} 1 & 0 & 0 \\ 0 & 1 & 0 \\ 0 & 0 & 1 \end{pmatrix} + \chi_a\, S \begin{pmatrix} \frac{2}{3} & 0 & 0 \\ 0 & -\frac{1}{3} & 0 \\ 0 & 0 & -\frac{1}{3} \end{pmatrix}, \qquad (9.55)$$

So $\Delta\chi = (\chi_s + \frac{2S}{3}\chi_a) - (\chi_s - \frac{S}{3}\chi_a) = \chi_a S$. Therefore, $\Delta\chi$ is proportional to S and its measurement clearly displays a first order phase transition at a temperature $T = 35°C$.

Cotton-Mouton effect

If the sample is immersed in an uniform magnetic field $\vec{H}$ parallel to the director $\vec{n}$, then the magnetic contribution (9.51) to the free energy reads $F_{mag} = -\frac{\mu_0}{2}V(\chi_s H^2 + \frac{2}{3}\chi_a S H^2)$. The presence of a magnetic field results in the addition to Expression (9.53) of $f_L(S,T)$ of a term $-h \times S$ with $h = \frac{1}{3}\mu_0\chi_a H^2$: $f_L(S,h,T) = f_L(S,T) - hS$. The extremisation $\partial f_L(S,h,T)/\partial S = 0$ can be written as:

$$\tilde{a}S - bS^2 + cS^3 = h\,. \qquad (9.56)$$

For $T > T_C$ and in the presence of a weak magnetic field, the value of S must be small ($S = 0$ when $\vec{H} = \vec{0}$) and Equation (9.56) can be linearised. The solution follows immediately and reads: $S \simeq h/\tilde{a} = h/(a \times (T - T^*)) \propto H^2/(T - T^*)$.

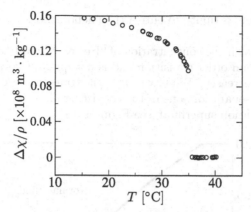

FIGURE 9.16 Magnetic anisotropy $\Delta\chi/\rho$ of the 5CB liquid crystal as a function of temperature T. ρ is the density. Experimental data were extracted from B. J. Frisken, *Nematic liquid crystals in electric and magnetic fields* (PhD thesis, The University of British Columbia, 1989, `http://hdl.handle.net/2429/29104`).

When $S \neq 0$, the anisotropy associated with the orientational order results in a optical birefringence of the liquid crystal. It can be shown that the difference Δn between the two refraction indices is proportional to S. The ratio $\Delta n/H^2$ is called the *Cotton-Mouton constant*. The approach proposed here predicts a Cotton-Mouton constant behaving as $1/(T - T^*)$ for $T \geq T_C$. This behaviour corresponds indeed to the experimental results reported in Figure 9.17 which displays the Cotton-Mouton constant plotted as a function of T for MBBA (MBBA is an organic compound with the following chemical structure: $C_4H_9 - \bigcirc - N = CH - \bigcirc - O - CH_3$).

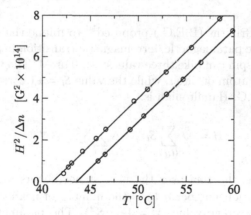

FIGURE 9.17 Inverse of the Cotton-Mouton constant as a function of temperature T, for two samples (with slightly different transition temperatures) of MBBA liquid crystal. The solid lines are fits of the experimental data (white dots). Data extracted from T. W. Stinson III and J. D. Litster, *Phys. Rev. Lett.* **25**, 503 (1970).

9.4 TRI-CRITICAL TRANSITION: BLUME-EMERY-GRIFFITHS MODEL

This section is devoted to the study of the impact of the addition of ^{3}He impurities on the superfluid phase transition of ^{4}He (see Section 7.3.1 for a discussion of this transition). It gives the opportunity to enrich the phenomenology of first and second order phase transitions.

Experiment shows that if the concentration of ^{3}He, x, in liquid ^{4}He is low, the superfluid transition remains a second order transition and is not qualitatively affected (however, the transition temperature decreases). When $x > 0.67$, the transition becomes discontinuous and goes along with a phase separation: one of the coexisting phases is ^{4}He rich and superfluid, the other is ^{3}He rich and non superfluid, see Figure 9.18.

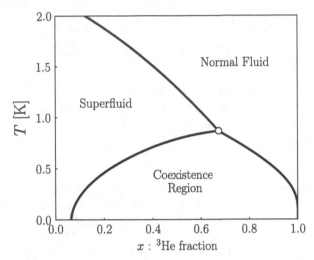

FIGURE 9.18 $T - x$ phase diagram of a ^{3}He + ^{4}He mixture. The data are extracted from G. Chaudhry, *Thermodynamic properties of liquid ^{3}He – ^{4}He mixtures between 0.15 K and 1.8 K* (PhD thesis, Massachusetts Institute of Technology, 2009, http://hdl.handle.net/1721.1/54873). The white circle is the "tricritical point" of coordinates $x_t = 0.674$, $T_t = 0.867$ K.

Blume, Emery and Griffiths (B.E.G.) proposed[16] to mimic the normal fluid/superfluid transition by a schematic paramagnetic/ferromagnetic transition in an Ising system. In the B.E.G. Ising model, the spin can take three values: +1, 0 or −1. The value $S_i = 0$ corresponds to the presence of a ^{3}He atom on site i, while the value $S_i = \pm 1$ corresponds to the presence of a ^{4}He atom. The B.E.G. Hamiltonian is:

$$H = -J \sum_{\langle i,j \rangle} S_i S_j + \Delta \sum_{i=1}^{N} S_i^2 - \Delta N \,, \tag{9.57}$$

with $J > 0$ and $\sum_{\langle i,j \rangle}$ is the sum over the nearest neighbours on a lattice with N sites ($N \gg 1$) and a coordination number q (q is the number of nearest neighbours). The ^{3}He concentration x is the average value: $x = 1 - \langle S_i^2 \rangle$. The parameter Δ controls the ^{3}He concentration: $x \to 1$ if $\Delta \to +\infty$ and $x \to 0$ if $\Delta \to -\infty$. The "superfluid" order parameter is $\phi = \langle S_i \rangle$, i.e. the appearance of superfluidity is schematically described by a spontaneous magnetisation of the Ising system.

Hamiltonian (9.57) can be studied within the framework of a mean field approximation. To this end, it is convenient to rewrite the interaction term by separating S_i into a mean and a fluctuating part: $S_i = \phi + \sigma_i$, where $\phi = \langle S_i \rangle$ plays the role of an order parameter and $\sigma_i = S_i - \langle S_i \rangle$ represents the fluctuation of the Ising spin around its mean value. Neglecting

---

[16]M. Blume, V. J. Emery, and R. B. Griffiths, Phys. Rev. A **4**, 1071 (1971).

the terms of order $O(\sigma^2)$, the interaction term can be written as:

$$\sum_{\langle i,j\rangle} S_i S_j \simeq -\frac{Nq\,\phi^2}{2} + q\,\phi \sum_{i=1}^{N} S_i \,. \tag{9.58}$$

With this approximation, H is separable into N independent Hamiltonians $H = \sum_{i=1}^{N} h(S_i)$ with $h(S_i) = \frac{1}{2}qJ\phi^2 - qJ\phi S_i + \Delta(S_i)^2 - \Delta$. The partition function of the Hamiltonian $h(S_i)$ is simply:

$$z = \sum_{S_i \in \{-1,0,1\}} \exp\{-\beta\,h(S_i)\} = e^{-\beta qJ\phi^2/2}e^{\beta\Delta}\left[1 + 2e^{-\beta\Delta}\cosh(q\phi\beta J)\right] \,. \tag{9.59}$$

The total partition function is $Z = z^N$ and, for a given value ϕ of the order parameter, the system's free energy can be written as:

$$\frac{F_L(\phi,T,\Delta)}{N} = -\Delta + \frac{1}{2}qJ\phi^2 - k_{\mathrm{B}}T\ln\left[1 + 2e^{-\beta\Delta}\cosh(\beta qJ\phi)\right] \,. \tag{9.60}$$

To obtain the expression of the ^{3}He concentration x as a function of T, ϕ and the other parameters of the problem, the separability of the approximated mean field Hamiltonian H can be used and only the single site Hamiltonian h is considered:

$$x = 1 - \langle S_i^2\rangle = 1 - \frac{1}{z}\times\sum_{S_i \in \{-1,0,1\}} S_i^2\exp\{-\beta\,h(S_i)\} = \frac{1}{1 + 2\,e^{-\beta\Delta}\cosh(q\phi\beta J)} \,. \tag{9.61}$$

In the superfluid phase, $\phi = 0$ and:

$$x = \frac{\lambda - 1}{\lambda} \,, \quad \text{where} \quad \lambda \equiv 1 + \frac{e^{\beta\Delta}}{2} \,. \tag{9.62}$$

A Taylor expansion of Equation (9.60) around $\phi = 0$ gives[17]:

$$\frac{F_L(\phi,T,\Delta)}{N} = a(T,\Delta) + b(T,\Delta)\phi^2 + c(T,\Delta)\phi^4 + d(T,\Delta)\phi^6 + \dots \tag{9.63}$$

where

$$b(T,\Delta) = \frac{qJ}{2}\left(1 - \frac{qJ}{\lambda\,k_{\mathrm{B}}T}\right) \quad \text{and} \quad c(T,\Delta) = \frac{qJ}{8\lambda^2}\left(\frac{qJ}{k_{\mathrm{B}}T}\right)^3\left(1 - \frac{\lambda}{3}\right) \,. \tag{9.64}$$

Within the mean field approach, the determination of the transition temperature is achieved by studying under which condition the self-consistent equation $\langle S_i\rangle = \phi$ admits non-zero solutions. An equivalent an simpler method consists in studying the behaviour of the free energy [Equation (9.63)] and determining below which temperature the extremum of F_L in $\phi = 0$ (which is a minimum at high-temperature) becomes a local maximum (as illustrated in Figure 9.9). This occurs when $b(T,\Delta)$ becomes negative, i.e., when $qJ > \lambda k_{\mathrm{B}}T$. Defining the reduced temperature as $\tau = k_{\mathrm{B}}T/(qJ)$, this condition reads $G(\tau) < 1$, with $G(\tau) = \tau[1 + \frac{1}{2}\exp(\Delta/(qJ\tau))]$. A graphical determination of the critical temperature can be achieved by plotting $G(\tau)$ (see Figure 9.19). For $\Delta/qJ > 0$, there are either two or no solutions to the equation $G(\tau) = 1$. For $\Delta/qJ < 0$ there is always a unique solution.

Note that at the critical temperature T_{C} the order parameter ϕ is equal to zero and thus $1/\lambda = 1 - x$ [see Equation (9.59)]. According to the relation $k_{\mathrm{B}}T_{\mathrm{C}}/(qJ) = 1/\lambda$, it yields $k_{\mathrm{B}}T_{\mathrm{C}} = qJ(1 - x)$. Thus, for $x = 0$ one gets[18] $k_{\mathrm{B}}T_{\mathrm{C}} = qJ$ and in the general case $T_{\mathrm{C}}(x)/T_{\mathrm{C}}(0) = 1 - x$.

---

[17]$\ln[1 + a\cosh(b\phi)] = \ln(1 + a) + \frac{ab^2}{1+a}\frac{\phi^2}{2} + \frac{ab^4}{(1+a)^2}(1 - 2a)\frac{\phi^4}{24} + O(\phi^6)$.

[18]This is a natural result. According to Equation (9.59), $x = 0$ corresponds to $\Delta = -\infty$; in this case $G(\tau) = \tau$ and $\tau_c = 1$ is the trivial solution of $G(\tau_c) = 1$.

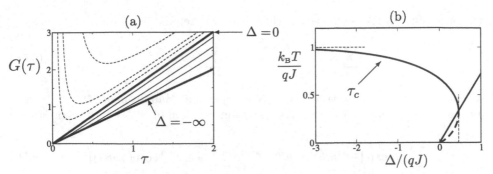

FIGURE 9.19 (a) Behaviour of $G(\tau)$ for different values of the parameter $\Delta/(qJ)$. The dashed curves are plotted for $\Delta/qJ > 0$ and the solid curves for $\Delta/qJ < 0$. The two thick straight lines correspond to the limiting cases $\Delta = 0$ and $\Delta/qJ = -\infty$ for which $G(\tau) = 3\tau/2$ and $G(\tau) = \tau$, respectively. The reduced critical temperature $\tau_c = k_BT_C/(qJ)$ is the solution of $G(\tau_c) = 1$. (b) τ_c as a function of $\Delta/(qJ)$. For $\Delta > 0$, there are two roots, one of which lies on the dashed line. On this plot the horizontal dashed line $\tau_c = 1$ is the asymptote for $\Delta \to -\infty$ and the vertical dashed line $(\Delta/(qJ) \simeq 0.462)$ is the value beyond which there is no transition. The straight line going through the origin corresponds to the equation $k_BT = \Delta/\ln(4)$; this is the position of the zeros of $c(T, \Delta)$ (see Exercise 9.9). Its intersection with the curve $T = T_C(\Delta)$ defines the position of the tricritical point of the B.E.G. model.

9.4.1 TRICRITICAL POINTS

Only the expansion of F_L up to fourth order in ϕ has been considered previously. A more precise discussion of the B.E.G. model requires to study a Landau theory of the form of Equation (9.63) taking into account the fact that $c(T, \Delta)$ can change sign. To simplify the discussion, only a model in which $a(T, \Delta) = 0$ and $d(T, \Delta)$ is a small positive constant (simply written d) is considered. d is supposed small enough so that the contribution $d\phi^6$ is not the dominant term in Equation (9.63) for the typical values of ϕ considered here[19]:

$$\frac{F_L(\phi, T, \Delta)}{N} = b(T, \Delta)\phi^2 + c(T, \Delta)\phi^4 + d\,\phi^6 \ . \tag{9.65}$$

Figure 9.20 shows two typical curves $b(T, \Delta) = 0$ and $c(T, \Delta) = 0$ in the plane (Δ, T). The intersection point of the two curves is called the *tricritical point*. Its coordinates are written (Δ_t, T_t).

Exercise 9.9. Determine the coordinates of the tricritical point in the B.E.G. model.

Solution: $c(\Delta, T)$ *cancels when* $\lambda = 3$, *i.e.* $k_BT = \Delta/\ln(4)$ *[see Equations (9.62) and (9.64)]. When* $\lambda = 3$, *the point at which* $b(T, \Delta) = 0$ *verifies* $qJ = 3k_BT$. *Thus, the coordinates of the tricritical point are* $k_BT_t/qJ = 1/3$ *and* $\Delta_t/(qJ) = \ln(4)/3 \simeq 0.462$. *The tricritical point corresponds to the crossing of the solid curves in Figure 9.19(b).*

F_L is plotted as a function of ϕ in Figure 9.21(a) for different values of temperature when $\Delta < \Delta_t$. The case $\Delta > \Delta_t$ is shown Figure 9.21(b). Different temperatures correspond to noticeably different behaviours because the coefficients c and b in Equation (9.65) are functions of T. In particular, the shape of the curves is strongly modified when these coefficients change sign.

---

[19]Its contribution shall nevertheless not be neglected, as it becomes dominant in Equation (9.63) when $\phi \to \pm\infty$.

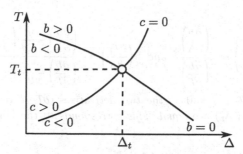

FIGURE 9.20 Plot of the curves $b(T,\Delta) = 0$ and $c(T,\Delta) = 0$ in the plane (Δ,T). When $\Delta < \Delta_t$, b changes sign at a higher temperature than c does. The opposite situation occurs when $\Delta > \Delta_t$. The intersection point (Δ_t, T_t) of the two curves is called the *tricritical point*.

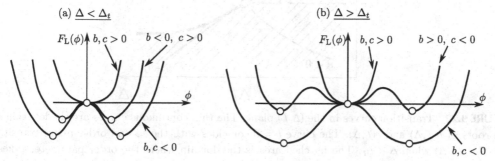

FIGURE 9.21 $F_L(\phi, T, \Delta)$ plotted as a function of ϕ for different values of T. (a) for $\Delta < \Delta_t$, (b) for $\Delta > \Delta_t$.

For $\Delta < \Delta_t$, the system undergoes a second-order phase transition when b changes sign. For $\Delta > \Delta_t$, this transition becomes a first-order transition between a $\phi = 0$ phase and a $\phi = \phi_1$ phase ($\phi_1 \neq 0$). The transition takes place at a temperature T_1 such that:

$$F_L(\phi_1, T_1, \Delta = F_L(0, T_1, \Delta) \quad \text{and} \quad \left.\frac{\partial F_L}{\partial \phi}\right|_{\phi_1, T_1, \Delta} = 0 . \tag{9.66}$$

This occurs when the two secondary minima of $F_L(\phi)$ that appear when $c < 0$ are exactly at the same level as the minimum at $\phi = 0$. In other words, $F_L(0) = 0 = F_L(\phi_1) = F_L(-\phi_1)$. Knowing that $\phi_1 \neq 0$, the two Equations (9.66) can be written as:

$$b + c\,\phi_1^2 + d\,\phi_1^4 = 0 \quad \text{and} \quad 2b + 4c\,\phi_1^2 + 6d\,\phi_1^4 = 0 . \tag{9.67}$$

This yields

$$\phi_1^2 = -c(T_1, \Delta)/(2d) \quad \text{and} \quad b(T_1, \Delta) = c^2(T_1, \Delta)/(4d) . \tag{9.68}$$

This last relation fixes the value of T_1.

Exercise 9.10. Plot the first and second order transition lines in the (Δ, T) plane.

Solution: For $\Delta < \Delta_t$, the critical temperature T_C is defined by $b(T_C, \Delta) = 0$, and for $\Delta > \Delta_t$, the temperature T_1 is defined by $b(T_1, \Delta) = c^2(T_1, \Delta)/(4d)$. The two curves meet in (T_t, Δ_t), since at that point $b = c = 0$.

The derivatives $\partial T_C/\partial \Delta$ and $\partial T_1/\partial \Delta$ can also be calculated. To do so, it should be noticed that the slope of a curve defined by the implicit equation $f(x, y) = 0$ is[20] $\partial y/\partial x = -(\partial f/\partial x)/(\partial f/\partial y)$.

---

[20]This can be proven the following way: for a point of the curve with coordinates (x, y), the relationship between dx and dy must be found so that the point with coordinates $(x + dx, y + dy)$ lies also on the curve.

Therefore:

$$\left(\frac{\partial T_C}{\partial \Delta}\right) = -\frac{\left(\frac{\partial b}{\partial \Delta}\right)}{\left(\frac{\partial b}{\partial T}\right)} \quad \text{and} \quad \left(\frac{\partial T_1}{\partial \Delta}\right) = -\frac{\left(\frac{\partial b}{\partial \Delta}\right) - \frac{c}{2d}\left(\frac{\partial c}{\partial \Delta}\right)}{\left(\frac{\partial b}{\partial T}\right) - \frac{c}{2d}\left(\frac{\partial c}{\partial T}\right)}. \tag{9.69}$$

At the tricritical point $c(T_t, \Delta_t) = 0$ so the two derivatives $(\partial T_C/\partial\Delta)_{\Delta_t^-}$ (at the left of Δ_t) and $(\partial T_1/\partial\Delta)_{\Delta_t^+}$ (at the right of Δ_t) are equal. This corresponds to the behaviour illustrated in Figure 9.22.

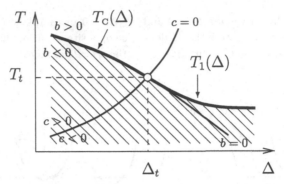

FIGURE 9.22 Transition curves in the (Δ, T) plane. The thin continuous curves are the locations of the roots of $b(T, \Delta)$ and $c(T, \Delta)$. The curve $b = 0$ coincides with the second order phase transition line $T = T_C(\Delta)$ when $\Delta < \Delta_t$. The hatched area is the domain where the order parameter takes a non-zero value.

Let $F(T, \Delta)$ be the system's free energy. Form (9.57) of the Hamiltonian shows that, in the framework of the B.E.G. model, the ^{3}He fraction $x = 1 - \langle S_i^2 \rangle$ can be calculated as $x = -\frac{1}{N}(\partial F/\partial\Delta)_T$. Let us determine the value of x in the different phases in the case of a first-order transition (when $\Delta > \Delta_t$).

In the mean field approximation, $F(T, \Delta)$ is equal to $F_L(\phi, T, \Delta)$ calculated for the value of ϕ such as $\partial F_L/\partial\phi = 0$; since this derivative is equal to zero, then $(\partial F/\partial\Delta)_T = (\partial F_L/\partial\Delta)_{T,\phi}$. In the phase where $\phi = 0$, it leads to $x = x_0 = -\frac{1}{N}(\partial F_L/\partial\Delta)_{T,0} = 0$, while in the case $\phi = \phi_1$, it leads to $x = x_1 = -\frac{1}{N}(\partial F_L/\partial\Delta)_{T,\phi_1} = -\phi_1^2(\partial b/\partial\Delta) - \phi_1^4(\partial c/\partial\Delta)$. Therefore, when $\Delta > \Delta_t$, the first order transition line is associated to a discontinuity of the ^{3}He concentration x which, taking into account Relations (9.68), can be written as:

$$x_1 - x_0 = \frac{c}{2d}\left(\frac{\partial b}{\partial \Delta}\right) - \frac{b}{d}\left(\frac{\partial c}{\partial \Delta}\right), \tag{9.70}$$

where every function is evaluated in $T = T_1(\Delta)$. One might object that this result is obtained from the simplified form of the free energy [Equation (9.65)] while the series expansion [Equation (9.63)] is less schematic: in particular it takes into account a ϕ-independent term $a(T, \Delta)$ which adds a contribution $\partial a/\partial\Delta$ to the expressions of x_0 and x_1, but this does not modify the difference (9.70).

At the tricritical point, $x_0 = x_1$ since $b = c = 0$. In the framework of the B.E.G. model, it corresponds to $\lambda = 3$ [see Equation (9.64)], i.e., using (9.62), $x = 2/3$, in very good agreement with experimental results giving $x \simeq 0.67$. The B.E.G. model is thus able to explain qualitatively and quantitatively the behaviour of the ^{3}He-^{4}He mixture in the vicinity of the tricritical point as illustrated in Figure 9.18 which shows the transition points in a (T, x) diagram.

Exercise 9.11. Determine the critical exponent δ characterising, in the vicinity of the tricritical point, how ϕ depends on an external field h.

Solution: A term $-h\phi$ must be added to the free energy [Equation (9.65)]. The functional extremisation around the critical point gives $6d\phi^5 = h$ i.e. $\delta = 5$.

Exercise 9.12. Determine the critical exponent[21] β characterising how ϕ goes to zero when approching the tricritical point while staying in the hatched area depicted in Figure 9.22.

Solution: The order parameter is solution of $\partial F_L/\partial \phi = 0$ i.e.

$$\phi^2 = \frac{1}{3d}\left[-c + \sqrt{c^2 - 3bd}\right] . \tag{9.71}$$

Note that one is not necessarily near the tricritical transition here. Let us assume that, in the vicinity of the curves $b = 0$ and $c = 0$, b and c linearly go to zero, i.e. $b(T,\Delta) \propto T - T_C(\Delta)$. If one approaches the tricritical point while remaining in the hatched area depicted in Figure 9.22, the order parameter will cancel as $|T - T_t|^\beta$, where the exponent β depends on the chosen path. If (Δ_t, T_t) is approached along the curve $b = 0$ in the region where $c < 0$, then, according to Equation (9.71), $\phi^2 = -2c/(3d)$ and thus $\beta = 1/2$. If, on the other hand, (Δ_t, T_t) is approached along the curve $c = 0$ in the region where $b < 0$, then $\phi^2 = \sqrt{-b/(3d)}$ and $\beta = 1/4$.

---

[21] Be careful with the notations: here $\beta \neq 1/k_B T$!

10 Phase Transition – Spatial Variations of the Order Parameter

10.1 GINZBURG-LANDAU THEORY

The Ginzburg-Landau theory is the version of Landau mean-field theory which applies to non-homogeneous systems. In this case, the order parameter is not uniform, but is instead a function $\phi(\vec{r})$ of the position $\vec{r}$. $\phi(\vec{r})$ is assumed to vary over a typical distance D much larger than the microscopic scale (for example, the lattice spacing a if the paradigm of the Ising model is considered). More precisely, $\phi(\vec{r})$ is defined as the average of the microscopic order parameter over a coarse grained cell centred in $\vec{r}$. The cell dimensions are large with respect to a and small with respect to D. Note that, in the case of the Ising system for example, the value $\phi(\vec{r})$ determined as an average of the spin on the coarse grained cell centred in $\vec{r}$ may correspond to several microscopic spin configurations within the cell.

The free energy F_L is then a functional of the spatially varying order parameter $\phi(\vec{r})$. In this chapter, a functional is a mapping from a space defined by a set of functions (here the set of all possible functions $\phi(\vec{r})$) onto $\mathbb{R}$. F_L must, of course, include a term $\int d^d r f_L(\phi)$ which is the immediate generalisation of Equation (9.41) to the case where ϕ depends on position. If there is no other contribution to the free energy, one speaks of "local density approximation"; but in all generality a part of the energy must account for the cost associated with the spatial variations of ϕ: in the integral over the system's volume, contributions from the derivatives of ϕ with respect to the coordinates must be considered. The simplest terms, of the form $g(\phi)\partial_i\phi$, should be discarded: they turn into surface contributions irrelevant in the thermodynamic limit (see Exercise 10.1). For the same reason, the terms proportional to $\partial_{ij}\phi$ are also irrelevant.

Exercise 10.1. Show that in the thermodynamic limit a term of the form $g(\phi)\partial_i\phi$ does not contribute to free energy.

Solution: The term can be rewritten as $g(\phi)\partial_i\phi = \vec{\nabla} \cdot [G(\phi)\vec{e}_i]$ where G is an antiderivative of g. Gauss-Ostrogradskii theorem[1] gives:

$$\int_\Omega d^d r \, g(\phi)\partial_i\phi = \int_{\partial\Omega} d^{d-1} r \, G(\phi)\,\vec{n}\cdot\vec{e}_i \, ,$$

where Ω is the region of the space $\mathbb{R}^d$ occupied by the system and whose boundary is denoted $\partial\Omega$. In the right-hand side of the above expression, $\vec{n}$ is the outward pointing normal vector. Thus, the volume integral of $g(\phi)\partial_i\phi$ yields a surface term that cannot be a relevant contribution to describe a system in the thermodynamic limit.

The simplest relevant contributions to the volume energy density associated with the spatial variation of the order parameter are therefore of the form $\phi\,\partial_{ij}\phi$ or $\partial_i\phi\,\partial_j\phi$. When integrating over the volume the former reduce to the latter, so the simplest contribution can be written as: $\int d^d r \sum_{i,j=1}^d g_{ij}\,\partial_i\phi\,\partial_j\phi$ where the g_{ij}'s are real constants. In the following, the

---

[1]see Formula (A.45) for a 3 dimensional formulation.

DOI: 10.1201/9781003272427-10

analysis is restricted to the very frequent case where[2]: $g_{ij} = g\,\delta_{ij}$: this occurs in the case of an invariance by rotation (but cubic symmetry is sufficient to ensure this behaviour). Also accounting for a possible position dependent external field $h(\vec{r})$, the free energy is [compare with Equations (9.41) and (9.43)]

$$F_L[\phi, h] = F_0(T) + \int \mathrm{d}^d r \left\{ g \left|\vec{\nabla}\phi\right|^2 + f_L(\phi(\vec{r}), h(\vec{r})) \right\} . \tag{10.1}$$

And the generalisation of Equation (9.44) becomes

$$\frac{\delta F_L}{\delta \phi(\vec{r})} = 0 , \tag{10.2}$$

where the symbol $\delta F_L/\delta\phi(\vec{r})$ is the functional derivative, a generalisation of the usual derivative, defined and studied in the Section 10.2. It will be seen that Equation (10.2) can be written as an equation for $\phi(\vec{r})$ [see Equation (10.9)]:

$$-2g\vec{\nabla}^2\phi + 2a(T - T_\mathrm{c})\phi + 4d\phi^3 = h(\vec{r}) . \tag{10.3}$$

Finally, note that, for now, Relation (10.2) is only motivated by its analogy with Equation (9.44). A precise justification will be given in the next chapter (Section 11.2).

10.2 FUNCTIONAL DERIVATIVE

Consider a functional $F[\phi(\vec{r})]$. When a small variation of ϕ occurs ($\phi(\vec{r}) \to \phi(\vec{r}) + \delta\phi(\vec{r})$), the value of F slightly changes: $F \to F + \delta F$. Under fairly general assumptions, it can be shown show (and this will be verified in every example studied below) that δF can be written as:

$$\delta F = \int \mathrm{d}^d r\, K_1(\vec{r})\,\delta\phi(\vec{r}) + \frac{1}{2} \int \mathrm{d}^d r\, \mathrm{d}^d r' K_2(\vec{r},\vec{r}')\delta\phi(\vec{r})\delta\phi(\vec{r}') + \cdots \tag{10.4}$$

Relation (10.4) defines the quantities K_1 and K_2. The quantity $K_1(\vec{r})$ will henceforth be written as:

$$K_1(\vec{r}) = \frac{\delta F}{\delta \phi(\vec{r})} , \tag{10.5}$$

and is called the *functional derivative* of F with respect to $\phi(\vec{r})$. $K_2(\vec{r},\vec{r}') = \delta^2 F/\delta\phi(\vec{r})\delta\phi(\vec{r}')$ is the second derivative. It verifies $K_2(\vec{r},\vec{r}') = K_2(\vec{r}',\vec{r})$, this can be seen by performing the change of variable $\vec{r} \leftrightarrow \vec{r}'$ in the last integral of the right-hand term of Equation (10.4). $K_1(\vec{r})$ can be seen as a functional of ϕ and it is then possible to show that $K_2(\vec{r},\vec{r}')$ is the functional derivative $\delta K_1(\vec{r})/\delta\phi(\vec{r}')$.

Of course, examples where Expansion (10.4) is not well defined mathematically can be imagined. However, in all cases studied in this book, it is clearly correct and allows direct calculation of the functional derivative. Let us now consider some representative cases and list some important properties of functional derivatives:

Properties

In the following, some properties of functional derivatives are listed. They will be of use later in the chapter.

- Consider the functional $F[\phi] = \int \mathrm{d}^d r\, \phi^2(\vec{r})$. If $\phi(\vec{r})$ is sightly modified then F becomes $F[\phi + \delta\phi] = \int \mathrm{d}^d r\, [\phi(\vec{r}) + \delta\phi(\vec{r})]^2 = F[\phi] + 2\int \mathrm{d}^d r\phi(\vec{r})\delta\phi(\vec{r}) + \cdots$. Thus, the functional derivative is here $\delta F/\delta\phi(\vec{r}) = 2\phi(\vec{r})$.

---

[2]The constant g is positive. Otherwise the system would be unstable with respect to spatial fluctuations of the order parameter.

- To slightly generalise the above point, consider now the functional $F[\phi] = \int d^d r f(\phi(\vec{r}))$ where $f(\phi)$ is a differentiable function with the derivative $f'(\phi)$. Following the same type of calculation:

$$\text{with}\quad F[\phi] = \int d^d r f(\phi(\vec{r})), \quad \text{the functional derivative is} \quad \frac{\delta F}{\delta \phi(\vec{r})} = f'(\phi(\vec{r})) . \quad (10.6)$$

- Consider the functional $F[\phi] = \int d^d r K(\vec{r} - \vec{r}_0)\phi(\vec{r})$. It is then clear that $\delta F/\delta \phi(\vec{r}) = K(\vec{r} - \vec{r}_0)$.

- $F[\phi] = \phi(\vec{r}_0)$. This is a special case of the previous example for which $K(\vec{r}-\vec{r}_0) = \delta^{(d)}(\vec{r}-\vec{r}_0)$, where $\delta^{(d)}$ is the Dirac distribution in $\mathbb{R}^d$. One immediately finds: $\delta F/\delta \phi(\vec{r}) = \delta^{(d)}(\vec{r} - \vec{r}_0)$.

- $F[\phi] = f(\phi(\vec{r}_0))$. Combining the previous results, the following useful formula is obtained:

$$\frac{\delta f(\phi(\vec{r}_0))}{\delta \phi(\vec{r})} = f'(\phi(\vec{r})) \, \delta^{(d)}(\vec{r} - \vec{r}_0) . \quad (10.7)$$

- Consider the functional F defined in Equation (10.6). Either by writing an expansion of the form of Equation (10.4) up to second order in $\delta\phi$ or by using Equation (10.7) to calculate the functional derivative of the right hand side expression in Equation (10.6), one gets: $\delta^2 F/\delta\phi(\vec{r})\delta\phi(\vec{r}') = f''(\phi(\vec{r}))\delta^{(d)}(\vec{r}' - \vec{r})$.

- $F[\phi] = \int d^d r' \, d^d r'' K(\vec{r}', \vec{r}'')\phi(\vec{r}')\phi(\vec{r}'')$. A simple calculation leads to $\delta F/\delta\phi(\vec{r}) = \int d^d r'[K(\vec{r}, \vec{r}') + K(\vec{r}', \vec{r})] \phi(\vec{r}')$ and $\delta^2 F/\delta\phi(\vec{r})\delta\phi(\vec{r}') = K(\vec{r}, \vec{r}') + K(\vec{r}', \vec{r})$.

- $F[\phi] = \int_\Omega d^d r |\vec{\nabla}\phi(\vec{r})|^2$. Here the integration is performed on a specified domain Ω of $\mathbb{R}^d$. Its boundary is a manifold $\partial\Omega$. Clearly, $\delta F = 2\int_\Omega d^d r \, \vec{\nabla}\phi(\vec{r}) \cdot \vec{\nabla}\delta\phi(\vec{r})$. This expression can be written in a form similar to the first right-hand term of Equation (10.4). Indeed, $\vec{\nabla}\phi \cdot \vec{\nabla}\delta\phi = \vec{\nabla}\cdot(\delta\phi\vec{\nabla}\phi)-\delta\phi\vec{\nabla}^2\phi$. Then, using Gauss-Ostrogradskii theorem [Equation (A.45)] yields $\delta F = 2\int_{\partial\Omega} d^{d-1}\sigma \, \delta\phi\vec{\nabla}\phi \cdot \vec{n} - 2\int_\Omega d^d r \, \delta\phi\vec{\nabla}^2\phi$, where $\vec{n}$ is the outward pointing normal of $\partial\Omega$. The surface term are arbitrarily set to zero and the final expression for δF lead to:

$$\text{with}\quad F[\phi] = \int_\Omega d^d r |\vec{\nabla}\phi(\vec{r})|^2, \quad \text{the functional derivative is} \quad \frac{\delta F}{\delta\phi(\vec{r})} = -2\vec{\nabla}^2\phi(\vec{r}) .$$
$$(10.8)$$

- Combining the different cases considered above yields the following expression for the derivative of Landau functional [Equation (10.1)]:

$$\frac{\delta F_L}{\delta\phi(\vec{r})} = -2g\vec{\nabla}^2\phi(\vec{r}) + \frac{\partial f_L}{\partial\phi} . \quad (10.9)$$

And thus Equation (10.3) is indeed an explicit form of Equation (10.2).

- Finally, suppose that for each point of space the value of a function $\phi(\vec{r})$ is a functional of a field $h(\vec{r})$: $\phi(\vec{r}) = \phi(\vec{r}, [h])$. Let $F[\phi]$ be a functional of ϕ. F can also be considered as a functional of h and the functional derivative $\delta F/\delta h(\vec{r})$ might be calculated. To do so, it is necessary to evaluate how F is modified if h varies: $h(\vec{r}) \to h(\vec{r}) + \delta h(\vec{r})$. The induced variation of ϕ is $\delta\phi(\vec{r}') = \int d^d r(\delta\phi(\vec{r}')/\delta h(\vec{r})) \, \delta h(\vec{r})$, which immediately leads to:

$$\frac{\delta F}{\delta h(\vec{r})} = \int d^d r' \, \frac{\delta F}{\delta\phi(\vec{r}')} \frac{\delta\phi(\vec{r}')}{\delta h(\vec{r})} . \quad (10.10)$$

This formula is a generalisation of the chain rule formula to functional derivatives.

10.3   SEVERAL NON-HOMOGENEOUS CONFIGURATIONS

The order parameter can be position-dependent, either in the presence of an external field or due to the boundary conditions imposed on the system. These two cases are studied in the present section.

10.3.1   EFFECT OF A WEAK NON-UNIFORM EXTERNAL FIELD

It is important to take into account spatial variations of the order parameter in the presence of a non-uniform external field. Suppose that the system is initially subjected to no external field. It is maintained at a temperature T which can be either larger or lower than the critical temperature T_C. It is clear that the solution of Equation (10.3) can be written as $\phi(\vec{r}) = \phi_0$ where ϕ_0 is the homogeneous solution given be Expression (9.37) (it cancels above T_C and has a finite value below T_C). In the presence of a weak external field $h(\vec{r})$, the order parameter becomes $\phi(\vec{r}) = \phi_0 + \delta\phi(\vec{r})$, where $\delta\phi$ is considered small in the sense that Expression (10.3) can be linearised as follows:

$$-2g\vec{\nabla}^2\delta\phi + \left[2a(T - T_C) + 12d\phi_0^2\right]\delta\phi = h(\vec{r}) . \tag{10.11}$$

Since Equation (10.11) is linear, it can be solved using the method of Green's functions. Let the function $\chi(\vec{r})$ be solution of:

$$\left[-\vec{\nabla}^2 + \xi^{-2}\right]\chi(\vec{r}) = \frac{1}{2g}\delta^{(d)}(\vec{r}) , \tag{10.12}$$

where the "correlation length" ξ is defined as

$$\xi^{-2} = \frac{2a(T - T_C) + 12d\phi_0^2}{2g} = \frac{1}{2g}\left.\frac{\partial^2 f_L}{\partial\phi^2}\right|_{\phi_0} , \tag{10.13}$$

or

$$\xi = \begin{cases} \sqrt{\dfrac{g}{a(T - T_C)}} & \text{when } T > T_C , \\[3ex] \sqrt{\dfrac{g}{2a(T_C - T)}} & \text{when } T < T_C . \end{cases} \tag{10.14}$$

$\chi(\vec{r})$ is the value of $\delta\phi(\vec{r})$ corresponding to a very peculiar external field: a Dirac distribution. Once Equation (10.12) is solved, it is easy to verify that the solution of Equation (10.11) for any field $h(\vec{r})$ can be written as: $\delta\phi(\vec{r}) = \int d^d r' \chi(\vec{r} - \vec{r}')h(\vec{r}')$.

The solution of Equation (10.12) in any dimension d will be discussed in details later (see Section 11.2.1), but to set ideas, the case $d = 1$ will be considered here. In one dimension, Equation (10.12) reads:

$$\left[-\frac{d^2}{d^2 x} + \xi^{-2}\right]\chi(x) = \frac{1}{2g}\delta(x) . \tag{10.15}$$

It is clear that the solution of Equation (10.15) is an increasing exponential for $x < 0$ and a decreasing one for $x > 0$ (in each region of space, the solution diverging to infinity is discarded). Since $\chi(x)$ must be continuous in zero, the solution reads $\chi(x) = \mathcal{A}\exp(-|x|/\xi)$. To determine the value of constant $\mathcal{A}$, Equation (10.15) is integrated between $-\epsilon$ and $+\epsilon$ (with $\epsilon > 0$). It gives:

$$\left[-\frac{d\chi}{dx}\right]_{-\epsilon}^{+\epsilon} + \frac{1}{\xi^2}\int_{-\epsilon}^{+\epsilon}\chi(x)dx = \frac{1}{2g} .$$

The limit $\epsilon \to 0$ gives $-\,\mathrm{d}\chi/\mathrm{d}x|_{0^+} + \mathrm{d}\chi/\mathrm{d}x|_{0^-} = 1/(2g)$, or $\mathcal{A} = \xi/(4g)$ and finally

$$\chi(x) = \frac{\xi}{4g} \exp\left(-\frac{|x|}{\xi}\right) . \tag{10.16}$$

The behaviour of $\phi(x)$ for the very peculiar external field $h(x) = \delta(x)$ is shown in Figure 10.1. It can be seen in Figure 10.1 that the influence of a localised perturbation extends only over

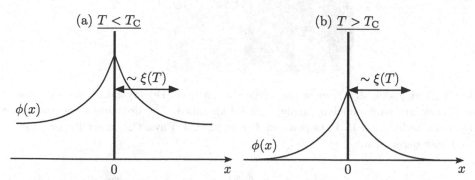

FIGURE 10.1 Behaviour of $\phi(x)$ in the presence of the external field $h(x) = \delta(x)$ (schematically represented as a bold vertical line) for (a) $T < T_C$ and (b) $T > T_C$. In both cases, $\phi(|x| \gg \xi) = \phi_0$: a localised external field modifies the order parameter only over a region of typical extend ξ.

a region of size ξ. This observation can be used to determine the order of magnitude of ξ in the ordered phase ($T < T_C$): ξ is the length over which the distortion of the order parameter has an energetic cost comparable to the "condensation energy", the energy gained by the system when it goes from a configuration $\phi = 0$ to $\phi = \phi_0$. Thus, $g|\vec{\nabla}\phi|^2 \sim |f_L(\phi_0)| = d\phi_0^4$ giving $g|\phi_0/\xi|^2 \sim d\phi_0^4$, or $\xi^2 \sim g/(d\phi_0^2)$, in qualitative agreement with Equations (10.13) and (10.14).

Formula (10.14) indicates that the correlation length diverges when $T \to T_C$. This divergence can be given a simple physical interpretation when considering temperatures T slightly higher than T_C: since for a temperature slightly lower than T the system spontaneously adopts a non-zero value of the order parameter[3] it will naturally tend to be very sensitive to an external field. This increased sensitivity is reflected by the increase of the size of the perturbed area (ξ increasing) and the amplitude of the response (the amplitude of χ also increases, see Equation (10.16)).

10.3.2 DOMAIN WALL

A very important type of inhomogeneous behaviour corresponds to the configuration where two regions with opposite values of the order parameter are in contact: in one region, the order parameter is mostly $+\phi_0$, while in the other, it is mostly $-\phi_0$. In magnetic systems, this configuration corresponds to the interface between two magnetic domains (or Weiss domains). It is illustrated in Figure 10.2 in a still quite simple but less schematic case than the scalar order parameter: the spin orientation has only two preferred directions directed along axes parallel to the edges of the figure.

The present section is devoted to the description of the transition zone between two domains. Consider a tri-dimensional ($d = 3$) configuration at $T < T_C$ in the absence of external field ($h = 0$) where ϕ depends on x only with $\phi(x \to \pm\infty) = \pm\phi_0$ (see Figure 10.3). The interface is thus planar, and perpendicular to the x direction. If the free energy density

---

[3]This corresponds to a spontaneous magnetisation in the case of the Ising model.

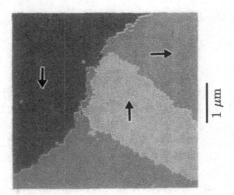

FIGURE 10.2 Magnetic structure of the surface of an iron (Fe) crystal. Three domains whose magnetisations are represented by arrows can be identified. The domains are separated by so-called *domain walls*. After H. P. Oepen and J. Kirschner, J. Phys. Colloques **49**, C8-1853 (1988), reprinted with permission.

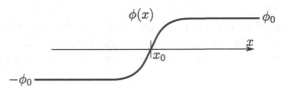

FIGURE 10.3 Sketch of the variation of the order parameter at the interface between two domains. Referring, for instance, to H. P. Oepen and J. Kirschner, J. Phys. Colloques **49**, C8-1853 (1988) it can checked that this type of profile provides a realistic description of the experimental observations.

of the homogeneous system is written as $f_L(\phi) = a(T - T_C)\phi^2 + d\phi^4$, $\phi(x)$ verifies a one-dimensional version of Equation(10.3) (with $h = 0$). This equation admits the following first integral:

$$g\left(\frac{\mathrm{d}\phi}{\mathrm{d}x}\right)^2 - f_L(\phi) = E_{\mathrm{cl}}, \tag{10.17}$$

Formula (10.17) is formally analogous to the energy conservation equation of a fictitious classical particle: this is why the integration constant is noted E_{cl}. ϕ plays the role of the position of the fictitious particle, x the role of time, and $-f_L(\phi)$ the role of the potential energy. The domain wall configuration to be described corresponds to the "motion" of a fictitious particle from $-\phi_0$ to $+\phi_0$ at an energy E_{cl} equal to the maximum $-f_L(\pm\phi_0) = d\phi_0^4$ of the fictitious potential. This behaviour is illustrated in Figure 10.4. Equation (10.17) can be written as $g(\mathrm{d}\phi/\mathrm{d}x)^2 = f_L(\phi) - f_L(\phi_0) = d(\phi^2 - \phi_0^2)^2$. Defining x_0 as the point where $\phi = 0$, one gets:

$$\int_{x_0}^{x} \mathrm{d}x = \int_0^{\phi} \frac{\mathrm{d}\phi}{\mathrm{d}\phi/\mathrm{d}x} = \sqrt{\frac{g}{d}} \int_0^{\phi} \frac{\mathrm{d}\phi}{\phi_0^2 - \phi^2} = \sqrt{\frac{g}{d}} \frac{1}{\phi_0} \operatorname{artanh}(\phi/\phi_0). \tag{10.18}$$

Noting that $\phi_0\sqrt{d/g} = (2\xi)^{-1}$ gives:

$$\phi(x) = \phi_0 \tanh\left[\frac{x - x_0}{2\xi}\right]. \tag{10.19}$$

Solution (10.19) describing the domain wall corresponds to a local minimum of the Landau free energy in the space of possible configurations $\phi(x)$. It is certainly not optimal: the

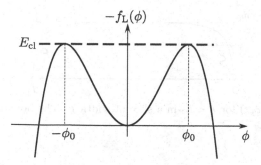

FIGURE 10.4 "Potential energy" as a function of ϕ as seen by the fictitious particle. The domain wall configuration corresponds to a "motion" from ϕ_0 to $+\phi_0$ performed at the constant "energy" E_{cl} shown as the horizontal dashed segment and equal to $-f_L(\phi_0)$.

corresponding free energy is greater than that of a uniform configuration. As a result, it can be associated with a surface tension, defined as the energy cost (per unit of transverse area) of the domain wall with respect to the uniform solution. Let L_y and L_z be the spatial dimensions of the sample in the y and z directions, respectively. The surface tension is:

$$\sigma = \frac{F_L[\phi] - F_L[\phi_0]}{L_y L_z} = \int_{\mathbb{R}} dx \left[g \left(\frac{d\phi}{dx}\right)^2 + f_L(\phi) - f_L(\phi_0) \right] = 2g \int_{\mathbb{R}} dx \left(\frac{d\phi}{dx}\right)^2 , \qquad (10.20)$$

where ϕ is the domain wall solution [Equation (10.19)]. In the last term of Equation (10.20), the first integral defined by Equation (10.17) was used. Doing the change of variable $\phi = \phi(x)$ allows to write (again thanks to the first integral) $dx(d\phi/dx)^2 = d\phi[f_L(\phi) - f_L(\phi_0)]^{1/2}/\sqrt{g}$ and thus the last term of Equation (10.20) can be expressed as:

$$\sigma = 2\sqrt{g\,d} \int_{-\phi_0}^{\phi_0} d\phi \left(\phi_0^2 - \phi^2\right) = \frac{4}{3} g \, \phi_0^2/\xi . \qquad (10.21)$$

It can be checked *a posteriori* that, apart from the factor $\frac{4}{3}$, this result could have been obtained by a quick order of magnitude calculation, similar to the one presented at the end of Section 10.3.1.

10.3.3 EFFECT OF BOUNDARY CONDITIONS

Another source of spatial variation of the order parameter are the boundary conditions imposed on the sample. To fix ideas, consider a slab infinite in the y and z directions and bounded in the x-direction with $x \in [0, L]$. For all y and z, the boundary condition $\phi(0, y, z) = 0 = \phi(L, y, z)$ is imposed. It is clear that the order parameter is a function of x only, noted $\phi(x)$. For a temperature $T > T_C$ and in the absence of external field ($h = 0$), the usual solution $\phi(x) = 0$ for an infinite system is compatible with the boundary conditions. By contrast, for $T < T_C$, the solution $\phi(x) = \phi_0$ is not.

However, as already observed in the previous subsection, the order parameter is able to deviate from its nominal value. The system is expected to adopt a configuration similar to the one depicted in Figure 10.5: The boundary conditions impose $\phi(0) = \phi(L) = 0$ but $\phi(x)$ tends towards its nominal value ϕ_0 at a distance ξ from the sample edges. Yet, this configuration is only possible if $L \gtrsim 2\xi$. According to Expression (10.14) of ξ, this cannot be realised if T is too close to T_C, i.e. $T \lesssim T_C - 2g/(aL^2)$. This means that the transition temperature in a slab is lower than the transition temperature in the bulk by a term of order $g/(aL^2)$. The following exercise gives a quantitative evaluation of this lowering of the critical temperature.

FIGURE 10.5 Shape of $\phi(x)$ for $T < T_C$ in a slice of width L. The boundary conditions are $\phi(0) = \phi(L) = 0$.

Exercise 10.2. Critical temperature in a thin film.
This exercise is dedicated to the quantitative study of the behaviour of the order parameter in a thin film of thickness L (a slab). On the film's boundary, the order parameter is set to zero.

1. Write the differential equation verified by ϕ and show that a first integral can be found. Give an intuitive interpretation of this equation.

   a. Show that if $T > T_C$ the only possible solution is $\phi = 0$.

   b. The case $T < T_C$ is now considered. Justify that except from the trivial solution $\phi = 0$, the minimum energy solution has most certainly the shape depicted in Figure 10.5.

2. Show that, the solution shown in Figure 10.5 is acceptable only if the width L of the thin film is greater than a value $L_{\min}$. Determine $L_{\min}$ and show that the phase transition temperature $T_C^{(L)}$ in the film is reduced compared to the transition temperature T_C in the bulk. Express $T_C^{(L)}$ as a function of T_C and the parameters of the problem.

Solution:
1. Equation (10.2) can be written here as:

$$-2g\frac{d^2\phi}{dx^2} + \frac{\partial f_L}{\partial \phi} = 0, \qquad (10.22)$$

and admits a first integral [Equation (10.17)]. The interpretation of this first integral is the same as the one given in Section 10.3.2: there is a formal analogy with the energy conservation of a fictitious classical particle of "mass" $2g$, "position" ϕ, at "time" x, moving in a "potential" $-f_L(\phi)$. As in Section 10.3.2, E_{cl} can be viewed as the energy of the fictitious classical particle.

a) If $T > T_C$, the effective potential is concave with $-f_L(0) = 0$. The only possible "trajectory" compatible with the boundary conditions $\phi(0) = \phi(L) = 0$ is $\phi(x) = 0$.

b) If $T < T_C$, the effective potential has the shape depicted in Figure 10.4. To obtain a solution $\phi(x)$ that obeys the boundary conditions in $x = 0$ and $x = L$, E_{cl} must lie between 0 (the local minimum of the effective potential) and $-f_L(\phi_0)$ (maximum of the effective potential). Typical "trajectories" are shown in Figure 10.6: they are functions oscillating between two opposite extremum values $\pm\phi_m$, where ϕ_m is the value of ϕ for which $f_L(\pm\phi_m) = E_{cl}$. For a given value of the film thickness L, not all values of E_{cl} are acceptable: the oscillation period of $\phi(x)$ must be a multiple of $L/2$ in order to satisfy boundary conditions in 0 and L.

2. It is clear that the profile of $\phi(x)$ showing the least oscillation (Figure 10.6(b)) corresponds to the solution minimising the free energy (oscillations have an energy cost due to the gradient term in the Ginzburg-Landau functional). For this profile, one obtains:

$$\frac{L}{2} = \int_0^{L/2} dx = \int_0^{\phi_m} \frac{d\phi}{d\phi/dx} = \sqrt{g}\int_0^{\phi_m} \frac{d\phi}{\sqrt{E_{cl} + f_L(\phi)}}. \qquad (10.23)$$

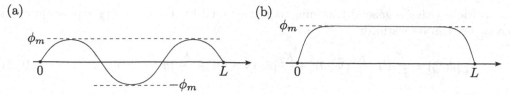

FIGURE 10.6 Possibles solutions $\phi(x)$ satisfying Equation (10.22) for $T < T_C$. The boundary conditions are $\phi(0) = \phi(L) = 0$. The two profiles have different values ϕ_m of the maximum of the order parameter.

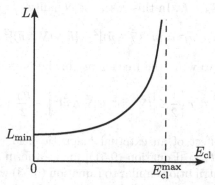

FIGURE 10.7 Graphical representation of Equation (10.23) relating L to E_{cl}. When $E_{cl} \to 0$, the oscillations of $\phi(x)$ become sinusoidal, and their period no longer depends on their amplitude. In this case, to obtain a solution of the form shown in Figure 10.6(b), L must be exactly equal to twice this period: this is the minimum value L_{min} allowed for L. When $E \to E_{cl}^{max} = -f_L(\phi_0)$, the period of nonlinear oscillations may diverge (depending on their amplitude). This ensures that a solution always exists when L becomes very large.

This equation relates the value of E_{cl} to the width L of the film. The corresponding plot is shown in figure 10.7.

To determine the smallest value L_{min} reached by L, let us consider the case $E_{cl} \to 0$ in Formula (10.23). In this limit, a quadratic approximation can be made: $f_L(\phi) \simeq -a(T_C - T)\phi^2$ and $E_{cl} = -f_L(\phi_m) \simeq a(T_C - T)\phi_m^2$ giving:

$$L_{min} = 2\sqrt{\frac{g}{a(T_C - T)}} \int_0^{\phi_m} \frac{d\phi}{\sqrt{\phi_m^2 - \phi^2}} = \pi\sqrt{\frac{g}{a(T_C - T)}} = \sqrt{2}\,\pi\,\xi\,.$$

The relation $L_{min} \leq L$ can be written as

$$T \leq T_C^{(L)} = T_C - \pi^2 g/(aL^2)\,.$$

$T_C^{(L)}$ *is the highest temperature for which a configuration of the system corresponding to a non-uniformly zero order parameter can be found. Note that the above exact expression of $T_C^{(L)}$ is in good agreement with the qualitative result obtained using the order of magnitude reasoning presented in the paragraph preceding this exercise.*

10.4 STATIC DEFORMATION IN A NEMATIC CRYSTAL

Consider a liquid crystal (see Section 9.3) in a uniaxial nematic configuration where the parameter S is homogeneous and equal to 1. In addition, the director $\vec{n}$ can now depend on the position. Using a general reasoning as in Section 10.1, the following expression of the free energy can be justified:

$$F_L\left[\vec{n}(\vec{r})\right] = \int d^3r \left\{\frac{K_1}{2}(\vec{\nabla}\cdot\vec{n})^2 + \frac{K_2}{2}[\vec{n}\cdot(\vec{\nabla}\wedge\vec{n})]^2 + \frac{K_3}{2}[\vec{n}\wedge(\vec{\nabla}\wedge\vec{n})]^2\right\} \quad . \tag{10.24}$$

Since $\vec{n}\cdot\vec{n} = 1$, the phenomenological constants K_1, K_2 and K_3 have the dimension of forces with a typical order of magnitude $K_i \sim 10^{-12}$ N. For example, for the MBBA nematic crystal[4] at 25 °C: $K_1 = 6\times 10^{-12}$ N, $K_2 = 2.9\times 10^{-12}$ N and $K_3 = 9.5\times 10^{-12}$ N. A common approximation leading to a qualitatively correct result is named the *single constant approximation*: $K_1 = K_2 = K_3 = K$. In this case, the equality[5]

$$(\vec{\nabla}\wedge\vec{n})^2 = [\vec{n}\cdot(\vec{\nabla}\wedge\vec{n})]^2 + [\vec{n}\wedge(\vec{\nabla}\wedge\vec{n})]^2 \, , \tag{10.25}$$

valid when $\vec{n}\cdot\vec{n} = 1$, allows to write the free energy in the presence of a magnetic field as:

$$F_L\left[\vec{n}(\vec{r}),\vec{H}(\vec{r})\right] = \int d^3r \left\{\frac{1}{2}K\left[(\vec{\nabla}\cdot\vec{n})^2 + |\vec{\nabla}\wedge\vec{n}|^2\right] - \frac{\mu_0}{2}\chi_a(\vec{H}\cdot\vec{n})^2\right\} \quad . \tag{10.26}$$

In Equation (10.26), the effect of an external magnetic field was taken into account by including the contribution F_{mag} [Equation (9.51)] under a form appropriate to the studied configuration ($S = 1$). A contribution similar to Equation (9.53) would here add an irrelevant constant term and is therefore not included.

Consider a device in which the liquid crystal occupies a space between two glass walls. The interaction between the nematic phase and the glass constrains the director to be perpendicular to the surface of the glass on each wall. A magnetic field parallel to the walls is applied (see Fig. 10.8). When the field strength exceeds a certain threshold H_{crit}, the optical properties of the system change abruptly. This effect, observed for the first time by Fredericksz and his collaborators (1927 and 1933), is described in the following.

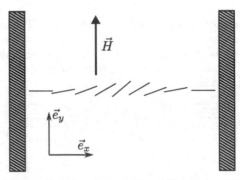

FIGURE 10.8 Schematic of the Fredericksz device. The rods orientations are depicted by the plain segments. The glass walls are the hashed rectangles.

The two walls confining the liquid crystal are parallel to the yz-plane and located at $x = 0$ and $x = L$. The surface area of the walls is $L_y L_z$ where L_y and L_z are the dimension of the glass walls in the y and z directions, respectively. The orientation of the director is given by the angle $\theta(x)$ such that $\vec{n} = \cos\theta(x)\,\vec{e}_x + \sin\theta(x)\,\vec{e}_y$ with the boundary conditions

---

[4]see I. Haller, J. Chem. Phys. **52**, 1400 (1972).

[5]This equality stems from the fact that, for any vector $\vec{A}$ making an angle θ with $\vec{n}$: $|\vec{n}\cdot\vec{A}|^2 = A^2\cos^2\theta$ and $|\vec{n}\wedge\vec{A}|^2 = A^2\sin^2\theta$, hence $A^2 = |\vec{n}\cdot\vec{A}|^2 + |\vec{n}\wedge\vec{A}|^2$.

$\theta(0) = \theta(L) = 0$. The magnetic field is uniform and reads $\vec{H} = H\,\vec{e}_y$. The free energy per unit area [Equation (10.26)] can be written as a functional of $\theta(x)$:

$$\frac{F_L[\theta]}{L_y L_z} = \int_0^L dx \left\{ \frac{K}{2} \left(\frac{d\theta}{dx} \right)^2 - \frac{\mu_0}{2} \chi_a H^2 \sin^2 \theta \right\}. \tag{10.27}$$

The extremisation of Equation (10.27) with respect to the variations of $\theta(x)$ leads to the equation:

$$\xi_H^2 \frac{d^2\theta}{dx^2} + \sin\theta \cos\theta = 0, \quad \text{where} \quad \xi_H = \sqrt{\frac{K}{\mu_0 \chi_a H^2}}. \tag{10.28}$$

The parameter ξ_H is called the *magnetic length*. It is the length over which the change of orientations of the molecules has an energy cost comparable to the energy of interaction with the magnetic field $[K(\theta/\xi_H)^2 \sim \mu_0 \chi_a H^2 \theta^2]$. The differential Equation (10.28) must be solved with the boundary conditions: $\theta(0) = 0 = \theta(L)$. $\theta(x) = 0$ is a trivial solution. If the magnetic field is strong enough, other solutions can be found by following an approach similar to the one described in Exercise 10.2. First, note that Equation (10.28) admits a first integral:

$$\xi_H^2 \left(\frac{d\theta}{dx} \right)^2 + \sin^2\theta = E_{cl}. \tag{10.29}$$

E_{cl} is the integration constant, as in Formula (10.17), and the same interpretation can be made here: Equation (10.29) is formally identical to the conservation of energy equation of a classical particle whose "position" θ varies as a function of "time" x. The term $\xi_H^2 (d\theta/dx)^2$ plays the role of a kinetic energy and the term $\sin^2\theta$ the role of a potential energy. A graphical interpretation of Equation (10.29) is given in Figure 10.9 which is the counterpart of Figure 10.4 for the case studied here. If it is possible to find a solution $\theta(x)$ of Equation (10.29) with a non-zero integration constant, then it should behave as illustrated in Figure 10.9(b). Note that this is the simplest non-trivial solution of Equation (10.29). Other solutions with a larger number of oscillations between 0 and L could be thought of but these solutions would not be physically acceptable (the liquid crystal molecules would be aligned with the magnetic field and then reverted to horizontal, etc). The solution represented in Figure 10.9(b) is the only physically acceptable non-trivial solution. If this solution is mathematically allowed, then it will automatically be energetically favoured over the trivial one ($\theta = 0$) since it represents a configuration where the molecules tend to align with the magnetic field (more or less strongly depending on the value θ_m).

Assume that a non-trivial solution of Equation (10.29) exists. The integration yields the evolution law $\theta(x)$. For $x \in [0, L/2]$, $x = \int_0^x dx = \int_0^{\theta_m} d\theta/(d\theta/dx)$, which leads to:

$$\frac{x}{\xi_H} = \int_0^\theta \frac{d\theta}{\sqrt{\sin^2\theta_m - \sin^2\theta}}, \tag{10.30}$$

and in particular when $x = L/2$:

$$\frac{L}{2\xi_H} = \int_0^{\theta_m} \frac{d\theta}{\sqrt{\sin^2\theta_m - \sin^2\theta}}. \tag{10.31}$$

If $L \gg \xi_H$ (i.e., in the limit of strong magnetic fields) A value of θ_m satisfying Equation (10.31) can always be found (it is sufficient to choose θ_m close enough to $\pi/2$). On the other hand, if L is small, this is not always possible. Indeed, suppose that $L \ll \xi_H$, then θ_m must be small and it is legitimate to use the small angle expansion of $\sin^2\theta \simeq \theta^2$

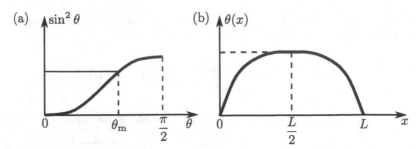

FIGURE 10.9 Illustration of the method of the fictitious classical particle. (a) Effective potential $\sin^2 \theta$ as a function of fictitious position. The horizontal line marks the integration constant in (10.29), corresponding to a maximum value θ_m of θ. (b) Corresponding profile $\theta(x)$. Note that the solution with opposite sign is also acceptable and corresponds to a configuration where the orientation of the molecules is symmetric with respect to the Ox axis to the one shown in Figure 10.8(b).

in Equation (10.31). The right-hand integral of Equation (10.31) is then easily calculated[6] leading to $L/\xi_H = \pi$. Since L is fixed, this relation is verified only if H is greater than a critical value H_{crit}:

$$H_{\mathrm{crit}} = \frac{\pi}{L} \sqrt{\frac{K}{\mu_0 \chi_a}} . \tag{10.32}$$

When $H > H_{\mathrm{crit}}$, Equation (10.30) can be solved and x is calculated as a function of θ through an elliptic integral[7].

Note that the order of magnitude of the critical field can be obtained by a simple physical reasoning based on the analysis of the different contributions to the energy [Equation (10.27)]: a non-trivial configuration ($\theta(x) \neq 0$) is only possible if the magnetic energy gain compensates the energy cost of the molecules orientation change, i.e. if $\frac{1}{2}\mu_0\chi_a H^2 L > \frac{K}{2}(\frac{\pi/2}{\xi_H})^2\xi_H$, which leads – up to a dimensionless multiplicative factor – to the critical value of the magnetic field [Formula (10.32)].

Exercise 10.3.   Give the expression of $\theta(x)$ when the distance L between the two walls tends to infinity.

Solution: In this case, the molecules align perfectly with the field when the distance to the wall is much greater than the magnetic length ξ_H. Thus, $\theta(x \gg \xi_H) = \theta_m = \pi/2$ and since the two walls are separated by a very large distance, the behaviour in the vicinity of a wall no longer depends on the distance to the other. It is sufficient to consider only the vicinity of the wall in $x = 0$ for example and calculate Integral (10.30) with $\theta_m = \pi/2$. This gives: $\frac{x}{\xi_H} = -\ln\tan\left(\frac{\pi}{4} - \frac{\theta}{2}\right)$, or

$$\theta(x) = \frac{\pi}{2} - 2\arctan\left(e^{-x/\xi_H}\right) . \tag{10.33}$$

---

[6]The change of variable $\theta = \theta_m \sin\alpha$ with $\alpha \in [0, \pi/2]$ is used.

[7]$x = \xi_H F(\varphi | \sin^2\theta_m)$, where $\varphi = \arcsin\left(\frac{\sin\theta}{\sin\theta_m}\right)$ and F is an elliptic integral of the first kind:

$$F(\varphi|k^2) = \int_0^\varphi d\theta(1 - k^2\sin^2\theta)^{-1/2} .$$

10.5 GINZBURG-LANDAU THEORY OF SUPERCONDUCTIVITY

At very low temperature, some materials become superconducting: they do not resist the flow of an electric current and they repel external magnetic fields (only a sufficiently intense field can penetrate inside a superconducting sample).

In 1950, Ginzburg and Landau proposed a phenomenological theory of superconductivity inspired by Landau theory of second order phase transitions (1937). They used a complex order parameter $\phi(\vec{r})$ whose physical meaning was unclear at the time, and which has since then been interpreted as describing a field of electron pairs. They wrote the system's free energy in the presence of a magnetic field by imposing gauge invariance. This theory enabled a detailed description of the rich effects caused by the mutual interaction of a superconductor and a static magnetic field. It is a remarkable illustration of the effectiveness of the mean effective field approach.

The superconducting phase is described by a complex order parameter $\phi(\vec{r})$ acting as an effective wave function. In the presence of a magnetic induction field $\vec{B}(\vec{r})$, the Ginzburg-Landau free energy is[8]:

$$F_L = F_0(T) + \int \mathrm{d}^3 r \left\{ \frac{\hbar^2}{2m} \left| \left(\vec{\nabla} - \frac{iq}{\hbar} \vec{A} \right) \phi \right|^2 + \tilde{a}|\phi|^2 + \frac{d}{2}|\phi|^4 + \frac{\vec{B}^2}{2\mu_0} \right\}, \qquad (10.34)$$

where m and q are two parameters of the model. The value of m is arbitrary, but not that of q. This point will be discussed at the end of Section 10.5.1. $\vec{A}(\vec{r})$ is the vector potential, related to the magnetic induction field ($\vec{B} = \vec{\nabla} \wedge \vec{A}$). d is a constant positive parameter, and $\tilde{a} = a \times (T - T_C)$ where $a > 0$ and T_C is the temperature of the superconducting transition. It varies, depending on the materials, from a few degrees Kelvin ($T_C = 4.1$ K for Hg and 9.2 K for Nb) to a value that can be higher than the boiling temperature of nitrogen ($T_C = 92$ K for $YBa_2Cu_3O_7$). The current records are held by hydrogen sulfide H_2S and LaH_{10} which, under enormous pressures, becomes superconducting at 203 K and 250 K, respectively[9].

The extremisation of F_L with respect to the variations of ϕ and $\vec{A}$ leads to two coupled equations describing the equilibrium configuration of the order parameter and the magnetic field. Since ϕ is complex, the extrimisation must be performed with respect to the variations of its real and imaginary parts. An equivalent method consists in extremising independently with respect to ϕ and ϕ^*. This amounts, for example, to writing $\partial \phi^2 / \partial \phi^* = \partial (\phi \phi^*) / \partial \phi^* = \phi$. Since Expression (10.34) of F_L is real, $\delta F_L / \delta \phi^* = (\delta F_L / \delta \phi)^*$ and it is sufficient to calculate one of these two quantities, $\delta F_L / \delta \phi^*$ for instance. The only difficulty comes from the gradient contribution in Expression (10.34). Writing

$$\left| \left(\vec{\nabla} - \frac{iq}{\hbar} \vec{A} \right) \phi \right|^2 = \left[\left(i\vec{\nabla} + \frac{q}{\hbar} \vec{A} \right) \phi \right] \cdot \left[\left(-i\vec{\nabla} + \frac{q}{\hbar} \vec{A} \right) \phi^* \right], \qquad (10.35)$$

one gets

$$\int \mathrm{d}^3 r \left| \left(\vec{\nabla} - \frac{iq}{\hbar} \vec{A} \right) \phi \right|^2 = \int \mathrm{d}^3 r \, \phi^* \left(i\vec{\nabla} + \frac{q}{\hbar} \vec{A} \right)^2 \phi + \text{surface term}. \qquad (10.36)$$

---

[8]The usual notations of the field of superconductivity are used here. They are slightly different from the conventions used elsewhere in the book. The coefficient of the ϕ^4 term in Equation (10.34) differs by a factor of 2 with respect to the one used in Equation (9.41), and the term in front of the gradient contribution is written as $\hbar^2/(2m)$ instead of g as in Equation (10.1). Later, a correlation length ξ is also defined (see Equation (10.46)). Its expression differs from Definition (10.14) by a factor $\sqrt{2}$.

[9]see A. P. Drozdov, M. I. Eremets, I. A. Troyan, V. Ksenofontov & S. I. Shylin, Nature **525**, 73 (2015) and A. P. Drozdov et al., Nature **569**, 528 (2019). Note that room temperature superconductivity might have been observed in carbonaceous sulfur hydride at very high pressure [E. Schneider et al., Nature **586**, 373 (2020)].

To obtain Equation (10.36) starting from Equation (10.35), the only somewhat difficult step is to use, for a vector field $\vec{V}$, the general relation[10] $(\vec{V} \cdot \vec{\nabla})\phi^* = \vec{\nabla} \cdot (\phi^* \vec{V}) - \phi^* \vec{\nabla} \cdot \vec{V}$, which, when $\vec{V}(\vec{r}) = -\mathrm{i}(\mathrm{i}\vec{\nabla} + \frac{q}{\hbar}\vec{A})\phi$ allows to write:

$$\left(\mathrm{i}\vec{\nabla} + \frac{q}{\hbar}\vec{A}\right)\phi \cdot (-\mathrm{i}\vec{\nabla}\phi^*) = \mathrm{i}\phi^* \vec{\nabla} \cdot (\mathrm{i}\vec{\nabla} + \frac{\mathrm{i}q}{\hbar}\vec{A})\phi - \mathrm{i}\vec{\nabla} \cdot \left[\phi^*(\mathrm{i}\vec{\nabla} + \frac{q}{\hbar}\vec{A})\phi\right] \ . \tag{10.37}$$

The last term of the right-hand side of Equation (10.37) is a divergence which, according to Gauss-Ostrogradskii theorem [Formula (A.45)], gives a surface contribution which can be ignored in the thermodynamic limit. Then, Expression (10.36) is easily obtained. The subsequent extremisation with respect to ϕ^* is not difficult and the equation $\delta F_L/\delta \phi^* = 0$ can be written as:

$$\frac{1}{2m}\left(\mathrm{i}\hbar\vec{\nabla} + q\vec{A}\right)^2 \phi + \tilde{a}\,\phi + d\,|\phi|^2 \phi = 0 \ . \tag{10.38}$$

The extremisation of Equation (10.34) with respect to $\vec{A}$ is done most simply by going back to the definition of the functional derivative given in Section 10.2, i.e. by evaluating the change of F_L due to a variation $\delta \vec{A}$ of $\vec{A}$:

$$\delta \left[\int \mathrm{d}^3 r\, \frac{\vec{B}^2}{2\mu_0}\right] = \frac{1}{\mu_0}\int \mathrm{d}^3 r\, \vec{B} \cdot \left(\vec{\nabla} \wedge \delta\vec{A}\right) = \frac{1}{\mu_0}\int \mathrm{d}^3 r\, \left[\delta\vec{A} \cdot \vec{\nabla} \wedge \vec{B} + \vec{\nabla} \cdot \left(\delta\vec{A} \wedge \vec{B}\right)\right] \ . \tag{10.39}$$

The last equality in Relation (10.39) can be verified by a direct calculation on the integrand[11]. The divergence in the last term of Equation (10.39) gives – using Gauss-Ostrogradskii theorem [Formula (A.45)] – a surface term which is irrelevant in the thermodynamics limit. One can also, without much difficulty, calculate the variation of the other term of Equation (10.34) involving the vector potential and finally obtain the total change of F_L due to a variation $\delta\vec{A}$:

$$\delta F_L = \int \mathrm{d}^3 r\, \delta\vec{A} \cdot \left(\frac{1}{\mu_0}\vec{\nabla} \wedge \vec{B} - \vec{J}_s\right) \ , \tag{10.40}$$

where

$$\vec{J}_s(\vec{r}) = -\frac{q^2}{m}|\phi|^2\vec{A} - \frac{\mathrm{i}q\hbar}{2m}\left(\phi^*\vec{\nabla}\phi - \phi\vec{\nabla}\phi^*\right) \ . \tag{10.41}$$

Hence, the extremisation of the Ginzburg-Landau free energy with respect to the vector potential leads to a Maxwell-Ampere type equation:

$$\vec{\nabla} \wedge \vec{B} = \mu_0 \vec{J}_s \ , \quad \text{where} \quad \vec{J}_s = \frac{q}{m}\rho\left[\hbar\vec{\nabla}\theta - q\vec{A}\right] \tag{10.42}$$

is called the *supercurrent density*. Its expression in Formula (10.42) is calculated from Relation (10.41) by writing the order parameter as $\phi = [\rho(\vec{r})]^{1/2}\exp\{\mathrm{i}\theta(\vec{r})\}$. Sometimes $\rho = |\phi|^2$ is called the *pair density*.

   Equations (10.38) and (10.42) are coupled equations governing the behaviour of the order parameter and the vector potential inside the superconductor. It can easily be checked that if a gauge transformation is performed on the vector potential $\vec{A} \to \vec{A}' = \vec{A} + \vec{\nabla}\chi$ (where $\chi(\vec{r})$ is any scalar field) then the new solution of Equation (10.38) is $\phi'(\vec{r}) = \phi(\vec{r})\exp\{\mathrm{i}\frac{q}{\hbar}\chi(\vec{r})\}$. Expression (10.34) of F_L is therefore gauge invariant (if the changes $\vec{A} \to \vec{A}'$ and $\phi \to \phi'$ are concomitantly made). The same holds true for the current density $\vec{J}_s$.

---

[10]This formula can be easily verified by direct computation, or, even better, using tensor notations with implicit summation over the repeated indices: $V_j \partial_j \phi^* = \partial_j(V_j \phi^*) - \phi^* \partial_j V_j$.

[11]The result can be obtained quickly by using tensor notations. Writing ϵ_{ijk} the fully antisymmetric tensor of rank 3, one gets: $\vec{B} \cdot (\vec{\nabla} \wedge \delta\vec{A}) = B_i \epsilon_{ijk}\partial_j \delta A_k = \partial_j(\epsilon_{jki}\delta A_k B_i) + \delta A_k \epsilon_{kji}\partial_j B_i = \vec{\nabla} \cdot (\delta\vec{A} \wedge \vec{B}) + \delta\vec{A} \cdot \vec{\nabla} \wedge \vec{B}$.

To gain acquaintance with the phenomenology of superconductivity, let us consider a uniform system in the absence of any magnetic field. In this case, the order parameter is a constant ϕ_0, solution of (10.38):

$$\phi_0 = \begin{cases} 0 & \text{if } T > T_C, \\ [a(T_C - T)/d]^{1/2} & \text{if } T < T_C. \end{cases} \tag{10.43}$$

A model nonuniform case at temperature below T_C can also be considered. In this situation, $\phi(\vec{r}) = 0$ is imposed in the half-space $x \leq 0$. For $x > 0$, according to Equation (10.38), $\phi(x)$ verifies:

$$-\frac{\hbar^2}{2m}\frac{d^2\phi}{dx^2} + \tilde{a}\phi + d\phi^3 = 0 . \tag{10.44}$$

One should look for the solution of Equation (10.44) that verifies $\phi(0) = 0$ and $\phi(x \to \infty) = \phi_0$, where ϕ_0 is given by Equation (10.43) (for the case $T < T_C$). This solution reads[12]:

$$\phi(x) = \phi_0 \tanh\left(\frac{x}{\sqrt{2}\,\xi}\right) , \tag{10.45}$$

where ξ is the "correlation length". At temperature below T_C, ξ is given by:

$$\frac{\hbar^2}{2m\xi^2} = -\tilde{a} = a(T_C - T) . \tag{10.46}$$

Apart from a factor 2, this definition is identical to Definition (10.13) and $\xi \propto (T_C - T)^{-1/2}$ as well.

10.5.1 MEISSNER EFFECT AND FLUXOID QUANTISATION

In this section, the interaction of the electromagnetic field with the superconductor is described in a simplified way: the pair density ($\rho = \phi^*\phi$) is supposed to be unaffected by the electromagnetic field and keeps the constant value ρ_0 determined in the previous section. This means that only weak electromagnetic fields are considered. While such fields do not affect the pair density, they can however induce currents in the superconductor, so that the pairing field can be written as $\phi(\vec{r}) = \sqrt{\rho_0}\exp\{i\theta(\vec{r})\}$. In the studied configuration, a field of this type exists in the half-space $x > 0$ which is occupied by the superconducting medium, while the half-space $x < 0$ is empty, and thus $\phi(x < 0) = 0$. A more accurate study should use a less schematic description of $\phi(x)$ [see the study leading to Equation (10.45)]. It will be seen in Section 10.5.4 that the situation considered here corresponds to an extreme type II superconductor.

Taking the curl of the Maxwell-Ampere Equation (10.42), it is found that inside the superconductor the magnetic field obeys:

$$\vec{\nabla}^2\vec{B} = \frac{1}{\lambda_L^2}\vec{B}, \quad \text{with} \quad \lambda_L = \sqrt{\frac{m}{\mu_0 q^2 \rho_0}} . \tag{10.47}$$

The system's characteristic length λ_L is called *London length*. If a uniform magnetic field $\vec{B}_0 = B_0\vec{e}_z$ is imposed in the half-space $x < 0$, then the magnetic field in the half-space $x > 0$ is $\vec{B}(x) = B_0\exp(-x/\lambda_L)\vec{e}_z$. The London length is also called the *penetration depth*. It means that a static magnetic field cannot penetrate a superconductor[13] deeper than λ_L:

---

[12]One can be convinced of the correctness of Equation (10.45) either by a direct verification or by using the first integral methode discussed in Section 10.3.2.

[13]It is worthwhile to recall that in a usual conductor, *oscillating fields* are exponentially damped.

this is the Meissner effect, observed experimentally by Meissner in 1933. It corresponds to ideal diamagnetism, as schematically illustrated in Figure 10.10: electric currents flow near the superconductor's surface and create a magnetic field perfectly balancing the imposed field $\vec{B}$ in the sample. In the schematic configuration considered here, the current density inducing this screening is $\vec{J}_s = B_0/(\mu_0 \lambda_L)\exp(-x/\lambda_L)\vec{e}_y$. The Meissner effect is often illustrated in scientific popularisation articles by a picture of a small magnet levitating above a superconductor.

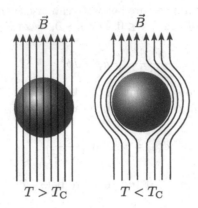

$$\vec{B} \qquad\qquad \vec{B}$$

$$T > T_C \qquad\qquad T < T_C$$

FIGURE 10.10 Schematic illustration of the Meissner effect in a superconductor: below the critical temperature, the superconductor sample expels magnetic field lines. Source: `https://en. wikipedia.org/wiki/Meissner_effect`.

In the interior of a superconducting sample immersed in a static magnetic field, far form the surface (more precisely at a depth larger than λ_L), $\vec{B} = 0$ and the Maxwell-Ampere Equation (10.42) imposes $\vec{J}_s = 0$ and $\vec{A} = \frac{\hbar}{q}\vec{\nabla}\theta$. Along a path connecting a point $\vec{r}_1$ to a point $\vec{r}_2$, the line integral of $\vec{A}$ is:

$$\int_{\vec{r}_1}^{\vec{r}_2} \vec{A}.\mathrm{d}\vec{\ell} = \frac{\hbar}{q}\Big[\theta(\vec{r}_2) - \theta(\vec{r}_1)\Big] . \qquad (10.48)$$

Let us apply the above formula along a closed curve $\mathscr{C}$ inside a toroidal superconductor, as represented in Figure 10.11(a). The field ϕ must be single-valued, but this constraint can be relaxed for the phase θ in a superconductor having a non-simply connected shape: it is sufficient for θ to vary by a multiple of 2π on a closed contour, such that $\oint_{\mathscr{C}} \vec{A}.\mathrm{d}\vec{\ell} = \frac{\hbar}{|q|}2\pi n$ with $n \in \mathbb{Z}$. Using Stokes' theorem [Formula (A.46)], this implies that the flux of $\vec{B}$ through the surface bounded by curve $\mathscr{C}$ is quantised in units of $\Phi_0 = h/|q|$. Experiments by Doll and Näbauer and by Deaver and Fairbank showed in 1961 that the flux is effectively quantised with a quantum Φ_0. The measured value is $\Phi_0 = 2.07 \times 10^{-7}$ gauss·cm^2 (see Figure 10.11). The corresponding value of the charge is $|q| = h/\Phi_0 = 1.998\,e$, where e is the elementary charge[14]. This value, equal to $2e$ within experimental uncertainties, confirms the interpretation of the order parameter as an effective wave function for electron pairs. However, nothing in the theory fixes the normalisation $\int \mathrm{d}r^3\,|\phi|^2$: the amplitude of the field ϕ is a free parameter which amounts to allowing a global scaling factor on $\hbar^2/(2m)$, a and d. It is the reason why it was stated at the beginning of the section that the value of m is not fixed in the theory.

---

[14]Recall that the S.I unit of magnetic field is the Tesla (T) and that 1 gauss$=10^{-4}$ T. The Planck constant is $h \simeq 6.62 \times 10^{-34}$ J·s and the elementary charge is $e \simeq 1.60 \times 10^{-19}$ C.

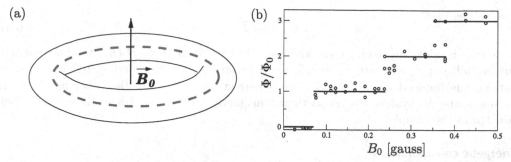

FIGURE 10.11 (a) Schematic representation of the experimental set-up: a toroidal superconductor is immersed in a static magnetic field. Curve $\mathscr{C}$ (dashed line) corresponds to the integration path in Formula (10.48). (b) Magnetic flux through a hollow superconducting tin cylinder as a function of the applied magnetic field. Figure adapted from B. S. Deaver and W. M. Fairbank, Phys. Rev. Lett. **7**, 43 (1961).

10.5.2 THERMODYNAMICS AND MAGNETISM

This section presents an advanced discussion of the thermodynamics of magnetic systems. It addresses the rich behaviour of a superconductor in the presence of magnetic fields. In particular, ideal paramagnetism is discussed anew and thermodynamic potentials are defined. These potentials are revealed instrumental to tackle the notion of critical magnetic field in Section 10.5.3.

Reminders: Magnetism in matter

The macroscopic magnetic induction $\vec{B}$ is the value of the microscopic field $\vec{b}$ averaged over a volume, small at the macroscopic scale: $\vec{B} = \langle \vec{b} \rangle$. Therefore, $\vec{\nabla} \cdot \vec{B} = 0$.

Let $\vec{\jmath}$ be the microscopic current density. If no current is injected into the system, the flux of $\langle \vec{\jmath} \rangle$ through any surface Σ bounding the system is equal to zero: $\int_{\Sigma} d^2 r \, \langle \vec{\jmath} \rangle \cdot \vec{n} = 0$. If it were otherwise, this would mean that current is indeed injected into the system. It thus is legitimate to assume the existence of a field $\vec{M}$ such as $\langle \vec{\jmath} \rangle = \vec{\nabla} \wedge \vec{M}$, with $\vec{M} = \vec{0}$ outside of the sample. Indeed, Stokes theorem [Formula (A.46)] leads to $\int_{\Sigma} d^2 r \langle \vec{\jmath} \rangle \cdot \vec{n} = \oint_{\Gamma} \vec{M} \cdot d\vec{r} = 0$ since the curve Γ bounding Σ never penetrates the sample and $\vec{M}$ is thus strictly equal to zero on Γ. Taking $\frac{1}{\mu_0} \vec{B} - \vec{M} = \vec{H}$, one gets $\vec{\nabla} \wedge \vec{H} = \vec{0}$ as an immediate consequence of $\vec{\nabla} \wedge \vec{B} = \mu_0 \langle \vec{\jmath} \rangle = \mu_0 \vec{\nabla} \wedge \vec{M}$. It can be shown that $\vec{M}$ is the magnetic moment density[15], i.e the magnetisation. Thus, when no current is injected into the system, the macroscopic current density is:

$$\vec{J} = \vec{\nabla} \wedge \vec{M} \, . \tag{10.49}$$

The approach presented in this section is general, and it is sufficient to replace $\vec{J}$ by $\vec{J}_s$ for the formulas to apply to the case of a superconductor. When a current is injected in the system, $\langle \vec{\jmath} \rangle = \vec{\nabla} \wedge \vec{M} + \vec{J}_{\text{cond}} = \vec{J} + \vec{J}_{\text{cond}}$, where $\vec{J}_{\text{cond}}$ is the "conduction current" density (equal to zero when no current is injected in the system or when considering a dielectric). The magnetic field is then defined as:

$$\vec{H} = \frac{1}{\mu_0} \vec{B} - \vec{M} \, , \tag{10.50}$$

---

[15]Indeed, the total magnetic moment of the sample $\mathscr{M}$ defined as $\mathscr{M} \equiv \frac{1}{2} \int_V \vec{r} \wedge \langle \vec{\jmath} \rangle = \frac{1}{2} \int_V \vec{r} \wedge (\vec{\nabla} \wedge \vec{M})$ can be written as $\mathscr{M} = \int_V d^3 r \, \vec{M}$. The physics of magnetic materials can be quickly reviewed by reading the first four chapters of *Electrodynamics of continuous media: Volume 8* by Landau and Lifshitz (Butterworth-Heinemann, 1984).

and verifies

$$\vec{\nabla} \wedge \vec{H} = \vec{J}_{\text{cond}} \, . \tag{10.51}$$

Note that inside a superconductor, Meissner effect imposes $\vec{M} = -\vec{H}$ ($\vec{B} = 0$). The magnetic susceptibility $\chi = \partial M / \partial H|_{H=0}$ is effectively equal to -1 corresponding to ideal diamagnetism: the induced current $\vec{J}_s$ leads to a perfect screening. On the other hand, in the normal state the system has no particular magnetic property and a static magnetic field penetrates the sample.

Energetic considerations

Consider a configuration where no current is injected in the system. The current $\vec{J}$ is induced by the external magnetic field and can therefore be written as in Equation (10.49). Outside the sample, there exists a magnetic field created by conduction currents (typically the sample is placed inside a coil) noted $\vec{J}_{\text{ext}}$ (no longer $\vec{J}_{\text{cond}}$ since this current is outside the sample). In this case, there is a clear distinction between $\vec{J}$ (which is non-zero only inside the considered object) and $\vec{J}_{\text{ext}}$ (which is non-zero only outside).

A situation is considered, where, at a fixed temperature, the magnetic induction varies slightly: $\vec{B}(\vec{r}) \to \vec{B}(\vec{r}) + \delta\vec{B}(\vec{r})$. This operation is performed over a time δt, and is assumed to be quasi-static and reversible. It induces an electric field $\vec{E}$ such as $\vec{\nabla} \wedge \vec{E} = -\delta\vec{B}/\delta t$. $\vec{E}$ exerts on the external currents a work $\delta t \int \vec{J}_{\text{ext}} \cdot \vec{E} \, d^3 r$. So, if the roles are now reversed, considering the physical situation where the modification of $\vec{B}$ is due to external currents, the work done by these external currents is:

$$\delta W = -\delta t \int \vec{E} . \vec{J}_{\text{ext}} d^3 r - \delta t \oint (\vec{E} \wedge \vec{H}) \cdot \vec{n} \, d^2 r \, . \tag{10.52}$$

The last term of the right-hand side of Equation (10.52) is the incoming flow of the Poynting vector $\vec{S} = \vec{E} \wedge \vec{H}$ ($\vec{n}$ is the outgoing normal). In the first right-hand term of Equation (10.52), the current can be written as $\vec{J}_{\text{ext}} = \vec{\nabla} \wedge \vec{H}$ [see Equation (10.51)] and the relation[16]:

$$\vec{E} \cdot \vec{\nabla} \wedge \vec{H} = -\vec{\nabla} \cdot (\vec{E} \wedge \vec{H}) + \vec{H} \cdot (\vec{\nabla} \wedge \vec{E}) \tag{10.53}$$

allows to rewrite Equation (10.52) in the form:

$$\delta W = -\delta t \int \vec{H} \cdot (\vec{\nabla} \wedge \vec{E}) \, d^3 r + \delta t \int \vec{\nabla} \cdot (\vec{E} \wedge \vec{H}) \, d^3 r - \delta t \oint (\vec{E} \wedge \vec{H}) \cdot \vec{n} \, d^2 r \, . \tag{10.54}$$

The last two terms of Equation (10.54) cancel out according to Gauss-Ostrogradskii theorem [Formula (A.45)] and thus:

$$\delta W = \int \vec{H} \cdot \delta\vec{B} \, d^3 r \, . \tag{10.55}$$

Since the process considered is reversible and isothermal, the work is equal to the variation of Helmoltz free energy: $\delta W = dF$. In the case where $\vec{B}$ and T can change concomitantly, the general relation is:

$$dF = \int \vec{H} \cdot \delta\vec{B} \, d^3 r - S \, dT \, . \tag{10.56}$$

If the equivalent of Gibbs free energy is defined by:

$$G = F - \int d^3 r \, \vec{B} . \vec{H} \, , \tag{10.57}$$

---

[16]Relation (10.53) can be easily demonstrated using tensor notations: $\vec{E} \cdot \vec{\nabla} \wedge \vec{H} = E_i \epsilon_{ijk} \partial_j H_k = \partial_j (\epsilon_{jki} H_k E_i) + H_k \epsilon_{kji} \partial_j E_i = \vec{\nabla} \cdot (\vec{H} \wedge \vec{E}) + \vec{H} \cdot (\vec{\nabla} \wedge \vec{E})$.

then $dG = -\int \vec{B} \cdot \delta \vec{H} \, d^3r - SdT$. An operation similar to the one leading to Equation (10.54) shows that $\int \vec{B} \cdot \delta \vec{H} \, d^3r = \int \vec{A} \cdot \vec{\nabla} \wedge \delta \vec{H} \, d^3r$. Now, according to Equation (10.51), $\vec{\nabla} \wedge \delta \vec{H} = \delta \vec{J}_{\text{ext}}$ such that

$$dG = -\int \vec{A} \cdot \delta \vec{J}_{\text{ext}} \, d^3r - S \, dT \, . \tag{10.58}$$

So G is extremal when $\vec{J}_{\text{ext}}$ and T are fixed. Working at a fixed $\vec{J}_{\text{ext}}$ means that the current flowing through the coils that create the magnetic field in which the sample is immersed is kept constant. This is the typical experimental situation.

Let us now focus on the case of a superconducting system. According to the relation established previously and to be consistent, $G_L = F_L - \int d^3r \, \vec{B}.\vec{H}$ must be extremised with respect to the variations[17] of $\vec{A}$ and no longer F_L. This operation is the subject of Exercise 10.4 and leads to the very natural equation:

$$\vec{\nabla} \wedge \vec{B} = \mu_0 (\vec{J}_s + \vec{J}_{\text{ext}}) \, . \tag{10.59}$$

Inside the superconductor $\vec{J}_{\text{ext}} = 0$, Relation (10.59) is reduced to Relation (10.42) and the superconducting system is still described by Equations (10.38) and (10.42).

Exercise 10.4. Extremise G_L with respect to variations of $\vec{A}$ for a superconducting system described by (10.34).

Solution: As already seen, $\delta F_L = -\int d^3r \delta \vec{A} \cdot \vec{J}_s$ and $\delta[\int d^3r \vec{B}^2/2\mu_0] = \frac{1}{\mu_0} \int d^3r \delta \vec{A} \cdot \vec{\nabla} \wedge \vec{B}$. Thus (neglecting surface terms):

$$\delta \left[-\int d^3r \, \vec{B} \cdot \vec{H} \right] = \delta \left[-\int d^3r \left(\vec{\nabla} \wedge \vec{A} \right) \cdot \vec{H} \right] = \delta \left[-\int d^3r \, \vec{A} \cdot \vec{\nabla} \wedge \vec{H} \right] \, .$$

The term at the right-hand side of the above equation is derived from the central term using a relation similar to Equation (10.53) and neglecting the contribution of the surface terms. Then $\vec{\nabla} \wedge \vec{H}$ can be replaced by $\vec{J}_{\text{ext}}$ and, at fixed $\vec{J}_{\text{ext}}$, one gets

$$\delta \left[-\int d^3r \, \vec{B} \cdot \vec{H} \right] = -\int d^3r \, \delta \vec{A} \cdot \vec{J}_{\text{ext}} \, .$$

It thus follows that

$$\delta G_L = \int d^3r \, \delta \vec{A} \cdot \left[\frac{1}{\mu_0} \vec{\nabla} \wedge \vec{B} - \vec{J}_s - \vec{J}_{\text{ext}} \right] \, . \tag{10.60}$$

Hence, the extremisation of G_L with respect to variations of $\vec{A}$ indeed leads to Equation (10.59).

Thermodynamic definition of magnetisation

So far, Relation (10.55) was used. It expresses the work in terms of the fields that actually prevail within the body. At this point, it is interesting to derive an expression of the work (not only valid for a superconductor) in terms of the applied external fields[18]. This yields a relation used in the rest of the book which defines magnetisation as a function of the variation in the sample's free energy induced by a change in the external magnetic induction. Let $\vec{B}_0$ and $\vec{H}_0 = \vec{B}_0/\mu_0$ be the "external" fields, i.e. the fields that exist in the absence of of the sample, for the same current distribution that creates $\vec{B}$ and $\vec{H}$ when the sample is present[19]. One can write:

$$\vec{H} \cdot \delta \vec{B} = \vec{H}_0 \cdot \delta \vec{B}_0 + (\vec{H}_0 \cdot \delta \vec{B} - \vec{B}_0 \cdot \delta \vec{H}) + (\vec{B}_0 \cdot \delta \vec{H}_0 - \vec{H}_0 \cdot \delta \vec{B}_0) +$$
$$\vec{B}_0 \cdot (\delta \vec{H} - \delta \vec{H}_0) + (\vec{H} - \vec{H}_0) \cdot \delta \vec{B} \, . \tag{10.61}$$

---

[17]and with respect to the variations of ϕ^*, but the calculations are the same for evaluating $\delta G_L/\delta \phi^*$ and $\delta F_L/\delta \phi^*$.

[18]V. Heine, Math. Proc. Cambridge Phil. Soc. **52**, 546 (1956).

[19]Remember that, in this section, only the case where no current is injected in the system is considered.

Noting that $\vec{H}_0 = \vec{B}_0/\mu_0$ and $\delta\vec{H} = \delta(\vec{B}/\mu_0 - \vec{M})$, the second term of the right-hand side of Equation (10.61) can be written as $\vec{B}_0 \cdot \delta\vec{M} = \delta(\vec{B}_0 \cdot \vec{M}) - \vec{M} \cdot \delta\vec{B}_0$. The next term is clearly zero. The last two terms cancel out when integrating over the entire space[20]. Expression (10.55) of the work can thus be rewritten as:

$$\delta W = \int d^3r\, \vec{H}_0 \cdot \delta\vec{B}_0 - \int_V d^3r\, \vec{M} \cdot \delta\vec{B}_0 \,. \tag{10.62}$$

The first term on the right-hand side of Equation (10.62) is equal to $\delta[\int d^3r\, \vec{B}_0^2/2\mu_0]$ and corresponds to the variation of the magnetic field energy when $\vec{B}_0 \rightarrow \vec{B}_0 + \delta\vec{B}_0$. The second term is an integral over the volume V of the sample (the only part of space where the magnetisation $\vec{M} \neq 0$). If only the free energy of the system is of interest, the first contribution of the right-hand side of Equation (10.62) can be discarded and so:

$$dF_{\text{syst}} = -S\, dT - \int_V d^3r\, \vec{M} \cdot \delta\vec{B}_0 \,. \tag{10.63}$$

This means that component i ($i = x$, y or z) of the magnetisation is the functional derivative $M_i(\vec{r}) = -\delta F_{\text{syst}}/\delta B_0|_i(\vec{r})$. For uniform fields, with the less precise notations used in the rest of the book, this relation can be written as:

$$M_i = -\frac{1}{V}\left(\frac{\partial F}{\partial B_i}\right)_{N,V,T} \,, \quad \text{where} \quad i = x, y \text{ or } z \,. \tag{10.64}$$

This relation was already used to calculate the magnetisation in the Landau diamagnetic configuration [see Equation (8.84)] which is the version using the grand potential. A relation equivalent to Equation (10.64) has also been derived in the particular case of the Ising system [see Equation (9.11)].

10.5.3   CRITICAL MAGNETIC FIELD

When a sample is placed in a coil with a magnetic field $\vec{H}$ at a temperature $T < T_C$, experiments show that superconductivity disappears if the intensity of the field is higher than a critical value $H_C(T)$: everything happens as if the sample cannot expel a too intense field and becomes "normal" again (in the sense of "non-superconducting"). As intuition suggests, the critical field depends on temperature: it is for instance clear that $H_C(T_C) = 0$. Experiments show that $H_C(T)$ varies approximately as $H_C(T) \simeq H_C(0)(1 - T^2/T_C^2)$. This behaviour is illustrated schematically in Figure 10.13(a).

A theoretical evaluation of the value of $H_C(T)$ can be obtained by comparing the Ginzburg-Landau free enthalpy (Gibbs free energy) for a superconducting and a normal system. Inside a uniform superconductor, the magnetic field is zero, the order parameter has a constant value, and according to Equation (10.57), the free enthalpy density is $g_s = G_L/V = \tilde{a}|\phi_0|^2 + \frac{d}{2}|\phi_0|^4$ with $|\phi_0|^2 = -\tilde{a}/d$ [see Equation (10.43)] and thus $g_s = -\tilde{a}^2/(2d)$.

---

[20]The integral $\int d^3r\, \vec{U} \cdot \vec{V}$ is equal to zero for real vector fields ($\vec{U} = \vec{U}^*$ and $\vec{V} = \vec{V}^*$) verifying the proprieties: (i) $\vec{\nabla} \cdot \vec{U} = 0$ and (ii) $\vec{\nabla} \wedge \vec{V} = 0$ (property (i) is verified by $\vec{B}_0$ and $\delta\vec{B}$ and property (ii) by $\delta\vec{H} - \delta\vec{H}_0$ and $\vec{H} - \vec{H}_0$). The nullity of the integral is proven by using the Parceval-Plancherel theorem [Formula (A.35)] which can be written here as:

$$\int \vec{U}^*(\vec{r}) \cdot \vec{V}(\vec{r})\, d^3r = \int \vec{\mathcal{U}}^*(\vec{k}) \cdot \vec{\mathcal{V}}(\vec{k}) \frac{d^3k}{(2\pi)^3} \,, \tag{10.65}$$

where $\vec{\mathcal{U}}$ and $\vec{\mathcal{V}}$ are the Fourier transformed fields of $\vec{U}$ and $\vec{V}$. The properties (i) of $\vec{U}$ and (ii) of $\vec{V}$ imply that $\vec{\mathcal{U}}(\vec{k}) \perp \vec{k}$ while $\vec{\mathcal{V}}(\vec{k}) \parallel \vec{k}$. $\vec{\mathcal{U}}$ and $\vec{\mathcal{V}}$ are mutually orthogonal and Integral (10.65) is equal to zero.

On the other hand, in the normal (non-superconducting) state $\phi_0 = 0$ and $\vec{B} = \mu_0 \vec{H}$, giving a free enthalpy density $g_n = -B^2/(2\mu_0) = -\frac{1}{2}\mu_0 H^2$. The superconducting state is thermodynamically favourable when $g_s < g_n$ which immediately translates into $H < H_C(T)$ with

$$H_C(T) = \frac{a(T_C - T)}{\sqrt{d\,\mu_0}} \ . \tag{10.66}$$

Ginzburg-Landau theory thus predicts a critical field whose value decreases linearly with temperature, which is not consistent with experimental observations for $T \in [0, T_C]$. This is not a serious problem if one remembers that mean field theory is designed to apply in the immediate vicinity of the critical temperature: in the vicinity of T_C, the experimental dependence shows the same behaviour as Equation (10.66) since, in this case $T_C^2 - T^2 \simeq 2T_C(T_C - T)$.

Similarly, it should be noted that, in the presence of an external field, the superconductor/normal transition is no longer a second order transition, but a first order transition since it is linked to a discontinuity of the free enthalpy and the order parameter. As in the case of the Ising system in the presence of an external magnetic field, it is outside of the general framework of the second order transitions for which the Landau method was conceived.

10.5.4 INTERFACE BETWEEN NORMAL AND SUPERCONDUCTING PHASES

Situations where the external magnetic field is just at the critical value correspond to the coexistence line of a first order phase transition, and the two phases, normal and superconducting, can coexist. In this subsection, the interface between these two phases is described and the associated surface energy is calculated.

Let us consider a planar interface within the sample between the normal and superconducting phases. The system is immersed in a magnetic field H_C. The interface is the yOz-plane and the x-axis points towards the superconducting phase. The system's configuration is sketched in Figure 10.12. The distribution of all quantities depends only on the x coordinate. The Coulomb gauge ($\vec{\nabla}.\vec{A} = 0$) is chosen to express the vector potential, i.e. $dA_x/dx = 0$, and it is therefore possible to set $A_x = 0$. Then the y-axis is chosen so that $\vec{e}_y$ is always parallel to $\vec{A}$: $\vec{A} = (0, A(x), 0)$ so $\vec{B} = (0, 0, dA/dx)$.

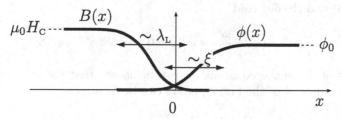

FIGURE 10.12 Schematic representation of the interface between the superconducting (right) and normal (left) phases. The magnetic field cancels over a distance of order λ_L in the superconducting region. The order parameter cancels in the normal part over a distance of the order of ξ.

It is instructive to calculate the expression of the field $\vec{H}$. In the configuration studied here, $\vec{J}_s$ can be written as $\vec{J}_s = J_s(x)\vec{e}_x$ and vector $\vec{M}$ defined by Equation (10.49) can be written as $\vec{M} = M(x)\vec{e}_z$. Thus, in total, $\vec{H}$ defined by Equation (10.50) has only a component along $\vec{e}_z$: $\vec{H} = H(x)\vec{e}_z$. Relations (10.42) and (10.49) can be written here as $\frac{1}{\mu_0}dB/dx = J_s(x)$ and $dM/dx = J_s(x)$. Then, the derivative of the $\vec{e}_z$ coordinate of $\vec{H}$ defined by Equation (10.50) can be written as $dH/dx = J_s - J_s = 0$: $\vec{H}$ is uniform and keeps the same value over the entire space. Note that this reasoning does not use the fact that $H = H_C$ but only the system's geometry and is thus valid for any field $\vec{H}$. In the configuration studied here, $\vec{H} = H_{\rm C}\,\vec{e}_z$.

With a reasoning similar to the one carried out for the study of domain walls in Section 10.3.2, it is legitimate to define the surface tension associated with the normal-superconductor interface as [see Equation (10.20)]:

$$
\begin{aligned}
\sigma &= \frac{G[\phi] - G_0}{L_y\,L_z} \\[2mm]
&= \int_{-\infty}^{\infty} {\rm d}x \left\{ \frac{\hbar^2}{2m} \left| \left(\vec{\nabla} - \frac{iq}{\hbar}\vec{A} \right)\phi \right|^2 + \tilde{a}|\phi|^2 + \frac{d}{2}|\phi|^4 + \frac{B^2}{2\,\mu_0} - B(x)H_{\rm C} - g_{nc} \right\},
\end{aligned}
\tag{10.67}
$$

where L_y and L_z are the spatial extents of the interface in the y and z directions, respectively. Note that in this formula, Expression (10.57) with a uniform field $H_{\rm C}$ was used. In Equation (10.67), g_{nc} is the free enthalpy density in the uniform normal phase in the presence of a field $H_{\rm C}$. It is exactly equal to the free enthalpy density g_{sc} in the uniform superconducting phase in the presence of a field $H_{\rm C}$, such that $G_0/V = g_{nc} = g_{sc} = -\frac{1}{2}\mu_0\,H_{\rm C}^2$.

In the geometry considered here, the coupled Equations (10.38) and (10.42) verified by the fields $\phi(x)$ and $A(x)$ read

$$
-\frac{\hbar^2}{2m}\frac{{\rm d}^2\phi}{{\rm d}x^2} + \left(\tilde{a} + \frac{q^2A^2(x)}{2m} + b\,|\phi(x)|^2 \right)\phi(x) = 0\,,
\tag{10.68}
$$

and

$$
\frac{{\rm d}^2A}{{\rm d}x^2} = \frac{\mu_0 q^2}{m}\,\phi^2(x)\,A(x)\,.
\tag{10.69}
$$

Equation (10.68) having real coefficients, ϕ can be set as real, which is done from now on. The following dimensionless variables are also be used: $X = x/\lambda_{\rm L}$, $\varphi = \phi\sqrt{b/|\tilde{a}|}$, $\mathscr{B} = B/B_{\rm C}$ (where $B_{\rm C} = \mu_0 H_{\rm C}$) and $\mathscr{A} = A/(B_{\rm C}\lambda_{\rm L})$. The derivative with respect to X is written with a prime (thus $\mathscr{B} = \mathscr{A}'$). The (dimensionless) Ginzburg-Landau parameter is also defined as:

$$
\kappa = \lambda_{\rm L}/\xi\,.
\tag{10.70}
$$

With the usual parametrisation, this reads $\kappa = m(\hbar|q|)^{-1}\sqrt{2d/\mu_0}$. With these new variables, Equations (10.68) and (10.69) read

$$
-\frac{1}{\kappa^2}\varphi'' + (-1 + \frac{\mathscr{A}^2}{2} + \varphi^2)\varphi = 0 \quad \text{and} \quad \mathscr{A}'' = \varphi^2\mathscr{A}\,.
\tag{10.71}
$$

With the second of the above equations, it can be shown that $(\mathscr{A}^2\varphi^2 - \mathscr{A}'^2)' = 2\mathscr{A}^2\varphi\varphi'$ from which a first integral of the first equation can be deduced:

$$
\frac{2}{\kappa^2}\varphi'^2 + (2 - \mathscr{A}^2)\varphi^2 - \varphi^4 + \mathscr{A}'^2 = 1\,.
\tag{10.72}
$$

The unit term in the right-hand side of Equation (10.72) is the integration constant that was evaluated by considering the situation deep into the normal phase ($\mathscr{A}' = \mathscr{B}$ and $\lim_{x\to-\infty}\mathscr{B} = 1$). This enables to write the surface tension as:

$$
\begin{aligned}
\sigma &= \frac{B_{\rm C}^2\lambda_{\rm L}}{2\,\mu_0} \int_{-\infty}^{\infty} {\rm d}X \left\{ \frac{2}{\kappa^2}\varphi'^2 + (\mathscr{A}^2 - 2)\varphi^2 + \varphi^4 + (\mathscr{A}' - 1)^2 \right\} \\[2mm]
&= \frac{B_{\rm C}^2\lambda_{\rm L}}{\mu_0} \int_{-\infty}^{\infty} {\rm d}X \left\{ \frac{2}{\kappa^2}\varphi'^2 + \mathscr{A}'(\mathscr{A}' - 1) \right\} \\[2mm]
&= \frac{B_{\rm C}^2\lambda_{\rm L}}{2\,\mu_0} \int_{-\infty}^{\infty} {\rm d}X \left\{ (\mathscr{A}' - 1)^2 - \varphi^4 \right\}\,.
\end{aligned}
\tag{10.73}
$$

The first equality in Relations (10.73) is the dimensionless form of Equation (10.67), the second follows from Equation (10.72). The last one is obtained from the first by writing $\int_{\mathbb{R}} dX \varphi'^2 = -\int_{\mathbb{R}} dX \varphi \varphi''$ (the integrated term is equal to zero) and then using the first equation in Equation (10.71). Expressions (10.73) enable the calculation of the surface tension in the following two limiting cases:

<u>$\kappa \ll 1$</u> It is the typical situation for pure metals. In that case, the London (penetration) length λ_L is much smaller than the coherence length ξ ($\lambda_L \ll \xi$) and, looking at Figure 10.12, it is clear that the approximation $\mathcal{B}(X) = \Theta(-X)$ holds (where Θ is the Heaviside function). For $X > 0$, Equation (10.72) then leads to $\varphi' = \frac{\kappa}{\sqrt{2}}(1 - \varphi^2)$ which is easily integrated into:

$$\varphi(X) = \Theta(X) \tanh\left(\frac{\kappa}{\sqrt{2}} X\right). \tag{10.74}$$

In that case, the second equation of Relations (10.73) enables an easy calculation of the surface tension yielding:

$$\sigma = \frac{B_C^2}{2\mu_0} \xi \frac{4\sqrt{2}}{3}. \tag{10.75}$$

<u>$\kappa \gg 1$</u> It is the limit $\xi \ll \lambda_L$. In that case, one can write $\varphi(X) = \Theta(X)$ and then:

$$\mathcal{B}(X) = \begin{cases} 1 & \text{if } X < 0, \\ \exp(-X) & \text{if } X > 0. \end{cases} \tag{10.76}$$

Here, the third equation of Relations (10.73) is useful and leads to

$$\sigma = -\frac{3}{2} \frac{B_C^2}{2\mu_0} \lambda_L. \tag{10.77}$$

It means that in the presence of a critical magnetic field, the surface tension between the normal and superconducting phases is negative: the two phases tend to interpenetrate in order to maximise their contact surface.

It can be shown that σ is exactly equal to zero when $\kappa = 1/\sqrt{2}$. Superconductors with $\kappa < 1/\sqrt{2}$ are called *type I superconductor* and the others ($\kappa > 1/\sqrt{2}$), *type II superconductors*. Lead, lanthanum, tantalum, mercury, tin, zinc are type I superconductor. Niobium, vanadium, many metallic alloys (Nb_3Ge, Nb_3Si, V_3Si...) as well as the so-called *high critical temperature superconductors* (such as mixed barium, copper and yttrium oxides discovered in 1986, or iron-based superconductors discovered in 2006) are type II superconductors. The two limiting cases considered above are called *extreme type I superconductor* ($\kappa \ll 1$) and *extreme type II superconductor* ($\kappa \gg 1$).

In a type I superconductor, $\sigma > 0$ opposes the formation of a normal core in the superconductor. This has little impact on the scenario of the breaking of superconductivity induced by a strong magnetic field described in Section 10.5.3. One must just be careful that, due to this positive surface tension, metastable superconductivity can exist even if $H > H_C(T)$.

For a type II superconductor, the surface tension of the normal/superconductor interface is negative: even when $H < H_C$ nuclei of normal phase can penetrate the superconducting zone. There are then two critical magnetic fields, $H_{C1}(T)$ (called the *lower critical field*) and $H_{C2}(T)$ (*upper critical field*) behaving as depicted in Figure 10.13(b). When $H < H_{C1}(T)$, the material is superconducting. When $H > H_{C2}(T)$, the material is normal. Between the two, one speaks of a mixed phase: small domains of normal phase (vortices that are filaments of transverse extension ξ) penetrate the sample and tend to form a regular structure in space:

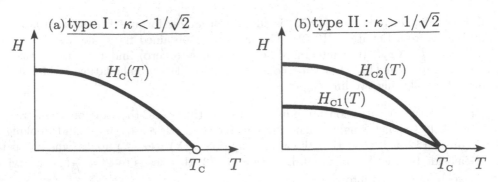

FIGURE 10.13   Critical fields in (a) type I superconductor and (b) type II superconductor.

an Abrikosov lattice. Figure 10.14 is a very nice example of such lattice. It is remarkable that the arrangement of vortices on a lattice was predicted by Abrikosov in 1957[21] based on Ginzburg-Landau theory before it was observed experimentally[22].

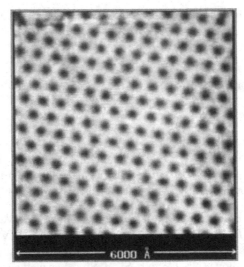

FIGURE 10.14   Abrikosov lattice due to a 1 T magnetic field applied to a NbSe$_2$ sample at 1.8 K. Figure reprinted with permission from H. F. Hess, R. B. Robinson, R. C. Dynes, J. M. Valles, Jr., and J. V. Waszczak Phys. Rev. Lett. **62**, 214 (1989). Copyright (1989) by the American Physical Society.

Exercise 10.5.   Determine the value of H_{C2} in a type II superconductor.

Solution: _Close to the disappearance of superconductivity, the order parameter_ ϕ _is small and screening is inefficient. In Equation (10.38), the term in_ ϕ^3 _can be ignored and_ $\vec{A}$ _is almost equal to the vector potential of the external field: this equation can be then written similarly to Expression (8.76) of the Schrödinger equation for a particle of charge q in a uniform magnetic field_

[21]A. A. Abrikosov, Sov. Phys. JETP **5**, 1174 (1957). The result was actually already obtained in 1953, but Landau approved its publication only after the publication by Feynman of an article about vortices in superfluid helium.

[22]D. Cribier, B. Jacrot, L. Madhav Rao and B. Farnoux, Phys. Letters **9**, 109 (1964); U. Essmann and H. Träuble, Phys. Lett. A **24**, 526 (1964).

$B = \mu_0 H$. The "eigenlevels" of this equation thus take a form similar to Relation (8.79) which can be here written as:

$$\frac{\hbar^2}{2m\xi^2} = -\tilde{a} = \frac{p_z^2}{2m} + (n + \tfrac{1}{2})\frac{|q|\hbar\mu_0 H}{m} .$$

The largest value of H, i.e. H_{C2}, is then obtained when $n = 0$ and $p_z = 0$. It reads:

$$H_{C2} = \frac{2m|\tilde{a}|}{|q|\hbar\mu_0} = \frac{1}{\mu_0}\frac{\Phi_0}{2\pi\xi^2} = \sqrt{2}\,\kappa\, H_C , \tag{10.78}$$

where Φ_0 is the flux quantum described in Section 10.5.1. The expressions for H_C and κ are given in Equations (10.66) and (10.70).

Note that while a type I superconductor experiences a first order phase transition when the field is higher that the critical value H_C (see Section 10.5.3), the transition is here a second order transition (the order parameter $\phi = 0$ when $H_{C2} > H_C$).

11 Phase Transitions – Validity of the Mean Field Theory – Scaling Laws

It has been seen in the previous chapters that Landau mean field approach relies on an incomplete partition function, the use of which is only justified if the fluctuations are Gaussian. This approach is manifestly invalid in the vicinity of a second order phase transition, since the second derivative of the Landau free energy with respect to the order parameter is equal to zero at the critical temperature T_C[1] and the magnitude of the fluctuations near the critical point become prohibitive for the use of the theory. This is a real paradox, since Landau theory is designed to be valid exactly in the vicinity of second order phase transition points (see Section 9.2.2). The present chapter focuses first on the discussion of this problem and the precise evaluation of the domain of validity of mean field theories. In a second part, mean field predictions are compared with experimental data and the notion of scale invariance in the vicinity of the critical point is discussed.

11.1 FLUCTUATION-DISSIPATION THEOREM

The canonical partition function is written as a sum over all configurations of an exponential of the system's energies, i.e. schematically:

$$Z = \sum_{\{\phi(\vec{r})\}} W[\phi] \exp\{-\beta E[\phi]\}. \tag{11.1}$$

In Equation (11.1), the sum is performed over all possible functions $\phi(\vec{r})$. For a given $\phi(\vec{r})$, $E[\phi]$ is the system's energy and $W[\phi]$ is the number of microscopic configurations[2] (microstates) which correspond to the same macroscopic function $\phi(\vec{r})$. Both $E[\phi]$ and $W[\phi]$ are functionals of ϕ and Equation (11.1) can be rewritten by introducing the exact functional $\mathscr{F}_L[\phi] = E[\phi] - \frac{1}{\beta} \ln W[\phi]$ ($\mathscr{F}_L$ is not the Landau functional, but the Landau functional is, of course, an approximation of the true $\mathscr{F}_L$, hence the similarity in notations). The canonical partition function can thus be written as

$$Z = \sum_{\{\phi(\vec{r})\}} \exp\{-\beta \mathscr{F}_L[\phi]\} = \int \mathscr{D}\phi \exp\{-\beta \mathscr{F}_L[\phi]\}. \tag{11.2}$$

A new notation has been introduced in the right-hand side of Equation (11.2). It corresponds to a well defined object: the functional integral. This notation will be preferred in the following as it is widely used because it can be given a rigorous mathematical meaning[3].

---

[1] See the discussion in Section 9.2.1 and the behaviour of the Landau free energy illustrated in Figure 9.9.

[2] Remember that a given value of the order parameter $\phi(\vec{r})$ corresponds to many microscopic configurations, because $\phi(\vec{r})$ is defined through a coarse grain averaging (see the discussion in Section 10.1).

[3] To this end, space is discretised into N_0 small cells of volume v centred on points $\vec{r}_i$ $i \in \{1, 2, \ldots, N_0\}$. The following approximation is introduced: the field is assumed to take a constant value $\phi_i = \phi(\vec{r}_i)$ in each cell (the approximation becomes exact when $v \to 0$). The integral of a given functional $F[\phi]$ is then $\int \mathscr{D}\phi \, F[\phi] = \int_{-\infty}^{\infty} d\phi_1 \int_{-\infty}^{\infty} d\phi_2 \cdots \int_{-\infty}^{\infty} d\phi_{N_0} \, F[\phi]$. According to the discretisation procedure, $F[\phi]$ is approximated in the last expression by a simple function of the ϕ_i's.

DOI: 10.1201/9781003272427-11

However, in the remainder of this chapter, the powerful concept of functional integration introduced in Equation (11.2) is merely used as a convenient notation.

In the presence of an external field, $\mathscr{F}_L[\phi]$ is replaced by

$$\mathscr{F}_L[\phi, h] = \mathscr{F}_L[\phi] - \int d^d r \, h(\vec{r}) \phi(\vec{r}), \tag{11.3}$$

and, in the following, the notation $\mathscr{F}_L[\phi, h]$ is used in order to encompass the most general situation.

Let us now determine the quantity $\langle \phi(\vec{r}) \rangle$, the value of $\phi(\vec{r})$ averaged over the statistical distribution (be careful, this is not a spatial average). $\langle \phi(\vec{r}) \rangle$ is defined as:

$$\langle \phi(\vec{r}) \rangle = \frac{1}{Z} \int \mathscr{D}\phi \, \phi(\vec{r}) \exp\{-\beta \mathscr{F}_L[\phi, h]\}. \tag{11.4}$$

This relationship can be written in another form using the following procedure: in the presence of an external field $h(\vec{r})$, the exact partition function [Formula (11.2)] can be seen as a functional $Z[h]$. When $h(\vec{r})$ varies by an amount $\delta h(\vec{r})$, the partition function becomes

$$Z[h + \delta h] = \int \mathscr{D}\phi \, \exp\{-\beta \mathscr{F}_L[\phi, h]\} \exp\left\{\beta \int d^d r \, \delta h(\vec{r}) \, \phi(\vec{r})\right\}. \tag{11.5}$$

An expansion of the second exponential at first order in δh gives $Z[h + \delta h] = Z[h] + \delta Z$ with

$$\delta Z \simeq \beta \int d^d r \, \delta h(\vec{r}) \int \mathscr{D}\phi \, \phi(\vec{r}) \exp\{-\beta \mathscr{F}_L[\phi, h]\}. \tag{11.6}$$

Comparing with the expression of the average value defined by Equation (11.4) and using Definition (10.5) of the functional derivative leads to the following exact relation:

$$\langle \phi(\vec{r}) \rangle = \frac{1}{\beta Z} \frac{\delta Z}{\delta h(\vec{r})} = -\frac{\delta F}{\delta h(\vec{r})}, \tag{11.7}$$

where $F = -k_B T \ln Z$.

The susceptibility of the system is defined by

$$\chi(\vec{r}, \vec{r}') = \frac{\delta \langle \phi(\vec{r}) \rangle}{\delta h(\vec{r}')} = -\frac{\delta^2 F}{\delta h(\vec{r}) \delta h(\vec{r}')}. \tag{11.8}$$

This definition is quite explicit: the function $\chi(\vec{r}, \vec{r}')$ describes how the average order parameter in $\vec{r}$ is affected by a modification of the external field in $\vec{r}'$. The right-hand side relation in Equation (11.8) follows directly from Expression (11.7).

Let G be the correlation function of the order parameter. It is defined as:

$$G(\vec{r}, \vec{r}') = \langle \phi(\vec{r}) \phi(\vec{r}') \rangle - \langle \phi(\vec{r}) \rangle \langle \phi(\vec{r}') \rangle. \tag{11.9}$$

Manipulations similar to those leading to Relation (11.7) allow to write the first contribution of the right-hand side of Equation (11.9) as a second order functional derivative:

$$\langle \phi(\vec{r}) \phi(\vec{r}') \rangle = \frac{1}{\beta^2 Z} \frac{\delta^2 Z}{\delta h(\vec{r}) \delta h(\vec{r}')}.$$

Thus, in total:

$$\begin{aligned} G(\vec{r}, \vec{r}') &= \frac{1}{\beta^2 Z} \frac{\delta^2 Z}{\delta h(\vec{r}) \delta h(\vec{r}')} - \frac{1}{\beta^2 Z^2} \frac{\delta Z}{\delta h(\vec{r})} \frac{\delta Z}{\delta h(\vec{r}')} \\ &= \frac{1}{\beta^2} \frac{\delta^2 \ln Z}{\delta h(\vec{r}) \delta h(\vec{r}')} = -\frac{1}{\beta} \frac{\delta^2 F}{\delta h(\vec{r}) \delta h(\vec{r}')}. \end{aligned} \tag{11.10}$$

A comparison with Equation (11.8) trivially leads to:

$$G(\vec{r}, \vec{r}') = \frac{1}{\beta} \chi(\vec{r}, \vec{r}').$$ (11.11)

Relation (11.11) is improperly called a *fluctuation-dissipation theorem* (one should rather speak of fluctuation-response). It links the correlations in the system (encoded in the function $G(\vec{r}, \vec{r}')$) to the way the system responds to an external field, i.e. to the susceptibility $\chi(\vec{r}, \vec{r}')$.

11.2 MEAN FIELD AS A SADDLE-POINT APPROXIMATION

The saddle-point approximation method (see Appendix A.3) is here generalised to the case of a functional integral and the field ϕ is expanded in the vicinity of $\phi_c(\vec{r})$ solution of:

$$\frac{\delta \mathscr{F}_L}{\delta \phi(\vec{r})} = 0.$$ (11.12)

In this section (and also at the beginning of Section 11.3) $\phi_c(\vec{r})$ is the solution of Equation (11.12) and should not be confused with the generic order parameter $\phi(\vec{r})$ which, in Expression (11.2) for instance, spans over the whole set of functions defined on $\mathbb{R}^d$. It is nonetheless always possible to write $\phi(\vec{r}) = \phi_c(\vec{r}) + \eta(\vec{r})$ and a series expansion to second order around ϕ_c gives $\mathscr{F}_L[\phi, h] = \mathscr{F}_L[\phi_c, h] + \Delta\mathscr{F}_L[\phi_c, \eta]$ with [see Equation (10.4)]:

$$\Delta\mathscr{F}_L[\phi_c, \eta] \simeq \frac{1}{2} \int d^d r \, d^d r' \frac{\delta^2 \mathscr{F}_L}{\delta\phi(\vec{r})\delta\phi(\vec{r})} \eta(\vec{r})\eta(\vec{r}').$$ (11.13)

The mean field approximation consists in writing:

$$
\begin{aligned}
F &= -k_{\mathrm{B}}T \ln\left(\int \mathscr{D}\phi \, \exp\{-\beta\mathscr{F}_L[\phi, h]\}\right) \\
&\simeq -k_{\mathrm{B}}T \ln\left(\exp\{-\beta\mathscr{F}_L[\phi_c, h]\}\int \mathscr{D}\eta \, \exp\{-\beta\Delta\mathscr{F}_L[\phi_c, \eta]\}\right) \simeq \mathscr{F}_L[\phi_c, h].
\end{aligned}
$$ (11.14)

In Equation (11.14), the final approximate equality amounts to simply neglecting the contribution of the functional integral over $\eta(\vec{r})$. This is exactly what was done in the uniform case [see Section 9.2.1, Formula (9.33)]. In this case, $F[T, h] \simeq \mathscr{F}_L[T, h, \psi_c]$. Similarly to what was done for the last term in Formula (9.33), the functional integral can be calculated over η to justify that, indeed, its contribution is typically negligibly small in the thermodynamic limit. However this would be a waste of time because – as in Formula (9.33) – this contribution is actually not negligible in the vicinity of a phase transition, i.e. in the region of primary interest to us. In the following, the last approximation in Equation (11.14)) is simply accepted and its validity will be verified *a posteriori* (see Section 11.3).

It is interesting to note that in the mean field approximation[4]:

$$\langle \phi(\vec{r}) \rangle = \phi_c(\vec{r}),$$ (11.15)

----

[4]This can be proven starting from the exact relation (11.7): $\langle \phi(\vec{r}) \rangle = -\delta F/\delta h(\vec{r})$. Then $\delta F/\delta h(\vec{r})$ is calculated using the saddle point approximation of F [Formula (11.14)]: $F = \mathscr{F}_L[\phi_c, h]$. This functional differentiation must be calulated carefully because the order parameter ϕ_c is also a functional of $h(\vec{r})$. Hence F depends on $h(\vec{r})$ in 2 ways: *via* its dependence in $\phi_c(\vec{r})$ and *via* the coupling $-\int d^d r h(\vec{r})\phi(\vec{r})$ (see Equation (11.3)). Thus, using Equation (10.10):

$$\frac{\delta F}{\delta h(\vec{r})} = \frac{\delta \mathscr{F}_L[\phi_c, h]}{\delta h(\vec{r})} = \int d^d r' \left[\frac{\delta \mathscr{F}_L}{\delta\phi(\vec{r}')}\bigg|_{\phi_c} \times \frac{\delta\phi_c(\vec{r}')}{\delta h(\vec{r})} \right] - \phi_c(\vec{r}).$$

The first term of the right-hand side of this expression is equal to zero according to Equation (11.12), and so $-\delta F/\delta h(\vec{r}) = \phi_c(\vec{r})$, which is Equation (11.15).

i.e. the average value of the order parameter is identified with its most probable value. This result was already hinted at the end of Section 9.1.3.

Formula (11.14) legitimises the mean field calculation of the order parameter by means of Equation (10.2), as presented in Section 10.1: within the framework of Ginzburg-Landau theory, the unknown functional $\mathscr{F}_L$ is approximated by the phenomenological functional F_L. Then, and this is at the heart of the mean field approximation, the system's free energy $F(T)$ is identified by means of the approximated formula (11.14) to the value of $F_L[\phi_c(\vec{r}); T]$ evaluated for an order parameter ϕ_c, solution of Equation (11.12), i.e. solution of Equation (10.2) in the framework of the Ginzburg-Landau approximation.

11.2.1 CORRELATION FUNCTION IN LANDAU THEORY

The fluctuation-dissipation theorem [Relation (11.11)] allows the evaluation of the mean field correlations of the order parameter by calculating $\chi(\vec{r}, \vec{r}') = \delta\phi(\vec{r})/\delta h(\vec{r}')$ and starting from any configuration where $\phi(\vec{r})$ is a solution of Equation (10.3)[5] which is rewritten here for readability as:

$$-2g\vec{\nabla}^2\phi + 2a(T - T_C)\phi + 4d\phi^3 = h(\vec{r}). \tag{11.16}$$

Equation (11.16) defines, in each point $\vec{r}$, $\phi(\vec{r})$ as a functional of the external field h. The corresponding functional derivative is calculated by determining the modification $\phi(\vec{r}) \to \phi(\vec{r}) + \delta\phi(\vec{r})$ of the order parameter resulting from a change $h(\vec{r}) \to h(\vec{r}) + \delta h(\vec{r})$ of the external field. $\delta\phi$ and δh are related through the linear equation

$$-2g\vec{\nabla}^2\delta\phi + \left[2a(T - T_C) + 12d\phi^2(\vec{r})\right]\delta\phi = \delta h(\vec{r}). \tag{11.17}$$

Let $\chi(\vec{r}, \vec{r}')$ be solution of:

$$\left\{-2g\vec{\nabla}_{\vec{r}}^2 + 2a(T - T_C) + 12d\phi^2(\vec{r})\right\}\chi(\vec{r}, \vec{r}') = \delta^{(d)}(\vec{r} - \vec{r}'). \tag{11.18}$$

It is easy to check by direct substitution that the solution of Equation (11.17) is:

$$\delta\phi(\vec{r}) = \int d^d r' \chi(\vec{r}, \vec{r}') \delta h(\vec{r}'). \tag{11.19}$$

Relation (11.19) shows that $\chi(\vec{r}, \vec{r}')$ is the functional derivative $\delta\phi(\vec{r})/\delta h(\vec{r}')$ (see Section 10.2 and the definition of the functional derivative). It was therefore legitimate to write the solution of Equation (11.18) as $\chi(\vec{r}, \vec{r}')$. The technique just exposed is indeed a simple way to determine the functional derivative $\delta\langle\phi(\vec{r})\rangle/\delta h(\vec{r}')$ [Equation (11.8)] within the mean field theory.

Exercise 11.1. Derive Equation (11.18) verified by $\chi(\vec{r}, \vec{r}')$ in the framework of Landau mean field theory using the technique of functional derivative.

Solution: _As shown by Equation (11.15), in the framework of the mean field theory, $\langle\phi(\vec{r})\rangle$ can be assimilated with the function $\phi(\vec{r})$ that extremises F_L. Taking the functional derivative $\delta/\delta h(\vec{r}')$ of both terms of (11.16) directly yields Equation (11.18). The only difficulty lies in the calculation of terms such as $\delta\phi^3(\vec{r})/\delta h(\vec{r}')$. Formulas (10.7) and (10.10) show that the result of this particular calculation is $3\phi^2(\vec{r})\chi(\vec{r}, \vec{r}')$._

---

[5] $\phi(\vec{r})$ is the quantity $\phi_c(\vec{r})$ in the previous section. To simplify notations, the index "c" is omitted henceforth.

Exercise 11.2. Recover Expression (9.46) considering an homogeneous system where ϕ, h and δh are independent of position.

Solution: For a homogeneous system χ does not depend separately on $\vec{r}$ and $\vec{r}\,'$, but is instead a function of the difference $\vec{r} - \vec{r}\,'$. Thus, Equation (11.19) can be written as $\delta\phi(\vec{r}) = \int d^d r' \chi(\vec{r} - \vec{r}\,') \delta h(\vec{r}\,')$. Since δh is here independent of position, the integration is trivial. $\delta\phi$ does not depend on position either and can be written as:

$$\delta\phi = \chi\,\delta h \quad \text{with} \quad \chi = \int d^d r' \chi(\vec{r}\,'). \tag{11.20}$$

Note that this relation was established using only the definition of the susceptibility and homogeneity assumptions. It is therefore very general and its validity is not limited to the mean field theory[6]. In the framework of the mean field theory, according to Equation (11.18) with $\phi(\vec{r}) = \phi_0$, Equation (11.20) leads to

$$\chi = \frac{1}{2a(T - T_C) + 12d\phi_0^2}.$$

Using Expression (9.37) for ϕ_0, Expression (9.46) of the susceptibility is naturally recovered.

The solution of Equation (11.18) allows the evaluation of the susceptibility $\chi(\vec{r},\vec{r}\,')$ and thus the determination of the correlation function $G(\vec{r},\vec{r}\,')$ thanks to the fluctuation-dissipation theorem [Formula (11.11)]. Let us apply this procedure in the simple case of zero external field ($h(\vec{r}) = 0$). In this case, $\phi(\vec{r})$ is a constant, given by Expression (9.37). Combining Equations (11.11) and (11.18) gives here:

$$\left[-\vec{\nabla}^2 + \xi^{-2}\right] G(\vec{r},\vec{r}\,') = \frac{k_B T}{2g}\,\delta^{(d)}(\vec{r} - \vec{r}\,'), \tag{11.21}$$

where ξ is the correlation length[7] [Equation (10.13)]. Equations (11.21) is solved by means of a Fourier transform. Since the system is homogeneous, the correlation function is a function of $\vec{r} - \vec{r}\,'$ only and the Fourier transform reads

$$G(\vec{r},\vec{r}\,') = G(\vec{r} - \vec{r}\,') = \int_{\mathbb{R}^d} \frac{d^d q}{(2\pi)^d}\,\hat{G}(\vec{q})\,\exp\{i\vec{q}\cdot(\vec{r} - \vec{r}\,')\}. \tag{11.22}$$

Carrying over in Equation (11.21) and remembering that the Fourier transform of the Dirac distribution is unity [a result which follows from Equation (A.32)] leads to

$$\hat{G}(\vec{q}) = \frac{k_B T}{2g}\,\frac{1}{q^2 + \xi^{-2}}. \tag{11.23}$$

Using the trick which consists in writing $D^{-1} = \int_0^\infty du\,\exp(-uD)$ (valid for any $D > 0$), and with $\vec{R} = \vec{r} - \vec{r}\,'$, the correlation function becomes:

$$G(\vec{R}) = \frac{k_B T}{2g} \int_0^\infty du \int_{\mathbb{R}^d} \frac{d^d q}{(2\pi)^d} \exp\left\{i\vec{q}\cdot\vec{R} - u(q^2 + \xi^{-2})\right\}, \tag{11.24}$$

where the integration over $\mathbb{R}^d$ can now be evaluated as a product of d independent integrals[8]. Thus,

$$G(\vec{R}) = \frac{k_B T}{2g}\,\frac{1}{\xi^{d-2}} \int_0^\infty \frac{dt}{(4\pi t)^{d/2}} \exp\left\{-t - \frac{X^2}{4t}\right\}, \quad \text{where} \quad X = \frac{|\vec{R}|}{\xi}. \tag{11.25}$$

---

[6]Relation (11.20) is used in a general context at paragraph "Behaviour of the correlation function", page 368.

[7]ξ diverges in the vicinity of the critical temperature as $\xi \propto |T - T_C|^\nu$ with $\nu = 1/2$ according to the mean field theory [Equation (10.13)], while in reality, for the Ising model for example, $\nu = 1$ in two dimensions and $\nu = 0.630(2)$ in 3 dimensions.

[8]If q_n and R_n are the Cartesian components of $\vec{q}$ and $\vec{R}$ respectively, then for $n \in \{1, \dots, d\}$: $\int_{\mathbb{R}} dq_n \exp\{iq_n R_n - uq_n^2\} = \sqrt{\frac{\pi}{u}} \exp(-R_n^2/4u)$ (see Equation (A.30)).

This can be written as

$$G(\vec{R}) = \frac{k_B T}{2g} \frac{\xi^{2-d}}{(4\pi)^{d/2}} g(X), \tag{11.26}$$

where

$$g(X) = \int_0^\infty dt \, \exp[-f(t)] \quad \text{with} \quad f(t) = t + \frac{X^2}{4t} + \frac{d}{2} \ln(t). \tag{11.27}$$

Expression (11.27) is well suited for an evaluation of $g(X)$ by means of the saddle point approximation (see Appendix A.3). According to this method, the behaviour of $g(X)$ is governed by the behaviour of $f(t)$ in the vicinity of its minimum. f' cancels at $t_c = \frac{1}{2}[-\frac{d}{2} + \sqrt{(d/2)^2 + X^2}]$ and $f''(t) = X^2/(2t^3) - d/(2t^2)$. Two cases must be distinguished:

- When $X \gg 1$: $t_c \simeq X/2$, $f(t_c) \simeq X + \frac{d}{2}\ln(X/2)$ and $f''(t_c) \simeq 4/X$. The saddle point method yields the approximate formula:

$$g(X) \simeq \sqrt{\pi} \left(\frac{2}{X}\right)^{(d-1)/2} \exp(-X), \quad \text{when} \quad X \gg 1. \tag{11.28}$$

- When $X \ll 1$: $t_c \simeq X^2/(2d)$ and $f''(t_c) \simeq 2d^3/X^4$. The saddle point method cannot be used in this case because the lower boundary of integration ($t = 0$) cannot be substituted by $-\infty$ since in Equation (11.27) the argument of the exponential at this point is not large and negative ($t_c^2 f''(t_c)$ is not large compared to 1, see the discussion in Section A.3). It it thus not possible to approximate G by a Gaussian integral over the whole real axis.

On the other hand, it remains true that the maximum of the integrand is reached for $t_c \simeq X^2/(2d)$. So, in the region near t_c, the first term (t) in Expression (11.27) for $f(t)$ is negligible with respect to the term $X^2/(4t)$. Accordingly, one can write $g(X) \simeq \int_0^\infty t^{-d/2} \exp\{-X^2/(4t)\}dt$. Note that this result is only valid when $d > 2$ (otherwise the integral diverges at infinity). The cases $d = 1$ and 2 must be treated separately (Case $d = 1$ is treated explicitly in Section 10.3.1, case $d = 2$ corresponds to a logarithmic divergence when $R \to 0$, see Exercise 11.4). Then, a simple change of variable for $d > 2$ leads to:

$$g(X) \simeq \Gamma\left(\frac{d}{2} - 1\right)\left(\frac{2}{X}\right)^{d-2}, \quad \text{when} \quad X \ll 1, \tag{11.29}$$

where the Γ function is defined by Formula (A.2).

The behaviour of $G(R = |\vec{r} - \vec{r}'|)$ is sketched in Figure 11.1. Correlations decrease as a power law when $R \ll \xi$ and exponentially when $R \gg \xi$. In the vicinity of the critical temperature, the correlation length diverges [$\xi \to \infty$ when $T \to T_C$, see Equation (10.13)] and the order parameter correlations then decrease as a power-law over the entire space without a characteristic length scale: a plot representing the behaviour of $\phi(\vec{r})$ in the entire space is thus scale invariant.

Exercise 11.3. Give the exact form of $G(R = |\vec{r} - \vec{r}'|)$ in dimension $d = 3$.

Solution: Formulas (11.22) and (11.23) read here:

$$G(R) = \frac{k_B T}{2g} \int_0^\infty \frac{dq}{4\pi^2} \frac{q^2}{q^2 + \xi^{-2}} \int_0^\pi d\theta \sin\theta \, \exp(iqR\cos\theta). \tag{11.30}$$

The value of the integral over θ is $\frac{1}{iqR}(e^{iqR} - e^{-iqR})$, which allows to write $G(R)$ as:

$$G(R) = \frac{k_B T}{2g} \frac{1}{iR} \int_{\mathbb{R}} \frac{dq}{4\pi^2} \frac{q \, e^{iqR}}{q^2 + \xi^{-2}}. \tag{11.31}$$

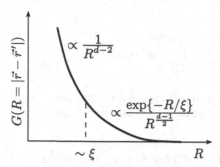

FIGURE 11.1 Schematic behaviour of $G(R = |\vec{r} - \vec{r}'|)$ in dimension $d > 2$. In dimension $d = 1$, G is a decreasing exponential as seen in Chapter 10 where the expression of $\chi(x) = \beta G(x)$ is given in Equation (10.16). In dimension $d = 2$, G shows a logarithmic divergence when $R \ll \xi$ [see Formula (11.36)].

This integral can be calculated by means of the residue theorem: the integration contour is closed in the plane of the complex variable q by a circle of infinite radius in the upper half-plane Im $q > 0$. The integral is then equal to $2i\pi$ times the residue in $q = i\xi^{-1}$ (the only pole in the upper half-plane) and:

$$G(R) = \frac{k_B T}{2g} \frac{\exp(-R/\xi)}{4\pi R}, \quad \text{for} \quad d = 3, \tag{11.32}$$

in agreement with both asymptotic estimates [Equations (11.28) and (11.29)].

Exercise 11.4. Starting directly from Equation (11.21), recover the limit expression of $G(R)$ corresponding to Relation (11.29) and valid for $R \ll \xi$ in dimension $d > 2$ and $d = 2$.

Solution : *Equation (11.21) imposes that G be singular at $R = 0$. Hence, when $R \ll \xi$, this equation can approximately be written as $-\vec{\nabla}^2 G = (k_B T/(2g)) \delta^{(d)}(\vec{r} - \vec{r}')$. A volume integral on both terms of this equation is then performed over a sphere of radius R centred in $\vec{r}'$. Writing the Laplacian as the divergence of the gradient of G, the Gauss-Ostrogradskii theorem [Formula (A.45)] can be used to replace the volume integral of $\vec{\nabla}^2 G$ by a surface integral yielding:*

$$\oint R^{d-1} d^{d-1}\Omega \, \vec{n}_s \cdot \vec{\nabla}G = -\frac{k_B T}{2g}, \tag{11.33}$$

where $d^{d-1}\Omega$ is the element on integration over the solid angles in $\mathbb{R}^d$ and $\vec{n}_s$ is the outgoing normal to the sphere of radius R. Denoting S_d the integral over solid angles [see Formulas (A.22) and (A.24)] allows to rewrite this equation as:

$$S_d R^{d-1} \frac{dG}{dR} = -\frac{k_B T}{2g}. \tag{11.34}$$

This expression can be easily integrated for $d > 2$ and one obtains (the integration constant shall not be considered since only the dominant contribution when $R \to 0$ is of interest here):

$$G(R) = \frac{k_B T}{2g} \frac{\Gamma(d/2)}{(d-2)2\pi^{d/2}} \frac{1}{R^{d-2}}, \quad \text{for} \quad R \ll \xi \quad \text{and} \quad d > 2. \tag{11.35}$$

Using the properties of the Γ function[9] it can be seen that Expression (11.35) is identical to Expression (11.26) in the limit (11.29).

Expression (11.34) can also be integrated when $d = 2$ yielding:

$$G(R) = -\frac{k_B T}{2g} \frac{1}{2\pi} \ln(R), \quad \text{for} \quad R \ll \xi \quad \text{and} \quad d = 2. \tag{11.36}$$

---

[9]$\Gamma(\frac{d}{2}) = (\frac{d}{2} - 1)\Gamma(\frac{d}{2} - 1)$ [see Equation (A.3)].

Exercise 11.5. Show that

$$\int_{|\vec{r}| \leq \xi} \mathrm{d}^d r\, G(\vec{r}, \vec{0}) \simeq k_\mathrm{B} T\, \frac{\xi^2}{2g}. \tag{11.37}$$

Solution: A first method consists in noticing that G decreases rapidly when $|\vec{r}| > \xi$ (see Figure 11.1) and that the integral over the domain $|\vec{r}| \leq \xi$ can accordingly be approximated by the integral over the entire space $\mathbb{R}^d$. Then:

$$\int_{|\vec{r}| \leq \xi} \mathrm{d}^d r\, G(\vec{r}, \vec{0}) \simeq \int_{\mathbb{R}^d} \mathrm{d}^d r\, G(\vec{r}, \vec{0}) = \hat{G}(\vec{0}), \tag{11.38}$$

where $\hat{G}$ is the Fourier transform of G, defined by Formula (11.22). Relation (11.23) leads directly to Formula (11.37).

An equivalent result can be obtained by evaluating the left-hand side term of Formula (11.37) using the approximation of G valid when $|\vec{r}| \ll \xi$ [Equations (11.26) and (11.29)]. Then

$$\int_{|\vec{r}| \leq \xi} \mathrm{d}^d r\, G(\vec{r}, \vec{0}) = \int \mathrm{d}^{d-1}\Omega \int_0^\xi r^{d-1} \mathrm{d}r\, G(\vec{r}, \vec{0}) \simeq k_\mathrm{B} T\, \frac{\xi^2}{2g}\, \frac{\Gamma(\frac{d}{2} - 1)}{(4\pi)^{d/2}}\, S_d \int_0^1 X\, \mathrm{d}X,$$

where S_d is the integral over angles in $\mathbb{R}^d$ (see Appendix A.4). Using the explicit expression (A.24) yields:

$$\int_{|\vec{r}| \leq \xi} \mathrm{d}^d r\, G(\vec{r}, \vec{0}) \simeq k_\mathrm{B} T\, \frac{\xi^2}{2g}\, \frac{1}{2^{d-1}(d-2)}, \tag{11.39}$$

which again corroborates Formula (11.37).

11.3 VALIDITY OF THE MEAN FIELD APPROXIMATION: GINZBURG CRITERION

This criterion – also known as the Levanyuk-Ginzburg criterion – allows to assess the validity of the mean field theory. As discussed in the introduction of the present chapter, strictly speaking, the large fluctuations at the critical point invalidate the mean field approach. Ginzburg criterion quantitatively evaluates the relative amplitude of these fluctuations which are described by the correlation function G defined by Equation (11.9) which can be written as $G(\vec{r}, \vec{r}') = \langle \eta(\vec{r})\eta(\vec{r}') \rangle$, where $\eta(\vec{r}) = \phi(\vec{r}) - \langle \phi(\vec{r}) \rangle$. The relative fluctuations' amplitude is evaluated at a temperature $T < T_\mathrm{C}$ by calculating the integral $\langle \eta(\vec{r})\eta(\vec{r}') \rangle$ over a domain Ω (of volume V) and comparing to the square of the integral of $\langle \phi(\vec{r}) \rangle$ over the same domain[10]. A scalar quantity, the so-called *Ginzburg parameter*, is thus defined as:

$$E_G = \frac{\displaystyle\int_{\Omega \times \Omega} \mathrm{d}^d r\, \mathrm{d}^d r'\, \langle \eta(\vec{r})\eta(\vec{r}') \rangle}{\left| \displaystyle\int_\Omega \mathrm{d}^d r\, \langle \phi(\vec{r}) \rangle \right|^2} = \frac{\displaystyle\int_{\Omega \times \Omega} \mathrm{d}^d r\, \mathrm{d}^d r'\, G(\vec{r}, \vec{r}')}{V^2 |\phi_c|^2} = \frac{\displaystyle\int_\Omega \mathrm{d}^d r\, G(\vec{r}, \vec{0})}{V |\phi_c|^2}. \tag{11.40}$$

In the two right-hand side terms of Equation (11.40) E_G is evaluated in the framework of the mean field theory [with $\langle \phi \rangle = \phi_c$, see Equation (11.15)].

How to choose the domain Ω? It cannot be the entire space since E_G would be then equal to zero: the denominator would diverge while the numerator would remain finite because the integral $\int_{\mathbb{R}^d} \mathrm{d}^d r\, G(\vec{r}, \vec{0})$ is finite (see Exercise 11.5). It is natural to take a volume with a typical extend ξ such that $V \simeq \xi^d$. In that case the numerator of the right-hand term of

---

[10]Why don't we use the integral of $\langle \phi(\vec{r}) \rangle^2$ over Ω instead? This can be justified by simple dimensional arguments (left as an exercise).

Equation (11.40) can be simply evaluated (see Exercise 11.5) and the Ginzburg parameter reads

$$E_G \simeq \frac{k_B T}{2g} \frac{\xi^{2-d}}{\phi_c^2} \simeq \frac{k_B T_C}{2g} \frac{\xi^{2-d}}{\phi_c^2} \propto \frac{1}{(T_C - T)^{2-d/2}}. \tag{11.41}$$

where Expression (10.14) was used. For $d < d_c = 4$, E_G diverges in the vicinity of the critical temperature and the mean field approximation is not strictly valid. On the other hand, for $d > d_c$, the relative amplitude of fluctuations cancel at the critical point and the phase transition is adequately described by the mean field approximation. d_c is called the *upper critical dimension*.

It is possible to perform a more quantitative analysis. Let the reduced temperature be defined as $t = (T - T_C)/T_C$. The Ginzburg parameter [Expression (11.41)] can then be written as $E_G = (t_G/t)^{2-d/2}$ with:

$$t_G^{2-d/2} = \left(\frac{2aT_C}{g}\right)^{d/2} \times \frac{2d}{a^2 T_C} \times \frac{k_B}{4} = \frac{1}{\xi_0^d} \times \frac{1}{\Delta c_V} \times \frac{k_B}{4}, \tag{11.42}$$

where ξ_0 is the correlation length at $T = 0$ [see Equation (10.12)] and $\Delta c_V = a^2 T_C/(2d)$ is the discontinuity of the volumetric heat capacity at the critical temperature T_C [see Equation (9.40)]. $\Delta c_V \times \xi_0^d$ is roughly of the order of the heat capacity of a region of size ξ_0 which, according to Dulong and Petit law (see Section 4.1.5), is equal to $k_B N_\ell(\xi_0)$ where $N_\ell(\xi_0)$ is the number of degree of freedom in a box of size ξ_0 ($N_\ell(\xi_0)$ is of the order of the number of particles inside the box). Eventually:

$$E_G = \left(\frac{t_G}{t}\right)^{2-d/2} \quad \text{with} \quad t_G \sim [N_\ell(\xi_0)]^{2/(d-4)}. \tag{11.43}$$

Even when $d < 4$, it is only when $|t| \ll t_G$ that the amplitude of fluctuations invalidate the mean field theory. Thus, below the upper critical dimension, far enough from the critical point (when $|t| \gg t_G$), the contribution of fluctuations do not significantly affect the mean field approximation which remains valid.

Exercise 11.6. Determine the upper critical dimension for a mean field theory characterised by the critical exponents β and ν.

Solution: According to Equation (11.41) showing that $E_G \propto \xi^{2-d}/\phi_c^2 \propto t^{\nu(d-2)-2\beta}$, the Ginzburg parameter does not diverge at the critical point if $d > d_c = 2(1 + \beta/\nu)$.

Exercise 11.7. Determine the upper critical dimension of the Blume-Emery-Griffiths model studied in Section 9.4.

Solution: It was seen that under relatively general conditions, the exponent β of the B.E.G. model is equal to $1/4$. The determination of exponent ν is obtained as follows: the correlation function is solution of

$$\left[-2g\vec{\nabla}^2 + \frac{\partial^2}{\partial\phi^2}\left(\frac{F_L}{N}\right)\right] G(\vec{r}, \vec{0}) = k_B T \delta^{(d)}(\vec{r}),$$

and thus $\xi^{-2} = (2g)^{-1} \partial^2 (F_L/N)/\partial\phi^2$ [see Equation (10.13)]. To calculate ξ, it is sufficient to work in the non-superfluid (normal) phase ($\phi = 0$) leading to $\xi^{-2} = b(T, \Delta)/g \propto (T - T_t)$, i.e. $\nu = 1/2$. The results of Exercise 11.6 then shows that the upper critical dimension of the B.E.G. model is $d_c = 3$.

11.4   SOME COMPARISONS

The validity of the mean field theory can be addressed experimentally by studying the behaviour of the heat capacity in the vicinity of the transition. Landau theory predicts a simple discontinuity of C_V at the critical point [see Formula (9.40) and Figure 9.10] whereas the singular behaviour of the Ginzburg parameter at T_C suggests that C_V can diverge at this point.

11.4.1   SUPERCONDUCTORS

In a superconductor, the correlation length is very large: $\xi_0 \sim 10^3$ Å, while the typical interatomic distance is about 2 Å, and electrical neutrality dictates that the electron density is equal to the atom density[11]. Therefore, it also corresponds to a typical electron spacing of about 2 Å. This gives $N_\ell(\xi_0) \sim (10^3/2)^3 \sim 10^8$ and in dimension $d = 3$, $t_G \sim 10^{-16}$ according to Equation (11.43). It is, of course, impossible to approach (or even define the value of) the critical point with such a precision, and a good agreement with mean field theory is found for all temperatures.

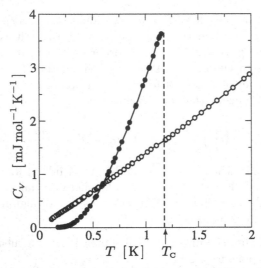

FIGURE 11.2   Heat capacity of aluminium (critical temperature T_C = 1.175 K). Figure adapted from N.E. Phillips, Phys. Rev. **114**, 676 (1959). The white (black) dots are the experimental data for the normal (superconducting) phase. Similar data for niobium were obtained by A. Brown, M. W. Zemansky, and H. A. Boorse, Phys. Rev. **92**, 52 (1953).

Figure 11.2 presents experimental data for aluminium (type I superconductor) with and without external magnetic field. When the external magnetic field is higher than $H_C(T = 0)$ the material remains in the normal phase (see Section 10.5.3), as reported in figure 11.2 (white dots, obtained for a 300 Gauss external magnetic field). In the absence of external magnetic field, aluminium becomes superconducting below T_C. At the transition, a discontinuity is observed in heat capacity as expected from mean field theory [see formula (9.40) and Fig. 9.10]: the very small value of t_G prevents the observation of a possible divergence in the vicinity of T_C.

---

[11] This is exactly true only if each atom provides one free electron. If one atom provide several electrons, there is a multiplicative term in the density (2 or 3 typically) which is insignificant for the typical inter-electronic distance.

11.4.2 ISING SYSTEM

For the two-dimensional Ising system on a square lattice [Hamiltonian (9.8) with $h = 0$], the mean field theory results (see Sections 9.1.2 and 9.1.3) can be directly compared with the exact Onsager solution [Phys. Rev. 65, 117 (1944)]. A first disagreement appears in the value of the critical temperature: $k_B T_C = 4J$ according to Bragg-Williams theory, while in reality $k_B T_C / J = 2 / \ln\left(1 + \sqrt{2}\right) \simeq 2.27$. But the real test of the validity of the mean field approach lies in the behaviour of thermodynamic quantities in the vicinity of the transition. The exact specific heat in the vicinity of T_C has a logarithmic divergence $(C_V / N k_B \simeq -\frac{2}{\pi}(2J/k_B T_C)^2 \ln |1 - T/T_C|)$ while the mean field theory[12] just predicts a discontinuity in T_C as shown in Figure 11.3.

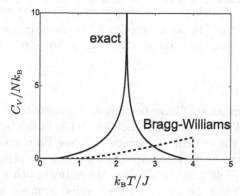

FIGURE 11.3 Comparison of the exact value of the heat capacity of a two-dimensional Ising system on a square lattice (solid line) with the Bragg-Williams result [Equation (11.44)] (dashed line).

11.4.3 LIQUID HELIUM

As already seen in Section 7.3, low-temperature ^{4}He undergoes a phase transition and becomes superfluid below temperature $T_\lambda = 2.17$ K. This is the temperature of the so called *lambda point*, named after the shape of the specific heat, which, plotted as a function of T, has roughly the shape of the greek letter λ in the vicinity of the transition (see Figure 11.4).

The behaviour of the heat capacity near the transition is not as predicted by the mean field theory, and seems to exhibit a self-similar structure (this point is discussed in the next section). It has long been questioned whether the singularity of the specific heat at T_λ is logarithmic, as for the Ising system (Figure 11.3). Very precise experiments (the most recent were conducted in the space shuttle[13]) showed that this is not the case. Instead, $C_V \propto |T - T_\lambda|^{-\alpha}$ with $\alpha = -0.01285 \pm 0.00038$.

11.5 BEHAVIOUR IN THE VICINITY OF THE CRITICAL POINT

According to Landau mean field theory, many thermodynamic quantities have, in the vicinity of the critical point, a power law behaviour characterised by a series of exponents called

---

[12]The mean field approach presented in Section 9.1.3 predicts $C_V(T > T_C) = 0$. Additionally, according to Equations (9.26) and (9.39):

$$\frac{C_V}{N k_B} = \frac{m^2 (1 - m^2) T_C^2}{T^2 - (1 - m^2) T T_C} \to \frac{3}{2} \quad \text{when} \quad T \to T_C^-, \tag{11.44}$$

where m is solution of Equation (9.28). The limit value when $T \to T_C^-$ in Equation (11.44) is obtained using the results of Exercise 9.1.

[13]see J. A. Lipa *et al.*, Phys. Rev. Lett. **76**, 944 (1996).

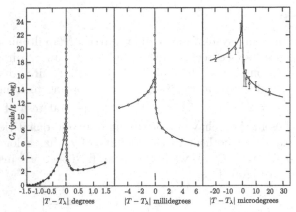

FIGURE 11.4 Heat capacity of ^{4}He as a function of temperature in the vicinity of T_λ. From Buckingam and Fairbank, *Prog. Low Temp. Phys.* III, p. 80 (1961). Copyright Springer 1961, reprinted with permission.

critical exponents. These exponents have been already encountered: for the liquid-gas transition using the van der Waals approach in Section 5.3.3, in the behaviour of the order parameter in the vicinity of the critical temperature T_C [see Formula (9.37) and the following discussion], in the behaviour of the heat capacity and the susceptibility at the critical point [see Equations (9.40) and (9.46)], in the shape of the critical isotherm [Equation (9.47)], etc. The definition of the critical exponents and their value in mean field theory (for a scalar order parameter) are summarised in Table 11.1.

TABLE 11.1 Definition of the critical exponents. $t = (T - T_C)/T_C$ is the reduced temperature and d is the dimension of the physical space. The rightmost column gives the value of each exponent predicted by the Landau mean field theory for a system with a scalar order parameter.

Heat capacity $(h = 0)$	$C \propto |t|^{-\alpha}$	$\alpha = 0$
Order parameter $(h = 0, T \lesssim T_C)$	$\phi \propto |t|^\beta$	$\beta = 1/2$
Susceptibility $(h = 0)$	$\chi \propto |t|^{-\gamma}$	$\gamma = 1$
Equation of state $(T = T_C, \text{weak } h)$	$\phi \propto \text{sgn}(h) \times |h|^{1/\delta}$	$\delta = 3$
Correlation function $(h = 0, T = T_C)$	$G(r) \propto 1/r^{d-2+\eta}$	$\eta = 0$
Correlation length $(h = 0)$	$\xi \propto |t|^{-\nu}$	$\nu = 1/2$

Experiments clearly demonstrate a power law behaviour of the thermodynamic quantities in the vicinity of the critical point, but the measured values of the critical exponents are in general not in agreement with the mean field theory predictions. This was already discussed in Sections 5.3.3 and 11.4.3. This is again evidenced in Figure 11.5 which shows the magnetisation of a ferromagnetic substance (EuS powder) as a function of temperature below the Curie temperature. The experimental data allows the determination the exponent β and point to a value $\beta = 1/3$, whereas mean field theory predicts $\beta = 1/2$.

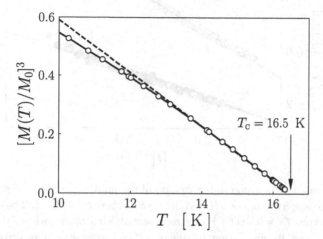

FIGURE 11.5　Third power of the magnetisation, $[M(T)/M(0)]^3$, of a EuS powder as a function of temperature T [adapted from P. Heller and G. Benedek, Phys. Rev. Lett. **14**, 71 (1965)]. The dots connected by a plain line correspond to experimental data. The dashed line corresponds to a fit by the empirical law $M(T)/M(0) = D(1 - T/T_C)^{1/3}$. As expected this fit looses accuracy away from the critical temperature ($T_C = 16.5$ K).

Another example of a scaling law is shown in Figure 11.6, which represents the heat capacity of argon near its critical point. Mean field theory predicts a discontinuity of the heat capacity [see Formula (9.40) and Figure 9.10], corresponding to an exponent $\alpha = 0$ in Table 11.1. Instead, the data are consistent with a value $\alpha = 0.13$ of the critical exponent. Figure 11.6 also illustrates a general feature: while the exponent of the power law is the same above and below the critical temperature, the multiplicative factors are different and the experimental data fall on two different curves depending on whether the temperature is larger or smaller than T_C. This type of behaviour should not come as a surprise, because it is also obtained in the framework of the mean field theory (see for example Expression (9.46) for susceptibility, or Expression (10.14) for the correlation length).

Note also that the exponents are universal, i.e. many transitions are characterised by identical exponents, and fall into the same so-called *universality class*. For instance, as Guggenheim noted in 1945, all liquid-gas transitions at the critical point appear to have the same exponent $\beta \simeq 1/3$ (see figure 5.7), and the same exponent is also obtained for the ferromagnetic transition, see results shown in figure 11.5.

11.6　SCALE INVARIANCE

The power law behaviour of certain quantities in the vicinity of the critical point corresponds to a scale invariance: "zoomed" curves have the same appearance as the original ones. This property is illustrated, for example, in Figure 11.4 which shows the behaviour of the heat capacity of ^{4}He in the vicinity of the superfluid transition. In the present section, it is first shown that any scale-invariant curve follows a power law, and then non-trivial relations between critical exponents are obtained.

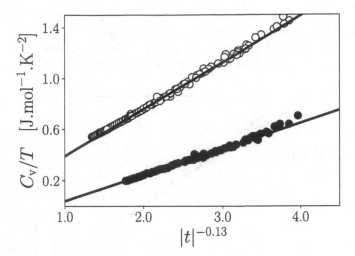

FIGURE 11.6 Heat capacity of argon in the vicinity of the liquid-vapor critical point. Experimental data are compared to empirical laws $C_V/T = A_\pm|t|^{-\alpha}+B_\pm$, where $t = T/T_C-1$. The data above (below) the critical temperature $T_C = 150.663$ K are represented with black (white) dots. The exponent α and constants $A_\pm$ and $B_\pm$ are fitting parameters. Constants with a positive (negative) index correspond to fits above (below) the critical temperature. The experimental data are compatible with $\alpha = 0.13$. Figure adapted from A.V. Voronel *et al.*, Sov. Phys. JETP **36**, 505 (1973).

11.6.1 POWER LAW

Power laws are special in that they are scale-invariant: for instance, by a change of scale the plots of $y = x^a$ for $x \in [0,1]$ and $x \in [0,\lambda]$ can be exactly superimposed, whereas nothing of the sort happens for the plot of $y = \exp(x) - 1$, for example.

The reverse property can be proven: any scale-invariant function obeys a power law. The concept of scale invariance can be defined as follows: let $f(x)$ be a function defined on $\mathbb{R}$. f is scale invariant if there exists a function g (defined on $\mathbb{R}^+$), such that for all $\lambda > 0$ and for all $x \in \mathbb{R}$:

$$f(\lambda x) = g(\lambda) \times f(x). \tag{11.45}$$

Considering the case $x = 1/\lambda$ enables to write $g(\lambda)$ as $g(\lambda) = f(1)/f(1/\lambda)$. Deriving Equation (11.45) with respect to x gives $\lambda f'(\lambda x) = g(\lambda)f'(x) = f(1)f'(x)/f(1/\lambda)$. This last expression is valid for all x in $\mathbb{R}$. If x is positive, taking $\lambda = 1/x$ one directly obtains $f'(1)/x = f(1)f'(x)/f(x)$. It immediately follows that, over $\mathbb{R}^+$, f is a power law of exponent $a = f'(1)/f(1)$ and can be written as: $f(x \geq 0) = K_{(+)}x^a$. To obtain a general expression of $f(x)$ over $\mathbb{R}^-$, Equation (11.45) is again used: rewriting this equality denoting $\lambda = X > 0$ and giving to x the values $x = \pm 1$ yields

$$g(X) = \frac{f(+X)}{f(+1)} = \frac{f(-X)}{f(-1)}. \tag{11.46}$$

This shows that $f(-X)/f(X)$ is a constant independent of X. Applying this property to $X = x$ allows to write f as:

$$f(x) = K_{(\pm)}|x|^a, \tag{11.47}$$

where $K_{(\pm)}$ is a constant which takes the value $K_{(+)}$ for $x > 0$ and $K_{(-)}$ for $x < 0$. Thus, roughly speaking, any scale invariant function f behaves as $|x|^a$. In the following, it is more convenient to adopt for such a behaviour a definition equivalent to Equations (11.45) and

(11.47): a scale-invariant function characterised by an exponent a is a homogeneous function verifying for $\lambda > 0$ and $x \in \mathbb{R}$:

$$f(\lambda^{1/a} x) = \lambda f(x). \tag{11.48}$$

Then, taking $\lambda = 1/|x|^a$, $f(x/|x|) = \frac{1}{|x|^a} f(x)$ i.e.,

$$f(x) = f(\pm 1)|x|^a. \tag{11.49}$$

This is the most general form of a homogeneous function of degree a. The term ± 1 in Expression (11.49) corresponds to the sign of x. Thus, a homogeneous function has the same exponent for all $x \in \mathbb{R}$, but the prefactor can be differently for $x > 0$ and $x < 0$. This is typically what happens in the vicinity of the critical point: many quantities behave as power laws of the quantity $t = (T - T_C)/T_C$, but the prefactors are different for $t > 0$ and $t < 0$ (see Figure 11.6).

The generalisation of Equation (11.48) to the case of functions of two variables $f(x, y)$ (generalised homogeneous function) is :

$$f(\lambda^{1/a} x, \lambda^{1/b} y) = \lambda f(x, y). \tag{11.50}$$

Taking $\lambda = 1/|x|^a$, one immediately finds $f(\pm 1, y/|x|^{a/b}) = |x|^{-a} f(x, y)$, which can be written as:

$$f(x, y) = |x|^a g_{\pm}\left(\frac{y}{|x|^{a/b}}\right), \quad \text{with} \quad g_{\pm}(Y) = f(\pm 1, Y), \tag{11.51}$$

where again the factor $\pm$ refers to the sign of x.

11.6.2 RELATIONSHIPS BETWEEN CRITICAL EXPONENTS

In the following, let $t = (T - T_C)/T_C$ be the reduced temperature. The volumetric free energy density $f(t, h)$ is written as $f(t, h) = f^{(r)}(t, h) + f^{(s)}(t, h)$, where $f^{(r)}$ is the part of the free energy which is regular in the vicinity of T_C ($t = 0$) and $f^{(s)}$ is the singular part. In agreement with numerous experimental results (some of which are presented in Section 11.5), it can be assumed that $f^{(s)}$ is a homogeneous function of t and h: there exists two exponents a and b such as

$$\frac{1}{\lambda} f^{(s)}(\lambda^{1/a} t, \lambda^{1/b} h) = f^{(s)}(t, h). \tag{11.52}$$

The exponents a and b are *a priori* unknown. In the following, they are related to the critical exponents defined in Table 11.1. Note that, by making the homogeneity assumption [Relation (11.52)], the free energy is not supposed to be analytic at the critical point, but only to be piece-wise analytic above and below the critical temperature as is, for the one-variable case, Function f in Equations (11.48) and (11.49).

In the following, let $f^{(s)}_{n,m}$ be the nth partial derivative of $f^{(s)}$ with respect to the first variable (t) and m^{th} with respect to the second variable (h). In other words, $f^{(s)}_{n,m} = \partial^{n+m} f^{(s)}/\partial t^n \partial h^m$.

Behaviour of the heat capacity

The aim here is not to specify the exact behaviour (with all the prefactors) of the quantities under scrutiny, but rather to determine the power law that governs the behaviour of the singular part at the transition. For the heat capacity, one thus writes $C(t, h) = -T(\partial^2 F/\partial T^2) \sim \partial^2 f^{(s)}/\partial t^2 = f^{(s)}_{2,0}$, meaning that the singular behaviour of the heat capacity at the critical point is dictated by the behaviour of $f^{(s)}_{2,0}$. According to Equation (11.52), it follows that the heat capacity behaves as $C(t, h) \sim \lambda^{\frac{2}{a}-1} f^{(s)}_{2,0}(\lambda^{\frac{1}{a}} t, \lambda^{\frac{1}{b}} h)$. For $h = 0$, taking $\lambda = |t|^{-a}$ leads to $C(t, 0) \sim |t|^{a-2} f^{(s)}_{2,0}(\frac{t}{|t|}, 0)$. Comparing to the behaviour given in Table 11.1, one gets $\alpha = 2 - a$.

Behaviour of the order parameter

One also has:

$$\phi(t,h) = -\partial f / \partial h \sim \lambda^{\frac{1}{b}-1} f_{0,1}^{(s)}\left(\lambda^{\frac{1}{a}}t, \lambda^{\frac{1}{b}}h\right). \tag{11.53}$$

At $h = 0$, taking again $\lambda = |t|^{-a}$ leads, for the order parameter near the transition, to a behaviour of the type $\phi(t,0) \sim |t|^{a(1-\frac{1}{b})} f_{0,1}^{(s)}(\pm 1, 0)$. Since $\phi(t > 0, 0) = 0$, this imposes $f_{0,1}^{(s)}(1,0) = 0$, and for $t < 0$ comparison with the definition of the critical exponent β is Table 11.1 yields $\beta = a(1 - 1/b)$.

Still considering ϕ, but staying now on the critical isotherm $t = 0$ and taking $\lambda = |h|^{-b}$, Equation (11.53) can be written as $\phi(0,h) \sim |h|^{b-1} f_{0,1}^{(s)}(0, \pm 1)$. This imposes $f_{0,1}^{(s)}(0, -1) = -f_{0,1}^{(s)}(0, 1)$ since ϕ is an odd function of h. As far as the critical exponent is concerned, one gets $b - 1 = 1/\delta$.

Behaviour of the susceptibility

Here, $\chi(t,h) = \partial \phi / \partial h \sim \lambda^{\frac{2}{b}-1} f_{0,2}^{(s)}(\lambda^{\frac{1}{a}}t, \lambda^{\frac{1}{b}}h)$. Taking $h = 0$ and $\lambda = |t|^{-a}$ yields $\chi(t,0) \sim |t|^{a(1-\frac{2}{b})} f_{0,2}^{(s)}(\pm 1, 0)$, which implies $a(1 - 2/b) = -\gamma$.

Summary

As it was just discussed, if the scale invariance assumption [Relation (11.52)] is valid, only two exponents (a and b) determine all the others (α, β, γ and δ). More explicitly, the analysis of the previous paragraphs "Behaviour of the heat capacity", "Behaviour of the order parameter" and "Behaviour of the susceptibility" imply that the following relations must be verified:

$$\begin{aligned} \beta\delta &= \beta + \gamma \quad \text{(Widom identity)}, \\ \alpha + 2\beta + \gamma &= 2 \quad \text{(Rushbrooke identity)}. \end{aligned} \tag{11.54}$$

There is another identity which is not independent of the others: $\alpha + \beta + \beta\delta = 2$ (Griffith identity). Of course, the mean field exponents verify these relations as do the exact exponents of the two-dimensional Ising ($\alpha = 0$, $\beta = \frac{1}{8}$, $\gamma = \frac{7}{4}$, $\delta = 15$, $\eta = \frac{1}{4}$ and $\nu = 1$).

Behaviour of the correlation function

The homogeneity assumption [Relation (11.52)] is not sufficient to relate the value of the exponent η governing the behaviour of the correlation function to the other critical exponents, since the knowledge of the free energy density alone does not permit the determination of $G(r)$. The fluctuation-dissipation theorem (11.11) combined with Relation (11.20) gives in a homogeneous system:

$$\chi = k_B T \int d^d r \, G(\vec{r} - \vec{r}'). \tag{11.55}$$

The behaviour of G given in Table 11.1 is valid only at the critical temperature ($t = 0$). In the vicinity of $t = 0$, the correlation function behaves as $G(r) \sim f(r/\xi)/r^{d-2+\eta}$, where $f(0) \sim 1$ and $f(x) \sim \exp(-x)$ when $x \gg 1$. Then, performing the change of variable $u = r/\xi$ in Equation (11.55) leads to $\chi \sim \xi^{2-\eta}$, i.e., as far as the critical exponents are concerned,

$$\gamma = \nu(2 - \eta). \tag{11.56}$$

11.6.3 EXPERIMENTAL DATA

Scale invariance in the vicinity of the critical point, and the resulting relationships between critical exponents, are confirmed by quite spectacular experimental tests. Consider for instance the equation of state (11.53). Taking $\lambda = |t|^{-a}$ yields $\phi(t, h) \sim |t|^{a(1-1/b)} f_{0,1}^{(s)}(t/|t|, h/|t|^{a/b})$. Then, expressing the scaling exponents a and b in terms of the critical exponents $[a(1 - 1/b) = \beta$ and $a/b = \beta + \gamma = \beta\delta]$ leads to:

$$\phi(t, h) \sim |t|^{\beta} \times F_{\pm}\left(\frac{h}{|t|^{\beta\delta}}\right), \qquad (11.57)$$

where the $\pm$ of F refers to sgn(t). Figure 11.7 shows the equation of state (11.57) for the ferromagnetic transition in Nickel obtained from Weiss and Forrer's data (1926) and reanalysed by Arrot and Noakes in 1967. This analysis gives $\beta = 0.4$, $\delta = 4.29$ and $\gamma = 1.31$, thus validating the Widom identity [Relation (11.54)] ($\beta\delta = 1.716$ and $\beta + \gamma = 1.71$). The same type of analysis has been performed by Ho and Lister for the magnetisation of CrBr$_3$ [Phys. Rev. Lett. **22**, 603 (1969)] and by Huang and Ho for EuO [Phys. Rev. B **12**, 5255 (1975)].

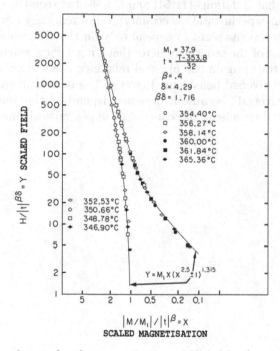

FIGURE 11.7 Equation of state for the magnetisation of Nickel in the vicinity of the Curie temperature $T_C = 353.8$ K. The leftmost curve corresponds to data for $T < T_C$, the rightmost one to data for $T > T_C$. The x-axis is the dimensionless magnetisation [ϕ in Equation (11.57)] divided by $|t|^{\beta}$. The y-axis shows $h/|t|^{\beta\delta}$. Figure reprinted with permission from A. Arrot and J. E. Noakes, Phys. Rev. Lett. **19**, 786 (1967). Copyright (1967) by the American Physical Society.

Experiments thus beautifully confirm scale invariance in the vicinity of the critical point and the corresponding relationships between critical exponents. It is also important to notice that transitions with very different physical characteristics seem to share the same critical exponents, as illustrated by the data of Table 11.2 giving the value of the critical exponents in dimension $d = 3$ for several transitions with scalar order parameter.

TABLE 11.2 Critical exponents for phase transitions in three-dimensional space characterised by a scalar order parameter.

	Liquid-gas	Binary mixture	Ferromagnetic material	Ising 3D
α	0.108±0.01	0.113±0.005	0.109	0.110±0.001
β	0.339±0.006	0.322±0.002	0.305	0.3265±0.0003
γ	1.20 ±0.02	1.239±0.002	1,22	1.2372±0.0005
δ	4.85 ±0.1	4.789 ±0.03	4.94	4.789±0.002
η	0.045 ±0.010	0.017±0.015		0.0364 ±0.0005
ν		0.625±0.006		0.6301 ±0.0004

In Table 11.2, the first column deals with the liquid/gas transition for different substances very near their critical points[14]; the second one deals with the mixing-demixing transition in a binary fluid mixture[15]; the third one deals with the ferromagnetic transition in CoF_2[16]; and the last one (the only non experimental data in the table) deals with the ferromagnetic transition in the three dimensional Ising model[17].

It can be checked that Relations (11.54) and (11.56) between the critical exponents are indeed satisfied (within experimental uncertainty), but a tendency towards universality can also be guessed: the exponents seem to depend only on the dimension of the space ($d = 3$ here) and the dimension of the order parameter (here, $n = 1$ for a scalar order parameter[18]). This universality illustrates again the physical relevance of the Landau mean field theory which also predicts a universal behaviour. However, the behaviour predicted by the mean field theory is "too universal" because it does not depend on the dimension of space, and the mean field theory thus misses the concept of upper critical dimension introduced in Section 11.3.

---

[14] The data are from M.A. Anismov *et al.*, Sov. Phys. JETP **49**, 844 (1979) and R. Hocken and M. R. Moldover, Phys. Rev. Lett. **37**, 29 (1976).

[15] Data taken from the book "The theory of critical phenomena" by J.J. Binney, N.J. Dowrick, A.J. Fisher and M.E.J. Newman, Clarendon Press, Oxford, 2002.

[16] Data from R.A. Cowley and E.K. Carneiro, J. Phys. C **13**, 3281 (1980) and C.A. Ramos, A.R. King and V. Jaccarino, Phys. Rev. B **40**, 7124 (1989).

[17] From A. Pelisseto and E. Vicari, Phys. Rep. **368**, 549 (2002) and H. Kleinert, Phys. Rev. D. **60**, 085001 (1999).

[18] Strictly speaking, the belonging or not to a certain universality class also depends on the interaction range.

12 Percolation

12.1 INTRODUCTION

The percolation transition is a geometrical phase transition (it is not associated to a Hamiltonian) allowing to illustrate on a simple model case the concepts developed in the previous chapters. One version of the problem is the following: consider a lattice in $\mathbb{R}^d$ comprising N sites, each site being occupied with a probability p. Figure 12.1 presents three different realisations (corresponding each to a different value of p) of percolation on a two-dimensional square lattice.

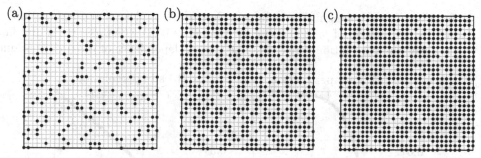

FIGURE 12.1 Site percolation on a 30×30 square lattice. Occupied sites are represented by black dots. (a) $p = 0.2$, (b) $p = 0.6$ (slightly above the critical probability $p_c = 0.592746\ldots$), and (c) $p = 0.8$.

The transition point is reached when the probability p is high enough for the existence of a "percolating path", i.e. a path enabling the crossing of the system by jumping from an occupied site to another occupied neighbouring site. Clearly, when $p = 0$, there is no percolating path, and when $p = 1$, all paths percolate. The value p_c of p at which the first percolating path appears is called the *critical probability*, or *percolation threshold*. In the thermodynamic limit, p_c depends only on the dimension of space and the type of lattice considered (square, triangular, cubic, etc.). The configuration with $p = 0.6$ in Figure 12.1(b) is slightly above the percolation threshold, and indeed several percolating paths can be identified.

The concept of percolation can be declined in many variants. In this chapter, two very common models will be studied: (i) the site percolation model in which each site is occupied with a probability p, this is the one used above to define the concept of percolation (see Figure 12.1); (ii) the bond percolation model where all sites are occupied and the bonds can be active with a probability p (see Figure 12.2).

An important concept in percolation theory is the notion of cluster: a cluster is defined as a set of nearest neighbour sites that are occupied (for site percolation) or connected by active bonds (for bond percolation). Different clusters are clearly visible in Figure 12.2. The order parameter of the percolation transition is:

$$\mathcal{P}(p) = \lim_{N \to \infty} \frac{\text{Number of sites of the largest cluster}}{N = \text{Number of sites in the lattice}}. \tag{12.1}$$

DOI: 10.1201/9781003272427-12

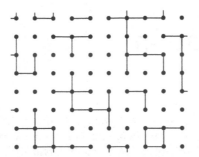

FIGURE 12.2 Bond percolation on a square lattice. Inactive links have been removed from the figure. Different clusters can be clearly identified (points connected by active bonds).

$\mathcal{P}$ is the fraction of sites belonging to the largest cluster[1]. It is clear that $\mathcal{P} \in [0,1]$ and that $\mathcal{P}$ is an increasing function of p. It is also clear that $\mathcal{P} \to 0$ (1) when $p \to 0$ (1) and finally that $\mathcal{P} = 0$ below the percolation threshold. The typical dependence of the order parameter

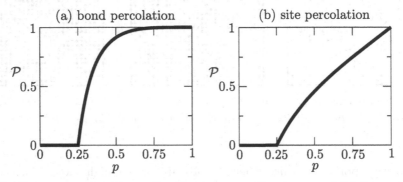

FIGURE 12.3 Order parameter $\mathcal{P}$ as a function of p according to mean field theory for (a) bond percolation and (b) site percolation on a lattice with a coordination number $z = 4$. The plots were obtained by numerically solving self-consistent equation (12.3) for bond percolation and Equation (12.13) (with $h = 0$) for site percolation.

on p is illustrated in Figure 12.3. This figure corresponds to a mean field approach on a lattice with a coordination number[2] $z = 4$ (the type of mean field approximation relevant to percolation is presented in Section 12.2.2). As one might, expect $\mathcal{P} < p$ for site percolation: this is an immediate consequence of Definition (12.1), since, for site percolation, p is the ratio of the number of occupied sites to the total number of sites. Note that Figure 12.3 presents mean field theory results (see Section 12.2.2) according to which the transition probability p_c is the same for site and bond percolation and equal to 1/4 on a square lattice. This is not correct in general (see some exact results in Section 12.2.1), and it can be shown that in all cases $p_c^{\text{lien}} \leq p_c^{\text{site}}$.

It thus appears that the percolation transition has all the properties of a second-order phase transition (continuous order parameter at the critical point) with the probability p (or rather $-p$ or even better $1/p$) playing the role of a temperature.

---

[1]Some authors define the order parameter as the fraction of *occupied* sites $\mathcal{P}_o$ belonging to the largest cluster. For bond percolation this definition is identical to Definition (12.1) since all sites are occupied: $\mathcal{P}^{\text{bond}} = \mathcal{P}_o^{\text{bond}}$. For site percolation, $\mathcal{P}^{\text{site}} = p\mathcal{P}_o^{\text{site}}$. An advantage of the use of $\mathcal{P}_o$ is that the curves $\mathcal{P}_o^{\text{bond}}(p)$ and $\mathcal{P}_o^{\text{site}}(p)$ behave very similarly. In particular, the mean field approximation (see Section 12.2.2) cannot distinguish between $\mathcal{P}_o^{\text{bond}}$ and $\mathcal{P}_o^{\text{site}}$. The proof is left as an exercise.

[2]In this chapter, the coordination number is written z and not q as in the rest of the book.

12.2 EVALUATION OF THE PERCOLATION THRESHOLD

In this section, the focus is put on the determination of the value of the percolation threshold p_c in different models, using exact or approximate methods.

12.2.1 BOND PERCOLATION ON A TWO-DIMENSIONAL SQUARE LATTICE

This is a case for which p_c can be determined exactly. Consider a square lattice and a given realisation of bond percolation. The dual configuration is defined as a realisation of percolation on the dual lattice (the lattice whose sites are at the centre of the original lattice's cells) for which a bond is active if and only if the bond of the original lattice that it intersects is inactive. A typical example is shown in Figure 12.4. It is clear that the dual configuration is also a realisation of bond percolation on a two-dimensional square lattice whose bonds are active with a probability $1 - p$ if the bonds of the initial configuration are active with a probability p. It is also clear that if there is a percolating path in the initial configuration, there is none in the dual configuration. On the other hand, if the initial configuration does not percolate, the dual configuration does (as is the case in Figure 12.4). Since the percolation threshold of the initial configuration is p_c, the dual configuration threshold is therefore $1 - p_c$. As these two configurations both correspond to a realisation of the same percolation model, it is obvious that $p_c = 1 - p_c$, i.e. $p_c = 1/2$ for bond percolation on a square lattice. Note that for site percolation on the same lattice, the threshold $p_c = 0.592746\ldots$ is only known numerically[3].

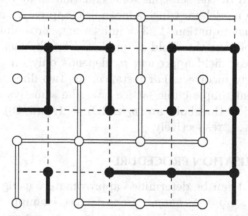

FIGURE 12.4 Bond percolation on a square lattice. The initial configuration is shown in black, the dual configuration in white. In the initial configuration, the active links are represented as thick solid lines, and the inactive ones as thin dotted lines. Only the active links are shown (as thick white lines) for the dual lattice.

Exercise 12.1. Use the dual lattice technique to show that the bond percolation threshold on a triangular lattice $p_c(\triangle)$ and on a hexagonal lattice $p_c(\bigcirc)$ are related by $p_c(\bigcirc) = 1 - p_c(\triangle)$.

Solution: It is obvious since the hexagonal and triangular lattice are dual one to another.

---

[3]In 3 dimensions, (numerical) results for bond and site percolation on a simple cubic lattice give the thresholds $p_c^{bond} = 0.248812\ldots$ and $p_c^{site} = 0.3116\ldots$, respectively. There exist a wikipedia page listing numerous information on p_c values for different lattices in different dimensions: http://en.wikipedia.org/wiki/Percolation_threshold .

12.2.2 MEAN FIELD APPROXIMATION

This section is devoted to a mean field study of bond percolation (site percolation is treated as an exercise in Section 12.3). Let $\mathcal{P}_i$ be the probability that site i belongs to the percolating cluster. An approximate value of the probability $1 - \mathcal{P}_i$ that site i is disconnected from the percolating can be calculated as follows. If site i is disconnected from the percolating cluster, it must not be connected to it through any of its z nearest neighbours. The probability of connection to the percolating cluster through a neighbour site j is $p\mathcal{P}_j$. So, the probability of not being connected to the percolating cluster through j is $1 - p\mathcal{P}_j$. If all these probabilities are assumed independent of each other (which is an approximation) then:

$$1 - \mathcal{P}_i = \prod_{j \in \text{nn}\{i\}} (1 - p\mathcal{P}_j), \tag{12.2}$$

where $\text{nn}\{i\}$ denotes the set of the z nearest neighbours of i. By homogeneity, in the thermodynamic limit, $\mathcal{P}_i$ does not depend on the considered site and thus $\mathcal{P}_i = \mathcal{P}$, the order parameter of the transition [Definition (12.1)]. Therefore, in the framework of the mean field approximation, the following self-consistent equation is obtained for the order parameter:

$$1 - \mathcal{P} = (1 - p\mathcal{P})^z. \tag{12.3}$$

Results of the numerical solution of this mean field equation are presented in Figure 12.3(a).

A graphical resolution of the self-consistent equation similar to the one presented in Section 9.1.2 for the Ising model [for which a mean field approach leads to the self-consistent equation (9.14)] shows that Equation (12.3) admits a non-zero solution only if $p > 1/z$: this is the value of the percolation threshold p_c predicted by the mean field theory for bond percolation[4]. In the mean field approach, p_c depends only on the lattice coordination number z, which is clearly incorrect. For instance, the two dimensional triangular lattice and the three dimensional simple cubic lattice have the same coordination number $z = 6$ but their bond percolation threshold are $p_c(\triangle, 2D) = 2\sin(\pi/18) = 0.347296355\ldots$ and $p_c(\text{cubic}, 3D) = 0.248812\ldots$, respectively.

12.2.3 THE RENORMALISATION PROCEDURE

The percolation threshold can be determined approximately using a simple version of the so-called *renormalisation group technique*. Consider the example of site percolation on a two dimensional square lattice with a lattice spacing a. A configuration percolates if there is a path formed by occupied neighbour sites going from the left side to the right side (or equivalently from the top to the bottom).

Scale invariance of the correlation function at the critical point (see Section 11.6) is invoked to perform a scaling of the system: 4 neighbour sites forming the vertices of a square cell are replaced by a "super-site" located at the centre of the cell. The super-site is occupied if the initial 4 sites allow a percolation path to propagate, say, horizontally. The lattice of super-sites is also a square lattice, with a lattice spacing $2a$. The procedure is illustrated in Figure 12.5. The bonds of the initial lattice are represented as dotted lines and those of the super-lattice as solid lines. For instance, the super-site on the top left of Figure 12.5 is considered unoccupied because it gathers 4 initial sites that do not allow propagation from left to right. Its right super-neighbour also contains 2 occupied and 2 empty sites, but it is considered occupied because it gathers 4 sites allowing left to right propagation. The renormalisation procedure is therefore not exact, as can be also verified from the following observation: in Figure 12.5, the super-lattice allows horizontal

---

[4]The same value is obtained for site percolation, see Exercise 12.3.

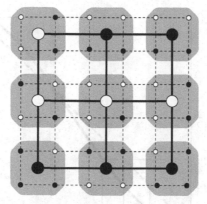

FIGURE 12.5 Illustration of the renormalisation procedure for site percolation on a square lattice. Each super-site groups the 4 sites located in the surrounding shaded area. The bonds of the initial lattice are represented as dotted lines and those of the super-lattice as solid lines.

propagation, whereas this is not the case for the initial lattice. It can also be seen that the initial lattice does not percolate from top to bottom, but it would be sufficient that its top right site be occupied for a vertical percolation to appear, whereas the super-lattice would remain unchanged (and still not percolate vertically). Of course, the renormalisation procedure would be more accurate if gathering more that 4 sites at each step. In the limit where *all* sites are considered at once, it is exact.

The occupation probability p' of the super-lattice is a function of the occupation probability p of the initial lattice: $p' = f(p)$. A super-site is occupied if it gathers 4 sites that also are (probability p^4), or if 3 of 4 sites are (corresponding to 4 configurations, each having a probability $p^3(1 - p)$), or if 2 of 4 sites are and allow a percolation path from left to right (corresponding to 2 configurations each, having a probability $p^2(1 - p)^2$). Thus $f(p) = p^4 + 4p^3(1 - p) + 2p^2(1 - p)^2 = p^2(2 - p^2)$. Repeating this procedure eventually leads to a unique site. If it is occupied the initial lattice is assumed percolating.

The function $f(p)$ has three fixed point: 0, 1 and $p^* = (-1 + \sqrt{5})/2$ (see Figure 12.6). If $p > p^*$, then $f(p) > p$. When $p_1 > p^*$, the iterative procedure described above generates a sequence $p_{n+1} = f(p_n)$ that converges to the stable fixed point 1. In this case, at the end of the renormalisation procedure, the initial configuration is found to be percolating. If, on the contrary, $p_1 < p^*$ the sequence converges to 0 – which is the other stable fixed point – and the initial lattice is found to be not percolating, as illustrated in Figure 12.6. This method thus predicts a site percolation threshold on a triangular lattice $p_c = p^* = 0.6180\ldots$, to be compared with the numerically determined value $p_c = 0.5927$ obtained by simulations on huge lattices. One can be surprised that it is the unstable fixed point that determines the behaviour of the system; but it is finally quite natural that the boundary between the two typical behaviours (percolating or non percolating) corresponds to an unstable point.

12.3 THERMODYNAMICS AND CRITICAL EXPONENTS

The value of the percolation threshold studied in Section 12.2 is far from being the only interesting characteristic of the percolation transition, just as the knowledge of the value of the critical temperature of a second order transition does not do justice to the richness of the subject. The behaviour of thermodynamic[5] quantities at the transition is also quite

---

[5] *Thermodynamic* is here a misnomer, since the physical quantities related to percolation are purely geometric.

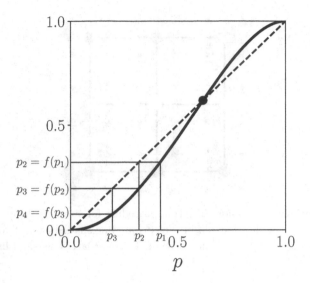

FIGURE 12.6 Graphical construction of the probabilities $p_1, p_2\ldots$ of each step of the renormalisation procedure described in the text. The solid line represents the function $f(p)$. The dashed line is the bisector. The horizontal and vertical thin lines serve to illustrate the iterative procedure: starting from p_1 located on the x-axis, $p_2 = f(p_1)$ is obtained on the y-axis which is transferred to the x-axis thanks to the bisector. The operation is then repeated. We start here with an occupation probability $p_1 = 0.42 < p^*$, the graphical construction illustrates the convergence of the p_n sequence to 0. The black point marks the unstable fixed point, with coordinates $(p^*, f(p^*) = p^*)$.

relevant. The study of second order transitions in Chapters 9, 10 and 11 has proven that this behaviour is all the more interesting as it is universal (see Section 11.6), which is verified again here: it will be seen that the study of the distribution of cluster sizes enables to characterise the percolation transition by critical exponents similar to those studied in the previous chapters. In contrast with the percolation threshold value which depends on the geometry of the lattice, the critical exponents are universal.

12.3.1 CLUSTER SIZE DISTRIBUTION

Let $\mathcal{N}(s)$ be the number of clusters comprising s sites. These are often referred to as clusters of "size" s. For site percolation, $\sum_{s=1}^{N} s \mathcal{N}(s) = pN$. For bond percolation, all sites are occupied, and thus $\sum_{s=1}^{N} s \mathcal{N}(s) = N$. Isolating in this sum the possible percolating cluster[6] (of size $\mathcal{P}N$) yields:

$$\sum_{s}^{*} s \mathcal{N}(s) + \mathcal{P}N = \begin{cases} pN & \text{for site percolation,} \\ N & \text{for bond percolation,} \end{cases} \tag{12.4}$$

where the notation $\sum_{s}^{*}$ means that the sum is restricted to non-percolating clusters (of "finite size"). Note that many authors use the notation $n(s) = \mathcal{N}(s)/N$. $n(s)$ is an intensive quantity in the thermodynamics limit.

---

[6]Mathematicians show that the percolating cluster, if it exists, is almost certainly unique. This corresponds to physical intuition: one can imagine that a configuration where there are several disjoint percolating clusters has a probability of zero in the thermodynamic limit.

- For site and bond percolation, the average size of finite size clusters is defined by the quantity:

$$S = \frac{\sum_s^* s^2 \mathcal{N}(s)}{\sum_s^* s \mathcal{N}(s)}. \tag{12.5}$$

Indeed, the total number of sites belonging to one of the clusters of size s is $s\,\mathcal{N}(s)$. There are thus $\sum_s^* s\,\mathcal{N}(s)$ sites in finite size clusters. Therefore, the probability that the cluster to which belong a site be of size s is $\pi(s) = s\,\mathcal{N}(s)/\sum_s^* s\,\mathcal{N}(s)$. And thus $S = \sum_s^* s\,\pi(s)$; this is Formula (12.5). Numerical simulations show that S diverges in the vicinity of p_c. This result will be later justified by relating S to the susceptibility χ defined by Equation (12.11). The associated critical exponent γ is defined as: $S \propto |p - p_c|^{-\gamma}$ when $p \to p_c$.

- Note that it could be tempting to define the cluster mean size as the quantity $\sum_s^* s\,\mathcal{N}(s)/\sum_s^* \mathcal{N}(s)$. But this would result in counting each cluster with an equal weight, independent of their size. As there are $\mathcal{N}(s)$ size s clusters, this would amount to say the probability of a size s cluster is $\mathcal{N}(s)/\sum_s^* \mathcal{N}(s)$.

12.3.2   FREE ENERGY

A pseudo-magnetic field h and a free energy $F(p, h)$ can be defined as now explained: A "ghost site" is added and connected to the occupied sites[7] of the lattice with a probability:

$$q = 1 - \exp\{-h\}, \tag{12.6}$$

with $h > 0$. This procedure is illustrated in Figure 12.7. The quantity h is roughly equivalent to a magnetic field because, as in the Ising model, if $h \neq 0$, the order parameter will be non-zero, even if $p \ll 1$. To see this point, let us consider the example of site percolation. If $p \ll 1$, in the absence of the ghost, there are no percolating clusters, only isolated ones of size 1. If a ghost is added, a macroscopic cluster formed by the (small, but finite) fraction of occupied sites connected to the ghost immediately appears. The resulting value of $\mathcal{P}$ is determined in Exercise 12.2.

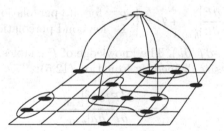

FIGURE 12.7 Schematic representation of the addition of a ghost for site percolation. The ghost site is represented by a white dot outside the physical space. The clusters pre-existing the ghost are enclosed by solid curves. After the addition of the ghost the largest cluster contains 9 sites.

Exercise 12.2.   Show that, for site percolation, the addition of a ghost leads to $\mathcal{P}(p \to 0) = p\,q$.

---

[7]Remember that for bond percolation all sites are occupied, while only a fraction p of the sites are occupied for site percolation.

Solution: When $p \to 0$, there are only size 1 clusters. The number of sites attached to the ghost is then $q\,p\,N$, hence the result. For bond percolation, all sites being occupied, the same reasoning leads to $\mathcal{P}(p \to 0) = q$.

In Figure 12.7, the largest cluster, which initially consisted of 5 sites, is increased to 9 sites after the addition of the ghost. The increase in size of the largest cluster, i.e., the variation of the order parameter caused by the ghost, can be precisely evaluated: If h goes from 0 to the non-zero value $\delta h \ll 1$, then $\mathcal{P} \to \mathcal{P} + \delta\mathcal{P}$ thanks to the connection through the ghost of new clusters to the percolating one. Indeed, the percolating cluster, if it exists (i.e. if $p > p_c$) will be connected to the ghost with a probability 1, as well as some other clusters, thus increasing the value of $\mathcal{P}$. If $p < p_c$, there is initially no percolating cluster, but when h becomes non-zero, the finite size clusters binding to the ghost create a macroscopic cluster. Thus, whether p is higher or lower than p_c, the addition of finite-sized clusters via the ghost changes the value of the order parameter. The probability that a cluster of size s joins the ghost when h goes from 0 to δh is $1 - (1 - q)^s \simeq s\,\delta h$ and this cluster adds s new sites to the percolating cluster. Summing over all cluster sizes gives:

$$\delta\mathcal{P} = \frac{1}{N} \sum_s^* \mathcal{N}(s) \times s\,\delta h \times s,$$

and thus

$$\left.\frac{\partial\mathcal{P}}{\partial h}\right|_{h=0} = \frac{1}{N} \sum_s^* s^2 \mathcal{N}(s). \tag{12.7}$$

The quantity determined in Relation (12.7), $\partial\mathcal{P}/\partial h$, is a susceptibility, similar to the quantity defined in Equation (9.45) for instance. It is useful to write it as the second derivative of a free energy with respect to h, as done in Equation (9.45). To do so, let us define the quantity:

$$F(p,h) \equiv \sum_s^* \mathcal{N}(s)\,e^{-sh}. \tag{12.8}$$

F plays indeed the role of a (dimensionless) free energy and Relation (12.4) immediately leads to

$$-\frac{1}{N}\left.\frac{\partial F}{\partial h}\right|_{h=0} + \mathcal{P} = \begin{cases} p & \text{for site percolation,} \\ 1 & \text{for bond percolation,} \end{cases} \tag{12.9}$$

(to be compared with $\phi = -\partial f/\partial h$). The knowledge of F enables, for example, the calculation of the average cluster size defined by Equation (12.5):

$$S = -\frac{\partial^2 F/\partial h^2\big|_{h=0}}{\partial F/\partial h\big|_{h=0}}. \tag{12.10}$$

Let us finally note that F allows the evaluation of the equivalent of a susceptibility, i.e. the quantity χ defined by:

$$\chi = \left.\frac{\partial\mathcal{P}}{\partial h}\right|_{h=0}. \tag{12.11}$$

Indeed, by comparing Equations (12.7) and (12.9), one immediately obtains:

$$\chi = \frac{1}{N}\left.\frac{\partial^2 F}{\partial h^2}\right|_{h=0}. \tag{12.12}$$

It is thus natural to denote γ the exponent associated with the divergence of χ in the vicinity of p_c (see Table 11.1). It is, of course, the same as the one associated with the divergence of S, as can be seen by comparing Expressions (12.4), (12.10) and (12.7).

An exponent β can also be defined by studying the behaviour of $\mathcal{P}$ in the vicinity of p_c and an exponent δ by studying the behaviour of $\mathcal{P}$ as a function of h at $p = p_c$. A scaling law assumption similar to the one presented in Section 11.6 leads to the Widom relation [Equation (11.54)]. It is, of course, verified in the framework of the mean field approximation[8] ($\beta = 1$, $\gamma = 1$ and $\delta = 2$) and also for the exact results in dimension 2 ($\beta = 5/36$, $\gamma = 43/18$ and $\delta = 91/5$).

Exercise 12.3.   Consider the site percolation model on a lattice with a coordination number z. p is the probability of occupation of a site. A ghost site is added and each occupied site can be connected to it with probability q. The ghost is associated with an effective external field h whose intensity is related to the binding probability with the ghost: $q = 1 - \exp\{-h\}$.

1/ Let $\mathcal{P}_i$ be the probability that site i belongs to the percolating cluster. In the thermodynamic limit where edge effect can be ignored, this probability does not depend on i and is written $\mathcal{P}$. Show that in the mean field framework, the probability is:

$$\mathcal{P} = p - p\,e^{-h}\left(1 - \mathcal{P}\right)^z. \tag{12.13}$$

2/ Let us first consider the case $h = 0$ (the ghost is "disconected").

(a) Propose a graphical solution of Equation (12.13) to determine $\mathcal{P}$ as a function of p. What is the value of the percolation threshold p_c predicted by mean field theory? Compare to the value obtained numerically for the two dimensional square lattice: $p_c = 0.592746\ldots$. Show that $\mathcal{P}(p) \leq p$. Plot the function $\mathcal{P}(p)$.

(b) Consider now the limit $p \to p_c^+$. Solve Equation (12.13) in a approximate way. Show that, in this limit, $\mathcal{P} \propto (p - p_c)^\beta$. What is the value of exponent β predicted by mean field theory?

3/ Consider now a non-zero field ($h \neq 0$).

(a) First consider the limit $p \to 0$. Determine how $\mathcal{P}$ depends on p and h within the mean field approach (hint: assume that $\mathcal{P}$ is small). Compare to the exact result.

(b) Plot $\mathcal{P}(p)$ for $h \neq 0$.

(c) Calculate $\partial \mathcal{P}/\partial h|_{h=0}$ for $p \in [0,1]$. Deduce from this result the value of exponent γ predicted by mean field theory.

(d) Consider now the case $p = p_c$. Determine the behaviour of $\mathcal{P}$ at low field. Deduce from this result the value of exponent δ predicted by mean field theory. Check that the Widom identity [Relation (11.54)] holds.

Solution: **1/** _Let us evaluate the probability_ $1 - \mathcal{P}_i$ _that site i does not belong to the percolating cluster. In this case, either site i is empty (probability $1 - p$), or is occupied (probability p) but is not connected to the ghost (probability $\exp(-h)$) and none of its neighbours is connected to the percolating cluster. This can be written as:_

$$1 - \mathcal{P}_i = (1 - p) + p\,e^{-h} \prod_{j \in \mathrm{nn}\{i\}} (1 - \mathcal{P}_j),$$

where $\mathrm{nn}\{i\}$ denotes the set of nearest neighbours of site i. In this formula, the probability that none of the neighbours of i is connected to the percolating cluster is written as a product of

---

[8]Mean field exponents for two-dimensional percolation are determined in Exercise 12.3.

probabilities, which is the same as assuming that these probabilities are independent: the mean field approximation is based on this simplifying assumption. In the thermodynamic limit, the probability $\mathcal{P}_i$ to be connected to the percolating cluster does not depend on the position and one writes $\mathcal{P}_i = \mathcal{P}$; the above relation then leads directly to (12.13).

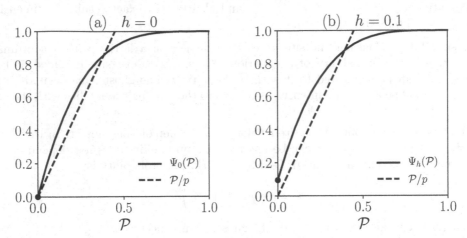

FIGURE 12.8 Graphical solution of Equation (12.13). (a): The solid curve represents $\Psi_0(\mathcal{P}) = 1 - (1 - \mathcal{P})^z$ for $z = 4$. The dashed line is $\mathcal{P}/p$ for $p = 0.45$. The intersection is located at $\mathcal{P} \simeq 0.386$. (b): The solid curve represents $\Psi_h(\mathcal{P}) = 1 - e^{-h}(1 - \mathcal{P})^z$ for $z = 4$ and $h = 0.1$. $\Psi_h(0)$ is non-zero (in the present case, $\Psi_h(0) \simeq 0.0951$). The dashed line is $\mathcal{P}/p$ for $p = 0.45$. The intersection is located at $\mathcal{P} \simeq 0.395$.

2/ When $h = 0$, the solution of Equation (12.13) corresponds to the intersection of the curve $\Psi_0(\mathcal{P}) = 1 - (1 - \mathcal{P})^z$ with the line $\mathcal{P}/p$. The corresponding curves are plotted in Figure 12.8(a). The curve $\Psi_0(\mathcal{P})$ behaves at the origin as $z\mathcal{P}$ and thus will cross the curve $\mathcal{P}/p$ if (and only if) $p > 1/z$, which sets the mean field site percolation threshold to $p_c = 1/z$. The curve $\mathcal{P}(p)$ is displayed in Figure 12.3. In the vicinity of $p = 1/z$, $\mathcal{P}$ remains small and the self-consistent equation can be solved by performing a series expansion of $\Psi_0(\mathcal{P})$. This yields $\mathcal{P} \simeq \frac{2z}{z-1}(p - \frac{1}{z})$, which implies that mean field theory predicts an exponent β equal to 1 for site percolation.

3/ When $h \neq 0$, the graphical solution of the self-consistent equation (12.13) is presented in Figure 12.8(b). In this case, there is always a crossing of the line $\mathcal{P}/p$ with the curve $\Psi_h(\mathcal{P}) = 1 - e^{-h}(1 - \mathcal{P})^z$. In the limit $p \to 0$, the line $\mathcal{P}/p$ is very steep and the intersection occurs for a very small value of $\mathcal{P}$. Then, Equation (12.13) can be solved by performing an expansion of $\Psi_h(\mathcal{P})$ around 0 and one gets $\mathcal{P} \simeq p(1 - e^{-h}) = p\,q$, in agreement with the small p limit of the exact result (see Exercise 12.2).

If the value of $\mathcal{P}$ for zero field is known, an increase $\delta\mathcal{P}$ associated to a small field δh can also be determined. To this end, one needs to solve the equation: $\frac{1}{p}(\mathcal{P} + \delta\mathcal{P}) = \Psi_{\delta h}(\mathcal{P} + \delta\mathcal{P})$. This yields

$$\frac{\delta\mathcal{P}}{\delta h} = \frac{(1 - \mathcal{P})^z}{1 - z\,p(1 - \mathcal{P})^{z-1}}.$$

In the vicinity of the critical point, when $p \to 1/z$, this relation is written as $\delta\mathcal{P}/\delta h = 1/(1 - z\,p)$, showing that the percolation exponent γ is equal to 1 in the framework of mean field theory.

When $p = p_c = 1/z$, the system is exactly at the percolation threshold. To make an analogy with thermodynamics, one could say that the system lies on the critical isotherm. In this case, the change of $\mathcal{P}$ for a low field h at constant p is obtained by solving $\mathcal{P}z = \Psi_h(\mathcal{P})$. This reads $z(z-1)\mathcal{P}^2 \simeq 2h$, and the exponent δ is thus equal to 2 in the framework of mean field theory. To summarise, mean field theory gives $\beta = \gamma = 1$ and $\delta = 2$, thus verifying Widom's identity.

Exercise 12.4. This exercise is dedicated to the study of site percolation on a two-dimensional triangular lattice by mean of the real space renormalisation technique.

The renormalisation technique is the following: A set of three neighbouring sites at the top of a triangle is replaced by a "super-site" (located in the centre of the triangle) that is occupied if and only if two (or more) of the three initial sites are occupied. This procedure results in a "super-lattice" as shown in Figure 12.9. If the initial lattice has a unit lattice spacing, the super-lattice has a lattice spacing $b = \sqrt{3}$.

1/ Let p' be the occupation probability of a super-site, and p be the occupation probability an initial site. Write p' as a function $p' = f(p)$. Plot this function and deduce the value of the critical probability p_c predicted by the renormalisation procedure. Compare to the exact value.

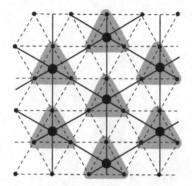

FIGURE 12.9 Illustration of the renormalisation procedure described in the text. The initial lattice is drawn in dotted lines and the super-lattice in solid lines. Each super-site groups the three sites located in the shaded region surrounding it. The possible occupation (or vacancy) of the sites is not represented in the figure.

2/ The critical exponent ν is related to the divergence of the correlation length ξ in the vicinity of the transition. For the initial lattice, $\xi = |p - p_c|^{-\nu}$ and for the super-lattice the same formula gives $\xi = b\,|p' - p_c|^{-\nu}$.

Determine the value of ν resulting from the renormalisation procedure. Compare to the exact result $\nu = 4/3$.

FIGURE 12.10 Configurations for which 3 initial neighbour sites lead to an occupied super-site. The configuration on the left has a probability p^3, the three configurations on the right (with 2 occupied sites) each occur with a probability $p^2(1-p)$.

Solution: 1/ _The renormalisation procedure leads to an occupied super-site in one of the 4 situations illustrated in Figure 12.10. According to this figure, the new probability is $p' = f(p) = p^3 + 3p^2(1-p)$. Function f has 3 fixed point (solutions of $f(p) = p$): $p = 0$, $p = 1$ and $p = 1/2$. The fixed point $p = 1/2$ is the unstable fixed point corresponding to the critical probability according to the analysis described in Section 12.2.3. The renormalisation procedure leads her to the exact result: $p_c = 1/2$._

2/ _To determine the exponent ν, one writes for the initial lattice $\xi = C|p - p_c|^{-\nu}$ and for the super-lattice $\xi = bC|p' - p_c|^{-\nu}$, where C is a constant. Eliminating ξ from these expressions yields:_

$$\nu = \frac{\ln(b)}{\ln \frac{p' - p_c}{p - p_c}}. \tag{12.14}$$

Actually, the relation $\xi \propto |p - p_c|^{-\nu}$ is valid only in the vicinity of p_c. Thus, the limit $p \to p_c$ should be taken in Expression (12.14). This yields

$$\nu = \frac{\ln(b)}{\ln[f'(p_c)]} = \frac{\frac{1}{2}\ln(3)}{\ln(3/2)} = 1.3548\ldots \tag{12.15}$$

which is very close to the exact result $\nu = 1.333\ldots$.

13 Dynamics of Phase Transitions

This chapter focuses on some aspects of the non-equilibrium dynamics of a system in the vicinity of a phase transition: how does a system behave when cooled below the critical temperature or when, at a fixed temperature, it is moved away from equilibrium? A first question deals with the appearance of order in an initially disordered system. A second one concerns the return to equilibrium after a perturbation. In the present chapter, these problems are addressed *via* an equation describing in a phenomenological way the weakly non-equilibrium dynamics of the order parameter characterising the system in the vicinity of a phase transition.

13.1 TIME-DEPENDENT GINZBURG-LANDAU EQUATION

Let us consider a non-equilibrium configuration. The order parameter should depend on time ($\partial_t \phi \neq 0$), until the system relaxes to the equilibrium configuration, solution of Equation (10.2). The quantities $\partial_t \phi$ and $\delta F_L / \delta \phi$ are thus related, in the sense that they must cancel together. In a configuration close to equilibrium, it is sound to assume that these two quantities are proportional so that:

$$\frac{\partial \phi}{\partial t} = -\Gamma \frac{\delta F_L}{\delta \phi} = -\Gamma \left\{ -2g \vec{\nabla}^2 \phi + \frac{\partial f_L}{\partial \phi} \right\}. \qquad (13.1)$$

where f_L is given by Equation (9.41) and Γ is a phenomenological parameter sometimes called the *Khalatnikov parameter*. In Section 13.2.1, it will be seen that Γ^{-1} can be interpreted as a viscosity. Γ must be positive so that the free energy minimum corresponds to a stable equilibrium. Indeed, during an interval of time dt, $\phi(\vec{r}, t)$ varies by $d\phi(\vec{r}, t) = \partial_t \phi\, dt$ and F_L varies by $dF_L = \int d^d r (\delta F_L / \delta \phi)\, d\phi$ (this is the definition of the functional derivative $\delta F_L / \delta \phi$, see Section 10.2). From Equation (13.1), it immediately follows that

$$\frac{dF_L}{dt} = -\frac{1}{\Gamma} \int \left(\frac{\partial \phi}{\partial t} \right)^2 d^d r. \qquad (13.2)$$

If $\Gamma > 0$ then $dF_L / dt < 0$ and the dynamics described by Equation (13.1) reflects the tendency of the system to spontaneously approach the free energy minimum, in agreement with the second law of thermodynamics. However, it should be noted that the mean field equation (13.1) describes a very schematic relaxation to equilibrium. In particular, during its time evolution, the system remains "stuck" in any local minimum of F_L. A possible way to improve the description of the return to equilibrium is to add to Equation (13.1) a random noise term $\eta(\vec{r}, t)$ effectively describing the thermal fluctuations inevitably disturbing the system. It is often assumed that η is an independent random variable in each point of space and time with zero mean $\langle \eta(\vec{r}, t) \rangle = 0$ and a white noise spectrum: $\langle \eta(\vec{r}', t')\eta(\vec{r}, t) \rangle \propto \delta^{(d)}(\vec{r}' - \vec{r})\delta(t' - t)$. Such improvements will not be considered here, and this chapter is limited to the description of the temporal phenomena induced by the effective equation (13.1). The rest of this section aims at familiarising the reader with the phenomenology of this equation through discussions of some of its main physical characteristics.

DOI: 10.1201/9781003272427-13

13.1.1 HOMOGENEOUS OUT-OF-EQUILIBRIUM SYSTEM

A first examination of the physics contained in Equation (13.1) can be carried on using the example of an unrealistic configuration where ϕ is out of equilibrium and thus evolves as a function of time t, while remaining spatially uniform (ϕ does not depend on position). In the limit of small deviations from equilibrium, one can write $\phi(t) = \phi_0 + \delta\phi(t)$. Then, Equation (13.1) is solved by $\delta\phi(t) = \delta\phi(0)\exp\{-t/\tau\}$ where:

$$\tau^{-1} = \Gamma \left.\frac{\partial^2 f_L}{\partial \phi^2}\right|_{\phi_0} = \begin{cases} 2\Gamma a\,(T - T_C) & \text{when} \quad T > T_C, \\ 4\Gamma a\,(T_C - T) & \text{when} \quad T < T_C. \end{cases} \tag{13.3}$$

Fluctuations are therefore damped on a time scale τ which diverges when $T \to T_C$. This is the phenomenon of "critical slowing down": In the vicinity of the critical temperature, the length scale of spatial fluctuation diverges and the corresponding equilibration time therefore become larger and larger. This divergence is characterised by a critical exponent z defined by:

$$\tau \propto \frac{1}{|\epsilon|^{\nu z}}, \tag{13.4}$$

where $\epsilon = (T - T_C)/T_C$ is the reduced temperature[1], and ν is the critical exponent associated with the divergence of the correlation length (see Table 11.1). In the framework of mean field theory, $\nu = 1/2$ and thus Expression (13.3) gives $z = 2$.

13.1.2 WEAKLY INHOMOGENEOUS OUT-OF-EQUILIBRIUM SYSTEM

Consider now a system where the order parameter varies as a function of time and position under the influence of an external field $h(\vec{r}, t)$. The system is assumed close enough to equilibrium and $h(\vec{r}, t)$ is considered small enough to write $\phi(\vec{r}, t) = \phi_0 + \delta\phi(\vec{r}, t)$ and linearise Equation (13.1). This leads to

$$\left(\frac{\partial}{\partial t} - 2g\Gamma\vec{\nabla}^2 + \frac{1}{\tau}\right)\delta\phi(\vec{r}, t) = \Gamma\,h(\vec{r}, t). \tag{13.5}$$

As already done in Sections 10.3.1 and 11.2.1, the susceptibility of the system is also defined here. Since the problem is time-dependent, a dynamical susceptibility $\chi(\vec{r}, t)$ (also called *linear response function*) must be considered. It is solution of:

$$\left(\frac{\partial}{\partial t} - 2g\Gamma\vec{\nabla}^2 + \frac{1}{\tau}\right)\chi(\vec{r}, t) = \Gamma\,\delta^{(d)}(\vec{r})\delta(t). \tag{13.6}$$

It is easy to check that the solution of Equation (13.5) can then be written as:

$$\delta\phi(\vec{r}, t) = \int \mathrm{d}t'\mathrm{d}^d r'\chi(\vec{r} - \vec{r}', t - t')h(\vec{r}', t'). \tag{13.7}$$

It is therefore sufficient to solve Equation (13.6) to determine, with the help of Equation (13.7), the expression of the fluctuations of the order parameter induced by the perturbation $h(\vec{r}, t)$. This is efficiently done by means of Fourier analysis [see Equation (A.36)]. Let $\hat{\chi}$ be the Fourier transform of the susceptibility. Both quantities are related via

$$\chi(\vec{r}, t) = \int \frac{\mathrm{d}^d k}{(2\pi)^d}\frac{\mathrm{d}\omega}{2\pi}\,\hat{\chi}(\vec{k}, \omega)\,\exp\{\mathrm{i}(\vec{k}.\vec{r} - \omega t)\}. \tag{13.8}$$

---

[1] In this chapter, since the letter t is already used for time, the reduced temperature is written as ϵ, in contrast with the convention used in Chapter 11.

The Fourier transform of Equation (13.6) leads to

$$\hat{\chi}(\vec{k},\omega) = \frac{\Gamma}{(\tau_k)^{-1} - \mathrm{i}\,\omega} \quad \text{with} \quad \frac{1}{\tau_k} = \frac{1}{\tau} + 2g\Gamma\vec{k}^2. \tag{13.9}$$

When $k = 0$ and $\omega = 0$ (i.e., for the homogeneous and time-independent component of the response), one gets $\hat{\chi}(\vec{0},0) = \tau\,\Gamma$ in agreement with Equation (9.46).

The computation of Integral (13.8) using Expression (13.9) for $\hat{\chi}(\vec{k},\omega)$ is performed as follows. First, the integral over ω is calculated using integration in the complex plane: the integrand has a simple pole in $\omega = -\mathrm{i}/\tau_k$; when $t > 0$, the integration contour should be closed by a path distant in the lower half-plane (Im $\omega < 0$) in order to be able to use the residue theorem (and then this pole contributes) whereas when $t < 0$, the contour must be closed from above (and then the integral is zero). This leads to:

$$\int_{\mathbb{R}} \frac{\mathrm{d}\omega}{2\pi} \frac{\exp\{-\mathrm{i}\omega t\}}{(\tau_k)^{-1} - \mathrm{i}\,\omega} = \begin{cases} \exp(-t/\tau_k) & \text{when } t > 0, \\ 0 & \text{when } t < 0. \end{cases} \tag{13.10}$$

Expression (13.9) of τ_k shows that the calculation of the integral over $\mathrm{d}^d k$ in Equation (13.8) corresponds to the evaluation of the inverse Fourier transform of a Gaussian function. The integral can be calculated as the product of d independent one dimensional integrals[2]. This yields:

$$\chi(\vec{r},t) = \Theta(t) \times \Gamma e^{-t/\tau} \times \frac{e^{-\vec{r}^2/(8g\Gamma t)}}{(8\pi g\Gamma t)^{d/2}}, \tag{13.11}$$

where Θ is the Heaviside function ($\Theta(t < 0) = 0$ and $\Theta(t > 0) = 1$). In Expression (13.11) for $\chi(\vec{r},t)$, the first right-hand side term (the one involving the Heaviside function) corresponds to causality: an external field induces, according to Equation (13.7), a fluctuation of the order parameter posterior to its cause. The second term describes a damping characterised by a typical time τ. The last term describes diffusion of the fluctuations of the order parameter induced by h: a fluctuation spreads in a time t over a distance $\sim \sqrt{g\Gamma t}$. This type of diffusive behaviour could have been anticipated because Equation (13.1) has roughly the same characteristics as a diffusion equation.

13.2 DOMAIN WALL MOTION

In this section, configurations further away from equilibrium than those studied in Sections 13.1.1 and 13.1.2 are considered. A particular attention is given to the dynamics of a domain wall separating two regions of space. In one of them (the region $x > 0$) the order parameter is mostly equal to $+\phi_0$, whereas in the other one ($x < 0$), it is mostly equal to $-\phi_0$ (see Section 10.3.2).

13.2.1 PLANAR INTERFACE

Let us first consider the simple configuration of a planar interface depicted in Figure 10.3. If, starting from this very inhomogeneous situation, a weak homogeneous magnetic field $h > 0$ is applied throughout the entire space, the system will try to relax to a state where ϕ takes the value $+\phi_0(h)$ everywhere. The interface will therefore start moving, for sure to the left, and one seeks to evaluate the speed v of this motion. The interface is assumed to move without deformation and a solution of the form $\phi(x,t) = \varphi(X = x - vt)$ is expected where v is yet unknown. In the following, we write $\varphi' = \mathrm{d}\varphi/\mathrm{d}X$. According to Equation (13.1), φ is

---

[2]If k_n and r_n are the Cartesian components of $\vec{k}$ and $\vec{r}$, then for $n \in \{1,\dots,d\}$: $\int_{\mathbb{R}} \mathrm{d}k_n \exp\{\mathrm{i}k_n r_n - 2g\Gamma t k_n^2\} = \sqrt{\frac{\pi}{2g\Gamma t}} \exp\{-r_n^2/(8g\Gamma t)\}$ [see Equation (A.30)].

solution of $2g\varphi'' - \partial_\varphi f_L + h = -\frac{v}{\Gamma}\varphi'$. Multiplying this equation by φ' and integrating over $\mathbb{R}$ yields:

$$\left[g(\varphi')^2 - f_L(\varphi) + h\varphi\right]_{-\infty}^{+\infty} = -\frac{v}{\Gamma}\int_\mathbb{R}(\varphi')^2 dX = -\frac{v}{\Gamma}\frac{\sigma}{2g}. \tag{13.12}$$

In the right-hand term of this equation, Expressions (10.20) and (10.21) of the surface tension have been used. This is legitimate since the shape of the interface is assumed to move without deformation. The left-hand term of Equation (13.12) is approximately $2h\phi_0$, so the speed v reads:

$$v = -4\Gamma g\,\phi_0\, h/\sigma. \tag{13.13}$$

This result can be interpreted by considering the interface as a classical particle subjected to two opposite forces: one deriving from an external potential (driven by h) and the other being a viscous force. Indeed, the domain wall is first subjected to a force induced by the external field h that reflects its tendency to move to the left in order to decrease the potential energy $U_{pot} = -\int \phi(x)h d^3r = -hL_yL_z\int\phi(x)dx$ which is the interaction potential energy of ϕ with h (as in Section 10.3.2, L_y and L_z are the spatial dimensions of the sample in the y and z directions, respectively). U_{pot} is a ill-defined quantity[3], but its variation upon a displacement δx_0 of the interface is well defined and reads $\delta U_{pot} = 2\phi_0 h L_y L_z \delta x_0$ (see Figure 13.1). Thus, a force per unit area can be defined:

$$\mathcal{F}_h = -\frac{1}{L_yL_z}\frac{\delta U_{pot}}{\delta x_0} = -2\phi_0 h. \tag{13.14}$$

This magnetic force is constant and directed to the left.

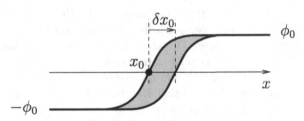

FIGURE 13.1 Sketch illustrating the calculation of the variation of potential energy resulting from a displacement $\delta x_0 > 0$ of the interface. δU_{pot} is equal to the area of the grey region multiplied by a factor hL_yL_z. The interface moving without deformation, the area of the grey region is $2\phi_0 \delta x_0$.

The additional dissipative force accounts for the fact that the system's free energy is not conserved. Using Equation (13.2) in the case of interest (where $\partial_t\phi = -v\varphi'$) and then Equation (13.12) allows to write the rate of dissipation of free energy as the power dissipated by a non-conservative force $\mathcal{F}_{vis}$:

$$\frac{1}{L_yL_z}\frac{dF_L}{dt} = \mathcal{F}_{vis} \times v, \quad \text{where} \quad \mathcal{F}_{vis} = -\frac{\sigma}{2g\Gamma}v. \tag{13.15}$$

Energy dissipation has been written in Equation (13.15) in order to evidence the role of a viscous-type force per unit area: $\mathcal{F}_{vis}$ is indeed proportional to the velocity. It should be noted that the analysis is here carried out by loosely speaking of force: since $\mathcal{F}_h$ and $\mathcal{F}_{vis}$ are forces per unit area, they are in fact pressures.

The interface moving under the effect of two forces of opposite directions, one constant and the other one viscous and opposite to the motion, the stable solution of the equation

---

[3]The integral $\int_\mathbb{R}\phi(x)dx$ does not converge.

of motion corresponds to a displacement at a constant velocity v for which the two forces balance each other exactly:

$$\mathcal{F}_{\text{vis}} + \mathcal{F}_h = 0, \quad \text{or} \quad -2h\phi_0 = \frac{\sigma}{2g\Gamma}v. \tag{13.16}$$

This corresponds indeed to expression (13.13) of the velocity of the interface and legitimates the above description of the domain wall as a classical particle subjected to the forces $\mathcal{F}_h$ and $\mathcal{F}_{\text{vis}}$.

Exercise 13.1. This exercise can be seen as a variation of the foregoing study. As previously, the system is at a temperature $T < T_C$ and $\pm\phi_0$ are the two equilibrium values of the order parameter (for which f_L reaches its minimum). Let us consider a two or three dimensional configuration where the order parameter is initially uniformly equal to ϕ_0 in the entire space except in a sphere of initial radius $R(0)$ inside which the order parameter is $-\phi_0$ (see Figure 13.2).

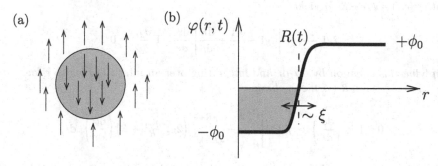

FIGURE 13.2 (a) Schematic representation of the initial state of the system. This describes for example an Ising system where a spherical domain contains spins downs (shaded area) while the remaining space is occupied by spins up. (b) solution of the time-dependent Ginzburg-Landau equation with a spherical symmetry and a rigid interface.

1/ One looks for a solution of Equation (13.2) with spherical symmetry: $\phi(\vec{r},t) = \varphi(r,t)$, with $r = |\vec{r}|$. It is further imposed that $\varphi(r,t)$ be of the form $\varphi(r,t) = \psi(r - R(t))$, where $R(t)$ is the radius of the sphere at time t (see Figure 13.2); i.e. the sphere contracts while always keeping the same interface profile. This is only possible in the limit where $R(t)$ is much larger than the thickness of the interface, which is admitted henceforth. Write the differential equation for which ψ is a solution. Deduce the evolution equation:

$$\frac{dR}{dt} + 2g\Gamma\frac{d-1}{R} = 0. \tag{13.17}$$

Solve this equation. How long does it take for the domain to disappear?

2/ The previous result can be recovered using a macroscopic approach in dimension $d = 3$. To do so, the analysis of Section 13.2.1 will be used. It was shown that a moving wall was experiencing a viscous force per unit area written here as $\mathcal{F}_{\text{vis}} = -\sigma/(2g\Gamma)\,dR/dt$ [compare to Equation (13.15)].

(a) Determine the expression of the surface tension force per unit area $\mathcal{F}_{\text{tens}}$ knowing that the potential energy associated with the surface tension is $U_{\text{tens}} = 4\pi R^2\sigma$.

(b) After a short period of time, the resorption speed of the sphere tends to a constant. In this case the acceleration is zero and the two forces $\mathcal{F}_{\text{vis}}$ and $\mathcal{F}_{\text{tens}}$ balance each other. Show that it leads to Equation (13.17).

3/ A slightly modified configuration is now considered. The system is still in dimension $d = 3$, but the precipitate (i.e., the sphere inside which the order parameter is $-\phi_0$) now evolves in the presence of a uniform external field $h(\vec{r}, t) = -h_0$ ($h_0 > 0$) which favours "down" spins.

Using the macroscopic approach of the previous question, write the equilibrium between $\mathcal{F}_{\text{vis}}$, $\mathcal{F}_{\text{tens}}$ and $\mathcal{F}_h$, the magnetic force per unit area (give the expression of $\mathcal{F}_h$). Show that there exists a critical radius R_{crit} beyond which the sphere does not collapse, but grows.

Solution : *1/ Using tensor notations with implicit summation over repeated indices, the Laplacian of $\phi = \varphi(r, t)$ is written (x_1, x_2 and x_3 are the Cartesian coordinated of $\vec{r}$ and $\partial_i = \partial/\partial x_i$):*

$$\vec{\nabla}^2 \phi = \partial_i \partial_i \varphi = \partial_i (\frac{\partial_r \varphi}{r}) x_i + \frac{\partial_r \varphi}{r} \partial_i x_i = \partial_r (\frac{\partial_r \varphi}{r}) \frac{x_i x_i}{r} + \frac{d}{r} \partial_r \varphi = \partial_r^2 \varphi + \frac{d-1}{r} \partial_r \varphi,$$

where $d = 2$ or 3 is the dimension of space. Carrying over into the time-dependent Ginzburg-Landau (13.1) with $\varphi(r, t) = \psi(r - R(t))$ yields

$$2g\Gamma \frac{d^2\psi}{dr^2} + \left[2g\Gamma \frac{d-1}{r} + \frac{dR}{dt} \right] \frac{d\psi}{dr} - \Gamma \frac{\partial f_L}{\partial \psi} = 0. \qquad (13.18)$$

By multiplying this equation by $d\psi/dr$ and integrating over an interval of width e located around $R(t)$ (i.e., for $R - \frac{e}{2} \leq r \leq R + \frac{e}{2}$), one obtains:

$$0 = \Gamma \left[g \left(\frac{d\psi}{dr} \right)^2 - f_L(\psi) \right]_{R-\frac{e}{2}}^{R+\frac{e}{2}} + \int_{R-\frac{e}{2}}^{R+\frac{e}{2}} \left(2g\Gamma \frac{d-1}{R} + \dot{R} \right) \left(\frac{d\psi}{dr} \right)^2 dr. \qquad (13.19)$$

If the value of e is chosen much larger than the thickness of the interface (i.e., $e \gg \xi$) but still small with respect to R, then, in Equation (13.19), (i) the integrated term is equal to zero and (ii) the factor $(2g\Gamma \frac{d-1}{R} + \dot{R})$ can be taken out of the integral. The remaining integral being strictly positive, this leads directly to Equation (13.17). The solution of this equation is $R^2(t) = R^2(0) - 4g\Gamma(d-1)t$. The sphere disappears in a time proportional to the square of its initial radius. As in Section 13.1.2, a diffusive result is obtained (which could have been anticipated).

2/ In three dimensions, when R varies by dR, U_{tens} varies by $dU_{\text{tens}} = 8\pi R\sigma dR$ and according to the definition of the force per unit area $dU_{\text{tens}} = -4\pi R^2 \mathcal{F}_{\text{tens}} dR$ where

$$\mathcal{F}_{\text{tens}} = -2\sigma/R. \qquad (13.20)$$

The sign corresponds to a centripetal force. A force per unit area is a pressure, and Equation (13.20) is the expression of the Laplace pressure experienced by a spherical drop[4].

The physics of relaxation contained in the time-dependent Ginzburg-Landau Equation (13.1) corresponds to a viscous dynamics where the Laplace pressure and the viscous force per unit area [of the type Equation (13.15)] are in equilibrium: indeed, the relation $\mathcal{F}_{\text{tens}} + \mathcal{F}_{\text{vis}} = 0$ results in Equation (13.17) which fixes the precipitate resorption dynamics.

3/ In the presence of an external magnetic field, an additional pressure $\mathcal{F}_h = -2h\phi_0$ appears [see Equation (13.14)] and the evolution equation becomes here ($h = -h_0$):

$$\frac{\sigma}{2g\Gamma} \dot{R} = \frac{2\sigma}{R} - 2h_0\phi_0. \qquad (13.21)$$

Beyond a critical radius $R_{\text{crit}} = \sigma/(h_0\phi_0)$ the magnetic force favouring the expansion of the precipitate exceeds the Laplace pressure and the sphere radius grows ($\dot{R} > 0$).

---

[4]see for example the website `https://en.wikipedia.org/wiki/Young-Laplace_equation`

13.2.2 INTERFACE OF ARBITRARY SHAPE, ALLEN-CAHN EQUATION

Here, a situation without external field ($h = 0$) is considered. It will be seen that the boundaries between domains deform in order to decrease the total surface area of the interfaces. This enables the system to relax by decreasing the free energy associated with surface tension. However, the equation that governs the dynamics of the interfaces does not explicitly involve surface tension, which came as a surprise when Allen and Cahn proposed their theory in 1975: the surfaces decrease by attacking the regions of high curvature.

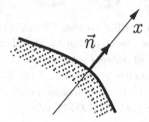

FIGURE 13.3 Local geometry of an interface. The normal vector $\vec{n}$ is oriented along the direction away the centre of curvature. The x-coordinate is the local normal coordinate.

Consider an interface between two domains. An x-axis can be defined locally along the normal vector $\vec{n}$ to the interface pointing in the direction opposite to the domain in which the centre of curvature is located (i.e. away from the centre of curvature, see Figure 13.3). In the vicinity of the interface, the order parameter ϕ varies rapidly in the x-direction so that the gradient of ϕ can be approximated by:

$$\vec{\nabla}\phi \simeq \partial_x\phi \, \vec{n}. \tag{13.22}$$

Also,

$$\vec{\nabla}^2\phi = \vec{\nabla} \cdot \left(\vec{\nabla}\phi\right) \simeq \partial_x^2\phi + \partial_x\phi \, \vec{\nabla}.\vec{n}. \tag{13.23}$$

The interface moves locally along the axis defined by $\vec{n}$ and, as in Section 13.2.1, the evolution of ϕ follows:

$$\partial_t\phi = -v \, \partial_x\phi, \tag{13.24}$$

where v is the *local* speed of deformation of the interface. Equation (13.1) can thus be written here as:

$$-v \, \partial_x\phi = \Gamma \left\{ 2g \, \partial_x^2\phi + 2g \, \partial_x\phi \, \vec{\nabla} \cdot \vec{n} - \partial f_L/\partial\phi \right\}. \tag{13.25}$$

Multiplying Equation (13.25) by $\partial_x\phi$ and then integrating[5] over an interval $[-e, e]$ where e is a length much larger that the interface thickness ξ and small with respect to the typical length scale over with the interface deforms (i.e., the typical radius of curvature), the Allen-Cahn equation is obtained:

$$v = -2g \, \Gamma \vec{\nabla} \cdot \vec{n} = -2g \, \Gamma K, \quad \text{where} \quad K \equiv \vec{\nabla} \cdot \vec{n}. \tag{13.26}$$

In two dimensions, $K = 1/R$ where R is the radius of curvature, and in three dimensions, $K = 1/R_1 + 1/R_2$ (R_1 and R_2 being the principal radii of curvature)[6]. The rate of deformation of the interface is therefore proportional to its curvature. Results of numerical simulations for the evolution of domains obeying the Allen-Cahn equation are shown in Figure 13.4.

---

[5]This procedure is identical to the one detailed in the solution of Exercise 13.1 [see Equation (13.19)].

[6]see for example the wikipedia pages https://en.wikipedia.org/wiki/Plane_curve for the two-dimensional version and https://en.wikipedia.org/wiki/Differential_geometry_of_surfaces for the three-dimensional version. To check these relations on an example, it is enough to consider the case of the sphere in $\mathbb{R}^d$ centred in O and of radius r. In this case, $\vec{n} = \vec{r}/r$ and $\vec{\nabla} \cdot \vec{n} = (d-1)/r$.

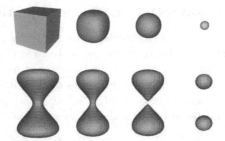

FIGURE 13.4 Domains evolving according to the Allen-Cahn equation (13.26) in a 3 dimensional space. The upper row shows the evolution of an initially cubic domain. The lower row shows the evolution of a domain of more exotic shape. It can be seen that the regions of strong curvature disappear faster (figure reproduced from M. Brassel, *Instabilités de forme en croissance cristalline* (PhD thesis, Université Joseph Fourier (Grenoble I), 2008), available at http://tel.archives-ouvertes.fr/tel-00379392/en/).

In the rest of this section, the space dimension is explicitly set to $d = 3$. The area $\mathcal{A}$ of a surface which undergoes at each of its points a normal displacement $dx = v dt$ is modified. The length elements $d\ell_1$ and $d\ell_2$ contained in the planes of the principal sections of the surface are subject to the increases $d\ell_1 dx/R_1$ and $d\ell_2 dx/R_2$, as illustrated in Figure 13.5. After displacement, the area of the surface element which was initially $d^2a = d\ell_1 d\ell_2$ is:

$$d\ell_1 \left(1 + \frac{dx}{R_1}\right) d\ell_2 \left(1 + \frac{dx}{R_2}\right) \simeq d^2a \left(1 + dx \left(\frac{1}{R_1} + \frac{1}{R_2}\right)\right). \tag{13.27}$$

Area $\mathcal{A}$ is is thus modified by a quantity

$$d\mathcal{A} = \int d^2a \, dx \left(\frac{1}{R_1} + \frac{1}{R_2}\right) = \int d^2a \, dx \, K. \tag{13.28}$$

In Equation (13.28), d^2a is the integration element along the surface and no integration over dx is done (but the displacement dx should not be taken out of the integral because it depends on the integration point). Since $dx = v \, dt$, Equation (13.26) allows the evaluation of the decrease of the interface area under the action of the Allen-Cahn dynamics:

$$\frac{d\mathcal{A}}{dt} = -2g\Gamma \int d^2a \, K^2 = -2g\Gamma \langle K^2 \rangle \mathcal{A}, \tag{13.29}$$

where $\langle K^2 \rangle$ is the average value of K^2, defined by the last equality in Equation (13.29).

The variation of free energy can also be calculated thanks to Formula (13.2). In the volume integral of Equation (13.2), two zones should be distinguished: one for the interior of the domains where the order parameter is constant and another for the volumes that contain the inhomogeneous regions where the interfaces are located. The first zone does not contribute to dissipation (since $\partial_x \phi = 0$ inside this region). A volume element of the second zone is written as $d^3r = d^2a \, dx$ where d^2a is the area element of the surface and x is the coordinate normal to the surface. The integral on variable x is performed over an interval $[-e, +e]$, where e is a length large with respect to the thickness ξ of the interface and small with respect to its typical radius of curvature. The time derivative of the free energy [Equation (13.2)] can be calculated using Equations (13.24) and (13.26):

$$\frac{dF_L}{dt} = -\frac{1}{\Gamma} \int_{-e}^{+e} dx \int d^2a \, v^2 (\partial_x \phi)^2 \simeq -4g^2 \Gamma \int_{-\infty}^{+\infty} dx \, (\partial_x \phi)^2 \int d^2a \, K^2. \tag{13.30}$$

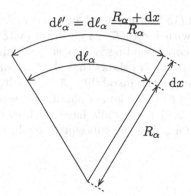

FIGURE 13.5 Increase of the length $d\ell_\alpha$ during a normal displacement dx of the surface in the principal section α ($\alpha \in \{1,2\}$). R_α is one of the two principal radii of curvature. The arc length after the increase is $d\ell'_\alpha$.

Relation (13.29) and Expression (10.20) of the surface tension eventually gives:

$$\frac{dF_L}{dt} = \sigma \frac{d\mathcal{A}}{dt}. \tag{13.31}$$

This relation shows that dissipation, i.e. the decay rate of of the free energy, is effectively obtained through a decrease of the area of the interfaces. This decrease is driven by the preferential alteration of the high curvature zones as specified by the Allen-Cahn equation (13.26).

13.3   QUENCH DYNAMICS

Consider a system above critical temperature, with $T > T_C$, in the absence of external field ($h = 0$). At time $t = 0$, the temperature is suddenly changed and takes a value $T < T_C$, fixed during the remaining evolution of the system. We aim here to describe the dynamics of the system after this quench. A simple guess would be to assume that the system will relax towards a homogeneous configuration in which ϕ randomly takes either the uniform value $+\phi_0$ or the uniform value $-\phi_0$. It will be seen that this simple scenario is incorrect and that an infinitely extended system never reaches a stationary state described by one of these two homogeneous configurations.

13.3.1   SHORT TIMES

Initially $\phi(\vec{r}, t = 0) = 0$, which at $T < T_C$ places the system in the unstable region of F_L (see Figure 9.9). For short times, it can be assumed that $\phi(\vec{r}, t)$ remains small. Therefore, the evolution equation (13.1) can be linearised around the initial value $\phi = 0$ giving:

$$\partial_t \phi = \Gamma \left[2g \vec{\nabla}^2 \phi + 2a(T_C - T)\phi \right]. \tag{13.32}$$

Performing a spatial Fourier decomposition: $\phi(\vec{r}, t) = \frac{1}{(2\pi)^d} \int d^d k \, \hat{\phi}(\vec{k}, t) \exp\{i\vec{k} \cdot \vec{r}\}$, it is clear that

$$\hat{\phi}(\vec{k}, t) = \hat{\phi}(\vec{k}, 0) \exp\{\gamma_k t\}, \tag{13.33}$$

with

$$\gamma_k = 2\Gamma \left[-gk^2 + a(T_C - T) \right] = \frac{1 - 2k^2\xi^2}{2\tau}, \tag{13.34}$$

where Expressions (10.13) and (13.3) where used for ξ and τ. According to Equations (13.33) and (13.34), the fluctuations with $k < (a(T_C - T)/g)^{1/2} = (\sqrt{2}\,\xi)^{-1}$ are unstable and diverge exponentially with t since the corresponding γ_k is positive. This perturbative analysis (valid for short times: $t < t_{\text{typ}} \sim (\gamma_{k=0})^{-1} = 2\tau$) shows that the long wave-length modes ($\lambda \gtrsim \xi$) are unstable: the larger λ, the greater the instability. As a result, domains form, with typical spatial extent much larger than ξ. These large domains are separated by well-defined walls, since, as seen in Section 10.3.2, domain walls have thickness $\sim \xi$. Within a domain, the order parameter is nearly uniform and the subsequent evolution of the system is governed by wall dynamics.

13.3.2 LONG TIME

It was just seen that after the initial instability has developed (just after quenching), the system is structured into domains whose subsequent evolution is governed by the dynamics of domain walls. This dynamics is driven by the Allen-Cahn equation (13.26).

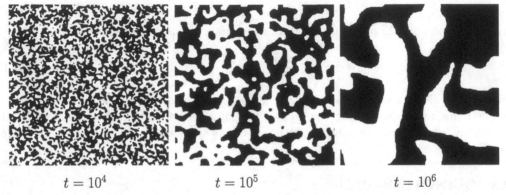

$$t = 10^4 \qquad\qquad t = 10^5 \qquad\qquad t = 10^6$$

FIGURE 13.6 Quenching of a two-dimensional Ising system with 512×512 sites. The system's temperature is initially above T_C. At $t = 0$, it is abruptly changed to $T < T_C$. The initial spin distribution corresponds to a completely disordered configuration. The 3 figures represent the magnetic domains for 3 post-quench times. The "up"("down") spin domains are shown in white (black). From A. J. Bray, S. N. Majumdar, G. Schehr, Adv. Phys. **62**, 225 (2013). Copyright Taylor and Francis 2013, reprinted with permission.

A simple analysis of the orders of magnitude in the Allen-Cahn equation (13.26) shows that an isolated domain of typical size L disappears in a typical time $\propto L^2$. This is what is observed, for example, in the schematic case of the circular precipitate studied in Exercise 13.1. It is therefore legitimate to assume that at a time t after a quench, only domains of typical size greater or equal to $t^{1/2}$ (in dimensionless units) remain. This is illustrated in Figure 13.6 by the quench of a two-dimensional Ising system.

A more quantitative description of the phenomenon can be obtained: as the size of the domains increases, the total number of interfaces decreases, and so does the average curvature, as exemplified in Figure 13.6. It is legitimate to assume that the domain structure undergoes an approximately self-similar change, i.e. it evolves while obeying a scaling law. In three dimensions, this implies that the surface is proportional to the average curvature $\langle K^2 \rangle$ defined by Equation (13.29). This result is not immediately intuitive but easy to confirm. Consider the simple case of the two-dimensional system shown in Figure 13.7. When the size of the domains is multiplied by a factor α ($\alpha = 2$ in Figure 13.7), their typical radius of curvature also changes: $R \to R' = \alpha R$, their number becomes $N' = N/\alpha^2$ and the total length L of the interfaces in the sample being proportional to NR becomes: $L \to L' = L/\alpha$.

In two dimensions, it can thus be seen that, for a sample of fixed size, the growth of the domains preserves the product LR. The same analysis in three dimensions (recommended as an exercise) shows that it is then the product $\mathcal{A}R$ that is conserved, where $\mathcal{A}$ is the total area of the interfaces in the system. As $\mathcal{A}$ is calculated for a sample of fixed volume, it is more appropriate to define $\mathcal{A}_v$, the area of domain walls per unit volume ($\mathcal{A}_v$ has the dimension of the inverse of a length) and then:

$$\langle K^2 \rangle = f\,\mathcal{A}_v^2, \tag{13.35}$$

where f is a dimensionless constant.

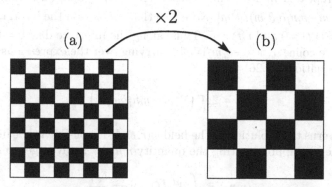

FIGURE 13.7 Schematic of the self-similar evolution of domains. During the evolution from structure (a) to structure (b), the area of the sample does not change, the size of the domains is increased by a factor of 2, their area by a factor of 4 and the total length of the interfaces is divided by 2.

Carrying over into Equation (13.29) gives $\dot{\mathcal{A}}_v = -2g\Gamma f\mathcal{A}_v^3$ leading to

$$\frac{1}{\mathcal{A}_v^2(t)} - \frac{1}{\mathcal{A}_v^2(0)} = 4\,g\,\Gamma f\,t. \tag{13.36}$$

This time dependence has been tested in a Fe-Al alloy. This system undergoes a second order phase transition between a disordered phase at high temperature and an ordered phase in which each of the two types of atoms forms a centred cubic lattice with an atom of one type at the centre of a cube whose vertices are occupied by an atom of the other type. Ordered domains appear when quenching through this transition. Domains with different lattice orientations are separated by domain walls. Allen and Cahn[7] were able to verify Relation (13.36): the initial contribution $1/\mathcal{A}_v^2(0)$ can legitimately be ignored in Equation (13.36) and then the product $\mathcal{A}_v^{-2}t^{-1}$ must be time independent. After a quench of an alloy containing 24% aluminum from a temperature $T = 1050$ K down to $T = 898$ K (while $T_C = 990$ K), at times $t = 6$ min and $t = 1$ hour, they obtained the same value $\mathcal{A}_v^{-2}t^{-1} = 1.1 \times 10^{-16}$ $m^2 \cdot s^{-1}$. They also verified that their data were compatible with Law (13.26) where the velocity does not depend on the surface tension σ (as one might have expected, see discussion at the beginning of Section 13.2.2).

13.3.3   O.J.K. THEORY

The approach proposed by Ohta, Jasnow, and Kawasaki[8] corresponds to a quantitative study of the correlation function of the order parameter during the domain growth as illustrated in Figure 13.6. Since almost everywhere in space $\phi(\vec{r},t) = \pm\phi_0$ except in the

[7]S.M. Allen and J.W. Cahn, Acta Metall. **27**, 1085 (1979).
[8]T. Ohta, D. Jasnow and K. Kawasaki, Phys. Rev. Lett. **49**, 1223 (1982).

vicinity of the interfaces, it can be postulated for simplicity that there exists an auxiliary field $m(\vec{r},t)$ that varies slowly in space such that $\phi(\vec{r},t) = \phi_0 \, \mathrm{sgn}[m(\vec{r},t)]$. The interfaces correspond to the manifolds defined by $m(\vec{r},t) = 0$. Then, from the Allen-Cahn equation (13.26), it can be shown (*modulo* some approximations) that the field m obeys a diffusion equation:

$$\partial_t m = D \vec{\nabla}^2 m, \quad \text{with} \quad D = 2 g \, \Gamma \, \frac{d-1}{d}, \tag{13.37}$$

where d is the dimension of space. Let us prove this result: Allen-Cahn equation (13.26) can be rewritten as a function of $m(\vec{r},t)$ since[9] $\vec{n} = \vec{\nabla}m/|\vec{\nabla}m|$. Using tensor notations with implicit summation over repeated indices, this relation reads: $n_i = \partial_i m / \sqrt{\partial_j m \partial_j m}$. Then, one gets $\vec{\nabla} \cdot \vec{n} = \partial_i n_i = (\partial_i \partial_i m - n_i n_j \partial_i \partial_j m)/|\vec{\nabla}m|$. Note also that a point on the interface has a position $\vec{r}(t)$ such that $m(\vec{r}(t),t) = 0$ and $\dot{\vec{r}}(t) = \vec{v}$. Thus, along the interface $\mathrm{d}m/\mathrm{d}t = 0 = \vec{v} \cdot \vec{\nabla}m + \partial_t m$. Since $\vec{v}$ and $\vec{\nabla}m$ are colinear, $v = -\partial_t m/|\vec{\nabla}m|$. Carrying over the expressions of v and $\vec{\nabla} \cdot \vec{n}$ in the Allen-Cahn equation (13.26) yields:

$$\partial_t m = 2g\Gamma \left(\vec{\nabla}^2 m - n_i n_j \partial_i \partial_j m \right). \tag{13.38}$$

This equation governs the evolution of the field $m(\vec{r},t)$. It is non-linear because $\vec{n} = \vec{\nabla}m/|\vec{\nabla}m|$. It can be simplified by approximating the quantity $n_i n_j$ by its average over solid angles[10]:

$$\langle n_i n_j \rangle = \int \mathrm{d}^{d-1}\Omega \, n_i n_j = \frac{\delta_{i,j}}{d}, \tag{13.39}$$

Thus, $n_i n_j \partial_i \partial_j m$ is replaced by the expression $\langle n_i n_j \rangle \partial_i \partial_j m = \frac{1}{d}\vec{\nabla}^2 m$: this approximation carried over into (13.38) leads directly to Equation (13.37).

The evolution equation (13.37) has been established for the field $m(\vec{r},t)$ without specifying its initial state $m(\vec{r},0)$. Since the initial temperature of the system is higher than T_C, in view of the short time instability described in Section 13.3.1, it is legitimate to assume that m is initially disordered with $\langle m(\vec{r},0) \rangle = 0$. To simplify the analysis, the size of the domains appearing just after the quench are ignored and it is assumed that, initially, the field m is an uncorrelated white noise: $\langle m(\vec{r}_1,0) \cdot m(\vec{r}_2,0) \rangle = \Delta \, \delta^{(d)}(\vec{r}_1 - \vec{r}_2)$, where Δ is a parameter whose value is irrelevant in the following. Then, the solution of Equation (13.37) for $t > 0$ can be written as:

$$m(\vec{r},t) = \int_{\mathbb{R}^d} \mathrm{d}^d r' \, K(\vec{r} - \vec{r}',t) \, m(\vec{r}',0) \quad \text{with} \quad K(\vec{R},t) = \frac{\exp\left[-\frac{R^2}{4Dt}\right]}{(4\pi Dt)^{d/2}}. \tag{13.40}$$

$K(\vec{R},t)$ is the propagator of the diffusion equation (13.37). Formula (13.40) can be verified by carrying over the resulting value of $m(\vec{r},t)$ into Equation (13.37). Then Equation (13.40) leads directly to

$$\langle m(\vec{x},t) \cdot m(\vec{y},t) \rangle = \Delta \int_{\mathbb{R}^d} \mathrm{d}^d r \, K(\vec{r},t) K(\vec{y} - \vec{x} + \vec{r},t)$$

$$= \frac{\Delta}{(8\pi Dt)^{d/2}} \exp\left\{ -\frac{|\vec{y} - \vec{x}|^2}{8Dt} \right\}. \tag{13.41}$$

---

[9]Remember that the gradient of a scalar field (here $\vec{\nabla}m$) is perpendicular to the "iso-field" surfaces, here defined by $m(\vec{r},t) = 0$.

[10]Equation (13.39) is easily obtained. A bit of thinking or a direct calculation shows that $\langle n_i n_j \rangle$ is a diagonal matrix. Space isotropy imposes that this matrix be proportional to identity. Since $\vec{n}$ is a unit vector, the trace $\langle n_i n_i \rangle$ is 1, from which (13.39) follows.

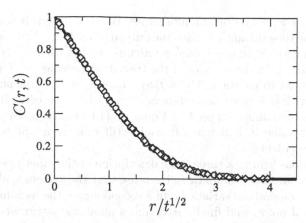

FIGURE 13.8 Comparison of the numerically determined correlation function with the result of the O.J.K. theory. The diamonds and white circles correspond to the results of numerical simulations for two different times; the continuous line corresponds to Formula (13.43). Figure adapted from K. Humayun and A. J. Bray, Phys. Rev. B **46**, 10594 (1992). The scattering coefficient D has been chosen to fit the numerical data.

The calculation of the value of the correlation function of the field m is only a first step towards the calculation of the correlation function of the physical field $\phi(\vec{r},t)$. The actual quantity of interest is

$$C(\vec{r},t) = \frac{1}{|\phi_0|^2} \langle \phi(\vec{x}+\vec{r},t) \cdot \phi(\vec{x},t) \rangle = \left\langle \mathrm{sgn}\left[m(\vec{x}+\vec{r},t)\right] \times \mathrm{sgn}\left[m(\vec{x},t)\right] \right\rangle. \qquad (13.42)$$

After a long calculation, one gets:

$$C(\vec{r},t) = \frac{2}{\pi} \arcsin\left[\exp\left\{-\frac{r^2}{8Dt}\right\}\right]. \qquad (13.43)$$

Expression (13.43) is in very good agreement with numerical simulations for the quenching of the Ising system, as can be seen in Figure 13.8.

13.4 KIBBLE-ZUREK MECHANISM

Section 13.3 discusses the kinetics of the appearance of ordered domains when a system is cooled very rapidly below its critical temperature. It is shown here that, for a second-order phase transition, any cooling is abrupt, in the sense that it takes place on time scales always shorter than the relaxation time scale. This is simply because the relaxation time diverges at the critical point, as seen in Relations (13.3) and (13.4). Any cooling below T_C thus shares some characteristics of a quench. In particular, the relaxation to the ordered state is not spatially uniform and leads to the formation of domain walls (or vortices in the case of a complex order parameter) whose number carries information about the characteristic cooling time and the behaviour of the equilibration time τ in the vicinity of T_C. This unavoidable formation of such "defects" is called the *Kibble-Zurek mechanism*[11].

Consider a configuration in which the temperature evolves with time as:

$$T(t) = T_C(1 - \epsilon(t)). \qquad (13.44)$$

---

[11] T. W. B. Kibble, J. Phys. A **9**, 1387 (1976); W. H. Zurek Nature **317**, 505 (1985).

In Equation (13.44), ϵ is the reduced temperature. By convention, it cancels at time $t = 0$ which is therefore the time instant at which the critical point is reached. Far from the critical point $\epsilon(t) = O(1)$ and the relaxation time τ remains small with respect to the time until the transition is reached. In the vicinity of the transition (reached at $T = T_C$), the reduced temperature is assumed to cancel as $\epsilon(t) = t/t_Q$ where t_Q is the "quenching time". In the vicinity of T_C, the critical slowing down (see Section 13.1.1) prevents the order parameter from adapting to the dynamics imposed by Equation (13.44): the order parameter remains "frozen" at the last value it had when it could still quickly adapt to the temperature constraint [Equation (13.44)].

These considerations lead to a three stage description of the kinetics of the transition: in the first stage, the dynamics is adiabatic in the sense that the system is able to adapt to the externally imposed temperature variations. In a second stage, the dynamics is frozen in the vicinity of the critical point, and finally it becomes adiabatic again when T is sufficiently lower than T_C. These three regimes correspond to the time intervals $(-\infty, -\hat{t}]$, $[-\hat{t}, \hat{t}]$ and $[\hat{t}, +\infty)$. The boundary $\hat{t}$ between the adiabatic and frozen periods can be estimated as the moment at which the relaxation time is equal to the time span until (or since) the transition: $\hat{t}$ is therefore a solution of $\tau(T(t)) \simeq t$. Writing:

$$\tau = \frac{\tau_0}{|\epsilon|^{\nu z}}, \tag{13.45}$$

[see Equation (13.4)] yields:

$$\hat{t} = (\tau_0 \, t_Q^{\nu z})^{1/(1+\nu z)}. \tag{13.46}$$

$\hat{t}$ is called the *freeze-out time*. According to Formula (13.45), the parameter τ_0 in Equation (13.46) is the value of τ at $T = 0$ K (when $\epsilon = 1$). However, Formula (13.45) is only valid when ϵ is close to 0, and it is better to consider τ_0 as a typical order of magnitude of the relaxation time (besides, strictly speaking, its value should depend on the sign of ϵ, see the discussion in Sections 11.5 and 11.6.1). The same remark applies to the parameter ξ_0 appearing in Equation (13.47) below.

Kibble-Zurek scenario sets the average domain size as equal to the value $\hat{\xi}$ of the relaxation distance at the freezing time, i.e., $\hat{\xi} = \xi(T(\hat{t}))$. Writing $\xi(\epsilon) = \xi_0 |\epsilon|^{-\nu}$ (see Table 11.1) yields

$$\hat{\xi} = \xi_0 \left(\frac{t_Q}{\tau_0}\right)^{\nu/(1+\nu z)}. \tag{13.47}$$

Estimate (13.47) of the domain size is often presented in terms of an estimate of the defects density. In the present case with a scalar order parameter, the defects are domain walls: hyper-surfaces of dimension $d - 1$ (where d is the dimension of space). Accordingly the density of domain walls is:

$$n \sim \frac{\hat{\xi}^{d-1}}{\hat{\xi}^d} = \frac{1}{\xi_0} \left(\frac{\tau_0}{t_Q}\right)^{\nu/(1+\nu z)}. \tag{13.48}$$

In one dimension, the density n corresponds to a number (of defects) per unit length, in dimension two to a length (of domain boundary) per unit area, etc.

Note that estimate (13.48) does not provide a precise value, but rather an order of magnitude of the density of defects. However, the power law behaviour must be experimentally observable. This requires the measurement of the number of defects for a given quench time t_Q and the repeat of the experiment for different values of t_Q. At the time of writing, numerical simulations confirm very well the validity of the Kibble-Zurek mechanism, and the power law behaviours [Equations (13.47) and (13.48)]. The comparison with experiments is, however, more delicate, because it is often difficult to vary t_Q over a large enough range to test the power law behaviour, and also because in many experiments the system is inhomogeneous and the theory must be enriched to be able to account for this aspect. However, it can be reasonably argued that several experimental studies obtained results consistent with the Kibble-Zurek mechanism[12].

13.5 DISAPPEARANCE OF A PHASE

Sections 13.3 and 13.4 considered a more or less sudden transition from a temperature $T_{\text{init}} > T_C$ to $T < T_C$ (at $h = 0$). In Figure 13.9, this corresponds to a transition schematically represented by the dashed white horizontal arrow: a quench through a second order phase transition. This section is devoted to the study of the dynamics of a first-order phase transition. In the framework of Landau mean field theory, this phenomenon is described by a transition (at fixed $T < T_C$) under the effect of an external field (see the discussion in Section 9.2.4). It is represented by the vertical dashed arrow in Figure 13.9.

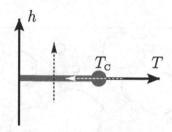

FIGURE 13.9 Phase diagram in the (T, h) plane. The thick line segment is the first order phase transition line located in $h = 0$ between $T = 0$ and $T = T_C$. It ends with the critical point in $T = T_C$ (see Figure 9.14). The quench considered in Section 13.3 is represented by the horizontal white dashed arrow. The type of first-order transition of interest here is represented by the vertical black dashed arrow.

Consider a generic first-order transition from an α-phase to a β-phase. In d dimensions, let I_d be the rate of appearance of new nucleation centres per unit "volume" and time (dimensions $[I_d] = L^{-d}T^{-1}$). These centres give rise to spherical domains of β-phase whose radius grows at a velocity v, assumed constant. In three dimensions, the KJMA theory (Kolmogorov 1937, Johnson and Mehl 1939, Avrami 1939) predicts that the fraction of volume occupied by the β-phase is given by

$$\frac{V_\beta}{V} = 1 - \exp\left\{-\frac{\pi}{3} v^3 I_3 t^4\right\}. \tag{13.49}$$

This result can be obtained as follows. During an interval $[t', t' + dt']$, $dn_\beta(t')$, new β-phase nucleation centres are created in the α-phase with:

$$dn_\beta(t') = V_\alpha(t')I_3 \, dt' = \left[V - V_\beta(t')\right]I_3 \, dt', \tag{13.50}$$

where $V_\alpha(t')$ [respectively $V_\beta(t')$] is the part of the total volume V occupied by the α-phase [respectively β] at time t'. Since a nucleation centre of β-phase cannot appear in the volume already occupied by this phase, Expression (13.50) takes into account only the nucleation centres appearing in the volume occupied by the α-phase: it is the reason why, in Equation (13.50), the multiplicative factor is not the total volume V but $V_\alpha(t')$.

[12]See for example in the framework of Bose–Einstein condensation: G. Lamporesi, S. Donadello, S. Serafini, F. Dalfovo, G. Ferrari, Nat. Phys. **9**, 656 (2013); N. Navon, A. L. Gaunt, R. P. Smith, Z. Hadzibabic, Science **347**, 167 (2015); L. Chomaz, L. Corman, T. Bienaimé, R. Desbuquois, C. Weitenberg, S. Nascimbène, J. Beugnon, J. Dalibard, Nature Comm. **6**, 6162 (2015).

Then, between times t and $t + \mathrm{d}t$, each bubble nucleated around a centre created at a previous time t' sees its volume increased by an amount $\mathrm{d}V = 4\pi R^2 v\, \mathrm{d}t$, where $R = v(t - t')$ is the bubble's radius at time t. Thus, in total, the volume V_β increases from V_β to $V_\beta + \mathrm{d}V_\beta$ with

$$\mathrm{d}V_\beta = \int_0^t \mathrm{d}n_\beta(t')\, 4\pi[v(t - t')]^2 v\, \mathrm{d}t, \tag{13.51}$$

or, using Equation (13.50),

$$\frac{\mathrm{d}V_\beta}{\mathrm{d}t} = 4\pi v^3 I_3 \int_0^t \mathrm{d}t' \left[V - V_\beta(t')\right](t - t')^2. \tag{13.52}$$

Equation (13.52) is an integro-differential equation fixing the evolution of $V_\beta(t)$. It is quite complicated although some of the possible domain overlaps were ignored: they are not considered in Equation (13.51), but taken into account in the expression of $\mathrm{d}n_\beta$ which is proportional to $V_\alpha(t')$.

FIGURE 13.10 Schematic representation of the growth of β-phase domains (white) in a system initially in α-phase (grey).

The following trick can be used to obtain a more precise formula: a volume V_β^* is defined by completely neglecting all overlaps. Then, one obtains $\mathrm{d}n_\beta^* = V I_3\, \mathrm{d}t'$ instead of Equation (13.50), i.e. even the nucleation centres appearing within domains occupied by β-phase are taken into account although they do not increase V_β. Then, one writes:

$$\frac{\mathrm{d}V_\beta^*}{\mathrm{d}t} = \int_0^t \mathrm{d}n^*\, 4\pi[v(t - t')]^2 v = \frac{4}{3}\pi(vt)^3 I_3\, V. \tag{13.53}$$

Then, the assumption $\mathrm{d}V_\beta = \mathrm{d}V_\beta^*(V - V_\beta)/V$ is made (certainly correct in the two limiting cases $V_\beta \to 0$ and $V_\beta \to V$). This allows the integration of Equation (13.53) and leads directly to the Equation (13.49).

Figure 13.11 [adapted from Y. Yamada *et al.*, Phys. Rev. Lett. **53**, 1665 (1984)] shows the evolution of the fraction of the favoured phase as a function of time. The system is an RbI ionic crystal, which above 3.5 kbar undergoes a first-order phase transition: it changes from a NaCl-like structure (cubic) to a CsCl-like structure (cubic-centred). The experimental data are well reproduced by a law of the form of Equation (13.49) with a single fitting parameter of the form $V_\beta/V = 1 - \exp\left(-\frac{\pi}{3}(t/\tau_0)^4\right)$. The good agreement with experiment is obtained by choosing $\tau_0 = 1.11 \times \tau_{1/2}$, where $\tau_{1/2}$ is the time at which half of the sample is filled by the new phase.

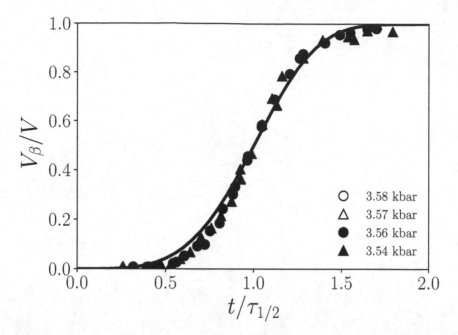

FIGURE 13.11 Experimental results for the fraction of the favoured phase as a function of time. The points are measurements performed at different pressures. For each pressure, $\tau_{1/2}$ is the time at which the new phase fills half of the sample. KJMA theory [Equation (13.49)] corresponds to the solid curve. Note that the KJMA law does not hold close to the transition pressure (experimental point at $P = 3.53$ kbar, not shown). In this case an empirical modification of Equation (13.49) accounting for the effects of a finite critical drop radius leads to a better fit [see Ishibashi and Takagi, J. Phys. Soc. Jpn. **31**, 506 (1971)].

Exercise 13.2. Show that in two dimensions, $S_\beta/S = 1 - \exp\{-\frac{\pi}{3}v^2 I_2 t^3\}$. What is the result in one dimension?

Solution: Using the method previously presented, the following expression is obtained: $\mathrm{d}S_\beta^*/\mathrm{d}t = 2\pi S I_2 v^2 \int_0^t (t - t')\mathrm{d}t'$, where $\mathrm{d}S_\beta/(S - S_\beta) = \pi I_2 v^2 t^2$, leading to the expected result. In one dimension, one obtains instead $L_\beta/L = 1 - \exp\{-I_1 v t^2\}$.

14 Bose–Einstein Condensates of Weakly Interacting Dilute Gases

Since the middle of the 90's, it has been known how to trap and cool vapours of bosonic atoms (typically rubidium ^{87}Rb or sodium ^{23}Na) below the Bose–Einstein condensation temperature (see Section 7.4). These systems are very cold ($T \sim 100$ nK) and exhibit strong quantum features. They are also inhomogeneous (because the gas is confined by an external potential) and subject to non-negligible interaction effects: the condensed atoms accumulate at the bottom of the trapping potential well until reaching relatively high densities[1] at which the physical properties of the system cannot always be described by the non-interacting model considered in Section 7.4. The system's low temperature is an asset for the theoretical treatment of the interactions because it is associated with a thermal wavelength that is larger than the typical distance between atoms[2]: in this regime, it is legitimate to ignore the details of the interaction and to use a schematic potential, designed only to correctly describe the low energy scattering between two atoms. The simplicity of the resulting model enables to account, in the framework of a mean field approach, for non-trivial interaction effects for which it is possible to obtain precise experimental information.

14.1 GROSS-PITAEVSKII EQUATION

The state of a Bose–Einstein condensate at low temperature is described by a complex order parameter $\phi(\vec{r},t)$ which acts as the "wave function of the condensate". As in Ginzburg-Landau theory of superconductivity (Section 10.5), a complex parameter enables to describe both the local density of the condensate (its modulus) and the current within the condensate [gradient of its phase, see Equation (14.70)]. It is customary to chose a normalisation such that $|\phi(\vec{r},t)|^2$ is equal to the local density $n(\vec{r},t)$ of condensed atoms (number of condensed atoms per unit volume):

$$|\phi(\vec{r},t)|^2 = n(\vec{r},t) \quad \text{with} \quad \int_{\mathbb{R}^3} d^3r \, n(\vec{r},t) = N_0(T), \qquad (14.1)$$

where $N_0(T)$ is the total number of atoms contained in the condensate (see Section 7.4). In a first stage (until Section 14.3), only the case at zero temperature is considered: $N_0(T) = N$, where N is the total number of atoms in the system, meaning that all atoms are assumed to belong to the condensate: depletion effects, caused by thermal fluctuations or by the interaction between atoms, are neglected.

In a Bose–Einstein condensate, dissipative effects are weak and the dynamics is richer than the one described by the time dependent Ginzburg-Landau equation which merely accounts for relaxation to equilibrium (see Section 13.1). A heuristic approach consists in pursuing as far as possible the interpretation of ϕ as the wave function of the condensate by

---

[1]These densities are, however, extremely low with respect to the density of ambient air for example: 10^{14} cm^{-3} compared to 10^{19} cm^{-3}.

[2]It is the degenerated gas regime, opposed to the Maxwell-Boltzmann limit [Relation (6.90), see Exercise 7.7].

DOI: 10.1201/9781003272427-14

postulating that the dynamics of the system is described by a time-dependant Schrödinger equation:

$$i\hbar \frac{\partial \phi}{\partial t} = -\frac{\hbar^2}{2m}\vec{\nabla}^2\phi + \left[V_{\text{trap}}(\vec{r}) + U_{\text{mf}}(\vec{r}, t)\right]\phi. \tag{14.2}$$

In this equation, V_{trap} is the trapping potential. Under usual experimental conditions, it is a harmonic well. When this well is isotropic, V_{trap} can be written as:

$$V_{\text{trap}}(\vec{r}) = \frac{1}{2}m\omega^2 r^2 = \frac{\hbar\omega}{2}\left(\frac{r}{a_{\text{ho}}}\right)^2, \quad \text{where} \quad a_{\text{ho}} = \sqrt{\frac{\hbar}{m\omega}} \tag{14.3}$$

is the typical extension of the ground state of Equation (14.2) when $U_{\text{mf}} \equiv 0$ (see Exercise 14.3). In Equation (14.2), U_{mf} is the mean field potential exerted by the condensate particles on one of them. If V_{int} is the actual interaction potential between two particles, then, intuitively:

$$U_{\text{mf}}(\vec{r}, t) = \int d^3r' \, V_{\text{int}}(|\vec{r} - \vec{r}'|)\, n(\vec{r}', t). \tag{14.4}$$

The potential V_{int} has a very repulsive core and Integral (14.4) diverges[3]. However, in a condensate, the relative energy of two particles is very low and it is legitimate to replace in Equation (14.4) the exact potential V_{int} by an effective potential V_{eff} for which the integral converges, provided that V_{eff} correctly describes the low energy scattering between two particles (i.e. it is not necessary to take into account the details of the two-body potential to describe the average interaction of a particle with its neighbours). In the dilute limit where the interparticle distance is large compared to the interaction range, the argument $\vec{r}'$ of $n = |\phi|^2$ in Equation (14.4) can be substituted for $\vec{r}$. This leads to the so-called *Gross-Pitaevskii equation*:

$$i\hbar\frac{\partial \phi}{\partial t} = -\frac{\hbar^2}{2m}\vec{\nabla}^2\phi + \left[V_{\text{trap}}(\vec{r}) + g\,|\phi|^2\right]\phi, \tag{14.5}$$

where

$$g = \int d^3r' \, V_{\text{eff}}(r'). \tag{14.6}$$

In the dilute regime, it can be shown that the low-energy scattering properties are expressed only as a function of the integral of V_{eff} (i.e. g): this is the Born approximation[4]. It allows to connect g to the "s-wave scattering length" which is a quantity, denoted a, which is homogeneous to a length and which parametrises the asymptotic behaviour of the two particles reduced wave function:

$$g = \frac{4\pi\hbar^2 a}{m}. \tag{14.7}$$

It will not be necessary here to get into the details of the two-body diffusion problem, and as far as we are concerned, Relation (14.7) is just a way to express the only parameter of the model (the interaction constant g) as a function of the scattering length a, which is characteristic of the low-energy interaction (for instance, the low-energy scattering cross section is $\sigma = 8\pi a^2$). In this chapter, only systems in which g and a are positive are considered, i.e. the effective interaction is repulsive (see Definition (14.6) of g).

---

[3]Or, at least, does not correctly describe the average force exerted by neighbour particles on a particle located at $\vec{r}$. A more accurate treatment should include the effect of short-range correlations, not taken taken into account in Equation (14.4), which regularise the integral.

[4]See for example, Chapter 11 of *Advanced Quantum Mechanics, Materials and Photons (Third edition)*, by R. Dick (Springer, 2016). The approach is consistent: according to Equation (14.5), all effective potentials are equally acceptable as long as integral (14.6) is the same, which, in the Born approximation, effectively corresponds to the same description of low energy scattering.

14.2   ZERO TEMPERATURE GROUND STATE OF A TRAPPED GAS

At $T = 0$, the system in equilibrium is in its ground state. This corresponds to a stationary configuration of the order parameter which takes the form $\phi(\vec{r}, t) = \phi(\vec{r}) \exp(-i\mu t/\hbar)$. Equation (14.5) can then be written as:

$$-\frac{\hbar^2}{2m} \vec{\nabla}^2 \phi + \left[V_{\text{trap}}(\vec{r}) + g \, |\phi|^2 \right] \phi = \mu \, \phi. \tag{14.8}$$

If g were zero, this equation would be exactly a stationary Schrödinger equation and μ would be the ground state energy. It is seen below (Exercise 14.2) that μ is is in fact the system's chemical potential.

14.2.1   VARIATIONAL APPROACH

Equation (14.8), which is the stationary version of the Gross-Pitaevskii equation, can be established through a procedure which calls upon less intuition than the one used to obtain Equation (14.5). To do so, a model is used where the effective interaction between two particles is described by the schematic potential $V_{\text{eff}} = g \, \delta^{(3)}(\vec{r} - \vec{r}')$. This is the simplest potential verifying Equation (14.6). The Hamiltonian of the system is then:

$$H = \sum_{i=1}^{N} \left(\frac{\vec{p}_i^2}{2m} + V_{\text{trap}}(\vec{r}_i) \right) + g \sum_{i=1}^{n} \sum_{j=i+1}^{n} \delta^{(3)}(\vec{r}_i - \vec{r}_j). \tag{14.9}$$

Since the system is a Bose–Einstein condensate in equilibrium, it is sound to use an approximation in which in all particles are assumed to occupy the same quantum state. This corresponds to the following ansatz for the system's N-body wave function:

$$\Psi(\vec{r}_1, \vec{r}_2, \cdots, \vec{r}_N) = \prod_{i=1}^{N} \frac{1}{\sqrt{N}} \, \phi(\vec{r}_i). \tag{14.10}$$

This is indeed a boson wave function (since it is totally symmetric upon a permutation of the particles). The function $\phi(\vec{r})$ is yet unspecified in Equation (14.10): it just has to satisfy the constraint (14.1) with $N_0(T) = N$; the factors $N^{-1/2}$ appearing in Equation (14.10) ensure the normalisation $\langle \Psi | \Psi \rangle = 1$. The system's energy evaluated in state Ψ is the quantity:

$$E[\phi] = \langle \Psi | H | \Psi \rangle = \int \mathrm{d}^3 r \left[\frac{\hbar^2}{2m} |\vec{\nabla}\phi|^2 + V_{\text{trap}}(\vec{r}) \, n(\vec{r}) + \frac{g}{2} \, n^2(\vec{r}) \right]. \tag{14.11}$$

In the integrand of the right-hand side of Equation (14.11), the first term is the kinetic energy, the second is the trapping potential energy, and the last one is the interaction potential energy.

Exercise 14.1.   Prove Formula (14.11).

<u>Solution:</u> *Hamiltonian (14.9) is the sum of one-body terms [of the form $\sum_{i=1}^{N} h_i$, with $h_i = \vec{p}_i^2/2m + V_{\text{trap}}(\vec{r}_i)$] and two-body interaction terms $\sum_{i<j} V_{\text{eff}}(\vec{r}_i - \vec{r}_j)$. The one-body terms give to $E[\phi]$ a contribution of the form*

$$E_1 = \sum_{i=1}^{N} \int \mathrm{d}^3 r_1 \cdots \mathrm{d}^3 r_N \frac{|\phi(\vec{r}_1)|^2}{N} \cdots \frac{|\phi(\vec{r}_{i-1})|^2}{N} \frac{\phi^*(\vec{r}_i) h_i \phi(\vec{r}_i)}{N} \frac{|\phi(\vec{r}_{i+1})|^2}{N} \cdots \frac{|\phi(\vec{r}_N)|^2}{N}. \tag{14.12}$$

The N terms of the above sum are equal, and in each of them, the integrals of the form
$\int \mathrm{d}^3 r_j |\phi(\vec{r}_j)|^2/N$ *are equal to unity. Thus, E_1 reads*

$$
\begin{aligned}
E_1 &= \int \mathrm{d}^3 r \, \phi^*(\vec{r}) h \, \phi(\vec{r}) = \int \mathrm{d}^3 r \, \phi^*(\vec{r}) \left[-\frac{\hbar^2}{2m} \vec{\nabla}^2 + V_{\mathrm{trap}}(\vec{r}) \right] \phi(\vec{r}) \\
&= \int \mathrm{d}^3 r \left[\frac{\hbar^2}{2m} |\vec{\nabla}\phi|^2 + V_{\mathrm{trap}} |\phi|^2 \right],
\end{aligned}
\tag{14.13}
$$

where in the last expression an integration by parts was performed[5].

The contribution to $E[\phi]$ of the interaction terms is denoted E_2. The same reasoning as above shows that it is made of $N(N-1)/2$ identical terms:

$$
E_2 = \frac{N(N-1)}{2N^2} \int \mathrm{d}^3 r \mathrm{d}^3 r' |\phi(\vec{r}\,')|^2 V_{\mathrm{eff}}(\vec{r} - \vec{r}\,') |\phi(\vec{r})|^2 \simeq \frac{g}{2} \int \mathrm{d}^3 r \, |\phi(\vec{r})|^4,
\tag{14.14}
$$

where the approximation $N - 1 \simeq N$ has been used (which is sound since in the considered system, the number of atoms is typically $N > 10^3$). The sum of Contributions (14.13) and (14.14) yields Expression (14.11) for $E[\phi] = E_1 + E_2$.

There is a theorem in quantum mechanics that states that for any normalised function Ψ, the quantity $\langle \Psi | H | \Psi \rangle$ is greater than the energy of the ground state[6]. Of course, this lower limit is reached if Ψ is the true ground state. This legitimates the use a "variational method": a family of ansätze [in our case, of the type Formula (14.10)] is chosen and the member of this family which minimises $E[\phi] = \langle \Psi | H | \Psi \rangle$ is identified with the (variational) ground state of the system. This technique only provides an upper bound of the ground state energy, but this is the best approximation that can be achieved within this approach.

$E[\phi]$ reaches an extremum under the constraint $\int \mathrm{d}^3 r |\phi(\vec{r})|^2 = N$ when the functional derivative of $E[\phi] - \mu \left(\int \mathrm{d}^3 r |\phi(\vec{r})|^2 - N \right)$ cancels, where μ is a Lagrangian parameter (see Section A.5). As already seen in Section 10.5, it is convenient for this purpose to consider ϕ and ϕ^* as independent functions. Moreover, since the quantity to extremize is real, the two functional derivatives $\delta E/\delta\phi$ and $\delta E/\delta\phi^*$ are complex conjugates of each other, and it is sufficient to impose:

$$
\frac{\delta}{\delta\phi^*} \left[E[\phi] - \mu \left(\int \phi^* \phi \, \mathrm{d}^3 r - N \right) \right] = 0, \quad \text{i.e.,} \quad \frac{\delta E[\phi]}{\delta\phi^*} = \mu \, \phi.
\tag{14.15}
$$

The calculation is straightforward (it is conducted according to the rules given in Section 10.2) and leads exactly to Equation (14.8) where the Lagrangian parameter μ can be interpreted as the chemical potential of the system, which is justified in exercise 14.2.

Exercise 14.2. Justify the interpretation of the constant μ appearing in Equations (14.8) and (14.15) as the system's chemical potential.

Solution: In the context of the above description, the system's energy E is the minimum value of $\langle \Psi | H | \Psi \rangle$ taken over the variational ansatz (14.10). This minimum is reached when ϕ verifies Equation (14.8) or, equivalently, Equation (14.15). If the number of particles in the system varies ($N \to N + \delta N$) then ϕ and E will also vary, by quantities δE and $\delta\phi$ respectively.

$$
\delta E = \int \mathrm{d}^3 r \left(\frac{\delta E[\phi]}{\delta\phi^*} \delta\phi^* + \frac{\delta E[\phi]}{\delta\phi} \delta\phi \right), \quad \text{and} \quad \delta N = \int \mathrm{d}^3 r \left(\phi \, \delta\phi^* + \phi^* \delta\phi \right).
\tag{14.16}
$$

---

[5]This method was already used in Equation (10.36), for instance. More precisely, one can write $\phi^* \vec{\nabla}^2 \phi = \vec{\nabla} \cdot (\phi^* \vec{\nabla}\phi) - \vec{\nabla}\phi^* \cdot \vec{\nabla}\phi$ and then use the Gauss-Ostrogradski theorem [Formula (A.45)].

[6]See for example, J. J. Sakurai, *Modern Quantum Mechanics* (Pearson, 1993).

In these expressions, one has to evaluate the functional derivatives for ϕ solution of (14.15). Then (14.16) immediately leads to $\delta E = \mu \delta N$, i.e. $\mu = \partial E / \partial N$, which is indeed the definition of the chemical potential[7].

14.2.2 THOMAS-FERMI APPROXIMATION

In order to evaluate the relative importance of the different terms of the stationary Gross-Pitaevskii Equation (14.8), it is useful to assess the order of magnitude of the different contributions to the system's energy E.

Let R be the typical value of the spatial extension of the condensate inside a harmonic trap. The kinetic contribution to the energy is $E_{\text{kin}} \sim N\hbar^2/(2mR^2) = \frac{1}{2}N\hbar\omega(a_{\text{ho}}/R)^2$. The trapping potential and the particles interactions contribute respectively for $E_{\text{trap}} \sim N\frac{1}{2}m\omega^2 R^2 = \frac{1}{2}N\hbar\omega(R/a_{\text{ho}})^2$ and[8] $E_{\text{int}} \sim 3N^2\hbar^2 a/(2mR^3)$.

In the ideal gas limit (no interactions), the condensate wavefunction is a Gaussian of typical width $R = a_{\text{ho}}$ (see Exercise 14.3). The system remains in a very weak interaction regime, with $R \sim a_{\text{ho}}$, as long as $E_{\text{kin}} \gg E_{\text{int}}$ which corresponds to $1 \gg Na/a_{\text{ho}}$. For a ^{87}Rb gas in a trap of angular frequency $\omega = 2\pi \times 350$ Hz, $a_{\text{ho}} \simeq 0.58$ μm, which is almost exactly 100 times the scattering length ($a \simeq 5.8$ nm). Thus, the ideal gas regime is irrelevant in usual experiments for which N is typically larger than 10^4 (in the previous example, for $N = 10^4$, $Na/a_{\text{ho}} = 10^2 \gg 1$).

Exercise 14.3. In the ideal gas regime, recover the order of magnitude $R = a_{\text{ho}}$ without explicitly solving Equation (14.8) (which is then a Schrödinger equation) but by minimising $E_{\text{kin}} + E_{\text{trap}}$.

Solution: it is sufficient to determine for which value of R the expression $E_{\text{kin}} + E_{\text{trap}} \propto (a_{\text{ho}}/R)^2 + (R/a_{\text{ho}})^2$ is at its minimum. One immediately gets $R = a_{\text{ho}}$. For the record, the exact solution of the Schrödinger equation in a harmonic trap is $\phi(\vec{r}) = \sqrt{N}(\pi a_{\text{ho}}^2)^{-3/4}\exp\left(-\frac{1}{2}r^2/a_{\text{ho}}^2\right)$, which leads to $\langle r^2 \rangle = N^{-1}\langle\phi|r^2|\phi\rangle = \frac{3}{2}a_{\text{ho}}^2$.

In the regime

$$\frac{aN}{a_{\text{ho}}} \gg 1, \tag{14.17}$$

which corresponds to many experimental situations, the interaction term dominates over the kinetic term in Equation (14.8). It is then legitimate to make the so-called *Thomas-Fermi* approximation which consists in simply neglecting the contribution in $\vec{\nabla}^2\phi$ in the Gross-Pitaevskii equation [Equation (14.8)]. The solution is then[9]:

$$n(\vec{r}) = |\phi(\vec{r})|^2 = \begin{cases} \frac{1}{g}(\mu - V_{\text{trap}}(\vec{r})) & \text{if } \mu > V_{\text{trap}}(\vec{r}), \\ 0 & \text{otherwise.} \end{cases} \tag{14.18}$$

When the trapping potential is the isotropic harmonic potential defined by Equation (14.3), the radius R of the condensate is defined by $n(r \geq R) = 0$, i.e. $V_{\text{trap}}(R) = \mu$. The chemical potential (or, equivalently, the radius R) is determined by the normalisation $\int \text{d}^3r|\phi|^2 = N$

---

[7]The derivative $\partial E/\partial N$ must be evaluated at constant entropy (see Equation (3.40)) but, here, the temperature is equal to zero, which implies $E = F$ and $\mu = (\partial F/\partial N)_T$ (see Equation (4.20)).

[8]To obtain the expression of E_{int}, $\frac{1}{2}g\int|\phi|^4 \text{d}^3r$ is evaluated using Equation (14.7) and, since $\int|\phi|^2\text{d}^3r = N$, $|\phi|^2 \simeq 3N/(4\pi R^3)\Theta(R - r)$.

[9]G. Baym and C. J. Pethick, Phys. Rev. Lett. **76**, 6 (1996).

of Expression (14.18). This leads to

$$R = \left(\frac{15\,g}{4\pi} \frac{N}{m\omega^2} \right)^{1/5} = a_{\mathrm{ho}} \left(\frac{15Na}{a_{\mathrm{ho}}} \right)^{1/5}, \quad \mu = \frac{m\omega^2 R^2}{2} = \frac{\hbar\omega}{2} \left(\frac{15Na}{a_{\mathrm{ho}}} \right)^{2/5}. \tag{14.19}$$

It can thus be seen that, in the regime (14.17), the condensate is much more spread out than the Gaussian solution of the ideal gas problem since $R \gg a_{\mathrm{ho}}$. The repulsion between the particles causes this large spread. Still, in the Thomas-Fermi regime, the kinetic energy is negligible in Equation (14.11) and the energy of the condensate is then $E \simeq E_{\mathrm{trap}} + E_{\mathrm{int}}$ with:

$$E_{\mathrm{trap}} = \int \mathrm{d}^3 r \, V_{\mathrm{trap}} \, |\phi|^2 = \frac{3}{7} \mu N, \quad \text{and} \quad E_{\mathrm{int}} = \frac{g}{2} \int \mathrm{d}^3 r \, |\phi|^4 = \frac{2}{7} \mu N, \tag{14.20}$$

where the integrals have been calculated using Expression (14.18). Here, interactions have also a major effect: Expressions (14.20) and (14.19) show that, in the regime (14.17), the energy is much larger than the value $\frac{3}{2} N \hbar \omega$ which would be the energy of an ideal Bose–Einstein condensate consisting in N particles occupying the ground state of a harmonic potential.

Exercise 14.4. Generalise the Thomas-Fermi approach to the case of an anisotropic harmonic trap of the form $V_{\mathrm{trap}}(\vec{r}) = \frac{1}{2} m (\omega_x^2 x^2 + \omega_y^2 y^2 + \omega_z^2 z^2)$.

Solution: Let us introduce the new coordinates $x' = \omega_x x / \bar{\omega}$, $y' = \omega_y y / \bar{\omega}$ and $z' = \omega_z z / \bar{\omega}$, where $\bar{\omega}^3 = \omega_x \omega_y \omega_z$. Writing $\mu = \frac{1}{2} m \bar{\omega}^2 \bar{R}^2$, the normalisation condition can be written as:

$$N = \frac{1}{g} \int \left(\mu - V_{\mathrm{trap}}(\vec{r}) \right) \mathrm{d}^3 r = \frac{m\bar{\omega}^2}{2g} \int_{r' < \bar{R}} (\bar{R}^2 - r'^2) \, \mathrm{d}^3 r', \tag{14.21}$$

where $r' = (x'^2 + y'^2 + z'^2)^{1/2}$ and use has been made of the fact that $\mathrm{d}^3 r' = \mathrm{d}^3 r$. The remaining calculation amounts to calculate a spherically symmetric integral similar to that leading to Formulae (14.19). These expressions therefore remain valid provided that ω is replaced by $\bar{\omega}$, R by $\bar{R}$ and the definition $a_{\mathrm{ho}} = \sqrt{\hbar / m \bar{\omega}}$ is used instead of its definition in Equation (14.3). With the same reasoning, it can be seen that Equations (14.20) still hold. However, one must be careful, the condensate is here anisotropic, with a maximum extension in direction α ($\alpha = x$, y or z) $R_\alpha = \bar{\omega}\bar{R}/\omega_\alpha$. $\bar{R}$ is the geometric mean $(R_x R_y R_z)^{1/3}$.

The relevance of the Thomas-Fermi approximation is illustrated in Figure 14.1 which represents the interaction energy plotted as a function of the number of atoms in a ^{23}Na Bose–Einstein condensate. This energy can be measured after a sudden opening of the trap, a process which cancels the contribution of E_{trap} to the total energy. Accordingly, in the Thomas-Fermi regime, only the $E_{\mathrm{int}} = \frac{2}{7} \mu N$ contribution should be taken into account in Equation (14.20), a procedure which reproduces well the experimental data[10]. The experiment was performed at different temperatures. In some cases (triangles in Figure 14.1), all atoms belong to the condensate. In others cases (circles), a fraction of non-condensed atoms[11] was present. The data shown in Figure 14.1 indicates that, under these experimental conditions, the non-condensed fraction has a minor contribution since the number of atoms in the condensate only is sufficient to determine the energy of the system.

---

[10]Data extracted from M.-O. Mewes, M. R. Andrews, N. J. van Druten, D. M. Kurn, D. S. Durfee, et W. Ketterle, *Phys. Rev. Lett.* **77**, 416 (1996).

[11]It shows that, in this case, the temperature T is not negligible with respect the the Bose–Einstein condensation temperature.

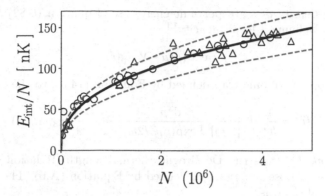

FIGURE 14.1 Energy per particle as a function of the number of atoms N contained in the condensate. The circles and triangles are experimental data. The black curve is Equation (14.20) evaluated for $\bar{\omega} = 2\pi \times 180$ Hz and $a = 65\,a_0$ where a_0 is the Bohr radius. The grey dashed curves correspond to $a = 90\,a_0$ and $40\,a_0 = (65 \pm 25)a_0$, reflecting the uncertainty in the exact value of a.

14.3 TRAPPED GAS AT FINITE TEMPERATURE

When the temperature is lower, but not small compared to the condensation temperature T_{BE}, the system is composed on the one hand of a condensed fraction ($N_0(T)$ atoms) and on the other hand of a "thermal cloud" corresponding to the remaining $N - N_0(T) = N_{th}(T)$ atoms, distributed on the excited levels of the harmonic trap. In order to develop an intuition of the system's phenomenology, it is useful to first treat the non-interacting case (ideal gas).

14.3.1 IDEAL GAS

It was seen in Section 7.4 that at a temperature below the Bose–Einstein condensation temperature T_{BE}, the chemical potential is zero: $N_0(T)$ atoms occupy the fundamental of the harmonic oscillator while the $N_{th}(T)$ other atoms are distributed over the excited levels, according to Law (7.50). Above the condensation temperature, the occupation of the ground state is not macroscopic: there is no condensate anymore and only a thermal cloud characterised by a chemical potential calculated using Relation (7.45).

It is interesting to determine the density profile of the thermal cloud and to compare it to the one of the condensate. Expression (14.22), given below, is valid whether T is higher or lower than the condensation temperature T_{BE}, but limited[12] to the regime $k_B T \gg \hbar\omega$. In this case, a semi-classical approach can be used to enumerate the states. States counting in phase space is performed imposing that the microstates are occupied according to the Bose–Einstein distribution [Equation (6.82)]. The density in real space is therefore:

$$n_{th}(\vec{r}) = \int d^3p\, dE\, \frac{\rho(\vec{r}, \vec{p}, E)}{\exp(\beta(E - \mu)) - 1},\qquad(14.22)$$

where

$$\rho(\vec{r}, \vec{p}, E) = \frac{1}{h^3}\delta\!\left(E - H(\vec{r}, \vec{p})\right)\qquad(14.23)$$

---

[12] Typical experiments are performed in the regime $k_B T \gg \hbar\omega$ since the condensation temperature T_{BE} is large with respect to $\hbar\omega/k_B$: the ratio between the two quantities is $(N/\zeta(3))^{1/3}$ which is approximately equal to 20 for $N = 10^4$.

is the phase space density of state per unit energy [see Equation (6.82)]. Here, $H(\vec{r}, \vec{p}) = p^2/(2m) + V_{\text{trap}}(\vec{r})$. Defining a local fugacity[13]

$$z(\vec{r}) = \exp\{\beta(\mu - V_{\text{trap}}(\vec{r}))\}, \tag{14.24}$$

leads, for the isotropic harmonic trap defined by Equation (14.3), to the expression:

$$n_{\text{th}}(\vec{r}) = \frac{1}{h^3} \int \frac{d^3 p}{[z(r)]^{-1} \exp(\beta p^2/2m) - 1} = \frac{1}{\lambda_T^3} g_{3/2}(z(r)). \tag{14.25}$$

where $\lambda_T = h(2\pi m k_{\text{B}} T)^{-1/2}$ is the De Broglie thermal length [Relation (3.48)] and $g_{3/2}$ belongs to the class of Bose functions g_ν defined by Equation (A.6). The total number of atoms in the thermal cloud is:

$$N_{\text{th}} = \int d^3 r\, n_{\text{th}}(r) = \left(\frac{k_{\text{B}} T}{\hbar \omega}\right)^3 g_3(e^{\beta \mu}), \tag{14.26}$$

see Equations (7.45) and (7.52) ($N_{\text{th}} = N$ and $\mu \leq 0$ when $T \geq T_{\text{BE}}$ whereas $\mu = 0$ and $N_{\text{th}} \leq N$ when $T \leq T_{\text{BE}}$).

The size of the non condensed cloud can be estimated using the energy equipartition theorem[14] seen in Section 4.1.5:

$$\langle \tfrac{1}{2} m \omega^2 r^2 \rangle_{\text{th}} \approx \tfrac{3}{2} k_{\text{B}} T, \quad \text{i.e.,} \quad \langle r^2 \rangle_{\text{th}} \approx 3 \frac{k_{\text{B}} T}{m \omega^2} = 3 a_{\text{ho}}^2 \frac{k_{\text{B}} T}{\hbar \omega}. \tag{14.27}$$

Exercise 14.5. Determine an expression for the mean square radius $\langle r^2 \rangle_{\text{th}}$ of the thermal cloud of an ideal gas valid at any temperature.

Solution: The quantity to be evaluated is $\langle r^2 \rangle_{\text{th}} = N_{\text{th}}^{-1} \int d^3 r\, r^2 n_{\text{th}}(r)$. The calculation is simpler if starting from Expression (14.22):

$$\langle r^2 \rangle_{\text{th}} = \int \frac{d^3 r\, d^3 p\, dE}{N_{\text{th}} h^3} \frac{r^2 \delta(E - H(\vec{r}, \vec{p}))}{\exp(\beta(E - \mu)) - 1}. \tag{14.28}$$

The integral over space can be calculated using the relation $\delta[f(r)] = \delta(r - a)/|f'(a)|$, valid when a is the only zero of the function $f(r)$. Thus:

$$\int d^3 r\, r^2 \delta(E - H(\vec{r}, \vec{p})) = \left(\frac{2E}{m}\right)^2 \frac{4\pi}{\omega^5 \sqrt{2mE}} \left(1 - \frac{p^2}{2mE}\right)^{3/2}. \tag{14.29}$$

and then (changing variable to $u = p/\sqrt{2mE}$):

$$\int d^3 p\, d^3 r\, r^2 \delta(E - H(\vec{r}, \vec{p})) = \frac{128 \pi^2 E^3}{m \omega^5} \int_0^1 u^2 (1 - u^2)^{3/2} du. \tag{14.30}$$

The integral of the right-hand side of Equation (14.30) is[15] $\pi/32$ and carrying over into Equation (14.28):

$$\langle \tfrac{1}{2} m \omega^2 r^2 \rangle_{\text{th}} = \frac{3}{2 N_{\text{th}}} \frac{(k_{\text{B}} T)^4}{(\hbar \omega)^3} g_4(e^{\beta \mu}), \tag{14.31}$$

where g_4 is a Bose function [Equation (A.6)]. The result expressed in this form corresponds to the potential energy per particle $E_{\text{trap}}/N_{\text{th}}$. If it is compared with the value of the total energy

---

[13] Compare to the usual definition (7.47).

[14] This estimation yields only an order of magnitude as the theorem is only valid in the classical limit.

[15] It is calculated thanks to the change of variable $u = \sin\theta$. Then: $\int_0^1 u^2 (1 - u^2)^{3/2} du = \int_0^{\pi/2} \sin^2\theta \cos^4\theta\, d\theta = \frac{1}{16} \int_0^{\pi/2} (1 + \frac{1}{2}\cos 2\theta - \cos 4\theta - \frac{1}{2}\cos 6\theta) d\theta = \pi/32$.

calculated (when $T < T_{BE}$, i.e. for $z = 1$) in Equation (7.58), it can be seen that $E_{trap} = \frac{1}{2}E$. This is a usual result of classical mechanics: in a harmonic well, the potential energy is half of the total energy. The present calculation shows that this result remains valid in quantum mechanics for a set of non-interacting bosons at non-zero temperature. The factor $\frac{1}{2}$ between E and E_{trap} is important when studying the energy of the gas after the trap has been opened (see Figure 7.15).

Using Equation (7.45) (or Equation (7.52) below the condensation temperature), Relation (14.31) can be written as:

$$\langle \tfrac{1}{2}m\omega^2 r^2 \rangle_{th} = \frac{3}{2}\,k_B T\, \frac{g_4(e^{\beta\mu})}{g_3(e^{\beta\mu})}. \tag{14.32}$$

At high temperatures, $\exp(\beta\mu)$ tends to zero [see Equation (7.48)] and the ratio $g_4(e^{\beta\mu})/g_3(e^{\beta\mu})$ approaches unity [$g_\nu(z) \simeq z$ when $z \to 0$, see Equation (A.7)]: the classical equipartition theorem is recovered. Below the condensation temperature, $\mu = 0$, $g_4(e^{\beta\mu})/g_3(e^{\beta\mu}) = \zeta(4)/\zeta(3) = O(1)$ and the order of magnitude (14.27) remains valid. The exact formula valid at any temperature is:

$$\langle r^2 \rangle_{th} = 3\,a_{ho}^2\, \frac{k_B T}{\hbar\omega}\, \frac{g_4(e^{\beta\mu})}{g_3(e^{\beta\mu})}. \tag{14.33}$$

At temperatures lower than the condensation temperature, the picture emerging from this analysis is that of a condensate in which N_0 atoms are concentrated on a domain of extension a_{ho} cohabiting with a much less dense thermal cloud which, according to Equation (14.33), spreads over a region of extend $\langle r^2 \rangle_{th}^{1/2}$ much larger than a_{ho} (since typically[12] $k_B T \gg \hbar\omega$).

14.3.2 INTERACTING GAS

In the non-ideal case, the picture of a system composed of a condensate co-existing with a dilute thermal cloud remains valid. At a temperature slightly above the transition temperature, since the gas is very dilute, interaction effects are small. Their order of magnitude can be estimated as $E_{int} \approx gN^2/(\frac{4}{3}\pi\langle r^2 \rangle^{3/2})$. Taking $T \approx T_{BE}$ and evaluating $\langle r^2 \rangle_{th}$ with Equation (14.27), one gets $E_{int} \approx (a/a_{ho})\hbar\omega$, which is a very small compared to the total energy of the gas at this temperature ($E \approx 3k_B T_{BE} \approx \hbar\omega N^{1/3}$). The Bose–Einstein condensation temperature is thus only slightly modified by the interactions. Here, it is simply assumed that it is unaffected and that Expression (7.49) still holds.

Throughout the rest of this section, only temperatures lower than T_{BE} are considered. In a first stage it will be assumed that the number of atoms in the condensate $N_0(T)$ remains given by law (7.53). As the atoms in the thermal cloud are very dilute, their contribution to the energy of the condensate atoms is negligible, and we can stick to the form (14.8) of the Gross-Pitaevskii[16]. In the rest of this section, the Thomas-Fermi regime (14.17) is considered, and the assumption[17] $N_0(T) \gg a_{ho}/a$ is made. Then, Formulae (14.18) and (14.19) remain valid as long as N is replaced by $N_0(T)$. To distinguish the case $T \neq 0$ from the $T = 0$ case studied in Section 14.2, we write $|\phi|^2 = n_0$, with thus $\int d^3r\, n_0(\vec{r}) = N_0(T)$. The Thomas-Fermi profile $n_0(r)$ of an isotropic condensate is shown in Figure 14.2.

The study of the effects of interaction on the thermal cloud is rather delicate. Since the thermal cloud is very dilute [the interparticle distance in the cloud is of the order of λ_T according to Equation (14.25)], the interactions within it[16] can be neglected in a first approximation. However, at the centre of the trap, the atoms belonging to the thermal cloud certainly interact with those of the condensate whose density is high. To determine

---

[16]See Section 14.4.2 for an improved approximation.

[17]When (14.17) holds, the relation $N_0 \gg a_{ho}/a$ is violated only very close to T_{BE} in a tiny interval of temperature of extension $\approx T_{BE} \times a_{ho}/(aN)$.

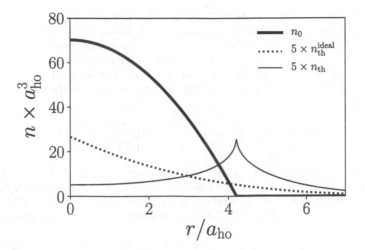

FIGURE 14.2 Density profile of a trapped vapour containing 10^4 atoms at $T = \frac{1}{2}T_{\mathrm{BE}}$ with $a/a_{\mathrm{ho}} = 10^{-2}$. The bold curve represents the Thomas-Fermi profile $n_0(r)$ of the condensate [Equation (14.18)]. The dashed curve represents $n_{\mathrm{th}}(r)$ in the non-interacting case [Equation (14.25)]. The thin line represents $n_{\mathrm{th}}(r)$ given by Equation (14.37) where the potential exerted by the condensate on the thermal cloud has been taken into account. The thermal cloud profiles have been multiplied by a factor of 5 for better visibility.

the interaction between an atom of the thermal cloud and the condensate, it is necessary to evaluate the matrix element of the interaction potential $V_{\mathrm{eff}} = g\,\delta^{(3)}(\vec{r}_1 - \vec{r}_2)$ between two bosons that do not occupy the same quantum state. The normalised spatial wavefunctions are denoted $\varphi_0(\vec{r})$ and $\varphi(\vec{r})$, one being that of the condensate ($\varphi_0 = \phi/\sqrt{N_0}$), the other ($\varphi$) corresponds to a thermal state and is orthogonal to φ_0. Indistinguishability prevents to ascribe a state (say φ_0) to an atom (say $\vec{r}_1$): as explained in Section 6.4, the symmetrised and normalised state of the two bosons is therefore of the form[18]:

$$\Phi(\vec{r}_1, \vec{r}_2) = \frac{1}{\sqrt{2}}\left[\varphi_0(\vec{r}_1)\varphi(\vec{r}_2) + \varphi_0(\vec{r}_2)\varphi(\vec{r}_1)\right]. \tag{14.34}$$

The contribution to the total energy of the interaction of a single atom of the thermal cloud with the whole condensate is then

$$\begin{aligned}
N_0\langle\Phi|V_{\mathrm{eff}}|\Phi\rangle &= g\,N_0\int \mathrm{d}^3r_1\mathrm{d}^3r_2\,\Phi(\vec{r}_1,\vec{r}_2)\delta(\vec{r}_1 - \vec{r}_2)\Phi(\vec{r}_1,\vec{r}_2) \\
&= \int \mathrm{d}^3r\,2gn_0(\vec{r})\,|\varphi(\vec{r})|^2,
\end{aligned} \tag{14.35}$$

where $n_0 = |\phi|^2 = N_0|\varphi_0|^2$. This implies that the condensate exerts an effective potential $2gn_0(\vec{r})$ on each atom of the thermal cloud. This potential is twice that experienced by the condensate atoms due to their mutual interaction [see for example the contribution $g|\phi|^2 = gn_0$ in Equation (14.8)]. This factor 2 comes from the two contributions to Equation (14.35) due to the symmetrised wavefunction [Equation (14.34)]: the first one is called *direct term* (or *Hartree term*), the second one is called *exchange term* (or *Fock term*). These two contributions are equal due to the special form of V_{eff} (a Dirac distribution), hence the factor 2.

---

[18]Compare to the Heitler-London *ansatz* for electrons [Relation (9.2)]. Here, no spin effects need to be taken into account and the normalisation is simpler as the considered configuration implies $\langle\varphi_0|\varphi\rangle = 0$.

The atoms in the thermal cloud therefore see a total potential $2gn_0 + V_{\text{trap}}$. In the Thomas-Fermi approximation, $n_0(\vec{r})$ is given by Equation (14.18) where, as already mentioned, one must evaluate μ replacing N by $N_0(T)$ in Formula (14.19). Following the semi-classical approach outlined in Section 14.3.1, for the thermal cloud atoms, the density of state per unit energy in phase space is given by Equation (14.23) with here $H(\vec{r}, \vec{p}) = p^2/2m + 2gn_0(\vec{r}) + V_{\text{trap}}(\vec{r})$ and the local fugacity is then [compare to the ideal case given by Formula (14.24)]

$$z(r) = \exp\left[\beta(\mu - V_{\text{trap}}(r) - 2gn_0(r))\right] = \exp(-\beta|\mu - V_{\text{trap}}(r)|). \tag{14.36}$$

In this formula, the chemical potential is given by Equation (14.19) in which N must be replaced by N_0. A result formally similar to Equation (14.25) is then obtained:

$$n_{\text{th}}(\vec{r}) = \frac{1}{\lambda_T^3} g_{3/2}(z(r)), \tag{14.37}$$

where here $z(r)$ is given by Equation (14.36) and not by Equation (14.24). Interaction effects have important implications for the density of the thermal cloud, as can be seen in Figure 14.2 by comparing the density profile of the thermal cloud of an ideal gas [Equation (14.25)] with the profile given by Equation (14.37). Interaction effects are of two kinds: (1) repulsion tends to expel the thermal cloud from the condensate region, and (2) the chemical potential of the thermal cloud in equilibrium with the condensate is not equal to zero as it is the case when $T \leq T_{\text{BE}}$ for the ideal gas profile given by Equation (14.25).

The total number of atoms in the thermal cloud is:

$$N_{\text{th}}(T) = \int d^3r\, n_{\text{th}}(\vec{r}) = \frac{1}{h^3} \int \frac{d^3r\, d^3p}{\exp(\beta|\mu - V_{\text{trap}}(r)|) \exp(\beta p^2/2m) - 1}. \tag{14.38}$$

Using the expansion $(\alpha - 1)^{-1} = \sum_{n=1}^{\infty} \alpha^{-n}$ (valid for $\alpha > 1$) allows to rewrite the above integral as:

$$\begin{aligned} N_{\text{th}}(T) &= \frac{1}{h^3} \sum_{n=1}^{\infty} \int d^3r\, d^3p\, \exp\left(-n\beta p^2/2m\right) \exp(-n\beta|\mu - V_{\text{trap}}(r)|) \\ &= \sum_{n=1}^{\infty} \left(\frac{2\pi m k_{\text{B}}T}{nh^2}\right)^{3/2} \int d^3r\, \exp(-n\beta|\mu - V_{\text{trap}}(r)|) \\ &= \left(\frac{k_{\text{B}}T}{\hbar\omega}\right)^3 \sum_{n=1}^{\infty} \frac{1}{n^3} F(n\beta\mu), \end{aligned} \tag{14.39}$$

where

$$\begin{aligned} F(x) &= \frac{4}{\sqrt{\pi}} \exp(-x) \int_0^{\sqrt{x}} t^2 \exp\left(t^2\right) dt + \frac{4}{\sqrt{\pi}} \exp(x) \int_{\sqrt{x}}^{\infty} t^2 \exp\left(-t^2\right) dt \\ &= \exp(x) + \frac{4}{\sqrt{\pi}} \int_0^x \sqrt{u} \sinh(u - x) du \simeq \begin{cases} 1 + x & \text{if } x \ll 1, \\ 4\sqrt{x/\pi} & \text{if } x \gg 1. \end{cases} \end{aligned} \tag{14.40}$$

Formula (14.39) is not of very convenient use[19] and it is suitable to cast it in an approximate, but more explicit form. This is possible in a regime where the condensate mean field only weakly perturbs the thermal cloud. The atoms of the cloud experience a potential due

---

[19]Indeed, the value of the chemical potential to be used in this formula is given by Relation (14.19) in which N must be replaced by $N_0 = N - N_{\text{th}}$. Relation (14.39) is thus an implicit equation for $N_{\text{th}}(T)$.

to the condensate that may be large: at $T \ll T_{BE}$, in the condensate region $(r < R)$, a cloud atom sees an effective field $2gn_0(\vec{r})$ created by the condensate, of the same order of magnitude as the trapping potential [see Equation (14.18)]. On the other hand, at $T \simeq T_{BE}$, the density of the condensate is very low and its interaction with the thermal cloud is negligible. There must therefore exist below T_{BE} a temperature range for which the effect of the condensate on the thermal cloud can effectively be considered as a small perturbation. This temperature range can be evaluated by requiring the interaction energy between the condensate and the thermal cloud $(2g \int n_0(\vec{r})n_{th}(\vec{r})\mathrm{d}^3r)$ to be small compared to the kinetic energy of the cloud $(\frac{3}{2}k_BTN_{th})$. As n_{th} varies over a much larger distance than n_0, one can write $\int n_0(\vec{r})n_{th}(\vec{r})\mathrm{d}^3r \simeq N_0\,n_{th}(0)$ and the condition of interest becomes[20]:

$$gN_0 \frac{N_{th}}{\frac{4}{3}\pi\langle r^2\rangle_{th}^{3/2}} \ll \frac{3}{2}k_BT\,N_{th}. \tag{14.41}$$

This can be written as:

$$k_BT \gg \hbar\omega\left(\frac{a\,N_0}{a_{ho}}\right)^{2/5} \approx \mu, \tag{14.42}$$

where Expression (14.19) was used, substituting N for $N_0(T)$, as it should at finite temperature. The regime of interest is therefore that in which the chemical potential of the condensate is small compared to k_BT. Inequality (14.42) can be rewritten as:

$$\frac{T}{T_{BE}} \gg \varepsilon\left(\frac{N_0}{N}\right)^{2/5}, \quad \text{where} \quad \varepsilon = \left(\frac{a\,N^{1/6}}{a_{ho}}\right)^{2/5} \tag{14.43}$$

plays the role of small parameters in this approach. For a gas of 10^5 atoms of ^{87}Rb in a trap with an angular frequency $\omega = 2\pi \times 350$ Hz, $a/a_{ho} \simeq 10^{-2}$ and one gets $\varepsilon = 0.34$. To sum up, Equations (14.44) and (14.45) below are obtained in a regime where $\varepsilon \ll 1 \ll \varepsilon N^{1/3}$. The first inequality corresponds to the regime where the perturbation of the thermal cloud by the condensate is small, the second corresponds to the Thomas-Fermi regime.

In the regime considered here, inequality (14.42) holds, or equivalently $\beta\mu \ll 1$. In the last term of Formula (14.39), the series expansion of $F(x)$ valid in the limit $x \ll 1$ can be used [see Equation (14.40)][21]. As $\sum_{n=1}^{\infty} n^{-\nu} = \zeta(\nu)$ (see Section A.1), this leads to the result:

$$N_{th}(T) \simeq \left(\frac{k_BT}{\hbar\omega}\right)^3 \left[\zeta(3) + \zeta(2)\frac{\mu}{k_BT}\right]. \tag{14.44}$$

For a non-interacting gas below the condensation temperature, $\mu = 0$ and the above expression is identical to Equation (7.52). For an interacting gas, Relation (14.42) ensures that the second term in the bracketed expression above is indeed a small correction to the main term. Expression (14.44) can be put into a form that involves the condensation temperature T_{BE} and the small parameter ε:

$$\frac{N_{th}(T)}{N} \simeq \left(\frac{T}{T_{BE}}\right)^3 + \frac{15^{2/5}\zeta(2)}{2\,\zeta(3)^{2/3}}\,\varepsilon\left(\frac{T}{T_{BE}}\right)^2\left(1 - \left(\frac{T}{T_{BE}}\right)^3\right)^{2/5}. \tag{14.45}$$

---

[20] The approximate expression $n_{th}(0) \approx N_{th}/(\frac{4}{3}\pi\langle r^2\rangle_{th}^{3/2})$ is used in Equation (14.41). Equation (14.25) could also be used.

[21] The summation over n in Expression (14.39) extends to infinity, and thus, in this formula, the argument of F is not always small compared to 1, even if $\beta\mu \ll 1$. It is thus not obvious that the small argument expansion of F can be used in Equation (14.39). However, it can be shown that the error introduced by replacing $F(n\beta\mu)$ by $1 + n\beta\mu$ plays a role only at order $O(\beta\mu)^2$, i.e. beyond the order of accuracy of (14.44).

This expression corresponds to an occupation of the thermal cloud larger than in the ideal case where N_{th} is given by Expression (7.52) [first term of the right-hand side of (14.45)]. As a result, the number $N_0(T) = N - N_{\text{th}}(T)$ of atoms in the condensate is decreased compared to the value given by Relation (7.53) valid for an ideal gas. The physical mechanism explaining this phenomenon is subtle, and is best understood if compared to the behaviour of a homogeneous gas (i.e. in the absence of a trap), studied in the next section.

14.4 INFINITE HOMOGENEOUS GAS AT FINITE TEMPERATURE

It is interesting to study the effects of interactions on the thermodynamics of a uniform density condensate, because the phenomenology is not exactly identical to that of a trapped condensate studied in Section 14.3.2. In addition, recent experiments have been able to measure the equation of state of interacting homogeneous Bose gas, which confirms the analysis underlying the treatment of interaction effects (see Section 14.4.2). In this section, a system of bosons contained in a box of volume V is considered. The non-interacting homogeneous Bose gas has been treated as an exercise in Chapter 7 (see Exercise 7.10). In the presence of interactions, we will follow the prescription of Section 14.3.2 and assume that T_{BE} has the same value than in the ideal case [Formula (7.69)].

14.4.1 FIRST APPROXIMATION

In a first stage, the approximate scheme of Section 14.3.2 is followed (and then refined in Section 14.4.2): at a temperature sufficiently below T_{BE}, the density of the thermal cloud is low, and so are its effects on the condensate. The Gross-Pitaevskii Equation (14.8) therefore remains valid (with here $V_{\text{trap}} \equiv 0$). For the condensate occupying the ground state, the kinetic term is negligible when $V \to \infty$ and Thomas-Fermi approximation is then exact. Therefore, one immediately obtains

$$\mu(T) = g\, n_0(T), \tag{14.46}$$

where $n_0(T)$ is given by Equation (7.70).

The atoms in the thermal cloud interacting with the condensate are subject to the Hamiltonian[22] $H(\vec{r}, \vec{p}) = \vec{p}^2/(2m) + 2g\, n_0$ and the "local" fugacity is then[23] $z = \exp(\beta(\mu - 2gn_0)) = \exp(-\beta\mu)$. The semi-classical approximation which, for the trapped gas, led to Expression (14.37) gives here:

$$n_{\text{th}} = \frac{1}{\lambda_T^3} g_{3/2}(e^{-\beta\mu}) = n\left(\frac{T}{T_{\text{BE}}}\right)^{3/2} \frac{g_{3/2}(e^{-\beta\mu})}{\zeta(3/2)}, \tag{14.47}$$

where T_{BE} is given by Equation (14.37) and μ by Equation (14.46). This relation is valid for $T < T_{\text{BE}}$, it is the counterpart for the homogeneous system of Expression (14.37) valid for trapped systems. In the absence of interactions, it leads, as it should, to a value of n_{th} identical to that given by Equation (7.70). In the presence of interactions the value of n_{th} given by Equation (14.47) is lower than in the case of the ideal gas[24]. The physical mechanism that explains this difference is the repulsion exerted by the condensate on the thermal cloud. In the trapped case (Section 14.3.2), this mechanism is also present, but leads to more subtle effects: the repulsion exerted by the condensate leads to an expulsion of the thermal cloud from the central region of the trap, towards a peripheral region (see

---

[22] As in Section 14.3.2, the interactions of the thermal cloud atoms with each other are neglected. The term $2gn_0$ is the Hartree-Fock term introduced in Equation (14.35) and the following discussion.

[23] Compare to Equation (14.36) valid in the same approximation regime for the trapped case.

[24] Because $g_{3/2}$ is an increasing function, when $z < 1$, $g_{3/2}(z) < g_{3/2}(1) = \zeta(3/2)$.

Figure 14.2), where the local fugacity given by Equation (14.36) is higher than in the ideal case, and in which the thermal fraction is therefore increased. In the trapped case, the net effect is an increase of N_{th} due to interactions. In the homogeneous case, thermal cloud and condensate are exactly superimposed. The expulsion mechanism which occurs in the trapped situation is impossible and instead there is a decrease in the thermal fraction compared to the ideal case which is caused by the uniform increase in chemical potential.

Relation (14.47) is an implicit equation determining the density of the thermal cloud in a self-consistent manner: n_{th} appears in the left-hand term and also in the right-hand term, via the quantity $\mu = g n_0 = g(n - n_{th})$. An explicit determination of n_{th} can be obtained perturbatively when $\mu \ll k_B T$ by using the expansion $g_{3/2}(1 - x) \simeq \zeta(3/2) - 2\sqrt{\pi x}$ (valid for $0 < x \ll 1$). This leads to:

$$
\begin{aligned}
\frac{n_{th}(T)}{n} &\simeq \left(\frac{T}{T_{BE}}\right)^{3/2} \left[1 - \frac{2\sqrt{\pi}}{\zeta(3/2)} \left(\frac{g n_0(T)}{k_B T}\right)^{1/2} \right] \\
&\simeq \left(\frac{T}{T_{BE}}\right)^{3/2} - \frac{\sqrt{8\pi a n^{1/3}}}{\zeta(3/2)^{2/3}} \frac{T}{T_{BE}} \left(1 - \left(\frac{T}{T_{BE}}\right)^{3/2}\right)^{1/2},
\end{aligned}
\tag{14.48}
$$

where Equation (14.7) was used to express g in terms of the s-wave scattering length. This relation is the counterpart, for a homogeneous system, of Expression (14.45) valid for a trapped system. The small parameter ε [defined in Equation (14.43)] relevant in the trapped case is replaced in the infinite and homogeneous case by the quantity $(a n^{1/3})^{1/2}$. These two parameters are not different in nature: in the framework of Thomas-Fermi approximation used in Section 14.3.2, it is easy to verify using Equations (14.18) and (14.19) that for a trapped gas at $T = 0$ K, $(a n^{1/3}(r = 0))^{1/2} = (15^{2/5}/8\pi)^{1/6} \varepsilon \simeq 0.7\,\varepsilon$. However, it is important to emphasise again that the sign of the correction to the main term in Equation (14.48) is the opposite of that in Equation (14.45).

14.4.2 EQUATION OF STATE. HARTREE-FOCK APPROXIMATION

For determining the equation of state of an interacting Bose gas, it is necessary to improve on the approximate treatment of interactions used so far. Up to now, only the interactions within the condensate and the force exerted by the condensate on the thermal cloud were accounted for. In the present section, every possible interaction channel is taken into account: the interaction effects within the thermal cloud and the force exerted by the thermal cloud on the condensate are also taken into account. For the schematic potential $V_{eff} = g\delta^{(3)}(\vec{r}-\vec{r}\,')$, the Gross-Pitaevkii equation (14.8) is thus modified into

$$
-\frac{\hbar^2}{2m}\vec{\nabla}^2\phi + \left[V_{trap}(\vec{r}) + g\,n_0(\vec{r}) + 2g\,n_{th}(\vec{r}) \right]\phi = \mu\,\phi,
\tag{14.49}
$$

where $n_0 = |\phi|^2$. As discussed above [Equation (14.34) and below], the factor 2 in the potential $2g n_{th}$ exerted by the thermal cloud on the condensate encompasses the contributions of the direct and exchange interaction terms. In the same line, the state of the thermal cloud atoms is described by a set of wave functions $\varphi_i(\vec{r})$ verifying

$$
-\frac{\hbar^2}{2m}\vec{\nabla}^2\varphi_i + \left[V_{trap}(\vec{r}) + 2g\,n_0(\vec{r}) + 2g\,n_{th}(\vec{r}) \right]\varphi_i = \epsilon_i\,\varphi_i,
\tag{14.50}
$$

with the normalisation $\int d^3r\,|\varphi_i|^2 = 1$ and[25] $n_{th}(\vec{r}) = \sum_{j\neq 0}|\varphi_j(\vec{r})|^2 n_B(\epsilon_j)$ where n_B is the Bose occupation factor (6.82). In the framework of this approach, the so-called *Hartree-Fock*

---

[25]Strictly speaking the term $j = i$ should be removed in the expression of n_{th} appearing in Equation (14.50). It is legitimate to overlook this correction, except when extremely close to $T = 0$, where a single state i can alone contribute significantly to n_{th}.

approximation, the energy of the system is:

$$E = \int \mathrm{d}^3r \left\{ \frac{\hbar^2}{2m}|\vec{\nabla}\phi|^2 + \sum_{i\neq0} \frac{\hbar^2}{2m} n_\mathrm{B}(\epsilon_i)|\vec{\nabla}\varphi_i|^2 \right.$$

$$\left. + V_\mathrm{trap}(n_0 + n_\mathrm{th}) + \tfrac{1}{2}gn_0^2 + 2gn_0n_\mathrm{th} + gn_\mathrm{th}^2 \right\}. \tag{14.51}$$

The second contribution in the right hand side of the above formula is the kinetic energy K_th of the thermal cloud. In the case of a uniform homogeneous system of volume V, $V_\mathrm{trap} \equiv 0$ and the densities n_0 and n_th of the condensate are position independent. Equation (14.49) leads to

$$\mu = gn_0 + 2gn_\mathrm{th}, \quad \text{with} \quad n = n_0 + n_\mathrm{th}, \tag{14.52}$$

[instead of Equation (14.46)] and Equation (14.51) simplifies to $E = K_\mathrm{th} + V[\tfrac{1}{2}gn_0^2 + 2gn_0n_\mathrm{th} + gn_\mathrm{th}^2]$. Multiplying (14.50) by $n_\mathrm{B}(\epsilon_i)\varphi_i^*(\vec{r})$, integrating over space and summing over i leads to $K_\mathrm{th} + 2g(n_0 + n_\mathrm{th})n_\mathrm{th}V = \sum_{i\neq0}\epsilon_i n_\mathrm{B}(\epsilon_i)$, which allows to rewrite the energy of the uniform system as

$$E = \sum_{i\neq0}\epsilon_i n_\mathrm{B}(\epsilon_i) + V\left[\tfrac{1}{2}gn_0^2 - gn_\mathrm{th}^2\right]. \tag{14.53}$$

Together with Equation (14.50), this expression shows that the energy of the homogeneous system is the sum of single particle energies of particles of the thermal cloud subjected to an external potential $2gn$ and a nonlinear term accounting for the remaining contributions of the interactions within the system. The grand potential of the system is then $J = E - TS - \mu N$ where $N = N_0 + N_\mathrm{th} = (n_0 + n_\mathrm{th})V$. Equation (14.53) allows to express J as

$$J = J_\mathrm{th} + V\left[\tfrac{1}{2}gn_0^2 - gn_\mathrm{th}^2\right] - \mu n_0 V, \quad \text{where} \quad J_\mathrm{th} = \sum_{i\neq0}\epsilon_i n_\mathrm{B}(\epsilon_i) - TS - \mu N_\mathrm{th}. \tag{14.54}$$

In the framework of the present approximation, the entropy S of the whole system is identical to that of a gas of N_th non-interacting bosons and the quantity J_th is thus the grand potential of a homogeneous ideal gas of N_th bosons subjected to an external potential $2gn$. The corresponding thermodynamic quantities are thus those determined in Exercise 7.10 where μ is replaced by $\mu - 2gn$. In particular [see Equations (7.68) and (7.72)]:

$$n_\mathrm{th} = \frac{1}{\lambda_T^3}g_{3/2}(\exp[\beta(\mu - 2gn)]), \tag{14.55a}$$

$$P_\mathrm{th} = \frac{k_\mathrm{B}T}{\lambda_T^3}g_{5/2}(\exp[\beta(\mu - 2gn)]). \tag{14.55b}$$

The total pressure of the system is $P = -J/V$ [see Equation (4.58)]. According to Equation (14.54), it reads

$$P = P_\mathrm{th} + gn_\mathrm{th}^2 - \tfrac{1}{2}gn_0^2 + \mu n_0. \tag{14.56}$$

For temperatures larger than T_BE, $n_0 = 0$, $n_\mathrm{th} = n$, the chemical potential μ is determined implicitly by Equation (14.55a) and Equation (14.56) reads

$$\text{for } T \geq T_\mathrm{BE}, \quad P = P_\mathrm{iBg}(T) + gn^2, \quad \text{with} \quad P_\mathrm{iBg}(T) = \frac{k_\mathrm{B}T}{\lambda_T^3}g_{5/2}\left(g_{3/2}^{-1}(n\lambda_T^3)\right). \tag{14.57}$$

The expression of P_iBg in the above formula is the same as the pressure of a homogeneous ideal Bose gas of density n above the transition temperature [see Equation (7.72)].

In the Bose–Einstein condensed phase, the chemical potential is instead given by Equation (14.52) and Equation (14.56) now reads

$$\text{for } T \leq T_{\text{BE}}, \quad P = \frac{k_{\text{B}}T}{\lambda_T^3} g_{5/2}(\exp[-\beta g n_0]) + g n^2 - \tfrac{1}{2} g n_0^2. \tag{14.58}$$

An explicit determination of P requires here to calculate n_0 which is achieved by self-consistently solving Equations (14.52) and (14.55a) for the three unknown quantities n_0, n_{th} and μ at fixed n and T. It is possible to circumvent the self-consistent procedure by deriving an approximate analytic expression for P valid in the regime $\beta g n_0 \ll 1$, i.e. close to the transition temperature (remaining in the region $T \leq T_{\text{BE}}$). Evaluating the pressure [Formula (14.58)] by using the approximate formula $g_{5/2}(1-x) \simeq \zeta(5/2) - x\zeta(3/2)$ (valid for $0 < x \ll 1$) leads to[26]

$$P \simeq \frac{k_{\text{B}}T}{\lambda_T^3} \zeta(5/2) - \frac{g n_0}{\lambda_T^3} \zeta(3/2) + g n^2 \left[1 - \tfrac{1}{2}(1 - n_{\text{th}}/n)^2\right]. \tag{14.59}$$

The first term in the right hand side of this expression is the ideal gas pressure below the condensation temperature [see Equation (7.72)]. It is convenient to rewrite the second term introducing the notation $n_c(T) = n \times (T/T_{\text{BE}})^{3/2} = \zeta(3/2)\lambda_T^{-3}$. Additionally, in the second and third term of Equation (14.59), it is legitimate to approximate n_{th} and n_0 by the non-interacting results ($n_{\text{th}} \simeq n_c$ and $n_0 \simeq n - n_c$), which eventually leads to

$$\text{for } T \leq T_{\text{BE}}, \quad P \simeq P_{\text{iBg}}(T) + \tfrac{1}{2}g(n^2 + n_c^2(T)), \quad \text{with} \quad P_{\text{iBg}}(T) = \frac{k_{\text{B}}T}{\lambda_T^3} \zeta(5/2). \tag{14.60}$$

Although this result has been derived by means of an approximation expected to hold only close to the condensation temperature (when $\beta g n_0 \ll 1$), it has the same asymptotic behaviour as Equation (14.58) when $T \to 0$. It is thus legitimate to expect that it gives a useful approximation of the Hartree-Fock result [Equation (14.58)] in the whole range of temperature $0 \leq T \leq T_{\text{BE}}$. Expressions (14.57) and (14.60) account for a modification of the ideal Bose gas result induced by interatomic repulsion. In the $T \to 0$ limit, P_{iBg} and n_c cancel as $T^{5/2}$ and $T^{3/2}$ respectively and only remains the interaction term $\tfrac{1}{2}g n^2$. It accounts for repulsion within the condensate and can be obtained directly from an analysis of Gross-Pitaevskii equation at $T = 0$ [see Equation (14.71) and below]. At finite temperature, the effect of interactions on the pressure is less simply expressed because of the direct and exchange contributions, already present in Equation (14.56), but, as expected, the net result is a nonlinear increase of the pressure with respect to the ideal gas value.

It has been recently possible to determine the canonical equation of state $P(v)$ of a weakly interacting Bose gas with homogeneous density $n = 1/v$ based on measurements on slices of a trapped system. The result of the Trento group[27] are presented in Figure 14.3 and compared with the results of Hartree-Fock approximation [Equations (14.57) and (14.60)]. The dashed line in the figure corresponds to the ideal Bose gas result $P_{\text{iBG}}(T)$. With the chosen rescaling, all isotherms of the ideal Bose gas fall on the same universal curve. This is no longer the case for the interacting system due to the addition of a new dimensionless parameter na^3 [a being the s-wave scattering length, see Equation (14.7)]. The discrepancy between the

---

[26]It is worth noticing that it is possible to solve Equation (14.55a) at the same order of accuracy by using the approximate formula $g_{3/2}(1 - x) \simeq \zeta(3/2) - 2\sqrt{\pi x}$ (valid for $0 < x \ll 1$). Since, below T_{BE}, Equation (14.55a) is formally equivalent to Equation (14.47), this procedure leads to a result identical to Equation (14.48), although derived here in the framework of a more accurate approximation than the one used in Section 14.4.1.

[27]C. Mordini, D. Trypogeorgos, A. Farolfi, L. Wolswijk, S. Stringari, G. Lamporesi, and G. Ferrari, Phys. Rev. Lett. **125**, 150404 (2020).

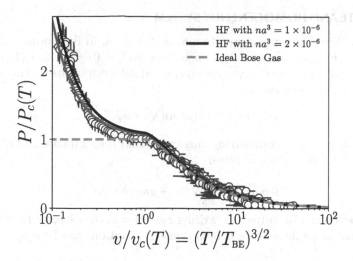

$$v/v_c(T) = (T/T_{\mathrm{BE}})^{3/2}$$

FIGURE 14.3 Equation of state of a homogeneous Bose gas of density $n = 1/v$. The volume per particle v is rescaled by a factor $v_c(T) = \lambda_T^3/\zeta(3/2)$ and the pressure by a factor $P_c(T) = k_{\mathrm{B}}T\zeta(5/2)/\lambda_T^{-3}$. The experimental data (points with error bars) are extracted from Mordini et al.[27] (courtesy of G. Lamporesi). The solid lines are plots of the Hartree-Fock results [Equations (14.57) and (14.60)] for two values of the interaction parameter corresponding to $na^3 = 1 \times 10^{-6}$ and 2×10^{-6}, where a is the s-wave scattering length (14.7).

experimental and theoretical results observed in Figure 14.3 does not point to a failure of the Hartree-Fock approximation, but is rather due to an oversimplification of the theoretical model: the experiments are performed on an elongated trapped system for which the density n varies from slice to slice whereas the interaction parameter (measured by a) remains constant. As a result, the product na^3 is not constant throughout the system contrarily to what has been assumed for simplicity in the theoretical model. Figure (14.3) shows the theoretical equation of state for two values of the interaction parameter $na^3 = 2\times10^{-6}$ which is the value at the centre of the trap, i.e. the largest value in the system and, for comparison, $na^3 = 1 \times 10^{-6}$. If instead the theory is conducted by taking, for each slice of the system (i.e. for each value of v), a parameter $na^{1/3}$ fixed experimentally, an excellent agreement between Hartree-Fock theory and the experimental results is obtained (see Note 27).

14.5 ELEMENTARY EXCITATIONS

Small spatial and temporal perturbations of the ground state are called *elementary excitations*. Their study is of major interest because they govern the system's low temperature thermodynamics. In the trapped case, however, their importance must be moderated, since at the usual experimental temperatures, it is the depletion effect of the condensate (studied in Section 14.3.2) that is the main phenomenon. It is only at very low temperatures that the elementary excitations contribute dominantly to the thermodynamic properties. However, their theoretical study is still interesting because it enables the description of a rich dynamics relevant to many non-equilibrium experimental situations. The study of the behaviour of the elementary excitations of the homogeneous system is first studied in Section 14.5.1. It is quite informative because it involves two similar but not identical phenomena, namely superfluidity and Bose–Einstein condensation.

14.5.1 INFINITE AND HOMOGENEOUS SYSTEM

Consider a Bose–Einstein condensate of uniform density n, in its ground state at $T = 0$ K. In accordance with Equations (14.46) (where here $n_0(T = 0) = n$) and (14.5) (where here $V_{\text{trap}} \equiv 0$), $\mu = gn$ and $\phi(\vec{r}, t) = \sqrt{n} \exp(-i\mu t/\hbar)$. Small excitations in the vicinity of this solution are sought in the form

$$\phi(\vec{r}, t) = \exp(-i\mu t/\hbar)\Big(\sqrt{n} + \delta\phi(\vec{r}, t)\Big), \tag{14.61}$$

assuming that $|\delta\phi| \ll \sqrt{n}$. Transferring this expression into Equation (14.5) and keeping only the linear terms in $\delta\phi$, one obtains:

$$i\hbar\,\partial_t \delta\phi = -\frac{\hbar^2}{2m}\vec{\nabla}^2 \delta\phi + gn\,(\delta\phi + \delta\phi^*). \tag{14.62}$$

This linear equation can be solved by writing $\delta\phi(\vec{r}, t) = \varphi_{\text{R}}(\vec{r}, t) + i\,\varphi_{\text{I}}(\vec{r}, t)$, where φ_{R} and φ_{I} are real functions. Since the system is homogeneous, solutions can be sought in the form of plane waves:

$$\begin{pmatrix} \varphi_{\text{R}}(\vec{r}, t) \\ \varphi_{\text{I}}(\vec{r}, t) \end{pmatrix} = \exp\left\{\frac{i}{\hbar}\big(\vec{p}\cdot\vec{r} - \epsilon\,t\big)\right\}\begin{pmatrix} \mathcal{A} \\ \mathcal{B} \end{pmatrix}, \tag{14.63}$$

where $\mathcal{A}$ and $\mathcal{B}$ are two complex constants[28]. Inserting this expression into Equation (14.62), one gets:

$$\begin{pmatrix} i\epsilon & p^2/(2m) \\ -p^2/(2m) - 2g\,n & i\epsilon \end{pmatrix}\begin{pmatrix} \mathcal{A} \\ \mathcal{B} \end{pmatrix} = \begin{pmatrix} 0 \\ 0 \end{pmatrix}. \tag{14.64}$$

This linear system has a non-trivial solution only if its determinant is zero, i.e.

$$\epsilon = \left[\frac{gn}{m}p^2 + \left(\frac{p^2}{2m}\right)^2\right]^{1/2} = c_s\,p\left[1 + \frac{1}{4}\left(\frac{p\xi}{\hbar}\right)^2\right]^{1/2}, \tag{14.65}$$

where

$$c_s = \left(\frac{gn}{m}\right)^{1/2} \quad \text{and} \quad \xi = \frac{\hbar}{mc_s}. \tag{14.66}$$

Relation (14.65) is the "Bogoliubov dispersion relation". It corresponds to an acoustic spectrum $\epsilon \simeq c_s p$ at long wavelengths ($p\xi \ll \hbar$, i.e. $\epsilon \ll gn = \mu$) and to a spectrum that resembles that of free particles $\epsilon \simeq p^2/(2m) + \mu$ at short wavelengths ($p\xi \gg \hbar$, i.e. $\epsilon \gg \mu$). This dispersion relation is represented in Figure 14.4. At low energies, the interpretation of elementary excitations as sound waves is corroborated by Expression (14.66) for the sound velocity c_s which corresponds to the generic expression at $T = 0$ K[29] : $mc_s^2 = \partial P/\partial n$ where[30] $P = -(\partial E/\partial V)_N = \frac{1}{2}gn^2$. Since a low energy excitation is a sound wave, it is a collective mode involving a large number of particles. At high energies, the physical interpretation of the dispersion relation [Equation (14.65)] is simpler because it involves only one particle: the relation $\epsilon \simeq p^2/(2m) + \mu$ is interpreted as the transfer of an atom from the condensate (energy $gn = \mu$) to an excited state belonging to the thermal fraction where the sum of the kinetic energy and the interaction with the condensate is[31] $p^2/(2m) + 2gn$. The energy to be supplied for this process is $p^2/(2m) + \mu$.

---

[28]It might seem odd to seek to represent the real functions φ_{R} and φ_{I} in the form of complex exponentials. This complex notation is, however, common. For example, one should read $\varphi_{\text{R}} = \text{Re}\,\{\mathcal{A}\exp[\frac{i}{\hbar}(\vec{p}\cdot\vec{r} - \epsilon\,t)]\}$. For the sake of legibility, the notation "real part" (Re), does not appear in Equations (14.63) and (14.64), although implicitly present.

[29]see https://en.wikipedia.org/wiki/Speed_of_sound

[30]In a homogeneous system at $T = 0$ K, the ground state energy given by Equation (14.11) is $E = \frac{1}{2}g\int d^3r\,n^2 = \frac{1}{2}gN^2/V$.

[31]The term $2gn$ is the Hartree-Fock interaction energy at $T = 0$ K, see Equation (14.35) and the following discussion.

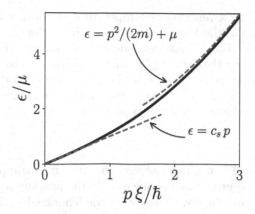

FIGURE 14.4 Solid curve: Bogoliubov dispersion relation given by Equation (14.65). The dashed curves are the low and high energy approximations.

The excitation spectrum of a homogeneous (weakly interacting) Bose gas was just determined. The analysis carried out in Section 7.3 for liquid helium can be applied in the present case also: the shape of the spectrum plotted in Figure 14.4 corresponds, like that of helium II in Figure 7.10, to a superfluid behaviour for which a Landau critical velocity [Equation (7.29)] and a normal (non-superfluid) fraction at non-zero temperature [Equation (7.34)] can be determined.

Relation (7.29) immediately gives the value of the critical velocity above which superfluidity disappears:

$$v_c = \min\left\{\frac{\epsilon(p)}{p}\right\} = c_s, \tag{14.67}$$

where c_s is given by Equation (14.66). This analysis shows that the non-interacting gas (for which $g \equiv 0$), although subject to Bose–Einstein condensation, is not superfluid. At variance, an interacting dilute gas displays superfluidity and its Landau critical velocity c_s is typically of the order of mm/s. The breaking of superfluidity according to Landau mechanism (by emission of elementary excitations) is, as is the case for superfluid helium, difficult to observe: nonlinear excitations (vortices or solitons) are emitted by the flow before the velocity c_s is reached.

In the superfluid regime, the density ρ_n of the normal fraction can be determined using Formula (7.34) which involves an integral whose integrand is peaked around $\epsilon \approx k_B T$. There are two temperature regimes in which an analytic result is easily obtained:

• $k_B T \gg \mu$: then one can write $\epsilon(p) \simeq p^2/(2m) + \mu$ in the integrand of Equation (7.34). With the change of variable $x = \beta p^2/(2m)$, one obtains:

$$\frac{\rho_n}{m} = \frac{4}{3\sqrt{\pi}} \frac{1}{\lambda_T^3} \int_0^\infty x^{3/2} \frac{e^{\beta\mu} e^x}{(e^{\beta\mu} e^x - 1)^2} dx = \frac{1}{\lambda_T^3} g_{3/2}(e^{-\beta\mu}), \tag{14.68}$$

where an integration by parts was performed and Definition (A.6) of the Bose integral was used. Comparing to Equation (14.47), one sees that $\rho_n = m\,n_{\text{th}}$: in this case, the normal fraction corresponds to the thermal cloud and behaves as $T^{3/2}$ (see Equation (14.48)).

• $k_B T \ll \mu$: in the integrand of Equation (7.34) the approximation $\epsilon(p) \simeq c_s\,p$ can be made and one exactly obtains Expression (7.35), with a normal fraction $\rho_n \propto T^4$. Then, in this regime, it can be shown that the thermal fraction is of the order of T^2.

The behaviour of ρ_n as a function of temperature is another indication of what was already noted in the case of the ideal Bose gas (which can form a Bose–Einstein condensate but not a superfluid): Bose–Einstein condensation and superfluidity are not identical phenomena. Here, it is seen that the depletion of the condensate and the non-superfluid part are not identical in the regime $k_B T \ll \mu$. They become so when $k_B T \gg \mu$ since in this case one is dealing with a region of the spectrum where elementary excitations lose their collective character and behave as individual particles.

14.5.2 TRAPPED GAS

When the system at $T = 0$ is not homogeneous because it is trapped in a harmonic well, when following the same approach as the one used in the previous section, one ends up with equations solvable only numerically. However, in the Thomas-Fermi limit characterised by Equation (14.17), an approximate approach can be used and leads to analytic expressions of the frequencies of the eigenmodes[32]. To do so, it is useful to put the Gross-Pitaevskii Equation (14.5) into a hydrodynamical form by writing: $\phi(\vec{r}, t) = A(\vec{r}, t) \exp(iS(\vec{r}, t))$, where A and S are two real-valued functions. Inserting this expression back into Equation (14.5) and considering the real and imaginary parts of this equation leads to two coupled equations governing the dynamics of the fields A and S:

$$\partial_t A^2 + \vec{\nabla} \cdot \left(A^2 \frac{\hbar}{m} \vec{\nabla} S \right) = 0, \quad \hbar \partial_t S = \frac{\hbar^2}{2m} \left(\frac{\vec{\nabla}^2 A}{A} - |\vec{\nabla} S|^2 \right) - V_{\text{trap}} - g A^2. \quad (14.69)$$

The first equation has the form of a conservation equation for the density $n = A^2 = |\phi|^2$ in the presence of a velocity field $\vec{v}$:

$$\partial_t n + \vec{\nabla} \cdot (n \vec{v}) = 0, \quad \text{where} \quad \vec{v}(\vec{r}, t) = \frac{\hbar}{m} \vec{\nabla} S. \quad (14.70)$$

If the gradient of the second equation is taken, after some manipulation it can be put in the form of an Euler equation[33] describing the dynamics of an inviscid fluid subjected to a potential V_{trap}:

$$m \left(\partial_t + \vec{v} \cdot \vec{\nabla} \right) \vec{v} = -\vec{\nabla} V_{\text{trap}} - \frac{1}{n} \vec{\nabla} \left(P + P_{\text{qu}} \right), \quad (14.71)$$

where $P = \frac{1}{2} g n^2$ is the pressure caused by interatomic repulsion [$T \to 0$ limit of Expression (14.60)] and P_{qu} is called the *quantum pressure*. Its origin is the quantum contribution to the kinetic energy of the particles. It reads

$$P_{\text{qu}} = \frac{\hbar^2}{4m} \left(\frac{|\vec{\nabla} n|^2}{n} - \vec{\nabla}^2 n \right) = \frac{\hbar^2}{2m} \left(|\vec{\nabla} A|^2 - A \vec{\nabla}^2 A \right). \quad (14.72)$$

The quantum pressure P_{qu} to (14.71) is small with respect to P when $\hbar^2 / m L_{\text{typ}}^2 \ll g \, n_{\text{typ}}$, where L_{typ} is the characteristic length over which typical variations of n occur and n_{typ} is the order of magnitude of n. This regime thus corresponds to the Thomas-Fermi limit where $E_{\text{kin}} \ll E_{\text{int}}$ (see Section 14.2.2), which can also be put in the form $L_{\text{typ}} \gg \xi_{\text{typ}}$, where ξ_{typ} is a typical value of the "relaxation length" ξ defined in Equation (14.66). In this regime, the quantum pressure term is negligible and Equation (14.71) can be written as:

$$m \partial_t \vec{v} + \vec{\nabla} \left(\frac{1}{2} m \vec{v}^2 + V_{\text{trap}} + g n \right) = 0. \quad (14.73)$$

---

[32] S. Stringari, Phys. Rev. Lett. **77**, 2360 (1996).

[33] see https://en.wikipedia.org/wiki/Euler_equations_(fluid_dynamics)

It is clear that the Thomas-Fermi profile [Equation (14.18)] is a stationary solution (with $\vec{v} = 0$) of Equations (14.70) and (14.73). To study the elementary excitations in the vicinity of this profile, one can write $n(\vec{r}, t) = n(\vec{r}) + \delta n(\vec{r}, t)$ and linearise the time-dependent equations by considering δn and $\vec{v}$ as small contributions. This leads to[32]:

$$\partial_t^2 \delta n - \vec{\nabla} \cdot \left(c_s^2(\vec{r}) \vec{\nabla} \delta n \right) = 0, \tag{14.74}$$

where $c_s(\vec{r}) = [gn(\vec{r})/m]^{1/2}$ can be interpreted as the local sound velocity[34]. Note that, in a system where the density is not homogeneous, this interpretation only makes sense in the limit $\xi_{\text{typ}} \ll L_{\text{typ}}$ where the density variations occur on a spatial scale much larger than the relaxation length, which corresponds exactly to the Thomas-Fermi regime.

Exercise 14.6. Use the approximate dynamics described by Equation (14.74) to study the dispersion relation of elementary excitations in a homogeneous condensate.

Solution: In a homogeneous condensate, $c_s = [gn/m]^{1/2}$ is independent on position. One can therefore look for plane wave solutions of Equation (14.74): $\delta n(\vec{r}, t) = \mathcal{A} \exp(i(\vec{p} \cdot \vec{r} - \epsilon t)/\hbar)$. This gives $\epsilon = c_s p$, which corresponds to the long-wavelength approximation of the exact dispersion relation (14.65), see Figure 14.4. This clarifies the domain of validity of the approach leading to Equation (14.74): Thomas-Fermi approximation is valid in the regime where the quantum pressure associated with the fluctuations δn (it is of order $(\hbar^2/m)\vec{\nabla}^2 \delta n \approx p^2 \delta n/m$) is small compared to the excess pressure associated with density fluctuations (which is itself of order $gn\delta n$). This imposes $p\xi \ll \hbar$ and $\epsilon \ll gn = \mu$, which is indeed the regime in which the Bogoliubov dispersion relation can be approximated by $\epsilon = c_s p$.

In a system where the density is not homogeneous, the wavelength λ must also be small compared to the typical size of the variations of $n(\vec{r})$. In a harmonic trap, this imposes: $\xi_{\text{typ}} \ll \lambda \ll R$, where R is the radius of the condensate given by Equation (14.19).

Equation (14.74) describes the dynamics of excitations of a trapped condensate. In the following, the case of an isotropic harmonic trap is considered. One possible mode of excitation is the "breathing mode" (or monopole) in which the density profile undergoes periodic isotropic expansion. This corresponds to a solution of Equation (14.74) in which $v(\vec{r}, t) \propto \vec{r} \exp(-i\Omega t)$ for $r < R$, where R is the radius of the condensate given by Equation (14.19). The linearised version of Equation (14.70) then enables the calculation of δn which is found in the form[35] $\delta n(\vec{r}, t) \propto (1 - 5r^2/3R^2) \exp(-i\Omega t)$. Inserting this expression into Equation (14.74) where, according to Equation (14.18), $c_s^2(\vec{r}) = \frac{1}{2}\omega^2(R^2 - r^2)$, one obtains $\Omega = \sqrt{5}\,\omega$.

It can be shown[32] that it is legitimate to look for generic harmonic solutions of Equation (14.74) in the form $\delta n \propto r^\ell P(r) Y_{\ell,m}(\theta, \varphi) \exp(-i\Omega t)$, where P is a polynomial of even order in r and $Y_{\ell,m}$ is a spherical harmonic. Among these solutions, those for which the polynomial P is a constant are particularly simple to deal with. They are called *surface modes* since the deformation is mainly localised on the surface of the condensate (and this more so as ℓ is large). Since $\vec{\nabla}^2(r^\ell Y_{\ell,m}) = 0$, for these modes, Equation (14.74) takes the form $-\Omega^2 \delta n = \vec{\nabla} c_s^2 \cdot \vec{\nabla} \delta n$. Since the gradient of c_s^2 has only a radial component ($\vec{\nabla} c_s^2 = -\omega^2 r\, \vec{e}_r$), one easily obtains $\Omega = \sqrt{\ell}\,\omega$.

The collective oscillations of trapped condensates have been studied in several experiments[36] which have highlighted the relevance of the approach relying on Equation (14.74) in the Thomas-Fermi limit [Relation (14.17)]. It should be noted that, in this limit, the

---

[34]Compare with Expression (14.66) valid in the uniform case.

[35]The divergence of a vector field of the form $\vec{V} = V\,\vec{e}_r$ is: $\vec{\nabla} \cdot \vec{V} = r^{-2}\partial_r(r^2 V)$.

[36]See for example, D. S. Jin, J. R. Ensher, M. R. Matthews, C. E. Wieman, and E. A. Cornell, Phys. Rev. Lett. **77**, 420 (1996).

eigenfrequencies do not depend on the interaction parameter g. However, they are clearly different from the angular frequencies of a non interacting condensate which are of the form $\Omega = (n_x + n_y + n_z)\omega = (2n_r + \ell)\omega$, where ℓ is the quantum number corresponding to the orbital angular momentum and n_r corresponds to the radial excitations[37] (e.g., $n_r = 0$ for the surface modes studied above).

---

[37]see for example, Claude Cohen-Tannoudji, Bernard Diu, Franck Laloë, *Quantum Mechanics* (volume I) (Wiley-VCH, 2019).

A Mathematical Complements

A.1 SPECIAL FUNCTIONS

- The Riemann zeta function is defined for $\nu > 1$ as:

$$\zeta(\nu) = \sum_{n=1}^{\infty} \frac{1}{n^{\nu}}. \tag{A.1}$$

Particular values of the function are given in Table A.1. The method used for the calculation of $\zeta(2)$ and $\zeta(4)$ is explained in Section A.6.2.

TABLE A.1 Useful values of the Riemann zeta function.

ν	3/2	2	5/2	3	7/2	4
$\zeta(\nu)$	2.61238	$\pi^2/6$	1.34149	1.20206	1.12673	$\pi^4/90$

- The function $\Gamma(x)$ is defined for $x > 0$ as:

$$\Gamma(x) = \int_0^{\infty} t^{x-1} \exp(-t)\,dt. \tag{A.2}$$

The Γ function is a generalisation of the factorial for a non-integer argument x. Indeed, it can be shown by means of an integration by parts that:

$$\Gamma(x + 1) = x\,\Gamma(x). \tag{A.3}$$

Since $\Gamma(1) = 1$, for $n \in \mathbb{N}^*$, one obtains by induction:

$$\Gamma(n) = (n - 1)! = (n - 1) \times (n - 2) \times \cdots \times 2. \tag{A.4}$$

It is useful to note that[1] $\Gamma(1/2) = \sqrt{\pi}$. Equation (A.3) then leads to the following result:

$$\Gamma(n + 1/2) = \sqrt{\pi} \left(n - \frac{1}{2}\right) \times \left(n - \frac{3}{2}\right) \times \cdots \times \frac{3}{2} \times \frac{1}{2} = \sqrt{\pi}\,\frac{(2n)!}{2^{2n}\,n!}. \tag{A.5}$$

- "Bose functions" g_{ν} are introduced in Chapter 7. They are defined for $z \in [-\infty, 1]$ and $\nu > 1$ as:

$$g_{\nu}(z) = \frac{1}{\Gamma(\nu)} \int_0^{\infty} \frac{x^{\nu-1}\,dx}{z^{-1}\exp(x) - 1}, \tag{A.6}$$

where $\Gamma(\nu)$ is given by Equation (A.2). When $z \in [0, 1]$, the series expansion $(a - 1)^{-1} = \sum_{n=1}^{\infty} a^{-n}$ valid for $a > 1$ leads to a series expansion of g_{ν}:

$$g_{\nu}(z) = \frac{1}{\Gamma(\nu)} \sum_{n=1}^{\infty} z^n \int_0^{\infty} x^{\nu-1} \exp(-nx)dx = \sum_{n=1}^{\infty} \frac{z^n}{n^{\nu}}, \tag{A.7}$$

---

[1] A change of variable $t = u^2$ in Definition (A.2) allows to write $\Gamma(1/2) = 2I_0(1)$, where I_0 is a Gaussian integral (A.12). The desired result then follows from (A.15).

DOI: 10.1201/9781003272427-A

where Equation (A.2) was used. For the particular value $z = 1$, one gets (see Equation (A.1))

$$g_\nu(1) = \zeta(\nu). \tag{A.8}$$

• Similarly to the Bose case, Fermi-Dirac integrals[2] are introduced in Chapter 8. They are defined for $z \in [-1, +\infty]$ and $\nu > 1$ by:

$$f_\nu(z) = \frac{1}{\Gamma(\nu)} \int_0^\infty \frac{x^{\nu-1}\,dx}{z^{-1}\exp(x) + 1}, \tag{A.9}$$

where $\Gamma(\nu)$ is defined by Equation (A.2). When $z \in [0, 1]$, the series expansion $(a + 1)^{-1} = -\sum_{n=1}^\infty (-a)^{-n}$, valid for $a > 1$, can be carried over into Equation (A.9) leading to:

$$f_\nu(z) = \frac{1}{\Gamma(\nu)} \sum_{n=1}^\infty (-1)^{n+1} z^n \int_0^\infty x^{\nu-1}\exp(-nx)dx = \sum_{n=1}^\infty (-1)^{n+1} \frac{z^n}{n^\nu}. \tag{A.10}$$

This series expansion allows to calculate the value of $f_\nu(1)$: separating the odd and even contributions to the series, one gets:

$$f_\nu(1) = \left(1 - \frac{1}{2^{\nu-1}}\right)\zeta(\nu), \tag{A.11}$$

where ζ is the Riemann zeta function defined by Equation (A.1).

A.2   GAUSSIAN INTEGRALS

Gaussian integrals of the form

$$I_n(\alpha) = \int_0^\infty x^n e^{-\alpha x^2}dx, \tag{A.12}$$

where $\alpha > 0$ and n is a positive integer, appear in many instances throughout this book. A first relation, important for their calulation, is[3]:

$$I_{n+2}(\alpha) = \frac{n+1}{2\alpha} I_n(\alpha). \tag{A.13}$$

The integral $I_1(\alpha)$ can be calculated without difficulties:

$$I_1(\alpha) = \int_0^\infty x\,e^{-\alpha x^2}dx = \left[-\frac{e^{-\alpha x^2}}{2\alpha}\right]_0^\infty = \frac{1}{2\alpha},$$

and then the recurrence relation (A.13) allows the calculation any Gaussian integral with odd subscript:

$$I_1(\alpha) = \frac{1}{2\alpha}, \quad I_3(\alpha) = \frac{1}{2\alpha^2}, \quad I_5(\alpha) = \frac{1}{\alpha^3} \cdots \tag{A.14}$$

---

[2] Bose integrals (A.6) and Fermi-Dirac integrals (A.9) are particular forms of the polylogarithmic function $\mathrm{Li}_\nu(z)$ defined for $|z| < 1$ and ν real as: $\mathrm{Li}_\nu(z) = \sum_{n=1}^\infty z^n/n^\nu$. Thus, $g_\nu(z) = \mathrm{Li}_\nu(z)$ and $f_\nu(z) = -\mathrm{Li}_\nu(-z)$, see https://fr.wikipedia.org/wiki/Polylogarithm.

[3] This can be demonstrated by differentiation of $I_{n+2}(\alpha)$ with respect to α or by by performing an integration by parts:

$$I_n(\alpha) = \int_0^\infty x^n e^{-\alpha x^2}dx = \left[\frac{x^{n+1}}{n+1}e^{-\alpha x^2}\right]_0^\infty + \frac{2\alpha}{n+1}\int_0^\infty x^{n+2}e^{-\alpha x^2}dx = \frac{2\alpha}{n+1}I_{n+2}(\alpha).$$

$I_0(\alpha)$ can be calculated by means of a trick which consists in writing its square as a double integral:

$$I_0^2(\alpha) = \left(\int_0^\infty e^{-\alpha x^2} dx \right) \left(\int_0^\infty e^{-\alpha y^2} dy \right) = \int_0^\infty \int_0^\infty e^{-\alpha(x^2+y^2)} dx dy.$$

This integral maybe computed using polar coordinates $x^2 + y^2 = r^2$, $0 \le \theta \le \pi/2$ and $dx dy = r dr d\theta$:

$$I_0^2(\alpha) = \int_0^{\pi/2} d\theta \int_0^\infty dr\, r\, e^{-\alpha r^2} = \frac{\pi}{2} I_1(\alpha) = \frac{\pi}{4\alpha}.$$

The knowledge of $I_0(\alpha)$ and relation (A.13) allows the calculation of every integral with an even subscript:

$$I_0(\alpha) = \frac{1}{2}\sqrt{\frac{\pi}{\alpha}}, \quad I_2(\alpha) = \frac{1}{4\alpha}\sqrt{\frac{\pi}{\alpha}}, \quad I_4(\alpha) = \frac{3}{8\alpha^2}\sqrt{\frac{\pi}{\alpha}} \dots \tag{A.15}$$

Note also that, thanks to a change of variable, $I_\nu(\alpha)$ can be written in terms of the Γ function defined by Equation (A.2):

$$I_\nu(\alpha) = \frac{\Gamma\left(\frac{\nu+1}{2}\right)}{2\,\alpha^{(\nu+1)/2}}, \quad \text{for all } \nu \in \mathbb{R}^+ . \tag{A.16}$$

A.3   SADDLE-POINT METHOD

In statistical physics, integrals of the type:

$$I = \int_a^b dt\, \exp\{-f(t)\}, \tag{A.17}$$

often need to be evaluated, where the function $f(t)$ has a minimum in t_c (a typical shape of $f(t)$ is shown in Figure A.1).

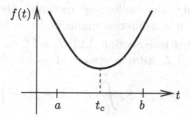

FIGURE A.1   Shape of the function $f(t)$ appearing in the definition of integral (A.17).

The dominant contribution to I comes from the vicinity of t_c which corresponds to the region where $\exp(-f)$ reaches its maximum value. This suggests an approximate method for evaluating I: in the vicinity of t_c, one can write $f(t) \simeq f(t_c) + \frac{1}{2}f''(t_c)(t-t_c)^2 + \dots$ This gives, by extending the integration domain to the whole real axis and using Expression (A.15) for $I_0(1)$

$$I \simeq e^{-f(t_c)} \sqrt{\frac{2\pi}{f''(t_c)}}. \tag{A.18}$$

The Gaussian function which approximates $\exp(-f)$ is integrated over $\mathbb{R}$ while, in I, the integration interval is limited to interval $[a, b]$. This approximation is valid only if $(b - t_c)^2$ and $(a - t_c)^2$ are large compared to $1/f''(t_c)$ (in which case the error is exponentially small).

As an illustration of the method, let us apply it for calculating an approximate value of $\Gamma(x+1)$ when $x \gg 1$. To this end, it is appropriate to write Equation (A.2) under the form $\Gamma(x+1) = \int_0^\infty \exp\{-f(t)\}\,dt$, where $f(t) = t - x\ln t$. The derivative f' cancels for $t_c = x$ and $f''(t_c) = 1/x$. Approximation (A.18) then leads to:

$$\Gamma(x+1) \simeq \sqrt{2\pi x}\,(x/e)^x \,. \tag{A.19}$$

In this approximate calculation, the lower bound of the integral ($b = 0$) was changed to $-\infty$. As discussed above, this is permissible when $(b - t_c)^2 = x^2 \gg 1/f''(t_c) = x$. Thus, Equation (A.19) is an approximation valid for large x.

Relation (A.19) is called *Stirling's formula*. It is often used to obtain an approximation of $N! = \Gamma(N+1)$ when $N \gg 1$. The results explicitly referred to in the main text is:

$$N! \simeq N^N e^{-N} \sqrt{2\pi N}. \tag{A.20}$$

The relative error induced by the use of Formula (A.20) is about 1% for $N = 10$, 0.1% for $N = 100$ and completely negligible for $N = 10^{23}$. In statistical physics, it is often $\ln N!$ which is evaluated approximately. In the thermodynamic limit, the term $\ln \sqrt{2\pi N}$ is completely negligible[4] compared to $N(\ln N - 1)$ which allows to write:

$$\ln N! \simeq N \ln N - N. \tag{A.21}$$

A.4   VOLUME OF A HYPERSPHERE

A hypersphere is a sphere in $\mathbb{R}^n$. In spherical coordinates, the infinitesimal volume element is $d^n r = r^{n-1}dr\,d^{n-1}\Omega$, where $d^{n-1}\Omega$ is the integration element over angles (in $\mathbb{R}^2$: $d\varphi$ and in $\mathbb{R}^3$: $d^2\Omega = \sin\theta d\theta d\varphi$). The angular integral S_n reads:

$$S_n = \int d^{n-1}\Omega. \tag{A.22}$$

It can be determined using the following method: the auxiliary integral $J_n = \int_{\mathbb{R}^n} d^n r \exp(-\vec{r}^{\,2})$ is calculated in two different manners. First, in a system of cartesian coordinates. In this case it is clear that (see Section A.2): $J_n = [\int_{\mathbb{R}} dx \exp(-x^2)]^n = [2I_0(1)]^n = \pi^{\frac{n}{2}}$. Computing instead the integral J_n using a system of spherical coordinates yields:

$$\pi^{\frac{n}{2}} = J_n = S_n \int_0^\infty r^{n-1}dr \exp(-r^2). \tag{A.23}$$

A change of variable in the right-hand term of Equation (A.23) leads to $J_n = \frac{1}{2}S_n\Gamma(\frac{n}{2})$, where the Γ function is defined by relation (A.2). Carrying over into Equation (A.23), one obtains:

$$S_n = \frac{2\pi^{\frac{n}{2}}}{\Gamma(\frac{n}{2})}. \tag{A.24}$$

Then, the volume of a hypersphere of radius $\mathcal{R}$ in $\mathbb{R}^n$ is

$$\mathcal{V}_n(\mathcal{R}) = S_n \int_0^{\mathcal{R}} r^{n-1}dr = \frac{\pi^{\frac{n}{2}}}{\Gamma(\frac{n}{2}+1)}\,\mathcal{R}^n = \frac{\pi^{\frac{n}{2}}}{(\frac{n}{2})!}\,\mathcal{R}^n, \tag{A.25}$$

where relation (A.3) was used: $\frac{n}{2}\Gamma(\frac{n}{2}) = \Gamma(\frac{n}{2}+1)$. This expression indeed matches well known results: $\mathcal{V}_1(\mathcal{R}) = 2\mathcal{R}$, $\mathcal{V}_2(\mathcal{R}) = \pi\mathcal{R}^2$ and $\mathcal{V}_3(\mathcal{R}) = \frac{4}{3}\pi\mathcal{R}^3$.

---

[4]For $N = 10^{23}$, this corresponds to neglecting 27 with respect to 10^{24}.

A.5   METHOD OF LAGRANGE MULTIPLIERS

The extremum (maximum or minimum) of a function $f(x_1, x_2, \ldots, x_n)$ of n variables is determined by

$$\mathrm{d}f = \sum_i \frac{\partial f}{\partial x_i} \mathrm{d}x_i = 0.$$

If variables $x_1, x_2, \ldots, x_n$ are independent, this condition is met when all the partial derivatives of f cancel:

$$\frac{\partial f}{\partial x_i} = 0, \quad \text{for} \quad i = 1, 2, \ldots, n. \tag{A.26}$$

If the variables are not independent, but linked together by a constraint that can be written as

$$g(x_1, x_2, \ldots, x_n) = 0, \tag{A.27}$$

then the extremum of f is no longer reached when the n Equations (A.26) are verified.

Let us take the example of a manufacturer who wants to produce cylindrical cans of fixed volume V_0, with the smallest possible surface area $\mathcal{A}$ (to minimise the amount of metal used). How to choose the height L and the radius R of the cylinder? In this very simple case, the function to minimise is the surface of the can $\mathcal{A} = f(L, R) = 2 \times \pi R^2 + 2\pi R \times L$, with the constraint $g(L, R) = \pi R^2 L - V_0 = 0$. Substituting $L = V_0/\pi R^2$ into $f(L, R)$, easily reduces to a one-variable function, which minimum is easily found to be reached for $R = (V_0/2\pi)^{1/3}$, that is for a ratio $L/R = 2$.

However, the approach just illustrated on a simple example is difficult to implement for a large number of variables. In this case, a very convenient technique known as the "method of Lagrange multipliers" allows the determination of the function extrema. A function $F(x_1, x_2, \ldots, x_n, \mu)$ is introduced which depends then of $n + 1$ *independent* variables:

$$F(x_1, x_2, \ldots, x_n, \mu) = f(x_1, x_2, \ldots, x_n) - \mu \underbrace{g(x_1, x_2, \ldots, x_n)}_{=0},$$

The extrema of this function are reached when (compare with Equations (A.26)):

$$\frac{\partial F}{\partial x_i} = \frac{\partial f}{\partial x_i} - \mu \frac{\partial g}{\partial x_i} = 0 \quad \text{for } i = 1, 2, \ldots, n, \tag{A.28}$$

$$\frac{\partial F}{\partial \mu} = -g(x_1, x_2, \ldots, x_n) = 0,$$

This last equation is none other than Constraint (A.27). The n first equations in Expression (A.28) thus relate the values of x_i which extremise F to the Lagrange multiplier μ, which is then calculated according to Constraint (A.27).

To illustrate this method, the above example of the can is used again. One looks for the minimum of $F(L, R, \mu) = f(L, R) - \mu g(L, R)$, such as

$$\frac{\partial F}{\partial L} = \frac{\partial f}{\partial L} - \mu \frac{\partial g}{\partial L} = 2\pi R - \mu \pi R^2 = 0$$

$$\frac{\partial F}{\partial R} = \frac{\partial f}{\partial R} - \mu \frac{\partial g}{\partial R} = 4\pi R + 2\pi L - 2\mu \pi R L = 0.$$

These two equations impose $R = 2/\mu$ and $L = 4/\mu$. From the constraint $g(L, R) = 0$, that is $V_0 = \pi R^2 L = 16\pi/\mu^3$, one deduces the value $\mu = (16\pi/V_0)^{1/3}$ and finally we have $R = L/2 = (V_0/2\pi)^{1/3}$.

In general, the extremum of a function $f(x_1, x_2, \ldots, x_n)$ with n variables linked together by M constraints $g_j(x_1, x_2, \ldots, x_n) = 0$, where $j = 1, 2, \ldots, M$ is given by the extremum of the function F with $n + M$ *independent* variables:

$$F(x_1, x_2, \ldots, x_n, \mu_1, \mu_2, \ldots, \mu_M) = f(x_1, x_2, \ldots, x_n) - \sum_{j=1}^{M} \mu_j \, g_j(x_1, x_2, \ldots, x_n),$$

where M Lagrange (μ_j for $j = 1, 2, \ldots, M$) have been introduced.

A.6   FOURIER ANALYSIS

A.6.1   FOURIER TRANSFORM

Consider a function $f(x)$ (with complex or real values) of a real variable x. Its Fourier transform $\hat{f}(k)$ is defined as:

$$\hat{f}(k) = \int_{\mathbb{R}} \mathrm{d}x \, f(x) \, \mathrm{e}^{-\mathrm{i}kx} \quad \text{and} \quad f(x) = \int_{\mathbb{R}} \frac{\mathrm{d}k}{2\pi} \, \hat{f}(k) \, \mathrm{e}^{\mathrm{i}kx}. \tag{A.29}$$

The first relation defines the Fourier transform $\hat{f}$. The second one deserves a proof, it is the inverse Fourier transform expressing f as a function of $\hat{f}$. Before providing a proof for this second relation, it is appropriate to evaluate the Fourier transform of a Gaussian function (in which α is a strictly positive real parameter):

$$\text{if} \quad f(x) = \exp\left(-\alpha \, x^2\right) \quad \text{then} \quad \hat{f}(k) = \sqrt{\frac{\pi}{\alpha}} \, \exp\left(-\frac{k^2}{4\,\alpha}\right). \tag{A.30}$$

This result can be obtained by writing $-\alpha x^2 - \mathrm{i}kx = -\alpha(x + \mathrm{i}k/2\alpha)^2 - k^2/4\alpha$ such that

$$\hat{f}(k) = \int_{\mathbb{R}} \mathrm{d}x \, \mathrm{e}^{-\mathrm{i}kx - \alpha x^2} = \mathrm{e}^{-k^2/4\alpha} \int_{\mathbb{R}} \mathrm{d}x \, \mathrm{e}^{-\alpha(x + \mathrm{i}k/2\alpha)^2}. \tag{A.31}$$

Then, the integration path of the integral of the right-hand term of Equation (A.31) can be shifted into the complex plane such that the integrand becomes real. Formula (A.15) can then be used to evaluate such a Gaussian integral, leading to Equation (A.30).

Another useful result shall be established:

$$\int_{\mathbb{R}} \mathrm{d}x \, \mathrm{e}^{\mathrm{i}(q-k)x} = 2\pi \, \delta(k - q), \tag{A.32}$$

where δ is the Dirac distribution. To do so, let us introduce the function $f_\alpha(x) = \mathrm{e}^{\mathrm{i}qx - \alpha x^2}$, where α is a positive real. Its Fourier transform can be easily calculated thanks to (A.30):

$$\int_{\mathbb{R}} \mathrm{d}x \, f_\alpha(x) \, \mathrm{e}^{-\mathrm{i}kx} = \int_{\mathbb{R}} \mathrm{d}x \, \mathrm{e}^{-\alpha x^2} \mathrm{e}^{\mathrm{i}(q-k)x} = 2\pi \, \frac{\mathrm{e}^{-(k-q)^2/4\alpha}}{\sqrt{4\pi\alpha}} \equiv 2\pi \, \delta_\alpha(k - q). \tag{A.33}$$

When $\alpha \to 0$ the function $\delta_\alpha(k - q)$ which is defined by the right-hand term of Equation (A.33) tends towards[5] $\delta(k - q)$, while, in the same limit, $f_\alpha(x)$ tends toward $\exp(\mathrm{i}qx)$, which demonstrates Formula (A.32), obtained as the limit $\alpha \to 0$ of Equation (A.33).

Once Formula (A.32) is proven, it is easy to show that the second relation of (A.29) is correct, and that it indeed enables the calculation of the inverse Fourier transform. This is achieved by writing

$$f(x) = \int_{\mathbb{R}} \mathrm{d}y \, \delta(x - y) \, f(y) = \int_{\mathbb{R}} \mathrm{d}y \, f(y) \int_{\mathbb{R}} \frac{\mathrm{d}k}{2\pi} \, \mathrm{e}^{\mathrm{i}k(x-y)} = \int_{\mathbb{R}} \frac{\mathrm{d}k}{2\pi} \, \hat{f}(k) \, \mathrm{e}^{\mathrm{i}kx}. \tag{A.34}$$

---

[5]Indeed, for all $\alpha > 0$: $\int_{\mathbb{R}} \mathrm{d}k \, \delta_\alpha(k - q) = 1$ and when $\alpha \to 0$, $\delta_\alpha(k - q)$ is an increasingly narrow function, with an increasingly sharp maximum reached when $k = q$.

In the second equality of relation (A.34), a form of Formula (A.32) was used, in which x and y play the role k and q. Then, in the third equality, the integrals were swapped and the first equation of Relations (A.29) was used.

The useful Parseval-Plancherel theorem is then easy to establish:

$$\int_{\mathbb{R}} \mathrm{d}x\, f^*(x)g(x) = \int_{\mathbb{R}} \frac{\mathrm{d}k}{2\pi}\, \hat{f}^*(k)\hat{g}(k), \qquad (A.35)$$

where $f^*(x)$ is the complex conjugate of $f(x)$. The result (A.35) can be proven by writing $f(x)$ and $g(x)$ as a function of their Fourier transform and then using (A.32).

All these formulas generalise directly to multidimensional and time-dependent situations. In such cases, Equation (A.29) becomes:

$$\hat{f}(\vec{k},\omega) = \int_{\mathbb{R}^3} \mathrm{d}^3r \int_{\mathbb{R}} \mathrm{d}t\, f(\vec{r},t)\, \mathrm{e}^{-\mathrm{i}(\vec{k}\cdot\vec{r}-\omega t)} \quad \text{and}$$
$$f(\vec{r},t) = \int_{\mathbb{R}^3} \frac{\mathrm{d}^3k}{(2\pi)^3} \int_{\mathbb{R}} \frac{\mathrm{d}\omega}{2\pi}\, \hat{f}(\vec{k},\omega)\, \mathrm{e}^{\mathrm{i}(\vec{k}\cdot\vec{r}-\omega t)}. \qquad (A.36)$$

A.6.2 FOURIER SERIES

A periodic function (of period a) can be decomposed into a Fourier series such that:

$$f(x) = \sum_{n\in\mathbb{Z}} f_n\, \mathrm{e}^{2\mathrm{i}\pi nx/a}, \quad \text{where} \quad f_n = \frac{1}{a}\int_0^a f(x)\, \mathrm{e}^{-2\mathrm{i}\pi nx/a}\mathrm{d}x. \qquad (A.37)$$

f_n are called *Fourier coefficients*. The right-hand side relation in Equation (A.37) can be easily proven by substituting in this formula $f(x)$ for its Fourier series expansion (left expression in Equation (A.37)).

A Parseval-Plancherel theorem similar to Formula (A.35) can here also be established:

$$\frac{1}{a}\int_0^a |f(x)|^2\mathrm{d}x = \sum_{n\in\mathbb{Z}} |f_n|^2. \qquad (A.38)$$

This formula enables the calculation of the value at certain points of the Riemann zeta function (A.1) which appears in quantum problems (in Chapters 7 and 8). In particular, it is possible[6] to calculate the value of $\zeta(2)$ and $\zeta(4)$ thanks to Formula (A.38). Let us perform a Fourier series expansion of the 2π-periodic function which is equal to x for $x \in\,]-\pi,\pi[$. According to Equation (A.37), its Fourier coefficient are $f_n = \frac{1}{2\pi}\int_{-\pi}^{\pi} x\,\mathrm{e}^{-\mathrm{i}nx}\mathrm{d}x = \mathrm{i}(-1)^n/n$ for $n \in \mathbb{N}^*$ and $f_0 = 0$. Parseval-Plancherel formula (A.38) gives $\sum_{n=-\infty}^{+\infty} |f_n|^2 = \frac{1}{2\pi}\int_{-\pi}^{\pi} x^2\mathrm{d}x = \pi^2/3$, and thus $\zeta(2) = \pi^2/6 = 1.644934\ldots$ The same technique can be used to calculate[7] $\zeta(4) = \pi^4/90 = 1.082323\ldots$

A.7 SEMI-CLASSICAL CALCULATIONS

The aim of this section is to establish a semi-classical formula used in Chapter 6 to obtain an evaluation of the density of state valid in the thermodynamic limit. To avoid confusion, quantum operators are written with a hat. To simplify the presentation, a one dimensional configuration is considered, but the generalisation to a higher number of dimensions does not present any difficulty.

---

[6]This calculation is known under the name of "Basel problem". It was solved by Euler in 1735, 90 years after it was first posed.

[7]In this case, one works with the 2π-periodic function which is equal to x^2 for $x \in\,]-\pi,\pi[$.

Let us consider a quantum system confined in a box: $x \in [-L/2, L/2]$. The size L of the box is supposed to be large with respect to any characteristic length of the system. A basis of the Hilbert space is built considering normalised plane waves verifying periodic boundary conditions:

$$\psi_n(x) = \frac{1}{\sqrt{L}} \exp\{ip_n x/\hbar\} \quad \text{where} \quad p_n = \frac{2\pi n \hbar}{L} = n\frac{h}{L} \quad \text{and} \quad n \in \mathbb{Z}. \qquad (A.39)$$

It is clear that $\psi_n(x + L) = \psi_n(x)$ and that $\int_{-L/2}^{L/2} \psi_m^*(x)\psi_n(x)\mathrm{d}x = \delta_{n,m}$.

Let $\hat{O}(\hat{x}, \hat{p})$ be an operator which classical equivalent is written as $O(x, p)$. $\hat{O}$ can be the position $\hat{x}$, the momentum $\hat{p}$, the Hamiltonian $\hat{H}$, etc. One wants to evaluate its trace. The calculation can be performed in the basis (A.39):

$$\mathrm{Tr}\,\hat{O} = \sum_{n \in \mathbb{Z}} \langle \psi_n | \hat{O} | \psi_n \rangle = \sum_{n \in \mathbb{Z}} \frac{1}{L} \int_{-L/2}^{L/2} \mathrm{d}x\, e^{-ip_n x/\hbar}\, \hat{O}\, e^{ip_n x/\hbar}, \qquad (A.40)$$

where, since the x representation was used to calculate the integral, $\hat{O}(\hat{x}, \hat{p}) = \hat{O}(x, \frac{\hbar}{i}\partial_x)$. The action of operator $\hat{O}$ on the function $\exp(ip_n x/\hbar)$ can be written at leading order in $\hbar$ as:

$$\hat{O}(\hat{x}, \hat{p})\, e^{ip_n x/\hbar} = O(x, p_n)\, e^{ip_n x/\hbar} + O(\hbar). \qquad (A.41)$$

To obtain Expression (A.41), all operators $\hat{x}$ and $\hat{p}$ were swapped in the expression of $\hat{O}$ in order to make the momentum operator act first on the wave function. These permutations of two non-commuting operators give rise to contributions of higher order in $\hbar$ which are neglected[8] in (A.41).

Then, letting $L \to \infty$, for every function $g(p)$

$$\frac{h}{L} \sum_{n \in \mathbb{Z}} g(p_n) \xrightarrow[L \to \infty]{} \int_{\mathbb{R}} \mathrm{d}p\, g(p). \qquad (A.42)$$

The p_n's being equidistant, separated by a spacing h/L, this formula simply amounts to write a Riemann sum (the left hand side member) for the integral of g over the whole real axis (the right hand side member). Relations (A.41) and (A.42) enable to express the trace (A.40) as:

$$\mathrm{Tr}\,\hat{O} = \sum_{n \in \mathbb{Z}} \frac{1}{L} \int_{-L/2}^{L/2} \mathrm{d}x\, O(x, p_n) + O(\hbar) \xrightarrow[L \to \infty]{} \frac{1}{h} \int_{\mathbb{R}^2} \mathrm{d}x\, \mathrm{d}p\, O(x, p) + O(\hbar). \qquad (A.43)$$

For a system with f degrees of freedom in the semi-classical limit where $\hbar$ is small compared to all quantities homogeneous to an action, (A.43) generalises to:

$$\mathrm{Tr}\,\hat{O} = \frac{1}{h^f} \int_{\mathbb{R}^{2f}} \mathrm{d}^f x\, \mathrm{d}^f p\, O(\boldsymbol{x}, \boldsymbol{p}). \qquad (A.44)$$

Equation (A.44) allows the calculation of the trace of a quantum operator as an integral over classical phase space. It was proven by implicitly assuming that the system under consideration can be described by the conjugate variables x and p. However, it is also valid in the general case of a system whose phase space is described by the conjugate variables q_j and p_j ($j \in \{1, \ldots, f\}$). Indeed, for any conjugate variables $[\hat{q}_j, \hat{p}_j] = i\hbar$, which amounts to say that in representation q, operator $\hat{p}_j$ is written as $\hat{p}_j = \frac{\hbar}{i}\partial/\partial q_j$, which is the key ingredient of our demonstration.

---

[8]The mechanism that brings about corrections of order $\hbar$ can be illustrated by considering the two model cases $\hat{O}_0 = \hat{x}\hat{p}$ and $\hat{O}_1 = \hat{p}\hat{x}$: $\hat{O}_0\psi_n(x) = xp_n\psi_n(x)$ whereas $\hat{O}_1\psi_n(x) = (xp_n + \hbar/i)\psi_n(x)$. $\hat{O}_0$ and $\hat{O}_1$ correspond to the same classical observable $O(x, p) = xp$. It is thus possible to write $\hat{O}_0\psi_n(x) = O(x, p_n)\psi_n(x)$ and $\hat{O}_1\psi_n(x) = O(x, p_n)\psi_n(x) + O(\hbar)$.

A.8 STOKES AND GAUSS-OSTROGRADSKI THEOREMS

In this brief section, we recall two theorems of vector analysis which are useful in several sections of the book.

Let $\vec{F}(\vec{r})$ be a vector field of $\mathbb{R}^3$ and let V be a domain of $\mathbb{R}^3$ bounded by a close surface S, then:

$$\int_V d^3r \, \vec{\nabla} \cdot \vec{F} = \oint_S d^2r \, \vec{F} \cdot \vec{n}, \tag{A.45}$$

where $\vec{n}$ is the unit normal vector to S, outward pointing with respect to V. This is the Gauss-Ostrogradski theorem.

Stokes theorem is also used troughout the book: Let S be a surface attached to a close curve $\mathscr{C}$, then:

$$\int_S d^2r \, \vec{n} \cdot (\vec{\nabla} \wedge \vec{F}) = \oint_{\mathscr{C}} d\vec{\ell} \cdot \vec{F}, \tag{A.46}$$

where $\vec{n}$ is the unit normal vector to S. The orientation of $\mathscr{C}$ is linked to that of $\vec{n}$ by the right-hand rule, see https://en.wikipedia.org/wiki/Right-hand_rule.

A.9 PHYSICAL CONSTANTS

Avogadro number: $\mathcal{N}_A = 6.022\ 140\ 76 \times 10^{23}$ mol^{-1}

Boltzmann constant: $k_B = 1.380\ 649 \times 10^{-23}$ J·K^{-1}

Ideal gas constant: $R = 8.314\ 462\ 618\ 153\ 24$ J·mol^{-1}·K^{-1}

Planck constant: $h = 6.626\ 070\ 15 \times 10^{-34}$ J·s

Gravitational constant: $G = 6.674\ 30(15) \times 10^{-11}$ m^3·kg^{-1}·s^{-2}

Speed of light in vacuum: $c = 2.997\ 924\ 58 \times 10^8$ m·s^{-1}

Elementary charge: $q_e = 1.602\ 176\ 634 \times 10^{-19}$ C

Electron mass: $m_e = 9.109\ 383\ 7015(28) \times 10^{-31}$ kg

Bohr magneton: $\mu_B = 9.274\ 010\ 0783(28) \times 10^{-24}$ J·T^{-1}

Index

Printed in the United States
by Baker & Taylor Publisher Services